T0390826

NANOPARTICLE THERAPEUTICS

NANOPARTICLE THERAPEUTICS
Production Technologies, Types of Nanoparticles, and Regulatory Aspects

Edited by

PRASHANT KESHARWANI

KAMALINDER K. SINGH

ELSEVIER

ACADEMIC PRESS

An imprint of Elsevier

Academic Press is an imprint of Elsevier
125 London Wall, London EC2Y 5AS, United Kingdom
525 B Street, Suite 1650, San Diego, CA 92101, United States
50 Hampshire Street, 5th Floor, Cambridge, MA 02139, United States
The Boulevard, Langford Lane, Kidlington, Oxford OX5 1GB, United Kingdom

Library of Congress Cataloging-in-Publication Data
A catalog record for this book is available from the Library of Congress

British Library Cataloguing-in-Publication Data
A catalogue record for this book is available from the British Library

ISBN 978-0-12-820757-4

For information on all Academic Press publications
visit our website at https://www.elsevier.com/books-and-journals

Publisher: Stacy Masucci
Acquisitions Editor: Andre Gerhard Wolff
Editorial Project Manager: Timothy Bennett
Production Project Manager: Omer Mukthar
Cover Designer: Mark Rogers

Typeset by STRAIVE, India

Contents

PART 2 Application of nanoparticles in drug delivery

8. Biodegradable self-assembled nanocarriers as the drug delivery vehicles — **293**

Charu Misra, Rakesh Kumar Paul, Nagarani Thotakura, and Kaisar Raza

9. Albumin nanoparticles—A versatile and a safe platform for drug delivery applications — **327**

Tamara Zwain, Neetika Taneja, Suha Zwayen, Aditi Shidhaye, Aparana Palshetkar, and Kamalinder K. Singh

17. Fate and potential hazards of nanoparticles in the environment   **581**

Govind Sharan Gupta and Alok Dhawan

18. Biological toxicity of nanoparticles   **603**

Violina Kakoty, Sarathlal K.C., Meghna Pandey, Sunil Kumar Dubey, Prashant Kesharwani, and Rajeev Taliyan

Contributors

May Azzawi
Cardiovascular Research Group, Department of Life Sciences, Faculty of Science and Engineering, Manchester Metropolitan University, Manchester, United Kingdom

Iara Baldim
CEB—Centre of Biological Engineering, University of Minho, Braga, Portugal; Faculty of Pharmaceutical Sciences of Ribeirão Preto, University of São Paulo, Ribeirão Preto, SP, Brazil

Arvind Kumar Bansal
Department of Pharmaceutics, Solid State Pharmaceutics Lab, National Institute of Pharmaceutical Education and Research (NIPER), Mohali, Punjab, India

Sarwar Beg
Department of Pharmaceutics, School of Pharmaceutical Education and Research, Jamia Hamdard (Hamdard University), New Delhi, India; School of Pharmacy and Biomedical Sciences, Faculty of Clinical and Biomedical Sciences, University of Central Lancashire, Preston, United Kingdom

Largee Biswas
Nanobiotech Lab, Department of Zoology, Kirori Mal College, University of Delhi, Delhi, India

Marco Cecchini
NEST, Scuola Normale Superiore and Istituto Nanoscienze-CNR, Pisa, Italy

Hira Choudhury
Department of Pharmaceutical Technology, School of Pharmacy, International Medical University, Kuala Lumpur, Malaysia

Carla Daruich de Souza
Laboratory for Radiation Therapy Sources Production, Nuclear and Energy Research Institute, National Nuclear Energy Commission, São Paulo, Brazil

Alok Dhawan
Nanomaterials Toxicology Group, CSIR-Indian Institute of Toxicology Research; Centre of Biomedical Research, Lucknow, Uttar Pradesh, India

Sunil Kumar Dubey
R&D Healthcare Division, Emami Ltd, Belgharia, Kolkata, India

Amal Ali Elkordy
Faculty of Health Sciences and Wellbeing, University of Sunderland, Sunderland, United Kingdom

Asima Farooq
Department of Natural Sciences, Faculty of Science and Engineering, Manchester Metropolitan University, Manchester, United Kingdom

Mariacristina Gagliardi
NEST, Scuola Normale Superiore and Istituto Nanoscienze-CNR, Pisa, Italy

Francisco M. Gama
CEB—Centre of Biological Engineering, University of Minho, Braga, Portugal

Honey Goel
Department of Pharmaceutics, University Institute of Pharmaceutical Sciences and Research, Baba Farid University of Health Sciences (BFUHS), Faridkot, India

Bapi Gorain
School of Pharmacy; Centre for Drug Delivery and Molecular Pharmacology, Faculty of Health and Medical Sciences, Taylor's University, Subang Jaya, Selangor, Malaysia

Govind Sharan Gupta
Unit of Molecular Toxicology, Institute of Environmental Medicine, Karolinska Institutet, Stockholm, Sweden

Ben Hodgson
Nano-biomaterials Research Group, School of Natural Sciences, Faculty of Science & Technology, University of Central Lancashire, Preston, United Kingdom

Ajaz Hussain
Private Practice Insight, Advice, and Solutions, Frederick, MD; Former President of the National Institute for Pharmaceutical Technology and Education, Minneapolis, MN; Deputy Director, Office of Pharmaceutical Science, US FDA, Silver Spring, MD, United States

Sanika Jadhav
Department of Pharmaceutics, Solid State Pharmaceutics Lab, National Institute of Pharmaceutical Education and Research (NIPER), Mohali, Punjab, India

Sarathlal K.C.
Department of Pharmacy, Birla Institute of Technology and Science, Pilani, Rajasthan, India

Violina Kakoty
Department of Pharmacy, Birla Institute of Technology and Science, Pilani, Rajasthan, India

Amanpreet Kaur
Department of Pharmaceutics, Solid State Pharmaceutics Lab, National Institute of Pharmaceutical Education and Research (NIPER), Mohali, Punjab, India

Prashant Kesharwani
Department of Pharmaceutics, School of Pharmaceutical Education and Research, Jamia Hamdard, New Delhi, India

Rajneet Kaur Khurana
Pharmaceutics Division, University Institute of Pharmaceutical Sciences, Panjab University, Chandigarh, India

Nisha Lamichhane
Nano-biomaterials Research Group, School of Natural Sciences, Faculty of Science & Technology, University of Central Lancashire, Preston, United Kingdom

Sheefali Mahant
Department of Pharmaceutical Sciences, Maharshi Dayanand University, Rohtak, Haryana, India

Asiya Mahtab
Department of Pharmaceutics, School of Pharmaceutical Education and Research, Jamia Hamdard, New Delhi, India

Tim Mercer
Nano-biomaterials Research Group, School of Natural Sciences, Faculty of Science & Technology, University of Central Lancashire, Preston, United Kingdom

Charu Misra
Department of Pharmacy, School of Chemical Sciences and Pharmacy, Central University of Rajasthan, Bandarsindri, Rajasthan, India

Sri Hari Raju Mulagapati
Analytical Science, BioPharmaceuticals Development, R&D, AstraZeneca, Gaithersburg, MD, United States

Jayabalan Nirmal
Translational Pharmaceutics Research Laboratory, Department of Pharmacy, Birla Institute of Technology and Science (BITS)-Pilani, Hyderabad, Telangana, India

Beatriz Ribeiro Nogueira
Laboratory for Radiation Therapy Sources Production, Nuclear and Energy Research Institute, National Nuclear Energy Commission, São Paulo, Brazil

Wanderley P. Oliveira
Faculty of Pharmaceutical Sciences of Ribeirão Preto, University of São Paulo, Ribeirão Preto, SP, Brazil

Aparana Palshetkar
C U Shah College of Pharmacy, SNDT Women's University; VES College of Pharmacy, Mumbai, India

Manisha Pandey
Department of Pharmaceutical Technology, School of Pharmacy, International Medical University, Kuala Lumpur, Malaysia

Meghna Pandey
Department of Pharmacy, Birla Institute of Technology and Science, Pilani, Rajasthan, India

Prashantkumar Khodabhai Parmar
Department of Pharmaceutics, Solid State Pharmaceutics Lab, National Institute of Pharmaceutical Education and Research (NIPER), Mohali, Punjab, India

Arun Parupudi
Dosage Form Design and Development, BioPharmaceutical Development, BioPharmaceuticals R&D, AstraZeneca, Gaithersburg, MD, United States

Rakesh Kumar Paul
Department of Pharmacy, School of Chemical Sciences and Pharmacy, Central University of Rajasthan, Bandarsindri, Rajasthan, India

Singh Raghuvir
Department of Pharmaceutical Sciences, Guru Jambheshwar University of Science and Technology, Hisar, Haryana, India

Vaikundamoorthy Ramalingam
Department of Animal Science, Jeonbuk National University, Jeonju, Republic of Korea; Centre for Natural Products & Traditional Knowledge, CSIR-Indian Institute of Chemical Technology, Hyderabad, India

Rekha Rao
Department of Pharmaceutical Sciences, Guru Jambheshwar University of Science and Technology, Hisar, Haryana, India

Kaisar Raza
Department of Pharmacy, School of Chemical Sciences and Pharmacy, Central University of Rajasthan, Bandarsindri, Rajasthan, India

Karan Razdan
Pharmaceutics Division, University Institute of Pharmaceutical Sciences, Panjab University, Chandigarh, India; School of Pharmacy and Biomedical Sciences, Faculty of Clinical and Biomedical Sciences, University of Central Lancashire, Preston, United Kingdom

Maria Elisa Chuery Martins Rostelato
Laboratory for Radiation Therapy Sources Production, Nuclear and Energy Research Institute, National Nuclear Energy Commission, São Paulo, Brazil

Komal Saini
Pharmaceutics Division, University Institute of Pharmaceutical Sciences, Panjab University, Chandigarh, India; School of Pharmacy and Biomedical Sciences, Faculty of Clinical and Biomedical Sciences, University of Central Lancashire, Preston, United Kingdom

Tapas Sen
Nano-biomaterials Research Group, School of Natural Sciences, Faculty of Science & Technology, University of Central Lancashire, Preston, United Kingdom

Maneea Eizadi Sharifabad
Nano-biomaterials Research Group, School of Natural Sciences, Faculty of Science & Technology, University of Central Lancashire, Preston, United Kingdom

Aditi Shidhaye
Colgate-Palmolive (India) Limited, Colgate Research Centre, Mumbai, India

Ali Shukur
Cardiovascular Research Group, Department of Life Sciences, Faculty of Science and Engineering, Manchester Metropolitan University, Manchester, United Kingdom

Lubna Siddiqui
Department of Pharmaceutics, School of Pharmaceutical Education and Research, Jamia Hamdard, New Delhi, India

Kamalinder K. Singh
School of Pharmacy and Biomedical Sciences, Faculty of Clinical and Biomedical Sciences; UCLan Research Centre for Smart Materials; UCLan Research Centre for Translational Biosciences & Behaviour, University of Central Lancashire, Preston, United Kingdom

Eliana B. Souto
CEB—Centre of Biological Engineering, University of Minho, Braga; Department of Pharmaceutical Technology, Faculty of Pharmacy, University of Coimbra, Coimbra, Portugal

J. Anand Subramony
Biologics Engineering, R&D, AstraZeneca, Gaithersburg, MD, United States

Sushama Talegaonkar
Department of Pharmaceutics, School of Pharmaceutical Sciences, Delhi Pharmaceutical Sciences and Research University (DPSRU), New Delhi, India

Rajeev Taliyan
Department of Pharmacy, Birla Institute of Technology and Science, Pilani, Rajasthan, India

Neetika Taneja
Sun Pharmaceutical Industries Ltd, Vadodara, Gujarat, India

Nagarani Thotakura
Department of Pharmacy, School of Chemical Sciences and Pharmacy, Central University of Rajasthan, Bandarsindri, Rajasthan, India

Anita K. Verma
Nanobiotech Lab, Department of Zoology, Kirori Mal College, University of Delhi, Delhi, India

Debra Whitehead
Department of Natural Sciences, Faculty of Science and Engineering, Manchester Metropolitan University, Manchester, United Kingdom

Azziza Zaabalawi
Cardiovascular Research Group, Department of Life Sciences, Faculty of Science and Engineering, Manchester Metropolitan University, Manchester, United Kingdom

Carlos Alberto Zeituni
Laboratory for Radiation Therapy Sources Production, Nuclear and Energy Research Institute, National Nuclear Energy Commission, São Paulo, Brazil

Tamara Zwain
School of Pharmacy and Biomedical Sciences, Faculty of Clinical and Biomedical Sciences; UCLan Research Centre for Smart Materials, University of Central Lancashire, Preston, United Kingdom

Suha Zwayen
School of Pharmacy and Biomedical Sciences, Faculty of Clinical and Biomedical Sciences, University of Central Lancashire, Preston, United Kingdom; Precision Nanosystems Inc., Vancouver, BC, Canada

Preface

Nanoparticle-based therapeutics have witnessed tremendous interest and advances in recent years, and this has been accelerated further with the introduction of nanotechnology-based COVID-19 vaccines. The intent of this book is to provide scientists and science interns with fundamental knowledge on the experiential and experimental aspects of nanoparticles and nanotechnology and to foster understanding of rational application in various fields, especially in differential diagnosis and treatment of diverse diseased states for improving and extending human life.

The book is a complete package in terms of providing a holistic understanding of the principle behind nanoparticles formation, process technologies employed for their fabrication, characterization, biofate, and pharmacokinetics along with different types of nanoparticles employed for a range of biomedical applications. It also covers the regulatory aspects, environmental hazards, and toxicity of the nanoparticles.

Chapter 1 introduces the advances in pharmaceutical nanotechnology and the exciting avenues for development and progress of nanoparticle research; it sets the scene for nanoparticle therapeutics with an emphasis on the recent state of the art and prospective challenges. The chapter focuses on dimensionality, particle types, and developmental methods, and details the nanoparticle classification and composition, with examples in each category. Emerging areas of nanoparticle therapy are reviewed, including the use of nanoparticles for gene delivery and RNA interference therapy, nanoparticles that can target dendritic cells, macrophages, and immune cells for cancer immunotherapy, and the use of nanoparticles in CRISPR technology and vaccine development, with an eye on recent events that have highlighted the need for vaccines to address global pandemics.

Chapter 2 compiles the technological advancements in the design of industrial-scale technologies for the pilot-scale production of nanoformulations such as nanoliposomes, nanocrystals, nanoparticles, nanosuspensions, and nanoemulsions, and address the pitfalls of small-scale research approaches to facilitate cost-effective design, maximum viability, stability, quality control, and therapeutic efficacy for better patient compliance. Chapter 3 elaborates key in vitro physicochemical parameters of nanoparticles in both dry and liquid states using specialized techniques that enable deeper insights to be gained into specific nanoplatform properties to tailor their in vitro and in vivo performance. The chapter discusses the various characterization methods, elaborating on their principles as well as investigating the advantages and limitations of these techniques.

Chapter 4 reviews the different methods for developing surface-engineered nanoparticles for treatment of different diseases, and diagnosis and theranostic approaches in different medical applications. Chapter 5 provides an insight into the biofate and cellular interactions of nanoparticles with an emphasis on lipid nanoparticles. It discusses distribution of nanoparticles and approaches for tumor targeting and brain targeting, and emphasizes their toxicity assessment. Chapter 6 focuses on current understanding of nanoparticle interactions with cells of the vessel wall, in particular endothelial cells and how these can be targeted by modifying nanoparticle surfaces via conjugation with tagging molecules that recognize surface endothelial determinants. Chapter 7 summarizes factors influencing pharmacokinetics, physiology-based pharmacokinetics, pharmacodynamics, and pharmacodynamics of nanoparticles for improvising release-kinetics, biodistribution, bioavailability, excretion, and toxicity with an emphasis on translation and therapy with future directions.

Chapter 8 focuses on self-assembled nanocarriers, and investigates various techniques for their fabrication, their classification, and clinical applications. It also discusses the chemistry behind the formation of supramolecular structure and their enhancing role in biomedical applications. Chapter 9 deliberates on the design and synthesis of nanoparticle-based drug delivery systems that use albumin as a drug carrier system and elaborates on their applications in various therapeutic conditions. Albumin-based nanoparticles are particularly interesting as attractive macromolecular carriers that are biodegradable and hold great value in the field of nanomedicine due to their safety, biocompatibility, high capacity for drug loading, and high binding efficiency with a range of drugs, providing a delivery vehicle for many small and large molecules alike.

Chapter 10 presents an overview of silver nanoparticles, highlighting the methods for synthesis and its mechanism of application in various fields such as antimicrobial, antimalarial, antibiofilm, antioxidant, and anticancer activity. Chapter 11 focuses on the potential use of silica nanoparticles as diagnostic tools and therapeutic intervention modalities, and provides an overview of the methods used to fabricate and characterize them. The chapter also highlights the strategies used to maximize the biocompatibility of silica nanoparticles to improve their applicability and bioavailability, minimize their off-target localization and effects, and promote their guided delivery.

Nanocrystals have emerged as a promising technology for addressing the challenges of poorly water-soluble drugs. Chapter 12 discusses the fundamentals of nanocrystals, the relation of crystal structure with nanonization, phase transformation during nanonization, and optimization of critical process parameters. Novel concepts in the field of nanocrystals such as miniaturization, IVIVC, functionalization, and applications in nonoral routes are also considered.

Chapter 13 systematically introduces the concept of magnetic nanomaterials, superparamagnetic materials such as SPIONs, various synthesis routes, various core-shell SPIONs, surface functionalization of core-shell SPIONs, and the main characterization techniques and application of SPIONs in medical diagnosis and therapy. Biomimicry is a novel discipline that explores innovative materials and techniques to imitate natural elements, molecules, mechanisms, and systems, to serve our technological society. Chapter 14 summarizes recent advances in the use of biomimetic and bioinspired nanoparticles for targeted drug delivery. Chapter 15 reviews advancements made in radioactive nanoparticles and new modalities of nanobrachotherapy, along with their benefits, challenges, and biomedical applications.

Chapter 16 gives an overview of regulatory pathways and federal perspectives pertinent to the complexity of nanoparticle systems and chaos precipitated by the COVID-19 pandemic. It elaborates on fundamental aspects of US FDA guidance on nanotechnology, providing a deeper understanding of "good practices" in the context of professional responses in chaotic, complex, complicated, and simple systems. Nanoparticles can enter the environmental matrices at different stages of their life cycle and their level of hazard regarding the environment depends on their fate after entering the environment, as they are most likely to behave differently after interactions with abundant biotic and abiotic components. Chapter 17 summarizes the current developments on the fate and impact of nanoparticles in the environment with special consideration at individual organism to food chain level. Nanoparticles, owing to their size, can penetrate the organs and the deepest layer of the skin easily. Chapter 18 provides a detailed review of the biological toxicity of nanoparticles and covers aspects on molecular, cellular, and tissue toxicity of nanomedicines and their effects on human health.

We are grateful to all contributing authors, who have worked strenuously on the chapters to provide up-to-date knowledge on pertinent topics. We sincerely hope that this book will serve as a comprehensive reference for everyone working in the broad area of nanoparticle therapeutics and that it will be an ever-ready source of helpful information.

Prashant Kesharwani
Kamalinder K. Singh

PART 1

Synthesis and characterization of NP for drug delivery

CHAPTER 1

Nanoparticle technologies: Recent state of the art and emerging opportunities

Arun Parupudi[a],*, Sri Hari Raju Mulagapati[b],*, and J. Anand Subramony[c],*

[a]Dosage Form Design and Development, BioPharmaceutical Development, BioPharmaceuticals R&D, AstraZeneca, Gaithersburg, MD, United States
[b]Analytical Science, BioPharmaceuticals Development, R&D, AstraZeneca, Gaithersburg, MD, United States
[c]Biologics Engineering, R&D, AstraZeneca, Gaithersburg, MD, United States

1. Introduction

Nanotechnology holds great promise for the development of pharmaceutical products and has been used successfully for three decades. Interest in this technology has been spurred by the ability of substances to behave differently in the "nano" ($\sim$1–200 nm) dimension and by the resulting implications of their physicochemical and biological properties. Higher surface area in the nano dimension has been a key driver of advances in this area, and unique physicochemical properties in the nano space have given rise to additional hypotheses with regard to the absorption-distribution, metabolism, and elimination of nanoparticles in physiological systems. The advances in and availability of process methodologies to create nanomaterials have added further impetus to this research.

An example of these advances is provided by the development of solid oral dosage forms. Nearly 40% of pharmaceutical molecules ultimately fail as therapeutic agents because of poor solubility and consequential low bioavailability. Nanotechnologies have made possible substantial improvements in the dissolution properties of active pharmaceutical ingredients in nano form, leading to the approval of various oral dosage forms (Table 1). In addition, nanotechnologies have made it possible to improve the dissolution of molecules for both oral and parenteral dosage forms.

Drug targeting is another area that has seen important advances owing to polymeric nanoparticle technologies. Nanoparticles have been postulated to escape the leaky vasculates of solid tumors, and hence, accumulate in tumor tissues, a phenomenon referred to as "enhanced permeation and retention" or "passive targeting." The surfaces of nanoparticles can be decorated with targeting ligands, small molecules, or antibodies via simple chemical reactions, potentially enabling the active targeting of drugs to specific tissues and organs via control of pharmacokinetic and intrinsic biodistribution properties.

---

* All authors contributed equally.

Table 1 Approved nanoparticles for drug delivery [1].

Nanoparticle class	Trade (generic) name	Material composition	Indication(s) (approval year, region)
Polymer-based nanoparticles	Adagen (pegademase)	PEGylated adenosine deaminase enzyme	Adenosine deaminase deficiency in patients with severe combined immunodeficiency disease (1990, US)
	Adynovate (antihemophilic factor)	Polymer-protein conjugate	Hemophilia (2015, US)
	Cimzia (certolizumab)	PEGylated antibody fragment	Crohn's disease (2008, US; 2009, EU)
	Copaxone, Glatopa (glatiramer)	Polymer composed of L-glutamate, L-alanine, L-lysine, and L-tyrosine	Multiple sclerosis (1996, US)
	Eligard (leuprolide)	Leuprolide acetate and polymer (PLGA)	Prostate cancer (2002, US)
	Krystexxa (pegloticase)	Polymer (PEGylated)-protein conjugate	Chronic gout (2010, US; 2013, EU)
	Macugen (pegaptanib)	PEGylated anti-VEGF aptamer	Neovascular age-related macular degeneration (2004, US; 2006, EU)
	Mircera (methoxy polyethylene glycol-epoetin beta)	Chemically synthesized erythropoiesis-stimulating agent	Anemia associated with chronic kidney disease (2007, US and EU)
	Neulasta (pegfilgrastim)	PEGylated GCSF	Chemotherapy-induced neutropenia (2002, US and EU)
	Oncaspar (pegaspargase)	PEGylated L-asparaginase	Acute lymphoblastic leukemia (1994, US; 2016, EU)
	Pegasys (peginterferon α-2a)	PEGylated IFN-α2a protein	Hepatitis B and C (2002, US and EU)
	PegIntron (peginterferon α-2b)	PEGylated IFNα-2b protein	Hepatitis C (2001, US; 2000, EU)
	Plegridy (peginterferon β-1a)	PEGylated IFNβ-1a	Multiple sclerosis (2014, US and EU)
	Renvela, Renagel (sevelamer)	Poly(allylamine hydrochloride)	Chronic kidney disease (2000, US and EU)
	Somavert (pegvisomant)	PEGylated hGH receptor antagonist	Acromegaly (2002, EU; 2003, US)
	Welchol (colesevelam)	Cross-linked poly(allylamine)	Type 2 diabetes (2000, US)

Table 1 Approved nanoparticles for drug delivery—cont'd

Nanoparticle class	Trade (generic) name	Material composition	Indication(s) (approval year, region)
Liposomal nanoparticles	Taxol (paclitaxel)	Paclitaxel	Ovarian cancer (1992, US)
	Taxotere (docetaxel)	Micelle of docetaxel	Breast cancer (1996, US; 1995, EU)
	Genexol-PM (paclitaxel)	Polymeric micelle of paclitaxel	Breast cancer (2001, South Korea)
	Abelcet (amphotericin B)	Liposomal amphotericin B lipid complex	Fungal infections (1995, US)
	AmBisome (amphotericin B)	Liposomal amphotericin B lipid complex	Fungal and protozoal infections (1997, US)
	Curosurf (poractant alfa)	Liposomal proteins SP-B and SP-C	Respiratory distress syndrome (1999, US)
	DaunoXome (daunorubicin)	Liposomal daunorubicin	Kaposi sarcoma (1996, US)
	DepoCyt (cytarabine)	Liposomal cytarabine	Lymphomatous meningitis (1999, US; 2001, EU)
	DepoDur (morphine sulfate extended-release liposome injection)	Liposomal verteporfin	Postoperative analgesia (2004, US)
	Doxil, Caelyx (doxorubicin)	Peglycated liposomal doxorubicin	Kaposi sarcoma (1995, US; 1996, EU)
	Marqibo (vincristine)	Liposomal vincristine	Acute lymphoblastic leukemia (2012, US)
	Mepact (mifamurtide)	Liposomal mifamurtide	Bone cancer (2009, EU)
	Myocet (doxorubicin)	Non-PEGylated liposomal doxorubicin	Breast cancer (2000, EU)
	Onivyde (irinotecan)	Liposomal irinotecan	Pancreatic cancer (2015, US; 2016, EU)
	Visudyne (verteporfin)	Liposomal verteporfin	Age-related macular degeneration (2000, US and EU)
Protein nanoparticles	Abraxane ABI-007 (paclitaxel)	Albumin-bound paclitaxel nanoparticle	Breast cancer (2005, US; 2008, EU)
	Ontak (denileukin)	Engineered protein combining IL-2 and diphtheria toxin	Cutaneous T-cell lymphoma (1999, US)

Continued

Table 1 Approved nanoparticles for drug delivery—cont'd

Nanoparticle class	Trade (generic) name	Material composition	Indication(s) (approval year, region)
Nanocrystals	Emend (aprepitant)	Aprepitant	Chemotherapy-induced emesis (2003, US and EU)
	EquivaBone (hydroxyapatite)	Nanocrystalline calcium phosphate	Bone voids and defects (2009, US)
	Focalin XR (dexamethylphenidate)	Dexamethylphenidate HCl	Attention deficit hyperactivity disorder (2005, US)
	Invega, Sustenna, Xeplion (paliperidone)	Paliperidone palmitate	Schizophrenia (2009 US; 2011, EU)
	Megace ES (megestrol)	Megestrol acetate	Anorexia, cachexia (2005, US)
	nanOss (hydroxyapatite)	Nanostructured calcium apatite	Bone loss (2005, US)
	OsSatura (hydroxyapatite)	Calcium apatite	Bone loss (2003, US)
	Ostim (alendronate)	Hydroxyapatite	Bone loss (2004, US)
	Rapamune (sirolimus)	mTOR-inhibiting macrolide	Prevention of kidney transplant rejection (1999, US; 2001, EU)
	Ritalin LA (methylphenidate)	Methylphenidate hydrochloride	Attention deficit hyperactivity disorder (2002, US)
	Ryanodex (dantrolene)	Dantrolene sodium	Malignant hypothermia (2014, US)
	TriCor (fenofibrate)	Fenofibrate	Hypercholesterolemia, hypertriglyceridemia (2004, US)
	Triglide (fenofibrate)	Fenofibrate	Hypercholesterolemia, hypertriglyceridemia (2005, US)
	Vitoss (calcium phosphate)	B-tricalcium phosphate	Bone loss (2003, US)
	Zanaflex (tizanadine)	Tizanidine HCl	Muscle spasticity (1996, US)
Inorganic and metallic nanoparticles	Nanotherm iron oxide	Iron oxide nanoparticles	Glioblastoma (2010, EU)
	Dexferrum, DexIron (iron dextran)	High-MW iron dextran	Iron deficiency (1996, US)
	Feraheme, Rienso (ferumoxytol)	Ferumoxytol SPION with polyglucose sorbitol carboxymethyl ether	Iron deficiency anemia in chronic kidney disease (2009, US; 2012, EU)

Table 1 Approved nanoparticles for drug delivery—cont'd

Nanoparticle class	Trade (generic) name	Material composition	Indication(s) (approval year, region)
	Feridex, Endorem (ferumoxides)	SPION coated with dextran	MRI evaluation of RES-associated liver lesions (1996, US)
	Ferrlecit (sodium ferric gluconate complex)	Sodium ferric gluconate	Iron deficiency anemia in chronic kidney disease (1999, US)
	GastroMARK, Lumirem (ferumoxsil)	SPION coated with silicone	MRI contrast (1996, US and EU)
	INFeD (iron dextran)	Low-MW iron dextran	Iron deficiency anemia in chronic kidney disease (1974, US)
Virus-based nanoparticles	Gendicine (recombinant human p53 AdV)	Recombinant AdV engineered to express wild-type p53	Head and neck cancer (2004, China)
	Luxturna (voretigene neparvovec-rzyl)	Recombinant AAV vector serotype 2	Inherited retinal disease (2017, US)
	Glybera (alipogene tiparvovec)	Recombinant AAV vector serotype 1	Lipoprotein lipase deficiency (2012, EU)
	Zolgensma (onasemnogene abeparvovec-xioi)	Recombinant AAV vector serotype 9	Spinal muscular atrophy (2019, US)
	Valrox (valoctocogene roxaparvovec)	Recombinant AAV vector serotype 5	Hemophilia A (2020, US)

Abbreviations: *AdV*, adenovirus; *GCSF*, granulocyte colony-stimulating factor; *hGH*, human growth hormone; *IFN*, interferon; *MRI*, magnetic resonance imaging; *MW*, molecular weight; *PEG*, polyethylene glycol; *PLGA*, poly(lactic-*co*-glycolic acid); *RES*, reticuloendothelial system; *SP-B, -C*, lung surfactant B, C; *SPION*, superparamagnetic iron oxide nanoparticle; *VEGF*, vascular endothelial growth factor.

Nanoparticles have also been investigated as agents for gene therapy. The emergence of RNA interference therapy and immuno-oncology has engendered new areas of research using nanoparticles. For example, lipid nanoparticles have been successfully advanced to clinical translation for messenger RNA (mRNA) delivery. The charge and tunability of nanoparticles have been explored to target immune cells in order to activate the collective immune response, rather than targeting solid tumors. Nanoparticles may also find utility in enabling the delivery of CRISPR (clustered regularly interspaced short palindromic repeats)-Cas9 and as virus vectors to target various organs.

In this chapter, we review the various nanoparticle types and their classifications based on dimensionality and processing technologies. Also discussed are widely used analytical methods for the characterization of nanoparticles and some of the key applications of nanoparticles in drug development and delivery, with an emphasis on emerging applications of nanoparticles as virus vectors, in gene therapy, and for CRISPR–Cas9-guided therapy.

2. Dimensional classification of nanoparticles

Nanomaterials can be classified on the basis of geometry and size in multiple dimensions. When based on dimensionality, the classification of nanoparticles is divided into zero-dimensional, one-dimensional, two-dimensional, and three-dimensional structures (Fig. 1).

Zero-dimensional nanoparticles have features or dimensions (x, y, z) that are all less than 100 nm and possess lengths that equal their widths (i.e., materials in which nanoparticles are isolated from each other). Generally, zero-dimensional nanoparticles may be atomic clusters, filaments, or cluster assemblies.

In a one-dimensional system, two dimensions (x, y) of one-dimensional nanostructures are in nanoscale, whereas other dimensions are outside the nanoscale. Electrons are within two dimensions, and hence, cannot move freely in this system. Generally, one-dimensional nanostructures may be nanorods, nanowires, nanotubes, nanofibers, or other forms.

In a two-dimensional system, two dimensions (x, y) are outside of the nanometric size range, whereas one dimension is in the nanoscale range. Two-dimensional nanomaterials

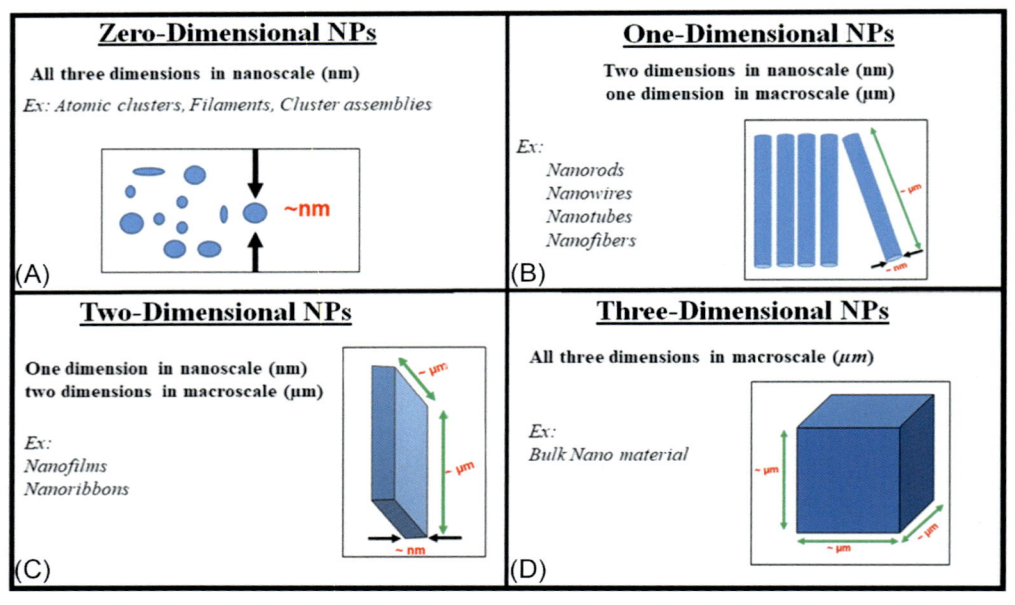

Fig. 1 Dimensional classification of nanostructures. (A) Zero-dimensional nanoparticles. (B) One-dimensional nanoparticles. (C) Two-dimensional nanoparticles. (D) Three-dimensional nanoparticles.

include nanofilms, nanosheets, and nanoribbons (ultrafine-grained overlayers or buried layers).

In a three-dimensional system, all three dimensions (x, y, z) are outside the nanoscale range. Electrons are fully delocalized and can move freely within all dimensions. This class includes bulk powders, dispersions of nanoparticles, bundles of nanowires, and nanotubes, as well as multiple nanolayers. Three-dimensional nanomaterials may be dendrimers, carbon nanotubes, or quantum dots [2, 3].

3. Nanoparticle material types

Nanoparticles can be also classified on the basis of their constituent materials. Broadly, they are classified as polymeric-based; lipid-based; inorganic, including metallic; and other macromolecular architecture. Fig. 2 lists these particle types with some examples of their constituent materials. Fig. 2 also lists other nanoparticle types not listed here and include composite nanoparticles that are composed of mixed metal oxides or other structures. Nanoparticles can also be made by unique solution chemistries, which is discussed later in this chapter (see Section 4).

3.1 Polymeric nanoparticles

Polymeric nanoparticles can be used as carriers or substrates to deliver vaccine antigens, proteins, and drugs to a desired target with lower cytotoxicity. They exist as two types, natural and synthetic.

Natural or biological polymers occur in nature and can be extracted from various natural materials. They are formed naturally during the life cycles of bacteria, fungi, plants, and animals. Some important natural polymers include dextrans, collagen, chitosan, gelatin, alginate, and hyaluronic acid. Most natural polymers are condensation polymers whose building blocks (monomers) are linked together due to release of a water molecule. Natural polymers have many advantages, including biocompatibility, biodegradability, reduced toxicity, and natural abundance. Limitations of natural polymers include high variability of the source material, complicated extraction process, and complex structures.

Synthetic polymer-based nanoparticles have gained attention because different monomers can be reacted in a controlled environment, resulting in a wide range of physical, chemical, and mechanical properties. Chemically synthesized nanoparticle materials include polyglycolic acid, polylactic acid, poly(lactide-*co*-glycolic acid) (PLGA), and poly (ε-caprolactone) [4, 5]. Advantages of synthetic polymers include biocompatibility, flexibility of material design, and ability to fine-tune various components to achieve a specific stoichiometry. The latter feature in particular has driven the broad utility of synthetic polymer-based nanoparticles; for example, charge, size, and aspect ratio can be controlled with various substituents and functionalization, and molecular weight and structure can be modified to achieve desired properties. Most synthetic polymers are derived from the

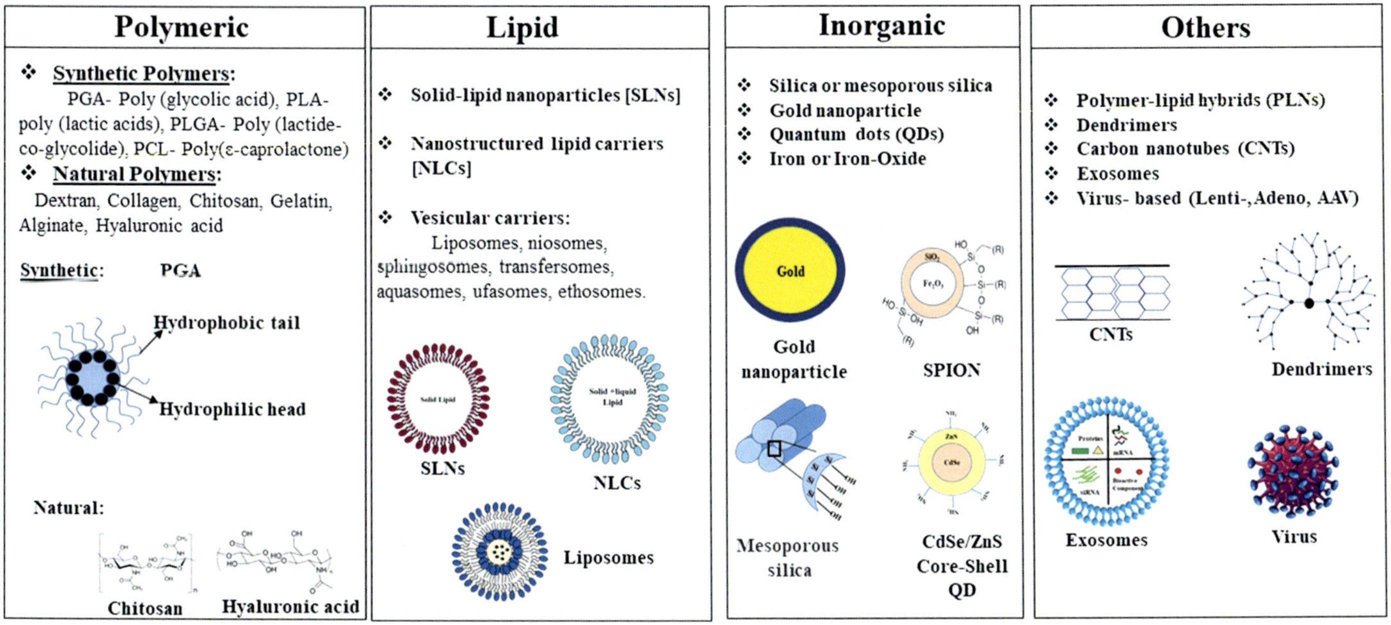

Fig. 2 Nanoparticle types and examples of their constituent materials.

ester family. They are considered to be the most promising drug carriers due to their bio-compatibility, which allows customization of key properties such as molecular weight and hydrophobicity [1, 6–9].

In the development of polymer nanocarriers, the drug is dissolved, entrapped, encapsulated, or covalently attached to a matrix or membrane polymer. Based on the method of preparation and drug load, these are divided into two types, nanospheres and nanocapsules. Nanospheres (matrix system) are solid-core structures in which the drug is encapsulated or adsorbed to the polymer matrix, whereas nanocapsules (reservoir system) are hollow-core structures in which the inner space is loaded with a drug of interest and surrounded by a unique polymer membrane [1, 6, 10]. Covalent attachments can also be made on spherically symmetrical nanoparticles with a variety of chemistries [11].

3.2 Lipid-based nanoparticles

Lipid-based nanoparticles are spherical particles composed of lipids with a lipidic core having a mean diameter of 50–1000 nm. The main advantages of lipid nanoparticles in drug delivery are due to their lower toxicity level and use of physiologically tolerated lipids, as well as the ability to achieve larger-scale manufacturing, protect the drug from degradation, control drug release, and achieve high transfection rate when direct membrane fusion is possible. Of the several types of lipid-based nanoparticles, important subclasses include solid lipid nanoparticles (SLNs), nanostructured lipid carriers (NLCs), and vesicular carriers such as liposomes, niosomes, sphingosomes, transfersomes, aquasomes, ufasomes, and ethosomes [12].

Introduced in 1991, SLNs are prepared from solid lipids that remain solid at room and body temperature and are mostly stabilized by suitable surfactants. SLNs contain purified triglycerides, diglycerides, complex glyceride mixtures, fatty acids, and waxes. Most SLN formulations contain 0.1%–30% lipid content, 0.5%–30% surfactant stabilizer (polysorbates, Tween, etc.), and 5% of the incorporated drug. These particles are less toxic due to the absence of organic solvents in the production process. Their disadvantages include insufficient drug-loading capacity, poor long-term stability, and drug expulsion caused by lipid polymorphic transition during storage. Usually, the drug-loading capacity of the lipid matrix for conventional SLNs is approximately 25%.

NLCs were introduced to overcome the limitations of SLNs. These carriers are composed of both solid- and liquid-phase lipids along with a surfactant layer. The major advantage of NLCs is their ability to incorporate more drug load than SLNs. Based on the structure of the lipid matrix used in production, three incorporation models have been proposed for NLCs: (1) imperfect type, in which NLCs cannot form highly ordered structures, resulting in structural imperfections; (2) multiple type, in which higher amounts of oil are mixed with the solid lipid layer, leading to phase separation and the resultant formation of oily nano compartments within the solid lipid matrix; and

(3) amorphous or structureless type, in which only a few lipid components recrystallize after homogenization and cooling of the nanoemulsion [12–14].

The liposome, one of the first vesicular systems that are spherical in structure, is composed of phospholipids and cholesterol with either one double layer (unilamellar vesicles) or multiple double layers (multilamellar vesicles). Liposomes are good candidate for drug delivery because their lipid composition is similar to that of the cell membrane, which may improve uptake. Other advantages include their ability to encapsulate the drug in both hydrophilic and lipophilic compounds in different phases, thereby protecting the drug from the external environment, as well as their low toxicity and lack of immunogenicity. Liposomes have been used to improve the therapeutic indexes of therapeutic agents by modifying drug absorption, reducing metabolism, prolonging biological half-life, and reducing toxicity. Drug distribution is thus controlled via the properties of the carrier rather than by the physicochemical characteristics of the drug substance. Modified liposomes are generally used for passive and active targeting of tumor sites. One of the most widely used and successful methods to obtain long-circulating, biologically stable liposomes is to coat the surface with inert polymers, such as polyethylene glycol (PEG). Doxorubicin (Doxil) was the first liposomal product approved by the U.S. Food and Drug Administration (FDA) which has a PEG polymer containing encapsulated doxorubicin [15]. Vesicular drug delivery systems have become one of the most widely used delivery systems due to their desirable properties and may offer potential for selective targeting [15–17].

3.3 Inorganic nanoparticles

Inorganic nanoparticles are composed of metal oxides or pure metals. Inorganic nanocarriers generally contain two regions, a core component with an inorganic substance and a shell region containing the oxide or the organic polymer layer (or metals) that protects the core region from the external biological microenvironment. Some examples of inorganic nanocarriers are gold, quantum dots (SiO_2), silica, and iron oxide [18, 19].

3.3.1 Silica and mesoporous silica nanoparticles

Silica and mesoporous nanoparticles are being increasingly used as an inorganic platform. This delivery system has a high surface-to-volume ratio that ensures a high loading capacity and homogeneous distribution of drug. To increase stability and protection from external stress or degradation, these nanoparticles are made up of a silica oxide framework. Silica-based nanoparticles have lower toxicity and higher biocompatibility than other metal oxides. Silica-based Cornell dots (C dots) received FDA approval in 2011 for phase 1 human trials [20, 21].

3.3.2 Gold nanoparticles

Gold nanoparticles have been used as a drug delivery system for therapeutic agents, in photodynamic therapy for cancer treatment, and as a diagnostic tool to detect biomarkers for various diseases. Gold nanoparticles can be used for both detection and direct cancer therapy with and without drug loading. These particles have received increased interest owing to their special properties in labeling biomolecules, delivery, and sensing. The photothermal properties of gold nanoparticles are used in photodynamic therapy, in which laser radiation with appropriate wavelength is shone on a tumor site containing gold nanoparticles, causing an increase in drug release near the tumor site, and resulting in localized therapy with lower nonspecific toxicity. Gold nanoparticles have several biomedical applications, including drug and gene delivery, cancer radiotherapy, and imaging of tumor cells. Clinical studies have used ultrafocal tissue ablation therapy with PEG-coated gold silica nanoshells, and gold nanoparticles containing active targeting ligands to target tumor sites have been tested in the clinic [18, 22].

3.3.3 Quantum dots

Quantum dots are optoelectronic materials with unique properties. They may be nanoparticles made from compound semiconductors, such as cadmium selenide or sulfide, or they may be made of silicon. Quantum dots can be used in cell culture to label live cells, prolonging stability compared with traditional organic dyes. Quantum dots are excellent tools for measuring the motility of cells via imaging. Compared with traditional organic dyes, quantum dots have better optical properties that have made a remarkable impact on cancer therapies. Because of the particle size of quantum dots, they are designed for simultaneous drug delivery, particularly for antitumor drug or gene delivery [18, 19], as well as in imaging and sensor applications [23].

3.3.4 Superparamagnetic iron oxide nanoparticles

Superparamagnetic iron oxide nanoparticles (SPIONs) have been successfully used as nanomedicines due to their superparamagnetic and ferromagnetic properties. The main advantages of SPIONs are minimal invasiveness, ability to access hidden tumors, and minimal side effects. SPIONs consist of an inner core of iron oxide (5–15 nm) and a silica outer shell (5 nm). Three types of iron oxide make up the inner core: magnetite (Fe_3O_4), maghemite (γ-Fe_2O_3), and hematite (α-Fe_2O_3). Each of these types of iron oxide has unique properties that are used to develop SPIONS with different molecular sizes and magnetic parameters. The outer shell is coated and coupled with different biocompatible functional groups, such as hydrophilic polymer (e.g., PEG) or polysaccharides. SPIONS are synthesized by coprecipitation, reverse microemulsion, and thermal decomposition methods. They can be used for cancer therapy with a variety of cargos, such as cytotoxic agents (chemotherapy), photosensibilizations (photodynamic therapy), and immune modulators (immunotherapy) [19, 24].

3.4 Other nanoparticle platforms

3.4.1 Lipid-polymer hybrid nanoparticles

Liposomes and polymeric-based systems have recently been integrated to introduce a novel drug delivery system known as lipid–polymer hybrid nanoparticles. These nanoparticles have additional advantages for cancer treatment owing to their unique features, which include a broad range of structural components, higher encapsulation, controlled drug release, biocompatibility, improved stability profile, and enhanced permeability and cellular uptake. Lipid–polymer hybrid nanoparticles can be used for nontargeted combination drug delivery, active targeted drug delivery, small interfering RNA (siRNA) delivery, and imaging agent delivery [25].

3.4.2 Dendrimers

Dendrimers are a type of polymeric, spherical macromolecule that have many branches originating from the center point in a spiderweb-like fashion. Self-assembled and formed by organic or inorganic hybrid nanoparticles, dendrimers can be attached to liposomes, nanoparticles, or carbon nanotubes through simple electrostatic interactions, encapsulations, and covalent conjugations. They are effective carriers for anticancer agents due to their small size (1–15 nm) and the fact that they are cleared through the kidneys, resulting in decreased in vivo toxicity. In recent years, polyamidoamine dendrimers have been widely used to combine liposomes and dendrimers into hybrid nanosystems for entrapping paclitaxel in the treatment of ovarian cancer [26]. A folate-PEGylated polyamidoamine dendrimer has been shown to enhance tumor selectivity for the delivery of doxorubicin [27]. In addition to these applications, dendrimers have also been used for gene delivery [28] and imaging technology for diagnostic applications, such as computed tomography imaging of cancer cells.

3.4.3 Carbon nanotubes

Carbon nanotubes can also be used as nanocarriers for the delivery of drugs and genes in cancer therapy. They are made by using carbon atoms from graphene sheets rolled into a seamless cylinder, which may be open-ended or capped. CNTs can be divided into single-walled nanotubes, which are made from a single sheet of graphene, and multiwalled nanotubes, made from multiple layers of graphene sheets. They are an ideal option for drug delivery because of their promising size, needle-like structure, and unique physicochemical properties. Some of the applications for CNT-drug interaction include absorption of the drug into CNT mesh, covalent or noncovalent conjugation of the drug onto the walls of CNTs, and catheters utilizing CNT channels. Limitations for CNTs applications include hydrophobicity and toxicity [29].

3.4.4 Exosomes

Exosomes comprise a subgroup of extracellular vehicles that enable intercellular communication between neighboring and distant cells. These particles are usually 30–150 nm in diameter, are secreted by cells into bodily fluids, and carry various biomacromolecules, including DNA, RNA, and proteins. Exosomes are surrounded by a phospholipid bilayer that is 30–200 nm in diameter and contains transmembrane proteins, lipid-anchored membrane proteins, peripherally associated membrane proteins, and soluble proteins of the exosome lumen. Exosomes are engineered at the cellular level through two main strategies: (1) manipulation of parent cells through genetics or metabolic engineering and (2) surface engineering of exosomes. Exosomes or exosome mimetics are currently being used as a promising drug carrier that offers advantages compared with liposomes and polymeric-based nanoparticles. Exosomes are nonimmunogenic and smaller in size, enabling penetration to deep tissues; have long circulation, the ability to target tissues, and strong biocompatibility; and may offer lower toxicity [30, 31].

3.4.5 Virus vectors

Virus vectors are commonly used by molecular biologists to deliver genetic material into cells. Of the several types of nonpathogenic virus vectors that have been used for gene therapy during clinical development, the most common are lentivirus, adenovirus (AdV), adeno-associated viruses (AAVs), and retroviruses. Although numerous virus vectors have been developed for clinical application, some general concerns surround their use, owing to their immunogenicity, oncogenicity, off-target effects, and potential for inflammation and insertional mutagenesis.

Lentiviruses make up a class of retroviruses (including HIV) carrying a transgene that integrates into the genome of the infected cell and permits stable expression in both dividing and nondividing cells. Generally, lentiviruses have been used for the production of induced pluripotent stem cells, as well as for direct gene therapeutics and as activating agents for immunotherapy. Recent years have seen an increase in preclinical investigations on in vivo and ex vivo delivery of gene therapies. Third-generation lentiviruses are becoming the vector of choice because of their high packaging capacity, stable gene expression in both dividing and nondividing cells, and low immunogenicity. Because lentiviruses can deliver large DNA molecules (approximately 8 kb), researchers are also interested in developing these vectors for CRISPR-Cas9 genome editing.

AdVs are often used as vectors for gene therapy because they can be grown in high titers, can accommodate large transgenes, have highly efficient transduction for most cell types, and do not integrate into the host genome. AdVs do have some disadvantages, however, including the high prevalence of preexisting immunity in humans, challenges in cloning due to larger genome size, and transient expression.

AAV vectors are a leading platform for gene therapy in many diseases because of their superior safety profiles and significantly longer and faster transgene expression compared with other virus vectors. Nine AAV serotypes have been identified with various levels of tropism for specific tissues and cell types. Due to their smaller size (approximately 20 nm) compared with lentiviruses and AdVs, AAVs have extraordinary diffusion capacity and thus are the best choice for tissues with small intercellular spaces, such as cancer cells, embryonic tissue, and tissue from brain regions, muscle, and kidney. The primary disadvantage of AAVs are their smaller packaging capacity, the difficulty of producing high titers for human clinical trials, and the risk of insertional mutagenesis due to vector integration. Currently, there are approximately 400 clinical trials using AAVs to treat a wide variety of diseases and disorders worldwide, and the number is steadily increasing.

4. Nanoparticle production technologies

Nanoparticles can be synthesized by a variety of methods, including bottom–up, top–down, and combination approaches. A broad classification of these approaches is summarized here. Fig. 3 illustrates approaches used for nanoparticle production.

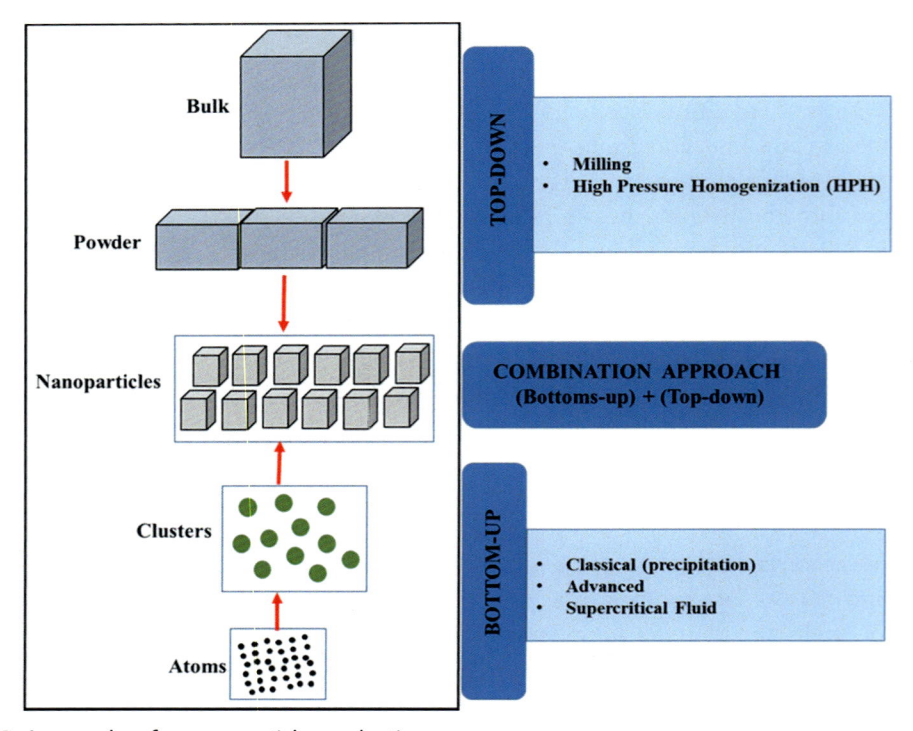

Fig. 3 Approaches for nanoparticle production.

4.1 Bottom-up approaches

In bottom-up procedures, nanomaterials are assembled from basic building blocks (molecules or nanoclusters). The molecule is dissolved in the solvent, and the conditions of the solution are changed so that the molecule starts to precipitate to form larger, stable structures. This approach makes it possible to control the size and shape of nanoparticles. Based on the phase type used during the production process, bottom-up approaches can be subdivided into liquid phase (precipitation process, sol-gel process, and hydrothermal process) and gas phase (aerosol process, flame hydrolysis, and spray hydrolysis).

The classical precipitation process (solvent-antisolvent approach) is a simple, cost-effective, and easy way to scale up the synthesis of nanomaterials. In the antisolvent precipitation method, the poorly soluble drug is first dissolved in a solvent, which is rapidly mixed with a solvent-miscible antisolvent, such as water. In recent decades, classical precipitation technologies have been further developed by academic and industry research groups to introduce advanced precipitation technologies. One interesting approach for advanced precipitation technology is evaporative precipitation into aqueous solution, in which the drug is dissolved in an organic solvent that is not miscible in water. The organic solvent (with drug) is later sprayed into heated water, resulting in immediate evaporation of the organic solvent and thus the instantaneous formation of drug nanoparticles. In the sol-gel process, material production involves the transition of a solution from a liquid solution ("sol") into a solid gel phase. The solution is usually prepared by using inorganic metal salts or metal-organic compounds (metal alkoxides).

Another bottom-up technique is based on supercritical fluid technology, which entails the use of supercritical fluids, i.e., gas converted to a supercritical fluid above its critical pressure and temperature. Among the number of gases used in this approach, carbon dioxide is considered to be the best choice because of its low critical point ($31.3°C$, $7.4\,MPa$), attractiveness for heat-sensitive materials, inertness, inexpensiveness, noninflammability, and safety. Supercritical carbon dioxide can be used as a solvent for active substances and excipients in techniques such as RESS (rapid expansion of supercritical solution), PGSS (particles from gas-saturated solution), RESOLV (rapid expansion of supercritical carbon dioxide solutions into a liquid solvent), RESAS (rapid expansion from supercritical to aqueous solution), and DELOS (depressurization of an expanded liquid organic solution). It can also be used as an antisolvent for the precipitation of active substances and excipients in an organic solvent, in techniques such as GAS (gaseous antisolvent), ASES (aerosol solvent extraction system), PCA (principal component analysis), SAS (supercritical antisolvent), ASAIS (atomization of supercritical antisolvent-induced suspension), and SEDS (solution-enhanced dispersion by supercritical fluids), or in methods based on assisted spray drying or aerosolization, such as CAN-BD (carbon dioxide–assisted nebulization with bubble drying) and SAA (supercritical assisted atomization). In the widely used SAS (supercritical antisolvent) method, the drug solution is sprayed through a nozzle in a closed pressurized system with supercritical carbon dioxide and undergoes precipitation as a result of rapid diffusion [32, 33].

4.2 Top-down approaches

In top-down approaches, nanomaterials are synthesized by breaking down larger drug crystals into smaller nano-sized crystals. These approaches include the techniques of ball (or wet) milling, high-pressure homogenization, ultrasonication, and laser fragmentation, all of which have been widely exploited for the synthesis of nanoparticles.

In the traditional approach of wet milling, a ball mill with a vessel or a vial filled with balls or rods is constructed from a variety of materials, such as ceramic, stainless steel, glass, or plastic polyamide. Using the principle of size reduction, a shear force reduces the particle size in the drug suspension. The size of the nanoparticles is usually controlled by coating their surfaces with stabilizers, such as surface modifiers (surfactants) or hydrocolloids (hydroxylpropyl methylcellulose), to prevent crystal aggregation. The first nanocrystal technology drug was approved by the FDA in 1999 for synthesizing sirolimus (Rapamune). Other drug products synthesized with wet milling techniques include aprepitant, fenofibrate, megestrol acetate, and paliperidone [34–36].

High-pressure homogenization (HPH), which combines both pressure and mechanical forces, is a disintegration method used to reduce the size of nanoparticles. In this technique, high pressure is used to reduce the size of drug particles in a liquid medium containing surfactants for stabilization. HPH methods can be classified into three types: (1) jet-stream homogenization (microfluidization), (2) piston-gap homogenization in water, and (3) piston-gap homogenization in water mixtures or in nonaqueous media.

Microfluidization is a high-pressure homogenization technique in which two fluid streams are made to collide under high pressure (up to 1.7×10^5 kPa). During the collision, small particles are generated as a result of particle, shear, and cavitation forces. Surfactants are added to the solution to stabilize the nanoparticle size during the process. In a study on particle size reduction for a BCS class II drug, particle size reduction depended on various process factors, including number of homogenization cycles, homogenization pressure, and stabilizer concentration. After 60 cycles, high homogenization pressure (30,000 psi [$\sim 2.1 \times 10^5$ kPa]) resulted in a significant reduction in particle size compared with that obtained using low-pressure homogenization (10,000 psi [$\sim 6.9 \times 10^5$ kPa]). Surfactant concentration (increase from 10 to 12 mg/mL) also played a key role in particle size reduction through particle stabilization by forming a thin layer around the newly formed surface. The major limitation of HPH is that a high number of cycles are required to obtain sufficient particle size reduction.

Piston-gap homogenization can be performed in water or in nonaqueous or water-reduced media. In aqueous-based piston-gap homogenization, drug is dispersed in an aqueous surfactant solution, and the suspension is forced by a piston through the tiny gap (5–20 μm, based on viscosity of the suspension and applied pressure). The aqueous phase moves at a very high streaming velocity, and as a result, the dynamic pressure rises and the static pressure falls below the vapor pressure of the liquid. Therefore, water begins boiling, leading to the formation of gas bubbles in a process known as "cavitation." The gas bubbles generate a high-power shock wave that reduces drug particle size. In non-aqueous piston-gap homogenization, nonaqueous media, such as oil or PEG, or water mixtures, such as glycerol-water or ethanol-water, are used. Because of the nonaqueous medium, cavitation is very limited due to the low vapor pressure. As a result, this homogenization process can be performed in a low-temperature setup, which could be an advantage for thermolabile drugs. For poorly soluble drugs, a process to blend the drug within a water-soluble excipient has been developed. On exposure to water, the blend (poorly soluble drug in water-soluble excipient) forms nanoparticles, which are stabilized by the soluble excipient [37, 38].

4.3 Combination approaches

The integration of top-down and bottom-up techniques is expected to provide a synergistic approach for the production of nanoparticles. While top-down approaches are better for producing long-range structures (macroscopic connections), bottom-up techniques are better for creating particles of less than 100 nm. The combination approach is a two-step process that begins with pretreatment (bottom-up) followed by HPH (top-down). Recently, several combination methods have been developed to improve particle size reduction, including Nanoedge (Baxter), which uses pretreatment with microprecipitation followed by HPH; H 69, pretreatment by microprecipitation immediately followed by HPH, or "cavi-precipitation"; H 42, pretreatment by spray drying followed by HPH; H 96, pretreatment by freeze-drying followed by HPH; and combinative technology, pretreatment by bead milling followed by HPH [39].

5. Characterization of nanoparticles

In contrast to the numerous types of nanoparticles, their characterization is been broadly divided into just two classifications: physiochemical and functional characterizations. Physiochemical characterization pertains to conformational and colloidal properties such as size, aggregation, self-assembly, charge (zeta potential), and entrapment efficiency, as well as surface topographic features such as surface area, morphology, and composition. Functional characterization is based on the fate of nanoparticles inside the human body, referring to factors such as toxicity and in vivo distribution. Because the chemistry of nanoparticles varies widely, ranging from polymeric to organic, inorganic, and liposomes, attempts to delineate similar physiochemical characteristics are difficult, apart from defining size, charge, and hydrodynamic diameter. For this reason, it may be more beneficial to identify the critical characteristics that may affect performance in each system or nanoparticle type. For example, physiochemical characterization of iron carbohydrate colloidal drug products requires an understanding of the magnetic properties of the iron core in order to measure overall uniformity and to detect impurities, which is not applicable to polymeric micelles. Similarly, the thermodynamic properties of lipids, such as phase transition temperatures, are evaluated for lipid nanoparticles, which is not essential for other types of nanoparticles. This section highlights the heterogeneity of various nanoparticle types, underscoring the difficulties in devising a common characterization method that can be used to compare and contrast various particle types.

5.1 Physiochemical characterization

The physiochemical characteristics of nanoparticles dictate their performance under physiological conditions. For example, toxicity is determined to some extent by particle size, surface area, charge, and composition of the nanoparticle. Owing to their small size and large surface area, the reactivity and catalytic activity of nanoparticles may be enhanced. With their high surface energy, nanoparticles given intravenously can interact with blood proteins to form nanoparticle protein complexes, which can affect pharmacokinetic parameters such as accumulation and biodistribution. The ability of nanoparticles with sizes smaller than 6 nm to penetrate nuclei of normal cells also poses some risk of toxicity [40]. Pan et al. [41] found that the toxicity of 1.4-nm gold nanoparticles is 60 times greater than that of 15-nm gold nanoparticles. Zhao et al. [42] demonstrated that mesoporous silica nanoparticles with surface areas of approximately 100 nm adsorbed to the surface of erythrocytes without disturbing the membrane or morphology, whereas adsorption of ~600-nm nanoparticles induced hemolysis.

There is growing evidence that the surface charge of nanoparticles directly influences toxicity, but may also have some benefits, such as increased tumor penetration and cellular uptake. For instance, it has been demonstrated that positively charged nanoparticles can bind more strongly to negatively charged DNA than negative nanoparticles, promoting cell death [43]. Surface composition and core composition are critical parameters because a nanoparticle's ability to interact with biological entities is largely dependent on these features. For example, metal ions such as silver and cadmium can be leaked from nanoparticles, damaging cellular pathways. On the other hand, the outer surface of a nanoparticle can be tailored to enhance the delivery of a drug to a target organ. Hung and coworkers [44] demonstrated that the uptake of doxorubicin by cancer cells when the drug was formulated in PLGA-coated nanoparticles with N-acetyl histidine glycol phosphate was greater compared with regular PLGA nanoparticles because of the increase in imidazole moiety, which caused attraction to the negatively charged cell membrane. In some cases, both size and structure of nanoparticles influence overall protein production. For example, Arteta et al. [45] demonstrated 130 nm size mRNA LNP's with nonhomogeneous lipid distribution showed higher protein expression in human adipocytes and hepatocytes compared to their counter parts.

Physiochemical characterization of nanoparticles is required during scale-up and manufacturing so that the final nanoparticle product is well-defined and can be reproducibly generated. Significant changes in physiochemical attributes upon storage may define the overall shelf-life of a nanoparticle-based product. In addition to final product tests, physiochemical characterization during early-phase development can help in identifying critical quality attributes, which may provide necessary information about robust manufacturing processes. For example, the FDA revised the liposomal draft guidance on Doxil (doxorubicin) and Ambisome (amphotericin B), highlighting the importance of the physiochemical attributes of "particle size" and "lipid degradants" as key areas of attention for performance and stability [46]. A series of analytical methods are usually necessary to perform collective characterization, because no single analytical technique can cover all types of physiochemical properties. Identifying and evaluating new analytical techniques for nanoparticle characterization is a continuously emerging field because of the variability and reproducibility of the measurements obtained with current technologies. An example is provided by a recent interlaboratory comparison of zeta potentials for surface charge measurements of silica, gold, polystyrene, and cerium oxide (inorganic nanoparticles mainly used in biomedical applications). The large variations found in nanoparticle measurements highlighted the need for new analytical techniques for measuring zeta potential [47].

The following sections describe standard analytical methods for characterizing the most common physiochemical properties of nanoparticles. Some of these methods provide information that complements that of other methods or that describes the unique characteristics of nanoparticles. Also discussed are novel analytical techniques that are required for each physiochemical attribute. Finally, keeping the diversity of nanoparticles in mind, recommended strategies for characterizing various nanoparticle types are presented in Table 2.

5.1.1 Particle size distribution

Particle size distribution is regarded as a key physiochemical attribute of nanoparticles. In an analysis of regulatory submissions, a published report from FDA's Center for Drug Evaluation and Research mentions that approximately 48% of FDA submissions on nanoparticles showed light-scattering methods (particularly static light scattering [SLS], dynamic light scattering [DLS], and nanoparticle tracking analysis) as the primary choice of sponsors for measurement of particle size distribution [48]. Laser diffraction and microscopy were the second major class of methods (approximately 44%) within FDA submissions for particle size distribution.

5.1.2 Dynamic light scattering

DLS is used to measure the diffusion coefficient of nanoparticles by analyzing fluctuations in the intensity of scattered light when a laser light is transmitted through a nanoparticle suspension. The diffusion coefficient is translated into hydrodynamic size by the Stokes-Einstein equation, using the correct viscosity of the buffer medium:

Table 2 General physiochemical attributes and corresponding standard and advanced characterization techniques.

Physiochemical attribute	Standard characterization techniques	Advanced characterization techniques
Size and size distribution	Dynamic light scattering Nanoparticle tracking analysis Asymmetric flow field fractionation-multiangle light scattering Laser diffraction	Nuclear magnetic resonance Optical tweezers
Shape	Transmission electron microscopy Atomic force microscopy	Cryogenic transmission electron microscopy
Surface morphology	Scanning electron microscopy Transmission electron microscopy Atomic force microscopy	High-resolution transmission electron microscopy
Surface charge/zeta potential	Electrophoretic mobility Laser Doppler velocimetry	Tunable resistive pulse sensing
Chemical composition/identity	Fourier-transform infrared spectroscopy Scanning electron microscopy-energy-dispersive X-ray	Microthermal gravimetric analysis
Dispersion of nanoparticles in matrices/supports	Scanning electron microscopy Atomic force microscopy Transmission electron microscopy	
Stability (aggregation, self-assembly)	Static light scattering Asymmetric flow field fractionation-multiangle light scattering Transmission electron microscopy	Cryogenic transmission electron microscopy

$$D = \frac{KT}{6\pi n R_h},$$

where n is the viscosity of the buffer, K is the Boltzmann constant (a physical constant that relates the average relative kinetic energy of particles in a gas with the temperature of the gas), R_h is the hydrodynamic radius, and T is the measurement temperature. Smaller particles will have a larger diffusion coefficient and vice versa. Two important parameters reported from this technique are (1) $Z_{average}$ (size) and (2) polydispersity index. Although some of the FDA submissions reported size as a cut-off range (upper or lower range), most preferred to use $Z_{average}$. The polydispersity index is used to measure the width of the distribution and is usually reported as a range from 0 to 1. If the sample is unaggregated and homogenous with a value of less than 0.1, it is regarded as monodisperse. If the sample is in an aggregated state with a value of greater than 0.1, it may be polydisperse and heterogenous.

5.1.3 Static light scattering

Static light scattering (SLS) is used to measure the intensity of scattered light to obtain the average molecular weight of a macromolecule. The average of the scattering intensity from a solution of nanoparticles is used to extract information about the molecular weight and radii of gyration of the nanoparticles. The scattering intensity from a series of concentrations of nanoparticles is measured for each concentration at numerous angles and extrapolated to zero angle and zero concentration to give a weighted average of the molecular weight and radius of gyration, using the Zimm or Debye equation. In the case of self-assembly, molecular weight increases with concentration. By combining SLS and DLS measurements, one may obtain the ratio of the hydrodynamic distance (R_g) to the hydrodynamic radius (R_h), which provides shape information. For spherical nanoparticles, the characteristic value is approximately 0.775, and more elongated structures are assumed when the $R_g{:}R_h$ ratio is greater than 0.775. SLS and DLS techniques are also coupled with separation methods, such as size exclusion and asymmetric flow field fractionation, to measure the size and shape of the monomeric and aggregated species in the solution [49].

5.1.4 Nanoparticle tracking analysis

Nanoparticle tracking analysis is used to measure the single-particle diffusion coefficient by tracking the Brownian motion of an individual particle in solution via high-resolution video acquisition and enhanced contrast microscopy. This technique is more appropriate than DLS for measuring polydisperse populations because size distributions obtained by DLS are biased toward larger components (intensity is proportional to [radius]^6). Despite these advantages, the applicability of nanoparticle tracking analysis is limited because it requires the sample concentration to be 10^8–10^9 particles per mL, which may differ from the sample concentration required for the intended application. This technique also poses some difficulties when used to characterize turbid samples because of the high background scattering from the medium, which masks the identity of the particles of interest.

5.1.5 Laser diffraction

Laser diffraction is used to measure particle size on the basis of the diffraction pattern that is generated by light scattering from particles at different angles. Raw light-scattering data are analyzed by Mie or Fraunhofer algorithms to generate volume distribution or mean particle size in terms of D10, D50, and D90 (i.e., the size, in microns, that results in a distribution of 10%, 50%, and 90%, respectively, of the measurements lying below that diameter). Although this technique has been used by many researchers, its major disadvantage is due to the fact that the conversion from volume distribution to number distribution is based on several assumptions that may give large erroneous results.

5.2 Advanced techniques for size characterization

Although most industrial applicants use the techniques just described, the use of optical tweezers is gaining momentum as an advanced technique for particle characterization. Optical tweezers are instruments used to trap particles with a strong beam of light in the Rayleigh or Mie region. Once the particles are trapped, the type and displacement of the particle is measured by laser intensity shifts, using high-quality numerical aperture optics. Recently, Curry et al. [50] from the National Institute of Standards and Technology developed a high-throughput system by using an array of optical tweezers and low-numerical-aperture optics to achieve a size resolution of 1 nm or less. In other research, Aziz et al. [51] studied gold nanoparticles by using a new optical tweezers design.

5.2.1 Transmission electron microscopy

Transmission electron microscopy (TEM) is a high-resolution technique used to reveal structural details, size distribution, and morphology of nanoparticles made up of lipids and proteins and hard nanoparticles composed of metallic particles, carbon, or plastics. TEM is based on the interaction between a high-energy electron beam and a thin sample. The amplitude and phase variations of the transmitted beam provide images of the nanoparticle. Based on the sample preparation, TEM can be classified as negative-stain, freeze-fractured, and cryogenic TEM. In negative-stain TEM, nanoparticles are adsorbed onto an electron microscope support and a heavy-metal stain is applied. Although this variation of TEM is easy to use, its application in nanoparticle research is limited by artifacts such as aggregation, particle collapse, and nonuniform deposition during sample preparation. Another limitation of this technique is due to difficulty in controlling the adsorption of nanoparticles to the electron microscope support.

A simple variation of the negative stain was developed by Hacker et al. [52], who embedded aqueous suspensions of nanoparticles within hydrophilic films composed of methylcellulose. This approach makes it possible to control aggregation and patchy distributions and hinders particle collapse while retaining the native structure of the nanoparticles. In addition, the use of methylcellulose allows stabilization of metal calibration particles in the hydrophilic films, facilitating estimations of nanoparticle numbers and concentrations with precision and sensitivity.

In freeze-fractured TEM, the sample is sandwiched between two copper holders, vitrified via rapid freezing, and visualized under a TEM microscope. Although this method has some advantages over negative-stain TEM (e.g., no need for drying), artifacts may result from insufficient freezing rates and redeposition of the solvent molecules.

In cryogenic TEM, a nanoparticle suspension is solidified by rapid freezing at cryogenic temperatures, and the specimen is visualized in the frozen state. Advantages of this approach include preservation of aggregation state and hydration, as well as removing the need for heavy-metal contrast. Cryogenic TEM is particularly useful in understanding the morphological features of a drug's inner liposomes, especially when regulatory agencies recommend controlling the morphology of the final product. One such example is demonstrating the "coffee bean"-like shape of doxorubicin inside liposomes [53]. Cryogenic TEM is also required to confirm the unusual behavior of self-assemblies due to sample preparation artifacts encountered with conventional TEM. For example, one study using negative-stain TEM showed that cellulose nanoparticles laterally self-assembled to form flat objects, but cryogenic TEM was required to confirm that self-assembly was not an artifact in the sample preparation [54]. Cryogenic TEM, thus, can provide a qualitative observation, although quantification remains challenging because the use of lower doses of electrons limits resolution.

5.2.2 Nuclear magnetic resonance

Nuclear magnetic resonance (NMR) is a spectroscopic technique used to characterize the spin state of active nuclei under the influence of a strong magnetic field. Excitation of the nuclei produces a spectrum of absorption intensity versus magnetic-field strength. NMR is a powerful tool for understanding the interaction of nanoparticles with their surrounding molecules. It has been extensively used to study the spectra of molecules with sizes in the range of 1–2 nm. Usage with larger nanoparticles (>2 nm), however, results in a slower tumbling motion of the molecules, leading to broadening of the overall peak line (overlapping resonances) of the NMR spectrum. This in turn causes signal losses over the course of NMR experiments. Significant improvements in NMR technology, as well as in labeling strategies for NMR active nuclei (^{15}N and ^{13}C), circumvent this issue by measuring heteronuclear resonances between ^{15}N and ^{1}H or ^{2}H nuclei and analyzing the data in two- or three-dimensional spectra. With these advancements, this technique can be used to understand two particular features of nanoparticles: (1) interaction of the molecules within themselves (self-assemblies), or with any partner molecules on their surfaces; and (2) the size properties of the molecules via measurement of the diffusion coefficient.

Calzolai et al. [55] studied protein ubiquitin interaction with gold nanoparticles by using two-dimensional ^{15}N-^{1}H heteronuclear single-quantum coherence (HSQC) NMR experiments. They compared HSQC NMR spectra of a ubiquitin gold nanoparticle complex with gold nanoparticle alone and observed chemical shift perturbations only in the ubiquitin group, confirming a specific interaction of gold nanoparticles with ubiquitin. Coelho and colleagues [56] studied the structure and physiochemical properties of gold nanoparticles modified with PEG and their interaction with the proteasome inhibitor bortezomib. Their results confirmed that bortezomib was included into nanoparticles, either by association with the PEG chains or by direct adsorption onto the surface of the gold nanoparticles. These findings provided insight into the overall structure of bortezomib-loaded, PEG-coated gold nanoparticles.

NMR, based on diffusion coefficient or diffusion-ordered spectroscopy NMR, can be useful for studying metallic nanoparticles, which are in the size range of 1–5 nm. By using incremental gradients of the magnetic pulse field, NMR signal amplitudes of the labeled nuclei change depending on their diffusion coefficients. Using Laplace inverse transformation, the signal decays are analyzed to construct the diffusion-ordered NMR spectrum. The diffusion coefficient can be converted into hydrodynamic size by using the Stokes-Einstein equation. Zhang et al. [57] studied the interaction of polyallylamine hydrochloride with diamond nanoparticles, using diffusion-ordered NMR, to identify the surface-tethered polymer in comparison with solution-phase polymer. Their results provided detailed information on molecular distribution, which led to future work on the interaction of nanoparticles with cellular membranes.

5.2.3 Zeta potential

Zeta potential is a measure of surface charge that is regarded as a valuable parameter by both the scientific community and regulatory agencies. It is a measure of the difference between a nanoparticle's electrostatic potential and that of the bulk solution. To obtain the zeta potential, a laser is passed through a nanoparticle solution under the influence of a varied electric field. The shift in frequency, or phase shift, of the scattered light caused by the Doppler effect is measured and related to electrophoretic mobility, which in turn is related to the zeta potential using Henry or Smolochuski approximations. Data analysis with existing methodologies becomes more problematic when characterizing highly aggregated samples or when measuring nanoparticles in high salt buffer conditions. A novel methodology, tunable resistive pulse sensing, circumvents these issues by simultaneously measuring zeta potential, size, particle shape, concentration, and anisotropy. Zeta potential can also be used to infer the stability of nanoparticles as well as changes in surface charge during formulations. For example, Vogel et al. [58] studied the zeta potential of magnetic nanoparticles and magnetic nanoparticles modified with RNA. They also measured the zeta potential of distearoyl phosphocholine (DSPC) and dimyristoyl phosphatidylglycerol (DMPG) liposomes in different molar ratios and found that the mixed liposomes became increasingly negative with an increase in DMPG content.

5.2.4 Chemical composition

Fourier-transform infrared spectroscopy is routinely used to determine the surface composition of nanoparticles. This technique is based on the absorption of mid-infrared radiation (4000–400 cm^{-1}) by the functional groups within the molecule, which provides a characteristic absorption spectrum that can be analyzed to determine molecular structure

Table 3 Recommended characterization strategies for classified nanoparticles.

Characterization tool	Polymer NP	Lipid NP	Inorganic	Virus vectors, exosomes
Dynamic light scattering	✓	✓	✓	✓
aF4–MALS	✓	✓	✓	✓
Nanoparticle tracking analysis	✓	✓	✓	✓
Optical tweezers	✓	✓	✓	✗
Transmission electron microscopy	✓	✓	✓	✓
Cryogenic transmission electron microscopy	✓	✓	✓	✓
Zeta potential	✓	✓	✓	✗
Fourier-transform infrared spectroscopy	✓	✓	✓	✗
SEM-EDX	✓	✓	✓	✗
Nuclear magnetic resonance	✓	✓	✓	✗
Microtransmission electron microscopy	✓	✓	✓	✗

Abbreviations: *aF4*, asymmetric flow field fractionation; *EDX*, energy-dispersive X-ray; *MALS*, multiangle light scattering; *NP*, nanoparticle; *SEM*, scanning electron microscopy.

and interactions. Another commonly used technique is microthermogravimetric analysis, in which the physical properties of the nanomaterials (mass change) are recorded as a function of temperature. Scientists at the National Institute of Standards and Technology have used this technique to obtain information about the purity of carbon nanotubes and the presence of PEG on silicon oxide nanoparticles [59].

In light of the diversity of the nanoparticles classified in the preceding sections, we recommend the characterization strategy shown in Table 3 for use during the initial stages of nanoparticle development.

6. Applications of nanotechnology

6.1 Oral and parenteral drug delivery

Pharmaceutical drug development has been the greatest benefactor of nanoparticle technologies in the life sciences. Reducing the particle size of a substance greatly improves the surface area, thereby allowing modulation of the dissolution properties of pharmaceutically active ingredients in the nano dimension. Table 4 summarizes the FDA-approved parenterally and orally administered drug products that were developed with nanotechnology approaches.

Absorption barriers pose one of the main challenges in drug delivery. The size of nanoparticles allows them to pass through the blood brain barrier [60] and enables their

Table 4 FDA-approved products based on nanosizing technologies.

Drug	Trade name	Indication	Company
Rapamycin, sirolimus	Rapamune	Prevention of kidney transplant rejection	Wyeth, Pharmaceuticals
Aprepitant	Emend	Chemotherapy-induced emesis	Merck
Fenofibrate	TriCor, Triglide	Hypercholesterolemia, mixed dyslipidemia	Abbott, Sciele Pharma
Megestrol	Megace ES	Anorexia, cachexia	Par Pharmaceutical
Paliperidone palmitate	Invega, Sustenna, Xeplion	Schizophrenia	Janssen
Griseofulvin	Gris-PEG	Ringworm infection	Valeant Pharmaceuticals
Nabilone	Cesamet	Chemotherapy-induced emesis	Eli Lilly, Valeant Pharmaceuticals

Modified from P. Tyagi, J.A. Subramony, Nanotherapeutics in oral and parenteral drug delivery: key learnings and future outlooks as we think small, J. Control. Release: Official Journal of the Controlled Release Society 272 (2018) 159–168.

use for retroorbital delivery [61] and tumor targeting. The latter use, one of the most-studied areas in nanotechnology research, is the focus of this section.

6.2 Tumor targeting

Nanotechnology techniques have improved the ability to develop parenteral dosage forms of these agents. Notable nanoparticle-based drugs are doxorubicin, paclitaxel, vincristine, daunorubicin, and irinotecan in the United States; Myocet, a liposome-encapsulated form of doxorubicin, and mifamurtide in Europe; Geneol-PM, a polymeric nanoparticle micelle formulation of paclitaxel, in Korea; and SMANCS [poly(stylene-*co*-maleic acid)-conjugated neocarzinostatin] in Japan [62].

Over the last two decades, considerable attention has been focused on using nanoparticle technology to target parenterally administered drugs to tumors to overcome off-target toxicity. The so-called "enhanced permeation and retention effect" is the ability of nanoparticles of approximately 100 nm in size to escape leaky vasculature and become localized in tumors while evading lymphatic drainage. In this route of drug targeting, therapeutically active molecules are encapsulated in a polymeric or lipid matrix and then injected systemically. The molecules can also be conjugated with releasable linkers. The amount of drug in the systemic circulation versus that in the tumor is measured to determine drug specificity and bioavailability to the tumor tissue. This method of passive targeting has been shown to considerably enhance drug uptake by the tumor, and several original research studies have documented this effect in animal models [35]. However, passive targeting with nanoparticle technology has met with only moderate success in translational clinical studies, due not only to the heterogeneity of the tumor vasculature

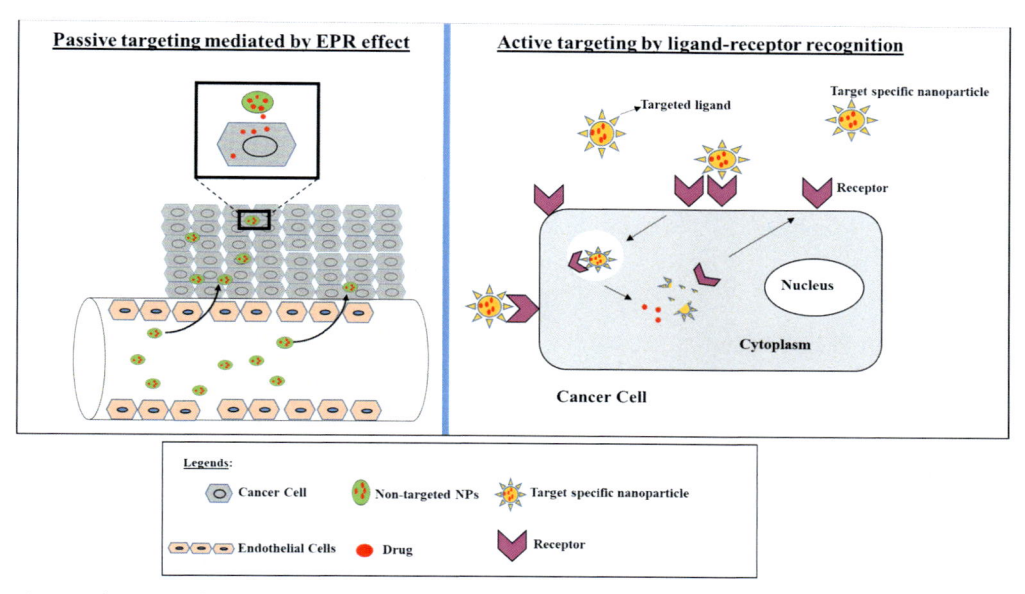

Fig. 4 Schematic showing active and passive targeting mechanisms.

but also to very low accumulation in tumor, which results in large off-target uptakes. Fig. 4 illustrates active and passive targeting mechanisms.

To overcome these limitations, an active targeting strategy has been introduced in which ligands such as proteins (antibody and its fragments) and nucleic acids, or other ligands such as peptides, carbohydrates, and vitamins, are used to target tumor tissues with nanoparticles containing active agents [63]. The ligands are attached to the surface of nanoparticles such that they can selectively recognize and bind to the receptors that are overexpressed on tumor targets. In principle, this method can promote accumulation of active molecules within tumor tissues, thereby reducing off-target toxicities.

Cellular targets that are being considered for active targeting strategies include human epidermal growth factor receptors (HER), prostate-specific membrane antigens, transferrin receptors, folate receptors, vascular endothelial growth factors, integrins, vascular cell adhesion molecule 1, and stromal cells (fibroblasts and macrophages). Table 5 lists some active targeting strategies that employ nanoparticle systems and the corresponding targets that are currently being investigated in preclinical and clinical studies. Although most of the novel nanoparticle drugs are based on the active targeting approach, only a small percentage have progressed from the preclinical to the clinical phase. The limited translation of these drugs to the clinic has been attributed to factors such as overexpressed receptors in normal healthy tissues and endosomal escape. Other factors, such as the hard parenchyma of the blood brain barrier [73] and adsorption of proteins to nanoparticles in the biological milieu, have also been found to be contributing factors in specific cases. To add

Table 5 Some actively targeted nanoparticles registered in clinical and preclinical studies.

Study and disease indication (clinical trail identifier)	Targeting ligand	Clinical Phases (present status)	Nanoparticle drug component
Phase I Study of TENPA in Advanced Solid Cancer (NCT02979392)	B-tubulin in microtubules	1 (terminated)	Paclitaxel encapsulated with human serum albumin–hemin complex stabilized by noncovalent interactions [63,64]
MesomiR 1: A Phase I Study of TargomiRs as 2nd- or 3rd-Line Treatment for Patients with Recurrent MPM and NSCLC (NCT02369198)	EGFR	1 (completed)	mIR-16 encapsulated with EngeneIC minivectors to mimic microRNA [65]
Safety study of CALAA-01 to treat Solid-Tumor Cancers (NCT00689065)	Transferrin receptor	1 (terminated)	siRNA encapsulated with a linear cationic cyclodextrin polymer with a hydrophilic polymer consisting of adamantane PEG [66]
Anti-EGFR Immunoliposomes in Solid Tumors (NCT01702129)	EGFR	1 (completed)	Doxorubicin encapsulated with anti-EGFR immunoliposomes constructed by covalently linking cetuximab Fab region to liposomal membrane [67]
Phase II Study of Combined Temozolamide and SGT-53 for Treatment of Recurrent Glioblastoma (NCT02340156)	Transferrin receptor	2 (terminated)	Cationic liposome encapsulating human wild-type p53 targeted with antitransferrin receptor single-chain antibody [68]
A Phase 2 Study to Determine the Safety and Efficacy of BIND-014 (Docetaxel Nanoparticles for Injectable Suspension), Administered to Patients with Metastatic Castration-Resistant Prostate Cancer (NCT01812746)	Prostate-specific membrane antigen	2 (completed)	Docetaxel polymeric nanoparticles [69]
Safety and Pharmacokinetic Study of MM-302 in Patients with Advanced Breast Cancer (NCT01304797)	HER2	1 (on-going)	HER2-targeted PEGylated antibody–liposomal doxorubicin conjugate [70]
R547 cyclin–dependent kinase ATP inhibitor targeted	Transferrin receptor	Preclinical	Polymeric PEG-PE micelles loaded with R547 [71]

to the complexity, poor correlation between in vitro and in vivo findings due to inaccurate predictive models in preclinical settings has also hindered reproducibility [74].

In addition to active and passive targeting approaches, targeted drug release from nanoparticles can also be achieved by leveraging the local environment of the tumor, including factors such as low pH, redox gradients, and external stimuli such as light, temperature, and magnetic fields. Encapsulating drugs with heat-sensitive lipids such as DSPC allow a drug to remain stable in normal physiological conditions, and upon heating, DSPC undergoes phase changes and becomes more permeable to the drug. ThermoDox (Celsion), a proprietary technology using a thermally sensitive liposome to encapsulate doxorubicin, was developed with this technique and has exhibited higher efficacy compared with its counterparts in hepatocellular carcinoma.

6.3 Nanoparticles as immunomodulating agents

Although conventional applications of nanotechnologies involve the delivery of therapeutic drugs to a target site, some nanoparticles can be engineered to possess immunomodulatory properties that mimic the biological interactions between antigen–presenting cells (APCs) and T cells. This is achieved by modifying nanomaterial composition or geometry or by docking the surface with appropriate immuno-oncology agents. For example, lipid nanoparticles can be engineered to form interbilayer-cross-linked multilamellar vesicles that can efficiently pack antigens and adjuvants to co-deliver the antigen load to APCs, thereby stimulating an antitumor T-cell response. These vesicles showed higher potency than did simple liposomes or multilamellar vesicles in antibody production [75]. In other research, Kuai and coworkers [76] developed a high-density lipoprotein-mimicking nanodisc platform personalized with patient-specific neoantigens to co-deliver antigens and adjuvants to APCs in order to enhance $CD8^+$ cytotoxic T-lymphocyte responses. They found that, when combined with anti-CTLA-4 and anti-PD-1 antibodies, nanodiscs completely eliminated B16-F10 and MC-38 tumors. The propensity of nanoparticles to selectively bind to serum protein albumin to form protein corona, and their preferential uptake by phagocytic cells for deposition in the spleen, enhanced the immune response.

Nanoparticles have also shown great potential for delivering chimeric antigen receptor (CAR) genes into T cells. Smith et al. [77] developed synthetic DNA nanocarriers by coupling anti-CD3 F(ab′)2 fragments with poly(β-amino ester) polymers, which facilitated T-cell uptake. To transport the DNA cargo into the nucleus, they functionalized the polymer with peptides containing microtubule-associated sequences and nuclear localization sequences. These nanoparticles showed minimal toxicity with shelf lives of more than 1 year.

During cancer progression, tumor cells are known to alter the microenvironment, secreting antiinflammatory cytokines such as tumor growth factor beta (TGF-β), by which they suppress the activity of immune-effective cells. Xu et al. [78] developed a system for co-delivery of liposome protamine hyaluronic acid containing siRNA nanoparticles for immune-suppressive cytokine TGF-β, using mannose-modified lipid calcium phosphate nanoparticles to deliver TRP-2 tumor antigen and CpG oligonucleotide adjuvant to the dendritic cells. Although lipid–core–peptide vaccine alone was ineffective during their initial studies, priming the tumor environment with lipid–polymer hybrid nanoparticles inhibited tumor growth by 52%.

Nanoparticle surfaces can also be engineered to activate tumor-associated macrophages to phagocytize cancer cells, thereby enhancing the antigen presentation process by macrophages. In one notable study, the cell surface of a carboxylated polymer nanoparticle was co-conjugated with anti-HER-2 antibody and calreticulin in a 1:3 ratio [79]. Once the nanoparticles bind to the tumor cells, calreticulin is thought to produce an "eat me" signal to induce phagocytosis of macrophages [79].

In another emerging area, nanoparticles are used to modulate specific steps in the immune activation cascade. For example, carboxymethyl dextran iron oxide nanoparticles, originally approved for iron deficiency, have been shown to polarize tumor-associated macrophages into proinflammatory M1 phenotypes, reducing tumor growth by 42% [80]. In another study, manganese dioxide nanoparticles encapsulated in hyaluronic acid and coated with mannan polysaccharide were shown to alter the tumor hypoxic environment [81]. This was achieved through the interaction of nanoparticles with H_2O_2 in the tumor tissues to produce oxygen, thereby alleviating hypoxia and increasing the pH of the tumor microenvironment. The authors also reported that the presence of a hyaluronic acid layer facilitated the differentiation of M2-type macrophages to the M1 type [81]. Tumor progression involves conversion of M1- to M2-type macrophages, which suppresses infiltration of $CD8^+$ T cells. This work also demonstrated a shift of tumor-associated macrophages to the M1 type, which can exert antitumor effects [81].

6.4 Nanoparticles in the evolving landscape of nonviral gene therapy

Nonviral gene delivery systems using DNA, mRNA, and siRNA represent an exciting approach for the treatment of cancer and other diseases. mRNA and siRNA have greater potential as carriers than DNA because they do not integrate into the nucleus, but they are also susceptible to degradation by endonucleases and therefore cannot be internalized by cells. One way to circumvent this issue is to envelop nucleic acid cargos with virus- or nonvirus-based platforms for endosomal escape. Viruses that are used as delivery platforms include AAV, lentivirus, and Sendai virus; nonvirus delivery platforms include lipid-based (solid lipid nanoparticles and liposomes), polymer-based, and hybrid lipid-polymer-based nanoparticles. Although both delivery platforms have shown efficacy in many preclinical and clinical studies, nonvirus platforms are preferred over virus-based platforms, which are known to induce immunogenicity. Cationic liposomes were among the earliest nonvirus platforms to be explored and showed some success due to their biocompatibility [82]. However, it has been demonstrated that cationic lipid heads cause toxicity and immunogenicity [83]. In addition, anionic serum proteins can adsorb onto positively charged liposomes, compromising the delivery of nucleic acid to cells.

To address these challenges, Cullis et al. [84] developed ionizable lipids as delivery vehicles for mRNA. Ionizable lipids possess pH-dependent charge profiles, acquiring a positive charge at low pH and a neutral charge at physiological pH. This feature enables them to condense with mRNA in acidic buffers and subsequently fuse with negatively charged endosomes, releasing nucleic acid into the cell. Ionizable lipids drew increased attention after the initial approval of the siRNA drug patisiran, which contains MC3 as an ionizable lipid component. Oberli et al. [85] engineered an ionizable lipid with other components, such as phospholipids, cholesterol, a PEG-containing lipid, and an additive, to deliver an mRNA vaccine and induce a cytotoxic T-cell response. Upon subcutaneous administration into mice, the nanoparticles not only showed transfection in dendritic cells, macrophages, B cells, and neutrophils, but also induced CD8 T-cell proliferation. Some researchers have also pointed out that the ratio of the ionizable lipid to RNA or to other components can be adjusted to modulate the entrapment efficiency of mRNA. For example, Ball et al. [86] studied five formulations with increasing ratios of ionizable lipids to mRNA to co-deliver siRNA and mRNA into HeLa cells and mice and to measure the delivery efficacy of factor VII and luciferase. When compared with siRNA and mRNA nanoparticles alone, co-delivered nanoparticles with the highest ratio of ionizable lipid to RNA showed greater expression for both factor VII and luciferase.

There has been increasing research activity by the pharmaceutical industry in the area of advancing RNA therapeutics (Table 6). For example, a biodegradable ionizable lipid has been developed for delivering mRNA to nonhuman primates. In a preclinical study, it was demonstrated that triplet mixtures of the lipid-delivered mRNA encoding interleukin 23 (IL-23), IL-36γ, and OX40 ligand produced tumor regression in various tumor environments [87]. Currently, this study is in phase 1 trials. In other work, an ionizable lipid was developed for loading both Cas9 mRNA and single-guide RNA (sgRNA) for transthyretin gene and other applications [88]. In addition to lipid nanoparticles, polymeric nanoparticles, dendrimers, cell-penetrating peptides (protamine active), and other materials, including combinations of cationic and zwitterionic amino lipids, are also being explored for mRNA delivery.

6.5 Hybrid nanoparticles for RNA delivery

Despite the remarkable clinical and commercial success of polymeric- and lipid-based nanoparticles in the field of oncology, these molecules also have certain limitations,

Table 6 Some current clinical trials for mRNA-based nonvirus anticancer gene therapy.

Study title (ClinicalTrials.gov ID)	Protein target(s)	Delivery vehicle	Disease
Dose Escalation and Efficacy Study of mRNA-2416 for Intratumoral Injection Alone and in Combination with Durvalumab for Patients with Advanced Malignancies (NCT03323398)	OX40 ligand	Lipid nanoparticles	Solid tumor of lymphoma
Study to Evaluate the Safety & Tolerability of MRT5005 Administered by Nebulization in Adults with Cystic Fibrosis (RESTORE-CF) (NCT03375047)	Cystic fibrosis transmembrane conductance regulator	Lipid nanoparticles	Cystic fibrosis
Safety, Tolerability, and Immunogenicity of mRNA-4157 Alone in Subjects with Resected Solid Tumors and in Combination with Pembrolizumab in Subjects with Unresectable Solid Tumors (KEYNOTE-603) (NCT03313778)	Neoantigen epitopes	Lipid nanoparticles	Solid tumors
Safety and Efficacy Trial of an RNActive-Derived Prostate Cancer Vaccine in Hormone Refractory Disease (NCT00831467; EudraCT 2008-003967-37)	Trimeric autotransporter adhesin Prostate-specific antigen Prostate stem cell antigen Prostate-specific membrane antigen STEAP1	RNA active protamine	Prostate carcinoma
HPV Anti-CD40 RNA Vaccine (HARE-40) (NCT03418480)	Trimeric autotransporter adhesin Prostate-specific antigen Prostate stem cell antigen Prostatic acid phosphatase Prostate-specific membrane antigen MUC1 STEAP1	RNA active protamine	Prostate carcinoma

Table 6 Some current clinical trials for mRNA-based nonvirus anticancer gene therapy—cont'd

Study title (ClinicalTrials.gov ID)	Protein target(s)	Delivery vehicle	Disease
LipoMerit (NCT02410733)	Trimeric autotransporter adhesin NY-ESO-1 antigen MAGE-A3 Tyrosine Putative tyrosine-protein phosphatase	Lipo-Merit DOTMA (DOTAP)/ DOPE lipoplex	Advanced melanoma
CT7, MAGE-A3, and WT1 mRNA-Electroporated Autologous Langerhans-Type Dendritic Cells as Consolidation for Multiple Myeloma Patients Undergoing Autologous Stem Cell Transplantation (NCT01995708)	CT7 MAGE-3 WT1 mRNA-electroporated Langerhans cells	Dendritic cell–loaded mRNA	Malignant melanoma

Abbreviations: *DOPE*, 1,2-dioleoyl-*sn*-glycerol-phosphoethanolamin; *DOTAP*, 1,2-dioleoyl 3-trimethylammonium propane; *DOTMA*, 1,2-di-*O*-octadecenyl-3-trimethylammonium propane; *MAGE*, melanoma antigen gene; *mRNA*, messenger RNA; *MUC1*, mucin 1; *NY-ESO-1*, New York esophageal squamous cell carcinoma 1; *STEAP1*, six transmembrane epithelial antigene of prostate 1; *WT*, wild type.
Modified from P.S. Kowalski, A. Rudra, L. Miao, D.G. Anderson, Delivering the messenger: advances in technologies for therapeutic mRNA delivery, Mol. Ther.: The Journal of the American Society of Gene Therapy 27(4) (2019) 710–728.

including high burst release and a short-duration release period during anticancer therapy. To overcome these limitations, a new hybrid particle, a polymer-lipid encapsulation matrix, has been designed that not only retains the benefits of polymeric- and lipid-based nanoparticles, but also offers additional advantages that preclude the shortcomings of lipid- and polymer-based nanoparticles. For example, Kaczmarek and colleagues [89] demonstrated that systemic delivery of mRNA to the lungs with hybrid lipid-polymer nanoparticles consisting of poly(β-amino esters) with lipid PEG produced greater in vitro potency when compared with delivery by lipid particles alone. In another preclinical study, Stadler et al. [90] reported that hybrid polymer-lipid nanoparticles used to deliver modified mRNA induced the expression of bispecific antibodies, resulting in tumor cell lysis in xenograft mice models.

6.6 Virus vectors as nanoviruses for nucleic acid delivery

The underlying strategy for virus vector–mediated gene therapy is the reengineering of a virus into a therapeutic gene. Of the virus vectors currently being explored for a variety of diseases, lentivirus and oncolytic viruses such as AdV and AAV are among the most notable. With the approval of voretigene neparvovec (Luxturna), AAV vector–based gene therapy is also gaining momentum for the treatment of a variety of diseases, although it remains in the preclinical stage for oncology applications. One study reported that a recombinant AAV vector expressing tumor necrosis factor–related apoptosis-inducing ligand (TRAIL/Apo2L) prevented tumor growth in mice by activating caspase-3 and cytochrome c from mitochondria [91]. Encouraging results were also reported in a study in which tumor vascularization was prevented by using AAV to express angiostatin [92].

Despite the high transduction efficiency of virus vectors, their application in cancer treatment has met with challenges in systemic delivery, such as rapid clearance from the bloodstream and insufficient accumulation of vectors in target tissue. In addition, administration of AAV and other virus vectors directly into tissue is known to result in diffusion from the target tissue to nontarget tissue, necessitating multiple doses to maintain therapeutic efficiency. Repeat dosing of virus vectors can also result in the development of neutralizing antibodies, attenuating therapeutic efficiency.

To address these issues, virus vectors have been encapsulated into magnetic nanoparticles, hydrogels, microspheres, and microparticles. Of these, magnetic nanoparticles are most advantageous because of their potential for use as contrasting agents for magnetic resonance imaging and the ability to use them for targeted delivery. Sapet et al. [93] developed a magnetic hybrid nanoparticle and complexed it to a replication-defective AdV for in vivo and in vitro cell transduction. The expression of green fluorescent protein in CAR-positive U251N cancer cells was taken to indicate the delivery of the AdV vector. This finding was confirmed by magnetic resonance contrast imaging, on which CAR-positive cells showed dark positive contrast. Kim et al. [94] encapsulated AAV vectors into magnetic nanoparticles and successfully delivered them into human neuronal stem cells. This approach holds promise for future delivery of virus vectors, despite its slow progress into human translation.

6.7 Future avenues: Nanoparticles for delivering CRISPR-Cas9 gene editing systems

Genome editing tools such as zinc finger nucleases, transcription activator factor-like nucleases (TALEN), and CRISPR-Cas9 systems offer versatile approaches for inducing the inactivation of functional genes in mammalian cells. Of these approaches, CRISPR-Cas9 systems offer the unique advantage of reducing off-target toxicity and can be used to introduce precise and specific incisions in a DNA molecule, allowing natural repair of the genes. The targeted CRISPR-Cas9 tool is made up of two distinct units, a Cas9 enzyme endonuclease and an sgRNA. Three delivery formats have been identified for CRISPR-Cas9 gene editing: (1) delivery of the CRISPR-Cas9 plasmid in one vector, (2) mixture of CRISPR-Cas9 mRNA and sgRNA, and (3) CRISPR-Cas9 ribonucleoprotein and sgRNA. These components can be directly delivered ex vivo to cells harvested from patients before being reintroduced into patients as CAR-T cells, hemopoietic cells, or germline cells. Direct in vivo delivery to tumor cells is not possible, however, because of the high molecular weight and complexity of the system, which is considered to be one of the major obstacles to its clinical translation. Currently, virus vectors such as AdV, AAV, and lentivirus, as well as nonvirus vectors such as liposomes, polymers, exosomes, and gold nanoparticles, offer promising approaches for delivering CRISPR–Cas9 machinery in vivo.

Lentivirus vectors were the first virus vectors to be explored for the delivery of CRIPSR–Cas9 to introduce genome-wide, targeted mutations into mouse embryonic stem cells and demonstrated exceptional efficiency and specificity over genetic screening with RNA interference technology [95, 96]. Although some studies showed that persistent expression of Cas9 protein produced some off-target effects, these were circumvented by the design of self-inactivating constructs in which one sgRNA codes for Cas9 and limits its expression while the another sgRNA acts against the target sequence [97]. The use of recombinant AAV vectors for gene editing has also increased, owing to their low immunogenicity, serotype-specific tissue targeting, and low toxicity in preclinical models. However, the gene coding for CRISPR–Cas9 ($\sim$4.2 kb) is close to the packaging capacity of AAVs (4.7 kb), leaving less room for coding other promoters [98]. This issue was addressed by splitting Cas9 and other regulatory elements into two AAV vectors and delivering them together. The two-vector approach was demonstrated by Wang et al. [99] in preclinical models for correcting gene mutations that cause deficiency in the enzyme ornithine transcarbamylase. Zhan et al. [100] recently developed CRIS-PReader, a technology in which promoter-independent genome editing is achieved by an RNA translation activator and, when combined with CRISPR–Cas9, allows the delivery of all genome editing components in a single AAV vector. In contrast to AAV and lentiviruses, AdV vectors have a high packaging capacity ($\sim$35 kb), making them a suitable vehicle for genome editing tools because all the components can be delivered into a single vector without any modifications. Many researchers have explored the use of AdV for CRIPSR–Cas9 delivery. Gao et al. [101] used a high-capacity AdV 5 vector for delivering tetracycline-controlled transcription activator–inducible CRISPR–Cas9 to correct a mutated canine *FIX* gene directed against hemophilia. They also demonstrated that the delivery of these genome editing tools with a single-vector AdV was superior in correcting the mutations when compared with a two-vector approach. Although virus vectors are highly efficient, inherent drawbacks such as high immune responses, difficulty in large-scale production, off-target effects, and challenges in systemic delivery have led researchers to explore nonvirus delivery platforms.

6.8 Nonvirus delivery platforms for CRISPR-Cas9

Synthetic materials have the advantage of lower immunogenicity compared with virus vectors and so are currently being explored for CRISPR–Cas9 delivery. These materials include ionizable lipid-, polymeric-, and gold-based nanoparticles. Finn et al. [88] first made significant progress in the in vivo delivery of CRISPR–Cas9 and PEG–dimyristoyl glycerol to hepatocytes by using ionizable lipids, which resulted in >97% of transthyretin knockout in healthy mice. Similarly, cationic polymeric nanoparticles such as poly(β-amino esters) and polyethylenimine can also serve as excellent carriers for CRISPR–Cas-9 delivery due to their ability to condense negatively charged nucleic acids. Zhu et al. [102] demonstrated that poly (β-amino ester)-loaded CRIPSR-Cas9 plasmid complexes inhibited tumor growth in transgenic mice. In another study, Sun et al. [103] found that CRISPR–Cas9 vectors coated with polyethylenimine injected into tumor cells expressing enhanced green fluorescent protein induced significant reduction in tumor growth. Chen et al. [97] synthesized a biodegradable nanocapsule by using glutathione-degradable cross-linker with acrylate methoxy-PEG polymer to encapsulate a CRISPR ribonucleoprotein complex for in vivo and in vitro gene editing. The nanocapsule demonstrated fewer safety concerns, and the polymeric shell can be bound with peptides to target the nanocapsule to a predetermined cell or tissue type.

The use of inorganic particles, particularly gold nanoparticles, for CRISPR–Cas9 delivery has gained attention in particular for genetic diseases that require homology-directed repair. For example, Lee et al. [104] developed a nanoparticle vehicle, called "CRISPR-Gold," by conjugating gold nanoparticles with Cas9, sgRNA, DNA, and glutathione and then coating them with the cationic polymer poly(N-(N-(2-aminoethyl)-2-aminoethyl) aspartimide) for easy penetration into cells. The CRISPR–Cas9-loaded nanoparticles were then used to correct the *CXCR4* gene with an efficiency of 3%–4% in stem cells. In a subsequent study [105], the same researchers observed a 5.4% correction in the mutant gene when the nanoparticles were injected in vivo in a mouse model of Duchenne muscular dystrophy. This research also provided an opportunity to deliver CRISPR–Cas9 and Cpf1 (another guided RNA nuclease) into mouse brain via intrathecal injection. Wang et al. [106] condensed Cas9-sgPlk-1 plasmids on lipid-coated, TAT peptide–modified gold nanoparticles. The lipid coating enhanced the stability of the nanoparticles, and the TAT peptide guided the plasmids to enter the nuclei and knock out the *Plk-1* gene after its release from the gold nanoparticles. In another research, a mixture of synthetic particles and virus vectors was exploited to deliver Cas9 mRNA by using lipid nanoparticles and sgRNA and a homology-directed repair template in a mouse model of tyrosinemia type 1 [107]. Other approaches, such as single-walled carbon nanotubes, are also currently being explored as a delivery platform for CRISPR–Cas9.

Although CRISPR delivery mediated by synthetic nanoparticles has shown promising in vitro and in vivo efficacy in preclinical phases, the ability of these nanoparticles to interact with the human innate immune system poses some safety concerns [108]. To address these issues, the use of nature-derived nanoparticles called "exosomes" has emerged as an alternative approach. As natural particles, exosomes are expected to be devoid of hypersensitivity reactions. Kim et al. [109] successfully delivered CRISPR–Cas9 expression plasmids by using cancer-derived exosomes and induced suppression of PARP-1 [poly

(ADP-ribose) polymerase 1] expression in ovarian cancer. Recently, Lin et al. [110] developed hybrid exosome-liposome nanoparticles to encapsulate CRISPR-Cas9-expressed plasmids to mesenchymal stem cells in vivo. The researchers proposed that this approach can be advantageous for delivering CRISPR-Cas9 to transfect resistant cells. Although the translation of endosomes as delivery platforms for CRISPR-Cas-9 is moving at fast pace, technology for endosomal production at large scale remains a major challenge.

7. Conclusions

Nanoparticle technologies and research using nanomaterials have made a tremendous impact in the last three decades on the areas of drug development, imaging, and diagnostics. Progress in characterization techniques, as well as an enhanced understanding of nanoparticle structure, has been a significant contributor to this progress. Novel nanomaterials have been created by leveraging the fundamentals of materials science and by studying structure-property relationships, thereby enabling the development of various classes of nanoparticles with desirable design features. The first era of nanotechnology research gave rise to the rapid implementation of nanoparticle sizing technologies for drug development through improved partitioning and dissolution in solvent media. This period was followed by the exploration of nanoparticle technologies for tissue, tumor, and organ targeting. Significant advances have been made in our understanding of how nanoparticles undergo absorption, distribution, metabolism, and excretion in physiological systems, as well as the steps needed to improve these pharmacokinetic properties. The ability to target ligands to localize nanoparticles in specific tissues has further improved the utility of these technologies.

Important preclinical studies using nanoparticles for drug targeting have been accompanied by equally significant research into the factors that govern their uptake into cells to produce pharmacodynamic effects. Another research has explored endocytosis, the active transport mechanism by which molecules are brought into cells, and its various components, such as receptor-mediated endocytosis, phagocytosis, and pinocytosis, in both normal and diseased cells.

At the same time, the clinical application of nanotechnologies has been hindered by the issue of endosomal escape and the fact that only a tiny fraction of injected particles reaches the target tissue due to the intrinsic biodistribution of the particles, which leads to high off-target toxicity. Other challenges to the successful clinical translation of nanotechnologies stem from the variability of tumor models, including tumor age and the variability of the vascularity index between preclinical species and humans, which have hindered the creation of targeted therapies.

Recent research has seen a shift in the trend of using nanotechnologies to target tumors to using these techniques to modulate the immune system and to target immune

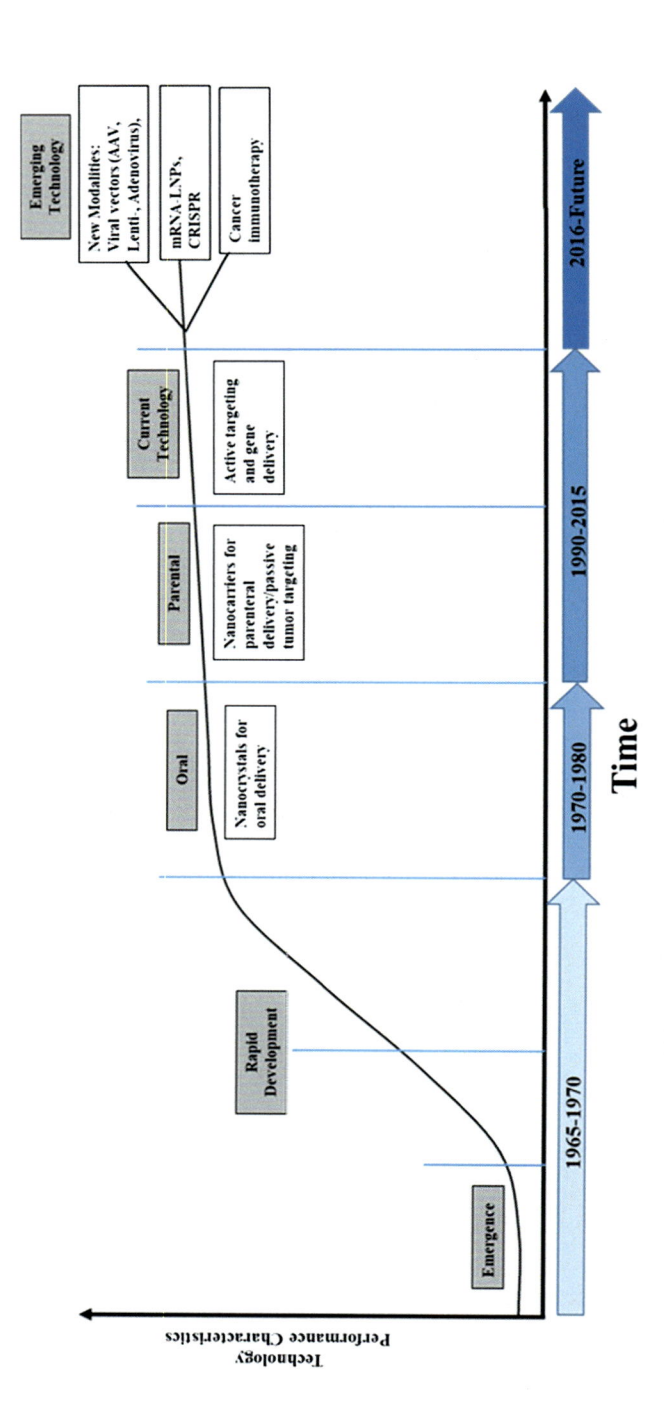

Fig. 5 Emergence and evolution of nanoparticle research.

1990-2010- Passive, active targeting and Gene delivery
2010- 2020- RNA delivery, emergence of immunotherapy, Novel vectors of gene delivery
2020- future – CRISPR mRNA, CRISPR- gene delivery (AAV),

cells. Immune targeting with nanoparticles, using the appropriate agents, can bring about an improved immune response that may, in principle, be superior to that produced by tumor-targeting therapy. New areas of research, such as the use of nanovirus vectors (e.g., AAV) to deliver genes, can be expected to see significant progress in the coming years. Another growing area is RNA therapy [111], which delivers ribo nucleic acids to cell. In fact, RNA therapy has made considerable progress in recent years relying on nanoparticle technologies to delivery RNAs compared to gene therapy that needs the use of electroporation equipment to delivery DNAs.

Nanoparticle technologies coupled with RNA therapy and other modalities such as virus are key to the development of vaccines for SARS-CoV-2 infection and/or for COVID-19 [112]. This will be a rapidly emerging area for the very near future.

Fig. 5 illustrates the emergence of nanoparticle research and how the technology has evolved. Nanoparticle research is branching into new areas from the understanding we have built thus far, drawing on and bringing together the cross-functional expertise of chemists, biologists, materials scientists, toxicologists, and translational scientists. Scale-up capabilities are becoming increasingly available due to advances in three-dimensional printing and the development of new synthetic schemes. These advances will improve the quality of nanoparticle production and make it amenable for mainstream therapies. Antibody-drug conjugates, a modality similar to targeted nanoparticles, have seen increasing success in recent years, and new insights into conjugation chemistry and disease biology through the successful use of antibody-drug conjugates will also contribute to advancements in nanoparticle technologies. Finally, the emergence of data and artificial intelligence will further the creation and design of successful endogenous nanoparticles to make these promising new therapies a reality.

Acknowledgment

The authors thank Deborah J. Shuman of AstraZeneca for editorial services and AstraZeneca researchers for their contributions to the development of nanoparticle tools and technologies.

References

[1] J.K. Patra, G. Das, L.F. Fraceto, E.V.R. Campos, M.D.P. Rodriguez-Torres, L.S. Acosta-Torres, et al., Nano based drug delivery systems: recent developments and future prospects, J. Nanobiotechnol. 16 (1) (2018) 71.

[2] R. Wahab, J. Ahmad, N. Ahmad, Application of multi-dimensional (0D, 1D, 2D) nanostructures for the cytological evaluation of cancer cells and their bacterial response, Colloids Surf. A Physicochem. Eng. Asp. 583 (2019) 123953.

[3] C. Zhang, B. Xie, Y. Zou, D. Zhu, L. Lei, D. Zhao, et al., Zero-dimensional, one-dimensional, two-dimensional and three-dimensional biomaterials for cell fate regulation, Adv. Drug Deliv. Rev. 132 (2018) 33–56.

[4] A. Bak, M. Ashford, D.J. Brayden, Local delivery of macromolecules to treat diseases associated with the colon, Adv. Drug Deliv. Rev. 136–137 (2018) 2–27.

[5] P. Tyagi, S. Pechenov, J. RiosDoria, L. Masterson, N.J. Dickinson, P. Howard, et al., Evaluation of pyrrolobenzodiazepine-loaded nanoparticles: a targeted drug delivery approach, J. Pharm. Sci. 108 (4) (2019) 1590–1597.

[6] B.L. Banik, P. Fattahi, J.L. Brown, Polymeric nanoparticles: the future of nanomedicine, Wiley Interdiscip. Rev. Nanomed. Nanobiotechnol. 8 (2) (2016) 271–299.

[7] B.V. Bonifacio, P.B. Silva, M.A. Ramos, K.M. Negri, T.M. Bauab, M. Chorilli, Nanotechnology-based drug delivery systems and herbal medicines: a review, Int. J. Nanomedicine 9 (2014) 1–15.

[8] Q. Tan, W. Liu, C. Guo, G. Zhai, Preparation and evaluation of quercetin-loaded lecithin-chitosan nanoparticles for topical delivery, Int. J. Nanomedicine 6 (2011) 1621–1630.

[9] R. Watkins, L. Wu, C. Zhang, R.M. Davis, B. Xu, Natural product-based nanomedicine: recent advances and issues, Int. J. Nanomedicine 10 (2015) 6055–6074.

[10] A. Mary Ealias, M.P. Saravanakumar, A review on the classification, characterisation, synthesis of nanoparticles and their application, IOP Conf. Ser. Mater. Sci. Eng. 263 (3) (2016) 032019.

[11] K.E. Sapsford, W.R. Algar, L. Berti, K.B. Gemmill, B.J. Casey, E. Oh, et al., Functionalizing nanoparticles with biological molecules: developing chemistries that facilitate nanotechnology, Chem. Rev. 113 (3) (2013) 1904–2074.

[12] A. Mukherjee, A.K. Waters, P. Kalyan, A.S. Achrol, S. Kesari, V.M. Yenugonda, Lipid-polymer hybrid nanoparticles as a next-generation drug delivery platform: state of the art, emerging technologies, and perspectives, Int. J. Nanomedicine 14 (2019) 1937–1952.

[13] P. Ghasemiyeh, S. Mohammadi-Samani, Solid lipid nanoparticles and nanostructured lipid carriers as novel drug delivery systems: applications, advantages and disadvantages, Res. Pharm. Sci. 13 (4) (2018) 288–303.

[14] S. Mukherjee, S. Ray, R.S. Thakur, Solid lipid nanoparticles: a modern formulation approach in drug delivery system, Indian J. Pharm. Sci. 71 (4) (2009) 349–358.

[15] Y. Barenholz, Doxil, the first FDA-approved nano-drug: lessons learned, J. Control. Release 160 (2) (2012) 117–134.

[16] J.C. Kraft, J.P. Freeling, Z. Wang, R.J. Ho, Emerging research and clinical development trends of liposome and lipid nanoparticle drug delivery systems, J. Pharm. Sci. 103 (1) (2014) 29–52.

[17] S.K. Sriraman, G. Salzano, C. Sarisozen, V. Torchilin, Anti-cancer activity of doxorubicin-loaded liposomes co-modified with transferrin and folic acid, Eur. J. Pharm. Biopharm. 105 (2016) 40–49.

[18] D. Lombardo, M.A. Kiselev, M.T. Caccamo, Smart nanoparticles for drug delivery application: development of versatile nanocarrier platforms in biotechnology and nanomedicine, J. Nanomater. 2019 (2019) 1–26.

[19] C. Martinelli, C. Pucci, G. Ciofani, Nanostructured carriers as innovative tools for cancer diagnosis and therapy, APL Bioeng. 3 (1) (2019) 011502.

[20] T.L. Nguyen, Y. Choi, J. Kim, Mesoporous silica as a versatile platform for cancer immunotherapy, Adv. Mater. 31 (34) (2019) e1803953.

[21] N. Poonia, V. Lather, D. Pandita, Mesoporous silica nanoparticles: a smart nanosystem for management of breast cancer, Drug Discov. Today 23 (2) (2018) 315–332.

[22] S. Jain, D.G. Hirst, J.M. O'Sullivan, Gold nanoparticles as novel agents for cancer therapy, Br. J. Radiol. 85 (1010) (2012) 101–113.

[23] C.T. Matea, T. Mocan, F. Tabaran, T. Pop, O. Mosteanu, C. Puia, et al., Quantum dots in imaging, drug delivery and sensor applications, Int. J. Nanomedicine 12 (2017) 5421–5431.

[24] M. Mahmoudi, S. Sant, B. Wang, S. Laurent, T. Sen, Superparamagnetic iron oxide nanoparticles (SPIONs): development, surface modification and applications in chemotherapy, Adv. Drug Deliv. Rev. 63 (1–2) (2011) 24–46.

[25] V. Dave, K. Tak, A. Sohgaura, A. Gupta, V. Sadhu, K.R. Reddy, Lipid-polymer hybrid nanoparticles: synthesis strategies and biomedical applications, J. Microbiol. Methods 160 (2019) 130–142.

[26] Y. Li, H. Wang, K. Wang, Q. Hu, Q. Yao, Y. Shen, et al., Targeted co-delivery of PTX and TR3 siRNA by PTP peptide modified dendrimer for the treatment of pancreatic cancer, Small 13 (2) (2017).

[27] Y. Gu, Y. Guo, C. Wang, J. Xu, J. Wu, T.B. Kirk, et al., A polyamidoamne dendrimer functionalized graphene oxide for DOX and MMP-9 shRNA plasmid co-delivery, Mater. Sci. Eng. C Mater. Biol. Appl. 70 (Pt 1) (2017) 572–585.

[28] J. Yang, Q. Zhang, H. Chang, Y. Cheng, Surface-engineered dendrimers in gene delivery, Chem. Rev. 115 (11) (2015) 5274–5300.

[29] A. Kavosi, S.H.G. Noei, S. Madani, S. Khalighfard, S. Khodayari, H. Khodayari, et al., The toxicity and therapeutic effects of single-and multi-wall carbon nanotubes on mice breast cancer, Sci. Rep. 8 (1) (2018) 8375.

[30] D. Ha, N. Yang, V. Nadithe, Exosomes as therapeutic drug carriers and delivery vehicles across biological membranes: current perspectives and future challenges, Acta Pharm. Sin. B 6 (4) (2016) 287–296.

[31] J.E. Pullan, M.I. Confeld, J.K. Osborn, J. Kim, K. Sarkar, S. Mallik, Exosomes as drug carriers for cancer therapy, Mol. Pharm. 16 (5) (2019) 1789–1798.

[32] R. Shegokar, R.H. Muller, Nanocrystals: industrially feasible multifunctional formulation technology for poorly soluble actives, Int. J. Pharm. 399 (1–2) (2010) 129–139.

[33] X. Ye, C.M. Wai, Making nanomaterials in supercritical fluids: a review, J. Chem. Educ. 80 (2) (2003) 198.

[34] R. Savla, J. Browne, V. Plassat, K.M. Wasan, E.K. Wasan, Review and analysis of FDA approved drugs using lipid-based formulations, Drug Dev. Ind. Pharm. 43 (11) (2017) 1743–1758.

[35] P. Tyagi, J.A. Subramony, Nanotherapeutics in oral and parenteral drug delivery: key learnings and future outlooks as we think small, J. Control. Release: Official Journal of the Controlled Release Society 272 (2018) 159–168.

[36] Z.H. Loh, A.K. Samanra, P.W. Sia Heng, Overview of milling techniques for improving the solubility of poorly water-soluble drugs, Asian J. Pharm. Sci. 10 (4) (2015) 255–274.

[37] R. Paliwal, R.J. Babu, S. Palakurthi, Nanomedicine scale-up technologies: feasibilities and challenges, AAPS PharmSciTech 15 (6) (2014) 1527–1534.

[38] D.R. Serrano, K.H. Gallagher, A.M. Healy, Emerging nanonisation technologies: tailoring crystalline versus amorphous nanomaterials, Curr. Top. Med. Chem. 15 (22) (2015) 2327–2340.

[39] J.U. Junghanns, R.H. Muller, Nanocrystal technology, drug delivery and clinical applications, Int. J. Nanomedicine 3 (3) (2008) 295–309.

[40] S. Huo, S. Jin, X. Ma, X. Xue, K. Yang, A. Kumar, et al., Ultrasmall gold nanoparticles as carriers for nucleus-based gene therapy due to size-dependent nuclear entry, ACS Nano 8 (6) (2014) 5852–5862.

[41] Y. Pan, S. Neuss, A. Leifert, M. Fischler, F. Wen, U. Simon, et al., Size-dependent cytotoxicity of gold nanoparticles, Small 3 (11) (2007) 1941–1949.

[42] Y. Zhao, X. Sun, G. Zhang, B.G. Trewyn, I.I. Slowing, V.S. Lin, Interaction of mesoporous silica nanoparticles with human red blood cell membranes: size and surface effects, ACS Nano 5 (2) (2011) 1366–1375.

[43] E. Frohlich, The role of surface charge in cellular uptake and cytotoxicity of medical nanoparticles, Int. J. Nanomedicine 7 (2012) 5577–5591.

[44] C.C. Hung, W.C. Huang, Y.W. Lin, T.W. Yu, H.H. Chen, S.C. Lin, et al., Active tumor permeation and uptake of surface charge-switchable theranostic nanoparticles for imaging-guided photothermal/chemo combinatorial therapy, Theranostics 6 (3) (2016) 302–317.

[45] M. Yanez Arteta, T. Kjellman, S. Bartesaghi, S. Wallin, X. Wu, A.J. Kvist, et al., Successful reprogramming of cellular protein production through mRNA delivered by functionalized lipid nanoparticles, Proc. Natl. Acad. Sci. U. S. A. 115 (15) (2018) E3351–e3360.

[46] M.L. Chen, M. John, S.L. Lee, K.M. Tyner, Development considerations for nanocrystal drug products, AAPS J. 19 (3) (2017) 642–651.

[47] G. Roebben, S. Ramirez-Garcia, V.A. Hackley, M. Roesslein, F. Klaessig, V. Kestens, I. Lynch, C.M. Garner, A. Rawle, A. Elder, V.L. Colvin, W. Kreyling, H.F. Krug, et al., Interlaboratory comparison of size and surface charge measurements on nanoparticles prior to biological impact assessment, J. Nanopart. Res. 13 (2011) 2675–2687.

[48] S.R. D'Mello, C.N. Cruz, M.L. Chen, M. Kapoor, S.L. Lee, K.M. Tyner, The evolving landscape of drug products containing nanomaterials in the United States, Nat. Nanotechnol. 12 (6) (2017) 523–529.

[49] B. Maherani, O. Wattraint, Liposomal structure: a comparative study on light scattering and chromatography techniques, J. Dispers. Sci. Technol. 38 (11) (2017).

[50] J.J. Curry, Z.H. Levine, Continuous-feed optical sorting of aerosol particles, Opt. Express 24 (13) (2016) 14100–14123.

[51] M.S. Aziz, N. Suwanpayak, M.A. Jalil, R. Jomtarak, T. Saktioto, J. Ali, et al., Gold nanoparticle trapping and delivery for therapeutic applications, Int. J. Nanomedicine 7 (2012) 11–17.

[52] C. Hacker, J. Asadi, C. Pliotas, S. Ferguson, L. Sherry, P. Marius, et al., Nanoparticle suspensions enclosed in methylcellulose: a new approach for quantifying nanoparticles in transmission electron microscopy, Sci. Rep. 6 (2016) 25275.

[53] P.J. Gaillard, C.C. Appeldoorn, R. Dorland, J. van Kregten, F. Manca, D.J. Vugts, et al., Pharmacokinetics, brain delivery, and efficacy in brain tumor-bearing mice of glutathione pegylated liposomal doxorubicin (2B3-101), PLoS One 9 (1) (2014) e82331.

[54] M.M. Modena, B. Ruhle, T.P. Burg, S. Wuttke, Nanoparticle characterization: what to measure? Adv. Mater. (Deerfield Beach, Fla.) 31 (32) (2019) e1901556.

[55] L. Calzolai, F. Franchini, D. Gilliland, F. Rossi, Protein–nanoparticle interaction: identification of the ubiquitin–gold nanoparticle interaction site, Nano Lett. 10 (8) (2010) 3101–3105.

[56] S.C. Coelho, S. Rocha, M.C. Pereira, P. Juzenas, M.A. Coelho, Enhancing proteasome-inhibitor effect by functionalized gold nanoparticles, J. Biomed. Nanotechnol. 10 (4) (2014) 717–723.

[57] Y. Zhang, C.G. Fry, J.A. Pedersen, R.J. Hamers, Dynamics and morphology of nanoparticle-linked polymers elucidated by nuclear magnetic resonance, Anal. Chem. 89 (22) (2017) 12399–12407.

[58] R. Vogel, A.K. Pal, S. Jambhrunkar, P. Patel, S.S. Thakur, E. Reategui, et al., High-resolution single particle zeta potential characterisation of biological nanoparticles using tunable resistive pulse sensing, Sci. Rep. 7 (1) (2017) 17479.

[59] E. Mansfield, K.M. Tyner, C.M. Poling, J.L. Blacklock, Determination of nanoparticle surface coatings and nanoparticle purity using microscale thermogravimetric analysis, Anal. Chem. 86 (3) (2014) 1478–1484.

[60] C. Saraiva, C. Praca, R. Ferreira, T. Santos, L. Ferreira, L. Bernardino, Nanoparticle-mediated brain drug delivery: overcoming blood-brain barrier to treat neurodegenerative diseases, J. Control. Release: Official Journal of the Controlled Release Society 235 (2016) 34–47.

[61] S. Jiang, Y.L. Franco, Y. Zhou, J. Chen, Nanotechnology in retinal drug delivery, Int. J. Ophthal. 11 (6) (2018) 1038–1044.

[62] D. Rosenblum, N. Joshi, W. Tao, J.M. Karp, D. Peer, Progress and challenges towards targeted delivery of cancer therapeutics, Nat. Commun. 9 (1) (2018) 1410.

[63] S. Biffi, R. Voltan, B. Bortot, G. Zauli, P. Secchiero, Actively targeted nanocarriers for drug delivery to cancer cells, Expert Opin. Drug Deliv. 16 (5) (2019) 481–496.

[64] H.J. Chung, H.J. Kim, S.T. Hong, Tumor-specific delivery of a paclitaxel-loading HSA-haemin nanoparticle for cancer treatment, Nanomed. Nanotechnol. Biol. Med. 23 (2020) 102089.

[65] N. van Zandwijk, N. Pavlakis, S.C. Kao, A. Linton, M.J. Boyer, S. Clarke, et al., Safety and activity of microRNA-loaded minicells in patients with recurrent malignant pleural mesothelioma: a first-in-man, phase 1, open-label, dose-escalation study, Lancet Oncol. 18 (10) (2017) 1386–1396.

[66] J.E. Zuckerman, I. Gritli, A. Tolcher, J.D. Heidel, D. Lim, R. Morgan, et al., Correlating animal and human phase Ia/Ib clinical data with CALAA-01, a targeted, polymer-based nanoparticle containing siRNA, Proc. Natl. Acad. Sci. U. S. A. 111 (31) (2014) 11449–11454.

[67] C. Mamot, R. Ritschard, A. Wicki, G. Stehle, T. Dieterle, L. Bubendorf, et al., Tolerability, safety, pharmacokinetics, and efficacy of doxorubicin-loaded anti-EGFR immunoliposomes in advanced solid tumours: a phase 1 dose-escalation study, Lancet Oncol. 13 (12) (2012) 1234–1241.

[68] S.S. Kim, A. Rait, E. Kim, K.F. Pirollo, E.H. Chang, A tumor-targeting p53 nanodelivery system limits chemoresistance to temozolomide prolonging survival in a mouse model of glioblastoma multiforme, Nanomed. Nanotechnol. Biol. Med. 11 (2) (2015) 301–311.

[69] K.A. Autio, R. Dreicer, J. Anderson, J.A. Garcia, A. Alva, L.L. Hart, et al., Safety and efficacy of BIND-014, a docetaxel nanoparticle targeting prostate-specific membrane antigen for patients with metastatic castration-resistant prostate cancer: a phase 2 clinical trial, JAMA Oncol. 4 (10) (2018) 1344–1351.

[70] P. Munster, I.E. Krop, P. LoRusso, C. Ma, B.A. Siegel, A.F. Shields, et al., Safety and pharmacokinetics of MM-302, a HER2-targeted antibody-liposomal doxorubicin conjugate, in patients with advanced HER2-positive breast cancer: a phase 1 dose-escalation study, Br. J. Cancer 119 (9) (2018) 1086–1093.

[71] R.R. Sawant, A.M. Jhaveri, A. Koshkaryev, L. Zhu, F. Qureshi, V.P. Torchilin, Targeted transferrin-modified polymeric micelles: enhanced efficacy in vitro and in vivo in ovarian carcinoma, Mol. Pharm. 11 (2) (2014) 375–381.

[72] S. Boondireke, M. Leonard, A. Durand, B. Thanomsub Wongsatayanon, Encapsulation of mono-myristin into polymeric nanoparticles improved its in vitro antiproliferative activity against cervical cancer cells, Colloids Surf. B Biointerfaces 176 (2019) 9–17.

[73] D. Raucher, S. Dragojevic, J. Ryu, Macromolecular drug carriers for targeted glioblastoma therapy: preclinical studies, challenges, and future perspectives, Front. Oncol. 8 (2018) 624.

[74] L. Belfiore, D.N. Saunders, M. Ranson, K.J. Thurecht, G. Storm, K.L. Vine, Towards clinical translation of ligand-functionalized liposomes in targeted cancer therapy: challenges and opportunities, J. Control. Release: Official Journal of the Controlled Release Society 277 (2018) 1–13.

[75] J.J. Moon, H. Suh, A. Bershteyn, M.T. Stephan, H. Liu, B. Huang, et al., Interbilayer-crosslinked multilamellar vesicles as synthetic vaccines for potent humoral and cellular immune responses, Nat. Mater. 10 (3) (2011) 243–251.

[76] R. Kuai, L.J. Ochyl, K.S. Bahjat, A. Schwendeman, J.J. Moon, Designer vaccine nanodiscs for personalized cancer immunotherapy, Nat. Mater. 16 (4) (2017) 489–496.

[77] T.T. Smith, S.B. Stephan, H.F. Moffett, L.E. McKnight, W. Ji, D. Reiman, et al., In situ programming of leukaemia-specific T cells using synthetic DNA nanocarriers, Nat. Nanotechnol. 12 (8) (2017) 813–820.

[78] Z. Xu, Y. Wang, L. Zhang, L. Huang, Nanoparticle-delivered transforming growth factor-beta siRNA enhances vaccination against advanced melanoma by modifying tumor microenvironment, ACS Nano 8 (4) (2014) 3636–3645.

[79] H. Yuan, W. Jiang, C.A. von Roemeling, Y. Qie, X. Liu, Y. Chen, et al., Multivalent bi-specific nanobioconjugate engager for targeted cancer immunotherapy, Nat. Nanotechnol. 12 (8) (2017) 763–769.

[80] S. Zanganeh, G. Hutter, R. Spitler, O. Lenkov, M. Mahmoudi, A. Shaw, et al., Iron oxide nanoparticles inhibit tumour growth by inducing pro-inflammatory macrophage polarization in tumour tissues, Nat. Nanotechnol. 11 (11) (2016) 986–994.

[81] M. Song, T. Liu, C. Shi, X. Zhang, X. Chen, Bioconjugated manganese dioxide nanoparticles enhance chemotherapy response by priming tumor-associated macrophages toward M1-like phenotype and attenuating tumor hypoxia, ACS Nano 10 (1) (2016) 633–647.

[82] M.A. Mintzer, E.E. Simanek, Nonviral vectors for gene delivery, Chem. Rev. 109 (2) (2009) 259–302.

[83] Z. Ma, J. Li, F. He, A. Wilson, B. Pitt, S. Li, Cationic lipids enhance siRNA-mediated interferon response in mice, Biochem. Biophys. Res. Commun. 330 (3) (2005) 755–759.

[84] P.R. Cullis, M.J. Hope, Lipid nanoparticle systems for enabling gene therapies, Mol. Ther.: : The Journal of the American Society of Gene Therapy 25 (7) (2017) 1467–1475.

[85] M.A. Oberli, A.M. Reichmuth, J.R. Dorkin, M.J. Mitchell, O.S. Fenton, A. Jaklenec, et al., Lipid nanoparticle assisted mRNA delivery for potent cancer immunotherapy, Nano Lett. 17 (3) (2017) 1326–1335.

[86] R.L. Ball, K.A. Hajj, J. Vizelman, P. Bajaj, K.A. Whitehead, Lipid nanoparticle formulations for enhanced co-delivery of siRNA and mRNA, Nano Lett. 18 (6) (2018) 3814–3822.

[87] S.L. Hewitt, A. Bai, D. Bailey, K. Ichikawa, J. Zielinski, R. Karp, et al., Durable anticancer immunity from intratumoral administration of IL-23, IL-36gamma, and OX40L mRNAs, Sci. Transl. Med. 11 (477) (2019).

[88] J.D. Finn, A.R. Smith, M.C. Patel, L. Shaw, M.R. Youniss, J. van Heteren, et al., A single administration of CRISPR/Cas9 lipid nanoparticles achieves robust and persistent in vivo genome editing, Cell Rep. 22 (9) (2018) 2227–2235.

[89] J.C. Kaczmarek, A.K. Patel, K.J. Kauffman, O.S. Fenton, M.J. Webber, M.W. Heartlein, et al., Polymer-lipid nanoparticles for systemic delivery of mRNA to the lungs, Angew. Chem. Int. Ed. Engl. 55 (44) (2016) 13808–13812.

[90] C.R. Stadler, H. Bahr-Mahmud, L. Celik, B. Hebich, A.S. Roth, R.P. Roth, et al., Elimination of large tumors in mice by mRNA-encoded bispecific antibodies, Nat. Med. 23 (7) (2017) 815–817.

[91] A. Mohr, G. Henderson, L. Dudus, I. Herr, T. Kuerschner, K.M. Debatin, et al., AAV-encoded expression of TRAIL in experimental human colorectal cancer leads to tumor regression, Gene Ther. 11 (6) (2004) 534–543.

[92] H.I. Ma, S.Z. Lin, Y.H. Chiang, J. Li, S.L. Chen, Y.P. Tsao, et al., Intratumoral gene therapy of malignant brain tumor in a rat model with angiostatin delivered by adeno-associated viral (AAV) vector, Gene Ther. 9 (1) (2002) 2–11.

[93] C. Sapet, C. Pellegrino, N. Laurent, F. Sicard, O. Zelphati, Magnetic nanoparticles enhance adenovirus transduction in vitro and in vivo, Pharm. Res. 29 (5) (2012) 1203–1218.

[94] E. Kim, J.S. Oh, I.S. Ahn, K.I. Park, J.H. Jang, Magnetically enhanced adeno-associated viral vector delivery for human neural stem cell infection, Biomaterials 32 (33) (2011) 8654–8662.

[95] H. Koike-Yusa, Y. Li, E.P. Tan, C. Velasco-Herrera Mdel, K. Yusa, Genome-wide recessive genetic screening in mammalian cells with a lentiviral CRISPR-guide RNA library, Nat. Biotechnol. 32 (3) (2014) 267–273.

[96] M.C. Milone, U. O'Doherty, Clinical use of lentiviral vectors, Leukemia 32 (7) (2018) 1529–1541.

[97] Y. Chen, X. Liu, Y. Zhang, H. Wang, H. Ying, M. Liu, et al., A self-restricted CRISPR system to reduce off-target effects, Mol. Ther.: The Journal of the American Society of Gene Therapy 24 (9) (2016) 1508–1510.

[98] D. Wilbie, J. Walther, E. Mastrobattista, Delivery aspects of CRISPR/Cas for in vivo genome editing, Acc. Chem. Res. 52 (6) (2019) 1555–1564.

[99] L. Wang, Y. Yang, C. Breton, P. Bell, M. Li, J. Zhang, et al., A mutation-independent CRISPR-Cas9-mediated gene targeting approach to treat a murine model of ornithine transcarbamylase deficiency, Sci. Adv. 6 (7) (2020) eaax5701.

[100] H. Zhan, Q. Zhou, Q. Gao, J. Li, W. Huang, Y. Liu, Multiplexed promoterless gene expression with CRISPReader, Genome Biol. 20 (1) (2019) 113.

[101] J. Gao, T. Bergmann, W. Zhang, M. Schiwon, E. Ehrke-Schulz, A. Ehrhardt, Viral vector-based delivery of CRISPR/Cas9 and donor DNA for homology-directed repair in an in vitro model for canine hemophilia B, Mol. Ther. Nucleic Acids 14 (2019) 364–376.

[102] D. Zhu, H. Shen, S. Tan, Z. Hu, L. Wang, L. Yu, et al., Nanoparticles based on poly (beta-amino ester) and HPV16-targeting CRISPR/shRNA as potential drugs for HPV16-related cervical malignancy, Mol. Ther.: The Journal of the American Society of Gene Therapy 26 (10) (2018) 2443–2455.

[103] W. Sun, W. Ji, J.M. Hall, Q. Hu, C. Wang, C.L. Beisel, et al., Self-assembled DNA nanoclews for the efficient delivery of CRISPR-Cas9 for genome editing, Angew. Chem. Int. Ed. Engl. 54 (41) (2015) 12029–12033.

[104] K. Lee, M. Conboy, H.M. Park, F. Jiang, H.J. Kim, M.A. Dewitt, et al., Nanoparticle delivery of Cas9 ribonucleoprotein and donor DNA in vivo induces homology-directed DNA repair, Nat. Biomed. Eng. 1 (2017) 889–901.

[105] B. Lee, K. Lee, S. Panda, R. Gonzales-Rojas, A. Chong, V. Bugay, et al., Nanoparticle delivery of CRISPR into the brain rescues a mouse model of fragile X syndrome from exaggerated repetitive behaviours, Nat. Biomed. Eng. 2 (7) (2018) 497–507.

[106] P. Wang, L. Zhang, W. Zheng, L. Cong, Z. Guo, Y. Xie, et al., Thermo-triggered release of CRISPR-Cas9 system by lipid-encapsulated gold nanoparticles for tumor therapy, Angew. Chem. Int. Ed. Engl. 57 (6) (2018) 1491–1496.

[107] H. Yin, C.Q. Song, J.R. Dorkin, L.J. Zhu, Y. Li, Q. Wu, et al., Therapeutic genome editing by combined viral and non-viral delivery of CRISPR system components in vivo, Nat. Biotechnol. 34 (3) (2016) 328–333.

[108] S.G. Antimisiaris, S. Mourtas, A. Marazioti, Exosomes and exosome-inspired vesicles for targeted drug delivery, Pharmaceutics 10 (4) (2018) 218.

[109] S.M. Kim, Y. Yang, S.J. Oh, Y. Hong, M. Seo, M. Jang, Cancer-derived exosomes as a delivery platform of CRISPR/Cas9 confer cancer cell tropism-dependent targeting, J. Control. Release: Official Journal of the Controlled Release Society 266 (2017) 8–16.

[110] Y. Lin, J. Wu, W. Gu, Y. Huang, Z. Tong, L. Huang, et al., Exosome-liposome hybrid nanoparticles deliver CRISPR/Cas9 system in MSCs, Adv. Sci. (Weinheim, Baden-Wurttemberg, Germany) 5 (4) (2018) 1700611.

[111] S. DeWeerdt, RNA Therapies Explained, 2019.

[112] T.T. Le, Z. Andreadakis, A. Kumar, R.G Roman, S. Tollefsen, M. Saville, S. Mayhew, The COVID-19 vaccine development landscape, Nat. Rev. Drug Discov. (2020) 305–306.

CHAPTER 2

Fabrication design, process technologies, and convolutions in the scale-up of nanotherapeutic delivery systems

Honey Goel[a], Lubna Siddiqui[b], Asiya Mahtab[b], and Sushama Talegaonkar[c]
[a]Department of Pharmaceutics, University Institute of Pharmaceutical Sciences and Research, Baba Farid University of Health Sciences (BFUHS), Faridkot, India
[b]Department of Pharmaceutics, School of Pharmaceutical Education and Research, Jamia Hamdard, New Delhi, India
[c]Department of Pharmaceutics, School of Pharmaceutical Sciences, Delhi Pharmaceutical Sciences and Research University (DPSRU), New Delhi, India

1. Introduction

The rise of nanotechnology-driven drug delivery therapeutics can be acclaimed as one of the most remarkable events of healthcare reform which has revolutionized the basics of medicine in the 21st century. Nanotechnology has dramatically transformed the field of medicine in the last 60 years. New drug delivery paradigms have evolved and reinvented the concepts of drug administration methods into patient-driven therapeutics. The global market is projected to grow by US$104.9 billion with compounded growth of 20.4% [1]. Fig. 1 displays the wide reach of nanotechnology into diverse sectors of pharmaceuticals, nanodevices, nanodiagnostics, and nanorobotics. In the last 5 years, FDA has approved first 3D print drug *Spritam* in 2015 [2], first gene therapy *Kymriah* in 2017 [3], and first digital drug Abilify *MyCite* in 2017 [4]. Advancement of nanotechnologies has provided huge impetus in the successful development of nanoformulations, viz., nanoliposomes, nanocrystals, nanoparticles, nano/microemulsions, nanosuspensions, and SMEDDS/SNEDDS, but their real impact has been quintessential in the cosmetic industry, with products including *Capture* (C. Dior), *Revitalift* (L'Oréal), *Revision* (Revision Skin Care), *Envirox* (Oxonica), *Lamesoft* (BASF Care), and *Emulgade* (BASF Care). Hence, the concept of modern medicine has been reinventing itself from traditional way of treating diseases to target the disease at cellular level employing tools of nanotherapeutics.

Several commercial nanotechnologies for nanocarriers (for example—pH/ion gradient method, spray drying technique, ethanol injection technique, media milling, high-pressure homogenization, $scCO_2$-supercritical anti-solvent (SAS) method, Microfluidization techniques) have shifted dynamics in favor of nanoplatforms to improve

47

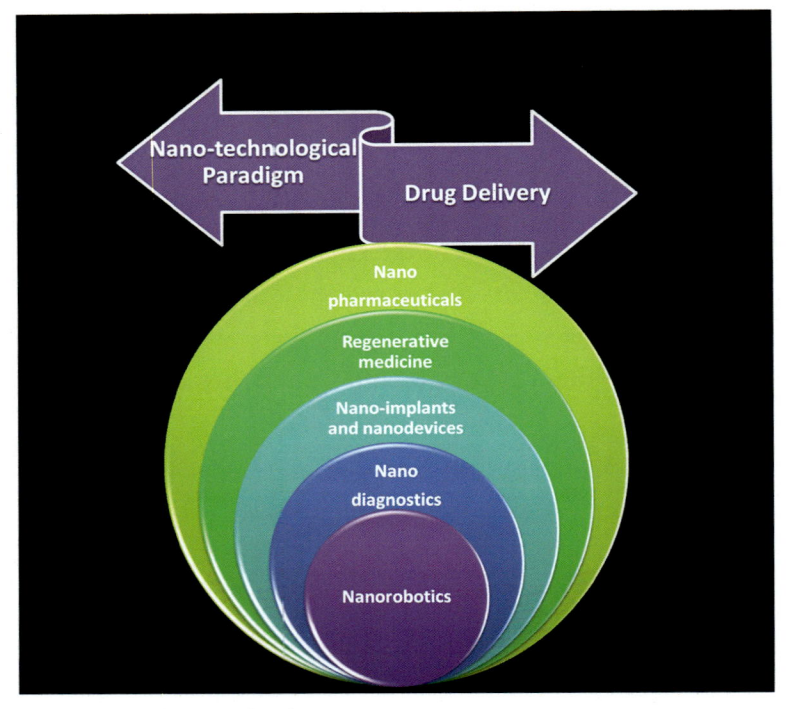

Fig. 1 Novel paradigms of nanotechnology.

therapeutic outcomes compared to traditional modes of drug delivery. A peek into the evolving role of industrially viable technologies has shown tremendous potential in enabling the low soluble drug candidates to effectively target the desired regions of the body. To keep abreast of the changing pulse of global therapeutic market, nanosized therapeutic systems have been capturing the space of conventional drug delivery systems and continue to gain momentum. Approximately 50 nanotechnology-based drug therapies have been approved for human use so far [5], out of which 20 nano drug therapies belong to the class of nanoliposomal drug delivery systems [6]. Since the last two decades, more than 1500 nanomedicine formulations have been filed for patents and have led to execution of vast number of clinical trials [7], with most of the clinical trials focussing on oncology-based products such as liposomal doxorubicin (DoX) or albumin bound paclitaxel (PTX) nanoparticles. Thus, it could be postulated that with the adoption of nanotechnology, fundamental changes in drug production and delivery designs are expected to affect nearly 50% of the total drug production in the coming next decade, expecting overall US$380 billion of revenue [8].

Drug delivery scientists have always envisioned to exploit the unrealized potential of nanoplatforms such as nanoliposomes, nanocrystals, nanoparticles, nanoemulsions, and nanosuspensions for improved delivery properties such as enhanced blood circulation time (via adding surface chemical modifications and controlling size), reduced cytotoxicity, sustained drug release, improving therapeutic window of drugs, targeting to selected tissues, medical imaging, gene delivery into cells (transfection), and analysis [9, 10]. The size of nanotherapeutic system plays a pivotal role in gauging where the nanoparticle will get accumulated inside the human body. As nanoparticles >200 nm are rapidly cleared by the spleen and liver, uniformity of size also becomes necessary to ensure drug accumulation at specific sites. During the last three decades, some of the nanotherapeutic products have enjoyed huge commercial success among diverse classes of drug delivery systems such as liposomes (*Doxil, AmBisome*), nanocrystals (*Triglide, Megace, Theodur*), nanoparticles (*Abraxane, Eligard*), nanoemulsions (*Diprivan, Restasis*), and SNEDDS/SMEDDS (*Neoral, Aptivus*). Among all of these nanoformulations, nanoliposomal systems are indisputably the dominant class which were also the first FDA-approved therapeutic nanosystems and still seizing a big chunk of clinical stage nanotherapeutic market.

Nevertheless, it is also a matter of fact that there is a huge gap in the clinical translation of nanoformulations from a lab scale approach to industrial scale-up owing to structural and physicochemical complexity of formulation itself, expensive methods of fabrication, and time-consuming processes [11]. Therefore, the essential element for the large scale cGMP production of nanotherapeutic systems is to reduce the scalability convolutions, numerous processing steps, or use of solvents in manufacturing process which will further assist in improving production yield, batch to batch reproducibility, consistency, and storage of the final product [12]. Nanoplatforms such as liposomes, nanoparticles, nanosuspensions, nanoemulsions, and SMEDDS have been successfully developed by addressing these aforementioned challenges. In order to ensure consistently similar high level of quality control with cost-effectiveness during the manufacturing of nanoformulations, methods must be simplistic, one-step processes with continuous flow of raw (feed) materials into final product. Further, validation of methods and technologies has to be evaluated for batch to batch variation (through characterization of formulation parameters such as surface charge, size, encapsulation efficiency, and drug release profile) and stability of the final product after the manufacturing during long-term storage and upon clinical administration (i.e., in order to prevent massive drug release or aggregation in the bloodstream en route to site of action).

In the light of the aforesaid complexities inherent in the nanocarriers, the mechanistic understanding of the phenomena involved, correlation among operational factors, physicochemical parameters and rational design associated with a scalable technological process able to construct commercially viable nanosystems (especially nanoliposomes, nanocrystals, nanoparticles nanosuspensions, and nanoemulsions) is warranted. Therefore, the aim of this comprehensive chapter is firstly to provide readers a prospective of experimental challenges associated with critical quality attributes and aptest process variables of the nanoformulations, secondly to enlighten with most recent developments in the field of nano drug delivery applications, and finally to discuss pathways for translational development, a hard road to commercialization.

2. Liposomes as nanoplatform

Doxil (Janssen Pharmaceuticals, PA, USA) and AmBisome (Astellas Pharma, IL, USA) are two most prominent liposomal nanoproducts which have carved a niche for liposomes as a nanoplatform in the diverse nanotherapeutic drug delivery sector by accomplishing commendable success commercially.

2.1 General characteristics

Liposomes (terms derived from Greek *lipo*—means fat and *soma*—means body) are the artificial membranes prepared by either natural or synthetic phospholipids (PHLs), typically comprising a unilamellar or multi-lamellar bilayer and a hydrophilic core. Drugs are encapsulated inside the lipid bilayer for slow release until the vesicle gets degraded. Since the pioneering discovery of liposomes by Alec D Bangham in 1965 [13], liposomes have been classified through various means, i.e., according to their structural properties [i.e., particle size and number of bilayers forming vesicles, namely, *multi-lamellar; MLVs—>0.5 µm, oligolamellar; OLVs—>0.1–1.0 µm, unilamellar; UVs (further classified into small unilamellar SUVs—20–40 nm, medium unilamellar MUVs—40–100 nm, large unilamellar; LUVs— 100–1000 nm, giant unilamellar GUVs > 1 µm) and multivesicular vesicles; MVVs—>1.0 µm* on the basis of their drug loading, namely, *active and passive loading*; on the basis of their composition or work and mechanism of intracellular delivery, namely, *conventional; pH sensitive; cationic; immune; magnetic; thermosensitive; and stealth liposomes* [14]; on the basis of manufacturing method, namely, *ethanol injection vesicle (EIV), frozen and thawed MLV (FATMLV), vesicles prepared by french press (FPV), vesicles prepared by extrusion method (VET), stable plurilamellar vesicles (SPLV), vesicles prepared by fusion (FUV), dehydration- rehydration vesicles (DRV), vesicles made by reverse-phase evaporation (REV), and bubblesomes*]. Further, carrier systems (such as emulsomes, enzymosomes, sphyngosomes, transfersomes, ethosomes, pharmacosomes, and virosomes and aquasomes, bilosomes, and niosomes) that are considered close native of liposome family are categorized under lipid vesicular and as nonlipid vesicular systems, respectively. Generally, liposomal formulations show huge variability between and within the batches. Therefore, FDA has issued revised draft guidelines in April 2018 for industry on liposome drug products and its characterization [15].

2.2 Properties of liposomes

The critical formulation aspects of liposomal drug delivery depend on lipid composition, size, charge, drug and lipid ratio, and method of delivery. A significant breakthrough in the field of liposomal drug delivery was comprehended by imparting steric stabilization via coating the vesicle surface using hydrophilic polymers such as polyethylene glycol (PEG) known as PEGylation (Fig. 2). For historical reasons, liposomes without any physical or chemical modification on its surface are termed as *conventional* in order to distinguish from the surface-modified liposomes. Surface-grafted PEG polymers enhanced circulation time in the blood and provided stability in general by reducing interaction with blood proteins.

Although such liposomes are generally termed as sterically stabilized liposomes, owing to their capability to escape immune system, sometimes these are also referred to as stealth liposomes (*Stealth liposome*, also a trademark of Sequus Pharmaceuticals). These stealthy lipid vesicles with prolonged circulation are eventually removed by macrophages in the tissues (compared to clearance of conventional liposomes in liver and spleen) with relatively at a slower rate with respect to conventional liposomes. Further, several target binding biological units such as vitamins, glycoproteins, peptides, oligonucleotides, and polysaccharides have been attached to the liposome surface as targeting ligands that

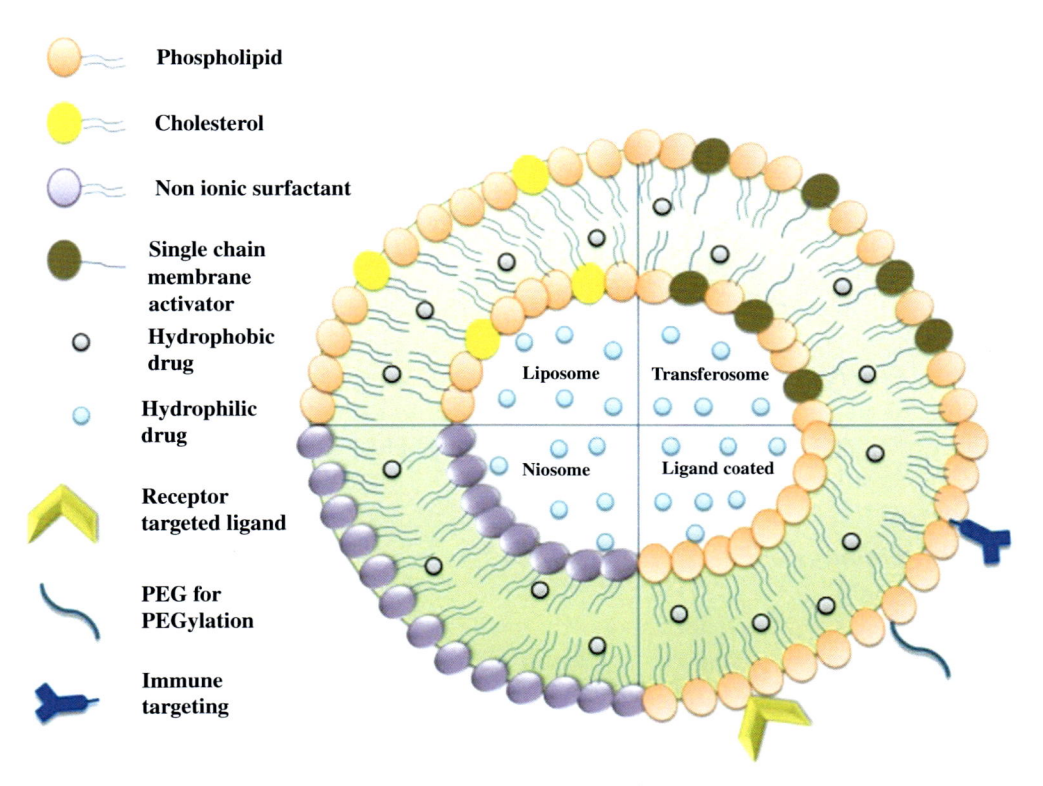

Fig. 2 Structure of a liposomal assembly and its possible applications.

exhibit specific affinity towards the receptors in order to exert therapeutic effect. Similarly, attachment of antibodies or antibody fragments to the liposomal surface as ligands has emerged as one of the most promising applications, also known as immune targeting or immunoliposomes [16]. Another advantage of liposomes is the thermosensitive feature, i.e., an increase of temperature (40–41°C) causes packing changes in the bilayer favoring the release of the encapsulated drug. These lipid-based thermo-devices succor in the specific release of cytotoxic agent to targeted site on the application of an external heat source [17].

2.3 Compositional architecture of liposomes

The major structural components of liposome moiety are Cholesterol (CHL), PHLs, and Stabilizing agents.

2.3.1 Cholesterol (CHL)

CHL does not construct a bilayer by itself, but tends to intercalate into PHL membranes (with its hydroxyl group oriented towards the aqueous surface and the aliphatic chain aligned parallel to the acyl chains in the center of the bilayer) at high concentrations (generally 1:1 or 2:1 M ratio of CHL to PHL). CHL molecules in the membrane increase separation between choline head groups which reduces the normal hydrogen bonding and electrostatic interaction. Although there is no distinct indication of arrangement of CHL in the bilayer, the high solubility of CHL in the PHL layer of liposome has been attributed to both hydrophobic and definite head group interactions.

2.3.2 Phospholipids (PHLs)

PHLs are the liposome-building lipids that comprise of phosphorus, a polar region, and nonpolar region in their structures. Various categories of PHLs offer diverse prospects to be used as excipients as mentioned in Table 1. Among the species diversity of these PHLs, selection of an appropriate PHL (critical for a stable and effective liposomal design) depends on the phase transition temperature (T_c), net charge, physical and chemical stability, and cost-effectiveness.

2.3.3 Stabilizing agents
Cryoprotectants

Circumvent fusion of vesicles and vesicle stability, cryoprotectants such as saccharides owing to its high glass transition temperature [18], oligosaccharides (mostly compatible to PEG), mucoadhesive polysaccharides, polar amino acids thermozeaxanthins, and organic solvents (as adjuvants to improve the sublimation rate) have been employed in the manufacturing process. Table 2 summarizes various classes of stabilizers which have been used for various types of PHLs or their combination ratios.

Table 1 List of some commonly employed phospholipids in liposome engineering along with their properties.

Phospholipids	Trade name/ Company	M.W (g/mol)	Fatty acid distribution	Molecular formula	T_c (°C)	Net charge (pH 7.4)	Commercial application
Soybean phosphatidylcholine (SPC)	Lipoid S100 (Lipoid GmbH)	758.1	18:2	$C_{42}H_{80}NO_8P$	−20 to −30	0	Oral, Topical, Inhalational
Egg sphingomyelin (ESM)	Coatsome NM-10 (NOF Corp.)	703	15:0, 17:0	$C_{39}H_{79}N_2O_6P$	40	0	Intravenous (IV)
Egg phosphatidylcholine (EPC)	Lipoid E80 (Lipoid GmbH); Coatsome NC-50 (NOF Corp.)	773	12:20	$C_{42}H_{82}NO_8P$	−5 to −15	0	Oral, IV
Dimyristoylphosphatidylcholine (DMPC)	Coatsome MC-4040 (NOF Corp.)	677.9	14:0	$C_{36}H_{72}NO_8P$	24	0	IV
Dipalmitoylphosphatidylcholine (DPPC)	Coatsome MC-6060 (NOF Corp.)	735.0	16:0	$C_{40}H_{80}NO_8P$	41.5	0	Inhalational, IV
Dioleoyl phosphatidylcholine (DOPC)	Coatsome MC-8181 (NOF Corp.)	786.1	18:1	$C_{44}H_{84}NO_8P$	−22	0	Epidural
Distearoyl phosphatidylcholine (DSPC)	Coatsome MC-8080 (NOF Corp.)	790.1	18:0	$C_{44}H_{89}NO_8P$	55	0	IV and Inhalational
Dimyristoyl phosphatidylglycerol (DMPG)	Coatsome MG-4040LS (NOF Corp.)	666.9	14:0	$C_{34}H_{67}O_{10}P$	23	−1	IV

Continued

Table 1 List of some commonly employed phospholipids in liposome engineering along with their properties—cont'd

Phospholipids	Trade name/ Company	M.W (g/mol)	Fatty acid distribution	Molecular formula	T_c (°C)	Net charge (pH 7.4)	Commercial application
Dioleoyl phosphatidylglycerol (DOPG)	Coatsome MG-8181LS (NOF Corp.)	775.0	18:1	$C_{42}H_{79}O_{10}P$	−18	−1	Epidural
Distearoyl phosphatidylglycerol (DSPG)	Coatsome MG-8080LS (NOF Corp.)	779.1	18:0	$C_{42}H_{83}O_{10}P$	55	−1	IV
Dimyristoyl phosphatidylethanolamine (DMPE)	Coatsome ME-4040 (NOF Corp.)	635.9	14:0	$C_{33}H_{66}NO_8P$	50	0	IV
Dimyristoyl phosphatidylserine (DMPS)	Coatsome MS-4040LS (NOF Corp.)	679.9	14:0	$C_{34}H_{66}NO_{10}P$	38	−1	IV
1,2-Dioleoyloxy-3-trimethylammonium propane (DOTAP)	Coatsome CL-8181TA (NOF Corp.)	698.5	18:1	$C_{43}H_{83}NO_8S$	−0	+1	IV
Polyethylene glycol 2000-distearoylphosphatidylethanolamine (PEG-2000-DSPE)	Sunbright DSPE-PTE020 (NOF Corp.)	2805.5	18:0	$C_{133}H_{267}N_2O_{55}P$	NA	−1	IV
Tetramyristoyl Cardiolipin•$(Na)_2$	Cardiolipin (Avanti Polar Lipids)	1285.6	14:0	$C_{65}H_{124}Na_2O_{17}P_2$	59	−2	IV, Topical

Table 2 Stabilizers commonly used in various types of liposomes.

Class	Choice of stabilizers	Phospholipids and its combinations (in molar ratio)
Saccharides	Trehalose	DMPC, DPPC, EPC, EPC:CHL (4:1), EPC:PS (9:1), POPC:PS (9:1), DDA:TDB (5:1), DOTAP:DOPE (1:1)
	Lactose	EPC:PS (9:1)
	Sucrose	DDA:TDB (5:1), EPC:PS (9:1), DOTAP:DOPE (1:1), EPC, DOPC:CHL (7:3)
	Glucose	DPPC, DMPC, EPC
	Sucrose/Raffinose	DPPC, DMPC
	Maltose	DPPC
	Raffinose	EPC
	Stachyose	EPC
	Verbascose	EPC
	Maltotriose-heptaose	DPPC
Oligosaccharide	Inulin	EPC, DOTAP:DOPE (1:1)
Polysaccharide	Dextran	EPC, DOTAP:DOPE (1:1)
Amino acids	Proline	DPPC, EPC
	Histidine	EPC:CHL (4:1)
	Lysine	EPC:CHL (4:1)
	Arginine	EPC:CHL (4:1)
Sugars	Mannitol	DOPC:CHL (7:3), DPPC:CHL (7:3)
	Sorbitol	EPC
Starches	Hydroxyethyl starch	EPC
Organic solvents	Dimethyl sulfoxide	DPPC
Fatty acids	Glycerol	DPPC
	2-O-(alfa-D-glucopyranosyl) glycerol	EPC
Osmolytes	Glycinebetaine	EPC

Edge activators

These are the single chain biocompatible surfactants which aid by softening the lipid bilayer, e.g., bile salts/anionic surfactants (sodium cholate, sodium deoxycholate, sodium taurocholate), or hydrophilic nonionic surfactants (e.g., Spans-20/40/60/80 or tweens 20/60).

Applications:

• Edge activators have the ability to encapsulate both hydrophilic and lipophilic moieties.
• It enhances deformability of bilayer by lowering the interfacial tension.

- It prolongs half-life of drug by increasing duration of systemic circulation due to encapsulation.
- It provides superior penetration properties.

2.4 Marketed liposomal formulations

In the field of nano drug delivery therapeutics, Doxil was the first FDA-approved nano-product in the year 1995, based on pegylated liposomal technology encapsulating DoX (Adriamycin; member of anthracycline group) for the therapeutic management of metastatic ovarian cancer and AIDS-related Kaposi's sarcoma. Presently, many liposome-based nanoformulations are under clinical use such as Caelyx (Janssen-Cilag, Europe), Evacet (Liposome Company Inc.), Doxil and Myocet (GP-Pharm, Barcelona, Spain), and Lipodox (Sun Pharma, India). Majority of the liposomal formulations have been approved for administration through intravenous (IV) route; however, other routes like intramuscular delivery (IM) for hepatitis A (Epaxal, Berna Biotech Ltd., Switzerland) and influenza virus (Inflexal V, Berna Biotech Ltd., Spain) have also been approved [19, 20]. Table 3 summarizes the list of various FDA-approved liposomal products commercialized for various conditions in the global pharmaceutical as well as cosmaceutical sector.

2.5 Methods of preparation

During the last five decades, several modes to classify liposomes have been documented in the literature. Among all, the best classified on the basis of drug loading can be passive loading and active (or remote) loading. These methods can be further extended into technologies, viz., mechanical dispersion methods, solvent dispersion methods, techniques based on detergent removal, novel methods, and gradient methods for the liposomal engineering as illustrated in Fig. 3.

Until August 2016, several liposome formulations were approved by regulatory agencies, mainly because of their improved biodistribution; however, production methods of these vesicles are still lagging in terms of industrial scale purposes, preventing innovative therapies from reaching patients worldwide. Most of the liposome manufacturing strategies lack batch to batch reproducibility, and therefore, are difficult to scale-up and have a low production rate and high fabrication costs. Hence, it could be postulated that many of these passive loading methods are practically unsuitable for large scale production of liposomes. Therefore, in order to aim for better understanding of the industrial scalable technologies, liposomes have been demarcated into two categories, viz., technologies for large scale production and technologies for small scale production (Fig. 4).

Both active and passing loading techniques are used for the small scale as well as large scale manufacturing of liposomes.

Table 3 Comprehensive study of marketed liposomal formulations approved by global regulatory bodies in chronological order.

Proprietary nano product/manufacturer/ approval year	Type of dosage form/delivery route/shelf life	Drug payload/technique	Composition/size (nm)	Therapeutic indication	References
Commercial Application in Pharmaceutical Sector					
Arikayce Insmed Inc. (Approved by FDA in September 2018)	Suspension; Oral Inhalational route; 60 days (4°C)	Amikacin sulfate/Cross-flow ethanol injection	DPPC: CHL (Molar ratio: 2:1) in ethanol; ~200 nm	*Mycobacterium avium* complex (MAC) lung disease	[21, 22]
CPX-351 (Vyxeos) (Jazz Pharmaceuticals plc) (Approved by FDA in 2017 and EMEA in 2018)	Powder for reconstituted Suspension; IV route; 24 months	Daunorubicin + cytarabine (daunorubicin, cytarabine 5:1)	Liposome; DSPC, DSPG, CHL (7:2:1); 100 nm	Acute myeloid leukemia	[23]
ONPATTRO (Alnylam Pharmaceuticals) (Approved by FDA in August 2018 and EMEA in 2018)	Solution; IV route; 24 months (2–8°C)	Patisiran (siRNA)	A lipid complex DL in-MC3-DMA, CHL, DSPC, PEG$_{2000}$-C-DMG	Hereditary transthyretin amyloidosis	[24]
Onivyde (MM-398) Merrimack Pharmaceuticals Inc. (Approved by FDA in 2015)	Suspension; IV route; 36 months	Irinotecan + fluorouracil + folinic acid	Pegylated liposome; DSPC:MPEG-2000:DSPE (3:2:0.015 M ratio); 80–140	Adjuvant therapy in pancreatic adenocarcinoma	[25]
Lipodox (Sun Pharmaceutical Industries Ltd.) (Approved by USFDA in 2012)	Suspension; IV route; 36 months	DoX HCl (Generic version of (Doxil/ Caelyx)/Ammonium ion gradient method	Pegylated liposome; DSPC:CHL:PEG$_{2000}$-DSPE (56:39:5 M ratio; 20 nm	Kaposi's sarcoma, ovarian and breast cancer	[26, 27]
Marqibo (formerly Onco-TCS) Talon Therapeutics, Inc. (Approved by FDA in 2012)	Suspension; IV route; 24 months (2–8°C)	Vincristine sulfate	Liposome Sphingomyelin: CHL (60:40 M ratio); <100 nm	Non-Hodgkin's lymphoma and leukemia	[28]

Continued

Table 3 Comprehensive study of marketed liposomal formulations approved by global regulatory bodies in chronological order—cont'd

Proprietary nano product/manufacturer/ approval year	Type of dosage form/delivery route/shelf life	Drug payload/technique	Composition/size (nm)	Therapeutic indication	References
Exparel Pacira Pharmaceuticals, Inc. (Approved by FDA in 2011)	Suspension; IV route; Up to 30 days (20–25°C)	Bupivacaine HCl	Liposome; DEPC, DPPG, CHL and Tricaprylin; 3000–30,000	Pain management	[29]
Inflexal V FDA approved in 1997; Approved by EMEA in 2008 (Crucell, Berna Biotech)	Suspension; IM route; 12 months	Inactivated hemaglutinine of Influenza virus strains A and B	Liposome; DOPC:DOPE (75:25 M ratio)	Influenza	[20]
Epaxal (FDA approved in 1993; EMEA approved in 2006 and is currently used in Switzerland and Argentina) (Crucell, Berna Biotech)	Suspension; IM route; 36 months	Inactivated hepatitis A virus (strain RG-SB)	Liposome; DOPC:DOPE (75:25 M ratio)	Hepatitis A	[19]
Mepact Takeda Pharmaceutical Ltd. (FDA denied approval in 2007; Authorized for use in European Union 2009)	Powder for reconstituted suspension; IV route; 30 months	Mifamurtide (Muramyl tripeptide PE)	Non-PEGylated liposome; DOPS:POPC (3:7 M ratio)	High-grade, resectable, nonmetastatic osteosarcoma	[30]
DepoCyt (DepoFoam technology) (Approved (1999/2007)	Suspension; Intraspinal route; 18 months	Cytabirine	CHL, Triolein, DOPC, and DPPG (11:1:7:1 M ratio); 20 nm	Lymphomatous meningitis; Neoplastic meningitis	[28]

Lipusu (Nanjing Luye Sike Pharmaceutical Co., Ltd.) Approved in China (2006); FDA in 2005	Suspension; IV route; 24 months (2–8°C)	PTX	Liposome; 72 g PC, 10.8 CHL in ethanol; 400 nm	Solid tumors of Gastric, ovarian, and lung cancer	[31]
DepoDur SkyPharma Inc. (Approved by FDA in 2004)	Suspension; Epidural route; 24 months	Morphine sulfate (Epidural route)	CHL, triolein, DOPC, DPPG (11:1:7:1); 17,000–23,000	Pain management	[32]
Visudyne Novartis (Approved by FDA in 2000)	Powder; IV route; 48 months	Verteporphin	Verteporphin:DMPC and EPG (1:7.5–15 M ratio); SUV ~ < 100 nm	Age-related macular degeneration; Ocular histoplasmosis	[33]
Myocet Elan Pharmaceuticals (Approved by EMEA in 2000)	Powder; IV route; 18 months	DoX + Cyclophosphamide/pH gradient method	Non-PEGylated liposome; EPC:CHL (55:45 M ratio); OLV-180 nm	Metastatic breast cancer	[34]
Curosurf (Chiesi USA) (Approved by FDA in 1999)	Suspension; Intratracheal use only 18 months	Poractant alfa (Drug:Lipid-1:0.96)	DPPC:SPB:SPC (30:0.45:5–11.6 M ratio);	Respiratory distress syndrome in premature infants	[5]
Ambisome Astellas Pharma (Approved by FDA in 1997)	Powder; IV route; 36 months	Amphotericin B (AmB) (Drug: Lipid −1:1 M ratio)/ Spray Drying Technique	Liposome; HSPC:DSPG: CHL: AmB (2:0.8:1:0.4 M ratio); SUV < 100 nm	Presumed fungal infections	[32]

Continued

Table 3 Comprehensive study of marketed liposomal formulations approved by global regulatory bodies in chronological order—cont'd

Proprietary nano product/manufacturer/ approval year	Type of dosage form/delivery route/shelf life	Drug payload/technique	Composition/size (nm)	Therapeutic indication	References
Amphotec (Approved by FDA in 1996; Status-discontinued)	Powder; IV route; 24 months	AmB	A Lipid Complex; Cholesteryl sulfate	Severe fungal infections	[35]
DaunoXome (Approved by FDA 1996; FDA status-discontinued) (Galen Pharma)	Emulsion; IV route; 12 months	Daunorubicin citrate lipid/drug 18.7:1 [w/w]	Non-Pegylated; DSPC/CHL, 2:1 M ratio; SUV ∼ 45 nm	HIV-associated Kaposi's sarcoma (primary) and Leukemia	[36]
Doxil Sequus Pharmaceuticals (Approved by FDA in 1995) Caelyx (Approved by EMEA in 1996) (Janssen Pharmaceuticals)	Suspension; IV route; 20 months	DoX/Ammonium ion gradient method	Pegylated; HSPC:CHL:PEG$_{2000}$-DSPE (56:39:5 M ratio); 100 nm	Ovarian, breast cancer, HIV-associated Kaposi's sarcoma	[37]
Abelcet (Approved by FDA in 1995)	Suspension; IV route; 24 months	AmB (Drug: Lipid −1:1 M ratio)	A Lipid Complex; DMPC:DMPG (7:3 M ratio); MLV < 5 μ	Severe fungal infections	[32]

Commercial Application in Cosmaceuticals and FMCG Sector

Product Name/ Company	Type of dosage form	Key ingredients	Application/use
Lancome Sleil Soft-Touch Antiwrinkle Sunscreen (SPF15) (L'Oreal)	Liposomal cream	Squaline, zinc oxide, xanthan gum, vitamin, vitamin D3	Antiaging cream
Revitalift (L'Oreal)	Nanosomal cream (very small liposomes)	Pro-Retinol A	Antiwrinkle and Firming face and neck contour cream
Fillderma Lips Lip Volumizer (Sesderma)	Liposomes	Hyaluronic acid, fermented black sweet tea, lactic acid, niacinamide, etc.	Increases volume of lips, fills wrinkles contour, and hydrates the skin

C-Vit Liposomal Serum (Sesderma)	Liposomal Serum	Vitamin C, Ascorbyl glucoside, Mulberry extract, Hyaluronic acid etc.	Hydration, boosts collagen synthesis, enhances skin's elasticity and firmness, and brightens the complexion
Liposome eye cream (Decorte)	Liposomal cream	*Panax ginseng* Root Extract, Tocopheryl Nicotinate Palmitoyl Pentapeptide-4, Glycerin, Hyaluronic acid, CHL	Moisturizes, firms, and brightens the delicate skin around the eyes
Face concentrate (Russell Organics Liposome Concentrate Skincare)	Liposomal Lotion	Vegetable oil, Floral water loaded with superoxide dismutase, beta glucans etc.	Hydrating and rejuvenating. Makes skin firmer, softer, and smoother
Face & Neck Lotion (Clinicians Complex)	Liposomal Lotion	Sunflower Oil, Squalane, Sodium Hylauronate, Tissue Respiratory Factors, Super Oxide Dismutase, Vitamin, etc.	Nourishes skin and prevents photoaging
Capture Totale (Christian Dior)	Liposomal Serum	Longoza, Limonene, Rye seed extract, rice protein, lecithin, and elastin peptides	Removes wrinkles and dark spots and has radiance effect with sunscreen
Dermosome (Angelift)	Liposomal Collagen Serum	Wheat Germ Extract, Seaweed Extract, Ferulic Acid Ferment Extract, Sunflower Seed Oil, Hyaluronic acid, Anthocyanins, Blueberry extract, etc.	Moisturizer
emerginC Hyper Vitalizer Face cream (emerginC)	Liposomal cream	Alpha-lipoic acid, Coenzyme Q10, Lutein, etc.	Antiaging, corrects uneven skin tone

Continued

Table 3 Comprehensive study of marketed liposomal formulations approved by global regulatory bodies in chronological order—cont'd

Proprietary nano product/manufacturer/ approval year	Type of dosage form/delivery route/shelf life	Drug payload/technique	Composition/size (nm)	Therapeutic indication	References
Decorte Moisture Liposome Face Cream (Decorte)	Liposomal cream	Squalane, hydrogenated lecithin, triethyl hexanoin, akebia trifoliata stem extract, *Asparagus officinalis* stem extract, *Fagus sylvatica* bud extract, *Impatiens balsamina* flower/leaf/stem extract, royal jelly extract, grape leaf extract, carbomer, CHL, xanthan gum, etc.	Moisturizer		
Natural Progesterone Liposomal Skin Cream (NOW Solutions)	Liposomal Cream	*Aloe barbadensis* Leaf Juice, Sunflower Seed Oil, Jojoba Seed Oil, Natural Progesterone, Xanthan Gum, Rosemary Leaf Extract, Vitamin A, Carrot Seed Oil, Vitamin E, Panax Ginseng Root Extract, *Serenoa serrulata* Fruit Extract, Matricaria Flower Extract, Lemongrass Oil.		Maintenance of healthy feminine balance	
Advanced Night Repair Protective Recovery Complex (Estee Lauder)	Liposomal Face serum	Bifida Ferment Lysate, *Cola acuminata* Seed Extract, Hydrolyzed Algin, Caffeine, Lecithin, Tripeptide-32, Bisabolol, Squalane, Sodium Hyaluronate, Lactobacillus Ferment, Chamomile, Yeast Extract, Hydrogenated Lecithin, Tocopheryl Acetate, Xanthan Gum,	Skin repair		

Product	Type	Ingredients	Application
Lumessence Eye Cream (Aubrey Organics)	Liposomal Cream	*Rosa rubiginosa* seed oil, Camelia *sinensis* leaf oil, oat extract, rye seed extract, shea butter, carageeaan	Antiwrinkle & firming cream
Kerstin Florian Rehydrating Liposome Day Creme (Kerstin Florian)	Liposomal ream	Shea butter, horse chestnut seed extract, xanthan gum, bisabolol, ruscus aculeatus root extract, panax ginseng root extract, bitter orange flower oil, glycyrrhetinic acid, *Sambucus nigra* flower extract, sodium hyaluronate, limonene, tocotrienols, palm oil, ascorbyl palmitate, etc.	Moisturizer
Mayu Niosome Base Cream (Laon Cosmetics)	Niosomal cream	Wild cultivated ginseng extract, horse oil, saponins, etc.	Antiwrinkle, whitening, and moisturizing cream
Simply Man Match Antiage Response Cream (Nouvelle)	Niosomal cream	Pomegranate seed oil, ribonucleotide monophosphate, Ginseng extract, avocado oil, mineral salts, etc.	Treatment of wrinkles, firming and moisturizer
Identik Masque Floral Repair (Identik)	Niosomal cream	Adenosine, *Punica granatum* seed extract, hydrolysed yeast extract	Hair repair masque
Identik Shampooing Floral Repair (Identik)	Niosomal cream	cocamidopropyl betaine, betaine, punica granatum seed extract, hydrolyzed yeast extract, adenosine phosphate, hydrolyzed wheat protein, hydrolyzed soy protein, polyglyceryl-10 dipalmitate, apricot kernel oil, etc.	Hair repair shampoo
Eusu Niosome Makam Pom Whitening Facial Cream (Eusu)	Niosomal cream	*Glycyrrhiza glabra* extract, Jojoba seed oil, Macadamia Nut oil, Amla Extract, Squalene, etc.	Skin whitening

Continued

Table 3 Comprehensive study of marketed liposomal formulations approved by global regulatory bodies in chronological order—cont'd

Proprietary nano product/manufacturer/ approval year	Type of dosage form/delivery route/shelf life	Drug payload/technique	Composition/size (nm)	Therapeutic indication	References
Effect du Soleil (L'Oréal)	Lipsomes	Nonionic lipid	Tanning agents in liposomes		
Liposome aktions gel (Madame Nanette Biocosmetic)	Liposomal Gel	*Aloe vera*, thymus extract	Antiwrinkle and moisturizing cream		
Future Perfect (Estée Lauder)	Liposomal Gel	Vitamin E, A, cerebroside, ceramide, TMF	Antiwrinkle radiance moisturizer		
Aquasome LA (Nikko Chemical Co.)	Liposomes	Liposomes with humectants like glycerine, sorbitol	Moisturizer		
Eye Perfector (Avon Solutions)	Liposomal cream	Soothing cream with peptides to reduce eye puffiness	Eye depuffing cream prevents irritation, sagging, and wrinkles		
Royal jelly lift concentrate (Jafra cosmetics)	A complex of jelly liposomes and cellspan	Sunflower Sprout Extract, Wintercherry and Lotus flowers, *Larrea divaricata* Extract	Reduce wrinkles, firmness, moisturizes, and improves skin radiance.		
Formule Liposome Gel (Payot (Ferdinand Muehlens))	Liposomal Gel	Thymoxin, hyaluronic acid	Moisturizing, antiwrinkle and brightening cream		
Symphatic 2000 (Biopharm GmbH)	Liposomal cream	Thymus extract, vitamin A palmitate	Moisturizing cream		
Natipide II (Nattermann PL)	Liposomal Gel	Preformed Liposomal gel	Gel For do it yourself cosmetics		
Inovita (Pharm/ Apotheke)	Liposomal cream	Thymus extract, hyaluronic acid, vitamin E	Moisturizing and Antiwrinkle cream		
Flawless finish (Elizabeth Arden)	Liquid make-up	Cocoyl Sarcosine, Lanolin Alcohol, Magnesium Aluminum Silicate, Triethanolamine, Lithium Stearate, Cellulose Gum, Alpha-Isomethyl Ionone, Benzyl Salicylate, Butylphenyl Methylpropional, Citronellol, Eugenol, Hexyl Cinnamal, Hydroxycitronellal, Limonene	To correct skin imperfections and dark circles		

Nactosomes Lancome (L'Oreal)	Liposomal cream	Vitamins, retinolacetate	Moisturizer
Niosomes Lancome (L'Oreal)	Liposomal serum	Glyceropolyether with moisturizers	For fine lines and wrinkles, dryness, and dullness and uneven texture
Isabella Liposomes Repair Gel (Isabella Skin care)	UV-A/UV-B liposome repair gel with Cellular Regen	*Eucalyptus dives* Leaf Water, Vitamins A, B3, B5, C & E, Liposome, Leuconostoc/ Radish Root Ferment Filtrate, Spotted Orchid Extract, *Orchis maculata* Flower Extract, Ginger Root Extract, Menthol Leaf Extract, Rice Extract, Watermelon Extract, *Citrullus lanatus* Fruit Extract, *Artocarpus heterophyllus* Seed Extract, *Centella asiatica* Extract, *Populus tremuloides* Bark Extract, *Camellia sinensis* Leaf Extract, Plankton Extract, Cocus Nucifera	To repair, stimulate and hydrate the skin with anti-inflammatory, antiaging gel for collagen production with antioxidant properties
Micro 2000 Complex Anti-Stress (Elizabeth Arden)	Face Serum	Hydrospheres	For skin repair

EPG, egg phosphatidyl glycerol; *HSPG*, hydrogenated soy phosphatidylcholine; PEG2000-C-DMG, (R)-methoxy-PEG2000-carbamoyl-di-O-myristyl-sn-glyceride; *DEPC*, dierucoyl-sn-glycero-3-phosphocholine; *DPPG*, dipalmitoyl-sn-glycero-3-phosphoglycerol.

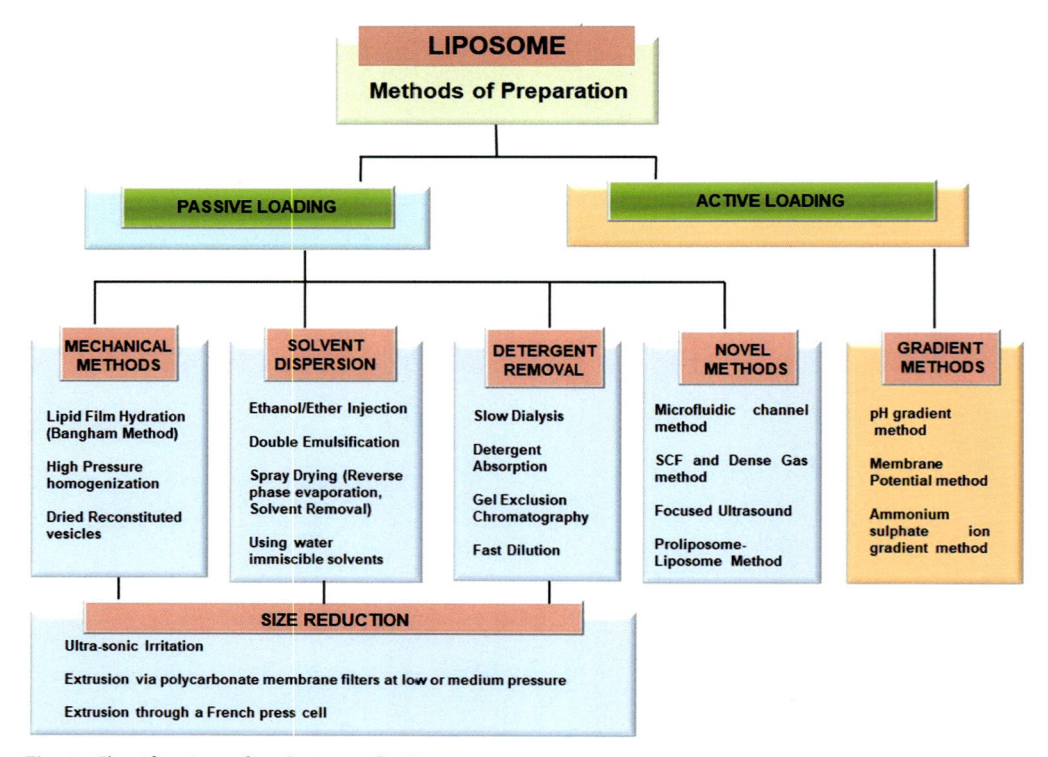

Fig. 3 Classification of techniques for liposome engineering.

2.5.1 Active (remote) loading methods

This method is based on the principle of transmembrane gradient (either pH gradient or ion gradient; i.e., the aqueous phases inside and outside the liposomes are different) which prompts the spontaneous transport of amphiphilic drugs from the extraliposomal aqueous medium across the lipid bilayer(s) into the intraliposomal aqueous phase followed by the interactions with a trapping agent in the core to lock in the drug (Fig. 5). Therefore, formation of drug aggregate with high energy of activation leads to stable complex with high and effective intraliposomal encapsulation.

Advantages

- A versatile method for amphipathic molecules (except drug structure should possess 1°, 2° or 3° amine groups).
- Allows independent variation of any liposomal parameter and higher drug to lipid ratios.
- High drug encapsulation and drug retention can be achieved for uncharged molecules (e.g., Docetaxel) where drug can be converted to prodrug and then could be actively loaded into the vesicle.

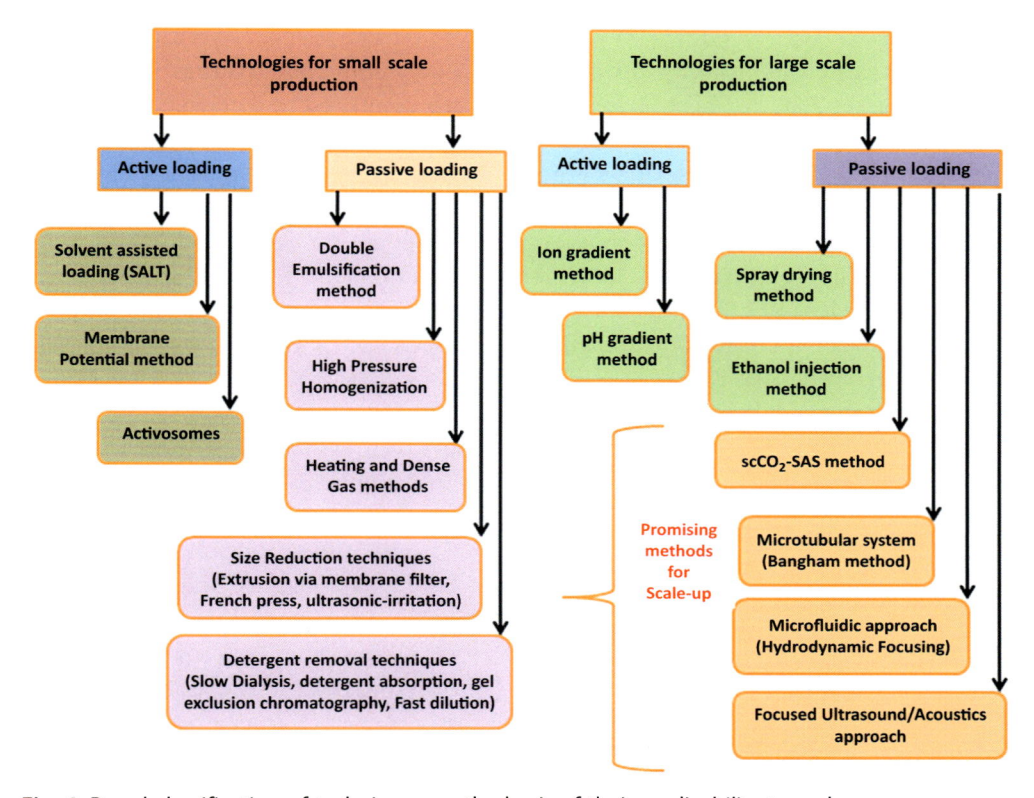

Fig. 4 Broad classification of techniques on the basis of their applicability to scale up.

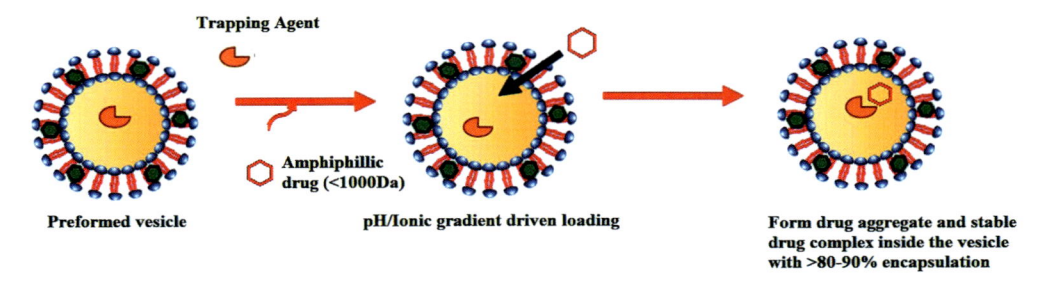

Fig. 5 Overview of active loading method. *(Adapted from G. Pauli, W.-L. Tang, S.-D. Li, Development and characterization of the solvent-assisted active loading technology (SALT) for liposomal loading of poorly water-soluble compounds, Pharmaceutics 11 (2019) 465, under creative common attribution license).*

Limitations
✓ Doesn't exhibit universality and applicable to only amphipathic drugs.
✓ Have the tendency of high leakage rate in vitro compared to in vivo.

Active loading methods were developed in late 1980s, yet they are one of the most employed methods to encapsulate a drug in preformed liposomal vesicles. The work of Nichols and Deamer (1976) conceptualized the idea of accumulation of weakly basic lipophilic molecules, e.g., catecholamines (many times greater than its solubility) inside the interior of liposome assembly in response to pH gradient [39]. Initially, liposomes were formed at low pH (pH $\sim$ 5), then these vesicles were exposed to alkaline solution of pH $\sim$ 8 in order to create 1000 fold difference in the H_3O^+ ions across the bilayer. The catecholamine molecules remained as free form in the basic exterior, as they freely permeated through the bilayer into low pH core; these catecholamines were subsequently protonated. These charged catecholamine molecules were no longer permeable and got locked in, which ultimately ended up in high accumulation of catecholamine molecules. Thereafter, active loading was studied using ion gradient (generated by internal acidic buffers or proton-generating dissociable salts, such as ammonium sulfate) for encapsulation of anthracycline molecules [40]. Briefly, concentrated solution was added to the liposomes and then loading was achieved after incubation at elevated temperature. The anthracycline molecules formed aggregates in the liposome core via precipitation with sulfate ions and obtained final loading of >90% without burst drug leakage. Various ion gradients (such as calcium acetate gradient, basified copper acetate gradients, phosphate ion gradients) have been employed for active loading of preformed vesicles.

Active loading can be achieved by both small scale (i.e., pH/ion gradients, membrane potential, and solvent-assisted techniques) as well as large scale methods (i.e., ion gradient and pH gradient). Another application of active loading technique is the employment of liposomes composed of cationic amphiphiles to form assemblies with polynucleotides and oligonucleotides (known as Activosomes) for gene delivery and modulation of its expression [41].

Modified applications—Solvent-assisted active loading technology (SALT)
In order to improve the encapsulation efficiency for poorly immiscible drugs and formulation stability in the liposome core, addition of small volume ($\sim$5% v/v) of solvent was carried out to the liposomal loading mixture. This simple approach resulted in rapid loading of poorly soluble drugs and attained high drug-to-lipid ratio with stable drug retention. Therefore, SALT emerged as a validated tool to improve the drug loading, retention, stability, palatability, and pharmacokinetic properties [38].

Commercial scale methods based on active loading
Industrial scale manufacturing of nanoformulations is quite challenging as the production processes are time-consuming and involve several complex stages. Hence, an effort has

been laid to compile standard procedures, validated adaptions, and examples of marketed products of liposomal formulations in order to develop a road map for the industrial scale-up of aforesaid nanoformulations with cost-effective design processes.

Transmembrane ion gradient technique: This technique is exclusively used for liposome design and touted as one of the best methods for producing efficient drug level per liposome with cost-effectiveness. Commercial liposomal technologies like *Doxil, Lipodox* (generic version of *Doxil*), and *Myocet* were developed through ion gradient and pH gradient techniques, respectively.

Principle: The principle of ammonium sulfate (ion) gradient $[(NH_4)_2SO_{4internal} \gg (NH_4)_2SO_{4external}]$ method is actually a base exchange of amphipathic weak base drug with the ammonium ions [42]. More than 90% of the DoX got encapsulated in the intraliposomal aqueous phase as precipitate of $(DoX-NH_3)_2SO_4$ as illustrated in Fig. 6.

High encapsulation in case of *Doxil* was attributed to the presence of sulfate as counter-anions of the ammonium cation, lipid composition, and temperature which affected the level of $(DoX-NH_3)_2SO_4$ precipitation. Further, the uniqueness of employing ammonium sulfate gradient over other ion gradients is that neither this approach requires fabrication of liposomes in acidic pH nor the alkalization of extra-liposome aqueous phase. The critical considerations domineering the efficient drug loading and stability are summarized as follows:

✓ Huge difference in the permeability coefficient ($\sim 10^{12}$) between neutral ammonia (10^{-1} cm/s) and SO_4^{2-} anion ($> 10^{-12}$ cm/s)

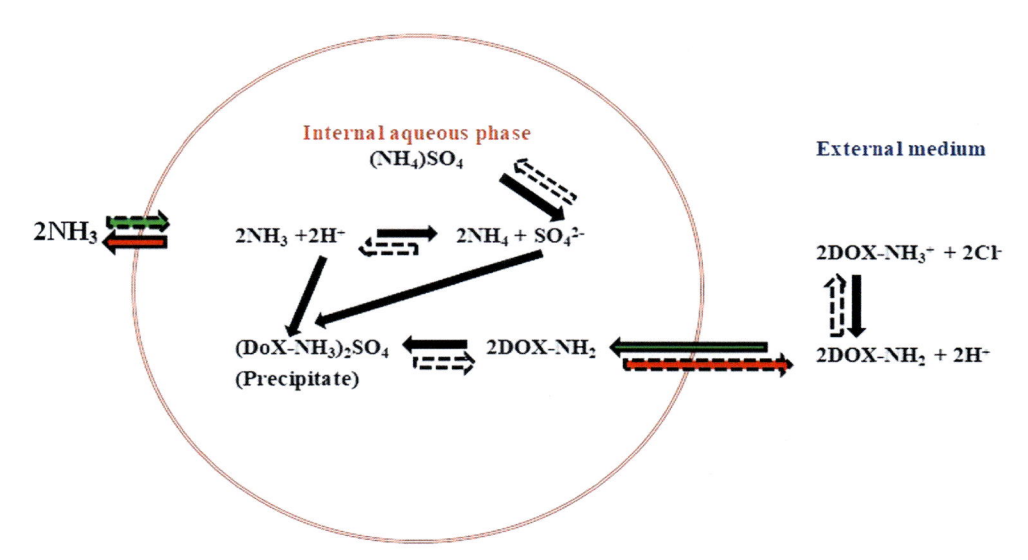

Fig. 6 Mechanism of transmembrane ammonium sulfate ion gradient in development of liposomal DoX.

✓ Initial pH gradient with $[H^+]_{internal} \gg [H^+]_{external}$
✓ Poor solubility profile of DoX sulfate precipitates (<2 mM) which reduce the osmotic pressure inside the vesicle internally and thereby assist in maintaining lipid bilayer integrity
✓ Asymmetry in partition coefficients of DoX within the inter- and intra-environments of the vesicle

Scale-up process for liposomal DoX (e.g., Doxil): A general multistep procedure for the development of liposomal DoX is given as follows [43]:
- Preparation of MLVs by lipid film hydration containing ammonium sulfate;
- Formation of LUVs from MLVs by freezing and thawing, followed by liposome extrusion through polycarbonate filters of a defined pore size that results in a homogeneous population of single bilayer vesicles of decreased particle size;
- Establishment of an ammonium sulfate gradient between the external aqueous phase and the interior aqueous compartment of the liposomal preparation (usually by size-exclusion column chromatography);
- Encapsulation of DoX by active loading.

The key common feature of commercial liposomal DoX manufactured by lipid film hydration (with subsequent downsizing of vesicles through freeze thawing/extrusion) followed by active loading obtains a physical form which could be either a liquid suspension or lyophilized powder having (with mean concentration of 8 mM of PHL and CHL of 0.1–0.2 mM in case of vaccines) vesicle size range of 80–100 nm. Commercialized products such as Doxil/Lipodox have been manufactured using ammonium ion gradient method to obtain the high loading efficiency of DoX, where the neutral form of DoX from a sucrose medium crosses the PHL bilayer (composed of hydrogenated soy phosphatidylcholine and CHL) and the protonated DoX formed inside the aqueous compartment with ammonium sulfate is retained [44]. The surface of liposomal moieties was sterically stabilized by anchoring distearoyl-phosphatidylethanolamine-PEG molecules to PHL bilayers. The step-wise procedure or flow chart of process is being summarized in Fig. 7.

2.5.2 Passive loading methods

It is worthy to emphasize that, for a small scale batch, conventional passive loading could possibly be achieved with variety of techniques for lipophilic drugs assuming partition coefficient of drug favors lipid phase, resulting in stable drug retention (Fig. 8).

However, passive drug methods employed for hydrophilic drugs often end up with low encapsulation efficiencies ($\sim$5%–20%) with remainder of the drug being in the bulk aqueous phase. Therefore, it can be implied that these methods require additional unit operations to eliminate nonencapsulated form of drug. Reverse-phase evaporation from an organic solvent has been used for encasing hydrophilic entities with encapsulation up

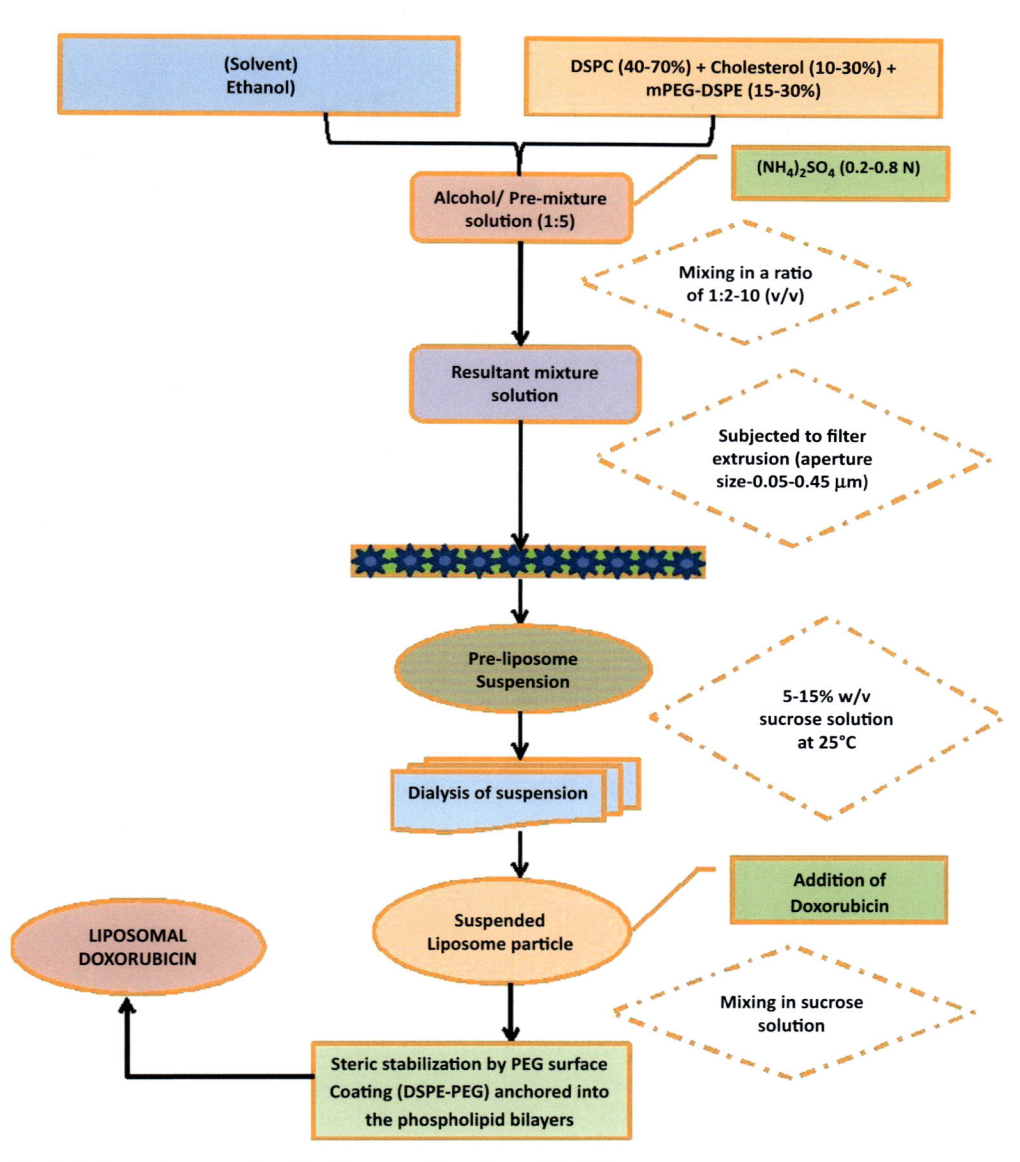

Fig. 7 Schematic procedure for the manufacturing of DoX liposomes.

to 50%. However, such methods are overly complex to regulate continuous solvent operation and thus it becomes very challenging to establish their sterile boundary [45].

Generally, passive loading of small drug molecules impinges upon their physical attributes and is often employed for the incorporation of hydrophobic drug candidates into the lipid bilayer as well as sequestration of hydrophilic drugs into aqueous core during

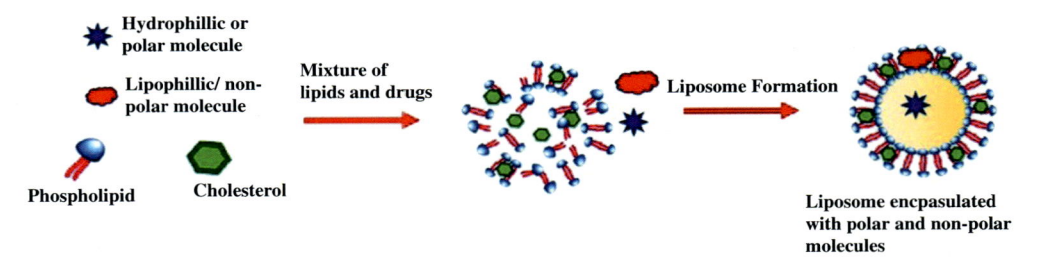

Fig. 8 Overview of passive loading. *(Adapted from G. Pauli, W.-L. Tang, S.-D. Li, Development and characterization of the solvent-assisted active loading technology (SALT) for liposomal loading of poorly water-soluble compounds, Pharmaceutics 11 (2019) 465, under creative common attribution license).*

liposome formation. Such drug entrapment techniques tend to produce low encapsulation efficiencies [38] or rapid leakage following systemic administration. The general overview of liposomal preparation based on passive drug loading methods and processes is depicted in Fig. 9.

Various shearing strategies for size reduction and polydispersity (such as ultrasonic-irritation, high-pressure mixing/extrusion by means of homogenizer, extrusion via French press or via membranes) are commonly employed in passive loading methods (i.e., lipid hydration or Bangham method, detergent depletion or emulsion methods, reverse-phase evaporation) and either fetch small batches with varying sizes or become counterproductive owing to very high pressures as well as difficulties associated with equipment sterilization [43].

Further, use of high-pressure homogenizer with intense energy often ends up in exposing the lipids and heat-sensitive molecules (proteins, peptides, and enzymes) to degradation. Furthermore, materials used in construction may also cause interference by adsorption of the lipid on the walls of the system. Additionally, increasing complexity of architecture of the systems and erosion of homogenizer's components (such as "O" rings) may result in contamination of liposomal mixtures. Therefore, these conventional passive loading methods (also regarded as top–down approaches) are discontinuous in nature and more conducive to small scale setup. On the other hand, novel technologies such as supercritical fluid methods and dense gas methods are still evolving. These methods involve processing steps of high temperature and high pressure for feed vessels which make resupply and continuous operation nonfeasible, respectively [46].

Ethanol injection technique

Ethanol injection technique is an interesting intervention for exclusively employed production of liposomes. In early 1970s, first report on fabrication of SUVs employing ethanol injection technique came into light as a substitute method to sonication [47]. Although the experimental results produced smaller size vesicles with poor encapsulation

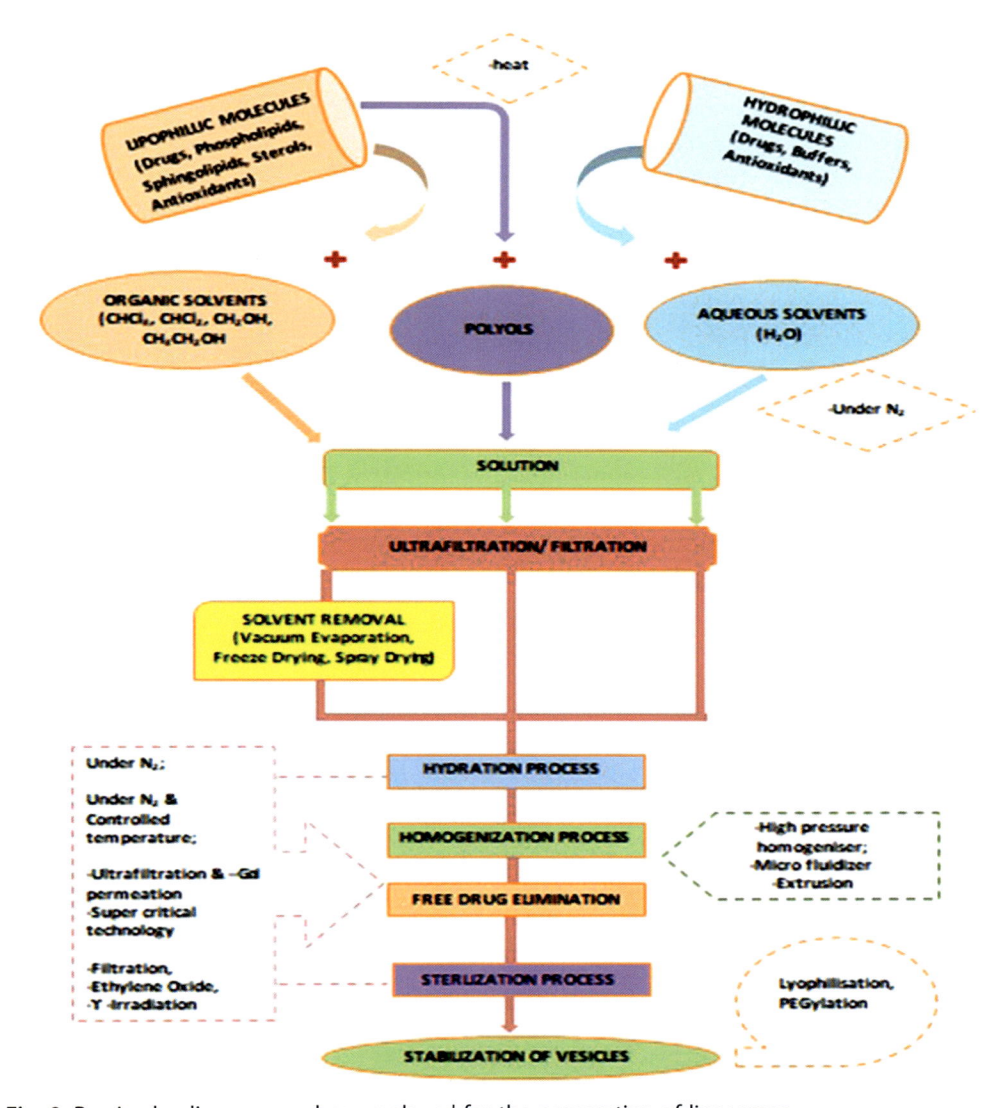

Fig. 9 Passive loading approaches employed for the preparation of liposomes.

(owing to low lipid concentration), it was later diversified into large range of vesicle size [48].

Advantages:
- Simple, highly reproducible (w.r.t. vesicular diameter and encapsulation rates), fast continuous process at pilot scale and doesn't allow lipid degradation or oxidative alterations.

- No mechanical forces are required to generate homogeneous and narrow distributed liposomes.
- Suitability for the entrapment of hydrophilic proteins by passive encapsulation, small amphiphilic drugs by a one-step remote loading technique, or membrane association of antigens for vaccination approaches.
- Modified methods such as membrane for injection improve micromixing of the organic phase into the aqueous phase and therefore optimize the preparation time.

Principle: In this technique, lipids dissolved in ethanol are rapidly injected using a thin needle into an excess of aqueous media (the lipid molecules precipitate and form bilayer planar fragments) while stirring, which leads to instantaneous generation of liposomes with diameter <300 nm [45]. The solvent is removed via dialysis to obtain liposomes (within acceptable limit of residual solvent). The curvature of lipid bilayer fragments obtained during this process becomes quasi-spherical, as lipid bilayers tend to reduce the exposure of hydrophobic part of their molecules to aqueous medium. Attempts have been followed to address such issues by using isopropanol (another water-soluble solvent) as a substitute to ethanol on the similar principle in ethanol injection technique [49]. The findings revealed better control of vesicle diameter, lesser toxicity profile than ethanol, and easy removal (with residual solvent content <0.5%).

Limitations: However, many limitations do exist along with these solvent-based processes, in which vesicle size increases with increase in lipid concentration.

Applications: This application has great potential for the commercial scale fabrication of liposomes for hydrophobic drugs. Further, this method can also be employed in conjunction with other processes such as high shear homogenization in certain cases.

Commercial scale methods based on ethanol injection technique

Among the passive loading methods, most successful scale-up/industrial techniques for liposome manufacture are either based on the principle of alcohol injection, where dissolved lipids are precipitated from an organic solvent into an aqueous solution (anti-solvent) by means of reciprocal diffusion of the alcohol and aqueous phases, or on the principle of spray drying, where organic solvents are used to dissolve the lipids and finally remove from the lipid-drug mixture to form a film via spray dryers.

The cross-flow injection (CFI) technique: For commercial scale-up of liposomes, cross-flow injection (CFI) techniques (modification of ethanol injection method) present the most suitable practical approaches for the continuous cGMP manufacturing of liposomes under ambient conditions (with production of 250 L of a batch in 1.5 h). These techniques guarantee batch to batch consistency and provide long-term stability of liposomes even at room temperature.

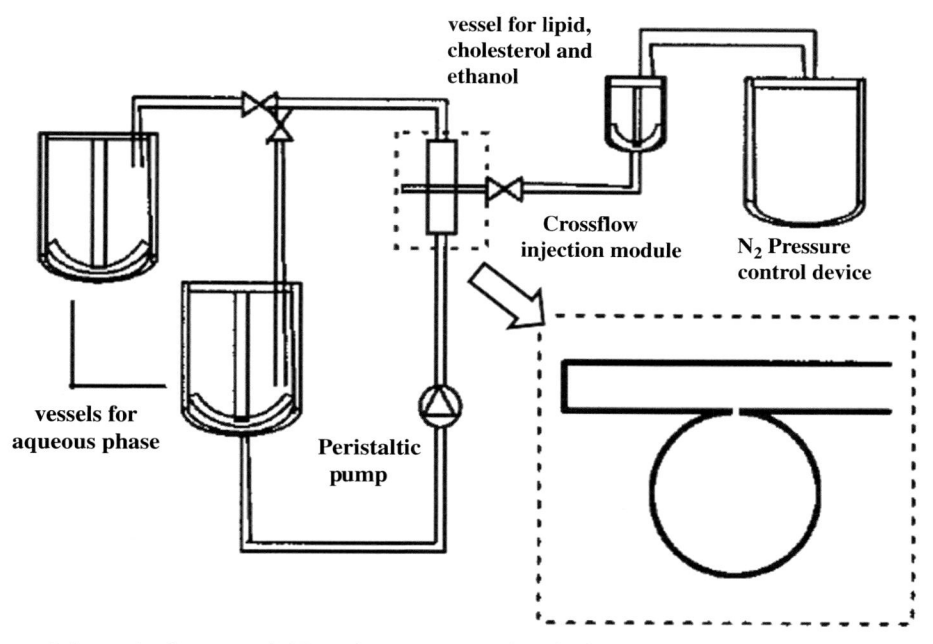

Fig. 10 Schematic diagram of CFI technique. *(Reproduced after licensed permission from publisher Taylor & Francis copyright@ 2002).*

Further, the CFI is a quite mild procedure with a closed system, one-step aseptic process for sensitive drugs (such as proteins, peptides, or thermolabile drugs). The CFI device comprises of two tubes welded together forming across and presenting an injecting hole at connecting point [50, 51] as shown in Fig. 10.

The organic phase (comprised of ethanol, PHLs, CHL, and drug) was injected into aqueous phase through this CFI device. Hence, this novel design enabled to control the vesicle size by manipulating the local lipid concentration at injection point (depending upon the process parameters such as injection pressure, lipid concentration, and injection rate) and assisted in easy scale-up by just changing the sterile vessels once the operational parameters are defined [52].

Modification of CFI: A modified approach employing microfluidic device for injection was suggested by Zhong et al. [53]. The technique comprises of three stages—first involves CFI utilizing special Y connector; second, ultrafiltration for removal of solvent; and third, high-pressure extrusion through polycarbonate membrane to manipulate liposome size. These injection techniques offer high end micromixing of both phases at injection point which results in bilayer planar fragments required prior to liposome formation for large scale production of batches.

Scale-up process of liposomal PTX (e.g., EndoTAG-1): The foundation of EndoTAG-1 (a highly promising liposomal formulation presently under Phase III trial for breast cancer) was derived from the serious concerns of earlier commercialized PTX-based formulation- *Taxol* (obtained from bark of *Taxus brevifolia*) used for the treatment of breast and ovarian cancer. The biggest obstacle in the formulation development of Taxol was the poor aqueous solubility of PTX ($\sim$1 µg/mL), which was improved by dissolving it in Cremophor EL (surfactant derived from polyethoxylated castor oil) and absolute alcohol as cosolvent. Further, prior to administration of formulation having improved solubility, a dilution step was a prerequisite which often resulted in precipitation and crystallization of PTX [54]. Therefore, in order to eliminate the irritation of precipitated drug, a filter was placed between the infusion bag and the injection port. Furthermore, apart from enhancing the solubility of PTX, addition of Cremophor EL also led to increased toxicity and hypersensitivity and thus failed to achieve the desired aims of the formulation.

Taking a lead from this matter, the concept of PTX-loaded cationic liposomes-based formulation, EndoTAG-1, for neovascular targeting of drug into the lipid bilayer was materialized by Michaelis and Haas in 2006 as well as Holvoet et al. in 2007 [55, 56]. The scale-up design of EndoTAG-1 via ethanol injection technique was feasible using combination of positively charged DOTAP-Cl and zwitterionic, DOPC as PHL base. The final formulation exhibited net positively charged which result in effective targeting of PTX to the negatively charged endothelial cells of the tumor blood vessels in breast and ovarian cancer. Although the viability of the liquid liposomal PTX formulation acquired through these methods was not suitable for long-term storage for several months (owing to crystallization tendency of PTX), with the introduction of lyophilization technique, these limiting issues were sorted and the product with storage shelf life of at least 24 months was obtained.

Membrane contactor technique: Literature studies have indicated several modified strategies adopted by researchers in ethanol injection for pilot scale operation such as membrane contactors in a tubular [57] or hollow fiber configuration [58] through SPG membranes (in which lipids are injected through membrane pore channels) [59] to improve the final yield (with volumes up to 3 L), vesicular stability, and robust mass scale development of sterile liposomes with cost-effectiveness (Fig. 11). The method has also shown huge potential for continuous, large scale liposome production [60].

Some studies have reported two diverse strategies (i.e., cross low recirculation of aqueous phase across the membrane surface and low frequency ($\sim$40 Hz) of the membrane for scale-up of the liposomes have been proposed).

Continuous process designs: Enormous amount of attention has been paid by the drug delivery scientists to evaluate the possibility of a single step scalable technique with cost-effective operational design, which involves programmable online flow-based

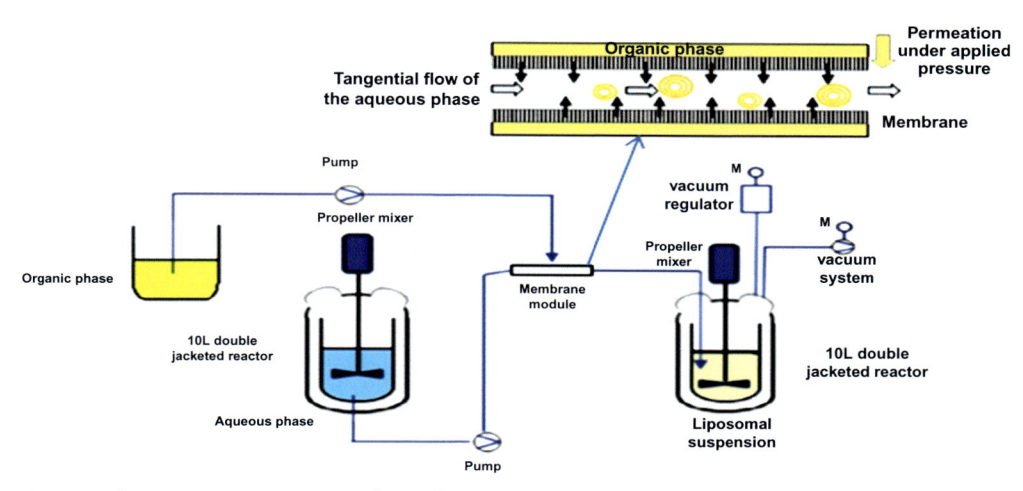

Fig. 11 Schematic representation of membrane contactor (hollow fiber configuration) experimental setup. *(Reproduced with licensed permission from Elsevier).*

strategies to realize controlled precipitation and self-assembly of PHL layers into uniform vesicular structures.

In order to aim for cost-effective scale-up design, Worsham et al. proposed a concept for the batch process design as well as continuous process design exploiting CFI technique where the liposome formulation can proceed indefinitely, if provided continuous supply of formulation of feed solutions [46]. The investigators proposed downstream unit operation of liposomes similar to biologics where unit operations such as tangential flow filtration (TFF) could be employed to remove the undesired elements like nonencapsulated drug or organic solvent and concentrate the final drug product to obtain a final desired strength.

Spray drying technique

This method is one of the cost-effective post-processing applications for the mass scale production of many nanoformulations, viz., liposomes, nanocrystals, nanosuspensions, and nanoparticles.

The application of spray drying technology in the field of liposome technology dates back to early 90s when the first study on the prospect of spray dried liposomes was published by [61]. However, Goldbach et al. investigated the preparation of spray dried liposomes of atropine (particle size of 200 nm) and α-tocopherol for pulmonary delivery employing spray drying technology [62]. Spray drying has also been widely implicated during the storage of liposomal products; however, it could be stated as the most cost-effective technologies for the mass development of liposomes.

Advantages:
• Able to operate in diverse applications and easily extrapolated to virtually any capacity.

- Adaptable to fully automated control that allows continuous processing of multiple process variables simultaneously.
- Can be used with both heat-resistant and heat-sensitive products without degradation.
- Offers high precision control over size, density, degree of crystallinity, volatile impurities, and residual solvents.
- Encapsulating solid and liquid particles.

The limitation of spray drying is the overall low thermal efficiency as the large volumes of heated air pass through the chamber without contacting a particle, thus not contributing directly to the drying. The schematic diagram represents a model of Nano/Mini spray dryer model (Buchi, Switzerland) employed to fabricate liposomes, nanoparticles, and nanocrystals (Fig. 12).

For commercial liposome design, Kukuchi et al. developed spray dried liposomes composed of DPPC:CH:DCP in a molar ratio of 7:3:1 (having lipid:core material in ratio of 39:30) using mannitol as core material and subsequently spray dried liposomes

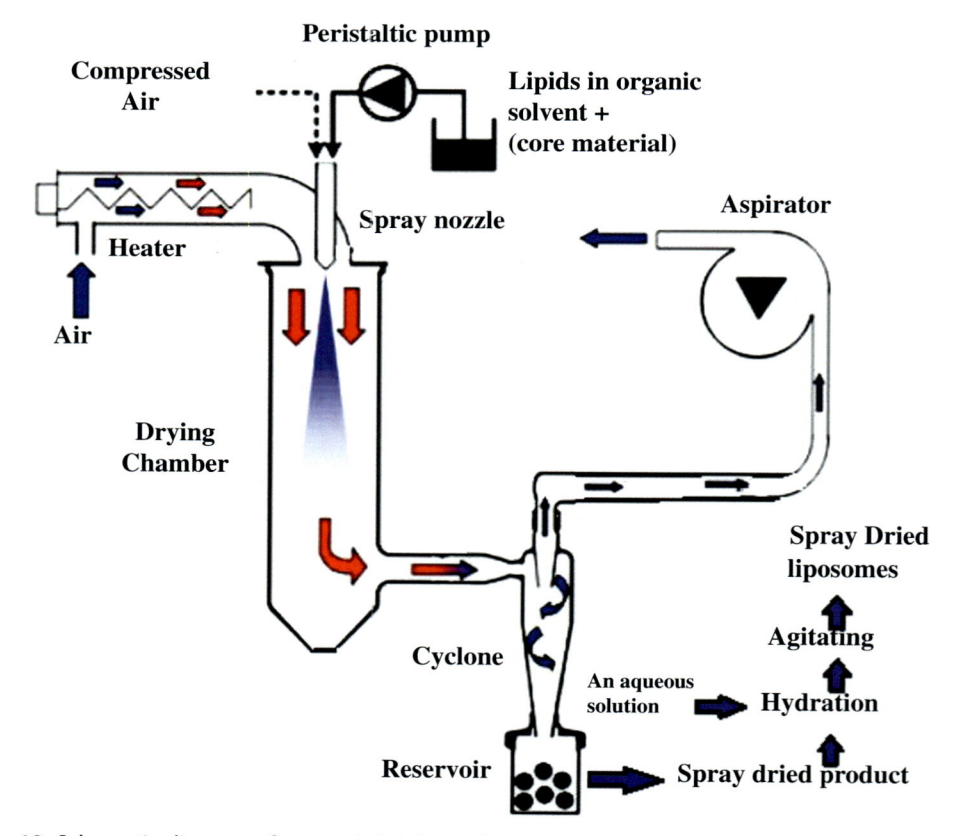

Fig. 12 Schematic diagram of a nano/mini Spray dryer system.

obtained were compared to conventional liposomes prepared with Bangham method [63]. It was reported that the use of mannitol as core material enhanced the final recovery of the product (by preventing the dry lipid mixture from adhesion to the wall of the chamber).

Another report investigated the development of AmB-EPC (using methanol as solvent) composite particles using spray dryer for the treatment of systemic fungal infections [64]. The study revealed that AmB-EPC composite particles (with size ranges of 192–505 nm) obtained as spray dried product were hydrated in phosphate buffer saline (pH 7.4) by gentle shaking at 25°C for 30 min. The final product was stored as dried form until hydration in order to prevent inherent instability issues.

Scale-up process of liposome AmB (e.g., AmBisome): The primary limitations associated with AmB liposomes were stability and toxicity issues, high concentration of residual solvents (such as methanol) required to dissolve lipid mixture, and time-consuming operations or additional processing steps (such as removal of the solvent from the mixture to form lipid–AmB film). In order to overcome these challenges, Proffitt et al. patented a novel method for industrial scale manufacturing of AmB-liposome which primarily addressed all the aforesaid concerns and showed marked reduction in toxicity [65]. The commercially blockbuster product *AmBisome* is also based on this spray drying technique (Fig. 13). The scale-up technological process for the development of liposomal AmB is discussed below.

Apart from volatile solvent-based methods, super critical processes have captured the attention of scientific community as a viable green alternative to fabricate nanoliposomes, nanocrystals, nanoparticles, and nanosuspensions for small scale as well as industrial scale production [66].

Supercritical fluid technology (SCFT)

Supercritical fluid technologies (SCFT) are the novel paradigms emerging in the field of nanotherapeutic drug delivery as a good green alternative to conventional preparation methods for nanoformulations design [67]. Supercritical fluid (SCF) is the "state of any compound, mixture or element above its critical temperature and critical pressure" exhibiting special physical properties, specifically liquid-like density, gas-like diffusivity and viscosity, and zero surface tension.

Principle: This green technology employs eco-friendly SCFs (such as supercritical carbon dioxide $scCO_2$, very dense at certain temperature and pressure beyond a critical value, i.e., $\sim$31°C and 1070 psig, respectively) in the development of nanoformulations in order to overcome the limitations of scaling up and requirement of large amounts of toxic organic solvents used in conventional methods.

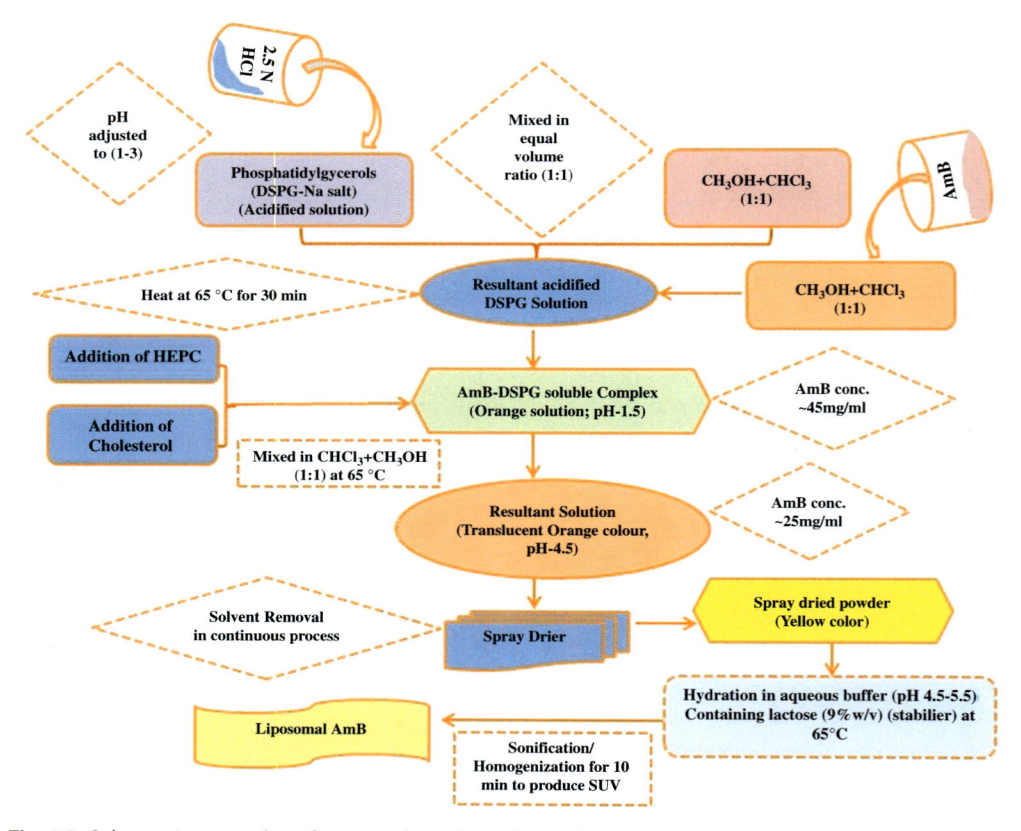

Fig. 13 Schematic procedure for manufacturing of AmB liposomes at industrial scale.

Applications: $scCO_2$ technique exhibits relatively low critical temperature and pressure and offers excellent prospect as an organic solvent-free method owing to its low cost, nontoxicity, noninflammability, recyclability and easy removal, and operation at moderate temperatures, preventing degradation of product in an inert atmosphere and dissolution properties analogous to those of nonpolar solvents [66]. Therefore, the application of SCFT in pharmaceutical industry includes nanoparticle and nanocrystal engineering, composite particle preparation, coating of solid dosage form, liposome preparation, extraction, and protein and peptide drying.

The first investigation of developing liposomes using $scCO_2$ was done in 1994 by Castor. Since then, several advances have been introduced to make this technology more viable for the large scale liposome production. AmB and cyclosporine-A (Cys-A) are the major two drugs which have been investigated for the industrially viable production of liposomes using supercritical fluid technologies [68–70].

The prime causative factor for the recent attention towards SCFT for liposome development is the negligible quantity of residual organic solvent present in the final material compared to conventional methods. The selection of processing technique with $scCO_2$ for biopolymers is governed by the interaction of the $scCO_2$ with the active ingredient, coating material of interest, and suitable solvent [71]. The advantages of SCFT are summarized as follows:

Advantages:
- Allow one-step preparation with a significant decrease or complete suppression of the use of organic solvent.
- Capable of producing pharmaceutically viable products of diverse size of unilamellar vesicles (100–1200 nm) with five-fold greater encapsulation efficiency compared to conventional methods.
- Offers the possibility of optimizing particle size, in situ sterilization, and large scale production in industrial settings.

Some of the limitations associated with this technique are low yield, use of high pressure (200–350 bar) during the process, requirement of special infrastructure, and restricted application in the liposomal drug development [45].

SCFT can be categorized into three broad classes on the basis of employment of $scCO_2$ discussed as follows:

(i) *As solvent for API and excipients*: e.g., Rapid expansion of supercritical solutions (RESS), Particle formation from gas saturated solutions (PGSS), Depressurization of an Expanded Solution into Aqueous Media (DESAM), Rapid expansion of a supercritical solution into a liquid solvent (RESOLV), Rapid expansion from supercritical to aqueous solutions (RESAS), Depressurization of an expanded liquid organic solution (DELOS), supercritical reverse-phase evaporation, and supercritical reverse-phase evaporation (SCRPE).

(ii) *As anti-solvent for API and excipients*: e.g., Gaseous Anti-solvent (GAS), Aerosol Solvent Extraction System (ASES), Particles by Compressed Anti-solvent (PCA), Supercritical Anti-solvent (SAS), Solution Enhanced Dispersion by Supercritical Fluids (SEDS), and Atomization of supercritical anti-solvent induced suspension (ASAIS).

(iii) *Use in aerosolization or spray drying techniques*: e.g., Carbon dioxide-Assisted Nebulization with Bubble Dryer (CAN-BD) and Supercritical Fluid-Assisted Atomization (SAA).

Commercial scale manufacturing of Nanoformulations (nanoliposomes/ nanocrystals/nanoparticles/nanosuspensions) using SCFT: SCFTs such as RESS employ minimal amount of organic solvents and reuse of SCFs; however; in case of liposome manufacturing, the major limitation is the immiscibility of PHLs in pure $scCO_2$. As PHLs assemble themselves only in aqueous medium, conventional RESS is applicable to other nanoformulations. Therefore, modified RESS approaches (such as RESS—coupled with vacuum-driven cargo loading; RESS—through a capillary tube; RESS—through a T-piece) have been introduced where PHLs and essential oils are dissolved in a mixture of $scCO_2$ and ethanol, and then the solution is sprayed into a buffer solution to form a liposome suspension which stated good prospect for scale-up production of liposomes. Techniques such as RESOLV involve atomization of $scCO_2$ solution dissolving drug into an aqueous medium which (causes rapid expansion of solution followed by quenching) ends up in particle formation. However, some of the issues related to particle recovery are still needed to be addressed.

Further, these green technologies are also capable of producing sterile, solvent-free, pharmaceutical-grade nanoformulations (nanocrystals, nanoparticles, and nanosuspensions). However, cost-effective production at large scale using SCFTs is yet evolving as majority of the nanoformulations are relatively less accessible due to high cost of manufacturing and convolutions in design for scale-up [72].

On the other hand, anti-solvent-based SCFTs (e.g., SAS) have been shown tremendous potential to fabricate low cost, commercial scale development of liposomes, nanocrystals, nanoparticle, or nanosuspensions.

Scale-up process of liposomes using SAS: Hwang and coworkers first studied the application of SAS process for the low cost manufacturing and scale-up design for industrial production of liposomal AmB, (based on the concept of proliposomes) using $scCO_2$ in which dried mixture of lipids, CHL, and drug of interest are coated with anhydrous lactose, a water-soluble carrier [73], as shown in Fig. 14.

In SAS operation, the resultant mixture was received into a sealed reaction vessel and $scCO_2$ was injected into the reaction vessel followed by stirring (at temperature ~45–65°C, pressure ~150–300 bar) at 500 rpm for 40 min. At the end of the operation, $scCO_2$ was discharged and vessel was decompressed to atmospheric pressure. Finally, lactose particles coated with drugs and lipids were obtained. These particles were hydrated by addition of an aqueous media (including water) in a supercritical process. Further, optimum size (~137 nm) was obtained using microfluidizer with drug loading efficiency of 89.2% and zeta potential of −42.49. Thereafter, freeze drying operation was carried out to stabilize the AmB liposomes.

The investigation was extended to study the effect of solvent, process temperature, pressure, surface area of the carrier particle, and hydration temperature in the preparation of liposomal AmB employing $scCO_2$ compared to conventional Bangham method [74]. Findings from the study showed $scCO_2$-mediated liposomes (size ~137 nm after homogenization and drug encapsulation efficiency of 90%) were bioequivalent to AmBisome nanoformulation pharmacokinetically and exhibited stability over a period of time.

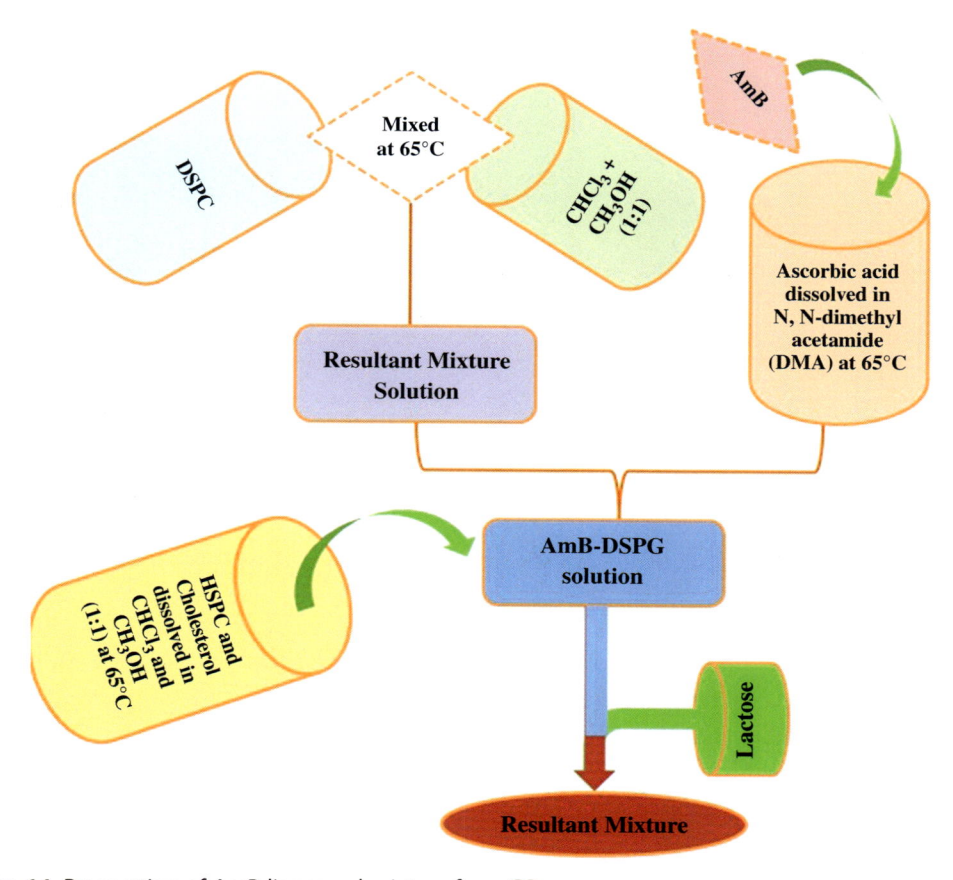

Fig. 14 Preparation of AmB liposomal mixture for scCO₂.

Latest developments

Microfluidic hydrodynamic focusing (MHF): Another potential application which enables control of nanoformulation size and size distribution during manufacturing, without additional postproduction steps [75, 76], has been reported in the literature. Confinement in space and controlled mixing in microfluidics make it an attractive approach for fabrication of nanoliposomes, nanocrystals, nanoparticles, and nanosuspensions with diverse size ranges. The self-assembly in microfluidic devices can be precisely controlled by varying flow rates, ratios of cross-flows, and the composition and concentration of lipids, resulting in tunable sizes and narrow size distributions.

Application of MHF in liposome design: In this technology, liposomes are spawned by hydrodynamically focusing a stream of lipids (by dissolving it in ethanol or isopropyl alcohol) into the central channel which is further intersected by the two-sided aqueous

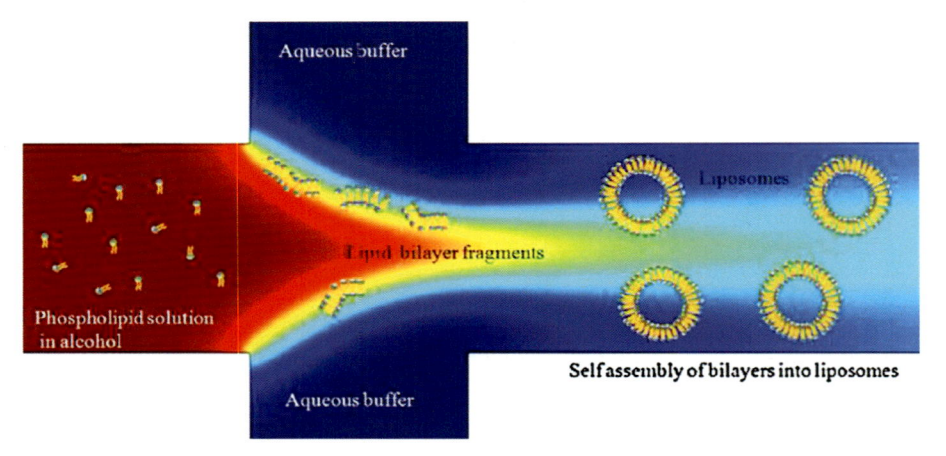

Fig. 15 Schematic diagram of MHF (diffusive mixing of PHL solution in alcohol and aqueous buffer are facilitated by micro channels). *(Reproduced with licensed permission copyright @Elsevier, 2014).*

streams (lead to controlled diffusive mixing), which ultimately ends up in producing liposomes due to varied shear forces generated at liquid interfaces by changing the flow rate ratio (Fig. 15).

In a new advancement to MHF technology, a thermoplastic microfabrication technique was devised based on a fully equipped pharmacy-on-a-chip model for liposomal production (including vesicle formation, drug encapsulation, and vesicle functionalization). The model works on "all-in-one" step process at low cost scale-up design for producing significantly high, clinically viable liposome volumes (approximately ~100 mg/h lipid) [77].

Electrostatic complexation of DNA using MHF: MHF technique has also been implicated for the complexation of DNA with cationic liposomes through electrostatic interactions. The principle of complex formation was based on two steps. Firstly, formation of multi-lamellar complex took place, followed by organization of DNA inside the lamellae. DNA solution and cationic liposomes (comprised of DOTAP: DOPC: 1:1) were channelized through the central stream and lateral position stream, respectively, at flow rate of 100 mm/s to obtain liposome size of 200 nm [78].

Focused ultrasound technology: A novel technology commercially marketed as Adaptive Focused Acoustics (AFA; a proprietary of Covaris, MA, USA) has been described in the literature [79]. The technology has been touted to fabricate efficient nanoliposome formulations at small scale as well as at pilot scale. Further, the technology has been validated to produce Doxil-like liposomes at 4°C with size of 200 nm with 30 min of processing time. It has been reported that the technology employs computer-guided process, which eliminates the need to heat the lipids or to dissolve them in a cosolvent during the formulation and ensures batch-to-batch repeatability. Further, the disposable closed flow-system prevents inter-batch contamination and alleviates the need for exhaustive wash cycles.

Microtubular system (based on Bangham method): This approach (based on the classical lipid film hydration method, also known as Bangham method) was investigated by Carneiro and Santana Andrade [80]. The study was carried out to demonstrate the feasibility of producing large scale production of liposomes having a mean diameter of 1500 nm via multitubular system (continuous adsorption of feeding lipid dispersion in ethanol in large tube-like columns, then film was dried by solvent evaporation, and finally hydrated using buffer solution).

3. Nanocrystals and nanosuspensions as nanoplatforms

3.1 General characteristics

Poorly water-soluble drugs take up about 60% of the commercially available active pharmaceutical ingredients (API), thus making formulation development a challenging task as poor water solubility leads to lesser absorption from gastrointestinal tract, ultimately leading to poor bioavailability [81]. Thus, it becomes essential to incorporate some methods to improve the water solubility of these drugs. Various methods are available like using cosolvents, surfactants, and β-cyclodextrin (β-CD) complexes, but each method usually has an indispensable disadvantage to it; to name a few, absence of solubility in GRAS organic solvents, toxicity issues related to surfactants (Cremophor EL in case of Taxol), and structural and conformational restrictions in case of encapsulation within β-CD. Micronization has also been proved to be insufficient in improving solubility of various biopharmaceutical classification system (BCS) class II and IV drugs. Hence, the concept of nanonization was put forward by Elan Nanosystems (CA, USA) in early 90s, where they developed nanocrystals of drug for oral administration and nanocrystal suspensions for parenteral drug delivery [82]. Nanocrystals, as the term suggest, are the crystalline form of the drug in nanosize range. In other words, nanocrystals are the nanoparticles (size below 1000 nm) of organized structure made up of 100% pure API along with some stabilizers (surfactants, polymeric stabilizer) when present as nanosuspensions in aqueous or nonaqueous media [83] (Fig. 16).

On the other hand, nanosuspensions are the submicron colloidal dispersions of poorly miscible drug particles (BCS class II and IV or Log $P > 2$) in an aqueous vehicle stabilized by surfactants with particle size distribution of generally less than 1 micron (having an average particle size range of 200–600 nm) for either oral, topical, parenteral, or pulmonary administration. Such nanosystems are employed to improve the erratic absorption profile and highly variable bioavailability as their performance is dissolution rate-limited (depending upon the fed/fasted state of the patient).

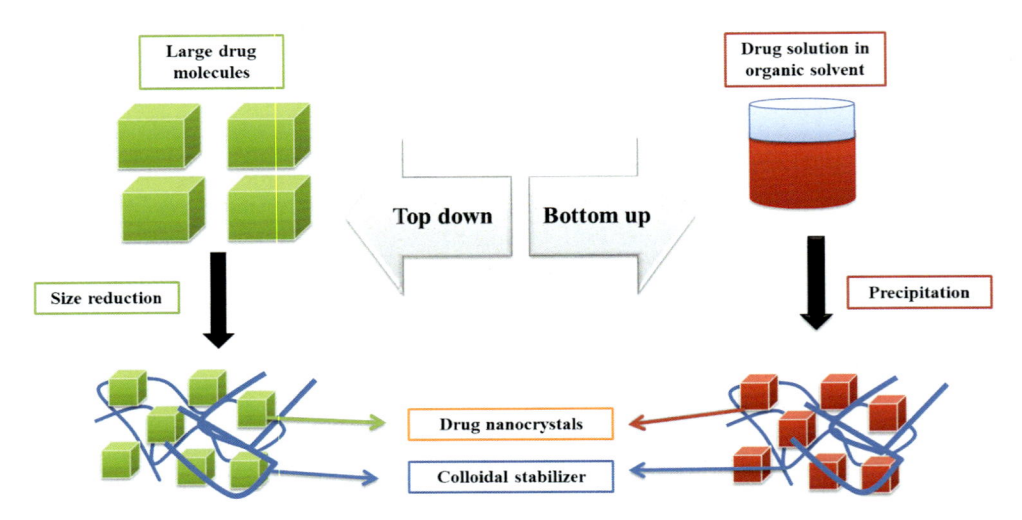

Fig. 16 Conversion of poorly water-soluble drug into nanocrystals.

In recent times, these nanosystems have received huge attention due to their cost-effectiveness and technical simplicity in manufacturing design compared to other delivery systems such as liposomes or other colloidal carriers. Nanosuspensions have been designed into various final dosage forms such as liquid, tablets, powders, pellets, capsules, and films.

3.1.1 Advantages of nanocrystals/nanosuspensions
- Most suitable of hydrophobic entities (as >40% of drug molecules are poorly soluble).
- Reduced tissue irritation in case of subcutaneous/intramuscular administration.
- Rapid dissolution and tissue targeting can be achieved by IV route of administration.
- Oral administration of nanosuspensions provides rapid and improved bioavailability.
- Rapid bioavailability and highly consistent dosing for ocular and inhalation delivery.
- Long-term physical stability and can be assimilated in tablets, pellets, hydrogels, and suppositories.

3.1.2 Limitations
In case of nanosuspensions, concerns are related to uniformity in dosing and sedimentation should be addressed while manufacturing.

3.2 Properties of nanocrystals
As described earlier, conversion of poorly water-soluble drug into nanocrystals leads to improved water solubility and bioavailability. The factors responsible for this revolutionized change are:

(i) Reduced size and increased surface area lead to increment in dissolution velocity of drug as per Noyes–Whitney equation (Eq. 1), where it can be seen that bioavailability of those drugs where dissolution velocity is a rate limiting step can be improved by size reduction.

(ii) Similarly, size reduction saturation solubility (C_s) of drug also improves which is directly proportional to dissolution velocity (dc/dt).

$$dc/dt = DAK_{w/o}(C_s - C_b)/Vh \qquad (1)$$

where, dc/dt is dissolution velocity, D is diffusion coefficient, A is surface area of drug, $K_{w/o}$ is partition coefficient, V is volume of dissolution medium, h is thickness of stagnant layer, and ($C_s - C_b$) is concentration gradient.

(iii) Amorphous state of a compound has the highest saturation solubility; thus, the developed nanocrystals need to be amorphous and within nano size range (20 to 50 nm) [84].

3.3 Properties of nanosuspensions

Nanosuspensions exhibit the potential to address the unique drug delivery issues associated with API by retaining it in a crystalline state, while enable them with increased drug loading during formulation development. Such nanosystems accommodating larger amount of drugs with minimum dose volume have additional benefits in parenteral and ophthalmic drug delivery system owing to the minimization of excessive use of harmful nonaqueous solvents and extreme pH. Some of the other advantages include increased stability, sustained release of drug, enhanced efficacy through tissue targeting, reduced first pass metabolism, and deep lung deposition.

3.4 Marketed nanocrystal and nanosuspension formulations

The production of nanocrystals/nanosuspensions has much to offer and thus their commercial production and clinical application bloomed in no time. Nanocrystallization not only improved the solubility and bioavailability profile of poorly soluble drugs, but also benefitted the absorption of drugs with low therapeutic index. Griseofulvin has a small absorption window mainly in duodenum; traditional therapy led to unpredictable and variable oral absorption (25%–70%). Griseofulvin nanocrystals with trade name GRIS-PEG by Novartis ensured improved gastric absorption. Nanocrystals/nanosuspensions ensured rapid release of drug for faster action, e.g., naproxen is required immediately to combat severe headaches. Nanocrystals/nanosuspensions have also improved patient compliance and ensured bioequivalency when drug is administered at fasted or fed state [85].

Nanocrystal or nanosuspension systems of various hydrophobic drug molecules overcame issues like addition of toxic surfactants like Cremaphor or other excipients.

Verapamil was earlier administered with organic acids (citric, fumaric, and itaconic acids) to improve its pH-dependent solubility; however, a revolutionary product by Lannett Company, Inc. under the trade name of Verelan PM was introduced in late 90s and composed of nanocrystals in the form of extended release capsules. The product successfully controlled and improved oral bioavailability of verapamil [86].

Cosmetic industry is also benefitted greatly by nanocrystal/nanosuspension technology. A commercially used antioxidant in cosmetics, Ascorbyl palmitate, faces hurdles like instability and inactivation due to speedy oxidation. A high-pressure homogenization-based technique, DissoCubes has been employed to synthesize nanocrystals using SDS and Tween 80 as stabilizers for improved stability and antioxidant activity [87].

Fluorophores or lumiscent semiconductors like zinc oxide, zinc sulphide, and cadmium sulphide are widely used as biological markers, staining agents, and contrast agents. Conventional semiconductors face issues like wide, uncontrollable, asymmetric emission spectrum and photochemical instability. Thus, the development of nanocrystals of fluorophores improved their properties as biological markers or diagnostic agents. Cesium lead bromide nanocrystals developed by Microfludizer technique using octylamine and octanoic acid as linear ligands and 2-ethylhexylamine and/or 2-ethylhexanoic acid as branched ligands significantly improved the photoluminescence and stability of semiconductor by inducing red shift in emission and by minimizing the bandwidth [88]. Henceforth, a wide range of nanocrystals and nanosuspensions (Table 4) in various dosage forms are commercially available in market for various indications after FDA approval.

3.5 Compositional architecture of nanocrystals/nanosuspensions

Composition of the nanomaterial is of utmost importance as the desired pharmacokinetics and pharmacodynamics of the final drug product greatly depend on its quality standards [89]. The final properties of nanocrystals/nanosuspensions depend on the method of production as well as various excipients used in product development.

Table 5 summarizes some common categories of excipients used in nanocrystals production along with the examples. GRAS category excipients are used commercially to ensure quality control, safety, and efficacy of the developed product, thus ensuring no toxic effect due to residual organic solvents or surfactants/stabilizers in the final formulation.

3.6 Methods of preparation

Nanocrystals/Nanosuspenions are prepared mainly by three approaches, namely, bottom-up, top-down, and combination technology, as shown in Fig. 17. Some of the post-production processes could be utilized in various top-down/bottom-up

Table 4 List of commercially available nanocrystal and nanosuspension preparations.

Marketed name	Drug	Company	Indication	Technology employed	Route of administration	FDA Approval
Rapamune	Rapamycin	Wyeth	Immunosuppressive	NanoCrystals	Oral tablets	2000
Emend	Aprepitant	Merck	Antiemetic	NanoCrystals	Oral capsules	2003
Triglide	Fenofibrate	SkyePharma	HyperCHLemia	IDD-P	Oral tablets	2005
MegaceES	Megestrole acetate	PAR Pharmaceuticals	Anorexia, cachexia	NanoCrystals	Oral suspension	2005
Invega Sustenna Xeplion	Paliperidone palmitate	Janssen	Antipsychosis	NanoCrystals	Parentral nanosuspension (i.m.)	2009
Nucryst	Silver	Nucryst Pharmaceuticals	Antibacterial	Silcryst (Self-developed; Reactive magnetron sputtering)	Wound dressings	2007
Cesamet	Nabilone	Lilly	Antiemetic	Coprecipitation	Oral capsules	2009
GRIS-PEG	Griseofluvin	Novartis	Antifungal	Coprecipitation	Oral tablets	1975
Avinza	Morphine sulfate	King Pharma	Anti-chronic pain	NanoCrystals	Oral tablets	2002
Zanaflex	Tizanidine HCl	Acorda	Muscle relaxant	NanoCrystals	Oral tablets	1996
Focalin XR	Dexmethyl-phenidate HCl	Novartis	Attention Deficit Hyperactivity Disorder (ADHD)	NanoCrystals	Oral capsules	2005
Naprelan	Naproxen sodium	Wyeth	Anti-inflammation	NanoCrystals	Oral tablets	1976
Herbesser	Diltiazem HCl	Mitsubishi Tanabe Pharmaceuticals	Antianginal	Media milling	IV	1998

Continued

Table 4 List of commercially available nanocrystal and nanosuspension preparations—cont'd

Marketed name	Drug	Company	Indication	Technology employed	Route of administration	FDA Approval
Theodur	Theophylline	Mitsubishi Tanabe Pharmaceuticals	Bronchodilator	Media milling	Oral extended release tablets	1979
RitalinLA	Methyl phenidate HCl	Novartis	Psycho-stimulant agent	Media milling	Oral sustained release tablets	1955
Ostim OsSatura NanOss Vitoss	Hydroxyapatite	Heraseus Kulzer Orthobiologics Rti Surgical Stryker	Bone substitute	–	Paste/Injections/ Foam packs/ Foam strips	–
Ryanodex	Dantrolene sodium	Eagle Pharmaceuticals	Malignant hypothermia	–	Freeze-dried powder for injection	–

Table 5 Excipients used for commercial production of nanocrystals/nanosuspensions.

Category	Examples of excipients
Solvents	Acetone, Ethanol, Methanol, Isopropanol, *N*-methylpyrrolidone, Olive oil, Cyclohexane, Benzene, Chloroform, Toluene, 1- octadecene, Trioctylphosphine (TOP), Oleylamine ethyl alcohol, acetone, butanol, ethyl formate, ethyl acetate, ethyl ether, methyl acetate, methyl ethyl ketone, triacetin
Cosolvents	Polyethylene glycols, propylene glycol, Ethylene glycol, Triethylene glycol, Thiourea, Dodecanethiol (1-DDT), *tert*-dodecyl mercaptan (*t*-DDT) Isopropyl alcohol
Stabilizers	
Surfactants	Sodium lauryl sulfate, Pluronics F68 and F127, Tween 80, sulfosuccinate, Lecithin, Tetronics 908 and 1107, cetyltrimethylammonium bromide
Polymers	Hydroxypropyl methylcellulose (HPMC), Polyvinyl alcohol, Carboxymethylcellulose sodium, PVP-K30, α-, β-, and γ-cyclodextrans, Polylactic acid, Poloxamer 188, Poloxamer 407, hydroxypropyl cellulose (HPC), hydroxyethyl cellulose (HEC), Soluplus (polyvinyl caprolactam-polyvinyl acetate-PEG copolymer, BASF)
Others	Vitamin E tocopherol, Hydrophobin, transferrin, immunoglobulin G, Human serum albumin and oleic acid cyclodextrin derivatives; polyethylene glycol succinate

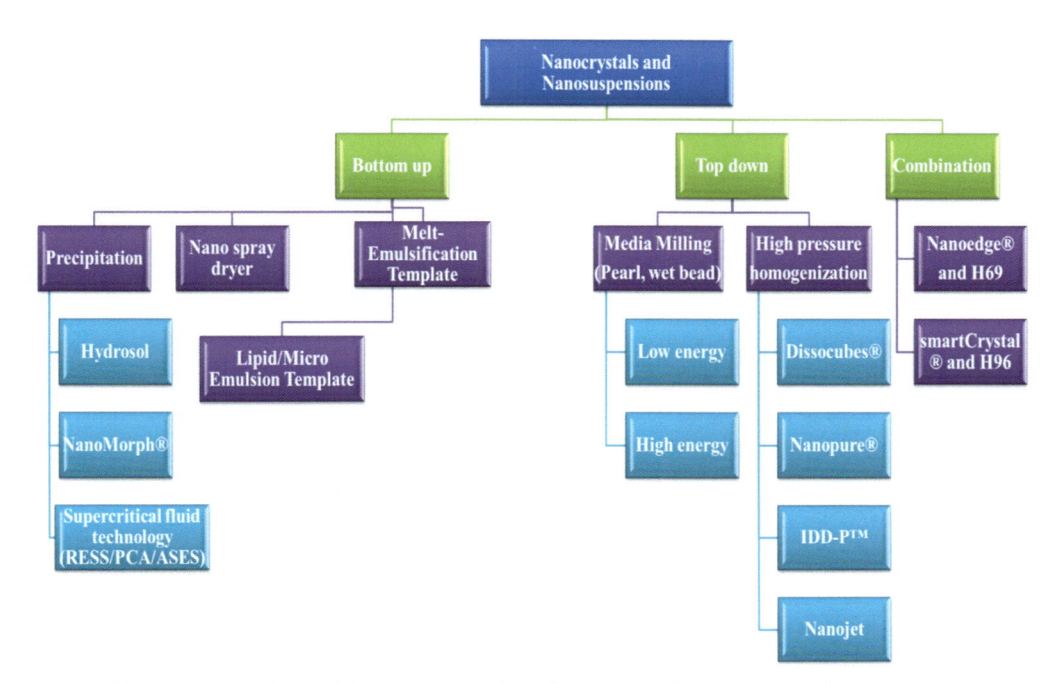

Fig. 17 Various methods used for the production of nanocrystals/nanosuspensions.

methods, viz., freeze drying, spray drying, and spray freezing. Nanosuspension/nano-crystal systems have achieved considerable success in improving the bioavailability of poorly soluble drug candidates for diverse routes of administration such as oral, parenteral, dermal, ocular, pulmonary, and targeted drug delivery applications. However, there is no "One fits all" approach which could be applicable to all the drugs. Scaling up, operational, and pharmacoeconomic factors mandate cost-effectiveness and simplicity in the manufacturing processes.

3.6.1 Bottom-up technologies

These approaches are based on the principle of dissolving the drug in a solvent and precipitating it in a controlled manner to nanoparticles through addition of an anti-solvent in the presence of surfactant.

The hydrosol approach resembles with emulsification-solvent evaporation method with respect to the use of solvent that has also been exploited. However, the only difference between the two techniques is that the drug solvent is miscible with the drug anti-solvent. Higher shear force prevents crystal growth and Ostwald ripening and ensures that the precipitates remain smaller in size. In case of nanocrystals, the first precipitation technique was developed by Sandoz (Novartis), where "hydrosol" was produced by "via humida paratum" (VHP) method in which nanocrystals were precipitated from drug solution in organic solvent on addition of anti-solvent in the presence of stabilizers to prevent aggregation beyond nano size. The reasons for its lack of industrial application are limited application for organic drugs, difficulty in removal of solvent residues, and poor solubility of drug in organic solvents [90].

NanoMorphs is another precipitation method where drug is precipitated out from its solution in water miscible organic solvent on addition of aqueous polymeric solution at specific temperature. The technique is used by Soliqs, Germany, and in food industry to make amorphous carotene nanoparticles like Lucarotin or Lucantin (BASF). O/W emulsion of carotenoids in digestible oil and surfactant suspended in suitable solvent and stabilized by protective colloid is prepared and finally amorphous carotene nanoparticles are obtained by lyophilization. Disadvantage of this technique is recrystallization of polymeric compounds and reduced bioavailability [91].

Further, these technologies also employ precipitation methods (supercritical fluids ($scCO_2$) function as solvent or anti-solvent). Some of the precipitation techniques with limited industrial application include sonoprecipitation, high gravity-controlled precipitation, evaporative methods (i.e., evaporative precipitation into aqueous solution (EPAS) and evaporative precipitation of nanosuspension (EPN)) and supercritical fluid techniques (e.g., RESS, RESOLV, RESAS, and SAS) and have been discussed in detail in Section "Scale-up process of liposomes using SAS." Another important technique is Nanospray Dryer developed by BUCHI, Switzerland, where freeze drying and spray drying techniques are combined and nanocrystals/nanosuspensions are produced by spraying atomized droplets of drug solution in cryogenic liquid like nitrogen [92].

Advantage—It produces small, porous, amorphous nanocrystals with high bioavailability.

Disadvantage—It is applicable only for drugs with low glass transition temperature.

Marketed product—Grisovin by BUCHI Switzerland is 1.5% *w*/w Griseofluvin nanocrystals dispersed in dichloromethane and 0.05% *w*/*v* Lutrol F127 as stabilizer.

Process variables—Conditions for nanocrystals synthesis are maintained as the following: gas used—nitrogen, gas flow rate—150 L/min, feed inlet temperature—50–60°C, product outlet temperature—30–40°C, spraying rate maintained at 100% at a pressure of 35–45 mbar. The detailed principle, working methodology, and process variables critical for this technique have been already discussed in Section "Spray drying technique."

Miscellaneous

Some of the nanosuspensions have been manufactured employing solvent evaporation of emulsions formed via HPH. Such techniques utilize partially water–soluble and volatile organic solvent (e.g., butyl lactate, benzyl alcohol, triacetin, and ethyl acetate) as the dispersed phase. The emulsion is made by dispersing the API in a mixture of organic solvent(s) and forming emulsion with water by HPH or other techniques. Dilution leads to formation of nanosuspensions by diffusion of the internal phase into the external phase when droplets convert into solid particles. The size of the emulsion droplets determines the particle size. Acyclovir nanosuspensions have been developed by this technique. The application of harmful solvents in the process leads to presence of residual amounts of these toxic solvents in the final product.

Another approach is the melt emulsification template, primarily employed for the fabrication design of solid lipid nanoparticles as well as nanosuspensions. In this method, API is dispersed in aqueous solution along with stabilizer, and then, the nanosuspension is heated above the melting point of the API and homogenized with a HPH to produce an emulsion. The final step of this technique is cooling off the emulsion to a room temperature or low temperature. Some of the factors affecting the particle size include concentration of API/stabilizer, type of stabilizer, and cooling condition. The major advantage of this method is the solvent-free operation for the fabrication of nanosuspensions.

Some approaches have also employed lipid/microemulsions as template for the design of nanosuspensions as these systems are thermodynamically stable and isotropically clear dispersions generally stabilized by an interfacial film of surfactant and cosurfactant. In this approach, first, the microemulsions are prepared; second, API solutions are mixed into the lipid/micro emulsion (used as a template); and finally, the drug loading efficiency is determined.

3.6.2 Top-down technologies

These approaches are based on the principle of mechanical attrition to render large crystalline particles into nanoparticles. Such technologies profess maximum industrial viability and are simply based on disintegration or size reduction of large drug crystals into nanocrystals by applying high-energy processes. These techniques are also used for synthesizing nanosuspensions where comminuted drug particles get homogenously suspended in the dispersion medium. The most common types of top-down methods are media milling and high-pressure homogenization (HPH) [93].

Media milling

One of the most commonly used top-down approaches for nanocrystals or nanosuspensions or nano/micro particle synthesis is media milling.

Principle—The principle behind media milling technique is simple mechanical attrition and collision between drug particles, milling media, and agitators, which lead to development of shear stress and cavitation and ultimate breaking of the drug particles.

Working—Drug along with stabilizers, dispersion medium, and milling media, i.e., pearls or wet beads (made up of ceramics, stainless steel, glass, and polystyrene-coated beads), are charged in to the milling chamber, which is either rotated completely or agitators move inside the chamber to cause attrition between milling media and drug, leading to size reduction of large drug crystals into nanocrystals dispersed as nanosuspensions in aqueous dispersion medium (Fig. 18). Continuous flow of coolant is necessary as heat is generated by attrition and collision of milling media and drug crystals. Media milling can be operated at two modes, i.e., low-energy (LE-MM) and high-energy media milling (HE-MM). NanoCrystals, a patented technique of Perrigo, operates on high-energy mode. Disadvantage with LE-MM is that it takes days to formulate nanosuspensions due to low-energy agitation.

Advantages—uniformity between batches, easy to scale-up, wide range of drug quantity can be processed, and narrow size distribution can be achieved.

Disadvantages—cleaning of machine is problematic, process is slow, drug wastage, and chances of contamination by milling media and milling chamber.

Process variables—The process variables that affect the quality of final product include:

Amount of feed—Ideally very low, 3%–20% volume of slurry.

Volume of milling media—Ideally high, 10%–50% volume of slurry.

Diameter of milling media—0.5 to 1.0 mm (Consistent for a batch).

Milling velocities and duration—80–90 rpm for 1–5 days or 1800–4800 rpm for 30–60 min.

Marketed product—Rapamune (FDA approval in 2000) was the first nanocrystals/nanosuspension product developed by Wyeth Pharmaceuticals (USA). It contains Rapamycin or Sirolimus as active drug agent which is obtained from *Streptomyces hygroscopicus* (actinomycetes) and is used as immunosuppressive agent. The product

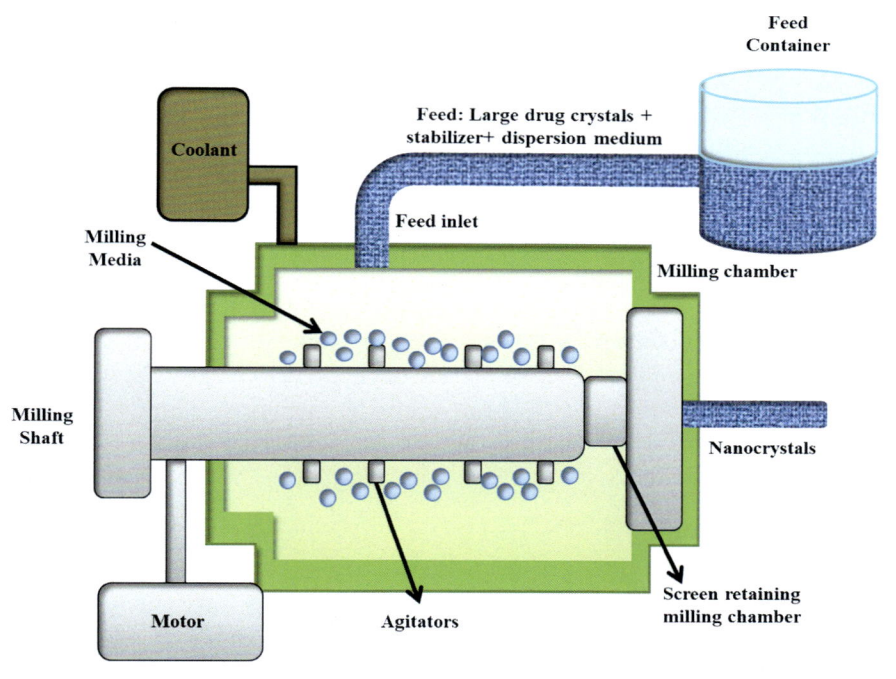

Fig. 18 Diagrammatic representation of media milling chamber for the production of drug nanocrystals or nanosuspensions.

has led to increase in the oral bioavailability of rapamycin by about 21% when compared to its oral solution [94].

Some other popular nanoproducts as nanosuspensions also emerged using media/wet milling technology are *Tricor, Emend*, and *Megace ES*. Few of the advantages of this method are the ease of scalability and batch to batch reproducibility. However, the major limitation is the erosion of pearls which could lead to contamination in the final product.

High-pressure homogenization (HPH)

Disintegration of drug is based on collision, cavitation, and shear forces between the piston and gap of homogenizer or when two streams of drug solution collide [95]. Different types of HPH techniques are used for commercial production of nanocrystals, nanosuspensions, nanoparticles, and nanoliposomes (as a part of size reduction step).

Microfluidizer technology: Y type or Z type Microfluidizer (Microfluidics Inc.) is used to ensure frontal collision between two jet streams of drug premix (stabilizers and dispersion medium) for collision and cavitation at a pressure of about 1700 bars. Insoluble Drug Delivery–Particles (IDD-P) technology of SkyePharma Canada Inc. is based on microfluidization. Disadvantage of this technique is that 50 to 100 cycles are required to achieve size reduction [96].

Piston-gap homogenization in water: Aqueous drug and surfactant suspension are forced through homogenization gaps (~5–20 µm diameter) by piston at a pressure of 1500 to 4000 bar, at room temperature. Disadvantages include not suitable for water-sensitive and thermolabile drugs as water may cause hydrolysis and drying of excess water may require expensive lyophilization techniques. *Dissocubes* technology of SkyePharma PLC utilizes this technique [97].

Piston-gap homogenization in nonaqueous media: Water has been replaced by nonaqueous dispersion medium (oils or hot melted polyethylene glycols) with low vapor pressure and homogenization is carried out at low temperatures. Thus, this method can be used for water-sensitive and thermolabile drugs. *Nanopure* technology developed by Pharma-Sol GmbH, Berlin, works on this principle [98].

Triglide (IDD-P technology): The oral tablets of Fenofibrate have been produced by IDD-P technology of SkyePharma, Canada, and are marketed by Sciele Pharma Inc. (Atlanta, USA) under the trade name of Triglide. These are oral tablets indicated for hypercholesterolemia and mixed dyslipidemia. Fenofibrate improves lipolysis and thus reduces plasma levels of low density lipoprotein CHL, very low density lipoprotein CHL, triglycerides, and apoproteins, and at the same time, increases high density lipo-protein CHL in plasma. Fenofibrate nanocrystals received FDA approval in year 2005 for marketing under the trade name *Triglide*. Need of nanocrystals of fenofibrate emerged when a difference of about 35% was observed in bioavailability of drug between fed and unfed patients. Nanonization ensured better drug uptake in unfed patients in the absence of fats, lipids and surfactants from food and gastrointestinal tract (Fig. 19) [99].

Other commercial nanoproducts based on homogenization processes are Nano-morph (Soligs/Abbot) and Nanocrystal (Elan Nanosystems).

Combinative technology

To overcome the disadvantages of conventional top-down and bottom-up techniques, various combinative technologies have been developed. These techniques can be cus-tomized as per the drug product in hand. First combination technique where micro-precipitation is followed by HPH, ultrasonication, or microfluidization is *Nanoedge* Technology (Baxter). *H69* technique (PharmaSol) is similar to *Nanoedge* where HPH is used for the cavitation of microprecipitate. Therefore, this technology relies on the precipitation of friable materials for subsequent fragmentation under conditions of high shear and/or thermal energy. On the other hand, *Nanojet* technology uses HPH followed by microfluidization for the downsizing of the particle size. Advantage of *Nanoedge* and H69 technique is the effective control on particle size as compared to individual

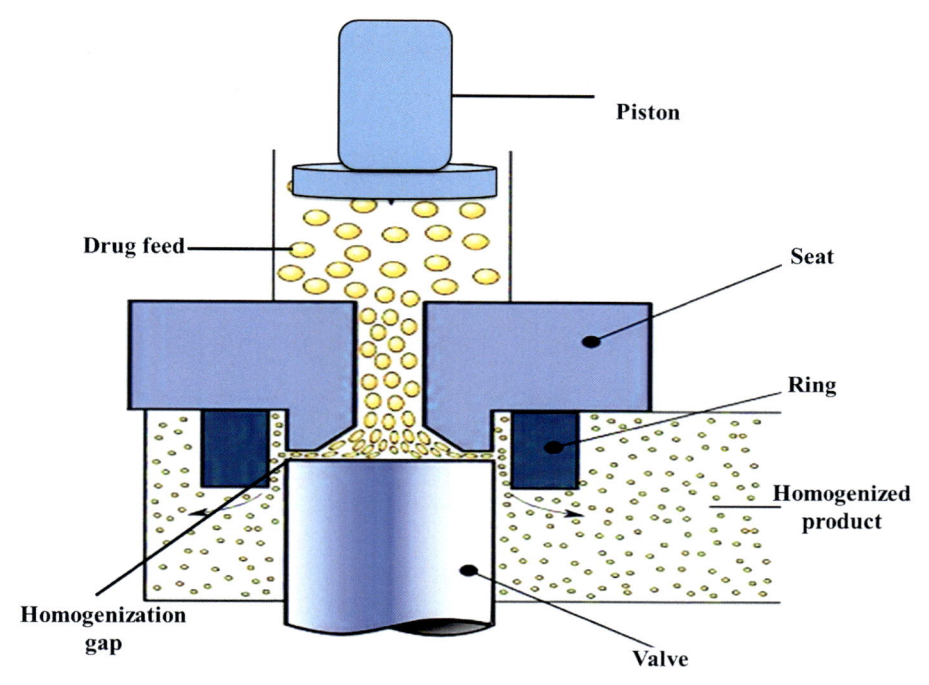

Fig. 19 Diagrammatic representation of high-pressure homogenizer for the production of nanocrystals or nanosuspensions (Parts from https://commons.wikimedia.org/wiki/File:Homogenizing_valve.svg).

conventional method; however, some disadvantages include organic solvent residues, larger particle size fraction, and expensive and complicated method [100]. In another commercial technique, anti–solvent precipitation takes place in high shear forces of homogenizer, thus nanocrystals as well as nanosuspensions are formed and simultaneously cavitation occurs and hence, better control on size can be achieved; this technique is *smartCrystal* technology (Abbott/Soliqs, Germany).

H96 technique (PharmaSol) is another variant of smartCrystal technique where precipitation is done by freeze drying. Absence of use of any organic solvent and suitability for water-sensitive and thermolabile drugs are the main advantages of these techniques. Another combinative technology is *H42* technique (PharmaSol) where spray drying is used for precipitation and size reduction is carried out by HPH [101].

4. Nanoparticles as nanoplatform

4.1 General characteristics

Particulate system having all three dimensions between 1 nm and 100 nm and capable of carrying and delivering a drug payload to desired tissue or organ site is in biomedical

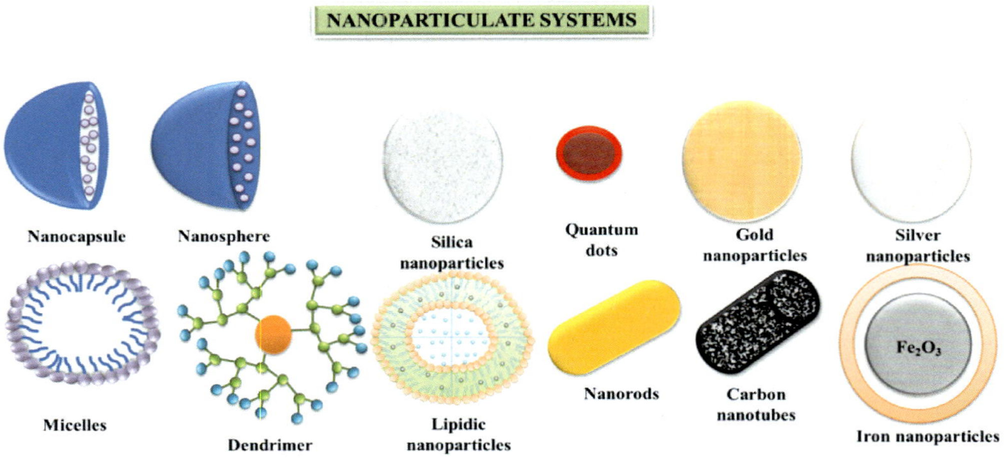

Fig. 20 Broad umbrella of nanoparticulate drug delivery systems.

terms addressed as nanoparticles [102]. This term is very broad and incorporates a wide variety of nanosized particulate systems like polymeric, lipidic, micellar nanoparticles, dendrimers, quantum dots, carbon nanotubes, nanorods, and silver/gold/iron nanoparticles [103] (Fig. 20).

Nanoparticles have become a field of interest for researchers from almost all walks of sciences as they present a plethora of advantages. However, nanoparticles are not only limited to research or lab level; in fact, a wide range of products have been approved by FDA for clinical application through almost all routes of administrations including oral, parenteral, and topical. Also, different commercially available nanoparticles are used in cosmetics as well as in diagnostic techniques.

Among the approximately 100 nanopharmaceuticals that have been approved by FDA so far, more than 36 nanoproducts belong to the category of various types of polymeric and inorganic nanoparticles [53]. Nanoparticles have become a choice of carrier for various reasons which primarily include—ease of manipulating size and surface chemistry for active/passive drug loading and targeting; controlled/sustained/targeted/localized release can be achieved; high drug payload can be achieved; specific receptor binding ligand can be decorated on nanoparticle surface; and nanoparticles can act as drug carriers, diagnostic agents, theranostic agents, in tissue engineering, and for developing scaffolds [104]. Some examples of the popular nanoparticle formulations which are commercially available in the global market are *Abraxane, Cosmofer, Atridox,* and *Eligard* (Table 6).

4.2 Classification of nanoparticles

The term nanoparticle is a wide umbrella which covers a variety of nanoparticulate systems which can be classified on the basis of (i) nature of material, (ii) source of material,

Table 6 List of commercially available nanoparticles.

Trade name	Drug	Type of nanoparticle	Company/ Manufacturer	Indication	Route of administration	FDA Approval year
Abraxane	PTX	Serum albumin nanoparticle	Abraxis Biosciences	Breast cancer, lung cancer, and pancreatic cancer	IV	2005
Cosmofer	Iron	Dextran colloids	Pharmacosmos UK limited	Anemia	IV injection and infusion	2001
Eligard	Leuprolide acetate	Thermoplastic polymeric (PLGA) nanoparticles	Tolmar	Prostate cancer, breast cancer, endometriosis, uterine fibroids, and early puberty	IV suspension	2002
Lupron Depot	Leuprolide acetate	PLGA nanoparticles	Abbvie endocrine Inc.	Endometriosis	Intramuscular suspension depot	1989
Optisol Envirox	–	Titanium dioxide nanoparticles modified by manganese	Oxonica	UV Sunscreen	Topical cream	2003
Atridox	Doxycycline Hyclate	PLA nanoparticles	Tolmar Inc.	Chronic Adult Periodontitis	Controlled released gel	2017
Feraheme	Ferumoxytol	Carbohydrate-coated iron oxide nanoparticle	AMAG Pharmaceuticals	Iron deficiency-related anemia	IV	2009
Somatuline Depot	Lanreotide acetate	Drug molecules arranged in nanotubes	Ipsen	Acromegaly, Carcinoid Syndrome and Neuroendocrine Carcinoma	IV	2007

Continued

Table 6 List of commercially available nanoparticles—cont'd

Trade name	Drug	Type of nanoparticle	Company/ Manufacturer	Indication	Route of administration	FDA Approval year
Somavert	Pegvisomant	Polyethylene glycol–Growth hormone receptor antagonist conjugate	Pfizer	Acromegaly	Subcutaneous injection (Reconstituted lyophilized powder)	2003
Sublocade	Buprenorphine	PLGA nanoparticles	Indivior	Opioid use disorder	Subcutaneous injection	2017
Feridex I.V.	Ferumoxides	Iron-dextran colloids	AMAG Pharmaceuticals	Liver lesions imaging agent	IV	2008
Lumason	Sulfur hexafluoride	Lipidic nano/microsphere	Bracco imaging	Ultrasound contrast agent	IV and intravesicle injectable suspension	2014
SonoVue	–	Phospholipid micro/nano bubble	Bracco imaging	Ultrasound contrast agent	Dispersion for injection	2001

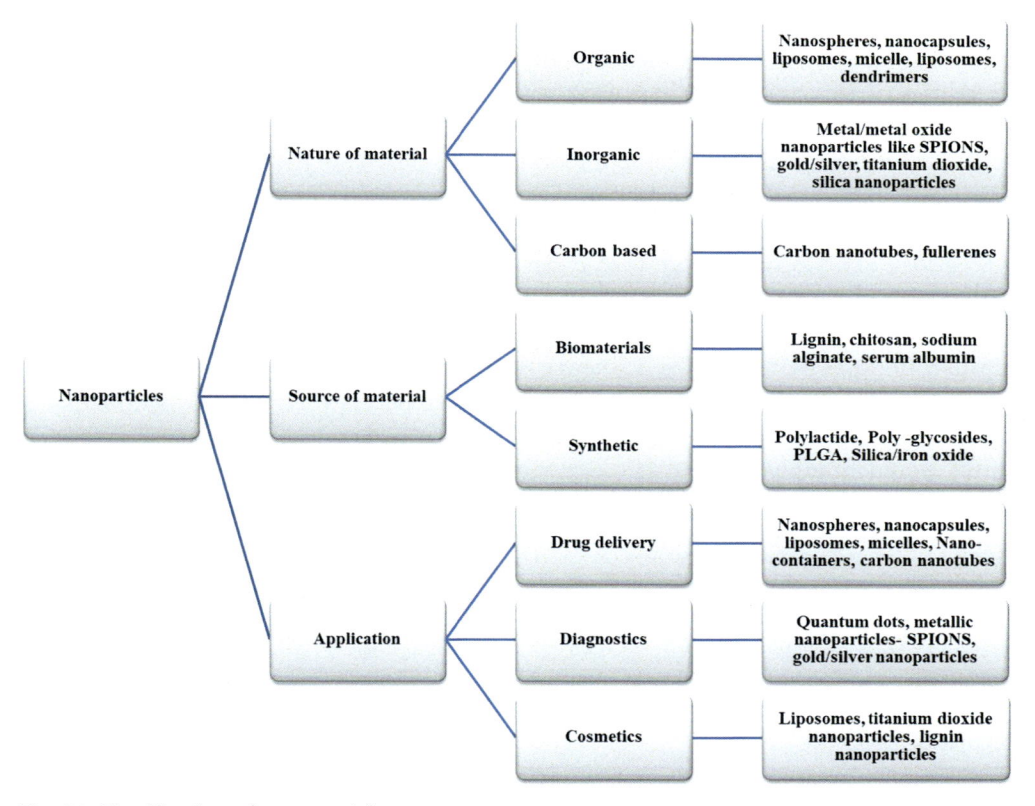

Fig. 21 Classification of nanoparticles.

and (iii) applications (Fig. 21). Broadly, nanoparticles can be classified as organic and inorganic. Nanoparticles made up of natural or synthetic polymers, lipids, or carbon are usually grouped into organic nanoparticles and are mainly used as drug carriers.

Nanoparticles made up of metal, metal oxides, or ceramics are known as inorganic nanoparticles and they found their major application in diagnostics and cosmetic industry (as UV protectant) [105].

4.2.1 Composition of nanoparticles

Composition of nanoparticles varies greatly and also affects the size, shape, and morphology of the nanoparticles [106]. Depending on the classification of nanoparticles, the basic structural units vary from polymeric (natural, synthetic and semisynthetic), metallic, metal oxide, semiconductor, carbon–based, lipids, mineral oils, to biomacromolecules, etc. Other excipients required for synthesis of nanoparticles vary depending on the method used for the preparation of nanoparticles.

Solvents (organic and inorganic) can be used for the dissolving polymers, drugs, or as dispersing agents in emulsion-based synthesis techniques or as anti-solvent for self-assembling of polymeric units in dialysis technique. Similarly, depending on the required storage time and compositional material, preservatives can be added to prevent spoilage of lipidic or oil-based nanoparticles like solid lipid nanoparticles (SLNs), liposomes, nano-structured lipidic carriers (NLCSs), etc. Inorganic nanoparticles like silver colloids itself act as preservatives and antimicrobial agents. Controlling the particle size and preventing aggregation of nanoparticles on storage can be achieved by addition of stabilizing agents to the dispersion of nanoparticulate systems. Coating of stabilizers sometimes even renders long circulating property to nanoparticulate system like PEGylated liposomes. Another important excipient is hardening agents which come into play when bridging between two polymeric subunits is required. This can be achieved by a bifunctional agent like glutaraldehyde as nanoparticles are usually composed of highly reactive constituents with exposed multiple functional groups which can easily react with these bridging agents and can lead to nucleation and condensation of nanoparticles. Table 7 summarizes the wide range of excipients used for the synthesis of nanoparticles.

Table 7 Composition of various nanoparticulate systems.

Nanoparticulate system	Category	Excipient
Polymeric nanoparticles	Natural polymers	Chitosan, gelatin, sodium alginate, albumin, lignin, collagen, pectin, starch
	Synthetic polymers	Poly(lactides) (PLA), poly(glycosides) (PGA), poly(lactide co-glycosides) (PLGA), polystyrene, poly(methacrylate) (PMMA)
	Semisynthetic	Ethyl cellulose (EC), hydroxy propyl methyl cellulose (HPMC), cellulose acetate phthalate (CAP), N,N,N-trimethyl chitosan chloride (TMC), pre-gelatinized starch, thiolated pectin
	Biodegradable	Chitosan, gelatin, sodium alginate, albumin, lignin, poly(lactide co-glycosides) (PLGA), poly cyanoacrylates
	Nonbiodegradable	Poly(methacrylate) (PMMA), polymethyl vinyl ether, Eudragit
Carbon-based nanoparticles	Carbon	Allotropic forms of carbon
Metallic nanoparticles	Metals	Gold, silver, copper, platinum

Table 7 Composition of various nanoparticulate systems—cont'd

Nanoparticulate system	Category	Excipient
Metal oxide nanoparticles	Metal oxide (semiconductor)	Iron oxide, titanium dioxide, silicon dioxide, zinc oxide, copper oxide
Micellar nanoparticles	Surfactants	Sorbitan esters, glycerol esters, polyethylene glycol esters, ethoxylated fatty esters
	Oil	Mineral oil, squalene, medium chain triglycerides, vegetable oils
Lipidic nanoparticles	Lipids	Soybean phosphatidylcholine (SPC), dioleoyl phosphatidylcholine (DOPC), dimyristoyl phosphatidylethanolamine (DMPE), egg phosphatidylcholine (EPC), egg sphingomyelin (ESM)
Quantum dots	Semiconductor	Silicon, cadmium sulphide/selenide, indium arsenide, grapheme
Dendrimers	Polymer	Pseudorotaxane, poly (propylene amine), polyamidoamine (PAMAM)
Nanoparticluate systems	Solvents	Water, dichloromethane, hexane, acetone, ethanol, methanol, dimethyl sulphoxide, ethylene glycol, dioxane
	Stabilizers	Polyethylene glycol, gum arabic, polylactic acid, poly(vinylpyrrolidone), polyaniline, poly(*N*-isopropylacrylamide-acrylamide-allylamine), polystyrene, polyisoprene
	Preservatives	Lactic acid, propionic acid, sodium sorbate, ethyl parabens, butlyparabens, sodium benzoates
	Hardening agents	Glutaraldehyde, formaldehyde, ammonium sulfate, hydrochloric acid, carbon tetrachloride, cyclohexane

4.2.2 Scale-up process for ABRAXANE

Systemic administration of water-insoluble PTX has been a challenge for a long time. However, development of protein shell-bound PTX nano/microparticles redispersed in biocompatible aqueous solvent revolutionized the therapy for breast cancer, lung cancer, and pancreatic cancer. The current assignee of the US Patent for Human Serum Albumin (HSA)-bound PTX is Abraxis BioScience Inc., formerly VivoRx pharmaceutical company, US. PTX is entrapped within the polymeric shell of HSA of cross-sectional dimensions of below 10 µm. HSA molecules were crosslinked by disulphide bond or by a bridging agent like glutaraldehyde using techniques like high-pressure homogenization, microfluidics, and ultrasonication (Fig. 22).

Abraxane is available as IV suspension which needs to be resuspended at time of administration and is designated as orphan drug and first-line treatment for metastatic

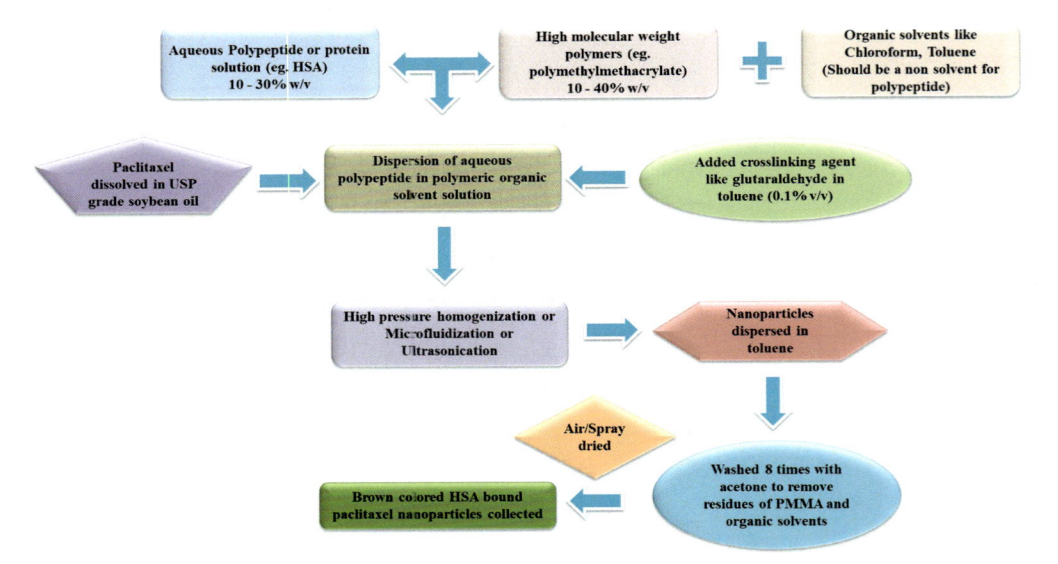

Fig. 22 Schematic representation of method of preparation of HSA shell-bound PTX nanoparticles.

breast cancer, nonsmall cell lung cancer (NSCLC), and pancreatic cancer along with gemcitabine in USA.

4.3 Method of preparation

Nanoparticle is a broad term as discussed earlier; for sake of clarity and ease of understanding, this section includes the various methods of preparation of polymeric and inorganic nanoparticles on small scale as well as on pilot scale.

4.3.1 On small scale

Exhaustive research has been done in the development and fabrication of nanoparticles and a wide range of methods are used at small scale (Fig. 23), ranging from simple dialysis to complex and sophisticated SCFT to a newer concept of green synthesis [107]. Our major focus of discussion is the industrial methods used for nanoparticle synthesis.

Polymeric nanoparticles

Three different principles are commonly employed for the synthesis of polymeric nanoparticles.

Dispersion of preformed polymer—Preformed nanoparticles dispersed in an organic solvent phase along with stabilizers/surfactants are assembled into nanoparticles by various techniques like salting out, dialysis, solvent evaporation, etc. Basic principle is to change (make hostile) the microenvironment of dispersed/dissolved preformed polymers in order to lead to assembling of polymeric units in the form of nanoparticles. These

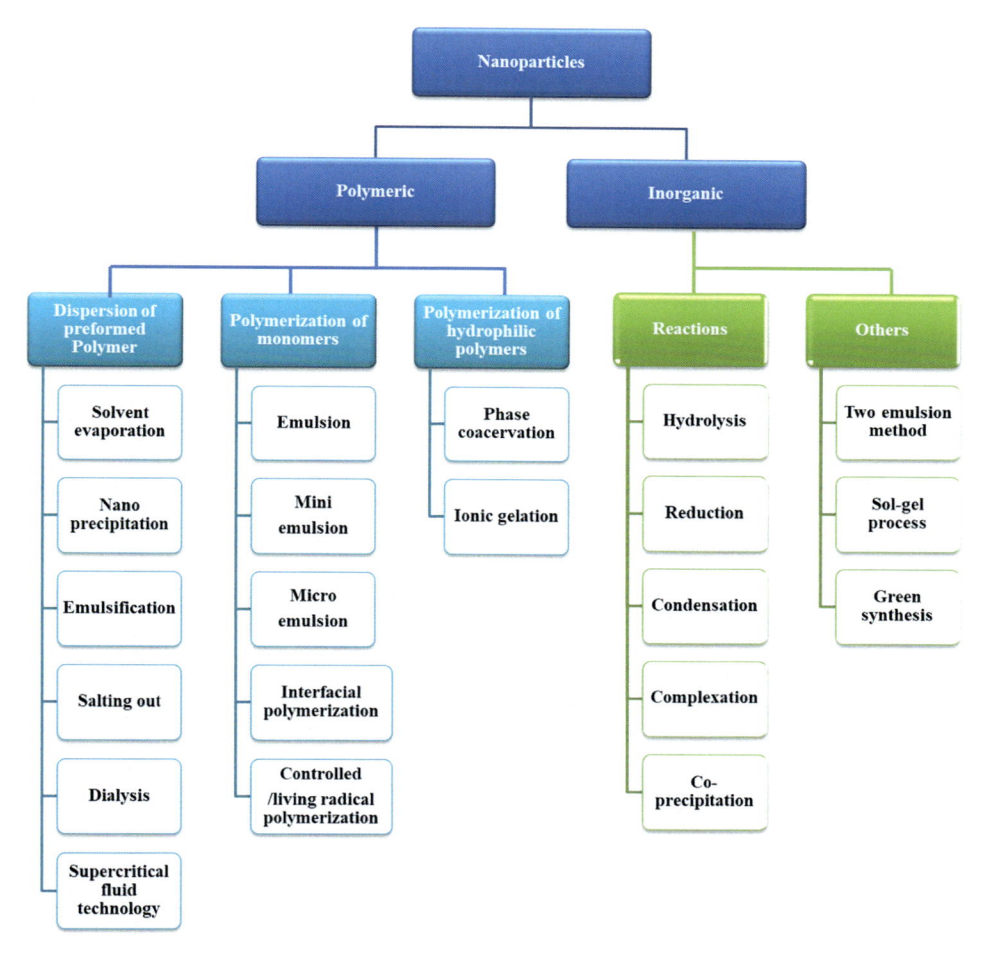

Fig. 23 Classification of methods used for the synthesis of polymeric and inorganic nanoparticles on small scale.

methods are mainly utilized for biodegradable polymers like poly(–lactic acid) PLA, poly (–glycolide acid) PLG, and polycyanoacrylates [108].

Polymerization of monomers—Monomeric units are polymerized during the development of nanoparticles and drug is either dissolved/dispersed in the solution of monomeric units and get entrapped during polymerization or drug can later be adsorbed on the surface of nanoparticles. Polymerization of monomers like butyl or alkyl cyanoacrylates in emulsion has also been reported in literature [109].

Coacervation and ionic gelation of hydrophilic polymers—Biodegradable hydrophilic polymers like chitosan, gelatin, sodium alginate, etc. can be developed into nanoparticles by employing the surface charge present on these polymers using ionic gelation/

coacervation method, where aqueous solution of oppositely charged polymers is mixed together to form nanoparticles due to electrostatic attraction or ionic bond formed between opposite charged polymers [110].

Inorganic nanoparticles

Inorganic nanoparticle synthesis usually involves various reactions like hydrolysis and complexation (iron oxide nanoparticles), reduction (gold and silver nanoparticles), and condensation (silica nanoparticles). Aqueous solutions/dispersions of metal/metal oxide ions are subjected to various chemical reactions in the presence of organic phase and stabilizers/surfactants (cetyltrimethylammonium bromide (CTAB), SDS, and PVP) to form nanoparticles [111]. New concept in development of inorganic nanoparticles is green synthesis, where aim is to reduce hazardous waste generation as well as improve economic and environmental benefits using sustainable methods of synthesis. In these techniques, use of organic solvents is minimized and plant extracts are used for isolation of nanoparticles or microorganisms–mediated synthesis is used to prepare biogenic nanoparticles [112].

4.3.2 On large/commercial scale

Commercial production of nanoparticles can be classified into top–down and bottom–up approaches. Polymeric nanoparticles available commercially are mainly in form of polymer-drug conjugates prepared by mixing of the polymer and drug under suitable conditions. However, a wide range of methods for producing inorganic nanoparticles commercially have evolved over a period of time [5, 24, 113]. Due to advancement in commercial production methods of nanoparticles, a wide range of nanoparticle-based biomedical, diagnostic, and cosmetic formulations are available in the market (Table 8).

Table 8 Commercial techniques employed for synthesizing diverse type of nanoparticles.

Approach	Technique	Type of nanoparticles prepared
Top-down	Milling	Metal/metal oxide and polymeric
	High-pressure homogenization (HPH)	Polymeric and organic
	Laser ablation synthesis in solution (LASiS)	Carbon-based and metal oxide
	Thermal decomposition	Carbon-based and metal oxide
	Sputtering	Metal
Bottom-up	Sol-gel process	Metal/metal oxide and carbon-based
	Spinning	Polymeric, micellar, lipidic
	Chemical vapor deposition (CVD)	Metal and carbon-based
	Pyrolysis	Metal oxide and carbon-based
	Biosynthesis	Organic, polymeric, metal-based

Top-down approaches

As the name suggests, the principle behind these techniques is the reduction of size of bulk material to nano range by using external forces.

Milling (mechanical/colloidal/pearl mills): This is the most commonly used technique to produce nanoparticles (polymeric as well as inorganic) and nanosuspensions at industrial level. Mechanical parts in form of agitators or milling media like pearls, beads, etc. are used to cause attrition. The reactants or excipients are mixed, comminuted, and annealed in controlled inert environment within the milling chamber to produce nanoparticles. Overall, plastic deformation forms the particle shape, attrition causes fracture and size reduction, and cold-welding controls the size distribution [114].

High-pressure homogenization (HPH): Another popular top-down technique is HPH, where size reduction is caused by collision, shear forces, and cavitation. Homogenizer comprises of a piston–gap assembly, where the piston forces liquid carrying solid particles (subjected to size reduction) into the homogenizing gap (valve). Gap is a small orifice between the valve and valve seat; when the liquid passes from the small orifice or gap, it is under high pressure and, on leaving the orifice, suddenly the pressure is reduced which causes cavitation and reduction in particle size [115]. This technique along with its applications has already been discussed in Section "High pressure homogenization (HPH)."

Laser ablation synthesis in solution (LASiS): Traditional chemical reduction methods have now been replaced by LASiS, where the material which has to be reduced to nano size is placed in a suitable solvent (water or inert gas) and then laser beam is focused on one point of the material surface leading to heating and evaporation of the material, forming plasma of vapors which on condensation forms clusters; then nucleation occurs and further condensation on the surface of nucleus forms nanoparticles. The synthesis occurs in water or organic solvent in the absence of stabilizer; thus it is considered as "green synthesis" and used for preparing quantum dots, carbon nanotubes and nanowires, and core shell structure nanoparticles [116].

Thermal decomposition: Metal and metal oxide nanoparticles are also prepared by thermal decomposition method which involves an endothermic reaction. The bulk material is subjected to ambient temperature, leading to chemical decomposition of the material as the chemical bonds between the molecules break. This particular temperature is called decomposition temperature. The technique produces stable and small size monodisperse metal nanoparticles [117].

Sputtering: This technique is used to deposit a thin film of metal nanoparticles on target surface. In the sputter chamber, sputtering material (cathode) is sputtered or bombarded by a beam of high-energy ions (noble gas ions like argon and helium produced by high voltage DC glow discharge), which leads to momentum transfer from ions to atoms of cathode which get released from cathode surface as nanoparticles and get deposited on the surface of the targeted material (anode) [118].

Bottom-up approaches

As opposed to top-down approaches, these are constructive techniques, where atoms are clustered and build into nanoparticles.

Sol-gel process: It is the most common bottom-up technique used for the synthesis of nanoparticles as it can be used for any type of nanoparticles. It is a wet chemical process where a chemical colloidal solution of precursor (sol) is treated to produce an integrated network (gel) which is later dried and calcinated to produce nanoparticles. The precursors are mainly metal alkoxides which when treated with acid lead to conversion of sol into a gel made up of liquid and solid phase where discrete units of metal oxides arrange into a 3D-network (gel). Later, the gel is separated by sedimentation, centrifugation, or filtration. Further removal of liquid or solvent is carried out by calcination to produce powdered nanoparticles [119].

Spinning: This method is most commonly used technique for preparing organic (polymeric, micellar, and lipidic) nanoparticles. Spinning disc reactor (SDR) consists of a chamber made up of suitable material (stainless steel), a disc spinning in the bottom, inlets for feed and aqueous phase, control panels for temperature, pH, etc. The head space of reactor is maintained with inert gases like nitrogen to remove oxygen and prevent side reactions. The precursor or feed is introduced into the spinning chamber along with stabilizers and the disc is allowed to spin or rotate at controlled speed, which leads to fusing of discrete units of precursors into nanoparticles. The characteristic properties like shape, size and size distribution, surface charge, and porosity of nanoparticles depend on various factors like rotation speed of disc, material and surface of disc, flow rate of feed, composition of feed, location of inlet of feed, and maintenance of temperature and pH [120].

Chemical vapor deposition (CVD): CVD is used for producing metal or carbon-based nanoparticles of uniform shape, monodispersity, high purity, hardness, and strength. In especially designed CVD reaction chamber, a layer of substrate (precursor for nanoparticles) is laid out and is maintained at ambient temperature. Deposition is then carried out, when gaseous molecules come in contact with the surface of the substrate. A thin film of nanoparticles gets deposited on the surface which is later recovered and collected. Temperature of the substrate is the main factor that controls the characteristics of the nanoparticles. CVD is not a popular technique as a highly sophisticated especially designed chamber is required and highly toxic gaseous by-products are formed during the production [121].

Pyrolysis: Pyrolysis is extensively used for large scale production of carbon and metal oxide nanoparticles in industries. In this technique, substrate for nanoparticle is either sprayed as aerosol (atomization of liquid state feed into droplets) or is available in volatile or gaseous state, which is introduced into pyrolysis furnace where feed is heated/burned/calcinated either by flame or by laser or by plasma to evaporate the solvents and produce nanoparticles. The nanoparticulate is released into the furnace air which is later collected and classified and nanoparticles are recovered and purified. This method is cost-effective, simple, efficient, and continuous high yielding method [122, 123].

Biosynthesis: Environment-friendly methods for developing nanoparticles have also become popular at industrial level. Biodegradable, biocompatible, and nontoxic nanoparticles are synthesized in industries by using precursors from natural sources like plant extracts (silver nanoparticles), bacteria (cellulose nanofibers), and fungi instead of using traditional organic solvents and synthetic monomers and polymers [124, 125].

5. Supramolecular drug delivery system as nanoplatform

5.1 General characteristics

In 1987, supramolecular chemistry received Nobel Prize for its contribution to principles of basic science as well as conceptual invention. However, 2 years later, practical and engineering aspects of supramolecular drug delivery system were also reconnoitered in the pharmaceutical field as a new paradigm for microfabrication.

Fig. 24 depicts various supramolecular systems such as surfactant-based systems (emulsions/microemulsions/nanoemulsions; polymeric micelles; liposomes; layer-by-layer assemblies) that can be used as drug carriers to modify physicochemical and

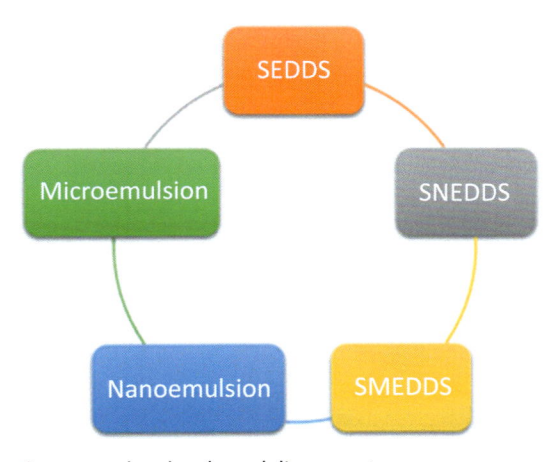

Fig. 24 Surfactant-based supramolecular drug delivery systems.

pharmacokinetic features of drugs. Notably, microemulsions have been used for oral drug delivery of poorly soluble drugs because of enhancements in bioavailability and expectable absorption behavior. *Neoral* is an example of supramolecular drug delivery system which acts as an immunosuppressant and used after transplant operations. It is one of the most well-known microemulsion-based drug delivery systems [126].

5.2 Self-emulsifying system (SES)/supramolecular drug delivery system (SDDS)

SES/SDDS is one of the most widespread and commercially possible oil-based methods for the delivery of drugs which have dissolution speed-limited absorption. SES is an isotropic mixture composed of oils, surfactants, cosurfactants, and sometimes cosolvents, which emulsify to produce oil/water or water/oil emulsion when come in contact with the gastrointestinal tract (GIT) [127]. After emulsification, they are further classified on the basis of the droplet size; they are self-emulsifying drug delivery systems (SEDDS), self-micro emulsifying drug delivery systems (SMEDDS), and self-nanoemulsifying drug delivery systems (SNEDDS) [128]. SEDDS are comparatively newer, lipid-based technological inventions with huge promise in improving the oral bioavailability of drugs. SEDDS formulation basically consists of lipidic and emulsifying excipients possessing an integral ability for drug solubilization [129].

5.2.1 Self-Nano and Micro-emulsifying drug delivery system (SNEDDS and SMEDDS)

SNEDDS are the anhydrous preconcentrate of nanoemulsions (Pronanoemulsion) designed by SEDDS. These are systems composed of anhydrous homogenous liquid mixtures comprising oil, surfactant, active pharmaceutical ingredient (API), and hydrophilic co-solubilizer [130] having exceptional property of forming oil in water nanoemulsion under gentle movement provided by digestive motility of the stomach and intestine [131, 132]. This system produces drug in solution inside nanosized oil droplets (>200 nm). These nanosized oil droplets would be emptied quickly from the stomach ensuing faster release of drug all over the GI tract.

SMEDDS, preconcentrate of microemulsion (Promicroemulsion), are isotropic mixtures composed of oils (natural or synthetic), solid, or liquid surfactants, or otherwise, one or more hydrophilic solvents in nature and cosolvents/surfactants having an exclusive ability of forming o/w microemulsions upon mild agitation in presence of aqueous media of GI fluids [133] (droplet size <200 nm). SMEDDS are stable, easy to produce, can be easily filled in soft gelatin capsules, and generate a drug comprising microemulsion having a large surface area upon gastrointestinal tract dispersion. The microemulsions will further enable the drug absorption due to a rapid digestion by gastrointestinal enzymes and successive transfer to mixed micelles or possible absorption straight from the emulsion particle, through partitioning of drug into the aqueous phase of intestinal fluids (Fig. 25).

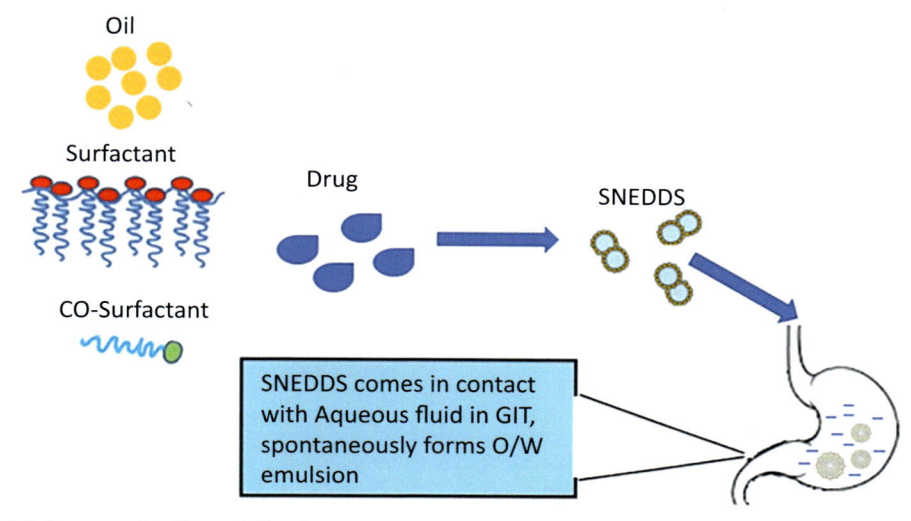

Fig. 25 Process of self-emulsification.

SMEDDS, as a technology, has significantly improved the formulation design of lipophilic drugs for enhancing oral bioavailability, enabling the formation of solubilized phases by reducing particle size to the molecular level, resulting in a solid state solution inside the carrier, altering drug uptake, and improving drug transport through intestinal lymphatic system to the systemic circulation [134]. Several hydrophobic drugs (lopinavir, saquinavir, ritonavir, tipranavir, and amprenavir) and immunosuppressants (Cys-A) have been encapsulated employing SEDDS formulations and have been successfully marketed. These commercial formulations include Gengraf soft gelatin capsule manufactured by Abbott, have Cys-A encapsulated, and is used as an immunosuppressant agent. It is composed of polyoxyl 35 castor oil and polysorbate 80. Fenogal is another marketed product manufactured by Genus having fenofibrate encapsulated in hard gelatin capsule composed of lauryl macrogol-glycerides (Gelucire 44/14). Fenogal is used to treat severe high triglyceride levels.

SNEDDS/SMEDDS are extremely effective in increasing the aqueous solubility, dissolution, and bioavailability of hydrophobic drugs (Fig. 26) [135]. SNEDDS having size in nanometric range provide a strong alternative to the conventional oral formulations of hydrophobic compounds [136]. There are various other properties of SNEDDS and SMEDDS that improve oral bioavailability such as reduction of cytochrome-P450 (CYP-450) breakdown in gut enterocytes, increase in lymphatic transport through payer-patches, and shielding against the first pass metabolism [137].

Two types of SNEDDS are conveyed: (a) liquid SNEDDS (L-SNEDDS) and (b) solid SNEDDS (S-SNEDDS). In most cases, the effectiveness of the SEDDS formulation is

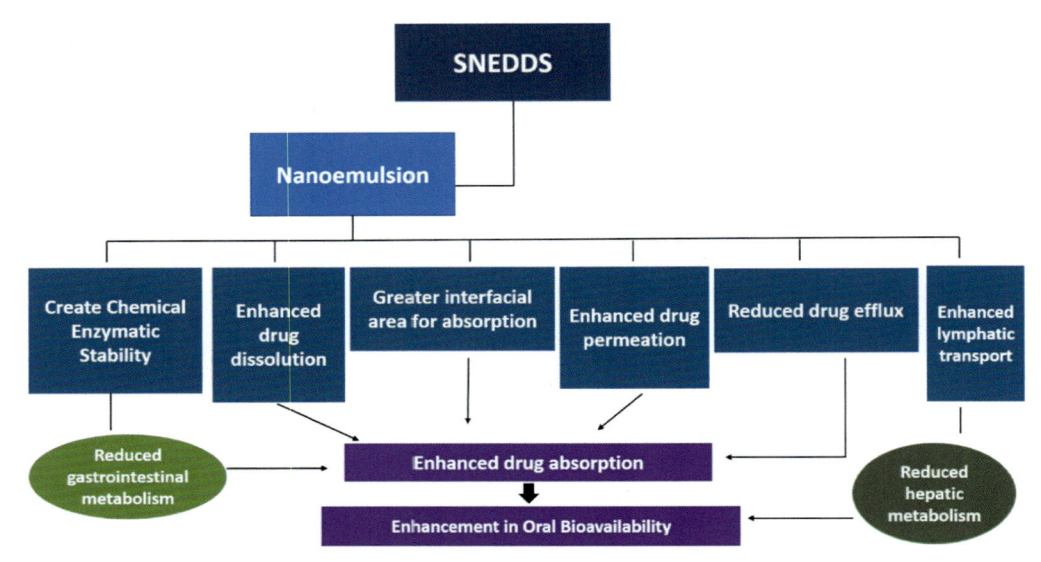

Fig. 26 Advantages of SNEDDS/SMEDDS.

drug-dependent. To reduce these complications, the liquid SEDDS are usually adsorbed onto inert carriers to modify them into solid SEDDS [138]. L-SNEDDS are converted to S-SNEDDS by spry drying, melt granulation, and inert solid adsorption such as aerosol, microcrystalline cellulose, and lactose. This method of converting L-SEDDS into S-SEDDS has various advantages such as stability, capability of manufacturing process, precision, and patient compliance. Thus, this approach of incorporation of L-SEDDS into solid dosage forms leads to the combination of advantages of lipid-based drug delivery systems with those of solid dosage forms [139].

5.2.2 Micro and nanoemulsions

Microemulsions are clear, isotropic liquid mixtures composed of oil, water, and surfactant, often in combination with a cosurfactant. Alternative names for these systems are often used, such as swollen micelle, transparent emulsion, solubilized oil, and micellar solution. Microemulsions are bicontinuous systems that are essentially composed of bulk phases of water and oil separated by a surfactant/cosurfactants-rich interfacial region. Microemulsions are further categorized into direct (o/w) and reversed (w/o). They have the advantage of spontaneous development, ease of manufacturing and scale-up, and enhanced drug solubilization and bioavailability [140, 141]. The high capacity of microemulsions for drugs makes them attractive formulations for pharmaceuticals.

Nanoemulsion is a kinetically stable clear dispersion of two immiscible phases, oil phase and water phase, in combination with surfactant molecules. Nanoemulsions are O/W or W/O dispersion of two immiscible liquids stabilized with the help of a proper surfactant. Nanoemulsions, despite having the same droplet size range as microemulsions, vary enormously in structural aspects [129]. Nanoemulsions are often mentioned as being translucent or transparent, rather than the characteristic opaque, milky white of traditional emulsions. The formation of nanoemulsions requires an input of energy. This energy can be supplied by either mechanical equipment or the chemical potential inherent within the component [142]. There are three types of nanoemulsion which are formed on the basis of the composition such as W/O nanoemulsion, O/W nanoemulsion, and bicontinuous nanoemulsions. In all of the above types of nanoemulsion, the interface becomes stable by addition of suitable blend of surfactants or cosurfactants [143]. There are various commercial products of nanoemulsion available for therapeutic use in market such as Liple manufactured by Mitsubishi pharmaceuticals as a vasodilator platelet inhibitor and Vitalipid manufactured by Fresenius kabi encapsulated with vitamins A, D, E, and K used as parenteral nutrition.

Self-emulsifying drug delivery systems, Nanoemulsion, Microemulsion, SNEDDS, and SMEDDS are composed of oil, surfactants, and cosurfactants, but differentiate from each other on the basis of droplet size, stability, and oil concentration as summarized in Table 9.

5.3 Compositional architecture of emulsifying drug delivery system

In view of pharmaceutical acceptability and the toxicity problems, the excipients selection is really difficult. The self-emulsification process is explicit to the concentration and nature of the oil/surfactant ratio, surfactant/cosurfactant ratio, and the appropriate temperature for self-emulsification. So, all these factors must be considered while excipients selection for these systems.

Major components used in fabrication of emulsifying drug delivery systems are oil, surfactant, and cosurfactant. Selection of oil phase plays an important role in the formulation as it regulates the quantity of drug that can be easily solubilized in the system [144]. Oils and essential oils resulting from natural sources are mainly used in microemulsion as oil phase. Oils/lipid phase is categorized into long chain triglycerides (LCTs, e.g., food grade fixed oils) and medium chain triglycerides (MCTs) [145]. Marketed preparation such as *Neoral* composed of olive oil is one of the LCTs that has revealed superior oral bioavailability [144]. Some of the SMEDDS composed of MCTs are less prone to oxidation and exhibit high solvent capacity compared to LCTs owing to their high ester group concentration. Labrafac CM 10, a MCT, has shown superior solubility for fenofibrate and produced wider microemulsion region at all surfactant/cosurfactant combinations than Maisine 35 (a LCT).

Semisynthetic derivatives obtained from natural oils are frequently being used as oil phase. Water is used as the polar solvent, whereas alcohol, protein, carbohydrates, and polyols are cosolvents [146]. Surfactant is an amphiphilic molecule composed of polar

Table 9 Overview of diverse emulsifying drug delivery systems.

Property	SEDDS	SNEDDS	SMEDDS	Nanoemulsion	Microemulsion
Components	Drug, oil, surfactant	Drug, oil, surfactant, cosurfactant/ hydrophilic cosolvent	Drug, oil, surfactant, cosurfactant/ hydrophilic cosolvent	Biphasic dispersion of two immiscible liquids	Oil, water, and surfactant
Size & appearance	Mean droplet size $\sim$200 nm to 5 µm; turbid	Mean droplet size $\sim$100 nm; Clear	Mean droplet size $\sim$<200 nm; clear–translucent	Mean droplet size $\sim$>500 nm	Mean droplet size $\sim$1–100 nm
Thermodynamic stability	Unstable	Stable	Stable	Unstable	Stable
Oil concentration	40%–80%	less as possible	less than 20%	5%–20%	up to 20%

group and nonpolar groups. The prime requirement to attain ultra–low interfacial tension at the oil water interface can be accomplished by selecting appropriate amphiphile. Cosurfactant (both type and concentration) plays an important role in the fabrication of these systems. It has been attained by incorporating an auxiliary constituent to surfactant called cosolvent or sometimes known as cosurfactant that leads to solubilization of surfactant at interface. There are various components used in self-emulsifying preparation as mentioned in Table 10.

Table 10 List of components employed in manufacturing nanoemulsions, microemulsions, and SNEDDS/SMEDDS.

Oil	Surfactants	Surfactant—HLB	Cosurfactants/solubilizers
Natural			
Soybean oil	POE–20–sorbitan monooleate	15	Ethanol
Castor oil	POE–20–sorbitan monolaurate	16.7	Benzyl alcohol
Teatree oil	Sorbitan monooleate	4.3	
Soyabean oil	Sorbitan monolaurate	8.6	Propylene glycol
Jojoba oil	Poloxamer 188 Poloxamer 407	29 21.5	Glycerol
Eucalyptus oil	POE–35–castor oil	12.5	PEG 400
Sesame oil	POE–40–hydrogenated castor oil POE–60–hydrogenated castor oil	15 14	Diethylene glycol monoethyl ether (Transcutol)
Babchi oil	PEG–660–12–hydroxystearate	15	Lauroglycol FCC
PG monocaprylate	Tocopheryl–PEG 1000–succinate	13.2	Lutrol E400
Synthetic			
Captex 355	Capryol 90		Transcutol P
Captex 8000	Gelucire 44/14	11	Glycerine
Witepsol	Cremophor RH 40	14–16	Ethylene glycol
Myritol 318	Imwitor 191, 742, 780k, 928, 988	3.7	Ethanol
Isopropyl myristate	Labrafil CS, M, 2125 CS	3–4	Propanol
Capryol 90	Lauroglycol 90	15	Ethanol

Continued

Table 10 List of components employed in manufacturing nanoemulsions, microemulsions, and SNEDDS/SMEDDS—cont'd

Oil	Surfactants	Surfactant—HLB	Cosurfactants/solubilizers
Sefsol-218	PEG MW	12–14	Isopropyl alcohol
Triacetin	Plurol Oleique CC		n-butanol
Isopropyl myristate	Poloxamer 124	12–18	PEG 400
Semisynthetic			
Isopropyl myristate	Tween 60	14.9	Cremophor RH40
Ethyl oleate	Tween 80	15.0	Plurololeique
Lauryl alcohol	Brij 58	16	
Tocopherol acetate	Soybean lecithin	7.0	Plurolisostearique
PEG-40 hydrogenated castor oil	Egg lecithin	7.5–8.5	Distearoylphosphatidyl ethanolamine-N-poly (ethyleneglucol)2000

5.4 Marketed emulsifying drug delivery formulations

Potential advantages of these systems include cost-effectiveness, as simple instrumentation for fabrication (such as simple mixers along with agitator and volumetric liquid filling equipment at large scale manufacturing). This enlightens the interest of pharmaceutical industry in manufacturing of such dosage forms. Nanoemulsions and microemulsion have amalgamated in various facets of drug delivery therapeutics such as cosmetics (such as deodorants, sunscreens, shampoos, lotions, nail enamels, conditioners, and hair serums) [147, 148] and transdermal drug delivery, cancer therapy, vaccines, nontoxic disinfectant cleaner, formulations for enhanced oral delivery of poorly soluble drug, ocular, intranasal, parenteral, and pulmonary drug delivery [149]. Table 11 summarizes the list of various FDA-approved SNEDDS/SMEDDS, nanoemulsion and microemulsion products, and cosmaceuticals commercialized worldwide.

5.5 Methods of preparation

There are various technologies employed for the preparation of self-emulsifying drug delivery systems as mentioned in Fig. 27.

Self-emulsifying drug delivery systems such as SNEDDS, SMEDDS, and microemulsions are prepared using high shear mixer at large scale. On the other hand, the fabrication of nanoemulsions is quite an expensive procedure (due to requirement of special type of instruments and development methods for droplets size reduction). Further, HPH,

Table 11 Marketed preparation of nanoemulsion, microemulsion, and SNEDDS/SMEDDS.

Drug name	Trade name	Indication	Manufacturer	Route of administration	FDA Approval
Application in pharmaceutical sector					
Propofol	Diprivan	General anesthesia	Zeneca Pharmaceuticals	I.V.	October 2, 1989
Diazepam	Diazemuls	Anxiety, tension, sedation, muscle spasm, convulsions, tetanus, and delirium	Kabi-Pharmacia	I.V.	July 29, 1997
Cyclosporin (Cys-A)	Restasis	Immunomodulator	Allergan, Inc.	Topical	October 10, 2003
Etomidate	Amidate	Anesthesia	Dumex	I.V.	February 26, 1999
Clofazimine	Lamprene	Leprosy	Geigy	Oral	December 15, 1986
Dronabionol	Marinol	Anorexia	Roxane	Oral	August 5, 1999
Ritonavir/ lopinavir	Kaletra	AIDS	Abbott	Oral	September 15, 2000
Progesterone	Prometrium	Endometrial hyperplasia	Solvay	Oral	May 14, 1998
Tritionoin	Vesanoid	Roche	Acne	Oral	February 14,2000
SNEDDS/SMEDDS					
Cyclosporine (Cys-A)	Neoral	Prophylaxis	Novartis	Oral	July 14, 1995
	Gengraf	Immunosuppressant, prophylaxis	Abbott laboratories	Oral	December 12, 2000
	Sandimmune	Immunosuppressant	Novartis	Oral	February 3, 1990

Continued

Table 11 Marketed preparation of nanoemulsion, microemulsion, and SNEDDS/SMEDDS—cont'd

Drug name	Trade name	Indication	Manufacturer	Route of administration	FDA Approval
Tipranavir	Aptivus	Antiretroviral	Boehringer Ingelheim Pharmaceuticals, Inc.	Oral	June 22, 2005
Isotretinoin	Accutane	Severe recalcitrant nodular acne	Roche Laboratories Inc.	Oral	July 5, 1982
Ritonavir	Norvir	HIV-1 infection, Antiviral	Abbott laboratories	Oral	June 29, 1999
Bexarotene	Targretin	Cutaneous T-Cell Lymphoma	Novartis	Oral	December 29, 1999
Calcitriol	Rocaltrol	Calcium regulator	Roche Laboratories Inc.	Oral	November 20, 1998
Glipizide	Glucotrol	Type 2 diabetes	Pfizer	Oral	April 26, 1994

Applications in cosmaceutical sector

Marketed product	Manufacturer	Indication
Nanocream	Sinerga	Wet wipes
Bepanthol-Protect Facial Cream Ultra	Bayer HealthCare	Moisturizing, antiaging, and antipollution
Korres Red Vine Hair Sun Protection	Korres	Prevents hair color from fading away
Precision-Solution Destressante Solution Nanoemulsion Peaux Sensitivit	Chane	Moisturizer
Phyto-Endorphin Hand Cream	Rhonda Allison	Softens and smoothes the skin
Nanovital Vitanics Crystal Moisture Cream	Vitacos Cosmetics	Skin moisturizing, elastic, and lightening effects
Vitacos Vita-Herb Nona-Vital Skin Toner	Vitacos Cosmetics	Moisturizer
Plantasil Micro	BASF Care Creations	Silicone free hair conditioning booster
Lamesoft PO 65	BASF Care Creations	Skin refatting agent in body washes
Emulgade CPE	BASF Care Creations	Preparation of creams and lotions, especially suitable for wet wipes.

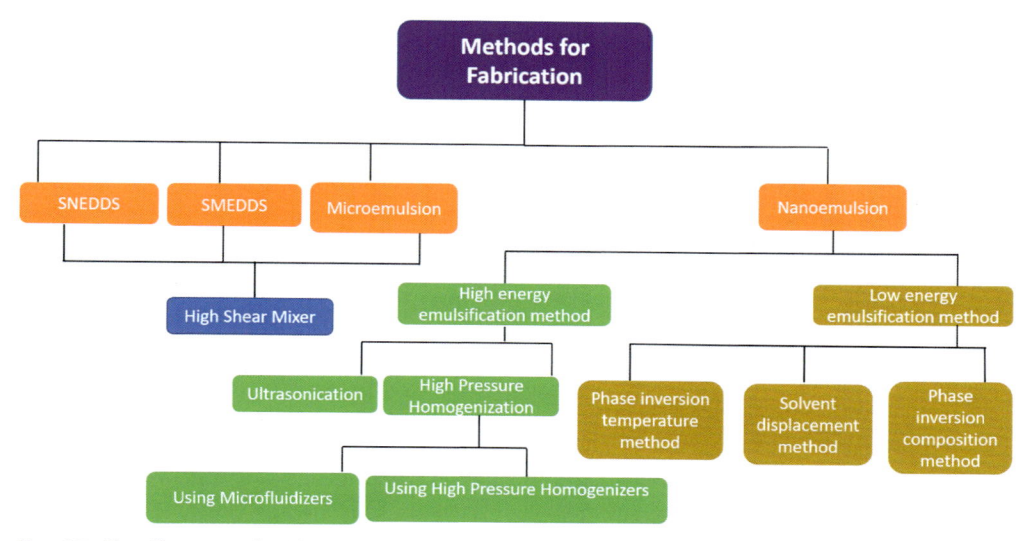

Fig. 27 Classification of self-emulsifying drug delivery systems on the basis of their methods of preparation.

microfluidization, and ultrasonication procedures are also highly expensive, which require huge amount of financial support. They also required massive amount of surfactant and cosurfactant essential for stabilizing the nano droplets. Microemulsion's use of excess amount of surfactant and cosurfactant increases cost [148].

SMEDDS and microemulsions preparation requires particular ratio of S_{mix} (Surfactant and cosurfactant) and Oil: S_{mix}, which are optimized with the help of pseudo ternary phase diagram. Optimized quantities of oil, surfactant, and cosurfactant are mixed using high shear mixer on large scale. Drug was taken and dissolved in this mixture and stored at room temperature [149].

However, high-energy method employs mechanical devices to generate intensely disruptive forces that break up the oil phase and water phase to yield droplets having nanosize. This can be achieved using high-energy devices such as ultrasonicators, microfluidizers, and high-pressure homogenizers [148].

The methods employed in fabrication of nanoformulations are diverse and show a great degree of overlapping [150]. There are various technologies available for the production of nanoformulations such as HPH, microfluidization, and ultrasonication.

5.5.1 High-energy methods
Microfluidization technique

Arthur D. Little Co. originally designed the microfluidizer, which was later carried out by the Microfluidics Corp. Microfluidizer has been used in the pharmaceutical industry to prepare pharmaceutical emulsions and has also been used to prepare flavor emulsions or homogenized milk recently [151]. *Microfluidization* has appeared as a novel method for the large scale development of delivery systems such as nanosuspensions, *solid lipid* nanocarriers, nanoemulsions, liposomes, and so on, having enhanced stability and bioavailability of encapsulated drug.

Principle: The active principle of a microfluidizer focused on microfluidization is a mixing technology that works at micro size level with the help of a device known as microfluidizer. In this technique, fluids are required to pass via the microchannels under high-pressure ranging from 500–20,000 psi [152] and dividing a pressure stream into two parts, passing each part via a fine orifice and guiding the flows at each part in the microfluidizer heart (interaction chamber) [153].

The aqueous and oil phases of macroemulsion are mixed and allowed to pass through the microfluidizer. The mixture of aqueous and oil phases is directed via microchannels under high pressure in the direction of the interaction chamber. At high velocity, two streams strike each other in the interaction chamber. Due to a sudden pressure drop, emulsification ensues because of turbulence, cavitation, and shear effects that lead to an impact, which yields stable nanoemulsions [153].

Instrumentation: Microfluidizer comprised of a pneumatic pump, an interaction chamber, and a filter. The pump could generate a pressure up to 500–30,000 psi from the compressed air supply. This instrument could be operated endlessly or recycled through a closed loop to have various cycles. The interaction chamber cooling was performed using tap water in order to improve the rise in temperature. A thermometer was positioned in the sample reservoir just afterwards the discharge port to monitor fluctuations in temperature throughout the microfluidization process. At the set pressure for each cycle, sample was prepared and guided via a microfluidizer. About one third of the microfluidized sample was used for size analysis, and the remaining amount of sample passed over microfluidizer again for the second cycle and then for the third cycle same process was followed [153]. A schematic illustration of a microfluidization system for preparation of formulation has been shown in Fig. 28.

Advantages: Microfluidization has superior advantages for its oral delivery in the medical field having greater bioavailability and the sustained release of added bioactive compounds. It has lately been used in the development of the nanoemulsion with a smaller particle size (<160 nm), higher stability, and higher encapsulation efficacy of incorporated bioactive compounds [154]. Microfluidizers formulate particle size distributions narrower and smaller compared to homogenizers. Such systems also yield stable nanoformulations using surfactant at low concentration [155]. Microfluidization also helps to extend shelf life of various products such as cream liqueurs and infant formulae by creating fine emulsions (particle size approx. 0.1 μm).

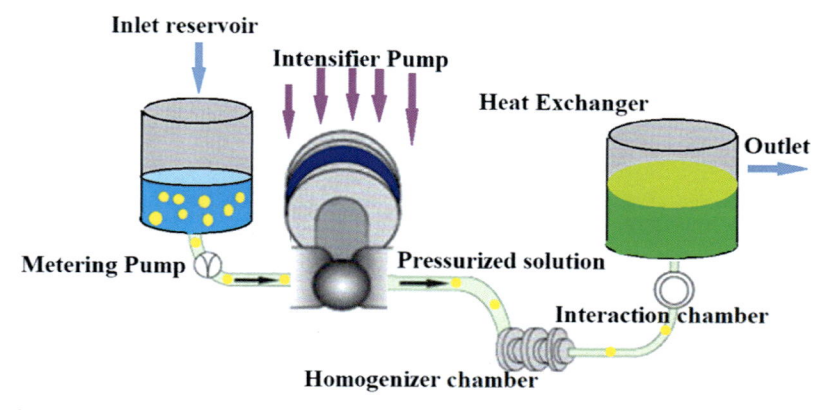

Fig. 28 Schematic representation of nanoemulsion development and size reduction using microfluidizer.

5.5.2 Low-energy methods

The low-energy methods utilize the energy contribution from component's chemical potential to produce nanoemulsions. Formation of nanoemulsions occurs spontaneously at oil and water phase interface via gentle mixing of the components. Nanoemulsion/ microemulsion development using low-energy methods depends on physicochemical parameters such as temperature, composition, and solubility. The low-energy methods involved in the nanoemulsion/microemulsion production are T_c, phase inversion composition (PIC), and solvent diffusion method. These methods include nominal energy generation, and thus avoid the degradation of heat labile compounds [156].

Phase inversion temperature method

This method involves phase invasion property of the molecules, where the emulsifiers alter their hydrophilicity or lipophilicity as a function of temperature having fixed composition. At low temperature, oil-in-water emulsion is formed at low temperature and water-in-oil emulsion is formed at high temperature as the solubility of emulsifier decreases in water with upsurge in temperature. The temperature at which there is change from oil-in-water to water-in-oil emulsion is known as phase inversion temperature. At a specific temperature, the curvature of emulsifier layer turns out to be zero and emulsifier solubility becomes almost equal in water and oil phase [159]. At this step, there is no propensity to form either O/W or W/O nanoemulsion and components form a bicontinuous or lamellar liquid crystalline system. Surfactant layer becomes concave at high temperature having negative curvature due to dehydration of hydrophilic nonionic surfactant. The solubility of surfactant is more in oil phase (lipophilic) than water phase and thus W/O nanoemulsion is produced. Temperature at which there is an alteration of O/W to W/O emulsion is known as phase inversion temperature [157].

For example, Cinnamon oil Nanoemulsion is prepared by using Phase inversion temperature method. Components such as cinnamon oil, nonionic surfactant, and water were allowed to heat at temperature above the phase inversion temperature of the system. It was quickly cooled by continuous stirring in order to ensure a spontaneous production of small oil droplets having mean droplet diameter of 101 nm. The cooling–dilution method augmented stability of the nanoemulsions at 4°C or 25°C for 31 days [158].

5.5.3 Scale-up process for nanoemulsion (e.g., Diprivan and Targretin)

Diprivan (injectable emulsion *formulation* of *Propofol*) is a stable and antimicrobial aqueous dispersion comprising a water–insoluble microdroplet matrix having a mean diameter of about 50 nm to 1000 nm and essentially propofol, propofol-soluble diluent, and amphiphilic agent for stabilizing surface [159]. The step–wise procedure for formation of Diprivan is being summarized in Fig. 29.

Targretin is an oral soft capsule which provides a new, easy, and only method for the treatment of intractable skin T–cell lymphoma. Bexarotene is a new synthetic retinoic acid analog and is in application for the intractable skin T–cell lymphoma treatment. The bexarotene soft capsule is made up of bexarotene, polysorbate, polyvidone, butylated hydroxyarisol, and PEG400. In this invention, PEG400 acts as a diluent, Polysorbate is polyoxyethylene sorbitan monoleate and acts as an emulsifying agent, Povidone K30 acts as a binding agent, and butylated hydroxyl anisole is used as an antioxidant. The step–wise procedure for formation of marketed product Targretin is being summarized in Fig. 30.

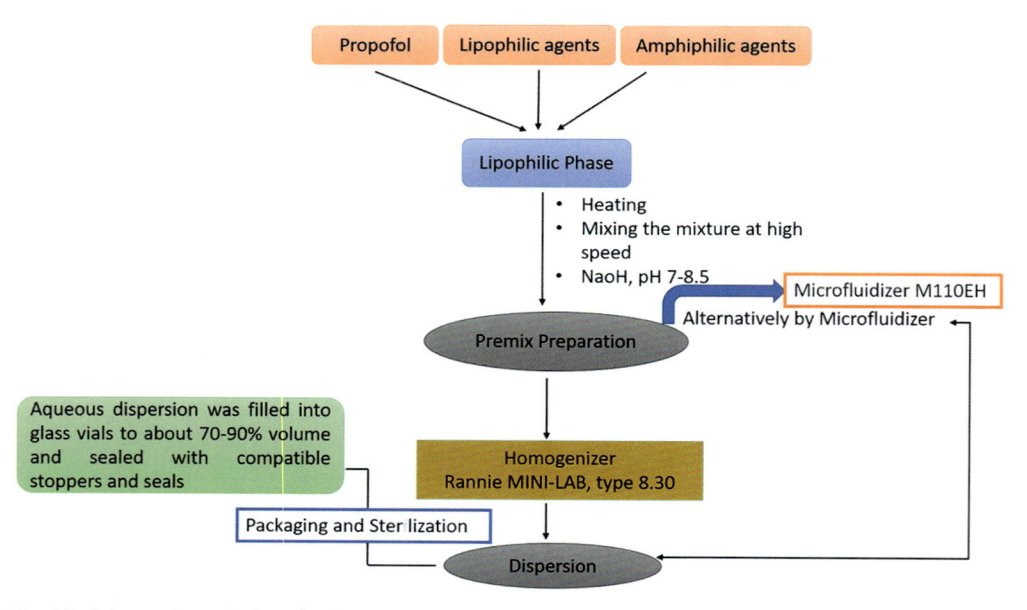

Fig. 29 Schematic procedure for the manufacturing of Diprivan Nanoemulsion by Zena Pharmaceuticals.

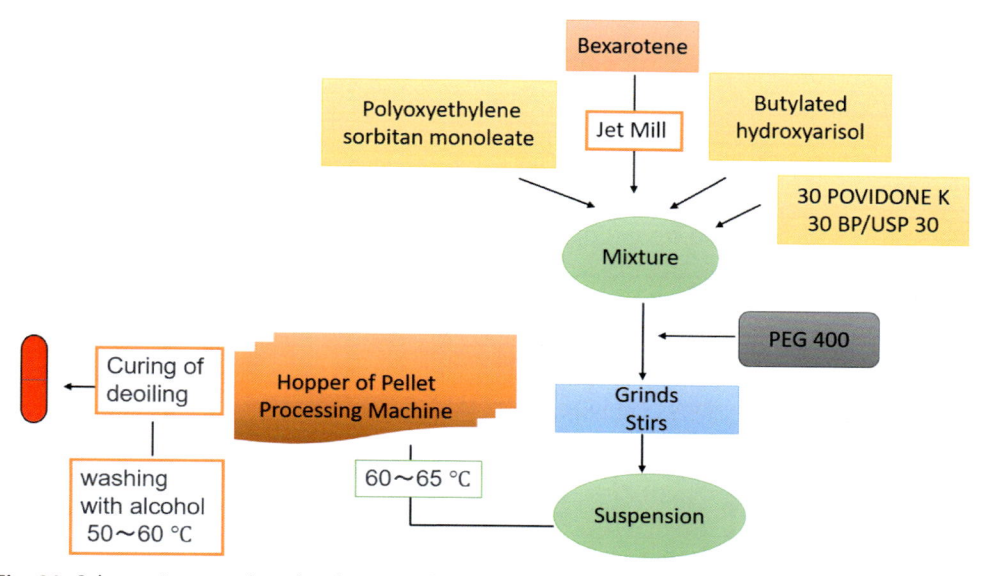

Fig. 30 Schematic procedure for the manufacturing of Targretin by Novartis.

6. Conclusions

Nanotechnology was earlier considered a science limited to researchers at laboratory scale with beaker and glass rod as their tools and a very little implementation at industrial level due to restrictions like stability concerns of nanoparticulate systems, difficulty in scaling up the synthesis techniques at commercial levels, obstacles in achieving and controlling aimed particle size, size distribution, and drug loading, requirement of expensive excipients and process equipments, and lack of technical skill at the end of technicians working in pilot scale–up and industrial plants.

But with the advanced engineering strategies and intelligent adaptations, nanotechnological architectures (especially liposomes, nanoparticles, nanocrystals, nanoemulsions, and nanosuspensions) have come a long way in the last five decades in terms of commercial implementation and expansion, continuously contributing to the disease management in the current pharmacological landscape, often limited by the concerns of drug delivery capabilities (encapsulation efficiencies and drug loading content), poor in vitro stability, and in vivo fate of the nanotherapeutic systems manufactured through conventional methods. Although exhaustive research is carried out daily in field of nanotechnology by researchers and experts throughout the globe, some specific techniques have only made their way to commercial level, thus understanding the principles and process parameters of these widely used techniques will benefit the future research and fulfill the gap between academic research and industrial demands. With the deep understanding of the structural, stability, and compositional attributes of the nanoplatforms, vehiculization of these potent therapeutic drugs or biomolecules or genes could be better facilitated for the futuristic design of next-generation therapeutic drug carriers. For instance, in case of liposomes, design of proper lipid composition and employment of active/passive loading targeting functionalizations into nanoparticle formulations may enhance the therapeutic efficacy, biodistribution, and lower toxicity of the drugs as well as the lipids. Further, it is a well-established avowal that commercial success of any dosage form hinges upon its ease of manufacturing (generally a one-step process) and cost-effective design for scale-up. Therefore, key parameters of any small scale technology or a novel process must be identified in order to optimize and promote scalability for mass production. Adoption of critical quality attributes and quality by design (QbD) approaches may provide a roadmap for the scalability and operatability in wider range of therapeutic application areas.

References

[1] Nanotechnology in Drug Delivery—GII, 2020. https://www.giiresearch.com/report/go297187-nanotechnology-drug-delivery.html. (Accessed 18 July 2020).

[2] S. Fitzgerald, FDA approves first 3D-printed epilepsy drug experts assess the benefits and caveats, Neurol. Today 15 (2015) 26–27.

[3] C. for B.E. and Research, KYMRIAH (tisagenlecleucel), FDA, 2019. https://www.fda.gov/vaccines-blood-biologics/cellular-gene-therapy-products/kymriah-tisagenlecleucel. (Accessed 19 July 2020).

[4] R.S. Plowman, T. Peters-Strickland, G.M. Savage, Digital medicines: clinical review on the safety of tablets with sensors, Expert Opin. Drug Saf. 17 (2018) 849–852.

[5] C.L. Ventola, Progress in nanomedicine: approved and investigational nanodrugs, Pharm. Ther. 42 (2017) 742.

[6] P. Skupin-Mrugalska, Liposome-based drug delivery for lung cancer, in: Nanotechnology-Based Target. Drug Delivery System for Lung Cancer, Elsevier, 2019, pp. 123–160.

[7] J.K. Patra, G. Das, L.F. Fraceto, E.V.R. Campos, M. del Pilar Rodriguez-Torres, L.S. Acosta-Torres, L.A. Diaz-Torres, R. Grillo, M.K. Swamy, S. Sharma, Nano based drug delivery systems: recent developments and future prospects, J. Nanobiotechnol. 16 (2018) 71.

[8] S. Bamrungsap, Z. Zhao, T. Chen, L. Wang, C. Li, T. Fu, W. Tan, Nanotechnology in therapeutics: a focus on nanoparticles as a drug delivery system, Nanomedicine (Lond.) 7 (2012) 1253–1271.

[9] U. Ruman, S. Fakurazi, M.J. Masarudin, M.Z. Hussein, Nanocarrier-based therapeutics and theranostics drug delivery systems for next generation of liver cancer nanodrug modalities, Int. J. Nanomedicine 15 (2020) 1437.

[10] S. Hossen, M.K. Hossain, M. Basher, M. Mia, M. Rahman, M.J. Uddin, Smart nanocarrier-based drug delivery systems for cancer therapy and toxicity studies: a review, J. Adv. Res. 15 (2019) 1–18.

[11] V. Agrahari, V. Agrahari, Facilitating the translation of nanomedicines to a clinical product: challenges and opportunities, Drug Discov. Today 23 (2018) 974–991.

[12] S. Hua, M.B. De Matos, J.M. Metselaar, G. Storm, Current trends and challenges in the clinical translation of nanoparticulate nanomedicines: pathways for translational development and commercialization, Front. Pharmacol. 9 (2018) 790.

[13] T.M. Allen, P.R. Cullis, Liposomal drug delivery systems: from concept to clinical applications, Adv. Drug Deliv. Rev. 65 (2013) 36–48.

[14] H. Daraee, A. Etemadi, M. Kouhi, S. Alimirzalu, A. Akbarzadeh, Application of liposomes in medicine and drug delivery, Artif. Cells Nanomed. Biotechnol. 44 (2016) 381–391.

[15] The United States Food and Drug Administration, 2020. https://www.fda.gov/. (Accessed 18 July 2020).

[16] A.A. Khan, K.S. Allemailem, S.A. Almatroodi, A. Almatroudi, A.H. Rahmani, Recent strategies towards the surface modification of liposomes: an innovative approach for different clinical applications, 3 Biotech 10 (2020) 163, https://doi.org/10.1007/s13205-020-2144-3.

[17] S. Nardecchia, P. Sánchez-Moreno, J. de Vicente, J.A. Marchal, H. Boulaiz, Clinical trials of thermosensitive nanomaterials: an overview, Nanomaterials (Basel) 9 (2019) 191.

[18] S. Franzé, F. Selmin, E. Samaritani, P. Minghetti, F. Cilurzo, Lyophilization of liposomal formulations: still necessary, still challenging, Pharmaceutics 10 (2018) 139.

[19] J. Lim, Y.-J. Song, W.-S. Park, H. Sohn, M.-S. Lee, D.-H. Shin, C.-B. Kim, H. Kim, G.-J. Oh, M. Ki, The immunogenicity of a single dose of hepatitis a virus vaccines (Havrix® and Epaxal®) in Korean young adults, Yonsei Med. J. 55 (2014) 126–131.

[20] R. Gasparini, D. Amicizia, P.L. Lai, S. Rossi, D. Panatto, Effectiveness of adjuvanted seasonal influenza vaccines (Inflexal V® and Fluad®) in preventing hospitalization for influenza and pneumonia in the elderly: a matched case-control study, Hum. Vaccin. Immunother. 9 (2013) 144–152.

[21] M. Shirley, Amikacin liposome inhalation suspension: a review in Mycobacterium avium complex lung disease, Drugs 79 (2019) 555–562.

[22] G. Eagle, R. Gupta, Methods for Treating Pulmonary Non-Tuberculous Mycobacterial Infections, 2019.

[23] M. Alfayez, H. Kantarjian, T. Kadia, F. Ravandi-Kashani, N. Daver, CPX-351 (vyxeos) in AML, Leuk. Lymphoma 61 (2020) 288–297.

[24] A.C. Anselmo, S. Mitragotri, Nanoparticles in the clinic, Bioeng. Transl. Med. 1 (2016) 10–29.

[25] U. Pelzer, J.-F. Blanc, D. Melisi, A. Cubillo, D.D. Von Hoff, A. Wang-Gillam, L.-T. Chen, J.T. Siveke, Y. Wan, C.T. Solem, Quality-adjusted survival with combination nal-IRI+ 5-FU/LV vs 5-FU/LV alone in metastatic pancreatic cancer patients previously treated with gemcitabine-based therapy: a Q-TWiST analysis, Br. J. Cancer 116 (2017) 1247–1253.

[26] V. Burade, S. Bhowmick, K. Maiti, R. Zalawadia, H. Ruan, R. Thennati, Lipodox®(generic doxorubicin hydrochloride liposome injection): in vivo efficacy and bioequivalence versus Caelyx®(doxorubicin hydrochloride liposome injection) in human mammary carcinoma (MX-1) xenograft and syngeneic fibrosarcoma (WEHI 164) mouse models, BMC Cancer 17 (2017) 405.

[27] J.A. Smith, L. Mathew, M. Burney, P. Nyshadham, R.L. Coleman, Equivalency challenge: evaluation of Lipodox® as the generic equivalent for Doxil® in a human ovarian cancer orthotropic mouse model, Gynecol. Oncol. 141 (2016) 357–363.

[28] G. Bozzuto, A. Molinari, Liposomes as nanomedical devices, Int. J. Nanomedicine 10 (2015) 975.

[29] J. Yeung, C.C. Crisp, D. Mazloomdoost, S.D. Kleeman, R.N. Pauls, Liposomal bupivacaine during robotic colpopexy and posterior repair: a randomized controlled trial, Obstet. Gynecol. 131 (2018) 39–46.

[30] J. Shi, P.W. Kantoff, R. Wooster, O.C. Farokhzad, Cancer nanomedicine: progress, challenges and opportunities, Nat. Rev. Cancer 17 (2017) 20.

[31] X. Xu, L. Wang, H.-Q. Xu, X.-E. Huang, Y.-D. Qian, J. Xiang, Clinical comparison between paclitaxel liposome (Lipusu®) and paclitaxel for treatment of patients with metastatic gastric cancer, Asian Pac. J. Cancer Prev. 14 (2013) 2591–2594.

[32] U. Bulbake, S. Doppalapudi, N. Kommineni, W. Khan, Liposomal formulations in clinical use: an updated review, Pharmaceutics 9 (2017) 12.

[33] M. Jain, M. Zellweger, A. Frobert, J. Valentin, H. van den Bergh, G. Wagnières, S. Cook, M.-N. Giraud, Intra-arterial drug and light delivery for photodynamic therapy using Visudyne®: implication for atherosclerotic plaque treatment, Front. Physiol. 7 (2016) 400.

[34] R. Eitan, A. Fishman, M. Meirovitz, H. Goldenberg, A. Amit, C. Koren, Y. Schneider, O. Rosengarten, A. Neuman, S. Keren-Rosenberg, Liposome-encapsulated doxorubicin citrate (Myocet) for treatment of recurrent epithelial ovarian cancer: a retrospective analysis, Anticancer Drugs 25 (2014) 101–105.

[35] K.V. Clemons, D.A. Stevens, Comparative efficacies of four amphotericin B formulations—Fungizone, Amphotec (Amphocil), AmBisome, and Abelcet—against systemic murine aspergillosis, Antimicrob. Agents Chemother. 48 (2004) 1047–1050.

[36] T.O. Olusanya, R.R. Haj Ahmad, D.M. Ibegbu, J.R. Smith, A.A. Elkordy, Liposomal drug delivery systems and anticancer drugs, Molecules 23 (2018) 907.

[37] A.G. Kohli, S. Kivimäe, M.R. Tiffany, F.C. Szoka, Improving the distribution of Doxil® in the tumor matrix by depletion of tumor hyaluronan, J. Control. Release 191 (2014) 105–114.

[38] G. Pauli, W.-L. Tang, S.-D. Li, Development and characterization of the solvent-assisted active loading technology (SALT) for liposomal loading of poorly water-soluble compounds, Pharmaceutics 11 (2019) 465.

[39] J.W. Nichols, D.W. Deamer, Catecholamine uptake and concentration by liposomes maintaining pH gradients, Biochim. Biophys. Acta Biomembr. 455 (1976) 269–271.

[40] G. Haran, R. Cohen, L.K. Bar, Y. Barenholz, Transmembrane ammonium sulfate gradients in liposomes produce efficient and stable entrapment of amphipathic weak bases, Biochim. Biophys. Acta Biomembr. 1151 (1993) 201–215.

[41] J.-S. Remy, C. Sirlin, J.-P. Behr, Gene transfer with cationic amphiphiles, in: Liposomes as Tools Basic in Research and Industry, CRC Press, Boca Raton, FL, 1994, pp. 159–170.

[42] Y.C. Barenholz, Doxil®—the first FDA-approved nano-drug: lessons learned, J. Control. Release 160 (2012) 117–134.

[43] D. Carugo, E. Bottaro, J. Owen, E. Stride, C. Nastruzzi, Liposome production by microfluidics: potential and limiting factors, Sci. Rep. 6 (2016) 25876.

[44] T.-J. Chen, S.-Y. Yang, C.-N. Liu, C.-Y. Huang, J.-C. Lin, Use and Manufacturing Process for Liposomal Doxorubicin Pharmaceutical Composition, 2005.

[45] L.A. Meure, N.R. Foster, F. Dehghani, Conventional and dense gas techniques for the production of liposomes: a review, AAPS PharmSciTech 9 (2008) 798.

[46] R.D. Worsham, V. Thomas, S.S. Farid, Potential of continuous manufacturing for liposomal drug products, Biotechnol. J. 14 (2019) 1700740, https://doi.org/10.1002/biot.201700740.

[47] S. Batzri, E.D. Korn, Single bilayer liposomes prepared without sonieation, Biochim. Biophys. Acta 298 (1973) 1015–1019.

[48] S. Stainmesse, H. Fessi, J.P. Devissaguet, F. Puisieux, Process for the Preparation of Dispersible Colloidal Systems of Amphiphilic Lipids in the Form of Oligolamellar Liposomes of Submicron Dimensions, US Patent 5174930A, 1992, p. 7.

[49] P. Gentine, A. Bubel, C. Crucifix, L. Bourel-Bonnet, B. Frisch, Manufacture of liposomes by isopropanol injection: characterization of the method, J. Liposome Res. 22 (2012) 18–30, https://doi.org/10.3109/08982104.2011.584318.

[50] A. Wagner, M. Platzgummer, G. Kreismayr, H. Quendler, G. Stiegler, B. Ferko, G. Vecera, K. Vorauer-Uhl, H. Katinger, GMP production of liposomes—a new industrial approach, J. Liposome Res. 16 (2006) 311–319.

[51] A. Wagner, K. Vorauer-Uhl, G. Kreismayr, H. Katinger, The crossflow injection technique: an improvement of the ethanol injection method, J. Liposome Res. 12 (2002) 259–270.

[52] A. Wagner, K. Vorauer-Uhl, Liposome technology for industrial purposes, J. Drug Deliv. 2011 (2011) 1–9, https://doi.org/10.1155/2011/591325.

[53] H. Zhong, G. Chan, Y. Hu, H. Hu, D. Ouyang, A comprehensive map of FDA-approved pharmaceutical products, Pharmaceutics 10 (2018) 263.

[54] Š. Koudelka, J. Turánek, Liposomal paclitaxel formulations, J. Control. Release 163 (2012) 322–334, https://doi.org/10.1016/j.jconrel.2012.09.006.

[55] U. Michaelis, H. Haas, Targeting of cationic liposomes to endothelial tissue, Liposome Technol. 3 (2006) 151–170.

[56] C. Holvoet, Y. Vander Heyden, G. Lories, J. Plaizier-Vercammen, Preparation and evaluation of paclitaxel-containing liposomes, Int. J. Pharm. Sci. 62 (2007) 126–132.

[57] C. Jaafar-Maalej, C. Charcosset, H. Fessi, A new method for liposome preparation using a membrane contactor, J. Liposome Res. 21 (2011) 213–220, https://doi.org/10.3109/08982104.2010.517537.

[58] A. Laouini, C. Jaafar-Maalej, S. Sfar, C. Charcosset, H. Fessi, Liposome preparation using a hollow fiber membrane contactor—application to spironolactone encapsulation, Int. J. Pharm. 415 (2011) 53–61, https://doi.org/10.1016/j.ijpharm.2011.05.034.

[59] K. Akamatsu, Y. Shimizu, R. Shimizu, S. Nakao, Facile method for preparing liposomes by permeation of lipid–alcohol solutions through Shirasu porous glass membranes, Ind. Eng. Chem. Res. 52 (2013) 10329–10332.

[60] N. Khayata, W. Abdelwahed, M.F. Chehna, C. Charcosset, H. Fessi, Preparation of vitamin E loaded nanocapsules by the nanoprecipitation method: from laboratory scale to large scale using a membrane contactor, Int. J. Pharm. 423 (2012) 419–427, https://doi.org/10.1016/j.ijpharm.2011.12.016.

[61] H. Hauser, G. Strauss, Stabilization of small unilamellar phospholipid vesicles during spray-drying, Biochim. Biophys. Acta Biomembr. 897 (1987) 331–334, https://doi.org/10.1016/0005-2736(87)90429-9.

[62] P. Goldbach, H. Brochart, A. Stamm, Spray-drying of liposomes for a pulmonary administration. II. Retention of encapsulated materials, Drug Dev. Ind. Pharm. 19 (1993) 2623–2636, https://doi.org/10.3109/03639049309047205.

[63] H. Kukuchi, H. Yamauchi, S. Hirota, A spray-drying method for mass production of liposomes, Chem. Pharm. Bull.(Tokyo) 39 (1991) 1522–1527, https://doi.org/10.1248/cpb.39.1522.

[64] J.-C.K. Kim Jong-Duk, Preparation by spray drying of amphotericin B-phospholipid composite particles and their Anticellular activity, Drug Deliv. 8 (2001) 143–147, https://doi.org/10.1080/107175401316906900.

[65] R.T. Proffitt, J. Alder-Moore, S.M. Chiang, Amphotericin B Liposome Preparation, US2004/0175417A1, 2004.

[66] P. Chakravarty, A. Famili, K. Nagapudi, M.A. Al-Sayah, Using supercritical fluid technology as a green alternative during the preparation of drug delivery systems, Pharmaceutics 11 (2019) 629, https://doi.org/10.3390/pharmaceutics11120629.

[67] B. William, P. Noémie, E. Brigitte, P. Géraldine, Supercritical fluid methods: an alternative to conventional methods to prepare liposomes, Chem. Eng. J. 383 (2020) 123106, https://doi.org/10.1016/j.cej. 2019.123106.

[68] S.-J. Hwang, P.R. Karn, W.K. Cho, H.J. Park, J.S. Park, Characterization and stability studies of a novel liposomal cyclosporin A prepared using the supercritical fluid method: comparison with the modified conventional Bangham method, Int. J. Nanomedicine (2013) 365, https://doi.org/10.2147/IJN.S39025.

[69] P.R. Karn, W. Cho, S.-J. Hwang, Liposomal drug products and recent advances in the synthesis of supercritical fluid-mediated liposomes, Nanomedicine 8 (2013) 1529–1548.

[70] P.R. Karn, H. Do Kim, H. Kang, B.K. Sun, S.-E. Jin, S.-J. Hwang, Supercritical fluid-mediated liposomes containing cyclosporin a for the treatment of dry eye syndrome in a rabbit model: comparative study with the conventional cyclosporin A emulsion, Int. J. Nanomedicine 9 (2014) 3791.

[71] S.H. Soh, L.Y. Lee, Microencapsulation and Nanoencapsulation using supercritical fluid (SCF) techniques, Pharmaceutics 11 (2019) 21, https://doi.org/10.3390/pharmaceutics11010021.

[72] B. Rivnay, J. Wakim, K. Avery, P. Petrochenko, J.H. Myung, D. Kozak, S. Yoon, N. Landrau, A. Nivorozhkin, Critical process parameters in manufacturing of liposomal formulations of amphotericin B, Int. J. Pharm. 565 (2019) 447–457, https://doi.org/10.1016/j.ijpharm.2019.04.052.

[73] S.J. Hwang, H.J. Park, W. Cho, K.-H. Cha, J. Park, C. Park, D. Gu, Method and Apparatus for Preparing Novel Liposome, 2013, US20130069261A1 https://patents.google.com/patent/US20130069261A1/en. (Accessed 19 July 2020).

[74] C. Lim, S.M. Abuzar, P.R. Karn, W. Cho, H.J. Park, C.-W. Cho, S.-J. Hwang, Preparation, characterization, and in vivo pharmacokinetic study of the supercritical fluid-processed liposomal amphotericin B, Pharmaceutics 11 (2019) 589.

[75] W.-Z.S. Lin, N. Malmstadt, Liposome production and concurrent loading of drug simulants by microfluidic hydrodynamic focusing, Eur. Biophys. J. 48 (2019) 549–558, https://doi.org/10.1007/s00249-019-01383-2.

[76] A. Zizzari, M. Bianco, L. Carbone, E. Perrone, F. Amato, G. Maruccio, F. Rendina, V. Arima, Continuous-flow production of injectable liposomes via a microfluidic approach, Materials (Basel) 10 (2017) 1411, https://doi.org/10.3390/ma10121411.

[77] R. Hood, D.L. DeVoe, Microfluidic Liposome Synthesis, Purification and Active Drug Loading, US2017/018972 A1, 2017.

[78] A. Otten, S. Köster, B. Struth, A. Snigirev, T. Pfohl, Microfluidics of soft matter investigated by small-angle X-ray scattering, J. Synchrotron Radiat. 12 (2005) 745–750, https://doi.org/10.1107/S0909049505013580.

[79] S. Kakumanu, A. Schroeder, Focused ultrasound—a novel tool for liposome formulation, Drug Dev. 12 (2012) 5.

[80] A.L. Carneiro, M.H.S. Andrade, Production of liposomes in a multitubular system useful for scaling up of processes, in: Progress in Colloid and Polymer Science, 2004, pp. 273–277.

[81] J. George, S. Sabapathi, Cellulose nanocrystals: synthesis, functional properties, and applications, Nanotechnol. Sci. Appl. 8 (2015) 45–54, https://doi.org/10.2147/NSA.S64386.

[82] S. Bansal, M. Bansal, R. Kumria, Nanocrystals: current strategies and trends, Int. J. Res. Pharmaceut. Biomed. Sci. 4 (2012) 10.

[83] L. Gao, G. Liu, J. Ma, X. Wang, L. Zhou, X. Li, Drug nanocrystals: in vivo performances, J. Control. Release 160 (2012) 418–430, https://doi.org/10.1016/j.jconrel.2012.03.013.

[84] J.-U.A.H. Junghanns, R.H. Müller, Nanocrystal technology, drug delivery and clinical applications, Int. J. Nanomedicine 3 (2008) 295, https://doi.org/10.2147/ijn.s595.

[85] H. Tao, N. Lavoine, F. Jiang, J. Tang, N. Lin, Reducing end modification on cellulose nanocrystals: strategy, characterization, applications and challenges, Nanoscale Horiz. 5 (2020) 607–627, https://doi.org/10.1039/D0NH00016G.

[86] K. Dvořáčková, P. Doležel, E. Mašková, J. Muselík, M. Kejdušová, D. Vetchý, The effect of acid pH modifiers on the release characteristics of weakly basic drug from Hydrophlilic–lipophilic matrices, AAPS PharmSciTech 14 (2013) 1341–1348, https://doi.org/10.1208/s12249-013-0019-1.

[87] V. Teeranachaideekul, V.B. Junyaprasert, E.B. Souto, R.H. Müller, Development of ascorbyl palmitate nanocrystals applying the nanosuspension technology, Int. J. Pharm. 354 (2008) 227–234, https://doi.org/10.1016/j.ijpharm.2007.11.062.

[88] S. Li, R.W. Baker, I. Lignos, Z. Yang, S. Stavrakis, P.D. Howes, A.J. de Mello, Automated microfluidic screening of ligand interactions during the synthesis of cesium lead bromide nanocrystals, Mol. Syst. Des. Eng. (2020), https://doi.org/10.1039/D0ME00008F.

[89] T.G. Meikle, B.P. Dyett, J.B. Strachan, J. White, C.J. Drummond, C.E. Conn, Preparation, characterization, and antimicrobial activity of Cubosome encapsulated metal nanocrystals, ACS Appl. Mater. Interfaces 12 (2020) 6944–6954, https://doi.org/10.1021/acsami.9b21783.

[90] B. Sinha, R.H. Müller, J.P. Möschwitzer, Bottom-up approaches for preparing drug nanocrystals: formulations and factors affecting particle size, Int. J. Pharm. 453 (2013) 126–141, https://doi.org/10.1016/j.ijpharm.2013.01.019.

[91] H.-K. Chan, P.C.L. Kwok, Production methods for nanodrug particles using the bottom-up approach, Adv. Drug Deliv. Rev. 63 (2011) 406–416, https://doi.org/10.1016/j.addr.2011.03.011.

[92] H. de Waard, H.W. Frijlink, W.L.J. Hinrichs, Bottom-up preparation techniques for nanocrystals of lipophilic drugs, Pharm. Res. 28 (2011) 1220–1223, https://doi.org/10.1007/s11095-010-0323-3.

[93] B. Van Eerdenbrugh, G. Van den Mooter, P. Augustijns, Top-down production of drug nanocrystals: nanosuspension stabilization, miniaturization and transformation into solid products, Int. J. Pharm. 364 (2008) 64–75, https://doi.org/10.1016/j.ijpharm.2008.07.023.

[94] K. Joshi, A. Chandra, K. Jain, S. Talegaonkar, Nanocrystalization: an emerging technology to enhance the bioavailability of poorly soluble drugs, Pharm. Nanotechnol. 7 (2019) 259–278, https://doi.org/10.2174/2211738507666190405182524.

[95] J.P. Möschwitzer, Drug nanocrystals in the commercial pharmaceutical development process, Int. J. Pharm. 453 (2013) 142–156, https://doi.org/10.1016/j.ijpharm.2012.09.034.

[96] T.W. Phillips, I.G. Lignos, R.M. Maceiczyk, A.J. de Mello, J.C. de Mello, Nanocrystal synthesis in microfluidic reactors: where next? Lab Chip 14 (2014) 3172–3180, https://doi.org/10.1039/C4LC00429A.

[97] C.M. Keck, R.H. Müller, Drug nanocrystals of poorly soluble drugs produced by high pressure homogenisation, Eur. J. Pharm. Biopharm. 62 (2006) 3–16, https://doi.org/10.1016/j.ejpb.2005.05.009.

[98] P. Patel Anita, J. Patel, S. Patel Khushbu, B. Deshmukh Aaishwarya, R. Mishra Bharat, A review on drug nanocrystal a carrier free drug delivery, Int. J. Res. Ayurveda Pharm. 2 (2011) 448–458.

[99] Y. Lu, Y. Li, W. Wu, Injected nanocrystals for targeted drug delivery, Acta Pharm. Sin. B 6 (2016) 106–113, https://doi.org/10.1016/j.apsb.2015.11.005.

[100] T. Chang, H. Zhan, D. Liang, J. Liang, Nanocrystal technology for drug formulation and delivery, Front. Chem. Sci. Eng. 9 (2015) 1–14, https://doi.org/10.1007/s11705-015-1509-3.

[101] T. Hatahet, M. Morille, A. Hommoss, C. Dorandeu, R.H. Müller, S. Bégu, Dermal quercetin smart-Crystals®: formulation development, antioxidant activity and cellular safety, Eur. J. Pharm. Biopharm. 102 (2016) 51–63, https://doi.org/10.1016/j.ejpb.2016.03.004.

[102] L. Siddiqui, J. Bag, D. Mittal, A. Leekha, H. Mishra, M. Mishra, A.K. Verma, P.K. Mishra, A. Ekielski, Z. Iqbal, Assessing the potential of lignin nanoparticles as drug carrier: synthesis, cytotoxicity and genotoxicity studies, Int. J. Biol. Macromol. 152 (2020).

[103] R. Gupta, H. Xie, Nanoparticles in daily life: applications, toxicity and regulations, J. Environ. Pathol. Toxicol. Oncol. 37 (2018) 209–230.

[104] V. Mohanraj, Y. Chen, Nanoparticles—a review, Trop. J. Pharm. Res. 5 (2006) 561–573.

[105] I. Khan, K. Saeed, I. Khan, Nanoparticles: properties, applications and toxicities, Arab. J. Chem. 12 (2019) 908–931.

[106] B. Nagavarma, H.K. Yadav, A. Ayaz, L. Vasudha, H. Shivakumar, Different techniques for preparation of polymeric nanoparticles-a review, Asian J. Pharm. Clin. Res. 5 (2012) 16–23.

[107] C. Vauthier, K. Bouchemal, Methods for the preparation and manufacture of polymeric nanoparticles, Pharm. Res. 26 (2009) 1025–1058.

[108] P.A. Grabnar, J. Kristl, The manufacturing techniques of drug-loaded polymeric nanoparticles from preformed polymers, J. Microencapsul. 28 (2011) 323–335, https://doi.org/10.3109/02652048.2011.569763.

[109] J.P. Rao, K.E. Geckeler, Polymer nanoparticles: preparation techniques and size-control parameters, Prog. Polym. Sci. 36 (2011) 887–913, https://doi.org/10.1016/j.progpolymsci.2011.01.001.

[110] U. Perera, N. Rajapakse, Chitosan nanoparticles: preparation, characterization, and applications, in: Seafood Processing By-Products, Springer, 2014, pp. 371–387.

[111] T. Kang, Y.G. Kim, D. Kim, T. Hyeon, Inorganic nanoparticles with enzyme-mimetic activities for biomedical applications, Coord. Chem. Rev. 403 (2020) 213092.

[112] Z.P. Xu, Q.H. Zeng, G.Q. Lu, A.B. Yu, Inorganic nanoparticles as carriers for efficient cellular delivery, Chem. Eng. Sci. 61 (2006) 1027–1040.

[113] O.V. Salata, Applications of nanoparticles in biology and medicine, J. Nanobiotechnol. 2 (2004) 3.

[114] M.S. Pereira, G.M.S. Mendes, T.S. Ribeiro, M.R. Silva, I.F. Vasconcelos, Influence of thermal-treatment effects on the structural and magnetic properties of $Sn_{1-x}Fe_xO_2$ Nanopowders produced by mechanical milling, J. Supercond. Nov. Magn. 33 (2020) 1–8.

[115] F. Patrignani, R. Lanciotti, Applications of high and ultra high pressure homogenization for food safety, Front. Microbiol. 7 (2016), https://doi.org/10.3389/fmicb.2016.01132.

[116] M. Kim, S. Osone, T. Kim, H. Higashi, T. Seto, Synthesis of nanoparticles by laser ablation: a review, KONA Powder Part. J. (2017), https://doi.org/10.14356/kona.2017009. advpub.

[117] A.T. Odularu, Metal nanoparticles: thermal decomposition, biomedicinal applications to cancer treatment, and future perspectives, Bioinorg. Chem. Appl. 2018 (2018), https://doi.org/10.1155/2018/9354708.

[118] P. Ayyub, R. Chandra, P. Taneja, A.K. Sharma, R. Pinto, Synthesis of nanocrystalline material by sputtering and laser ablation at low temperatures, Appl. Phys. A Mater. Sci. Process. 73 (2001) 67–73, https://doi.org/10.1007/s003390100833.

[119] J. Xu, H. Yang, W. Fu, K. Du, Y. Sui, J. Chen, Y. Zeng, M. Li, G. Zou, Preparation and magnetic properties of magnetite nanoparticles by sol–gel method, J. Magn. Magn. Mater. 309 (2007) 307–311, https://doi.org/10.1016/j.jmmm.2006.07.037.

[120] C.Y. Tai, Y.-H. Wang, H.-S. Liu, A green process for preparing silver nanoparticles using spinning disk reactor, AICHE J. 54 (2008) 445–452, https://doi.org/10.1002/aic.11396.

[121] J. Ding, S. Chen, N. Han, Y. Shi, P. Hu, H. Li, J. Wang, Aerosol assisted chemical vapour deposition of nanostructured ZnO thin films for NO_2 and ethanol monitoring, Ceram. Int. 46 (2020) 15152–15158.

[122] H. Li, S. Pokhrel, M. Schowalter, A. Rosenauer, J. Kiefer, L. Mädler, The gas-phase formation of tin dioxide nanoparticles in single droplet combustion and flame spray pyrolysis, Combust. Flame 215 (2020) 389–400.

[123] R. Mueller, L. Mädler, S.E. Pratsinis, Nanoparticle synthesis at high production rates by flame spray pyrolysis, Chem. Eng. Sci. 58 (2003) 1969–1976, https://doi.org/10.1016/S0009-2509(03)00022-8.

[124] A.M. Awwad, N.M. Salem, M.M. Aqarbeh, F.M. Abdulaziz, Green synthesis, characterization of silver sulfide nanoparticles and antibacterial activity evaluation, Chem. Int. 6 (2020) 42–48.

[125] S. Iravani, R.S. Varma, Greener synthesis of lignin nanoparticles and their applications, Green Chem. 22 (2020) 612–636, https://doi.org/10.1039/C9GC02835H.

[126] K. Kawakami, M. Ebara, H. Izawa, N.M. Sanchez-Ballester, J.P. Hill, K. Ariga, Supramolecular approaches for drug development, Curr. Med. Chem. 19 (2012) 2388–2398, https://doi.org/10.2174/092986712800269254.

[127] S.P. Kovvasu, P. Kunamaneni, R. Joshi, G.V. Betageri, Self-emulsifying drug delivery systems and their marketed products: a review, Asian J. Pharm. 13 (2019) 73–84.

[128] N.H. Shah, M.T. Carvajal, C.I. Patel, M.H. Infeld, A.W. Malick, Self-emulsifying drug delivery systems (SEDDS) with polyglycolyzed glycerides for improving in vitro dissolution and oral absorption of lipophilic drugs, Int. J. Pharm. 106 (1994) 15–23, https://doi.org/10.1016/0378-5173(94)90271-2.

[129] B. Singh, S. Bandopadhyay, R. Kapil, R. Singh, O.P. Katare, Self-emulsifying drug delivery systems (SEDDS): formulation development, characterization, and applications, Crit. Rev. Ther. Drug Carrier Syst. 26 (2009) 427–521.

[130] F.U. Rehman, K.U. Shah, S.U. Shah, I.U. Khan, G.M. Khan, A. Khan, From nanoemulsions to self-nanoemulsions, with recent advances in self-nanoemulsifying drug delivery systems (SNEDDS), Expert Opin. Drug Deliv. 14 (2017) 1325–1340, https://doi.org/10.1080/17425247.2016.1218462.

[131] A.A. Date, N. Desai, R. Dixit, M. Nagarsenker, Self-nanoemulsifying drug delivery systems: formulation insights, applications and advances, Nanomedicine 5 (2010) 1595–1616, https://doi.org/10.2217/nnm.10.126.

[132] M. Tanya, Smedds/Snedds: an emerging technique to solubility enhancement for the pharmaceutical industry, World J. Pharm. Pharm. Sci. (2017) 317–336, https://doi.org/10.20959/wjpps20177-9493.

[133] S. Reddy, T. Katyayani, A. Navatha, G. Ramya, Review on self micro emulsifying drug delivery systems, Int. J. Res. Pharm. Sci. 2 (2011) 382–392.

[134] S.D. Maurya, R.K. Arya, G. Rajpal, R.C. Dhakar, Self-micro emulsifying drug delivery systems (smedds): a review on physico-chemical and biopharmaceutical aspects, J. Drug Deliv. Ther. 7 (2017) 55–65, https://doi.org/10.22270/jddt.v7i3.1453.

[135] M. Kazi, A.A. Shahba, S. Alrashoud, M. Alwadei, A.Y. Sherif, F.K. Alanazi, Bioactive self-nanoemulsifying drug delivery systems (bio-SNEDDS) for combined Oral delivery of curcumin and Piperine, Molecules 25 (2020) 1703, https://doi.org/10.3390/molecules25071703.

[136] A.A.-W. Shahba, K. Mohsin, F.K. Alanazi, Novel self-nanoemulsifying drug delivery systems (SNEDDS) for oral delivery of cinnarizine: design, optimization, and in-vitro assessment, AAPS PharmSciTech 13 (2012) 967–977, https://doi.org/10.1208/s12249-012-9821-4.

[137] S. Gupta, S. Chavhan, K.K. Sawant, Self-nanoemulsifying drug delivery system for adefovir dipivoxil: design, characterization, in vitro and ex vivo evaluation, Colloids Surf. A Physicochem. Eng. Asp. 392 (2011) 145–155, https://doi.org/10.1016/j.colsurfa.2011.09.048.

[138] H. Chavda, J. Patel, G. Chavada, S. Dave, A. Patel, C. Patel, Self-nanoemulsifying powder of isotretinoin: preparation and characterization, J. Powder Technol. 2013 (2013) 1–9, https://doi.org/10.1155/2013/108569.

[139] B. Sanghai, G. Aggarwal, S.L. HariKumar, Solid self microemulsifying drug deliviry system: a review, J. Drug Deliv. Ther. 3 (2013) 168–174, https://doi.org/10.22270/jddt.v3i3.476.

[140] S. Madhav, D. Gupta, A review on microemulsion based system, Int. J. Pharm. Sci. Res. 2 (2011) 1888–1899.

[141] Y. Singh, J.G. Meher, K. Raval, F.A. Khan, M. Chaurasia, N.K. Jain, M.K. Chourasia, Nanoemulsion: concepts, development and applications in drug delivery, J. Control. Release 252 (2017) 28–49, https://doi.org/10.1016/j.jconrel.2017.03.008.

[142] J.P. Fast, S. Mecozzi, Nanoemulsions for intravenous drug delivery, in: M.M. de Villiers, P. Aramwit, G.S. Kwon (Eds.), Nanotechnology in Drug Delivery, Springer, New York, NY, 2009, pp. 461–489, https://doi.org/10.1007/978-0-387-77668-2_15.

[143] M. Jaiswal, R. Dudhe, P.K. Sharma, Nanoemulsion: an advanced mode of drug delivery system, 3 Biotech 5 (2015) 123–127, https://doi.org/10.1007/s13205-014-0214-0.

[144] A. Gurram, P. Deshpande, S. Kar, U. Nayak, N. Udupa, M. Reddy, Role of components in the formation of self-microemulsifying drug delivery systems, Indian J. Pharm. Sci. 77 (2015) 249, https://doi.org/10.4103/0250-474X.159596.

[145] C.W. Pouton, C.J.H. Porter, Formulation of lipid-based delivery systems for oral administration: materials, methods and strategies, Adv. Drug Deliv. Rev. 60 (2008) 625–637, https://doi.org/10.1016/j.addr.2007.10.010.

[146] J.B. Aswathanarayan, R.R. Vittal, Nanoemulsions and their potential applications in food industry, Front. Sustain. Food Syst. 3 (2019) 95, https://doi.org/10.3389/fsufs.2019.00095.

[147] S. Kaul, N. Gulati, D. Verma, S. Mukherjee, U. Nagaich, Role of nanotechnology in cosmeceuticals: a review of recent advances, J. Pharm. 2018 (2018) 1–19, https://doi.org/10.1155/2018/3420204.

[148] S.A. Chime, F.C. Kenechukwu, A.A. Attama, Nanoemulsions—advances in formulation, characterization and applications in drug delivery, in: A.D. Sezer (Ed.), Application of Nanotechnology in Drug Delivery, InTech, 2014, https://doi.org/10.5772/58673.

[149] S.K. Savale, A review—self nanoemulsifying drug delivery system (snedds), Int. J. Res. Pharm. Nano Sci. 4 (2015) 385–397.

[150] R. Paliwal, R.J. Babu, S. Palakurthi, Nanomedicine scale-up technologies: feasibilities and challenges, AAPS PharmSciTech 15 (2014) 1527–1534, https://doi.org/10.1208/s12249-014-0177-9.

[151] M. Kumar, R.S. Bishnoi, A.K. Shukla, C.P. Jain, Techniques for formulation of nanoemulsion drug delivery system: a review, Prev. Nutr. Food Sci. 24 (2019) 225–234, https://doi.org/10.3746/pnf.2019.24.3.225.

[152] A. Qadir, M.D. Faiyazuddin, M.D. Talib Hussain, T.M. Alshammari, F. Shakeel, Critical steps and energetics involved in a successful development of a stable nanoemulsion, J. Mol. Liq. 214 (2016) 7–18, https://doi.org/10.1016/j.molliq.2015.11.050.

[153] S. Mahdi Jafari, Y. He, B. Bhandari, Nano-emulsion production by sonication and microfluidization—a comparison, Int. J. Food Prop. 9 (2006) 475–485, https://doi.org/10.1080/10942910600596464.

[154] P. Ganesan, G. Karthivashan, S.Y. Park, J. Kim, D.-K. Choi, Microfluidization trends in the development of nanodelivery systems and applications in chronic disease treatments, Int. J. Nanomedicine 13 (2018) 6109–6121, https://doi.org/10.2147/IJN.S178077.

[155] T.J. Wooster, M. Golding, P. Sanguansri, Impact of oil type on Nanoemulsion formation and Ostwald ripening stability, Langmuir 24 (2008) 12758–12765, https://doi.org/10.1021/la801685v.

[156] J.B. Aswathanarayan, R.R. Vittal, Nanoemulsions and their potential applications in food industry, Front. Sustain. Food Syst. 3 (2019) 1–21, https://doi.org/10.3389/fsufs.2019.00095.

[157] N. Anton, T.F. Vandamme, The universality of low-energy nano-emulsification, Int. J. Pharm. 377 (2009) 142–147, https://doi.org/10.1016/j.ijpharm.2009.05.014.

[158] P. Chuesiang, U. Siripatrawan, R. Sanguandeekul, L. McLandsborough, D. Julian McClements, Optimization of cinnamon oil nanoemulsions using phase inversion temperature method: impact of oil phase composition and surfactant concentration, J. Colloid Interface Sci. 514 (2018) 208–216, https://doi.org/10.1016/j.jcis.2017.11.084.

[159] M.G. Vachon, (75) Inventors: Awadhesh K. Mishra, Verdun (CA); Gary W. Pace, Raleigh, NC (US), 2002, p. 11.

CHAPTER 3

In vitro physicochemical characterization of nanocarriers: a road to optimization

Honey Goel[a],*, Komal Saini[b,c,]*, Karan Razdan[b,c,]*, Rajneet Kaur Khurana[b], Amal Ali Elkordy[d], and Kamalinder K. Singh[c,e,f]

[a]Department of Pharmaceutics, University Institute of Pharmaceutical Sciences and Research, Baba Farid University of Health Sciences (BFUHS), Faridkot, India
[b]Pharmaceutics Division, University Institute of Pharmaceutical Sciences, Panjab University, Chandigarh, India
[c]School of Pharmacy and Biomedical Sciences, Faculty of Clinical and Biomedical Sciences, University of Central Lancashire, Preston, United Kingdom
[d]Faculty of Health Sciences and Wellbeing, University of Sunderland, Sunderland, United Kingdom
[e]UCLan Research Centre for Smart Materials, University of Central Lancashire, Preston, United Kingdom
[f]UCLan Research Centre for Translational Biosciences & Behaviour, University of Central Lancashire, Preston, United Kingdom

1. Introduction

Recent years have seen an increased interest in obtaining the information regarding the physical and chemical characteristics of new pharmaceutical preparations, at the level of both industry and academics. The attention on understanding such properties is driven by both the regulatory requirements and appreciation that understanding the properties can help in the design of "better" formulations or in nanotailoring the desired form. On nanoscale, atoms and molecules exhibit distinctive properties and present an opportunity for diverse range of useful and attractive applications. Materials with nanometric dimensions have different physicochemical characteristics in terms of size, surface properties, shape, composition, molecular weight, stability, and solubility, which are essential determinants of their biofate and physiological activities.

In recent years, research and development activities have taken the advantages of nanotechnology for targeting therapy in terms of selective delivery of protein, vaccine, and genes [1]. The term nanocarrier (NC) includes a variety of nanostructures with different characteristics, such as nanoparticles, nanospheres, nanocapsules, nanoemulsion, dendrimers, and nanosized vesicular carriers, that is, liposomes and niosomes [2]. Because of small size, the nanocarriers have the capability to reach the unreachable places, which basically consist of the tumor cells and inflamed tissues. Nanocarriers exhibit enhanced permeability and retention (EPR) effect combined with altered lymphatic drainage at the diseased site, leading to benefit of selective delivery of drugs, proteins, and vaccines

---

* All contributed equally as co-first authors.

Nanoparticle Therapeutics
https://doi.org/10.1016/B978-0-12-820757-4.00018-1

[3]. Mostly, these carriers can facilitate improvement in bioavailability aspects, solubility enhancement, extending the duration of action and protection from enzymatic degradation, as well as producing the optimal therapeutic efficacy. The designed nanocarrier structure for the sustained drug release and targeted delivery of the pharmaceutical drugs offers various advantages over the plain drug administration, i.e., improvement in drug quantity that reaches the targeted place, prevention of drug degradation, and reduction of off target side effects. In spite of advantages of nanocarriers, they possess limitations that involve stability issues because of their modified physical properties permitting particle-particle aggregation plus difficulty of handling nanoparticles both in liquid and dry conditions because of their small size and large surface area. Furthermore, the small size of the carriers limits drug loading and may result in burst release of encapsulated molecules. These types of investigational difficulties should be countered early on to permit successful clinical study and commercial application of nanoparticles [4]. The key aspect of engineered nanocarriers as a drug carrier is to analyze their particle size, surface morphology, surface charge, drug release, and other characteristics so that they fulfill their specific objectives.

The characterization of drug nanocarriers is vital to understand and tailor their desired actions in vitro and in vivo as well. This chapter will focus on the characterization methods applied to nanocarriers elaborating on the principles of these methods as well as investigating the advantages of various techniques.

2. Classification of nanocarriers

Nanocarriers (NCs) are colloidal particles (usually in the range of 10–100 nm) that are fabricated from biocompatible and biodegradable materials such as polymers or lipids. They have been studied extensively as targeted delivery systems in last 30 years due to their many potential advantages. The large surface area-to-volume ratio provides large contact area for the drug to interact with body, allowing for reduction in drug dosage and subsequent side effects. Moreover, the surface chemistry of NCs can be fine-tuned for drug targeting to desired site and prolonged drug circulation. Incorporation of drug in NCs can protect it from undue degradation. Finally, NCs can lend flexibility in more diverse formulation approaches and routes of administration [5]. A schematic diagram of various types of NCs used for drug delivery is depicted in Fig. 1.

2.1 Vesicular nanocarriers

2.1.1 Liposomes

Alec Bangham discovered liposomes more than five decades ago [6], and the term is derived from Greek words—*lipos* (fat) and *soma* (body). One or more phospholipid bilayers encasing an aqueous compartment(s) make up the liposomes. Cholesterol is added to confer rigidity to the vesicle structure. Delivery of both hydrophilic and hydrophobic drugs can be accomplished. They have several advantages as a drug carrier such as biomimetic nature, high biocompatibility, low immunogenic potential, and toxicity. More

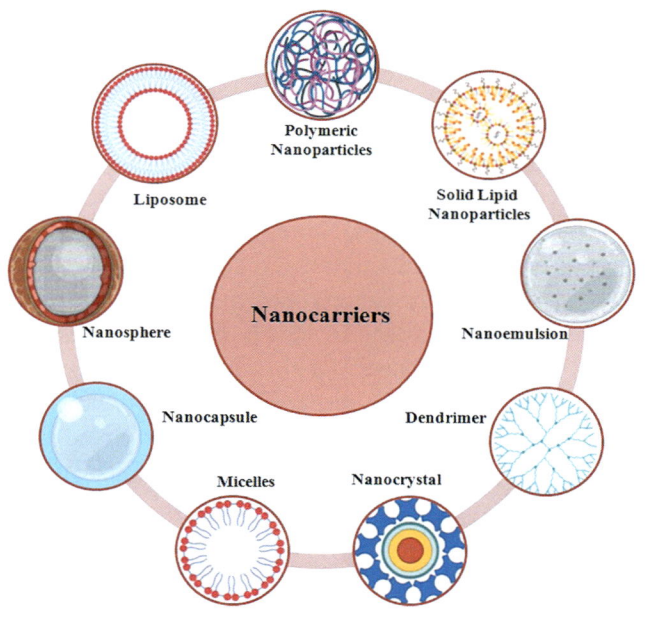

Fig. 1 Classification of nanocarriers for drug delivery.

than 40 liposome-based drug formulations are in market or in different stages of clinical research. Some of the significant examples include *Doxil, Depocyt, Ambisome*, and *Epaxal* [7].

Liposomes can be modified to get desired characteristics and actions. Incorporation of ethanol (up to 50%) in the phospholipid structure makes them flexible and malleable and are termed as ethosomes. Ethosomes are excellent carriers for delivery of drugs through the human skin. Ethanol fluidizes the skin lipids by lowering their transition temperature, allowing the delivery of therapeutic agents to deeper layers of skin. Incorporation of an edge activator (usually a surfactant) to liposomes gave rise to an ultradeformable structure called transferosomes. Routinely used edge activators are sodium cholate, polysorbates, sorbitan esters, and dipotassium glycyrrhizinate. The deformability of vesicle is increased owing to destabilization of lipid bilayer by reduction in interfacial tension [8].

2.1.2 Niosomes

L'Oreal was the first to develop and patent niosomes in the 1970s. These are vesicular structures composed of nonionic surfactants, cholesterol, and fatty alcohols. Surfactants such as alkyl ethers, alkyl esters, polysorbates, and sorbitan esters are most commonly used. Nontoxic and nonimmunogenic composition of niosomes makes them advantageous for drug delivery. They exhibit higher physical and chemical stability in comparison to liposomes and are cost-effective [8].

2.2 Particulate carriers

2.2.1 Polymeric nanoparticles

Polymeric nanoparticle (PNPs) are extensively used nanocarriers owing to their suitable characteristics in terms of design and biocompatibility. Their small size allows permeation through biological barriers and capillaries. PNP surface can be modified to enhance the duration of systemic circulation and find application in cancer therapy. Nanocapsules and nanospheres are two major types of PNPs. Nanospheres are the type of matrix–based particles in which drugs is embedded in a uniform manner, while in case of nanocapsules, drug is dispersed and surrounded by a special polymeric coat. Nanocapsules act as drug reservoirs owing to their vesicular form, in which the drug is present in an aqueous or nonaqueous core surrounded by polymeric shell, whereas nanospheres are spherical mass of polymers in which the drug is either dispersed in the polymeric matrix or adsorbed onto the surface. Both natural and synthetic polymers, such as albumin, chitosan, and polylactides (PLA), and poly(lactide-*co*-glycolides) (PLGA), can be used for the formation of PNPs [9].

2.2.2 Solid lipid nanoparticles and nanostructured lipid carriers

Developed as a substitute drug delivery system to PNPs and liposomes in 1990s [10], solid lipid nanoparticle (SLNs) are colloidal systems fabricated using of physiological lipids (which remain solid at body temperature), stabilized by suitable surfactant. Lipids used usually include triglycerides, partial glycerides, fatty acids, waxes, or mixtures thereof. Typical concentrations of solid lipid and surfactants used in SLN formulations are 0.1%–30% w/w and 0.5%–5% w/w, respectively. SLNs are spherical in shape with mean particle size in range of 50–1000 nm. Good biocompatibility, low toxicity, feasibility for scale up, and solvent-free preparation are some of the many advantages that SLN offer. However, the nanocarrier suffers from the drawback of low drug loading and drug expulsion during storage. To overcome these limitations, Muller et al. [10] put forth second-generation lipid carriers, the nanostructured lipid carriers (NLCs), which are prepared by precise blending of solid and liquid lipids, leading to nanostructures with enhanced drug loading and release profiles. Mixing of lipid molecules of various conformations and size leads to the formation of imperfect crystalline matrix, which can accommodate higher amount of drug. This improves drug loading capacity and reduces expulsion of drug during storage. NLC preparation entails mixing of solid and liquid lipid in a ratio of 70:30 to a ratio of 99.9:0.1, and surfactant concentration varies from 0.5% to 5% w/v [8].

2.2.3 Dendrimers

Dendrimers belong to family of synthetic polymers, which have well-defined, highly branched, three-dimensional structure. Poly(amidoamine) dendrimers (PAMAM) are most extensively used for drug delivery. Drug can be entrapped via three main sites in dendrimer structure, viz., empty spaces (molecular entrapment), branching points (hydrogen bonding), and external surface groups (charge-charge interaction). Dendrimers can serve a wide range of applications such as enhancing drug solubility, stability, and bioavailability. However, conventional dendrimers suffer from the problem of non-biodegradability and cytotoxicity. Hence, biodegradable dendrimers such as polyester and polyacetal dendrimers have gained importance [11, 12].

2.3 Inorganic nanocarriers

2.3.1 Silica nanoparticles

Mesoporous silica nanoparticles (MSNs) are nonbiodegradable NCs that have been investigated for drug delivery applications. Chemically, MSNs have active surfaces and honeycomb-like structures. By modifying the components used for the preparation of MSNs, their properties such as pore size, mesoporous nature, loading efficiency, etc., can be altered to meet the needs for drug delivery applications [13].

2.3.2 Gold nanoparticles

Unique physicochemical and optical properties of gold nanoparticles lend them to diagnostic and drug delivery applications. Their small size allows them to easily penetrate cells and deliver drugs. The inert nature of metallic gold inherently bestows it with low toxicity, and it can be modified easily by linking ligand for cell-specific delivery [14].

3. Characterization of nanocarriers

The most commonly faced issue during the characterization of nanoformulations is the reproducibility of the nanoformulations owing to broad distribution of discrete particle sizes with diverse dimensions (e.g., spherical structures are generally defined by a single size parameter whereas nonspherical structures entail multidimensional analysis), shapes, and defects. Furthermore, nanoparticle functionalization capacity, fluid drag, diffusion, optical properties, and cellular uptake are affected by shape and size [15]. The characterization investigations that are essential to corroborate the quality control of the nanotherapeutic systems have been classified on the basis of their internal physical state, i.e., dry phase or liquid phase. Fig. 2 depicts a framework that shall enable the readers to organize the most significant techniques suitable on the basis of various physicochemical parameters, viz., size distribution and morphological characteristics (shape, surface topography, and ultrastructure), surface charge, porosity, polymorphism and crystallinity, encapsulation efficiency, and drug release. Table 1 summarizes the various parameters of nanoplatforms that are characterized employing diverse technologies and methods.

3.1 Characterization in liquid state

Some of the commonly used techniques that are primarily employed for the analysis of samples in liquid state or dosage systems (such as nanoemulsions, microemulsion, and

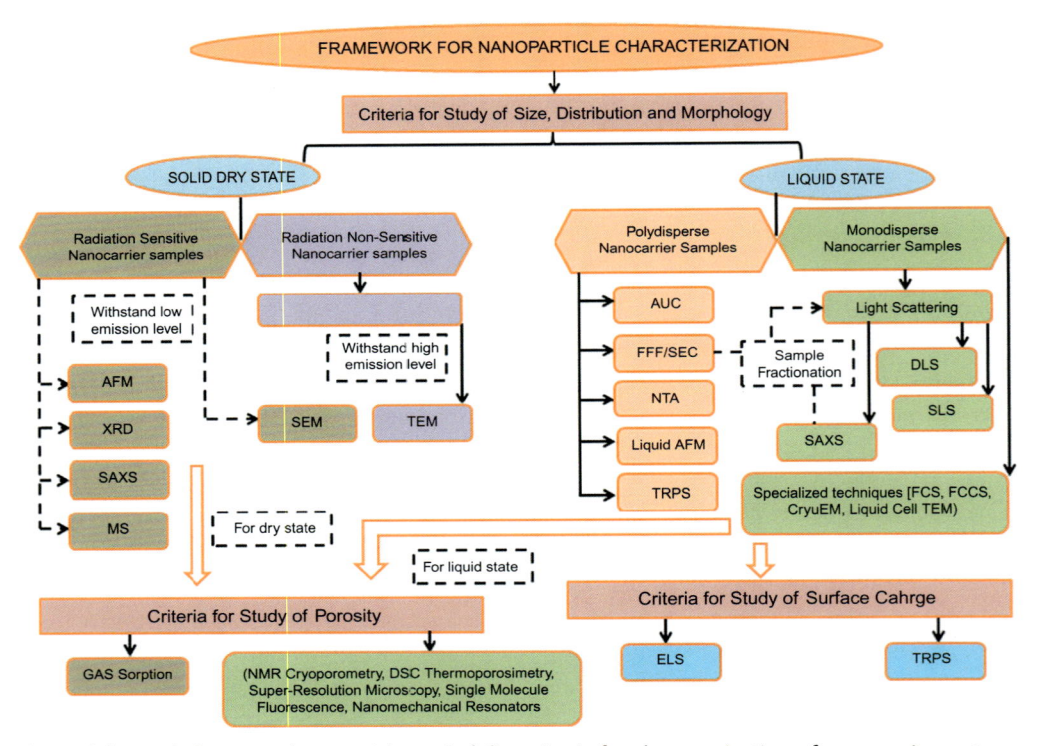

Fig. 2 Schematic framework comprising suitability criteria for characterization of nanocarrier systems on the basis of their physical states. *AFM*, atomic force microscopy; *AUC*, analytical ultracentrifugation; *DCS*, differential centrifugal sedimentation; *DLS*, dynamic light scattering; *DSC*, differential scanning calorimetry; *ELS*, electrophoretic light scattering; *FCS*, fluorescence correlation spectroscopy; *FFF*, field flow fractionation; *MS*, mass spectrometry; *NTA*, nanoparticle tracking analysis; *SAXS*, small angle X-ray scattering; *SEC*, size exclusion chromatography; *SLS*, static laser light scattering; *TEM*, transmission electron microscopy; *TRPS*, tunable resistive pulse sensing; *XRD*, X-ray diffraction.

nanosuspensions) for the characterization of particle size distribution include dynamic light scattering (DLS), static laser light scattering (SLS), differential centrifugal sedimentation (DCS), analytical ultracentrifugation (AUC), field flow fractionation (FFF), nanoparticle tracking analysis (NTA), electrophoretic light scattering (ELS), size exclusion chromatography (SEC), fluorescence correlation spectroscopy (FCS), and tunable resistive pulse sensing (TRPS). Entrapment efficiency or loading capacity utilize techniques such as UV–visible spectroscopic studies, high pressure liquid chromatography (HPLC), and drug release analysis or elemental analysis use inductively coupled plasma mass spectrometry (ICP-MS), respectively.

Table 1 Summary of characterization of diverse properties of various nanoplatforms.

Nano platforms	Characterization	
	Parameters	Methods/instrumentation
Physical		
Liposomes; SNEDDS; SMEDDS; Nanoemulsions; Microemulsions; Nanocrystals; Nanoparticles	Vesicle (size, shape, surface morphology, and size distribution)	Freeze fracture electron microscopy, polymerase chain reaction (PCR), gel permeation, dynamic light scattering (DLS), size exclusion chromatography (SEC), field–flow fractionation (FFF), transmission electron microscopy (TEM), cryogenic-TEM (cryo-TEM), and atomic force microscopy (AFM)
	Surface charge	Free flow electrophoresis
	Electric surface potential and pH	Zeta potential measurement, pH probes, electrophoretic mobility
	Phase behavior	Freeze fracture electron microscopy, X-ray diffraction (XRD), differential scanning calorimetry (DSC), and thermogravimetric analysis (TGA)
	Entrapment efficiency	Mini column centrifugation, gel exclusion, ion exchange, protamine aggregation, radiolabeling, centrifugation, dialysis, or column separation for liposomes isolation, followed by drug content determination
	Drug release	Diffusion cell/dialysis or centrifugation, followed by drug quantification using analytical method, such as UV–VIS spectrophotometry, fluorescence spectrometry, enzyme or proteinbased assays, gel electrophoresis, High pressure liquid chromatography (HPLC), and Liquid chromatography and mass spectrometry (LCMS)

Continued

Table 1 Summary of characterization of diverse properties of various nanoplatforms—cont'd

Nano platforms	Characterization	
	Parameters	**Methods/instrumentation**
Unique properties		
Liposomes	Lamellarity	Small angle X-ray scattering (SAXS), freeze fracture EM, CryoTEM, and ^{31}P NMR
SNEDDS; SMEDDS; Nanoemulsions; Microemulsions	Viscosity and pH	Brookfield viscometer, pH probes, potentiometer, electrophoretic mobility
	Dilution test	
	Optical birefringence	Abbe's refractometer
	Centrifugation test	Centrifuge
	Turbidity measurement	Turbidimeter
	Limpidity test	UV-VIS spectrophotometery
Nanosuspensions	Density	Hydrometer
	Viscosity and pH	Rotary viscometer
Nanocrystals	Polymorphic state	XRD, DSC, thermogravimetry, vibrational spectroscopy (infrared and Raman)
	Permeation studies	Franz diffusion cell
Nanoparticles	Optical absorption, transmittance, and reflectance	UV/VIS-diffuse reflectance spectrometer (DRS)
	Surface area	BET isotherm, titrimetric methods and NMR, scanning mobility particle sizer (SMPS), differential mobility analyzer (DMA)
	Composition	X-ray photoelectron spectroscopy (XPS), chemical digestion followed by mass spectrometry (MS), atomic emission spectroscopy (AES), and ion chromatography (IC)
Chemical		
Liposomes	PHL:CHL:drug	Barlett/Stewart assay, cholesterol oxidase assay, HPLC method as in individual monograph

Table 1 Summary of characterization of diverse properties of various nanoplatforms—cont'd

Nano platforms	Characterization	
	Parameters	**Methods/instrumentation**
	PHL: per oxidation hydrolysis	UV absorbance, TBA, iodometric, GLC, HPLC, TLC, fatty acid concentration
	CHL auto-oxidation	HPLC, TLC
	Antioxidant degradation	HPLC, TLC
	pH	pH meter
Liposomes; *Nanosuspensions SNEDDS*; *SMEDDS*; *Nanoemulsion*; *Microemulsion*	Osmolality	Osmometer
	Emulsification time and precipitation assessment	USP II dissolution apparatus
Nanocrystals	Apparent and supersaturated solubility	UV spectrophotometer, HPLC, LCMS
	Drug-stabilizer ratio	Dilution and temperature stability of particle size by DLS, Zetasizer
Nanoparticles	Absorption shift in doping and composites	UV/Vis-diffuse reflectance spectrometer (DRS)
	Refractive index and extinction coefficient	Spectroscopic ellipsometry
	Dielectric properties	Localized surface plasma resonance (LSPR) spectrum
	Concentration	Condensation particle counter
	Pyrogenicity	Pyrogen test (LAL test)
	Animal toxicity	Monitoring survival rates, histopathology

3.1.1 Particle size determination

Size is the spatial extent of any spherical/nonspherical object, unambiguously described either by single or multidimensional analysis. It is an essential element that transforms nanoparticles from alternate medicament dispensing system, including bulk powder. Although numerous techniques are available for measuring the size of nanocarriers, it is imperative to proclaim here that each technique has its own advantages and disadvantages, which should be assessed conscientiously while evaluating the nanocarriers. The selection of an explicit method depends on different criteria, for example, expected size and the nanoparticles population. Moreover, particles diameter shows an impact on the extrication of entrapped materials; therefore, minute particles offer the extensive surface area [16, 17].

As a result, most encapsulated agents show faster release in release media, while therapeutics gradually permeates out from the core layer of macroparticles. On the other hand, finer particles have a tendency to cluster while the delivery and vaulting of nanoparticles dispersion. Therefore, a concession is evident between a finer size and enhanced stability of nanoparticles. The size of nanoparticles is the significant parameter establishing their properties as well as safety in the biological systems, particularly when administered parenterally.

ynamic light scattering

DLS, also called as quasielastic light scattering or photon correlation spectroscopy (PCS), is one of the methods rampantly used to determine the particle size of solutions and size-distribution studies. Basic principle of DLS is envisaged around the Brownian diffusion of spherical particles, whose movement is associated with an equivalent hydrodynamic diameter. The instrument focuses a light of laser beam into the nanoparticle dispersion, and a photon detector detects the intensity of Doppler shift of the incident radiation, based on the detection of time-dependent fluctuations. The instruments contain three crucial components: laser source, sample holder, and light detector (Fig. 3). The Stokes-Einstein equation is used to measure particle size, where the particle diffusion timescale gets associated with the equivalent sphere hydrodynamic diameter of particles. This relation is based on both the viscosity and temperature of the solution upon which the scattered light is collected [18, 19].

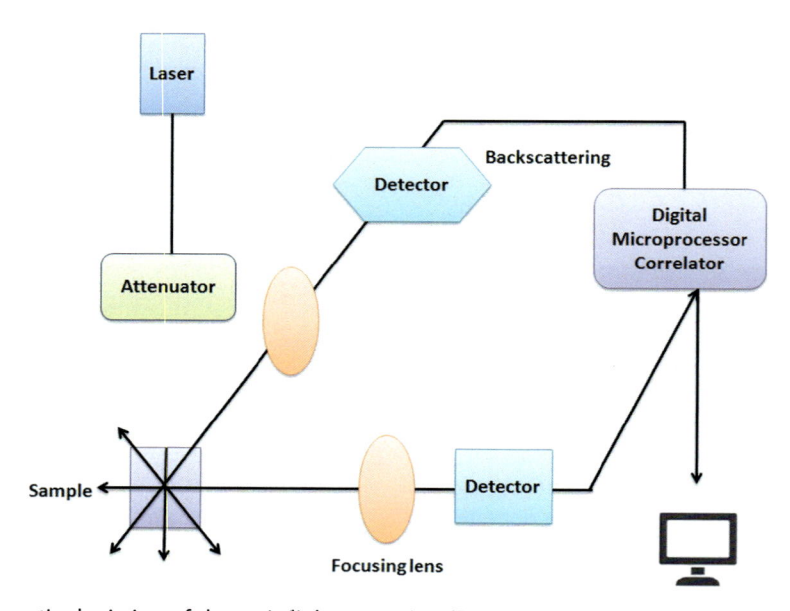

Fig. 3 Schematic depiction of dynamic light scattering (DLS) instrumentation.

Several features of DLS method are available concomitant to the equipment, collection of data, and the treatment. It is impossible to measure the size of particles that absorb at the similar wavelength as that of the laser used in the equipment. DLS determines the equivalent sphere hydrodynamic diameter and the actual particle size in the suspension can be underrated. DLS does not give information related to shape of nanoparticles. Moreover, laser radiation scattering of larger particles was found to be more efficient in comparison to smaller particles, as the light intensity scattered from small spherical particles is proportional to 6th power of particles radius; thus, the presence of a dust particle or traces of agglomerates can interfere in the results of DLS. It cannot differentiate between equally sized populations not including the coupling to a separation/fractionation process [20].

It is a well-documented fact that nanomaterials have a tendency to cluster in water, varying their size and surface properties, thus owing to different interactions with the water molecules that bound the nanoparticles. Consequently, the obtained size by DLS may be overrated and size distribution may be changed because of environmental dependence. Alternatively, DLS permits the measurements simulating the physiological conditions and imitating the in vivo behavior. The size of dried nanoparticles probably differs from when present in water, stomach, and intestines. Moreover, the size may also differ in the presence of plasma proteins and in ionic medium. As a result, DLS is an important technique for carrying out the hydrodynamic sizes of nanoparticles in the biological system [21].

Advantages
• Provides information about hydrodynamic radius (diffusion behavior of particles).
Limitations
• Can provide overrated size distribution due to environmental dependence.
Therefore, it could be postulated that DLS mainly deals with the development and confirmation of different types of nanoparticles, including polymeric nanoparticles, micelles, or liposomes, with the advantage of being possible to shape their properties to increase the efficacy in targeting or drug delivering.

DLS is generally used first to study the formulations stability in accordance with time and/or variations of temperature, second to discover the presence of aggregation in formulations prepared by different processes, and finally for fast determination of particle size of monodisperse samples.

Static laser light scattering or laser diffraction

SLS is an important approach that shows a much broader range of detection (20–2000 µm). SLS is also referred as laser light scattering, that can be used alone and in combination with the PCS to achieve a range of total population size from minute to macroparticles. Also, SLS is mainly used for emulsion, nanoemulsion, and colloids.

The principle of SLS is that when the laser beam is passed via liquid containing the suspended particles, the large particles scatter the light at narrow angles while the small particles show the scattering of light at broader angles. The obtained results using SLS can be used to gauge the correspondent spherical radius of particles as indicated by the Mie scattering solution (also called as the "Mie theory"). Large particles and small particles being mainly estimated by red lasers and blue lasers, respectively, although it is not suggested for colloidal suspensions containing particularly lesser diameter in comparison to laser wavelength.

Although laser diffractometer suggests a detection of particles polydispersity as it shows a broader size range from nanometer to 100 microns. However, the uses of this technique are limited to multicomponents nanocarriers because the available information for refractive index at the measurement wavelength is significant as the distribution of particle size is enormously based on these optical factors. The development of polarization intensity differential scattering (PIDS) noticeably augmented the SLS sensitivity to the finer particles [22]. This technology combines the wavelength dependence and polarization effects jointly therefore significantly amplified the SLS sensitivity towards the small particles. Yet, despite of technology advancement, it is significantly suggested to use PCS and SLS parallelly.

Advantages
- Provides analysis of mass-weighted size.
- Aids in determining molecular weight.

Limitations
- Nonspherical particles cause more diffused scatter patterns and are difficult to accurately interpret.

Differential centrifugal sedimentation

The underlying postulate behind this technique is that bigger particles settle rapidly in comparison to smaller particles if they have the same density. However, most nanocarriers do not sediment under the effect of gravity itself (due to their small size); therefore, sedimentation can be encouraged by centrifugation. It consists of a centrifuge tube with a hollow disk that can be clearly observed, and the operational speed of centrifuge ranges from 600 to 24,000 rpm. Disk compartment is filled to some extent with the fluid, which permits the rings of liquid to fall against density difference. The sample is administered at the center of a rotating disk for measurement. DCS gives a high-quality resolution and various nanocarriers having of size variation with <5% can be resolved completely. This technique has previously been employed to determine the size distribution, sedimentation coefficients and hydrodynamic radii of nanoscale polymer particles, functionalized quantum dots, inorganic nanoparticles including Pt and ZnO, FePt, TiO_2, MnO, ZnS, as well as modified gold nanoparticles and nanocrystals [23].

Recently, variation in the size of gold (Au) nanoparticles after the surface modification typically was examined by using DCS technique, and it demonstrated a 0.5 nm change in the nanoparticle size after the modifications and this shift amounted to 2.1 nm after modification in nanoparticles with an entity having high molecular weight, for instance, single-stranded DNA. The utilization of DCS approach for estimating the size of PEG-alkane thiol-modified Au nanoparticles was also described by Krpetic and colleagues [23].

Analytical ultracentrifugation

This methodology assists in determining size, its distribution and particle density characteristics in a dispersion medium/liquid phase containing nanoparticles, based on the principle of detection of the sedimentation properties of the nanoparticles when a centrifugal force is applied. The instrument comprises of an ultraspeed centrifuge equipped with a transparent cell and a detector to track the progression of the dynamic concentration profile of the particles along the cell axis during centrifugation, and the final thermodynamic equilibrium.

Advantages
- Highly sensitive to variations in density and mass of the nanoparticles.
- This technique facilitates multimodal sample distribution during the sedimentation process, where diverse particle sizes and its molecular weights can be eluted.
- Coupling the integrating multiwavelength optical characterization may expand its horizon to analyze nanoparticle-protein interactions, its shape, and optical behavior.

Limitations
- Data interpretation and translation of sedimentation gets affected by factors such as sample concentration or sample interactions.
- Involves high instrumental costs and time-consuming data analysis.

Field flow fractionation

FFF method is used for calculating the particle sizes of pigments, fillers, and lattices used in coating systems. Flow field-flow fractionation (F4) is increasingly employed as a mature separation technique able to size-sort and segregate the nanoparticles for their further analysis by online, disparate detection techniques. F4 belongs to the FFF family, flow-assisted methods preferably suitable to isolate the dispersed analytes up to the wide range of size, from nanometer to micrometer sized analytes. F4 principle for separation is totally dependent upon the particle diffusion coefficient [24]. Food and drug administration (FDA) approved the F4 as a technique that complements size exclusion chromatography (SEC) in the protein drug products validation, due to the higher selectivity in the high molar mass and nanometer-size range, and stationary phase absence makes the F4 suited to analysis of delicate, nanosized samples.

Advantages
The advantages of F4 approach for nanoparticle characterization are as follows:

- Involves sample fractionation.
- Enables accurate, high-resolution nanoparticle size distribution analysis.
- Analysis of nanoparticle aggregation in native states (important characteristic in any nanoparticle application).
- Isolation of the unbound constituents of the functional nanoparticles, which can be a key step for the optimization of nanoparticle synthesis.
- Establishment of optical aspects of the nanoparticles, separated from other dispersion components containing free chromophores, and of the correlation of spectroscopic characterization with size of nanoparticles.

Limitations
- Absolute size quantification might be challenging.

Asymmetrical flow field fractionation Asymmetrical flow field fractionation (AF4) is the elective size-separation method for the particles. AF4 strategy not only plays pivotal role in enhancing the resolution of size dimensions but also enables the analysis of complex multimodal samples and provides compatibility with complex media [25]. With the help of multidetectors, this approach could serve as a potential application to overcome the pitfalls of conventional dynamic light scattering and could assist in broadening the application for elucidation of particle characterization (Fig. 4) to drug delivery nanocarriers such as liposomes, solid lipid nanoparticles, and nanostructured lipid complexes.

Primarily, coupling the detectors with AF4 systems may generate significant data information about the particle dimensions and its distribution pattern, molar mass, mass concentration, particle shape or shape factor, and compositional study of various elements. In this approach, mobile liquid phase moves through a thin channel in which analyte diffusion is counterbalanced by a downward force provided by cross flow via an underlying porous membrane. This process leads to differential in particle velocities across the channel owing to its analyte size (as smaller ones leave the channel before the larger ones). The individual size fractions are traced in the form of separate bands as depicted in Fig. 5.

In one study, Li et al. coupled AF4 to differential refractive index (dRI) detector to estimate the liposome entrapment efficiency of actin by separating free actin from entrapped actin [26]. Shape/size distribution and conformational changes were examined, and actin-containing liposomes were obtained with larger mean size in comparison to pure liposomes.

Nanoparticle tracking analysis

From the above-mentioned techniques for particle size determination, DLS has become the potent and accessible tool to usually establish the size of nanoparticles. DLS shows the intensity of light scattering (due to Brownian movement of particles) is directly proportional to the 6th power of diameter of particles and found to be sensitive to large particles.

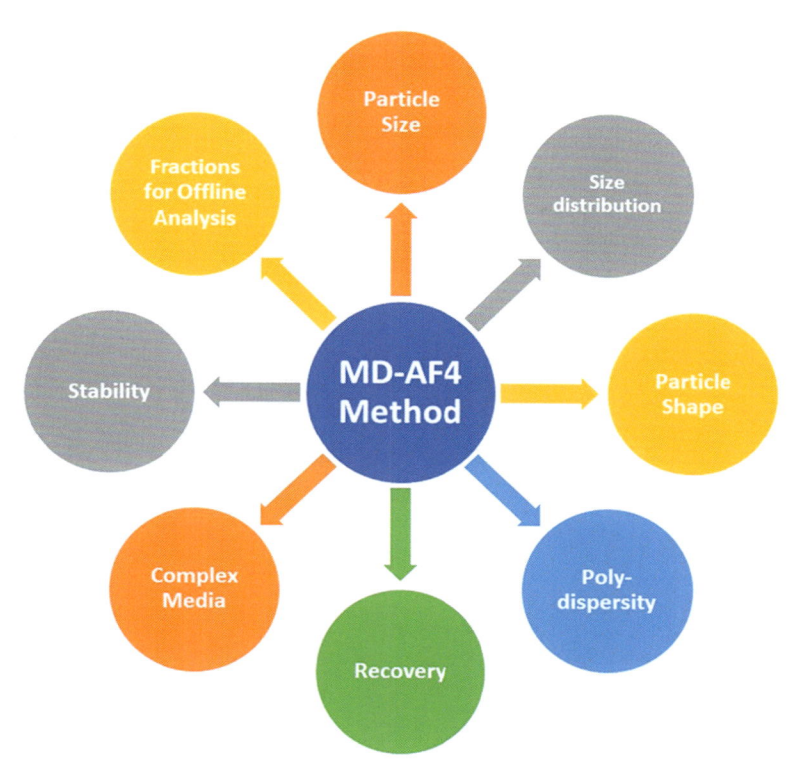

Fig. 4 Multidimensional aspects of multi detector-AF4 systems. *(Reproduced under Creative Commons CC-BY license from J. Parot, F. Caputo, D. Mehn, V.A. Hackley, L. Calzolai, Physical characterization of liposomal drug formulations using multi-detector asymmetrical-flow field flow fractionation, J. Control. Release 320 (2020) 495–510, by Elsevier.)*

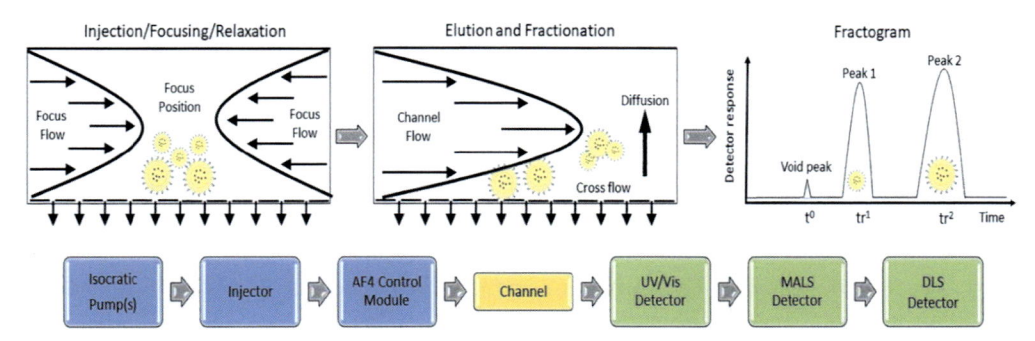

Fig. 5 Schematic representation of AF4 systems coupled with multi detectors. *(Reproduced under Creative Commons CC-BY license from J. Parot, F. Caputo, D. Mehn, V.A. Hackley, L. Calzolai, Physical characterization of liposomal drug formulations using multi-detector asymmetrical-flow field flow fractionation, J. Control. Release 320 (2020) 495–510, by Elsevier.)*

However, DLS is also known to suffer from drawbacks associated with the accurate size estimation. NTA is a method for visualizing and analyzing particles in liquids that relates the rate of Brownian motion to particle size. NTA was introduced in 2006 as commercial set-up for estimation of size and concentration colloidal suspensions [27]. It was also used in different research and development, as well as industry fields with diverse sample compositions, e.g., lipid nanoparticles, nanolipid carriers, cellular vesicles, virus particle, microvesicle and exosome, gold nanoparticle conjugation, fullerenes, or proteins [28–30].

This innovative system measures the particle size ranging from 30 to 1000 nm, with the lower limit of detection being based on the dioptric index of nanoparticles. The said method combines a laser light scattering microscopy with a charge-coupled device (CCD) camera, and then the particles can be visualized and tracked by specific image tracking and recording the nanoparticles in solution. The NTA software is then able to identify and track individual nanoparticles moving under Brownian motion and relates the movement to a particle size according to the following formula derived from the Stokes-Einstein Eq. (1):

$$\frac{(x, y)^2}{4} Dt = \frac{Kb \cdot T}{3\pi\eta\, d} \tag{1}$$

where Dt is the diffusion constant, a product of diffusion coefficient D and time t.

Kb is Boltzmann's constant, T is the absolute temperature, η is viscosity, and d is the diameter of the spherical particle.

These obtained tracks are further used to draw the histogram for estimation of the size distribution and the concentration of particles. Nanosight Ltd. developed the commercial NTA system and overtures assorted options as well as challenges for the operator, such as different capture modes or variable measurement and estimation parameters that need to be set for each measurement. Video recording of the tracks is the pioneering step of the measurement process. The second step, the data analysis and evaluation process, involves another set of parameters. Some of the merits and demerits of this application have been summarized as follows:

Advantages
- Enables single particle detection.
- Useful in the study of protein aggregates and provides information about the kinetics of protein aggregation.

Limitations
- Requires high scattering or fluorescent probe coupled NPs.
- Requires careful sample preparation in order to prevent aggregate distribution in measurement cell.

Ohlsson et al. [31] reported the use of NTA to check liposome stability and integrity by determining the solute transport in the sub-100 ms time scale across the lipid bilayer membrane of individual proteoliposomes. In research on nanoparticles, the application of NTA for the measurement of particle size and its distribution in the field of gene delivery has also been investigated. In a study a novel assay was devised to quantify the number of plasmids encapsulated by polymeric nanoparticles, and NTA has been used to establish the number density of plasmids per 100 nm nanoparticles [32]. Several other examples for measuring the size and enumerating nanoparticulate drug delivery systems by NTA have also been reported by authors [33–35]. Bolat et al. [36] have studied the particle concentration determination of emulsome formulations by using the NTA technique.

Size exclusion chromatography

This technique also termed as molecular sieve chromatography is employed for the separation and analysis of macromolecules (i.e., proteins, large industrial biopolymers, and macromolecular complexes) among the polydisperse populations of nanoparticles. This chromatographic method comprises a column packed with porous microparticles (with size range of up to nm) known as solid phase. On the other hand, liquid phase is inserted into the column under pressure in order to exploit the differential diffusion (i.e., biomolecules with small radius tend to diffuse more whereas larger ones undergo faster diffusion owing to shorter diffusion paths) of nanoparticles in accordance to their hydrodynamic radius, thus resulting in effective sample fractionation. Some of the merits and demerits are described as follows:

Advantages
- Commonly employed in both research and industrial settings for separation and analysis of proteins and viruses.
- Aids in purification of nanoparticles.
- Quantitative estimation and characterization of size and molecular weight.

Limitations

Size analysis using these techniques may not provide absolute information challenges involved in the calibration of relationship between elution times and particle size owing to different surface charge and shape characteristics.

Fluorescence correlation spectroscopy

FCS assists in the facilitation of gap between classical ensemble and contemporary single molecule measurements by assessing the spatial and temporal correlation of discrete molecules with themselves [37]. The principle of FCS is based on the detection of emitted light rather than scattered light. Moreover, it also provides data on concentration and molecular number fluctuations for nonlinear reaction systems that complement single-molecule measurements. As fluorophores diffuse in and out of the confined volume of excitation light in a confocal microscopy setup, they emit fluorescent light, and the resulting intensity fluctuations of the emitted light are recorded by a detector. Furthermore, the intensity of the fluctuations depends on the number of fluorophores present in excitation volume and could be used to produce autocorrelation, which in turn allows for extracting translational diffusion coefficients.

Advantages

- Assist in evaluation of chemical rate constants, molecular concentrations, or especially in the case of FCCS-binding events in complex systems.
- Useful for characterizing small, dynamic systems with low concentration at given time.

Tunable resistive pulse sensing

This methodology works on the principle of coulter counter, which measures the variation of ionic current through an orifice due to partial blocking of the orifice aperture by the passage of the particle. The signal amplitude scales linearly with respect to particle volume. This technique is quite useful for the quantification of size and charge of colloidal dispersion systems with single particle resolution. Furthermore, this approach allows the tunability of the aperture (i.e., enables clearing of the pore in the case of clogging, as well as to adapt the measurement sensitivity during acquisition) by stretching the pore containing membrane. Furthermore, precise prediction of size and charge becomes challenging owing to unpredictable calibration of aperture shape and size.

Advantages

- Characterizes the fate and interactions of nanoparticles in complex physiological media.
- Provides single particle resolution.
- Enables simultaneous characterization of size and surface charge.
- Possess large dynamic range for measurement of size.

Limitations

- Require careful calibration.
- Require conductive solutions for size determination.

3.1.2 Dispersibility and surface charge
Polydispersity index

PDI is the most critical parameter that gives the heterogeneity of a sample based on size. Polydispersity can take place because of the size distribution in a sample or agglomeration or aggregation during isolation or analysis. If the sample shows PDI equal to zero, called as monodisperse, while the PDI is near to one, then it said to be polydisperse. However, if found less than 0.3, these are generally considered as the homogeneous. Lipid–based drug delivery system or lipid carriers, for example, liposomes, nanoliposomes, vesicular phospholipid gels, solid lipid nanoparticles, transferosomes, etc., are employed for the enhancement of bioavailability and are also characterized for PDI for determining the size distribution of particles.

Zeta potential

Another vital quality control aspect for the evaluation of stability of cationic, anionic, or neutral nanoparticle systems is zeta potential (ZP), which can be defined as the magnitude of the overall charge acquired by any particle in a particular medium. In other words, it could be implied as the potential at hydrodynamic boundary layer. Higher is the potential (magnitude of the repulsive forces between intraparticles), more will be the stability of the dispersion. Generally, stable nanocarrier dispersions exhibit ZPs between $> +30$ or < -30 mV. Hence, it provides valuable information on surface character and functionality of the nanocarriers. Table 2 represents the information about the magnitude of the ZP with respect to the stability characteristics (i.e., characterization of electrostatic repulsions) of the nanoparticles.

Advantages

- Provide insights regarding their stability, circulation times, interaction between proteins, particle cell permeability, and biocompatibility as well for the drug nanocarrier systems.
- Assist in improving the biological performance of the nanotherapeutic carrier systems by circumventing surface charge related toxicities.

Among the zeta potential measurements, solvent ionic strength viscosity, pH, and temperature are some of the most critical parameters that significantly impact the outcome of zeta potential distribution. Therefore, in order to obtain precise data information related to batch-to-batch reproducibility, these parameters must be carefully controlled in the formulation process [38]. Some of the mechanistic approaches that are the principally used as functionality in the ZP instruments are summarized as follows [39].

Based on electrophoresis approach

Electrophoretic light scattering (ELS) technique employs electric field across the dispersion to evaluate the ZP by measuring the migrating rate of particles (as velocity is directly proportional to ZP) toward the electrode of opposite charge. This technique aids in characterizing the solid lipid nanoparticles and offers better resolution with reliable outcomes in comparison to others. Because of small size of lipidic nanoparticles and higher surface area with free energy, they have the tendency to strongly aggregate and form the secondary structures. Solid lipid nanoparticles having high ZP between ± 20 and ± 40 mV are found with high physical stability and less susceptible to form the aggregates or show

Table 2 Summary of zeta potential and its stability characteristics.

S. no	Zeta potential (mV)	Stability information of the nanoparticles
1.	0–5	Tend to agglomerate or aggregate
2	5–20	Minimally stable
3.	20–40	Moderately stable
4.	40+	Highly stable

enhancement in particle size. Nanoparticles with sufficient charge can repel each other and enhance the physical stability of system. This technique is used for lipid nanoparticles, polymeric nanoparticles, nanolipid carriers, liposomes, emulsion, and all the colloidal suspension.

Advantages
- A rapid approach for surface charge measurement.
- Typically combined with DLS/SLS.

Limitations
- Involves indirect estimation of ZP.

Based on electroacoustic approach

This technique exploits electroacoustic effects such as colloid vibration current and electric sonic amplitude that further aids in the determination of dynamic electrophoretic mobility in the commercial ZP instruments. Further, these approaches allow the measurement of intact samples without any alteration.

ZP is generally established by the laser Doppler anemometry and works on the principle of Doppler shift. When an electric field is applied through the diluted suspension of nanoparticles then the scattered light is observed for a frequency shift, i.e., used for the determination of electrophoretic mobility (μ, particle velocity/strength of electric field). The values used for ZP are not an absolute estimation of stability of nanoparticles, so direct conclusions without accomplishing the stability tests are not recommended. The surface charge on nanoparticles has a bearing on the in vivo fate of solid lipid nanoparticles. It is generally seen that solid lipid nanoparticles with positive ZP show longer circulating half-life. Fonte et al. showed that SLNs developed for delivery of insulin when coated with chitosan demonstrated increased circulating half-life [40]. This is related to the absorption of solid lipid nanoparticles on to protein components in the blood. On the other hand, ZP with positive charge is also shown to be related to the cell destabilizing effect because of the interaction between charges. In one study, it was observed that SLNs bearing negative ZP were found to be cleared by reticular endothelial system (RES) [41]. Similarly, Mozafari and Mortazavi showed the usefulness of ZP of nanoliposome surface like surface coating by polymers to improve the blood circulation life. ZP measurement is also significant for controlling the accumulation, fusion, and illuviation of nanoliposomes, which are the crucial parameters and affects the physical stability of nanoliposomal products [42, 43].

3.1.3 Entrapment efficiency and loading capacity

Ideally, NCs should have high entrapment efficiency (EE) and loading capacity (LC). Loading of drugs in NCs is usually carried out by two methods: (a) drug is added at the time of preparation and (b) absorption/adsorption of drug into preformed NCs. EE depends on various factors such as type of NC, method of preparation, and specific chemical and physical properties of drug. It can be determined by using direct or indirect method. In the indirect method, separation of free drug is achieved by using techniques such as centrifugation, ultrafiltration, dialysis of NCs, and unentrapped drug is estimated in the supernatant. EE is calculated as difference between the initial amounts of drug added and unentrapped drug in supernatant with respect to the total amount added initially for NC preparation [44]. However, in the direct method, NCs are solubilized in a suitable solvent and analyzed after filtration and suitable dilution. The %EE is calculated by using following equation:

% Entrapment efficiency = Amount of drug in NCs/initial amount of drug × 100

% Drug loading is calculated as *ratio of actual amount of drug loaded into NCs to total amount of carrier and drug taken* initially.

The analysis of drug can be carried out using methodologies such as UV-spectroscopy and high-performance liquid chromatography (HPLC), mass spectrometry and its various applications (Inductively Coupled Plasma Mass Spectrometry (ICP-MS), and gas sorption.

Spectroscopic analysis (UV-visible spectroscopy)

The application of UV/visible spectroscopy (spectral region from 200 to 1100 nm) is quite exquisite for identifying, characterizing, and studying nanomaterials, which primarily assists in quantifying the sum of absorbed and scattered light by the sample (as small as 60–100 μL using a microcell with a path length of 1 cm), when placed between a UV light source and a photodetector. The intensity of the UV beam is quantified before and after passing through the sample. The intensity measured is then compared at each wavelength to quantify the sample's wavelength dependent extinction spectrum. The data obtained is plotted by employing molar extinction as a function of length. Each spectrum is background corrected using a buffer blank to avoid spectral interference from the buffer in the sample extinction spectrum [45]. Nanoparticles exhibit optical properties that are sensitive to size, shape, concentration, agglomeration state, and refractive index. These properties are characterized with the help of UV-vis spectral systems to elucidate the size and concentration dependent optical properties.

Inductively coupled plasma mass spectrometry

This technique is highly sensitive approach to identify the trace elements in biological fluids and quantify the elemental analysis of metallic and nonmetallic samples (having an atomic mass ranges from 7 to 250) [46]. Since the last decade, researchers have shown more inclination towards contemporary tools such as ICP-MS for study of elemental composition in pharmaceutical nanosystems or biological samples rather than conventional technologies like atomic absorption spectroscopy (AAS) with the aim of achieving better sensitivity (with detection limits of parts per trillion or parts per billion), simultaneous and multiple analyte detection, higher throughput, and requirement of low sample volumes. ICP-MS allows analysis of only liquid samples that are aerosolized via nebulizer and subjected to ionization (through a high temperature heat source such as plasma from argon gas) by exciting the electrons of individual atoms. These ionized atoms get separated from neutral particles in vacuum and detected by mass spectrophotometer. The mass-to-charge ratio of ions is used to separate the elements and the concentration of each element is determined based on the ion signal proportion relative to an internal calibration standard.

Advantages
- Allows isotope comparison.
- Provides quick analysis with large analytical ranges up to eight figures.
- High sensitivity analysis—lower detection limits of most elements are in parts per trillion to parts per quadrillionorder.
- Enable simultaneous multielement analysis possible.
- High sample throughput and possess low detection limit.
- Low sample volume and simple sample preparation.
- High-resolution and tandem mass spectrometry (triple-quadrupole) instruments offer a very high level of interference control.

Limitations
- High equipment costs and operational costs (e.g., multiple high purity gases required such as argon plasma, lab set-up costs, e.g., air-conditioning, HEPA filters, pipe work, dust reduction measures).
- Required professionally trained expert to handle this sophisticated instrument.
- Interference needs to be controlled.

3.1.4 Drug release

An appropriately designed in vitro release experiment can shed light on behavior of the NCs as well as provide information about release mechanism and kinetics. There are several methods to evaluate drug release from NCs such as equilibrium dialysis, ultrafiltration, ultracentrifugation, and diffusion cells [47]. Dialysis is the preferred method to ascertain in vitro drug release from NCs. This involves suspending the drug encapsulated NCs alone or in a buffer inside a dialysis bag, which has a certain molecular weight cut-off (MWCO). Hence, to avoid drug transport as a limiting factor, membranes with high MWCO are selected for release studies. The dialysis bag is then suspended in release medium (maintained usually at 37°C under constant stirring) and assayed at different time intervals using appropriate analytical techniques [48]. Various factors influence drug release from NCs such as method of preparation, composition, drug to carrier ratio, and physical/chemical interactions among the components. Mechanisms of drug release from the NCs can be diffusion controlled, solvent controlled, stimuli controlled, and degradation controlled.

3.2 Characterization in dry state

The methodologies and techniques which are primarily employed for the analysis of samples in dry state (powdered form) for the characterization of parameters such as surface morphology and ultrastructure that encompasses microscopies, viz., electron microscopies, which further include scanning electron microscopy (SEM), transmission electron microscopy (TEM), high-resolution transmission electron microscopy (HRTEM), scanning tunneling microscopy (STM), and polarized light microscopy (PLM). Internal pore size, pore volume could be determined by surface area (porosimetry), while crystalline behavior and polymorphism are analyzed employing differential scanning calorimetry (DSC) and X-ray diffraction (XRD) respectively.

3.2.1 Surface morphology and ultrastructure
Electron microscopy

Generally, nanoparticle's morphology is established by employing the electron microscopy (EM) methods. As diffraction effects restrict the optical microscopy resolution, it is very difficult to study the small size particles having diameter >1 mm with light. Therefore, the better resolution is preferred, for instance, by electromagnetic radiation of shorter wavelength.

Advantages
- Provide information up to single particle resolution.
- Operational range at sub (nm) size and morphology.
- Can be coupled with elemental analysis.

Limitations
- Limited throughput.

Scanning electron microscopy

SEM is the most useful technique for the examination and analysis of micro- and nanoparticle imaging under high vacuum or low-pressure depiction of solid objects. The reason behind the use of SEM study is that it is used to determine the particle size owing to its resolution of 10 nm (100 Å). The components of this instrument include an electron gun, condenser lenses, and a vacuum system. This instrument usually generates three kinds of principal illustrations: external X-ray maps, backscattered electron and secondary electron images. The emittance of secondary electrons from the surface of sample is employed to attain the surface characteristics of the particles. Nanocarriers have the tendency to hold on vacuum, as the electron rays can destroy the particles. Although SEM produces the average diameter similar to the DLS method, these analysis techniques are time consuming, costly, and continuously supplemented with the size distribution pattern.

Zhang et al. prepared the poly(dimethylsiloxane)–gold nanoparticles loaded into the film and obtained scanning electron micrographs of the nanoparticles encapsulated inside the film [49]. Logeswari et al. developed the silver nanoparticles for their antibacterial activities in a formulation that incorporated the plant extracts [50]. Characterization of the developed silver nanoparticle was done using the X-ray diffraction (XRD), atomic force microscope (AFM), and SEM. Zang and Tang enhanced the imaging presentation of the colloidal nanoparticle and further characterized using the XRD, SEM, and TEM [51].

ZnO nanoparticles exhibited a flower-like nanostructure that was examined through the SEM study. Khosravi and co-workers reported the SEM study of paromomycin loaded nanoparticles and they found a noncrescent shape with abnormal surface, while the control group was observed with normal shape and size with a smooth surface [52].

Advantages

- Operate at low beam energies and hence suitable for radiation sensitive biological materials.
- Requires simple preparation of sample and easy to operate.

Limitations

- Application only limited to surface characterization.
- Requires conductive samples.
- Lower resolution than TEM.

Nonetheless, it is pertinent to note here that nanomaterials topography can be preserved by new microscopy techniques that does not require the sample drying (environmental or wet SEM) or by careful freezing of sample (cryo-SEM). The environmental SEM technique that permits the analyses of hydrated materials to be carried out without fixing, drying, freezing, or coating of specimen. The use of wet SEM is, as yet, mainly restricted to the characterization of microspheres and microcapsules, while cryo-SEM has been employed for characterizing the microspheres and nanoemulsions.

Transmission electron microscopy

TEM has been used for significant measurement of nanoparticle size and size distribution. This is the first-choice method that has the ability to measure the particles from 1 nm up to a few micrometers. TEM is used to observe the morphology of nanoparticles owing to its ultrahigh resolution tendency in contrast to SEM image that agree to investigating the granularity of the surface of particles or even crystalline lattice structure [53]. An incident beam of electrons is transmitted through a thin layer of the sample, converted into the unscattered electrons, elastically or in-elastically scattered electrons. The ratio of distance between the specimen and image plane of objective lens establishes the resolution power of TEM.

The physicochemical condition of the nanomaterials may be affected by the specimen fixation and drying as the TEM examination requires high level vacuum and thin layer of specimen for the penetration of electron-beam through the sample. Nanocarriers are fixed with negative staining solution (phosphotungstic acid) or derivatives (e.g., uranyl acetate) to resist the vacuum pressure of microscope [54]. After fixation, the particles are dried under a mercury lamp and then examined under a monochromatic beam of electrons that transmit through the sample and produce a micrograph.

Advantages

- Generate information data up to atomic resolution.
- Provides information on internal structure of the small NPs (such as crystallinity, core/shell).

Limitations
- Requires high beam energies which makes it nonconductive for the radiation sensitive biological materials.
- High-cost equipment and huge operational/handling expenses.

Nik et al. observed the mean particle size of pegylated liposomal galbanic acid liposomal formulation of 100 nm with narrow particle size. The small particle size, morphology, and, to some degree, the narrow particle size distribution were confirmed by the TEM [55]. Similarly, Yao et al. also studied the surface morphology of liposomes by TEM study and confirmed the spherical vesicular structure that could easily be detected with a particle size of approximately 120 nm, smaller than that measured by the DLS [49]. Also, Tianyu et al. prepared the gold nanoparticles and characterized them by using DLS technique along with complementary TEM study [56].

With the aim of TEM imaging, biological samples that have the large amount of water require dehydration and are then accumulated on a metal mesh and treated with electron stains. Every sample type has specific requirement due to their composition. It is possible to prepare the sample with careful freezing (cryo-TEM), which preserves its morphological character. Margulis-Goshen et al. showed the cryo-TEM studies on a celecoxib microemulsion (formulated with a volatile organic phase) that was utilized to produce the drug-containing nanoassemblies by spray-drying and re-dispersion in water [57].

High-resolution transmission electron microscopy

The image mode in high-resolution transmission electron microscopy (HRTEM) allows the direct imaging of crystallography of the atomic structure of the sample [58]. It is a specialized TEM technique to study the material properties on the atomic scale, for example, semiconductors, metals, nanoparticles, and sp2-bonded carbon (e.g., graphene, carbonnanotubes). HRTEM is also known as high resolution scanning TEM (HRSTEM, mostly in high angle annular dark field mode), and for disambiguation, the technique is also known as the phase contrast TEM. For three-dimensional crystal structure, it may involve the combination of several views, taken from different positions, into a 3D map. This process is known to as electron crystallography (EC).

The HRTEM image is originated by the interference in the image plane of electron wave with itself. Amplitude in the image plane is recorded because of the inability to record an electron wave. However, the information about sample structure is contained in electron wave phase. The microscopic aberrations (like defocus) are detected, which convert the wave phase into amplitudes in the image plane at the exit plane of specimen. The qualitative interaction can readily be achieved even if the crystallographic structure of the sample is found to be complex with the electron wave interaction leading to Bragg diffraction. After penetration of electron wave into the sample, it is attracted by the positive atomic potentials of atom cores and channels with the atom columns of crystallographic lattice (s-state model). In a sample, dynamical scattering of electrons not satisfying the weak phase object approximation , which is about all real samples, still remains the holy grail of EM. However, the physics behind the electron scattering and EM image formation are well known to allow the accurate simulation of electron micrographs. Conventionally, HRTEM has been mainly employed for the imaging, diffraction, and chemical analysis of solid materials. Carbon nanotubes were first discovered by HRTEM technique.

Scanning tunneling microscopy

STM belongs to the family of scanning probe microscopy (SPM), which uses the quantum tunneling current to generate the electron density illustrations for conductive or semiconductive surfaces and biomolecules attached on conductive substrates at the atomic scale [59, 60]. After following the generic principle for all SPM methods, i.e., bringing a susceptible probe in close proximity to the object surface determined to monitor the reactions of the probe, the vital STM components that comprises a sharp scanning tip, a xyz-piezo scanner for controlling the lateral and vertical movement of tip, a coarse control unit placing the tip close to sample within the tunneling range, a vibration isolation stage, and feedback regulation electronics.

If the tip moves across the sample in x–y plane, then the changes in surface height and density of states are detected and cause the changes in current while the mapped images are measured [61]. Although keeping the responding current unchanged by regulating the height of tip via the use of feedback electronics can generate an image of tip topography across the sample, STM or EM methods are used for characterization of biomolecules, and the samples are typically encapsulated into a matrix to retain their original conformations, followed by samples coating with a thin metallic layer, for instance, gold, before acquiring the images [62]. It is impossible to image these biomolecules in their native positions using traditional EM techniques that usually accompany a time-consuming process for the sample preparation. After reducing the disadvantages of EM methods, STM give an image with atomic scale resolution by, for example, using a Platinum-Iridium tip with a very sharp end. However, the high spatial resolution of STM provides the advantages for characterization of nanoscale biomaterials, for instance, size, shape, structure, and states of dispersion and aggregation; only few studies have been given using gold or carbon as the substrates [63]. The practical problems are mostly due to requirement of conductive surface of sample and detection of surface electronic structure. Unfortunately, the majority of biomaterials are insulating, and a simple connection of surface electronic structure of sample with its surface topography may not essentially exist. However, STM is a preferred tool for studying the conductive atomic structures like carbon nanotubes, fullerenes, and graphene [63].

Polarized light microscopy

Polarized light microscopy (PLM) is used for preliminary detection of lyotropic liquid crystalline structures (except cubic mesophases) [64]. Anisotropic systems deviate the plane of polarized light (birefringence, analogous to real crystals), typical black and white images, or colored textures when using an additional λ-plate. According to their textures, they can be classified as follows: (1) lamellar liquid crystalline phase, with the micrograph showing oily streaks with "Maltese crosses" and (2) hexagonal liquid crystalline structure (fan-like texture) [65]. In the case of lamellar phase, the Maltese crosses result from concentric rearrangement of plane layers and are of dominant texture. A micrograph of a lamellar liquid crystal confirms the several Maltese crosses (white color). A dark field is detected in isotropic systems because of lack of light deviation [66]. PLM can be estimated in the micron or submicron ranges as well. In case of liquid crystal particles (smaller dimensions), TEM can give the adequate resolution.

3.2.2 Porosity determination
Porosimetry

Porosimetry is a robust method for the characterization of porous materials, giving abundant information containing the pore size, pore volume, and surface area of a sample [67]. The experimental methodology basically depends upon the mass determination of a specific substance adsorbed at a particular pressure. A microbalance is used for weighing the sample before and after the adsorption of an adsorbate, for example N_2 or Hg. Brunauer, Emmett, and Teller (BET) equation and isotherm curve patterns are used to classify the pores, for instance: (a) micropores (with <2 nm pore size), (b) mesopores (with >5 nm to <50 nm pore size), and (c) macropores (with >50 nm pore size). Moreover, the adsorption isotherms can also be classified as the Types I–VI, as per their forms. Microporous materials are basically examined by the Type I isotherms. Types II and III isotherms are mainly explained the high and low affinity of macroporous materials for the adsorbate, respectively. For mesoporous materials, Types IV and V isotherms used with high and low interaction with the adsorbate, respectively.

Finally, the nonporous materials are evaluated by Type VI isotherms with the uniform surfaces. An optimal porous material can adsorb the massive amount of adsorbate gas at the lowest possible gas pressure, and then type I isotherm best explains this substrate, with its higher adsorption curve suggesting the adsorption at a lower pressure [68].

One drawback of porosimetry is that it can determine the entrance only but not the actual inner size of the pore. The closed pores cannot be determined by porosimetry because the gas cannot enter into the pores [67]. Surface area and porosity are the crucial parameter of porous drug carriers for estimating the drug loading and release studies. Porosimetry provides the information about the drug-storage capacity, which is associated with pore size as well as volume [69].

3.2.3 Crystallinity and polymorphism

Components of NCs can exist either in crystalline or amorphous state. In the crystalline state, the molecules are arranged in a three-dimensional repeating pattern or a unit cell whereas molecules are arranged randomly with high degree of disorder in the amorphous form. The amorphous form is usually less stable but has higher solubility than the crystalline form. Moreover, crystalline solids exhibit polymorphism, i.e., they can exist in more than one crystalline form. They are chemically identical but show distinct physical properties. For instance, the lipids (such as triglycerides could exist in various polymorphic forms, viz., α, β, and β' [70] used for preparation of SLNs and NLCs display a complex crystalline behavior. DSC and XRD are two useful techniques to furnish information pertaining to the physical state and crystal structure of NCs.

Differential scanning calorimetry

DSC is a versatile method that is used to determine alterations in structural properties of a sample as a function of time and temperature. As the temperature changes, heat quantity, which is radiated or absorbed excessively by the sample on the basis of temperature difference between sample and reference material, is determined by the instrument [71]. Two types of DSCs are used based on the mechanism of operation: heat-flux DSC and power compensated DSC. In the former type, sample pan and empty reference pan are placed on a thermoelectric disk enclosed in a furnace which is heated at a linear rate, transferring heat to the pans through disk. The temperature difference between both the pans is measured by thermocouples and is converted into a heat flow signal via a calibration procedure [72]. In the latter type, separate furnaces house the sample and reference pans and maintain them at same temperature. The difference in thermal power needed to keep them at the same temperature is determined and plotted as a function of time or temperature [73]. By convention, endothermic events are displayed as upwards signals in power compensated DSC and as downwards signals in heat flux DSC. Nitrogen or helium gases are used to purge the instrument to avoid reactions with the atmosphere. Typical scan rates are between 1°C/min and 10°C/min. For the purpose of peak resolution and sample investigation close to equilibrium, low scan rates are preferred, while sensitivity of measurement is enhanced at high scan rates as it leads to heat exchange within short period of time [74].

Houacine et al. [75] prepared NLCs of resveratrol and evaluated the effect of six different liquid lipids on various attributes of NLCs such as particle size, drug loading, and encapsulation efficiency. An extensive DSC study was employed to determine the interactions between solid lipid (trimyristin) and liquid lipids, and liquid lipids and resveratrol. Depression in melting point of trimyristin and resveratrol was observed with addition of liquid lipids. The study aided in selection of optimum liquid lipid, which led to the formation of crystalline nanoparticles with high drug solubility and stability.

X-ray diffraction

This versatile technique (operated in dry state) is quite useful in characterizing the crystallite sizes of ranges (1–100 nm) and provides information about the wide range of structural aspects (e.g., arrangement of the crystal components, and macroscopic information like mean shape and size) in crystalline samples.

Although minute thermal events in sample can be monitored and computed by DSC, direct revelation of the cause of the events is not possible. The precise nature of thermal transitions and deep analysis of crystal structure can be determined by XRD analysis. The basis for X-ray diffraction is constructive interference of monochromatic X-rays and sample. X-rays generated by cathode ray tube heat the filament to produce electrons which are accelerated and bombarded towards target. X-rays are filtered to produce monochromatic rays and collimated onto the sample. Intensity of scattered X-rays is recorded as the sample and detectors are rotated [76]. This information can be attained by examining the full width at half maximum (FWHM) of the Bragg reflections. The characteristics of the scattered X-rays indicate the arrangement of the crystalline material, using Bragg's Law:

$$n\lambda = 2d \cdot \sin\theta$$

where n is an integer, λ is the wavelength, θ is the scattering angle, and d is the interplanar distance.

Typical X-ray diffractograms are obtained at 2θ from 5 to 70 degrees. XRD can be used to identify the type of crystalline phase, crystallinity degree, orientation and chemical nature of the compound (in comparison to a known standard). Crystalline materials show sharp diffraction peaks while broad diffraction peaks are characteristic of amorphous materials [77]. Furthermore, in order to obtain mean crystal size of polycrystalline sample using the Scherrer equation (also termed as Debye-Scherrer equation) considering the ideal condition of a perfectly parallel, infinitely narrow and monochromatic X-ray beam incident on a monodisperse powder of cube-shaped crystallites. The equation is given as follows:

$$D_{hkl} = \frac{K\lambda}{B_{hkl \times \cos\theta}}$$

where D_{hkl} is the crystallite size in the path perpendicular to the lattice planes, hkl are the Miller indices of the planes belonging to the peak under study, K denotes as the crystallite-shape factor, λ, wavelength of the X-rays, B_{hkl} refers to FWHM of the diffraction peak in radians, and θ is denoted as Bragg angle.

Advantages
- Provides data information about crystallographic structure and size.

Limitations
- Provides information only about size of crystal domains.

Liu et al prepared SLNs loaded with diclofenac sodium using the phospholipid complexes (PCs) technology to improve the solubility of diclofenac sodium. XRD results showed the incorporation of diclofenac sodium in the PCs (either molecularly dispersed or in an amorphous form) [78].

3.3 Characterization in both (liquid/dry) states (specialized techniques)

3.3.1 Size analysis

Aerodynamic particle sizer

This technique is a vital tool for real time monitoring and high resolution-based characterization of solid aerosolized powder particles or liquid aerosolized droplets for the study of their particle aerodynamic diameters (ranges from 0.5 to 20 μm) based on particle time of flight as well as optical diameters (ranges from 0.3 to 20 μm) on scattered light intensity. The principle of aerodynamic particle sizer (APS) technique is primarily centered on the measurement of changes in the terminal velocity (that depends upon its particle size and calculated by analyzing the time taken by the particle in an accelerating flow field to pass between two spatially separated laser beams) of the aerosolized particles in an aerosol characterization chamber with the help of high-pressure air gun and a venturi pump (for powdered samples) or nebulizer (for liquid samples) [79].

Advantages
- Enable real time monitoring of size distributions and data storage.
- High size resolution, 1 s sampling, and precise measurement.

Limitations
- APS is a complicated device that requires complex optics, in conjunction with sample dilution, to avoid particle coincidence errors.
- Some sources of bias, such as droplet distortion and phantom particles, should also be considered.

Scanning mobility particle sizer

Another broadly used technique as the standard for measuring airborne particle size distributions of sub-0.5 μm aerosol particles of both powdered (aerosolized with the help of a high pressure air gun) and liquid samples (aerosolized with the help of nebulizer or fog generator nozzle) is scanning mobility particle sizer (SMPS) [80]. The basic principle of this technique is the mobility of a charged particle in an electric field. SMPS comprises of two additional devices, namely, differential mobility analyzer (works in response to high voltage electric field) and condensation particle counter (enlarge the particle to make them detectable by common optical counters) to assess the particle size by measuring electromobility diameter of aerosolized nanoparticles [81].

Particles enter through differential mobility analyzer and flow down a cylindrical tube with a high voltage center electrode and outer tube at ground. A small orifice allows those particles with electrical mobility to pass. A very fine range of particles pass through the exit port when the voltage is altered. The size of these particles is enhanced by condensation particle counter (as such particles are too small to be accurately counted by light scattering) that condenses butanol on the surface. After condensation, the particles are individually counted at each voltage and a particle size distribution of 10–500 nm is obtained.

Advantages

- Ease of portability and significantly lower power consumption.
- Able to produce aerosol spectra with a much higher temporal resolution (up to 192 channels) with fast measurements of <15 s scans.

Limitations

Although it is useful for measuring nanoparticles of spherical shape, there is lack of information on their performance on nanoparticles of anisometric shapes.

3.3.2 Molecular weight analysis
Matrix-assisted laser desorption/ionization mass spectrometry (MALDI-MS)

In mass spectroscopy, MALDI-TOF-MS is an ionization technique that employs laser energy absorbing matrix for ionizing small fragile molecules to larger organic analytes/biomolecules into gas phase with minimal fragmentation. It is widely used technique for tissue imaging, protein/peptide, and polymer mass analysis. In this approach, sample is first mixed with suitable matrix, which are selected on the basis of specific molecular design considerations (e.g., sinapinic acid, α-cyano-4-hydroxycinnamic acid, alpha-cyano or alpha-matrix and 2,5-dihydroxybenzoic acid) and applied to the metal plate, secondly, a pulsed laser irradiate the sample triggering ablation and desorption of the sample and matrix material and finally the analyte molecules are ionized by being protonated or deprotonated in the hot plume of ablated gases and then accelerated into time-of-flight mass spectrometer for their subsequent detection. The major application of this technology is the rapid diagnostic tool, i.e.,rapid turnaround time (<10 min) and an overall 95% accuracy for the identification at species level. However, some of the limitations include inability to discriminate between the related microbial species [82].

Atomic absorption spectroscopy

This approach can evaluate the mass concentration of element in both liquid and solid samples. The principle of AAS is based on the fact that free atoms of the element in gas phase generated in an atomizer can absorb radiation at specific wavelength that offers the technique excellent specificity and detection limits. The total amount of absorption depends on the number of free atoms present and the degree to which the free atoms absorb the radiation in the light path; the sample mass concentrations can be quantified by comparison of signal to calibration standards at known concentrations [83].

Advantages

- More elements in rapid sequence from one sample can be analyzed.

- Simplified operation and sample preparation.
- Completely sealed optics with quartz over coated mirrors offer protection in dusty or corrosive environment.
- The air purge system eliminates the chance of corrosion in rugged or corrosive environments.

Limitations

Compared to ICP-MS, the detection limits are relatively high, and only 2–5 mL of sample is required for analysis.

3.3.3 Lamellarity

Among the plethora of nanocarriers, liposome lamellarity (a term used in reference to the number of lipid bilayers of liposomes) is of utmost significance and a nontrivial matter in the process of characterization. This parameter that hugely vary with different types of preparation methods play a crucial role in defining the prominent characteristics of the liposome moiety such as affecting encapsulation efficiency, mediating the diffusion rate or efflux rate of drug from the vesicle, and finally, intracellular fate of the drugs delivered by the liposomes after they were taken up or processed in the cell [84]. Chiba et al. [85] quantitatively studied the comparative effect of using different methods of liposome preparation (i.e., hydration and inverted emulsion methods) on the lamellarity of the vesicles (as depicted in Fig. 6).

The results from the findings revealed significant difference in the unilamellarity of liposomes prepared by inverted emulsion (98%) compared to hydration method (49.8%). Some of the predominant techniques, which are employed are ^{31}P-nuclear magnetic resonance (^{31}P-NMR), Small angle X-ray scattering (SAXS), label-free differential interference contrast (DIC) microscopy, epifluorescence microscopy, cryoelectron microscopy, and electron microscopy.

^{31}P-nuclear magnetic resonance

Size and lamellarity are the most conspicuous quality control aspect of any liposomal system, which is determined by ^{31}P NMR in combination with the use of chemical shift reagents (primarily paramagnetic ions such as Mn^{2+}, Co^{2+}, and Pr^{3+}). These ions interact with phospholipids located in the outermost of monolayer resulting in perturbations of the nuclear spin relaxation time and are reflected on the spectrum as broadened peaks and reduced signal strength (generally multilamellar vesicles produce very broad peaks on the spectra owing to restricted anisotropic motion whereas the small unilamellar vesicles are less affected). Furthermore, the metal ion induced signal loss can be quantified after data acquisition to quantify the lamellarity. Investigations have shown that this technique does not provide the absolute information about the lamellarity depending on the experimental settings and the shape of the liposomes [86].

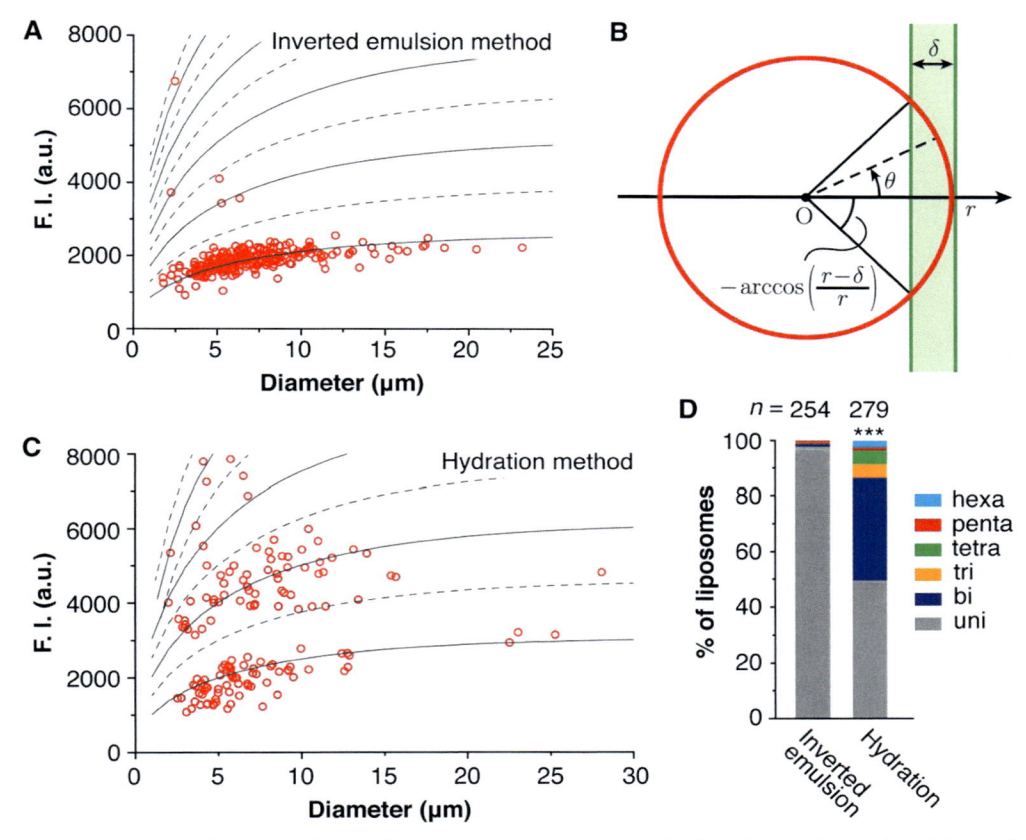

Fig. 6 Effect of different methods of liposome preparation on the lamellarity. *(Reproduced from M. Chiba, M. Miyazaki, S. Ishiwata, Quantitative analysis of the lamellarity of giant liposomes prepared by the inverted emulsion method, Biophys. J. 107 (2014) 346–354.)*

Small angle X-ray scattering

Another widely known application for the most accurate evaluation of lamellarity among the nanolipid carriers is the SAXS. In this technique, lipid vesicle is placed in a glass capillary tube and then subjected to X-ray radiation in order to cause scattering curves of both the sample and the blank, which are recorded by a camera with single dimensional position sensitive detector. The indirect Fourier transformation provides the electron distance distribution $p(r)$, which is calculated from scattered intensity in the measured sample. $p(r)$ gives the radial contrast profile of $\Delta p(r)$ in electron density relative to the mean value, which could be used to solve the internal structure of the scattering particles.

Advantages

- Provides excellent resolution of the Bragg peaks.
- Allows fast kinetics measurements.

- More readily accessible in a laboratory setting and does not require large scale facilities (such as accelerator driven or reactor-based sources, deuterated solvents and involved analysis procedures).

Limitations

The major limitation of SAXS is the radiation induced damage to the sample, which could be resolved with the use of other techniques such as small-angle neutron scattering.

Small angle neutron scattering

This potential nondestructive approach is based on the interaction with neutrons that provides vesicle structural information for bulk measurements (primarily membrane bilayer thickness, vesicle diameter, lamellarity) offering better contrast without any risk of damage to sample compared to SAXS. This technique is also suited for contrast matching measurements.

3.3.4 Surface topography and size analysis
Atomic force microscopy

AFM technique is used to examine the nanomechanical properties of every molecule and particle under closer physiological states. It can physically scan the particles through the probe tip of atomic scale at submicron level and particle size estimation can obtain with extremely higher resolution. Based on forces between tip and sample surface, AFM gives the topographical chart of objects. Afterwards, the scans are obtained in contact or non-contact form depending on their properties. AFM is the powerful characterization tool over the TEM/SEM due to its ability to scan the nonconducting nanocarriers without any special preparation of sample; however, delicate biological and polymeric nano- and microstructures could also be described [87].

AFM provides the opportunity of 3D visualization of nanocarriers with qualitative as well as quantitative properties about physical behaviors such as size, surface texture morphology, and roughness. Therefore, it can be implied that AFM can be used for nanomaterials characterization in different mediums such as ambient air, controlled environments and even liquid dispersions versus AFM study can be performed in liquid as well as gas medium. The AFM scans take more time in comparison to SEM but an entire measurement session inclusive of specimen treatment, image acquisition, and image analysis consumes much reduced time (approximately 1/4th time to generate the data as compared to SEM/TEM investigations). Furthermore, an AFM is an exceptionally cost-effective instrument accessible to nanocarriers imaging in comparison to other EM techniques. Additionally, the AFM requires considerable smaller laboratory space than the SEM/TEM studies (requires a specially trained operator) and much simpler to run.

Advantages

- Provide single particle resolution.
- Gives information about particle morphology.

Limitations
- Require NPs on hard surfaces.
- Limited throughput.

3.3.5 Study of internal structure and interaction mechanism
In situ liquid-cell and atomic-resolution TEM

This prime reason for the designing of this technique was to overcome the challenges related to TEM (restricted for analysis of samples in dry state only). LC-TEM allows the assessment of internal mechanism and interaction of the processes of gold nanoparticles like nucleation rates and growth of nanoparticles. Further, atomic-resolution transmission electron microscopy (AR-TEM) facilitates the dynamic processes and provides the real-time surveillance of fluctuations with atomic sensitivity at the molecular level. This approach has fostered several novel avenues in the field of nanoparticle-protein interactions.

Electron cryogenic transmission electron microscopy

Cryo-TEM has evolved into an essential tool for the characterization of colloidal drug delivery systems. The application of this technique is not only confined to size analysis, but also the shape and internal structure of nanoparticulate carrier systems as well as the overall colloidal composition of corresponding dispersions. Cryo-TEM, generally called as cryogenic electron microscopy (cryo-EM), is a type of TEM where the sample is examined at the cryogenic temperatures (usually liquid-nitrogen temperatures). Cryo-EM technique is gaining a popularity in structural biology system. The utility of Cryo-TEM allows the observation of specimens that does not require any staining and fixation in the native environment. Comparatively, X-ray crystallography requires crystallizing the specimen in comparison to Cryo-TEM, which can be complex, and putting them in nonphysiological atmosphere, which can in turn occasionally cause the functionally irrelevant conformational transforms.

Cryo-TEM after plunge freezing [88–90] permits direct analysis of colloids in the vitrified and frozen-hydrated phase, i.e., nearly the native state. Furthermore, freeze-fracture TEM also provides the detail about internal form of colloidal particles in addition to their phases (such as multilamellar vesicles, nanoparticles based on cubic or hexagonal phases). Cryo-TEM has been mainly used for analysis of biologically originated samples, for example, viruses, bacteria and thin cross sections of complex tissues [91–93]. For all step preparation of samples (e.g., environmental control during sample vitrification [89, 94]) and data processing (e.g., digitalization of the data processing, CCD cameras [95]), cryo-TEM has been extensively used. The information about three-dimensional structure of particles of interest (by three-dimensional reconstitution [96] or cryoelectron tomography [97, 98]) is also possible with this technique. However, mostly organic materials provide the poor contrast in cryo-TEM, but if combined with staining, it can increase the contrast. [99, 100]. There is another interesting cryo-TEM method that is associated with freeze-fracture, called as freeze-fracture direct imaging [101].

Margulis-Goshen et al. showed the cryo-TEM study on a celecoxib microemulsion prepared by using the volatile organic phase that was used to prepare the drug-containing nanoassemblies by spray-drying process and re-dispersion in the water [57]. Some authors have also used the cryo-TEM technique to deal with the questions associated with biopharmaceutics. For instance, the colloidal structure of simulated intestinal fluids and their processes during the lipolytic digestion of drug carrier systems were explained by comprehensive cryo-TEM analysis [102, 103]. Szebeni et al. [104] showed the impressive outcome of cryo-TEM on predicted potential of paclitaxel to precipitate from its micellar solution after dilution of a Cremophor EL/ethanol concentrate with phosphate buffer saline and on the effects of Taxol dilutions on the ultrastructure of human plasma.

Cryo-TEMs with high resolution giving the access to advanced technologies like cryo-tomography, will further expand the chances to obtain the three-dimensional information on the structures of interest. It may be wondered that, under favorable conditions, they might even permit to study the nanoparticles interaction with cells in more detail in the near future.

Electron tomography

This tomography technique (originally developed for X-ray study) is basically an extension of traditional TEM that facilitates to obtain detailed 3D structures of subcellular macromolecular objects. The principle is working on the acquisition of multiple 2D projections of the same sample at different angles associated with the incident beam and on the reconstruction of 3D depiction from these partial photographs [105]. This technique is quite suited to elucidate the 3D structure of a nanoparticle, properties of nanocrystals in liquid phase.

Limitations
- Poor statistics, very low throughput, and a high level of complexity.
- Particle drift and beam damage caused the modifications of the object during the analysis.

Super resolution microscopy

It is a well-documented fact that the light microscopy resolution is restricted as light because of its wave nature, which is related to diffraction. *Diffraction limit:* In the imaging process of an optical microscope, light rays from each point on the object converge to a one point at the image plane. However, diffraction of light on the object hinders the exact ray's convergence, resulting in a sharp point on the object to blur into a fixed sized patch on the image. A point object image with 3D intensity distribution is called as point spread function (PSF). The size of PSF is the primary factor for determination of microscopic resolution.

Conventionally researchers have employed electron microscopes to evidently comprehend the subcellular structures to discern all the living cells. However, advanced microscopies like super resolution microscopy (SRM) have made a colossal sway in investigating the synthetic materials owing to its nanometric resolution to ~10 nm, multicolor ability, minimal invasiveness, and includes the multiple techniques that achieve 20 times greater resolution as compared to traditional light microscopy (which is limited to only 200 nm owing to diffraction light) [106].

Types of SRM: These include three types as follows.

Stimulated emission depletion microscopy

This microscopy employs spatially patterned excitation technique. In this case, two lasers-excitation laser and stimulated emission depletion (STED) laser, which when used together on the focal plane results in diminishing the effective point spread function (PSF), thereby enhancing the resolution. STED laser terminates the excited–state fluorophores, which is positioned next to the excitation focal point with stimulated emission. Furthermore, a donut-shaped ring of fluorophores around the center of the image field gets suppressed instead of center. This leads to PSF lower than the diffraction limit [107]. Afterwards, incredibly sensitive single-photon detectors then pick up the signal from the central fluorophores.

Structured illumination microscopy

This form of SRM enhances the spatial resolution of light microscopy. In this case, the fluorescent sample is excited multiple times using striped illumination patterns and each time the stripe's orientation and position are changed and the obtained images are examined by using the computer software. The stripes fired at sample interact with the high frequency light generated from the sample and results in creating a third pattern that can be studied without difficulties. Using several images, additional aspects are found, and an image is remodeled with approximately twice the resolution as conventional light microscopy.

Advantages
- Enables live cell imaging.
- Generate 3D imaging data.
- Allows thick section imaging.

Stochastic optical reconstruction microscopy/photoactivation localization microscopy

This technique is the most recent application of SRM that activate or excite just a few fluorophores at a time, thereby reduce spatial overlap that allows images with low resolution (5 nm). In this application, photoswitchable dyes are combined with lower power activation laser, changing from dark to emission states. By imaging the photoswitchable fluorophores over time, precise location of molecules can be traced. Furthermore, it is of two types: In the first type of stochastic optical reconstruction microscopy (STORM), it uses an activator dye that switches the fluorophore on whereas the reporter dye, results in signal. In contrast, another type of STORM, known as direct STORM (dSTORM) does not want activator dye; rather, fluorophores are joined with specialized buffers and lasers to produce the photoswitching.

Therefore, limitation of conventional optical microscopy has rendered SRM as a viable tool in material science for giving the superior optical characterization as well as examining the spatial resolution below the diffraction limit by adopting schemes to improve the localization of single emitting fluorophores or dropping the size of point spread function of emitted light. Higher spatial resolution can be typically attained by EM or SPM: however, fluorescence-based characterization provides the enhanced selectivity, higher contrast, and nonsuperficial characterization of various nanosystems such as lipid coated nanocapsules, internal structure, and swelling mechanism of microgels, penetration and adsorption of different proteins into the framework of mesoporous silica nanoparticles.

Single molecule fluorescence microscopy

With the advancement in detector and dye technologies coupled to image analysis, single molecule fluorescence microscopy (SMFM) has become another indispensable tool in biosciences, which not only offers exceptional contrast (i.e., high signal–to–noise ratio for visualization) with minimal perturbation to biomolecules in the physiological context, but also provides useful information regarding the porous system (i.e., dynamics of molecules inside porous materials) and host guest interaction of nanomaterials intended for targeted application [108]. Such advance systems enable faithful detection and analysis of single fluorescent molecules used as reported tags in biological samples. Furthermore, these technologies have led to the discovery of green fluorescent protein (GFP), which have progressed the fluorescent microscopies to complete another new level. Some of the useful applications of this technology have been summarized as follows:

Advantages
- Aids in mechanistic understanding with mapping of internal porous structure and the defects related to nanomaterials.
- Assist in dynamic study, i.e., diffusional movement of fluorescent dyes into the pores or quantification of diffusional coefficients of dyes and oligonucleotides.
- Provides information about the occurrence and density of the functional groups on porous walls, which can affect the diffusional movement of guest molecules through the electrostatic interactions.

Nanomechanical resonators

With the emergence of nanoelectromechanical systems such as nanomechanical resonators, early prognosis of disease physiology and labeled free detection of bio/chemical-molecules at single-molecule (or atomic) resolution have turn out to be a veracity. Furthermore, such nanosystems also exhibit the ability to detect parameters such as molecular weight, elastic stiffness, surface stress, and elastic stiffness for the adsorbed molecules on the surface. The principle of molecular detection employing this method is working on molecular adsorption of particles onto the surface of resonator that result in the detection of shift in resonance frequency, i.e., proportional to the mass of adsorbed particle.

Advantages

- NMRs coupled with mass spectrometry enables mass analysis of neutral single nanoparticles regardless of its ionization state.
- This approach is extendable to colloidal dispersion systems by entrapping a microfluidic channel into the resonator via devices known as suspended microchannel resonators.
- Allows multiparticle flow analysis.
- Aids in the study of complex partitioning of binary solvent mixtures.

3.3.6 Porosity determination and pore size distribution analysis

Two techniques, i.e., thermoporometry and cryoporometry, are extensively used to measure the porosity, pore size distribution (depending upon their pore sizes, i.e., microporous (2 nm); mesoporous (2–50 nm); and macroporous (>50 nm), respectively) determinations [109] and study of crosslinking behavior in polymeric gel networks, e.g., cellulose nanofiber aerogels [110] for both liquid and solid samples. A small part of solid melts at lower temperature than the bulk solid as per Gibbs–Thomson equation. Therefore, a liquid is imbibed into a porous material and the sample is cooled until the liquid becomes frozen, and then heated until all the liquid is again melted. Hence, the melting temperature will convey the details of pore size distribution.

DSC thermoporometry

This calorimetric method facilitates the characterization of pore structure from the melting or depression in freezing point of a liquid confined in a pore, by reason of the added contribution of surface curvature to the phase-transition free energy (by sensing the transient heat flows at the period of phase transitions). Measuring the differences of phase temperature changes gives the direct information on the volume and the pores dimension accessible to the liquid. Although the changes with both the phases, i.e., freezing and melting, can potentially be employed for such a study, detection of melting temperatures is usually preferred to avoid uncertainties arising from supercooling effects.

Advantages

- Measure the internal pore size in the 1.5–150 nm range, whereas the conventional techniques only give information sizes of the pore-openings.
- Relatively shorter time and well suited for wet porous samples.
- Aids in determining the shape of pores by comparing the data of freezing and melting transitions.
- Helps in quantitative evaluation of pore size distributions.

NMR cryoporometry

This approach measures the quantity of mobile liquid employing NMR technique. ^{1}H NMR Cryoporometry has been employed effectively as a tool (even in wet state, unlike conventional pore characterization techniques) for the determination the pore size distribution of ultrafiltration membranes [111] and also for the characterization studies of the relation between drug release profile and pore structural evolution of polymeric nanoparticles [112]. These investigations proved direct relationship of different synthesis routes for nanoparticles that leads to different pathways in the evolution of the pore structure which significantly impact the drug release profiles among nanoscale polymeric nanoparticles. The suitability of cryoporometric liquids as probe materials significantly depends upon the pore diameter and the nature of the pore materials, e.g., water for <2 nm pore sizes, t-butanol for sizes of 10–60 nm, menthol for pore sizes of >60 nm [113].

Advantages
- NMR cryoporometry measures the true volume of pores for liquids.
- Pore size calibration is in good colinear agreement with gas adsorption measurements.
- Facilitate measurement of wider pore size range than gas adsorption or thermoporometry.

4. Regulatory and safety aspects of nanocarriers

The complexity in the development of nanosized carriers for bioactive molecules have been increased day by day [114]. Safety assessment is necessary not only for individual components but also for all the components used in the form of nanosized entity [115, 116]. The nanoencapsulated entity shows changes in terms of chemical composition, purity, concentration, as well as physicochemical properties, stability, and dermal penetration. Furthermore, safety assessment also involves the potential toxicological effects and exposure estimates underestimated use conditions for every component and nanoencapsulated entity as well. The appropriate assays for safety evaluation should be recognized and there is no need for strict regulations for the nanotherapeutics approval as compared to conventional drugs. The characterization process for nanoparticles is probably more difficult as compared to the conventional drugs prior to drug approval because they display the further physicochemical characterizations, like surface area, shape, as well as drug release. Although the physical characterization of nanoparticles is necessary to understand the toxicity mechanism and therapeutic efficacy that will eventually determine the suitability of nanotherapeutic for further clinical studies. There is a challenge for nanotherapeutics or small molecule drugs specifically to identify the toxic side effects. When a small molecule enters the blood brain barrier, it causes the neurological effects and nanoparticles can accumulate in brain tissues. On the other hand, several small molecules drugs can produce the adverse immune reactions because of the presence of problem for nanoparticles due to immunological activation [117].

It may not be acceptable to stick to a stricter approval method for nanotherapeutics as compared to the conventional drugs. Mostly biological constituents like cellular organelles, proteins, and nucleic acids are present in the nanosize range. Furthermore, variations in the regulatory practices from one region to another region involving nanotherapeutics (different guidelines for "nanosimilars") could impede the timely approvals. While the generic small molecule drugs can moderately easily be reproduced, biologics and nanoparticles exhibit the complexity enhancement that is not easy to replicate, unless the same materials and manufacturing processes are used. It can produce the small changes in nanoparticle properties could modify the biological impact; it is doubtful whether nanoparticles should be allowed to undertake the accelerated approval depend upon their bioequivalence [118]. Moreover, it is expected that precision medicine will participate in future for the approval of all therapeutics, despite category. The genome, epigenome, transcriptome, proteome, and metabolome could be employed to foresee the drug responses, so as to avoid the hypersensitivity responses and select patients that are expected to benefit from therapy [119].

Nanotechnological application may result in product attributes that differ from those of conventionally manufactured products, and therefore safety or effectiveness of FDA-regulated products that comprise of nanomaterials or else to involve the nanotechnological use should consider the unique properties and behaviors that nanomaterials may display. FDA will regulate nanotechnological products under existing statutory authorities, in accordance to specific legal standards applicable to every product under its jurisdiction [120].

5. Conclusions and perspectives

It is needless to argue why commercial scalability of any nanotherapeutic design contingents upon its reproducibility and quality control, the key issues hindering the manufacturing process to achieve a consistent product with the intended physicochemical characteristics, biological behaviors, and pharmacological profiles. Hence, standardized protocols for both accurate and precise nanotechnological characterization (method specific settings for each batch with full description of metadata) become quintessential to fully comprehend the origin of nanoparticle behavior and subsequently translate their performance benefits from the laboratories into real world applications. Furthermore, majority of the diverse characterization techniques impinges on single physical aspects or functionality at one time, revealing only a fractional interpretation of the nanotherapeutic system. Thus, it poses a big challenge to determine the physicochemical properties of a nanoformulation, comparing and evaluating the data with previous literature studies and exploring their structure-function relationships. Therefore, approaches/models should be adopted to characterize the same quantity employing more than one technique to assure the needs of application and to meet statistical requirements in the quality control for the pharmaceutical products with a reasonable statistical confidence.

Acknowledgment

One of the authors, Karan Razdan, acknowledges the Department of Science and Technology, New Delhi, India, for financial assistance as DST INSPIRE JRF (IF170172). Karan Razdan and Komal Saini are grateful to Commonwealth Scholarship Commission, United Kingdom, for their financial support.

References

[1] J.K. Patra, G. Das, L.F. Fraceto, E.V.R. Campos, M.D.P. Rodriguez-Torres, L.S. Acosta-Torres, et al., Nano based drug delivery systems: recent developments and future prospects, J. Nanobiotechnol. 16 (1) (2018) 71.

[2] Z. Liu, S. Tabakman, K. Welsher, H. Dai, Carbon nanotubes in biology andmedicine: in vitro and in vivo detection, imaging and drug delivery, Nano Res. 2 (2009) 85–120.

[3] W.H.D. Jong, P.J.A. Borm, Drug delivery and nanoparticles: applications and hazards, Int. J. Nanomedicine 3 (2008) 133–149.

[4] V.J. Mohanraj, Y. Chen, Nanoparticles a review, Trop. J. Pharm. Res. 5 (2006) 561–573.

[5] C. Ding, Z. Li, A review of drug release mechanisms from nanocarrier systems, Mater. Sci. Eng. C 76 (2017) 1440–1453.

[6] A.D. Bangham, R.W. Horne, Negative staining of phospholipids and their structural modification by surface-active agents as observed in the electron microscope, J. Mol. Biol. 8 (5) (1964) 660–668.

[7] H.I. Chang, M.K. Yeh, Clinical development of liposome-based drugs: formulation, characterization, and therapeutic efficacy, Int. J. Nanomedicine 7 (2012) 49–60.

[8] K. Razdan, V.R. Sinha, K.K. Singh, New paradigms in the treatment of skin infections: lipid nanocarriers to the rescue, in: Nanomedicine for Bioactives, Springer, 2020, pp. 317–339.

[9] M. Khalid, H.S. El-Sawy, Polymeric nanoparticles: promising platform for drug delivery, Int. J. Pharm. 528 (1–2) (2017) 675–691.

[10] R.H. Müller, K. Mäder, S. Gohla, Solid lipid nanoparticles (SLN) for controlled drug delivery–a review of the state of the art, Eur. J. Pharm. Biopharm. 50 (1) (2000) 161–177.

[11] A.S. Chauhan, Dendrimers for drug delivery, Molecules 23 (4) (2018) 938.

[12] D. Huang, D. Wu, Biodegradable dendrimers for drug delivery, Mater. Sci. Eng. C 90 (2018) 713–727.

[13] C. Bharti, U. Nagaich, A.K. Pal, N. Gulati, Mesoporous silica nanoparticles in target drug delivery system: a review, Int. J. Pharm. Investig. 5 (3) (2015) 124–133.

[14] W. Cai, T. Gao, H. Hong, J. Sun, Applications of gold nanoparticles in cancer nanotechnology, Nanotechnol. Sci. Appl. 1 (2008) 17–32.

[15] M.M. Modena, B. Rühle, T.P. Burg, S. Wuttke, Nanoparticle characterization: what to measure? Adv. Mater. 31 (32) (2019) 1901556.

[16] R. Konwar, A.B. Ahmed, An overview of preparation, characterization and application, Int. Res. J. Pharm. 4 (2016) 47–57.

[17] P. Ghosh, X. Yang, R. Arvizo, Z.J. Zhu, S.S. Agasti, Z. Mo, et al., Intracellular delivery of a membrane-impermeable enzyme in active form using functionalized gold nanoparticles, J. Am. Chem. Soc. 132 (8) (2010) 2642–2645.

[18] P.C. Lin, S. Lin, P.C. Wang, R. Sridhar, Techniques for physicochemical characterization of nanomaterials, Biotechnol. Adv. 32 (4) (2014) 711–726.

[19] S.K. Brar, M. Verma, Measurement of nanoparticles by light-scattering techniques, Trends Anal. Chem. 30 (1) (2011) 4–17.

[20] S. Bhattacharjee, DLS and zeta potential—what they are and what they are not? J. Control. Release 235 (2016) 337–351.

[21] J.B. Hall, M.A. Dobrovolskaia, A.K. Patri, S.E. McNeil, Characterization of nanoparticles for therapeutics, Nanomedicine (Lond.) 2 (6) (2007) 789–803.

[22] K. Jores, W. Mehnert, M. Drechsler, H. Bunjes, C. Johann, K. Mäder, Investigations on the structure of solid lipid nanoparticles (SLN) and oil-loaded solid lipid nanoparticles by photon correlation spectroscopy, field-flow fractionation and transmission electron microscopy, J. Control. Release 95 (2004) 217–227.

[23] Z. Krpetić, I. Singh, W. Su, L. Guerrini, K. Faulds, G.A. Burley, D. Graham, Directed assembly of DNA- functionalised gold nanoparticles using pyrol- imidazole polyamides, J. Am. Chem. Soc. 134 (2012) 8356–8359.

[24] T. Kowalkowski, B. Buszewski, C. Cantado, F. Dondi, Field-flow fractionation: theory, techniques, applications and the challenges, Crit. Rev. Anal. Chem. 36 (2006) 129–135.

[25] J. Parot, F. Caputo, D. Mehn, V.A. Hackley, L. Calzolai, Physical characterization of liposomal drug formulations using multi-detector asymmetrical-flow field flow fractionation, J. Control. Release 320 (2020) 495–510.

[26] S. Li, J. Nickels, A.F. Palmer, Liposome-encapsulated actin-hemoglobin (LEAcHb) artificial blood substitutes, Biomaterials 26 (17) (2005) 3759–3769.

[27] V. Filipe, A. Hawe, W. Jiskoot, Critical evaluation of nanoparticle tracking analysis (NTA) by NanoSight for the measurement of nanoparticles and protein aggregates, Pharm. Res. 27 (5) (2010) 796–810.

[28] R.A. Dragovic, C. Gardiner, A.S. Brooks, D.S. Tannetta, D.J. Ferguson, P. Hole, et al., Sizing and phenotyping of cellular vesicles using nanoparticle tracking analysis, Nanomedicine 7 (2011) 780–788.

[29] P. Kramberger, M. Ciringer, A. Strancar, M. Peterka, Evaluation of nanoparticle tracking analysis for total virus particle determination, Virol. J. 9 (2012) 265.

[30] C.Y. Soo, Y. Song, Y. Zheng, E.C. Campbell, A.C. Riches, F. Gunn-Moore, et al., Nanoparticle tracking analysis monitors microvesicle and exosome secretion from immune cells, Immunology 136 (2012) 192–197.

[31] G. Ohlsson, S. Tabaei, J.P. Beech, J. Kvassman, U. Johansson, P. Kjellbom, et al., Solute transport on the sub 100 ms scale across the lipid bilayer membrane of individual proteoliposomes, Lab Chip 12 (2012) 4635–4643.

[32] N.S. Bhise, R.B. Shmueli, J. Gonzalez, J.J. Green, A novel assay for quantifying the number of plasmids encapsulated by polymer nanoparticles, Small 8 (3) (2012) 367–373.

[33] J. Hsu, D. Serrano, T. Bhowmick, K. Kumar, Y. Shen, Y.C. Kuo, Enhanced endothelial delivery and biochemical effects of α-galactosidase by ICAM-1-targeted nanocarriers for Fabry disease, J. Control. Release 149 (3) (2011) 323–331.

[34] J. Park, W. Gao, R. Whiston, T. Strom, S. Metcalfe, T.M. Fahmy, Modulation of CD4+ T lymphocyte lineage outcomes with targeted, nanoparticle-mediated cytokine delivery, Mol. Pharm. 8 (1) (2011) 143–152.

[35] A.D. Tagalakis, S.M. Grosse, Q.-H. Meng, M.F.M. Mustapa, A. Kwok, S.E. Salehi, et al., Integrin-targeted nanocomplexes for tumour specific delivery and therapy by systemic administration, Biomaterials 32 (5) (2011) 1370–1376.

[36] Z.B. Bolat, Z. Islek, B.N. Demir, E.N. Yilmaz, F. Sahin, M.H. Ucisik, Curcumin- and Piperine-loaded Emulsomes as combinational treatment approach enhance the anticancer activity of curcumin on HCT116 colorectal cancer model, Front. Bioeng. Biotechnol. 8 (2020) 1–21.

[37] E.L. Elson, Fluorescence correlation spectroscopy: past, present, future, Biophys. J. 101 (12) (2011) 2855–2870.

[38] M.C. Smith, R.M. Crist, J.D. Clogston, Zeta potential: a case study of cationic, anionic, and neutral liposomes, Anal. Bioanal. Chem. 409 (2017) 5779–5787.

[39] J.D. Clogston, A.K. Patri, Zeta potential measurement, in: S McNeil (Ed.), Characterization of Nanoparticles Intended for Drug Delivery, in: S McNeil (Ed.), Methods in Molecular Biology, 697, Springer Nature, Switzerland, 2011, pp. 63–70.

[40] P. Fonte, F. Andrade, F. Araújo, C. Andrade, J.D. Neves, B. Sarmento, Chitosan-coated solid lipid nanoparticles for insulin delivery, Methods Enzymol. 508 (2012) 295–314.

[41] R. Sinha, G.J. Kim, S. Nie, D.M. Shin, Nanotechnology in cancer therapeutics: bioconjugated nanoparticles for drug delivery, Mol. Cancer Ther. 5 (2006) 1909–1917.

[42] W. Endreas, J. Brußler, D. Vornicescu, M. Keusgen, U. Bakowsky, T. Steinmetzer, Thrombin-inhibiting anticoagulant liposomes: development and characterization, ChemMedChem 11 (2016) 340–349.

[43] M.R. Mozafari, Liposomes: an overview of manufacturing techniques, Cell. Mol. Biol. Lett. 10 (2005) 711.

[44] K. Razdan, N.S. Sahajpal, K. Singh, H. Singh, S.K. Jain, Formulation of sustained-release microspheres of cefixime with enhanced oral bioavailability and antibacterial potential, Ther. Deliv. 10 (12) (2019) 769–782.

[45] M. Picollo, M. Aceto, T. Vitorino, UV-Vis spectroscopy, Phys. Sci. Rev. 4 (4) (2019).

[46] S.C. Wilschefski, M.R. Baxter, Inductively coupled plasma mass spectrometry: introduction to analytical aspects, Clin. Biochem. Rev. 40 (3) (2019) 115–133.

[47] R. Singh, J.W. Lillard Jr., Nanoparticle-based targeted drug delivery, Exp. Mol. Pathol. 86 (3) (2009) 215–223.

[48] Y. Zhou, C. He, K. Chen, J. Ni, Y. Cai, X. Guo, et al., A new method for evaluating actual drug release kinetics of nanoparticles inside dialysis devices via numerical deconvolution, J. Control. Release 243 (2016) 11–20.

[49] H. Yao, H. Lu, J. Zhang, X. Xue, C. Yin, J. Hu, et al., Preparation of prolonged-circulating galangin-loaded liposomes and evaluation of antitumor efficacy in vitro and pharmacokinetics in vivo, J. Nanomater. 2019 (2019) 1–10.

[50] P. Logeswari, S. Silambarasan, J. Abraham, Synthesis of silver nanoparticles using plants extract and analysis of their antimicrobial property, J. Saudi Chem. Soc. 48 (3) (2012) 1–8.

[51] Z. Zang, X. Tang, Enhanced fluorescence imaging performance of hydrophobic colloidal ZnO nanoparticles by a facile method, J. Alloys Compd. 619 (2015) 98–101.

[52] M. Khosravi, H.M. Rahimi, D. Doroud, E.S. Mirsamadi, H. Mirjalali, M.R. Zali, In vitro evaluation of mannosylated paromomycin-loaded solid lipid nanoparticles on acute toxoplasmosis, Front. Cell. Infect. Microbiol. 10 (2020) 1–10.

[53] A. Patri, M. Dobrovolskaia, S. Stern, Preclinical characterization of engineered nanoparticles intended for cancer therapeutics, in: M. Amiji (Ed.), Nanotechnology for Cancer Therapy, CRC Press, Boca Raton, FL, 2006, pp. 105–138.

[54] S.L. Pal, U. Jana, P.K. Manna, Nanoparticle: an overview of preparation and characterization, J. Appl. Pharm. Sci. 01 (2011) 6228–6234.

[55] M.E. Nik, B. Malaekeh-Nikouei, M. Amin, Liposomal formulation of galbanic acid improved therapeutic efficacy of pegylated liposomal doxorubicin in mouse colon carcinoma, Sci. Rep. 9 (2019) 9527.

[56] Z. Tianyu, B. Steven, H. Qun, Techniques for accurate sizing of gold nanoparticles using dynamic light scattering with particular application to chemical and biological sensing based on aggregate formation, ACS Appl. Mater. Interfaces 8 (2016) 21585–21594.

[57] K. Margulis-Goshen, E. Kesselman, D. Danino, S. Magdassi, Formation of celecoxib nanoparticles from volatile microemulsions, Int. J. Pharm. 393 (1–2) (2010) 230–237.

[58] J.C.H. Spence, H.R. Kolar, G. Hembree, C.J. Humphreys, J. Barnard, R. Datta, et al., Imaging dislocation cores—the way forward, Philos. Mag. 86 (29–31) (2006) 4781–4796.

[59] T.R. Albrecht, M.M. Dovek, C.A. Lang, P. GrUtter, C.F. Quate, S.W.J. Kuan, et al., Imaging and modification of polymers by scanning tunneling and atomic force microscopy, J. Appl. Phys. 64 (1988) 1178–1184.

[60] P. Avouris, Atom-resolved surface chemistry using the scanning tunneling microscope, J. Phys. Chem. 94 (1990) 2246–2256.

[61] D. Bonnell, Scanning Probemicroscopy and Spectroscopy: Theory, Techniques, and Applications, Wiley-VCH, NewYork, 2001.

[62] C. Kocum, E.K. Cimen, E. Piskin, Imaging of poly(NIPA-co-MAH)-HIgG conjugate with scanning tunneling microscopy, J. Biomater. Sci. Polym. Ed. 15 (2004) 1513–1520.

[63] H. Wang, P.K. Chu, Chapter 4—surface characterization of biomaterials, in: B. Amit, B. Susmita (Eds.), Characterization of Biomaterials, Academic Press, Oxford, 2013, pp. 105–174.

[64] N.K. Gaisin, O.I. Gnezdilov, T.N. Pashirova, E.P. Zhil'tsova, S.S. Lukashenko, L.Y. Zakharova, et al., Micellar and liquidcrystalline properties of bicyclic fragment-containing cationic surfactant, Colloid J. 72 (6) (2010) 764–770.

[65] F.B. Rosevear, The microscopy of the liquid crystalline neat and middle phases of soaps and synthetic detergents, J. Am. Oil Chem. Soc. 31 (12) (1954) 628–639.

[66] C.C. Müller-Goymann, Physicochemical characterization of colloidal drug delivery systems such as reverse micelles, vesicles, liquid crystals and nanoparticles for topical administration, Eur. J. Pharm. Biopharm. 58 (2) (2004) 343–356.

[67] H. Giesche, Mercury porosimetry: a general (practical) overview, Part. Part. Syst. Charact. 23 (1) (2006) 9–19.

[68] J. Van Brakel, S. Modrý, M. Svatá, Mercury porosimetry: state of the art, Powder Technol. 29 (1) (1981) 1–12.

[69] P. Sher, G. Ingavle, S. Ponrathnam, P. Poddar, A.P. Pawar, Modulation and optimization of drug release from uncoated low density porous carrier based delivery system, AAPS PharmSciTech 10 (2) (2009) 547–558.

[70] J.W. Hagemann, J.A. Rothfus, Polymorphism and transformation energetics of saturated monoacid triglycerides from differential scanning calorimetry and theoretical modeling, J. Am. Oil Chem. Soc. 60 (6) (1983) 1123–1131.

[71] D.T. Haynie, Biological Thermodynamics, second ed., Cambridge University Press, Cambridge, 2001.

[72] R.L. Danley, New heat flux DSC measurement technique, Thermochim. Acta 395 (1–2) (2002) 201–208.

[73] N. Zucca, G. Erriu, S. Onnis, A. Longoni, An analytical expression of the output of a power-compensated DSC in a wide temperature range, Thermochim. Acta 413 (1–2) (2004) 117–125.

[74] H. Bunjes, T. Unruh, Characterization of lipid nanoparticles by differential scanning calorimetry, X-ray and neutron scattering, Adv. Drug Deliv. Rev. 59 (6) (2007) 379–402.

[75] C. Houacine, D. Adams, K.K. Singh, Impact of liquid lipid on development and stability of trimyristin nanostructured lipid carriers for oral delivery of resveratrol, J. Mol. Liq. 316 (2020) 113734.

[76] A.A. Bunaciu, E.G. UdriŞTioiu, H.Y. Aboul-Enein, X-ray diffraction: instrumentation and applications, Crit. Rev. Anal. Chem. 45 (4) (2015) 289–299.

[77] L. Sawyer, D.T. Grubb, G.F. Meyers, Polymer Microscopy, Springer Science & Business Media, New York, 2008.

[78] D. Liu, L. Chen, S. Jiang, S. Zhu, Y. Qian, F. Wang, et al., Formulation and characterization of hydrophilic drug diclofenac sodium-loaded solid lipid nanoparticles based on phospholipid complexes technology, J. Liposome Res. 24 (1) (2014) 17–26.

[79] R. Fishler, J. Sznitman, A novel aerodynamic sizing method for pharmaceutical aerosols using image-based analysis of settling velocities, Inhalation 11 (3) (2017) 21–25.

[80] M.R. Stolzenburg, P.H. McMurry, Method to assess performance of scanning mobility particle sizer (SMPS) instruments and software, Aerosol Sci. Tech. 52 (2018) 609–613.

[81] L. Coquelin, N. Fischer, C. Motzkus, T. Mace, F. Gensdarmes, L.L. Brusquet, et al., Aerosol size distribution estimation and associated uncertainty for measurement with a Scanning Mobility Particle Sizer (SMPS), J. Phys.: Conf. Ser. 429 (2012). Nanosafe 2012: International Conferences on Safe Production and Use of Nanomaterials 13–15 November, Grenoble, France.

[82] C. Jurinke, P. Oeth, D. van den Boom, MALDI-TOF mass spectrometry, Mol. Biotechnol. 26 (2) (2004) 147–163.

[83] N.R. Bader, Sample preparation for flame atomic absorption spectroscopy: an overview, Rasayan. J. Chem. 4 (1) (2011) 49–55.

[84] V. Nele, M.N. Holme, U. Kauscher, M.R. Thomas, J.J. Doutch, M.M. Stevens, Effect of formulation method, lipid composition, and PEGylation on vesicle lamellarity: a small-angle neutron scattering study, Langmuir 35 (18) (2019) 6064–6074.

[85] M. Chiba, M. Miyazaki, S. Ishiwata, Quantitative analysis of the lamellarity of giant liposomes prepared by the inverted emulsion method, Biophys. J. 107 (2014) 346–354.

[86] M. Fröhlich, V. Brecht, R. Peschka-Süss, Parameters influencing the determination of liposome lamellarity by 31P-NMR, Chem. Phys. Lipids 109 (1) (2001) 103–112.

[87] H.G. Shi, L. Farber, J.N. Michaels, Characterization of crystalline drug nanoparticles using atomic force microscopy and complementary techniques, Pharm. Res. 20 (2003) 479–484.

[88] M.J. Costello, Cryo-electron microscopy of biological samples, Ultrastruct. Pathol. 30 (5) (2006) 361–371.

[89] S.U. Egelhaaf, P. Schurtenberger, M. Müller, New controlled environment vitrification system for cryo-transmission electron microscopy: design and application to surfactant solutions, J. Microsc. 200 (2) (2000) 128–139.

[90] H. Friedrich, P.M. Frederik, G. de With, N.A. Sommerdijk, Imaging of self-assembled structures: interpretation of TEM and Cryo-TEM images, Angew. Chem. Int. Ed. 49 (43) (2010) 7850–7858.

[91] F. Guo, W. Jiang, Single particle cryo-electron microscopy and 3-D reconstruction of viruses, in: Electron Microscopy, Humana Press, Totowa, NJ, 2014, pp. 401–443.

[92] M. Marko, C. Hsieh, R. Schalek, J. Frank, C. Mannella, Focused-ion-beam thinning of frozen-hydrated biological specimens for cryo-electron microscopy, Nat. Methods 4 (3) (2007) 215–217.

[93] L. Norlén, Nanostructure of the stratum corneum extracellular lipid matrix as observed by cryo-electron microscopy of vitreous skin sections, Int. J. Cosmet. Sci. 29 (5) (2007) 335–352.

[94] P.M. Frederik, D.H. Hubert, Cryoelectron microscopy of liposomes, Methods Enzymol. 391 (2005) 431–448.

[95] V.M. Unger, Electron cryomicroscopy methods, Curr. Opin. Struct. Biol. 11 (5) (2001) 548–554.

[96] J.L. Jimenez, J.I. Guijarro, E. Orlova, J. Zurdo, C.M. Dobson, M. Sunde, H.R. Saibil, Cryo-electron microscopy structure of an SH3 amyloid fibril and model of the molecular packing, EMBO J. 18 (4) (1999) 815–821.

[97] R.I. Koning, A.J. Koster, T.H. Sharp, Advances in cryo-electron tomography for biology and medicine, Ann. Anat. 217 (2018) 82–96.

[98] J.L. Milne, S. Subramaniam, Cryo-electron tomography of bacteria: progress, challenges and future prospects, Nat. Rev. Microbiol. 7 (9) (2009) 666–675.

[99] M. Adrian, J. Dubochet, S.D. Fuller, J.R. Harris, Cryo-negative staining, Micron 29 (2–3) (1998) 145–160.

[100] A. Wittemann, M. Drechsler, Y. Talmon, M. Ballauff, High elongation of polyelectrolyte chains in the osmotic limit of spherical polyelectrolyte brushes: a study by cryogenic transmission electron microscopy, J. Am. Chem. Soc. 127 (27) (2005) 9688–9689.

[101] L. Belkoura, C. Stubenrauch, R. Strey, Freeze fracture direct imaging: a new freeze fracture method for specimen preparation in cryo-transmission electron microscopy, Langmuir 20 (11) (2004) 4391–4399.

[102] D.G. Fatouros, B. Bergenstahl, A. Mullertz, Morphological observations on a lipid-based drug delivery system during in vitro digestion, Eur. J. Pharm. Sci. 31 (2) (2007) 85–94.

[103] K. Kleberg, F. Jacobsen, D.G. Fatouros, A. Müllertz, Biorelevant media simulating fed state intestinal fluids: colloid phase characterization and impact on solubilization capacity, J. Pharm. Sci. 99 (8) (2010) 3522–3532.

[104] J. Szebeni, C.R. Alving, S. Savay, Y. Barenholz, A. Priev, D. Danino, Y. Talmon, Formation of complement-activating particles in aqueous solutions of Taxol: possible role in hypersensitivity reactions, Int. Immunopharmacol. 1 (4) (2001) 721–735.

[105] P. Ercius, O. Alaidi, M.J. Rames, G. Ren, Electron tomography: a three-dimensional analytic tool for hard and soft materials research, Adv. Mater. 27 (38) (2015) 5638–5663.

[106] S. Pujals, N. Feiner-Gracia, P. Delcanale, I. Voets, A. Lorenzo, Super-resolution microscopy as a powerful tool to study complex synthetic materials, Nat. Rev. Chem. 3 (2019) 68–84.

[107] B. Huang, H. Babcock, X. Zhuang, Breaking the diffraction barrier: super-resolution imaging of cells, Cell 143 (7) (2010) 1047–1058.

[108] S. Shashkova, M.C. Leake, Single-molecule fluorescence microscopy review: shedding new light on old problems, Biosci. Rep. 37 (4) (2017), BSR20170031.

[109] T.J. Rottreau, C.M.A. Parlett, A.F. Lee, R. Evans, Extending the range of liquids available for NMR cryoporometry studies of porous materials, Micropor. Mesopor. Mat. 274 (2019) 198–202.

[110] Y. Kharbanda, M. Urbańczyk, O. Laitinen, K. Kling, S. Pallaspuro, S. Komulainen, et al., Comprehensive NMR analysis of pore structures in Superabsorbing cellulose nanofiber aerogels, J. Phys. Chem. C. Nanomater. Interfaces 123 (51) (2019) 30986–30995.

[111] J.D. Jeon, S.J. Kim, S.Y. Kwak, ^{1}H nuclear magnetic resonance (NMR) cryoporometry as a tool to determine the pore size distribution of ultrafiltration membranes, J. Membr. Sci. 309 (1–2) (2008) 233–238.

[112] N. Gopinathan, B. Yang, J.P. Lowe, K.J. Edler, S.P. Rigby, NMR cryoporometry characterisation studies of the relation between drug release profile and pore structural evolution of polymeric nanoparticles, Int. J. Pharm. 469 (1) (2014) 146–158.

[113] T.J. Rottreau, G.E. Parkes, M. Schirru, J.L. Harries, M.G. Mesa, R. Evans, NMR cryoporometry of polymers: cross-linking, porosity and the importance of probe liquid, Colloid Surf. A Physicochem. Eng. Asp. 575 (2019) 256–263.

[114] C. Sabliov, H. Chen, R. Yada, Nano- and Micro-scale Vehicles for Effective Delivery of Bioactive Ingredients in Functional Foods, Wiley-Blackwell, 2015, pp. 1–408.

[115] Q. Chaudhry, L. Castle, Safety assessment of nano- and micro-scale delivery vehicles for bioactive ingredients, in: C. Sabliov, H. Chen, R. Yada (Eds.), Nanotechnology and Functional Foods: Effective Delivery of Bioactive Ingredients, Wiley-Blackwell, 2015, pp. 348–357.

[116] EFSA, EFSA Guidance on risk assessment of the application of nanoscience and nanotechnologies in the food and feed chain: Part 1, human and animal health, EFSA J. 16 (7) (2018) 5327. www.efsa.europa.eu/en/efsajournal/pub/5327.

[117] A. Nel, Y. Zhao, L. Madler, Environmental health and safety considerations for nanotechnology, Acc. Chem. Res. 46 (3) (2013) 605–606.

[118] S.M. Moghimi, F.Z. Shadi, Defining and characterizing nonbiological complex drugs (NBCDs)—is size enough? The case for liposomal doxorubicin generics ('liposomal nanosimilars') for injection, GaBI J. 3 (2) (2014) 56–62.

[119] D. Rosenblum, D. Peer, Omics-based nanomedicine: the future of personalized oncology, Cancer Lett. 352 (1) (2014) 126–136.

[120] FDA's Approach to Regulation of Nanotechnology Products, 2018. https://www.fda.gov/science-research/nanotechnology-programs-fda/fdas-approach-regulation-nanotechnology-products.

CHAPTER 4

Surface engineering of nanoparticles for imparting multifunctionality

Hira Choudhury[a], Bapi Gorain[b,c], Manisha Pandey[a], Jayabalan Nirmal[d], and Prashant Kesharwani[e]

[a]Department of Pharmaceutical Technology, School of Pharmacy, International Medical University, Kuala Lumpur, Malaysia
[b]School of Pharmacy, Faculty of Health and Medical Sciences, Taylor's University, Subang Jaya, Selangor, Malaysia
[c]Centre for Drug Delivery and Molecular Pharmacology, Faculty of Health and Medical Sciences, Taylor's University, Subang Jaya, Selangor, Malaysia
[d]Translational Pharmaceutics Research Laboratory, Department of Pharmacy, Birla Institute of Technology and Science (BITS)-Pilani, Hyderabad, Telangana, India
[e]Department of Pharmaceutics, School of Pharmaceutical Education and Research, Jamia Hamdard, New Delhi, India

1. Introduction

To exert their therapeutic efficacy, the drug molecules must reach the site of action and need to interact with the target site. Therefore, the drugs need to cross the biological barriers to reach the site of action. Whatever the dosage form used to deliver the therapeutics, it should be capable to deliver the therapeutic agents across the biological barrier [1, 2]. These complex biological barriers are composed of different components (enzymatic barrier, physicochemical, or mechanical barrier) and different elements (cellular, endothelial, or epithelial membrane) to resist the passage of any foreign materials. The delivery of therapeutics to the specific site of the cell, tissue, or organ is of great challenge in the treatment of complex diseases, such as cancer, infections, or genetic disorders [1–4].

Increasing percentage of hydrophobic new chemical entities in the pipeline has brought several complications in drug delivery to attain desired physiological functions, because of their low aqueous solubility, low bioavailability, fast biological metabolism, and unavoidable toxic manifestations [5–8]. Novel and advanced drug delivery systems are enriching the delivery options with novel formulations to efficiently deliver the therapeutics via different routes of administration to effectively achieve desired therapeutic efficacy.

The recent advancement of nanotechnology has brought several technologies to fabricate a number of nanocarriers in the application in the medical field. These nanocarriers are effectively dealing with the challenges of several therapeutic agents, which are unable to deliver at therapeutic concentrations to the site of action using the conventional dosage forms [9–11]. Hence, the unique structure of the nanocarriers is known to overcome the limitations of delivering therapeutic agents, particularly those which are highly hydrophobic, or metabolized rapidly or may experience some efflux mechanism while

Nanoparticle Therapeutics
https://doi.org/10.1016/B978-0-12-820757-4.00001-6

absorption from the gastrointestinal tract. The improvement of the drug properties by the use of nanocarrier might be obtained by encapsulating the therapeutics (either hydrophilic or hydrophobic) within the core of the carrier, providing controlled release of the entrapped therapeutics, enhanced transportation of drugs from the site of administration to the site of action, preventing immature degradation by the metabolizing enzymes or by the phagocytic cells, and by decreased unnecessary systemic exposure of the therapeutics to reduce side effects [1, 12].

Researchers over the world are engaged in developing these miniaturized particles in the form of lipidic (liposome, noisome, solid lipid nanoparticles, etc.) or polymeric (polymeric micelles, nanoparticles, dendrimers, etc.) nanocarriers. Among these nanocarriers, nanoparticles are gaining tremendous research interest. These nanoparticles are known to increase the solubility of the hydrophobic agents largely by increasing enormous surface area, where these therapeutic agents are entrapped into polymeric or lipid environments to prevent further coagulation. Such increased solubility of the therapeutics improves the pharmacokinetic profiles of the entrapped drugs [13–17]. Thus, formulation scientists over the world are in the process of fabricating nanoparticles with diverse compositions and biological properties to deliver the therapeutic agents to the site of action to obtain a desired therapeutic effect. These formulations are modifying several properties of the entrapped drug component(s), thereby alter the release pattern, stability, biological interaction, and targeting behavior of the therapeutics [9, 18].

2. Need of surface engineering in drug delivery

The nanocarriers within the size range of 10–200 nm are advantageous in delivering therapeutics as these carriers are not recognized by the endogenous reticuloendothelial system (RES), and also not filtered by glomerular filtration, thereby retain within the system for a longer period of time [19–21]. Furthermore, in the cancerous microenvironment, the number of blood vessels is increased. This highly fenestrated structure meets the increasing demand for nutrition and oxygen by the fast-growing cancer cells [22]. Thus, the nanocarriers with an appropriate range of size can travel within the biological system and will reach the tumor environment at higher concentrations. This phenomenon is known as the enhanced permeability and retention (EPR) effect of nanocarriers. Further, such escape of recognition decreases systemic exposure of the drug, thereby decrease in side effects, which helps in increasing the tolerated dose of the therapeutics [23–28]. However, these approaches are lacking the direction to deliver the therapeutics to the particular site of action for desired therapeutic action [29].

Targeted delivery in any diseased condition refers to direct the drug carrier for dominant accumulation to the diseased site to exert its therapeutic role. Thus, to obtain effective targeted delivery of therapeutic agents, the carrier must retain within the biological system for a longer period of time and escape from the immunological system, so that it can target the specific organ, tissue, or cell to release the entrapped therapeutic agent [30, 31].

It is worth mentioning that the delivery of therapeutics, genes, DNA, or small-interfering RNA to the specific location is necessary to obtain desired therapeutic effect with minimal side effects. These factors act as a driving force for researchers to explore new strategies for drug delivery. Active targeting of the nanocarriers could significantly improve the delivery of entrapped drugs to the site of action as compared to the nontargeted deliveries [11, 32–34]. Thus, the EPR lead prolonged retention of the carriers is further aided in targeting to the site of delivery by the technique called active targeting. The active targeting of nanocarriers can be achieved by decorating the surface of the nanocarriers with specific ligands, which have an affinity for the specific receptors overexpressed at the desired diseased site. Therefore, many types of nanoparticles have been investigated with unique architectures to serve as delivery systems to treat different chronic diseases. Surface engineered nanocarriers have been investigated as an alternative approach to resolve these challenges [35, 36]. In this context, several biological ligands have been identified for surface engineering of nanoparticles to improve therapeutic efficacy (Fig. 1) [35, 37]. This strategy enhances the binding of the carriers to the surface of the cells at the targeted site and improves the transportation of the drug. This concept is not new; it was first proposed in 1980 where the liposomal deliveries were coupled to monoclonal antibodies for active targeting purposes [38]. Later, various kinds of ligands has been introduced in the targeting of nanocarriers, such as antibodies (e.g., Herceptin, CD19, Rituxan), peptides (e.g., RGD, NGR), proteins (e.g., transferrin, LHRH), aptamers (e.g., pegaptanib), and small molecules (e.g., folate, galactose) [1].

The selection of the ligand for the surface engineering is largely dependent on the overexpressed receptors on the targeted cells. An enormous number of receptors have been identified on the cell surface of different disease areas, which help in identifying the particular ligand. Numerous studies had been conducted in vitro and in vivo to establish the targetability of the carriers when a ligand is attached to the surface of the nanoparticles [6, 39–42]. The following section of the chapter will emphasize different ligands, which are successfully decorated onto the nanoparticles for their active targeting and improved efficacy.

3. Types of targeting moieties

Traditional approaches for drug delivery often lead to uncontrolled drug release with concentration spikes in plasma, which ends with harming the nontargeted organs. In the earlier section, it has been discussed how these nanocarriers circumvent these challenges due to their diminutive size and penetration ability to biological barriers [43]. Furthermore, targeting of the nanocarriers to the site of action could be enhanced by the surface engineering techniques, where different ligands can be attached to the surface

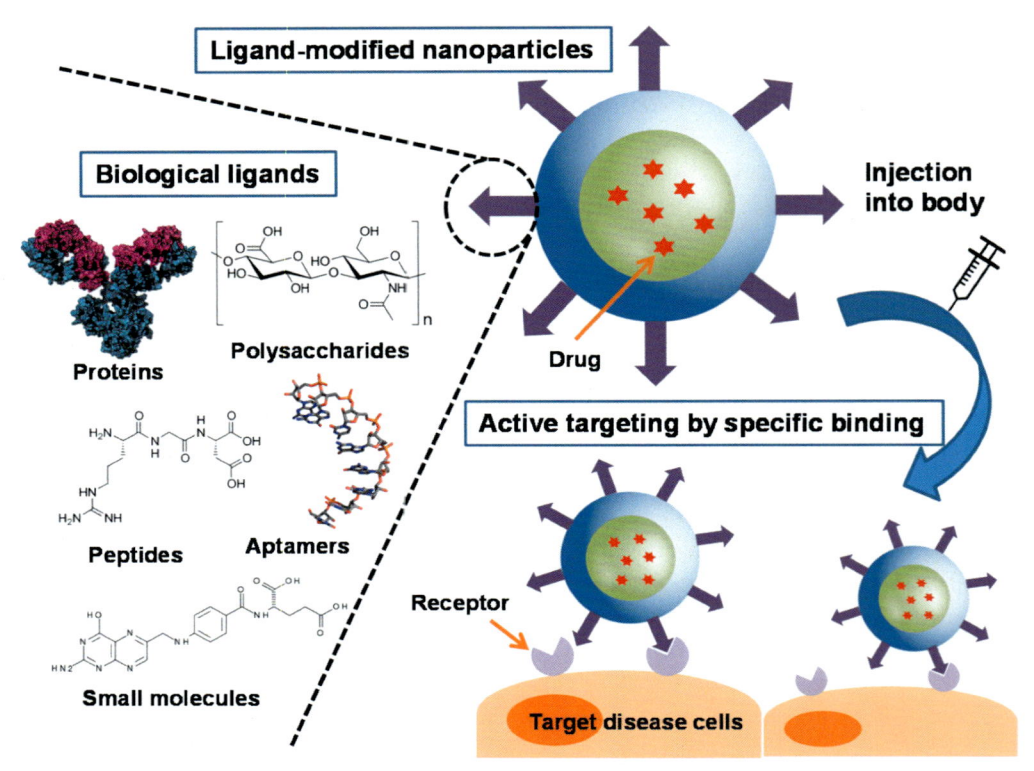

Fig. 1 Illustration of biological ligands for active targeting of nanoparticle drug carriers. *(Adapted with permission from J. Yoo, C. Park, G. Yi, D. Lee, H. Koo, Active targeting strategies using biological ligands for nanoparticle drug delivery systems, Cancers (Basel). 11 (2019) 640. https://doi.org/10.3390/cancers11050640.)*

of the nanocarriers for active targeting. A list of different types of ligands used in the active targeting of nanocarriers is provided in Table 1. This part will elaborate on various types of ligands for surface functionalization and their promising outcomes.

3.1 Proteins and polysaccharides

There has been a revolution in treatment efficacy due to target specificity by the conjugation of ligand to the drug carrier. Among the biological ligands, antibodies and their fragments have exquisite specificity and reduced off-target effects. However, their large sizes limit their number on the surface of miniaturized drug delivery. Their use is not limited to targeting; antibodies are also explored for therapeutic and theranostic purposes [44, 45].

Another goal of protein and polysaccharide-targeted nanocarriers is the safe delivery of labile therapeutic molecules to the site of action. Nascimento and team investigated that the silencing capability of Mad2 siRNA was enhanced by loading in epidermal

Table 1 Different targeting ligands and respective targets on the specific cells.

Category	Ligand	Specific target
Proteins and polysaccharides	Anti-annexin A2 antibody	Annexin A2 receptor
	Transferrin	Transferrin receptor
	CD19	CD20 or CD19 antigen
	Trastuzumab	HER-2 receptor
	Hyaluronic acid	CD44 receptors
	Hyaluronic acid ceramide	CD44 receptors
Peptides	cRGD	$\alpha_v\beta_3$ Integrin receptors
	H2009.I peptide	$\alpha_v\beta_6$ Integrin receptors
	IL-13 peptide	IL-13-Rα2 receptor
	AP-1 peptide	IL-4 receptor
	Bombesin peptide	Gastrin releasing peptide receptor
	Asparagine-glycine-arginine (NGR) peptide	Aminopeptidase N
	NR7 peptide	EGFR
	CVKTPAQSC peptide	CD133 + receptor
	Luteinizing hormone-releasing hormone LHRH peptide	LHRH receptor
Small molecules	(3-Aminomethylphenyl)-boronic acid	CD44 receptors
	Galactose	Asialoglycoprotein receptor
	Folic acid	Folate receptor
Aptamer	EpCAM aptamer	EpCAM protein
	AS-1411 aptamer	Nucleolin-targeted DNA aptamer
	Pegaptanib	VEGF receptor
	CD133 aptamer	CD133 protein

growth factor receptor (EGFR)-targeted nanoparticles for the treatment of nonsmall-cell lung cancer model [46, 47]. Moreover, they have performed biodistribution and pharmacokinetics of the above-stated nanoparticles in A549-DDP and A549-WT tumor-bearing mice. Results are in coherence to in vitro data that accumulation and targeting efficiency significantly improved compared to nontargeted nanoparticles [48].

Another protein-based ligand used for surface engineering is transferrin (Tf), an iron-binding glycoprotein. Tf is responsible for iron transport in the body and expressed on hepatocytes, intestinal cells, blood-brain barrier (BBB), epithelial cells of choroid plexus, and neurons [49]. The abundant expression of Tf is due to the need for iron for DNA synthesis and differentiation and regeneration of cells. Literature showed excessive expression (up to 100 folds) of Tf receptors on cancerous cells, which may lead to a significant role of Tf for surface engineering [50].

The current strategy on active targeting is reflected by the surface modification of the nanocarriers using antibody fragments, i.e., instead of attaching the whole antibody, active targeting could be achieved by attachment of single-chain fragment variables (scFv) and fragment antigen-binding (Fab) [51]. This novel technique has shown the advantage of avoiding antibody inactivation during the functionalization of the developed nanocarriers. Furthermore, this decrease in the size of the surface attachment helps in reducing the final size of the nanocarriers, where the carriers prevent the possibility of initiating immune response [52]. Simultaneously, Lin and the team had developed a triptolide-loaded liposomal delivery system with anti-CA-IX antibody functionalization. Thiolation reaction was performed in the presence of dithiothreitol, where the reduction reaction from the cleaved antibody by cleaving at the hinge. Finally, based on the results, the authors concluded that this CA-IX decorated carrier could be a promising platform for lung cancer therapy [53].

Active targeting can also be achieved by hyaluronic acid, which is a polysaccharide and is present in the extracellular matrix along with collagen. Hyaluronic acid is one of the representative markers for cancer stem cells and able to bind to CD44, usually overexpressed on cancer cells [54]. In the series of re-inventing the potential drug, Kumar et al. developed hyaluronic acid-dihydroartemisinin conjugate, which in presence of water formulate self-assembled nanoparticles. These nanoparticles were found to be cytotoxic on the lung cancer (A549) cell line. This data was further confirmed by reactive oxygen species, mitochondrial membrane potential, and apoptosis assay (Fig. 2) [55]. Conclusively, the above literature indicates the potential of hyaluronic acid in improving therapeutic outcomes of anticancer drugs.

3.2 Peptides

To achieve a better clinical outcome, active targeting is an emerging strategy, which utilizes overexpressed receptors on disease tissue. In this regard, peptide ligands having some advantage over others, such as low cost, ease of synthesis, low immunogenicity, high affinity, and specificity [56]. Several peptide ligands, such as p160 (VPWMEPAYQRFL), arginine-glycine-aspartic acid (RGD), and asparagine-glycine-arginine (NGR) are used to target keratin 1, integrin, and aminopeptidase N (or CD13), respectively, on different targeted cells, e.g., breast cancer cells. Hanieh et al. developed a peptide 1 (GE11) ligand to target EGFR specifically expressed on triple-negative breast cancer cells [57]. On the other hand, Chi et al. used an IL-4R-binding peptide-1 (sequence CRKRLDRNC) to target interleukin-4 receptor (IL-4R) overexpressed on tumor endothelial cells and lung cancer cells. IL-4R-binding peptide-1 was labeled with liposome loaded with doxorubicin (DOX) and results indicate efficient internalization in H226 tumor cells. Incoherence with in vitro data, in vivo studies showed higher accumulation of labeled liposomes in tumors compared to unlabeled [58].

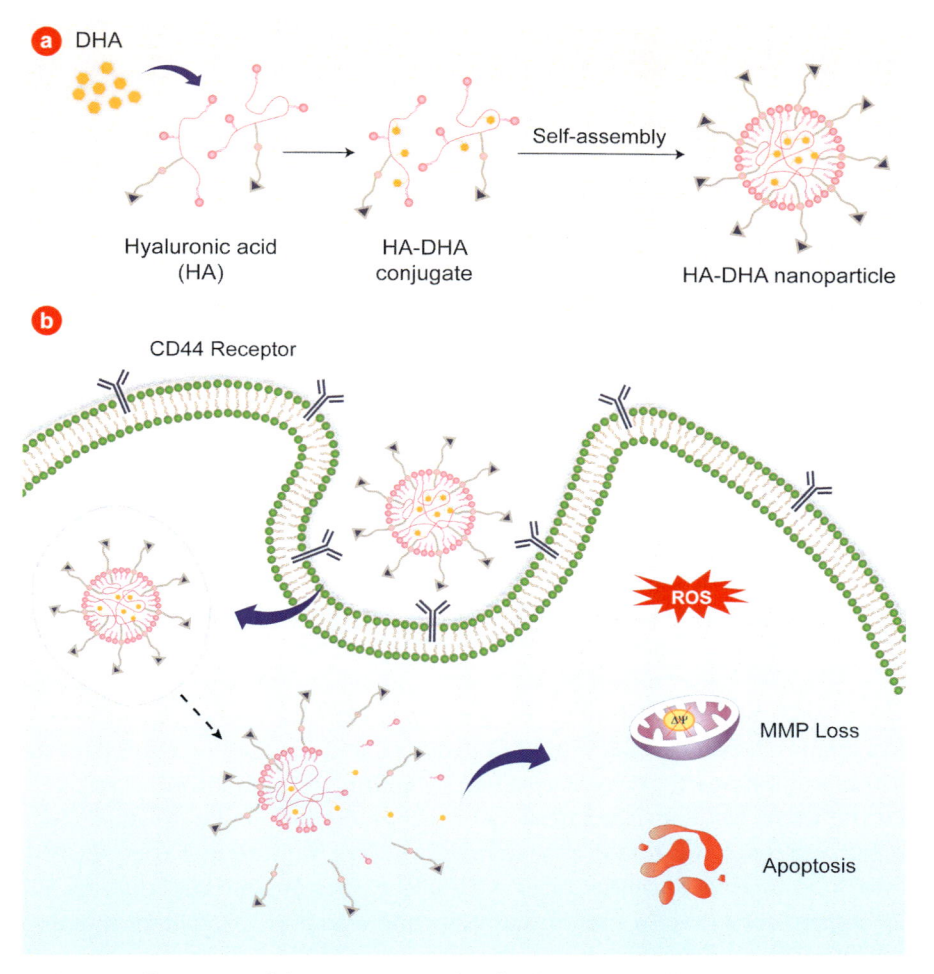

Fig. 2 Schematic illustration of the preparation of (A) hyaluronic acid-dihydroartemisinin conjugated nanoparticles. (B) The mechanistic insight into the antiproliferative potential of developed nanoparticles was investigated using apoptosis, reactive oxygen species (ROS) generation, and mitochondrial membrane potential (MMP) loss. *(Adapted with permission from R. Kumar, M. Singh, J. Meena, P. Singhvi, D. Thiyagarajan, A. Saneja, A.K. Panda, Hyaluronic acid—dihydroartemisinin conjugate: synthesis, characterization and in vitro evaluation in lung cancer cells, Int. J. Biol. Macromol. 133 (2019) 495—502. https://doi.org/10.1016/j.ijbiomac.2019.04.124.)*

In another research, RGD was used to target the integrin receptors. Thus, RGD was labeled on size-shrinkable nanoparticles loaded with DOX and metformin for combination therapy. Release of metformin and DOX occurs after breakage of imine bonds only in acidic conditions, which is mainly available in tumor tissue. Inflammation at the tumor site induced by nuclear factor-κB (NF-κB) was suppressed by metformin; however,

DOX kills the cancer cells. The presence of RGD enhanced accumulation in 4T1 and CT26 tumors. This indicated that the combination therapy has added advantage over conventional therapy [59]. Similarly, combination therapy for liver fibrosis by miR-29b and germacrone was investigated by Ji et al. Poly(ethylene glycol)-*block*-poly(lactide-*co*-glycolide) (PEG-PLGA) nanoparticles coloaded with miR-29b and germacrone was developed and labeled with cyclic RGD peptides (cRGDfK). cRGDfK was used to target integrin $\alpha v\beta 3$ overexpressed on the fibrotic liver cells. The results showed that the nanoparticles possess high cytotoxicity on hepatic stellate cells. Furthermore, nanoparticle accumulation was achieved in liver fibrotic mice treated with RGD labeled nanoparticles [60]. These studies indicated the efficiency of peptide-based ligand in targeting drug delivery.

3.3 Aptamer

Aptamers are sensitive, small, have immunogenicity, and are biodegradable. They are composed of several nucleotides and are classified as short nucleic acid. Because of these properties, aptamer gains considerable attention in research as a potential ligand for active targeting [61, 62]. In this context, Duo et al. fabricated mesoporous silica nanoparticles labeled with AS-1411 aptamer, coated with PEG and polydopamine to deliver CX-5461. CX-5461 is an inhibitor of rRNA synthesis and induces the death of tumor cells via prodeath autophagy. In vitro assay results revealed a significant enhancement of nucleolar accumulation of CX-5461. Similarly, in animal imaging assay, data showed aptamer labeled nanoparticles have a higher inhibition effect and distribution in cancer cells. Furthermore, histology analysis revealed that no significant toxicity of nanoparticles tagged with AS-1411 was found on other organs. This indicates the safety of AS-1411 tagged mesoporous silica nanoparticles with effective treatment targeting the nucleus of the cancerous cells. [63]. In another study, aptamer was used to target the nanoparticles for silencing of P-glycoprotein (P-gp) expression by siRNA and site-specific release of DOX for breast cancer treatment. The nanoparticles containing siRNA and DOX labeled with aptamer enhanced the knockdown of P-gp and concurrently cellular uptake of DOX in 4T1-R breast cancer cells [64]. Gui et al. fabricated polymer-lipid nanoparticles tagged with CD133 aptamers for effective and targeted delivery of all-trans retinoic acid (ATRA) to osteosarcoma initiating cells. CD133 is one of the primary markers overexpressed in osteosarcoma initiating cells, which was targeted by using aptamer. Results from tumorsphere formation assay, cytotoxicity assay, and flow cytometry revealed the therapeutic efficacy of ATAR was improved by tagging CD133 aptamers [65]. These studies represent an up-and-coming approach for the targeted and potential treatment of diseases.

3.4 Small molecules

Small molecule ligands act as intra- and extra-cellular signals for making regulation networks and are also involved in the various enzymatic reaction [66]. Small molecules are specific and have a high affinity toward receptors abundantly present on the cell surface [67]. Folic acid is one of the best examples of small-molecule ligand as folate receptor is highly expressed in some of the cancer cells. Basically, these cancer cells need a high amount of folate for DNA repair during rapid duplication. Folate as a ligand can reduce off-target organ toxicity and lessen the side effect of the therapeutic agent [68–72]. A stimuli-responsive nanocarrier loaded with DOX and tagged with folate was developed to target breast cancer. The maximum release of the drug was observed at pH 5.5 compared to pH 7.4, which limits the release at the tumor microenvironment. Cellular uptake was higher in nanoparticles labeled with folate as well as higher apoptosis was observed in the MCF-7 cell line compared to non-targeted nanoparticles. In coherence with in vitro data, in vivo results revealed high tumor regression (91%), and no significant off-target toxicity was observed when compared to free DOX-treated animals [73]. Similarly, phenylboronic acid (PBA) a small molecule ligand, was used to target sialylated glycans. Sialylated glycans are broadly overexpressed on cancer cells and selectively recognized by PBA. Deshayes et al. fabricated PBA labeled micelles loaded with oxaliplatin. Results confirmed improve cellular recognition and higher cellular uptake in B16F10 murine melanoma cells. Moreover, PBA-installed micelles inhibit tumor growth in lung metastasis models of melanoma. This suggests the improved therapeutic efficacy of cancer treatment [74].

4. Methods of surface engineering of nanoparticles

There are several approaches available, which are used to conjugate the aforementioned ligands to the surface of the nanoparticles. The primary objective of this technique is to attach the targeting ligand to the surface of the nanoparticles without compromising the functionality of the nanoparticles. This conjugation can be achieved by bioconjugation technique, click chemistry, and hybridization techniques. A brief description of the methods has been given in the connecting section.

4.1 Bioconjugation method

This conventional bioconjugation can be achieved by three different methods, such as physical interactions, linker chemistry, and direct conjugation.

4.1.1 Physical interaction

This interaction between the ligand moiety and nanoparticles can be achieved by hydrophobic, electrostatic, and affinity interactions. This technique can be useful in conjugating therapeutic moiety to the surface of the nanoparticles. The hydrophobic nature of most of the anticancer agents helped in interacting with the nanoparticles physically via hydrophobic interaction. Such interaction results in adsorption of the hydrophobic therapeutics under the hydrophobic coating layer [75]. Upon administration, the entrapped drug was released within the cells when the outer coat was digested. Simultaneously, the electrostatic physical interaction strategy was utilized to load small-interfering RNA on the cationic polyethyleneimine-coated nanoparticles [76]. These techniques possess different advantages of lacking if any adjustment steps and binding process is rapid. However, it is challenging to control or maintain the molecular alignment of the physically bound ligands to the nanoparticles. Thus, the mode of binding for the two methods under interaction is not proper. Alternatively, an effective bioconjugation of the ligands to the nanoparticles could be achieved by the third method, affinity interactions. An example of affinity interaction was documented between biotin and streptavidin, which forms a stable and strong linkage [77].

4.1.2 Direct conjugation

During the preparation method of nanoparticles, the functional groups are available while applying the coating layer on the surface of the nanoparticles. Depending on the need for surface functionalization, aldehyde, amine, or active hydrogen groups can be introduced on the surface of the nanoparticles. This technique has been found to properly conjugate the nanocarriers with different therapeutic agents, chelators for nuclear imaging, or any fluorescence dyes based on the purpose. The unmodified proteins or peptides could not be conjugated to the surface of the nanoparticles, as these are not reactive [78]. However, surface maleimides containing nanocarriers can be directly conjugated with proteins or peptides biomolecules containing thiol groups.

4.1.3 Linker chemistry

The binding orientation of ligands to the surface of the nanoparticles can be controlled by the linker molecules. Thus, this linker chemistry is favored in conjugating ligand molecules to the nanocarriers over the previous two methods. There are a number of linkers that have been identified and synthesized, which can conjugate proteins, peptides, carbohydrates, other small molecules, etc. Among the availability, the most common one is the reaction among sulfhydryl groups on the surface of the ligand to the anime-modified nanocarriers. Thus, the sulfhydryl moiety can be introduced by the attachment of cysteine residue to peptide/protein-based ligands, which facilitates conjugation by the linker chemistry to the amine group of the nanoparticles.

There are verities of linker molecules incorporated for the purpose of bioconjugation, such as heterobifunctional PEG molecules (succinimidyl ester-PEG-maleimide), N-succinimidyl-3-(2- pyridyldithio)-propionate, succinimidyl-4-(N-maleimidomethyl) cyclohexane-1-carboxylate, and N-succinimidyl iodoacetate. The orientation of functional groups in heterobifunctional PEG molecules facilitates proper orientation, where it contains succinimidyl esters (reactive to amine) at one end, with pyridyldithio, maleimide, or iodoacetate groups (reactive to thiols) at the other end. Such reactions could form complexes with the covalent linkage between the carrier and ligand; thus, purification is highly needed at each step of the modification [79].

4.2 Click chemistry

The concept of click chemistry has been brought into the pharmaceutical field by Kolb and team almost two decades ago [80]. The cycloaddition reactions in click chemistry are facilitated by the presence of Cu(I) catalyst. A stable triazole linkage is formed by the reaction of azide and alkyne groups. This triazole linkage facilitates aqueous solubility and even biocompatible. These highly specific reactions are producing high yields in aqueous media under a mild condition of reaction. Additionally, the specificity of the reaction prevents any other reactions (unfavorable) to occur, thereby finally, the purification step would be simpler and the conjugation will produce the desired compounds. These highly oriented linkages formed during click reaction make it suitable for conjugating with targeted ligands onto the surface of the nanoparticles. However, the use of Cu(I) in the reaction as a catalyst to facilitate the reaction could be responsible for severe toxicities to users, including kidney disease, liver disorder (hepatitis), and neurological disorder (Alzheimer's disease) [81].

Another approach of click reaction has been proposed by Devaraj and group, where the researchers did not use Cu(I) catalyst. They tried a reaction between 1,2,4,5-tetrazine and a trans-cyclooctene to establish a covalent biorthogonal reaction [82]. Avoiding Cu(I) in the reaction helps to minimize associated toxicities; thus, it is gaining attracted researchers' attention for application in molecular imaging at the cellular level.

4.3 Hybridization techniques

A novel approach of conjugating ligand to the nanoparticle surface had been proposed by Javier and team a decade ago [83]. In this method, the researchers extended their aptamer to offer a hybridization site, which could be complementary for the nanoparticles coated with oligonucleotide. The authors used gold nanoparticles, which were coated with oligonucleotide, a spacer (hexa(ethylene glycol)), and a sequence complementary to the aptamer extension. The hybridized aptamer was reported to capture the oligonucleotide-coated gold nanoparticles by maintaining the reaction condition. The resulted conjugate of aptamer and nanoparticle was further reported to detect the specific target of cells as evidenced by the reported reflectance imaging [83].

There are number of advantages reported for this hybridization method in conjugating ligand moiety to the nanoparticle surface:

(a)   The formulation is experiencing stability through the electrostatic repulsive force between the particles due to the presence of negatively charged phosphate groups of oligonucleotide.

(b)   This charge is stable even in the increased salt environment.

(c)   The conjugation of aptamers to the complementary hybridized sequence on the nanoparticle surface is formed easily.

(d)   The complementary sequence on a particular oligonucleotide surface conserved the integrity and stability of the bound aptamers during and after conjugation.

(e)   The numbers of aptamers needed for conjugation are small.

(f)   This method can be employed for conjugating different components on the nanoparticle surface, such as imaging, delivery, targeting, or therapeutic agents.

5. Surface engineered nanoparticles in targeted therapy

The application of surface-engineered nanoparticles has been extended in the treatment of various complex diseases. This section of the chapter will emphasize on application of surface-engineered nanoparticles in the treatment of cancer, autoimmune diseases (rheumatoid arthritis, AIDS), cardiovascular diseases, neurodegenerative diseases, ocular diseases, pulmonary diseases, and regenerative therapy.

5.1 Application of surface-engineered nanoparticles in cancer

One of the foremost causes of death among the worldwide human population is cancer. Conventional chemotherapies of cancer treatment limit its efficacy due to the associated side effects or due to the development of resistance. Progress of nanotechnology-based researches trying to overcome the issues of conventional therapies, however, to specifically target the site of disease is established to be more advantageous [2, 84, 85]. Thus, there are number of research studies ongoing in delivering chemotherapeutics, particularly to the cancer microenvironment to minimize systemic exposure and side effects.

A recent report by Martinelli and team summarized that antiepidermal growth factor (EGRF) monoclonal antibodies with anticancer drugs are more effective for the treatment of wild-type metastatic colorectal cancer [42]. Similarly, Roncato et al. fabricated avidin-nucleic-acid nanoassemblies (ANANAS) decorated with EGFR for the treatment of cancer. They compared the effectivity of EGFR antibody(cetuximab)-targeted ANANAS over antibody-drug conjugates (ADC). In the current scenario, ADC is frequently used as personalized therapy. In vitro results revealed that antibody-targeted ANANAS have significantly more internalization in MDA-MB-231 cells as well as more cytotoxic in vitro. Additionally, tumor–bearing mice treated with antibody(cetuximab)–targeted ANANAS showed better therapeutic efficacy compared to ADC treated group [86]. On the other hand, one study highlighted the use of cetuximab (CTX) decorated PLGA nanoparticles in CTX-resistant cancers. These nanoparticles were loaded with camptothecin (CPT) and the effectivity of nanoparticles was measured in KRAS mutant CTX-resistant cancer cells. CTX-targeted nanoparticles showed effective targeting with high cytotoxicity to the cancer cells in vitro as well as significant retardation in the growth of the PANC-1 tumor in vivo. This indicated that antibody repositioning can be relevant in the case of other antibodies restricted by resistance [41, 42].

Likewise, Tsai et al. developed chitosan nanoparticles conjugated with EGFR (CENP) for curcumin delivery to treat cancer. Cellular uptake study in MKN45 revealed the more than two folds enhancement of curcumin internalization was associated with CENP compared to nonconjugated chitosan nanoparticles. This may attribute to the positive charge of CENP which improved the interaction of nanoparticles with cell membrane contains negative charge. Moreover, this ionic interaction enhanced the recognition of EGF on CENP and lead to receptor-mediated endocytosis internalization of nanoparticles [40–42].

In this context, Tf conjugated PLGA nanoparticles for doxorubicin DOX delivery were fabricated to treat brain cancer. Cytotoxicity and cellular uptake data revealed the highest cell toxicity and uptake of Tf conjugated nanoparticles compared to nonconjugated nanoparticles by cancer cells. These results were further confirmed by in vivo experiments, showed the strongest antitumor activity and tumor growth inhibition by Tf-conjugated nanoparticles in mice. This finding suggests the Tf as an important ligand for surface engineering of nanoparticles [87]. Similarly, pH–sensitive PLGA nanoparticles were fabricated for enhancement of DOX accumulation inside the cancerous cells. pH sensitivity was imposed by the use of 77KS (surfactant) and active targeting was achieved by incorporation of Tf conjugation. The in vitro results showed pH–sensitive release of DOX in an acidic environment, which was further confirmed by hemolysis assay. Finally, cell toxicity studies were performed on HaCaT and HeLa, nontumor, and tumor cell lines. Tf conjugated nanoparticles showed notable protection of HaCaT cells and a significant reduction in HeLa cell growth [88]. Jain et al. adopted a similar concept, they prepared pH-sensitive PLGA nanoparticles coated with Tween 80 and conjugated with Tf on the surface for methotrexate (MTX) delivery tool. Tf-conjugated nanoparticles shown to have higher cytotoxicity on C6 glioma cells as well as enhance cellular uptake to brain cancer cells via endocytosis due to change in plasma membrane fluidity, facilitated by Tween 80 [39, 41]. These all studies revealed the Tf as a promising ligand for surface engineering to overcome the side effects of conventional delivery of cytotoxic drugs.

Choi et al. evaluated tumor targetability of hyaluronic acid nanoparticle coated with polyethylene glycol (PEG) (CPT-HA-NPs) for CPT delivery in cancer cells. The internalization of nanoparticles into SCC7 and MDA-MB-231 cells was reported via receptor-mediated endocytosis. Rapid release of CPT was observed when the enzyme Hyal-1 was used as release media. Moreover, dose-dependent cytotoxicity was also notified on different cancer cells (HCT 116, SCC7, and MDA-MB-231 cells) in contrast to normal cells. In vivo data of the study indicated selective accumulation and prolong circulation of the drug in the blood. It is also showed higher antitumor activity in mice [89]. In another study, combination chemotherapy was proposed for the treatment of colon cancer. Polymeric nanoparticles functionalized with hyaluronic acid were fabricated for the delivery of curcumin/CPT. In vitro release data revealed the sustained release profile and surface engineering with HA endowed nanoparticles with increased cellular uptake and cancer-targeting capability [90].

There are enormous examples available in the literature, a few have been incorporated in Table 2. These reports are suggesting the superiority of surface-engineered nanoparticles in the treatment of cancer with increased efficacy and decreased systemic exposure and toxicity.

5.2 Application of surface-engineered nanoparticles in autoimmune diseases

Interaction of nanocarriers to the immune system components is one of the interesting areas of treating the immune system. It had been established that the nanomaterials adsorb proteins on its surface from the circulatory blood when it is administered to form "proteins corona," which are then recognized by various immune cells [97]. Such components can be modified through surface engineering, through the incorporation of antigen into the immune system. Engineering to the surface of the nanoparticles can lead to enhance or inhibit the immune responses. Nowadays, targeted nanoparticles are widely in the investigation to target treatment or prevention of noninfectious and infectious diseases. Immunostimulatory molecules can be improved by localized nanoimmunotherapy, thereby reduce systemic exposure and toxicity [98, 99].

Researches since the last two decades have brought several nanotechnology-based types of research; however, efforts toward targeting immune-based diseases are limited. The literature says that these nanocarriers have been used to target rheumatoid arthritis and acquired immune deficiency syndrome (AIDS).

Although the exact mechanism of rheumatoid arthritis is not known, it has been said that the damage of bone and cartilage results due to the complex interaction between immune mediators [100]. To target the chemophotothermal treatment in rheumatoid arthritis, a group of researchers formulated MTX-loaded PLGA nanoparticles, which were then deposited on the nanometric gold sheet and conjugated with RGD peptide [101]. In the next study, the same group replaced the Au film with a complex film of Au/Fe/Au [102]. The accumulation of gold was measured in a collagen-induced arthritis animal model, where the Au was found to be concentrated within the inflamed joints of

Table 2 Use of surface engineered technology in the delivery of nanoparticles to the cancerous cells.

Nanocarrier	Ligand	Targeting receptor/cells	Research outcome	Source
Nanoparticle	Galactose	HeLa cells	• Tetraphenylethene-chitosan conjugate was tagged with galactose. • Size was recorded between 50 and 200 nm. • Fluorescence at the cancerous site remains intact.	[91]
PLGA nanoparticles	Trastuzumab	Human epidermal growth factor receptor 2 (HER2) positive	• Yield and entrapment efficiency was >90%. • Size of unconjugated nanoparticles was 74.66 ± 9.29 nm, whereas the conjugated had 141 ± 58.41 nm. • More cytotoxicity was reported with conjugated nanoparticles. • Drug uptake by the HER2-positive cells was higher.	[92]
Solid lipid nanoparticles	Cyclic Arg-Gly-Asp (cRGD)	$\alpha_v\beta_3$ Integrin receptors	• Active targeted nanoparticles were reported to accumulate in the cancer cells, whereas the nontargeted nanoparticles were accumulated significantly higher in the liver, kidney, and spleen. • The EPR and active targeting might be advantageous in treating cancers.	[93]
Stealth nanoparticle	Transferrin	Transferrin receptor on cancer cells	• Poly(ethylene) glycol-hydroxycamptothecin conjugate was conjugated with transferrin with 110 nm diameter.	[94]

Continued

Table 2 Use of surface engineered technology in the delivery of nanoparticles to the cancerous cells—cont'd

Nanocarrier	Ligand	Targeting receptor/cells	Research outcome	Source
Protein-inorganic nanoparticles	cRGD	cRGD overexpressed tumor cells	• Conjugate was reported to have high solubility and stability. • It produces cytotoxicity similar to free drugs. • Conjugate showed longer retention (8.94-fold) within the circulation and tumor accumulation (9.03-fold) when compared to unconjugated counterpart. • Theranostic approach of delivery was made by using disulfide-bond rich proteins with the quantum dots and PTX. • This hybrid nanoparticle was reported to possess a higher accumulation to the tumor site. • The cytotoxicity was enhanced due to higher accumulation. • This photo-induced assembly could be used for versatile applications in the biomedical field.	[95]
Multifunctional nanoparticles	EpCAM aptamer	ZR751 (human breast cancer) cell line	• Theranostic approach was made with the EpCAM aptamer conjugated n-loaded PLGA nanoparticles. • Superior therapeutic activity was reported	[96]

Table 2 Use of surface engineered technology in the delivery of nanoparticles to the cancerous cells—cont'd

Nanocarrier	Ligand	Targeting receptor/cells	Research outcome	Source
			by different in vitro cell line assays. • Superlative bioimaging modality was reported in different imaging models (tumor spheroid model and 2D monolayer culture)	

the animals. This increased concentration of the targeted carrier was explained by the binding of RGD to the $\alpha_v\beta_3$ receptors at the inflamed joints [102]. Furthermore, to target overexpressed scavenger receptors at the macrophage surface, a team had formulated the carrier with dextran sulfate ligand. The formulated nanoparticle with the dextran sulfate ligand was reported to accumulate to the synovial fluid of the inflamed joints in collagen-induced arthritis animals [103].

On the other hand, several nanotechnology-based studies had been performed toward effective control of AIDS. In this context, a poly(propylene imine) dendritic nanoparticle was formulated to load efavirenz. Later, the surface of the nanocarrier was decorated with Tuftsin. Finally, the surface-engineered nanocarrier was found to be recognized by the phagocytic cells due to the presence of Tuftsin, thereby resulted in enhanced uptake of the drug-loaded nanocarriers by the macrophages when compared with uninfected cells [104].

5.3 Application of surface-engineered nanoparticles in neurodegenerative diseases

Progressive loss of neuronal functions subsequently leads to the death of the neurons in neurodegenerative disease conditions. Due to the loss of neurons, patients may be diagnosed with multiple sclerosis, Parkinson's disease, and Alzheimer's disease, which might result in loss of memory, problems in movement, demented condition, and finally loss in quality of life [105–107]. Treatment of these neurological disorders is a huge challenge due to the presence of a rigid biological barrier, the BBB. This highly selective semipermeable membrane of BBB limits the transportation of most of the foreign components to the central unit of our body to maintain the body's homeostasis [108]. Therefore, a small fraction of the drug can permeate the brain; however, the increased dose can lead to increased brain concentration. Nevertheless, the associated side effects will be pronounced. Thus, targeted therapies of nanoparticles have been seen to cross the BBB effectively without exposing the systemic circulation much with the therapeutic agent [109, 110].

A peptide (Mimotopes) functionalized nanoparticle was developed by Führmann and the team to target multiple sclerosis. The polymeric nanoparticles functionalized with peptide were found to be accumulated at the site of injury [111]. Similarly, the experiments were also performed in Parkinson's disease for improved response. A glial cell-derived neurotrophic factor-loaded PEGylated nanoparticles had been shown to prevent the loss of dopaminergic neurons, thereby enhanced the level of dopamine level [112, 113]. Another research by Huang and team developed angiopep-conjugated dendrigraft poly-L-lysine for delivering genes in Parkinson's disease [114]. Here, angiopep was serving as a targeting ligand, which is specific to low-density lipoprotein receptor-related protein. As a result, these angiopep-conjugated nanoparticles reported increased cellular uptake and also exerted increased gene expression in the brain cells. Finally, the locomotor activity of the experimental animals was reported to be improved following 5 injections of the formulated nanoparticles [114].

Similarly, progress has also been made in the treatment of Alzheimer's disease. In this context, a PLGA nanoparticle loaded with Huperzine A was developed by Meng and team, where the authors modified the surface with lactoferrin-conjugated N-trimethylated chitosan. This novel approach had shown to provide a controlled release of entrapped therapeutic agent, enhanced bioadhesion, and superior targetability to the brain when applied through the intranasal routes of administration [115].

5.4 Application of surface-engineered nanoparticles in ocular diseases

Recent advancement of nanotechnology has also been progressed to deliver therapeutics to the eyes. These nanocarriers could cross several barriers of eyes (blood-retinal layer, corneal epithelium, and the uppermost mucoaqueous tear layer) to deliver the therapeutics effectively to the site of action [113, 116]. The targeted nanocarriers are also advantageous in producing less irritation to the eyes when applied, lowering the loss of the therapeutics and better bioavailability [117].

A micellar formulation of dexamethasone was developed for topical ocular delivery using chitosan oligosaccharide-valylvaline-stearic acid, where the researchers functionalized the surface of the nanocarrier with peptide transporter-1. Subsequent analysis revealed the penetration of the functionalized carrier to the posterior segment of the eye through the conjunctival route. Further studies revealed that the formulation can release the entrapped drug for a prolonged period, suggesting it as a promising tool for future development [118]. In another report, cell-penetrating peptides were used to surface decorate developed PEG-PLGA nanoparticles loaded with fluorometholone. These surface-engineered nanoparticles were reported to possess suitable characteristics for ocular administration without any cytotoxic manifestation. Additionally, this nanoparticular delivery was internalized into the mouse eye and HCE-2 cells and also showed anti-inflammatory activity [119].

5.5 Application of surface-engineered nanoparticles in pulmonary disease

There are several pulmonary disorders, such as pulmonary tuberculosis, chronic obstructive pulmonary disease, asthma, pulmonary fibrosis, etc., had gained the attention of the formulation scientists to formulate site-specific delivery of therapeutics [120]. These diseases sometimes become fatal because of not gaining attention or proper treatment. There are no effective therapies available currently to restore lung function; however, conventional therapies are used to deliver the therapeutics systematically or via a local application using inhalers [121].

In a report by Maretti et al., the researchers had developed solid–lipid nanoparticles to deliver rifampicin. This mannosylated solid–lipid nanoparticle was intended to be administered through an inhalational route for patients with tuberculosis. This mannose residue on the surface of the drug-loaded solid–lipid nanoparticles was hypothesized to be recognized by the infected alveolar macrophages. Such recognition had resulted in quick recognition and internalization of the carrier through phagocytosis. This evidence of surface engineering on nanoparticles provides superiority in treatment options in pulmonary diseases [122].

5.6 Application of surface-engineered nanoparticles in tissue engineering

This field of research is emerging, and researchers are trying their best to get some breakthroughs in this platform. Several nanocarriers are at different stages of the experiment for the generation of tissues through the creation of an externally applied scaffold with all the necessary ingredients for damaged tissue regrowth [123, 124]. This technology had been successfully progressed in bone-grafting in the generation of bone and dental tissue regeneration. The example of EquivaBone® is appropriate in this case. This is a nanotherapeutics consisting of demineralized bone matrix, carboxymethyl cellulose, and hydroxyapatite. Food and Drug Administration (FDA) approved this product in 2009 for the treatment of osteoinductive bone graft substitute [113]. This tissue engineering technology, with the help of nanocarriers, had been progressed for the treatment of ischemic stroke. In this context, the research output of Tian et al. would be applicable. The authors used exosomes for the brain delivery due to associated advantages of crossing ability to the BBB, high drug-delivery efficiency, innate stability, and low immunogenicity. However, to target the BBB, the researchers conjugated c(RGDyK) peptide to target brain lesions of the transient middle cerebral artery occlusion animals. Authors loaded curcumin within the surface-modified nanocarrier to exert its anti-inflammatory efficacy, and their satisfactory results proved the superiority of this formulation strategy [125, 126].

Overall, it can be said that the surface engineering of the nanoparticles helps the nano-carriers to deliver specifically to the target site to achieve more pronounced activity with minimal exposure to the systemic circulation, thereby minimizing the associated toxicities.

6. Surface engineered nanoparticles in diagnosis

Early detection of disease is a milestone of successful treatment. In the case of cancer, magnetic resonance imaging, x-ray, endoscopy, computed tomography (CT), etc., used to detect cancer upon visual changes in tissues; however, at this time point of detection, cancer cells start proliferation. Moreover, current diagnostic tools are unable to distinguish malignant to benign tumors. Hence, early detection is hindered by the use of old detection methods [127]. Needless to say, there is the crucial requirement of the emergence of new diagnostic methods to detect cancer at an early stage. In this regard, significant research has been carried out to identify various biomarkers, which have a close association with tumorigenesis and progression. Furthermore, the development of new technology is on the path to identify and target these biomarkers effectively for early detection [128]. From the 1990s, a plethora of nanoparticle research was pooled in the field of therapeutics and diagnosis. Regardless of their potential application as a diagnostic tool in several types of research, only iron oxide nanoparticles reached up to clinical trials. This may attribute to the reproducibility of the manufacturing process, variable pharmacokinetic properties, toxicity, and biodegradability. However, taking advantage of the unique features of nanoparticles coupled with a specific ligand can improve its sensitivity and efficacy as a diagnostic tool [129, 130]. Generation of smart nanoprobe with targeting ligand can sense the changes at a molecular level. So, it was identified that the microRNA biomarker is a promising target for tumor diagnosis in prostate cancer. Jou et al. developed a two-step sensing platform for the specific detection of miR-141. A semiconductor of CdSe/ZnS quantum dots (QDs) upgraded with FRET quencher-functionalized nucleic acid was used in the first step. The first step platform is nonsensitive to miR-141. However, the second step includes telomerase, which specifically provides a luminescence signal for miR-141. This platform was evaluated on the serum samples collected from prostate cancer patients as well as healthy volunteers. Impressively, the tool was able to discriminate the serum samples among cancer patients and healthy volunteers. The mechanism behind this activity is the cleavage of the covalent bond of nucleic acid-functionalized CdSe/ZnS QDs and FRET quencher due to hybridization of miR-141 with the above-mentioned unit. This cleavage activated the fluorescence of the QDs for detection [131].

On the other hand, the potential of magnetic nanoparticles conjugated with antibodies for magnetic resonance imaging capability was evaluated by Lee et al. The researchers did surface engineering of the nanoparticles with Herceptin and tested its capability on ovarian and breast tumor model. Results indicated higher magnetic resonance imaging sensitivity for cancer detection compared to the current diagnostic tool [132].

7. Surface engineered nanoparticles in theranostic approaches

In 1946, Seidlin and team treated metastasized thyroid adenocarcinoma patients with radioactive iodine and first introduced "Theranostic" concept in biomedical application [133]. Followed by, the first phase 1 clinical trial with 90Y-CYT-356 monoclonal antibody was projected to treat the hormone-refractory metastatic prostate cancer patients [134]. Although the approach was not successful, might be due to issues related to cellular entry to cancer cells of a larger molecule and hematological toxicity. Subsequently, with the development of nanotechnology, various attempts have been made with small molecules to achieve simultaneous targeted diagnosis and therapy to treat different cancers. Integration of imaging and therapeutic agent in same nanocarrier and further surface engineering by attaching targeting ligand has the potential to achieve simultaneous targeted diagnosis and therapy against various cancers at their curable stages. Their nanometric size range lead to enhance permeation and retention at the cancer microenvironment through leaky vasculature in cancer cells and facilitate passive targeting. Furthermore, surface engineering with targeting ligand owing to the surface functionalization properties of nanocarrier facilitate active targeting the cancer cells [3]. Wang et al. reported the potential targeted theranostic approach of superparamagnetic iron oxide nanoparticle platform [135]. Surface functionalization with anti-EGFR monoclonal antibodies for targeting overexpressed EGFR receptors in lung cancers proved to have better targeting ability to human lung cancer (H460) with improved magnetic resonance imaging at the cancer site. Additionally, targeted nanoparticles act synergistically and significantly improved in vivo ultrasonic energy deposition in magnetic resonance-guided focused ultrasound surgery [135]. Further to say, this multifunctional platform could be a potential tool to deliver gene, stimuli-responsive drug release, particularly when the surface of the nanocarriers are functionalized with targeting ligand for active delivery. Thus, it could provide all-in-one in a single tool, to achieve a successful theranostic approach against cancer therapy.

Schleich and team compared the passive targeting, active targeting by means of RGD grafting to $\alpha v\beta 3$ integrin, magnetic targeting, and the combination of targeting to $\alpha v\beta 3$ integrin and magnetic targeting using PLGA nanoparticle loaded with a chemotherapeutic drug, paclitaxel and supramagnetic iron oxide [136]. The group reported that the combination approach of active targeting and magnetic targeting drastically improved the accumulation of nanoparticles in targeted cancer sites further enhanced imaging and anticancer activity compared to single targeting and passive targeting nanoparticles

[136]. Therefore, it can be inferred that the theranostic approach using double targeting can be further explored to treat cancer.

Xiao et al. reported enhanced tumor accumulation in both in vitro and in vivo glioblastoma models of the developed pH stimuli-responsive drug release micelles loaded with DOX and labeled with 64Cu (for in vivo PET imaging) and cRGD peptide (functionalized for active targeting). Active targeting of theranostic approach has been made by PET imaging and chemotherapy using loaded DOX [137]. Although it is a challenging task to develop a stable nanocarrier along with active and passive targeting and theranostic approach; however, various attempts are in progress for efficacious diagnosis and therapy against various cancers via these different surface engineered nanotechnology platforms [138].

8. Conclusion

Presently, the thrust area of formulation research is nanotechnology, where the nanocarriers can deliver the entrapped therapeutics particularly to the site of action via passive and active targeting. This active targeting of nanoparticles is usually achieved by the conjugation of specific surface components to the nanoparticles. These specific components are helping the nanocarriers to accumulate and release the therapeutics at the disease site overexpressed with the receptor. Varied research in this field is progressing; however, only a few of them have reached the bedside of the patients. Researchers are trying to address the challenges of the scaling-up process, minimizing the complexity of the conjugation process, trying to minimize associated side effects through reduction of possibilities of cytotoxicity to the normal cells and immunogenicity. Moreover, it is expected to approach an era soon when diagnostic, therapeutic, and theranostic approaches can be easily used to target the diseased area through active targeting using surface-engineered nanoparticles.

Acknowledgments

Dr. Gorain would like to acknowledge the School of Pharmacy, Taylor's University, Selangor, Malaysia, and Dr. Choudhury and Dr. Pandey would like to acknowledge the School of Pharmacy, International Medical University, Malaysia for providing resources and support in completing this work. Dr. Kesharwani acknowledges the financial support from the University Grants Commission, New Delhi, India, through Start-Up Research Grant and Indian Council of Medical Research, New Delhi, India, through Extramural Research Grant.

References

[1] M.F. Attia, N. Anton, J. Wallyn, Z. Omran, T.F. Vandamme, An overview of active and passive targeting strategies to improve the nanocarriers efficiency to tumour sites, J. Pharm. Pharmacol. 71 (2019) 1185–1198, https://doi.org/10.1111/jphp.13098.

[2] H. Choudhury, B. Gorain, M. Pandey, R.K. Khurana, P. Kesharwani, Strategizing biodegradable polymeric nanoparticles to cross the biological barriers for cancer targeting, Int. J. Pharm. 565 (2019) 509–522, https://doi.org/10.1016/J.IJPHARM.2019.05.042.

[3] B. Gorain, H. Choudhury, A.B. Nair, S.K. Dubey, P. Kesharwani, Theranostic application of nanoemulsions in chemotherapy, Drug Discov. Today (2020), https://doi.org/10.1016/j.drudis.2020.04.013.

[4] P.V. Thanikachalam, S. Ramamurthy, Z.W. Wong, B.J. Koo, J.-Y. Wong, M.F. Abdullah, Y.H. Chin, C.H. Chia, J.Y. Tan, W.T. Neo, B. Sen Tan, W.F. Khan, P. Kesharwani, Current attempts to implement microRNA-based diagnostics and therapy in cardiovascular and metabolic disease: a promising future, Drug Discov. Today 23 (2018) 460–480, https://doi.org/10.1016/J.DRUDIS.2017.10.020.

[5] B. Gorain, H. Choudhury, E. Biswas, A. Barik, P. Jaisankar, T.K.T.K. Pal, A novel approach for nanoemulsion components screening and nanoemulsion assay of olmesartan medoxomil through a developed and validated HPLC method, RSC Adv. 3 (2013) 10887–10893, https://doi.org/10.1039/c3ra41452c.

[6] S. Singh, D. Hassan, H.M. Aldawsari, N. Molugulu, R. Shukla, P. Kesharwani, Immune checkpoint inhibitors: a promising anticancer therapy, Drug Discov. Today 25 (2020) 223–229, https://doi.org/10.1016/j.drudis.2019.11.003.

[7] P. Kesharwani, S. Banerjee, U. Gupta, M.C.I. Mohd Amin, S. Padhye, F.H. Sarkar, A.K. Iyer, PAMAM dendrimers as promising nanocarriers for RNAi therapeutics, Mater. Today 18 (2015) 565–572, https://doi.org/10.1016/j.mattod.2015.06.003.

[8] D. Pandey, P. Kesharwani, D. Jain, Entrapment of drug-sorbate complex in submicron emulsion: a potential approach to improve antimicrobial activity in bacterial corneal infection, J. Drug Deliv. Sci. Technol. 49 (2019) 455–462, https://doi.org/10.1016/J.JDDST.2018.12.006.

[9] B. Gorain, H. Choudhury, M. Pandey, P. Kesharwani, Paclitaxel loaded vitamin E-TPGS nanoparticles for cancer therapy, Mater. Sci. Eng. C 91 (2018) 868–880, https://doi.org/10.1016/j.msec.2018.05.054.

[10] H. Choudhury, B. Gorain, M. Pandey, S.A.S.A. Kumbhar, R.K. Tekade, A.K.A.K. Iyer, P. Kesharwani, Recent advances in TPGS-based nanoparticles of docetaxel for improved chemotherapy, Int. J. Pharm. 529 (2017) 506–522, https://doi.org/10.1016/j.ijpharm.2017.07.018.

[11] P. Kesharwani, H. Choudhury, J.G. Meher, M. Pandey, B. Gorain, Dendrimer-entrapped gold nanoparticles as promising nanocarriers for anticancer therapeutics and imaging, Prog. Mater. Sci. 103 (2019) 484–508, https://doi.org/10.1016/J.PMATSCI.2019.03.003.

[12] P. Kumar Singh, B. Gorain, H. Choudhury, S. Kumar Singh, P. Whadwa, Shilpa, S. Sahu, M. Gulati, P. Kesharwani, Macrophage targeted amphotericin B nanodelivery systems against visceral leishmaniasis, Mater. Sci. Eng. B Solid-State Mater. Adv. Technol. 258 (2020) 114571, https://doi.org/10.1016/j.mseb.2020.114571.

[13] H. Choudhury, B. Gorain, S. Karmakar, E. Biswas, G. Dey, R. Barik, M. Mandal, T.K. Pal, Improvement of cellular uptake, in vitro antitumor activity and sustained release profile with increased bioavailability from a nanoemulsion platform, Int. J. Pharm. 460 (2014) 131–143, https://doi.org/10.1016/j.ijpharm.2013.10.055.

[14] B. Gorain, H. Choudhury, A. Kundu, L. Sarkar, S. Karmakar, P. Jaisankar, T.K. Pal, Nanoemulsion strategy for olmesartan medoxomil improves oral absorption and extended antihypertensive activity in hypertensive rats, Colloids Surf. B Biointerfaces 115 (2014) 286–294, https://doi.org/10.1016/j.colsurfb.2013.12.016.

[15] R.A. Bapat, S. Dharmadhikari, T.V. Chaubal, M.C.I.M. Amin, P. Bapat, B. Gorain, H. Choudhury, C. Vincent, P. Kesharwani, The potential of dendrimer in delivery of therapeutics for dentistry, Heliyon 5 (2019), https://doi.org/10.1016/j.heliyon.2019.e02544.

[16] H. Choudhury, M. Pandey, Y.Q. Lim, C.Y. Low, C.T. Lee, T.C.L. Marilyn, H.S. Loh, Y.P. Lim, C.F. Lee, S.K. Bhattamishra, P. Kesharwani, B. Gorain, Silver nanoparticles: advanced and promising technology in diabetic wound therapy, Mater. Sci. Eng. C 112 (2020) 110925, https://doi.org/10.1016/j.msec.2020.110925.

[17] P. Kesharwani, K. Jain, N.K. Jain, Dendrimer as nanocarrier for drug delivery, Prog. Polym. Sci. 39 (2014) 268–307.

[18] S. Bozdağ Pehlivan, Nanotechnology-based drug delivery systems for targeting, imaging and diagnosis of neurodegenerative diseases, Pharm. Res. 30 (2013) 2499–2511, https://doi.org/10.1007/s11095-013-1156-7.

[19] M.J. Ernsting, M. Murakami, A. Roy, S.D. Li, Factors controlling the pharmacokinetics, biodistribution and intratumoral penetration of nanoparticles, J. Control. Release 172 (2013) 782–794, https://doi.org/10.1016/j.jconrel.2013.09.013.

[20] W. Wu, L. Luo, Y. Wang, Q. Wu, H. Bin Dai, J.S. Li, C. Durkan, N. Wang, G.X. Wang, Endogenous pH-responsive nanoparticles with programmable size changes for targeted tumor therapy and imaging applications, Theranostics 8 (2018) 3038–3058, https://doi.org/10.7150/thno.23459.

[21] R.A. Bapat, T.V. Chaubal, S. Dharmadhikari, A.M. Abdulla, P. Bapat, A. Alexander, S.K. Dubey, P. Kesharwani, Recent advances of gold nanoparticles as biomaterial in dentistry, Int. J. Pharm. 586 (2020) 119596, https://doi.org/10.1016/j.ijpharm.2020.119596.

[22] H. Choudhury, M. Pandey, T.H. Yin, T. Kaur, G.W. Jia, S.Q.L. Tan, H. Weijie, E.K.S. Yang, C.G. Keat, S.K. Bhattamisra, P. Kesharwani, S. Md, N. Molugulu, M.R. Pichika, B. Gorain, Rising horizon in circumventing multidrug resistance in chemotherapy with nanotechnology, Mater. Sci. Eng. C 101 (2019) 596–613, https://doi.org/10.1016/j.msec.2019.04.005.

[23] U. Prabhakar, H. Maeda, R.K. Jain, E.M. Sevick-Muraca, W. Zamboni, O.C. Farokhzad, S.T. Barry, A. Gabizon, P. Grodzinski, D.C. Blakey, Challenges and key considerations of the enhanced permeability and retention effect for nanomedicine drug delivery in oncology, Cancer Res. 73 (2013) 2412–2417.

[24] P. Kesharwani, R. Ghanghoria, N.K. Jain, Carbon nanotube exploration in cancer cell lines, Drug Discov. Today 17 (2012) 1023–1030.

[25] P. Kesharwani, R.K. Tekade, N.K. Jain, Generation dependent safety and efficacy of folic acid conjugated dendrimer based anticancer drug formulations, Pharm. Res. 32 (2015) 1438–1450, https://doi.org/10.1007/s11095-014-1549-2.

[26] P. Ma, R.J. Mumper, Paclitaxel nano-delivery systems: a comprehensive review, J. Nanomed. Nanotechnol. 4 (2013) 1000164, https://doi.org/10.4172/2157-7439.1000164.

[27] S. Singh, A. Numan, N. Agrawal, M.M. Tambuwala, V. Singh, P. Kesharwani, Role of immune checkpoint inhibitors in the revolutionization of advanced melanoma care, Int. Immunopharmacol. 83 (2020), https://doi.org/10.1016/j.intimp.2020.106417.

[28] S. Singh, M.M. Alrobaian, N. Molugulu, N. Agrawal, A. Numan, P. Kesharwani, Pyramid-shaped PEG-PCL-PEG polymeric-based model systems for site-specific drug delivery of vancomycin with enhance antibacterial efficacy, ACS Omega (2020), https://doi.org/10.1021/acsomega.9b04064.

[29] B. Gorain, S.K. Bhattamisra, H. Choudhury, U. Nandi, M. Pandey, Overexpressed receptors and proteins in lung Cancer, nanotechnology-based target, Drug Deliv. Syst. Lung Cancer. (2019) 39–75, https://doi.org/10.1016/B978-0-12-815720-6.00003-4.

[30] M.E. Davis, Z. Chen, D.M. Shin, Nanoparticle therapeutics: An emerging treatment modality for cancer, Nat. Rev. Drug Discov. 7 (2008) 771–782, https://doi.org/10.1038/nrd2614.

[31] L. Devi, R. Gupta, S.K. Jain, S. Singh, P. Kesharwani, Synthesis, characterization and in vitro assessment of colloidal gold nanoparticles of Gemcitabine with natural polysaccharides for treatment of breast cancer, J. Drug Deliv. Sci. Technol. 56 (2020), https://doi.org/10.1016/j.jddst.2020.101565.

[32] H. Choudhury, R. Maheshwari, M. Pandey, M. Tekade, B. Gorain, R.K. Tekade, Advanced nanoscale carrier-based approaches to overcome biopharmaceutical issues associated with anticancer drug 'Etoposide,', Mater. Sci. Eng. C 106 (2020) 110275, https://doi.org/10.1016/J.MSEC.2019.110275.

[33] P.K. Tripathi, S. Gupta, S. Rai, A. Shrivatava, S. Tripathi, S. Singh, A.J. Khopade, P. Kesharwani, Curcumin loaded poly (amidoamine) dendrimer-plamitic acid core-shell nanoparticles as anti-stress therapeutics, Drug Dev. Ind. Pharm. (2020) 1–46, https://doi.org/10.1080/03639045.2020.1724132.

[34] S. Nallamolu, V.R. Jayanti, M. Chitneni, L.Y. Khoon, P. Kesharwani, Self-micro emulsifying drug delivery system "SMEDDS" for efficient Oral delivery of Andrographolide, Drug Deliv. Lett. 10 (2019) 38–53, https://doi.org/10.2174/2210303109666190723145209.

[35] J. Yoo, C. Park, G. Yi, D. Lee, H. Koo, Active targeting strategies using biological ligands for nanoparticle drug delivery systems, Cancers (Basel) 11 (2019) 640, https://doi.org/10.3390/cancers11050640.

[36] P. Tagde, G. Kulkarni, D.K. Mishra, P. Kesharwani, Recent advances in folic acid engineered nano-carriers for treatment of breast cancer, J. Drug Deliv. Sci. Technol. 56 (2020) 101613, https://doi.org/10.1016/j.jddst.2020.101613.

[37] K. Kobayashi, J. Wei, R. Iida, K. Ijiro, K. Niikura, Surface engineering of nanoparticles for therapeutic applications, Polym. J. 46 (2014) 460–468, https://doi.org/10.1038/pj.2014.40.

[38] L.D. Leserman, J. Barbet, F. Kourilsky, J.N. Weinstein, Targeting to cells of fluorescent liposomes covalently coupled with monoclonal antibody or protein a, Nature 288 (1980) 602–604, https://doi.org/10.1038/288602a0.

[39] A. Jain, A. Jain, N.K. Garg, R.K. Tyagi, B. Singh, O.P. Katare, T.J. Webster, V. Soni, Surface engineered polymeric nanocarriers mediate the delivery of transferrin–methotrexate conjugates for an improved understanding of brain cancer, Acta Biomater. 24 (2015) 140–151, https://doi.org/10.1016/j.actbio.2015.06.027.

[40] W.H. Tsai, K.H. Yu, Y.C. Huang, C.I. Lee, EGFR-targeted photodynamic therapy by curcumin-encapsulated chitosan/TPP nanoparticles, Int. J. Nanomedicine 13 (2018) 903–916, https://doi.org/10.2147/IJN.S148305.

[41] W.J. McDaid, M.K. Greene, M.C. Johnston, E. Pollheimer, P. Smyth, K. McLaughlin, S. Van Schaeybroeck, R.M. Straubinger, D.B. Longley, C.J. Scott, Repurposing of Cetuximab in antibody-directed chemotherapy-loaded nanoparticles in EGFR therapy-resistant pancreatic tumours, Nanoscale 11 (2019) 20261–20273, https://doi.org/10.1039/c9nr07257h.

[42] E. Martinelli, D. Ciardiello, G. Martini, T. Troiani, C. Cardone, P.P. Vitiello, N. Normanno, A.M. Rachiglio, E. Maiello, T. Latiano, F. De Vita, F. Ciardiello, Implementing anti-epidermal growth factor receptor (EGFR) therapy in metastatic colorectal cancer: challenges and future perspectives, Ann. Oncol. 31 (2020) 30–40, https://doi.org/10.1016/j.annonc.2019.10.007.

[43] A.A. Yaqoob, H. Ahmad, T. Parveen, A. Ahmad, M. Oves, I.M.I. Ismail, H.A. Qari, K. Umar, M.N. Mohamad Ibrahim, Recent advances in metal decorated nanomaterials and their various biological applications: a review, Front. Chem. 8 (2020) 345, https://doi.org/10.3389/fchem.2020.00341.

[44] S. Ussar, S.G. Vienberg, C.R. Kahn, Metabolic disease: receptor antibodies as novel therapeutics for diabetes, Sci. Transl. Med. 3 (2011) 113–115, https://doi.org/10.1126/scitranslmed.3003447.

[45] H.M. Shepard, G.L. Phillips, C.D. Thanos, M. Feldmann, Developments in therapy with monoclonal antibodies and related proteins, Clin. Med. J. R. Coll. Physicians London 17 (2017) 220–232, https://doi.org/10.7861/clinmedicine.17-3-220.

[46] A.V. Nascimento, A. Singh, H. Bousbaa, D. Ferreira, B. Sarmento, M.M. Amiji, Mad2 checkpoint gene silencing using epidermal growth factor receptor-targeted chitosan nanoparticles in non-small cell lung cancer model, Mol. Pharm. 11 (2014) 3515–3527, https://doi.org/10.1021/mp5002894.

[47] F. Zeeshan, M. Tabbassum, P. Kesharwani, Investigation on secondary structure alterations of protein drugs as an Indicator of their biological activity upon thermal exposure, Protein J. 38 (2019) 551–564, https://doi.org/10.1007/s10930-019-09837-4.

[48] A.V. Nascimento, F. Gattacceca, A. Singh, H. Bousbaa, D. Ferreira, B. Sarmento, M.M. Amiji, Bio-distribution and pharmacokinetics of Mad2 siRNA-loaded EGFR-targeted chitosan nanoparticles in cisplatin sensitive and resistant lung cancer models, Nanomedicine 11 (2016) 767–781, https://doi.org/10.2217/nnm.16.14.

[49] G. Sharma, S. Lakkadwala, A. Modgil, J. Singh, The role of cell-penetrating peptide and transferrin on enhanced delivery of drug to brain, Int. J. Mol. Sci. 17 (2016) 806, https://doi.org/10.3390/ijms17060806.

[50] F. Danhier, O. Feron, V. Préat, To exploit the tumor microenvironment: passive and active tumor targeting of nanocarriers for anti-cancer drug delivery, J. Control. Release 148 (2010) 135–146, https://doi.org/10.1016/j.jconrel.2010.08.027.

[51] M.K. Riaz, M.A. Riaz, X. Zhang, C. Lin, K.H. Wong, X. Chen, G. Zhang, A. Lu, Z. Yang, Surface functionalization and targeting strategies of liposomes in solid tumor therapy: A review, Int. J. Mol. Sci. 19 (2018), https://doi.org/10.3390/ijms19010195.

[52] L. Feng, R.J. Mumper, A critical review of lipid-based nanoparticles for taxane delivery, Cancer Lett. 334 (2013) 157–175, https://doi.org/10.1016/j.canlet.2012.07.006.

[53] C. Lin, B.C.K. Wong, H. Chen, Z. Bian, G. Zhang, X. Zhang, M. Kashif Riaz, D. Tyagi, G. Lin, Y. Zhang, J. Wang, A. Lu, Z. Yang, Pulmonary delivery of triptolide-loaded liposomes decorated with anti-carbonic anhydrase IX antibody for lung cancer therapy, Sci. Rep. 7 (2017), https://doi.org/10.1038/s41598-017-00957-4.

[54] S. Misra, V.C. Hascall, R.R. Markwald, S. Ghatak, Interactions between hyaluronan and its receptors (CD44, RHAMM) regulate the activities of inflammation and cancer, Front. Immunol. 6 (2015) 205, https://doi.org/10.3389/fimmu.2015.00201.

[55] R. Kumar, M. Singh, J. Meena, P. Singhvi, D. Thiyagarajan, A. Saneja, A.K. Panda, Hyaluronic acid - dihydroartemisinin conjugate: synthesis, characterization and in vitro evaluation in lung cancer cells, Int. J. Biol. Macromol. 133 (2019) 495–502, https://doi.org/10.1016/j.ijbiomac.2019.04.124.

[56] Z. Jiang, J. Guan, J. Qian, C. Zhan, Peptide ligand-mediated targeted drug delivery of nanomedicines, Biomater. Sci. 7 (2019) 461–471, https://doi.org/10.1039/c8bm01340c.

[57] H. Hossein-Nejad-Ariani, E. Althagafi, K. Kaur, Small peptide ligands for targeting EGFR in triple negative breast cancer cells, Sci. Rep. 9 (2019) 1–10, https://doi.org/10.1038/s41598-019-38574-y.

[58] L. Chi, M.H. Na, H.K. Jung, S.M.P. Vadevoo, C.W. Kim, G. Padmanaban, T.I. Park, J.Y. Park, I. Hwang, K.U. Park, F. Liang, M. Lu, J. Park, I.S. Kim, B.H. Lee, Enhanced delivery of liposomes to lung tumor through targeting interleukin-4 receptor on both tumor cells and tumor endothelial cells, J. Control. Release 209 (2015) 327–336, https://doi.org/10.1016/j.jconrel.2015.05.260.

[59] Z. Lu, Y. Long, X. Cun, X. Wang, J. Li, L. Mei, Y. Yang, M. Li, Z. Zhang, Q. He, A size-shrinkable nanoparticle-based combined anti-tumor and anti-inflammatory strategy for enhanced cancer therapy, Nanoscale 10 (2018) 9957–9970, https://doi.org/10.1039/c8nr01184b.

[60] D. Ji, Q. Wang, Q. Zhao, Q. Zhao, H. Tong, M. Yu, M. Wang, T. Lu, C. Jiang, C. Jiang, Co-delivery of miR-29b and germacrone based on cyclic RGD-modified nanoparticles for liver fibrosis therapy, J. Nanobiotechnol. 18 (2020) 90, https://doi.org/10.1186/s12951-020-00645-y.

[61] H. Jo, C. Ban, Aptamer–nanoparticle complexes as powerful diagnostic and therapeutic tools, Exp. Mol. Med. 48 (2016), https://doi.org/10.1038/emm.2016.44, e230.

[62] P. Kesharwani, V. Gajbhiye, N.K. Jain, A review of nanocarriers for the delivery of small interfering RNA, Biomaterials 33 (2012) 7138–7150, https://doi.org/10.1016/j.biomaterials.2012.06.068.

[63] Y. Duo, M. Yang, Z. Du, C. Feng, C. Xing, Y. Wu, Z. Xie, F. Zhang, L. Huang, X. Zeng, H. Chen, CX-5461-loaded nucleolus-targeting nanoplatform for cancer therapy through induction of pro-death autophagy, Acta Biomater. 79 (2018) 317–330, https://doi.org/10.1016/j.actbio.2018.08.035.

[64] S. Chandra, H.M. Nguyen, K. Wiltz, N. Hall, S. Chaudhry, G. Olverson, T. Mandal, S. Dash, A. Kundu, A.K. Kundu, Aptamer-functionalized hybrid nanoparticles to enhance the delivery of doxorubicin into breast Cancer cells by silencing P-glycoprotein HHS public access, J. Cancer Treat. Diagnosis. 4 (2020) 1–13.

[65] K. Gui, X. Zhang, F. Chen, Z. Ge, S. Zhang, X. Qi, J. Sun, Z. Yu, Lipid-polymer nanoparticles with CD133 aptamers for targeted delivery of all-trans retinoic acid to osteosarcoma initiating cells, Biomed. Pharmacother. 111 (2019) 751–764, https://doi.org/10.1016/j.biopha.2018.11.118.

[66] H.F. Ji, D.X. Kong, L. Shen, L.L. Chen, B.G. Ma, H.Y. Zhang, Distribution patterns of small-molecule ligands in the protein universe and implications for origin of life and drug discovery, Genome Biol. 8 (2007) R176, https://doi.org/10.1186/gb-2007-8-8-r176.

[67] M. Toporkiewicz, J. Meissner, L. Matusewicz, A. Czogalla, A.F. Sikorski, Toward a magic or imaginary bullet? Ligands for drug targeting to cancer cells: principles, hopes, and challenges, Int. J. Nanomedicine 10 (2015) 1399–1414, https://doi.org/10.2147/IJN.S74514.

[68] L. Xu, Q. Bai, X. Zhang, H. Yang, Folate-mediated chemotherapy and diagnostics: An updated review and outlook, J. Control. Release 252 (2017) 73–82, https://doi.org/10.1016/j.jconrel.2017.02.023.

[69] Y.G. Assaraf, C.P. Leamon, J.A. Reddy, The folate receptor as a rational therapeutic target for personalized cancer treatment, Drug Resist. Updat. 17 (2014) 89–95, https://doi.org/10.1016/j.drup.2014.10.002.

[70] L. Ramzy, M. Nasr, A.A. Metwally, G.A.S. Awad, Cancer nanotheranostics: a review of the role of conjugated ligands for overexpressed receptors, Eur. J. Pharm. Sci. 104 (2017) 273–292, https://doi.org/10.1016/j.ejps.2017.04.005.

[71] A.K. Bakrania, B.C. Variya, S.S. Patel, Novel targets for paclitaxel nano formulations: hopes and hypes in triple negative breast cancer, Pharmacol. Res. 111 (2016) 577–591, https://doi.org/10.1016/j.phrs.2016.07.023.

[72] C. Chen, J. Ke, X.E. Zhou, W. Yi, J.S. Brunzelle, J. Li, E.-L. Yong, H.E. Xu, K. Melcher, Structural basis for molecular recognition of folic acid by folate receptors. SI, Nature 500 (2013) 486–489, https://doi.org/10.1038/nature12327.

[73] A. Kumar, S.V. Lale, M.R. Aji Alex, V. Choudhary, V. Koul, Folic acid and trastuzumab conjugated redox responsive random multiblock copolymeric nanocarriers for breast cancer therapy: in-vitro and in-vivo studies, Colloids Surf. B Biointerfaces 149 (2017) 369–378, https://doi.org/10.1016/j.colsurfb.2016.10.044.

[74] S. Deshayes, H. Cabral, T. Ishii, Y. Miura, S. Kobayashi, T. Yamashita, A. Matsumoto, Y. Miyahara, N. Nishiyama, K. Kataoka, Phenylboronic acid-installed polymeric micelles for targeting sialylated epitopes in solid tumors, J. Am. Chem. Soc. 135 (2013) 15501–15507, https://doi.org/10.1021/ja406406h.

[75] D.B. Shenoy, M.M. Amiji, Poly(ethylene oxide)-modified poly(ε-caprolactone) nanoparticles for targeted delivery of tamoxifen in breast cancer, Int. J. Pharm. 293 (2005) 261–270, https://doi.org/10.1016/j.ijpharm.2004.12.010.

[76] K. Singha, R. Namgung, W.J. Kim, Polymers in small-interfering RNA delivery, Nucleic Acid Ther. 21 (2011) 133–147, https://doi.org/10.1089/nat.2011.0293.

[77] N. Michael Green, Avidin and streptavidin, Methods Enzymol. 184 (1990) 51–67, https://doi.org/10.1016/0076-6879(90)84259-J.

[78] M.K. Yu, J. Park, S. Jon, Targeting strategies for multifunctional nanoparticles in Cancer imaging and therapy, Theranostics 2 (2012) 3–44, https://doi.org/10.7150/thno.3463.

[79] O. Veiseh, J.W. Gunn, M. Zhang, Design and fabrication of magnetic nanoparticles for targeted drug delivery and imaging, Adv. Drug Deliv. Rev. 62 (2010) 284–304, https://doi.org/10.1016/j.addr.2009.11.002.

[80] H. Kolb, M. Finn, K. Sharpless, Click chemistry: diverse chemical function from a few good reactions, Angew. Chemie. 40 (2001) 2004–2021, https://doi.org/10.1002/1521-3773(20010601)40:11<2004::AID-ANIE2004>3.3.CO;2-X.

[81] C.D. Hein, X.M. Liu, D. Wang, Click chemistry, a powerful tool for pharmaceutical sciences, Pharm. Res. 25 (2008) 2216–2230, https://doi.org/10.1007/s11095-008-9616-1.

[82] N.K. Devaraj, R. Upadhyay, J.B. Haun, S.A. Hilderbrand, R. Weissleder, Fast and sensitive pretargeted labeling of cancer cells through a tetrazine/trans-cyclooctene cycloaddition, Angew. Chem. Int. Ed. 48 (2009) 7013–7016, https://doi.org/10.1002/anie.200903233.

[83] D.J. Javier, N. Nitin, M. Levy, A. Ellington, R. Richards-Kortum, Aptamer-targeted gold nanoparticles as molecular-specific contrast agents for reflectance imaging, Bioconjug. Chem. 19 (2008) 1309–1312, https://doi.org/10.1021/bc8001248.

[84] N. Muhamad, T. Plengsuriyakarn, K. Na-Bangchang, Application of active targeting nanoparticle delivery system for chemotherapeutic drugs and traditional/herbal medicines in cancer therapy: a systematic review, Int. J. Nanomedicine 13 (2018) 3921–3935, https://doi.org/10.2147/IJN.S165210.

[85] H. Choudhury, B. Gorain, R.K. Tekade, M. Pandey, S. Karmakar, T.K. Pal, Safety against nephrotoxicity in paclitaxel treatment: Oral nanocarrier as an effective tool in preclinical evaluation with marked in vivo antitumor activity, Regul. Toxicol. Pharmacol. 91 (2017), https://doi.org/10.1016/j.yrtph.2017.10.023.

[86] F. Roncato, F. Rruga, E. Porcù, E. Casarin, R. Ronca, F. Maccarinelli, N. Realdon, G. Basso, R. Alon, G. Viola, M. Morpurgo, Improvement and extension of anti-EGFR targeting in breast cancer therapy by integration with the Avidin-nucleic-acid-Nano-assemblies, Nat. Commun. 9 (2018) 1–11, https://doi.org/10.1038/s41467-018-06602-6.

[87] Y. Cui, Q. Xu, P.K.H. Chow, D. Wang, C.H. Wang, Transferrin-conjugated magnetic silica PLGA nanoparticles loaded with doxorubicin and paclitaxel for brain glioma treatment, Biomaterials 34 (2013) 8511–8520, https://doi.org/10.1016/j.biomaterials.2013.07.075.

[88] L.E. Scheeren, D.R. Nogueira-Librelotto, L.B. Macedo, J.M. de Vargas, M. Mitjans, M.P. Vinardell, C.M.B. Rolim, Transferrin-conjugated doxorubicin-loaded PLGA nanoparticles with pH-responsive behavior: a synergistic approach for cancer therapy, J. Nanopart. Res. 22 (2020) 1–18, https://doi.org/10.1007/s11051-020-04798-7.

[89] K.Y. Choi, H.Y. Yoon, J.H. Kim, S.M. Bae, R.W. Park, Y.M. Kang, I.S. Kim, I.C. Kwon, K. Choi, S.Y. Jeong, K. Kim, J.H. Park, Smart nanocarrier based on PEGylated hyaluronic acid for cancer therapy, ACS Nano 5 (2011) 8591–8599, https://doi.org/10.1021/nn202070n.

[90] B. Xiao, M.K. Han, E. Viennois, L. Wang, M. Zhang, X. Si, D. Merlin, Hyaluronic acid-functionalized polymeric nanoparticles for colon cancer-targeted combination chemotherapy, Nanoscale 7 (2015) 17745–17755, https://doi.org/10.1039/c5nr04831a.

[91] K. Mandal, N.R. Jana, Galactose-functionalized, colloidal-fluorescent nanoparticle from aggregation-induced emission active molecule via polydopamine coating for cancer cell targeting, ACS Appl. Nano Mater. 1 (2018) 3531–3540, https://doi.org/10.1021/acsanm.8b00673.

[92] F.F. Kolahkaj, K. Derakhshandeh, F. Khaleseh, A.H. Azandaryani, K. Mansouri, M. Khazaei, Active targeting carrier for breast cancer treatment: Monoclonal antibody conjugated epirubicin loaded nanoparticle, J. Drug Deliv. Sci. Technol. 53 (2019) 101136, https://doi.org/10.1016/j.jddst.2019.101136.

[93] A.J. Shuhendler, P. Prasad, M. Leung, A.M. Rauth, R.S. DaCosta, X.Y. Wu, A novel solid lipid nanoparticle formulation for active targeting to tumor $\alpha_v\beta_3$ integrin receptors reveals cyclic RGD as a double-edged sword, Adv. Healthc. Mater. 1 (2012) 600–608, https://doi.org/10.1002/adhm.201200006.

[94] M. Hong, S. Zhu, Y. Jiang, G. Tang, C. Sun, C. Fang, B. Shi, Y. Pei, Novel anti-tumor strategy: PEG-hydroxycamptothecin conjugate loaded transferrin-PEG-nanoparticles, J. Control. Release 141 (2010) 22–29, https://doi.org/10.1016/j.jconrel.2009.08.024.

[95] J. Xie, L. Mei, K. Huang, Y. Sun, A. Iris, B. Ma, Y. Qiu, J. Li, G. Han, A photo-inducible protein-inorganic nanoparticle assembly for active targeted tumour theranostics, Nanoscale 11 (2019) 6136–6144, https://doi.org/10.1039/C9NR01120J.

[96] M. Das, W. Duan, S.K. Sahoo, Multifunctional nanoparticle-EpCAM aptamer bioconjugates: a paradigm for targeted drug delivery and imaging in cancer therapy, nanomedicine nanotechnology, Biol. Med. 11 (2015) 379–389, https://doi.org/10.1016/j.nano.2014.09.002.

[97] Z. Hussain, S. Khan, M. Imran, M. Sohail, S.W.A. Shah, M. de Matas, PEGylation: a promising strategy to overcome challenges to cancer-targeted nanomedicines: a review of challenges to clinical transition and promising resolution, Drug Deliv. Transl. Res. 9 (2019) 721–734, https://doi.org/10.1007/s13346-019-00631-4.

[98] R. Rezaei, M. Safaei, H.R. Mozaffari, H. Moradpoor, S. Karami, A. Golshah, B. Salimi, H. Karami, The role of nanomaterials in the treatment of diseases and their effects on the immune system, Open Access Maced. J. Med. Sci. 7 (2019) 1884–1890, https://doi.org/10.3889/oamjms.2019.486.

[99] B. Kwong, H. Liu, D.J. Irvine, Induction of potent anti-tumor responses while eliminating systemic side effects via liposome-anchored combinatorial immunotherapy, Biomaterials 32 (2011) 5134–5147, https://doi.org/10.1016/j.biomaterials.2011.03.067.

[100] J. Falconer, A.N. Murphy, S.P. Young, A.R. Clark, S. Tiziani, M. Guma, C.D. Buckley, Review: synovial cell metabolism and chronic inflammation in rheumatoid arthritis, Arthritis Rheumatol. 70 (2018) 984–999, https://doi.org/10.1002/art.40504.

[101] S. Lee, H. Kim, Y. Ha, Y. Park, S. Lee, Y. Park, K. Yoo, Targeted chemo-photothermal treatments of rheumatoid arthritis using gold half-shell multifunctional nanoparticles, ACS Nano 7 (2013) 50–57, https://doi.org/10.1021/NN301215Q.

[102] H.J. Kim, S.M. Lee, K.H. Park, C.H. Mun, Y.B. Park, K.H. Yoo, Drug-loaded gold/iron/gold plasmonic nanoparticles for magnetic targeted chemo-photothermal treatment of rheumatoid arthritis, Biomaterials 61 (2015) 95–102, https://doi.org/10.1016/j.biomaterials.2015.05.018.

[103] S.H. Kim, J.H. Kim, D.G. You, G. Saravanakumar, H.Y. Yoon, K.Y. Choi, T. Thambi, V.G. Deepagan, D.G. Jo, J.H. Park, Self-assembled dextran sulphate nanoparticles for targeting rheumatoid arthritis, Chem. Commun. 49 (2013) 10349–10351, https://doi.org/10.1039/c3cc44260h.

[104] T. Dutta, M. Garg, N.K. Jain, Targeting of efavirenz loaded tuftsin conjugated poly(propyleneimine) dendrimers to HIV infected macrophages in vitro, Eur. J. Pharm. Sci. 34 (2008) 181–189, https://doi.org/10.1016/j.ejps.2008.04.002.

[105] S. Md, N.A. Alhakamy, H.M. Aldawsari, H.Z. Asfour, Neuroprotective and antioxidant effect of naringenin-loaded nanoparticles for nose-to-brain delivery, Brain Sci. 9 (2019), https://doi.org/10.3390/brainsci9100275, E275.

[106] B. Gorain, H. Choudhury, M. Pandey, H.L.B. Mohd Zaki, N.I.B. Bakar, N.H. Hamdhan, N.A.B. Musa, N.F.B. Mustafa, J.S. Sangkari, N.A.A.B. Azmi, L. Park, P. Kesharwani, Mechanistic description of natural herbs in the treatment of dementia: a systematic review, Curr. Psychopharmacol. 07 (2018) 149–164, https://doi.org/10.2174/2211556007666180420124544.

[107] B. Gorain, D.C. Rajeswary, M. Pandey, P. Kesharwani, S.A. Kumbhar, H. Choudhury, Nose to brain delivery of nanocarriers towards attenuation of demented condition, Curr. Pharm. Des. 26 (2020), https://doi.org/10.2174/1381612826666200313125613.

[108] W.M. Pardridge, Drug transport across the blood-brain barrier, J. Cereb. Blood Flow Metab. 32 (2012) 1959–1972, https://doi.org/10.1038/jcbfm.2012.126.

[109] H.L. Liu, C.H. Fan, C.Y. Ting, C.K. Yeh, Combining microbubbles and ultrasound for drug delivery to brain tumors: current progress and overview, Theranostics 4 (2014) 432–444, https://doi.org/10.7150/thno.8074.

[110] S. Wohlfart, S. Gelperina, J. Kreuter, Transport of drugs across the blood-brain barrier by nanoparticles, J. Control. Release 161 (2012) 264–273, https://doi.org/10.1016/j.jconrel.2011.08.017.

[111] T. Führmann, M. Ghosh, A. Otero, B. Goss, T.R. Dargaville, D.D. Pearse, P.D. Dalton, Peptide-functionalized polymeric nanoparticles for active targeting of damaged tissue in animals with experimental autoimmune encephalomyelitis, Neurosci. Lett. 602 (2015) 126–132, https://doi.org/10.1016/j.neulet.2015.06.049.

[112] R. Huang, W. Ke, Y. Liu, D. Wu, L. Feng, C. Jiang, Y. Pei, Gene therapy using lactoferrin-modified nanoparticles in a rotenone-induced chronic Parkinson model, J. Neurol. Sci. 290 (2010) 123–130, https://doi.org/10.1016/j.jns.2009.09.032.

[113] A.A. Yetisgin, S. Cetinel, M. Zuvin, A. Kosar, O. Kutlu, Therapeutic nanoparticles and their targeted delivery applications, Molecules 25 (2020), https://doi.org/10.3390/molecules25092193.

[114] R. Huang, H. Ma, Y. Guo, S. Liu, Y. Kuang, K. Shao, J. Li, Y. Liu, L. Han, S. Huang, S. An, L. Ye, J. Lou, C. Jiang, Angiopep-conjugated nanoparticles for targeted long-term gene therapy of parkinson's disease, Pharm. Res. 30 (2013) 2549–2559, https://doi.org/10.1007/s11095-013-1005-8.

[115] Q. Meng, A. Wang, H. Hua, Y. Jiang, Y. Wang, H. Mu, Z. Wu, K. Sun, Intranasal delivery of Huperzine a to the brain using lactoferrin-conjugated N-trimethylated chitosan surface-modified PLGA nanoparticles for treatment of Alzheimer's disease, Int. J. Nanomed. 13 (2018) 705–718, https://doi.org/10.2147/IJN.S151474.

[116] M.G. Sharaf, S. Cetinel, L. Heckler, K. Damji, L. Unsworth, C. Montemagno, Nanotechnology-based approaches for ophthalmology applications: therapeutic and diagnostic strategies, Asia-Pacific J. Ophthalmol. 3 (2014) 172–180, https://doi.org/10.1097/apo.0000000000000059.

[117] Y. Diebold, M. Calonge, Applications of nanoparticles in ophthalmology, Prog. Retin. Eye Res. 29 (2010) 596–609, https://doi.org/10.1016/j.preteyeres.2010.08.002.

[118] X. Xu, L. Sun, L. Zhou, Y. Cheng, F. Cao, Functional chitosan oligosaccharide nanomicelles for topical ocular drug delivery of dexamethasone, Carbohydr. Polym. 227 (2020) 115356, https://doi.org/10.1016/j.carbpol.2019.115356.

[119] R. Gonzalez-Pizarro, G. Parrotta, R. Vera, E. Sánchez-López, R. Galindo, F. Kjeldsen, J. Badia, L. Baldoma, M. Espina, M.L. García, Ocular penetration of fluorometholone-loaded PEG-PLGA nanoparticles functionalized with cell-penetrating peptides, Nanomedicine 14 (2019) 3089–3104, https://doi.org/10.2217/nnm-2019-0201.

[120] A.L. Byrne, B.J. Marais, C.D. Mitnick, L. Lecca, G.B. Marks, Tuberculosis and chronic respiratory disease: a systematic review, Int. J. Infect. Dis. 32 (2015) 138–146, https://doi.org/10.1016/j.ijid.2014.12.016.

[121] J.S. Patil, S. Sarasija, Pulmonary drug delivery strategies: a concise, systematic review, Lung India 29 (2012) 44–49, https://doi.org/10.4103/0970-2113.92361.

[122] E. Maretti, L. Costantino, C. Rustichelli, E. Leo, M.A. Croce, F. Buttini, E. Truzzi, V. Iannuccelli, Surface engineering of solid lipid nanoparticle assemblies by methyl α-D-mannopyranoside for the

active targeting to macrophages in anti-tuberculosis inhalation therapy, Int. J. Pharm. 528 (2017) 440–451, https://doi.org/10.1016/j.ijpharm.2017.06.045.

[123] B. Gorain, M. Tekade, P. Kesharwani, A.K. Iyer, K. Kalia, R.K. Tekade, The use of nanoscaffolds and dendrimers in tissue engineering, Drug Discov. Today 22 (2017) 652–664, https://doi.org/10.1016/j.drudis.2016.12.007.

[124] C.M. Agrawal, R.B. Ray, Biodegradable polymeric scaffolds for musculoskeletal tissue engineering, J. Biomed. Mater. Res. 55 (2001) 141–150, https://doi.org/10.1002/1097-4636(200105)55:2<141::AID-JBM1000>3.0.CO;2-J.

[125] Y. Qi, L. Guo, Y. Jiang, Y. Shi, H. Sui, L. Zhao, Brain delivery of quercetin-loaded exosomes improved cognitive function in AD mice by inhibiting phosphorylated tau-mediated neurofibrillary tangles, Drug Deliv. 27 (2020) 745–755, https://doi.org/10.1080/10717544.2020.1762262.

[126] T. Tian, H.X. Zhang, C.P. He, S. Fan, Y.L. Zhu, C. Qi, N.P. Huang, Z.D. Xiao, Z.H. Lu, B.A. Tannous, J. Gao, Surface functionalized exosomes as targeted drug delivery vehicles for cerebral ischemia therapy, Biomaterials 150 (2018) 137–149, https://doi.org/10.1016/j.biomaterials.2017.10.012.

[127] Y. Zhang, M. Li, X. Gao, Y. Chen, T. Liu, Nanotechnology in cancer diagnosis: progress, challenges and opportunities, J. Hematol. Oncol. 12 (2019) 141, https://doi.org/10.1186/s13045-019-0833-3.

[128] H. Chen, Z. Zhen, T. Todd, P.K. Chu, J. Xie, Nanoparticles for improving cancer diagnosis, Mater. Sci. Eng. R Rep. 74 (2013) 35–69, https://doi.org/10.1016/j.mser.2013.03.001.

[129] S.C. Baetke, T. Lammers, F. Kiessling, Applications of nanoparticles for diagnosis and therapy of cancer, Br. J. Radiol. 88 (2015), https://doi.org/10.1259/bjr.20150207, e20150207.

[130] X. Chi, D. Huang, Z. Zhao, Z. Zhou, Z. Yin, J. Gao, Nanoprobes for in vitro diagnostics of cancer and infectious diseases, Biomaterials 33 (2012) 189–206, https://doi.org/10.1016/j.biomaterials.2011.09.032.

[131] A.F.J. Jou, C.H. Lu, Y.C. Ou, S.S. Wang, S.L. Hsu, I. Willner, J.A.A. Ho, Diagnosing the miR-141 prostate cancer biomarker using nucleic acid-functionalized CdSe/ZnS QDs and telomerase, Chem. Sci. 6 (2015) 659–665, https://doi.org/10.1039/c4sc02104e.

[132] J.H. Lee, Y.M. Huh, Y.W. Jun, J.W. Seo, J.T. Jang, H.T. Song, S. Kim, E.J. Cho, H.G. Yoon, J.S. Suh, J. Cheon, Artificially engineered magnetic nanoparticles for ultra-sensitive molecular imaging, Nat. Med. 13 (2007) 95–99, https://doi.org/10.1038/nm1467.

[133] S.M. Seidlin, L.D. Marinelli, E. Oshry, Radioactive iodine therapy; effect on functioning metastases of adenocarcinoma of the thyroid, JAMA 132 (1946) 838–847, https://doi.org/10.1001/jama.1946.02870490016004.

[134] N. Deb, M. Goris, K. Trisler, S. Fowler, J. Saal, S. Ning, M. Becker, C. Marquez, S. Knox, Treatment of hormone-refractory prostate cancer with 90Y-CYT-356 monoclonal antibody, Clin. Cancer Res. 2 (1996) 1289–1297.

[135] Z. Wang, R. Qiao, N. Tang, Z. Lu, H. Wang, Z. Zhang, X. Xue, Z. Huang, S. Zhang, G. Zhang, Y. Li, Active targeting theranostic iron oxide nanoparticles for MRI and magnetic resonance-guided focused ultrasound ablation of lung cancer, Biomaterials 127 (2017) 25–35, https://doi.org/10.1016/J.BIOMATERIALS.2017.02.037.

[136] N. Schleich, C. Po, D. Jacobs, B. Ucakar, B. Gallez, F. Danhier, V. Préat, Comparison of active, passive and magnetic targeting to tumors of multifunctional paclitaxel/SPIO-loaded nanoparticles for tumor imaging and therapy, J. Control. Release 194 (2014) 82–91, https://doi.org/10.1016/J.JCONREL.2014.07.059.

[137] Y. Xiao, H. Hong, A. Javadi, J.W. Engle, W. Xu, Y. Yang, Y. Zhang, T.E. Barnhart, W. Cai, S. Gong, Multifunctional unimolecular micelles for cancer-targeted drug delivery and positron emission tomography imaging, Biomaterials 33 (2012) 3071–3082, https://doi.org/10.1016/J.BIOMATERIALS.2011.12.030.

[138] V.J. Yao, S. D'Angelo, K.S. Butler, C. Theron, T.L. Smith, S. Marchiò, J.G. Gelovani, R.L. Sidman, A.S. Dobroff, C.J. Brinker, A.R.M. Bradbury, W. Arap, R. Pasqualini, Ligand-targeted theranostic nanomedicines against cancer, J. Control. Release 240 (2016) 267–286, https://doi.org/10.1016/J.JCONREL.2016.01.002.

CHAPTER 5

Biofate and cellular interactions of lipid nanoparticles

Iara Baldim[a,b], Wanderley P. Oliveira[b], Rekha Rao[c], Singh Raghuvir[c],
Sheefali Mahant[e], Francisco M. Gama[a], and Eliana B. Souto[a,d]
[a]CEB—Centre of Biological Engineering, University of Minho, Braga, Portugal
[b]Faculty of Pharmaceutical Sciences of Ribeirão Preto, University of São Paulo, Ribeirão Preto, SP, Brazil
[c]Department of Pharmaceutical Sciences, Guru Jambheshwar University of Science and Technology, Hisar, Haryana, India
[d]Department of Pharmaceutical Technology, Faculty of Pharmacy, University of Coimbra, Coimbra, Portugal
[e]Department of Pharmaceutical Sciences, Maharshi Dayanand University, Rohtak, Haryana, India

1. Introduction

By definition, drug targeting and delivery is the process of delivering a drug molecule to a specific site where the pharmacological effect is desired [1]. A targeted release may simplify the drug administration protocols and reduce the amount of drug necessary, thus improving the cost–effectiveness [2, 3] and reducing the side effects. A drug administered by intravenous injection distributes in the whole body, reaching organs, tissues and cells where it's action is not required, potentially causing harmful effects [1]. Therefore targeting is essential for the drug to reach the necessary concentration for therapeutic efficacy, while minimizing the toxic and unwanted effects to other organs caused by random and uncontrolled bioactive action. Nowadays, most of the targetable controlled drug delivery systems are based on nanotechnology [3].

High selectivity and specificity ensure the physical contact of the bioactive molecule with the physiological target only at the desired location of the body. Indeed, the development of highly selective and site-specific delivery systems capable of maintaining the optimal effect in a suitable time frame [3] is one of the main challenges in nanotherapy. Desirable features in nanocarriers include biodegradability and biocompatibility, nonimmunogenicity and stability in the bloodstream [4]. Prolonged circulation time and moderated degradation rate are also important parameters to be considered in the design of a suitable nanocarrier delivery system, once they provide enough time for the system to reach the targeted tissue [5]. These features can be found in different nanosystems available, such as liposomes, solid lipid nanoparticles (SLN), and nanostructured lipid carriers (NLC), nanogels, micelles, dendrimers, and so on. The surface area of a particulate system increases several orders of magnitude with the decrease of its size to nanometric scale, promoting higher interactions with the biological environment [6]. The particle size plays an important role in determining the biofate of nanoparticles upon administration to the body. In addition, parameters such as nanoparticle surface chemistry (charge, hydrophobicity/hydrophilicity) as well as the composition of the wall material are also prominent factors affecting nanoparticle circulation time and the biodegradation rate [5].

Nanosystems are already used in the treatment of various chronic diseases, particularly malignant tumors, where the use of cytotoxic drugs as chemotherapeutic agents remains as the primary treatment choice. Cytotoxic agents are poorly selective, highly toxic to healthy cells, and can even stimulate cellular drug resistance, a defense mechanism by which the tumor cell expels the drug out of its cytoplasm, reducing the exposure to the cytotoxic agent [7].

Another challenge of nanotechnology in the pharmaceutical field is the brain drug delivery. This is possibly the most challenging site of action in the human body, as the blood–brain barrier (BBB) prevents the action of more than 98% of the new chemical entities discovered [8]. The BBB, whose function is to maintain brain homeostasis, is a complex network of brain micro vessels, composed of endothelial cells that form tight junctions, allowing the passage of a few types of molecules (e.g., low molecular weight lipophilic molecules, nutrients and some peptides) [9].

Lipid nanoparticles have shown high loading capacity and ability to target drugs to a specific site providing a prolonged release profile [10–12]. In addition, these delivery systems have also shown to be able to penetrate the BBB [13–20]. Thus these systems attracted an expanding scientific and industrial attention in the last years, creating new solutions for therapeutic applications. After the intravenous administration, nanocarriers may be cleared from the systemic circulation by the reticuloendothelial system (RES) [3, 7]. RES is a part of the immune system consisting of phagocytic cells that rapidly remove from bloodstream (seconds to minutes) drug carriers identified as foreign elements, particularly in the spleen, lymph nodes and liver [7, 21]. This process, and consequently the biodistribution, is determined by the stability, size and surface characteristics of the nanoparticles [3, 22]. A drug targeting system requires vectors with enough circulation time in the body as to reach their site of action [3] as well as the ability to penetrate the endothelium and reach the target tissue [23]. The physicochemical and biological factors involved in the nanoparticles biodistribution are interrelated [24] and are depicted in Fig. 1.

2. Passive and active targeting

The goal of drug targeting is to maximize the local effect of the drug together with low systemic exposure. The targeting mechanism can be classified into two pathways: passive versus active (Fig. 2). Passive targeting does not rely on any site-specific targeting strategy. It is a passive process, in which targeting is guided by both the physicochemical properties of carriers, and the physiological and histological characteristics of the target organ and the interaction between them, resulting in the accumulation of therapeutic nanoparticles in this target organ [22, 41]. Characteristics such as hydrophobicity and surface charge as

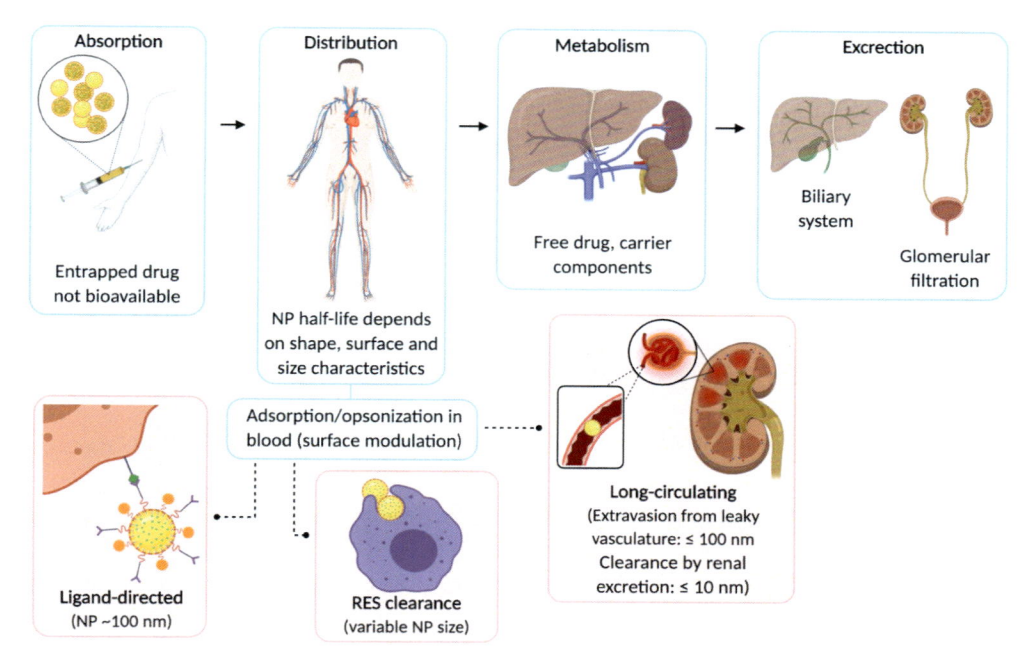

Fig. 1 Overview of nanoparticles biodistribution and the influence of particle size on this process. NP, nanoparticles; RES, reticuloendothelial system.

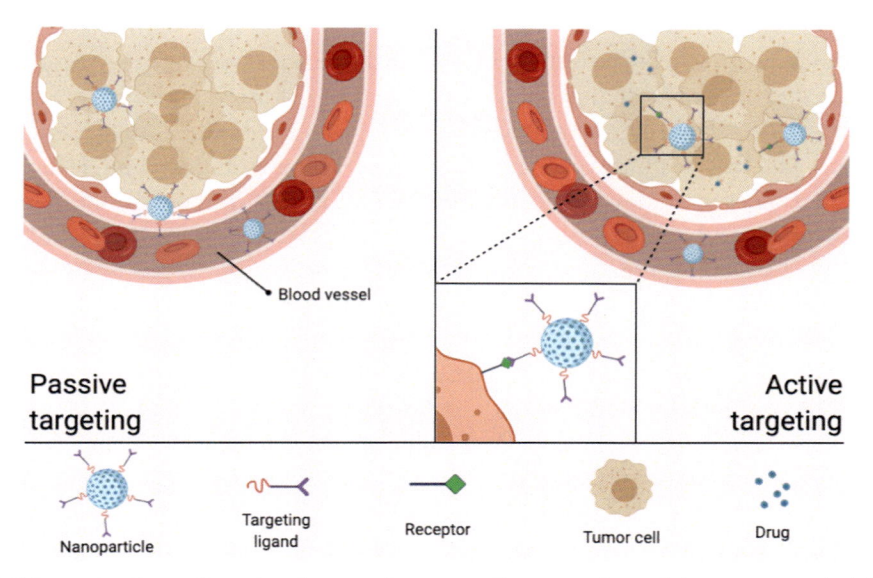

Fig. 2 Schematic illustration showing passive vs. active targeting of intravenous-administered nanoparticles to reach tumor cells.

well as size and mass of carrier particles directly influence this type of targeting [1]. It also involves the use of innate characteristics of the nanoparticles to induce the targeting, such as surface charge. Cationic liposomes, for example, can bind to the negatively charged phospholipids of tumor endothelial cells by electrostatic interactions [42]. Another example of the use of passive targeting is the delivery of antibiotics to treat diseases such as leishmaniosis, tuberculosis, and listeriosis. In these cases the carrier must be phagocytized by the infected macrophages, the drug being released intracellularly [3]. Thus, differently from other applications, in this case, the carrier nanoparticles must be recognized by the macrophages and the RES system.

Passive targeting also has special relevance to oncological applications, where it is known as enhanced retention and permeability effect. Local environment of tumor is quite different from that of healthy tissues. The vascular endothelium of tumors is usually leaky and not uniform, presenting enhanced vascular permeability. The pore diameter in the tumor capillaries may vary from 100 nm to almost 1 μm, depending on the type of tumor [43]. Therefore nanoparticles of a certain size range (between 10 and 500 nm) can penetrate the endothelium and accumulate inside the tumor interstitial space through a passive and nonselective process [22, 23]. However, this type of targeting requires a long blood circulation time, avoiding RES and namely internalization by macrophages, able to maintain both postcirculation therapeutic activity and drug retention until target release [3].

Active targeting, on the other hand, is a site-specific delivery strategy. This approach consists in decorating the surface of nanocarriers to provide selective and specific recognition of the target site, where a drug's pharmacological activity is required. Such interactions may be achieved by antigen–antibody binding, ligand–receptor binding, or even by physical signals, such as changes in pH, temperature, ultrasound, magnetic field application, etc. [1, 3]. Active targeting may occur after passive accumulation of nanoparticles at the target site, that is, after passive targeting [22]. Examples of ligands used for this purpose are antibodies, peptides, or molecules such as folic acid and sugars [43], which will bind to plasma membrane receptors such as low density lipoprotein (LDL), folate, and transferrin receptors [3]. One of the key challenges in designing active targeting mechanisms in drug delivery has been to find highly specific and nonimmunogenic ligands [9]. As a result, the high specificity will provide undeniable increase in therapeutic effect, reduced exposure time of the carrier system to plasma circulation and consequently, lower side effects and drug toxicity.

Lipid-based nanoparticles such as nanoemulsions, liposomes, SLN, and NLC have received significant attention in the treatment of various chronic diseases, including cancer, cardiovascular and central nervous system (CNS) diseases. These nanosystems exhibit several advantages, such as low toxicity compared to polymeric and inorganic nanoparticle systems, biocompatibility, capacity to transport both hydrophilic and hydrophobic compounds, prolonged half-life, controlled drug release, and easy production scalability

[44]. In addition, triglycerides, fatty acids, and waxes release natural occurring byproducts upon degradation.

Lipid nanoparticles are a viable strategy to formulate pharmaceuticals for oral, topical, pulmonary, and parenteral drug delivery [45]. The route taken by the drug and the barriers to be overcome depend on the route of administration used. Many active pharmaceutical ingredients are physicochemically instable in the gastrointestinal tract, showing low oral bioavailability and narrow therapeutic index, thus requiring parenteral administration. Although SLN and NLC formulations for parenteral use are still under development stage, lipid emulsions composed by fatty acids are widely used in parenteral nutrition since the 1960s [46]. The design of site-specific lipid systems involves the correct choice of the lipids and surfactant materials. The use of GRAS (generally recognized as safe) lipids and surfactants is highly recommended to obtain easy approval of the drug regulatory authorities and good in vivo tolerance [47]. The functionalization of these systems has been investigated for active and passive targeting, using different types of ligands, as peptides, saccharides, and therapeutic molecules, as proteins and antibodies [48]. Surface modifications of lipid-based nanoparticles allow adjusting their properties as well as overcoming the limitations of different administration routes [49]. Many studies highlight the advantages and characteristics of the functionalization and target of lipid nanoparticles, as listed in Table 1.

3. Lipid-based nanoparticles for drug targeting and delivery

There is an increased momentum in the pace of nanomedicine discovery. The development of lipid-based drug carriers has been a crescent area of interest in research over the past years. These carriers constitute a broad and diversified group of nanoparticles that are particularly relevant for drug targeting. Fig. 3 exemplifies the structural differences presented by the four main types of lipid-based nanoparticles: liposomes, nanoemulsions and both lipid nanoparticles, SLN and NLC.

Because of the small particle size and minimized risk of blood clotting and aggregation, lipid-based nanosystems can be used for all parenteral applications [46]. Moreover, they are formed by physiologically well-tolerated components, being relatively safe [50–52]. Once accumulated in a tissue, the lipid content degrades faster than biodegradable polymeric materials, producing less toxic end products after degradation [22, 53].

Liposomes arguably are the most studied and easily synthesized class of nanoparticles and become the first nanomedicines in Food and Drug Administration clinical trials [54]. They are self-assembling vesicles composed of a phospholipid bilayer membrane with the potential of carrying not only water-soluble drugs, but also lipophilic molecules entrapped within the bilayer. Their biocompatibility, biodegradability, and versatility are some of the advantages. The active targeting of liposomes is achieved by conjugating cell surface receptor ligands to the liposome membrane. The disadvantage of having rapid

Table 1 Examples of parenteral delivered lipid-based nanoparticles for specific drug targeting.

Target	Encapsulated drug	Lipid matrix	Nanosystem	Production method	Ligand	Application	Ref.
Targeted gene therapy	Delivery of genes	Compritol ATO 888 and stearylamine	Cationic SLN	Microemulsion	SLN:DNA: streptavidin complex	Production of a targeted nonviral gene therapy vector	[25]
Solid tumors	Doxorubicin	Stearic acid	SLN	Microemulsion	PEG 2000	Coated SLN with hydrophilic molecules carrying anticancer drugs can reach cancer cells in solid tumors more effectively than healthy tissues	[26]
Tumor	Camptothecin	Stearic acid	SLN	Hot high-pressure homogenization	–	Compared to the free drug, camptothecin SLN presented higher uptake in RES organs and good targeting to the liver, lung, spleen, heart, blood, and, particularly to the brain, which confirms its efficiency in delivering drugs across the blood–brain barrier	[27]

Solid tumors	Dexamethasone	Stearic acid	SLN	Solvent displacement	Folate coating	Folate and dexamethasone improved cell targeting and nuclear targeting ability of SLN, forming a promise active targeting for drug/gene delivery	[28]
Brain (treatment of central nervous system fungal infection)	Itraconazole	Precirol ATO 5 and Transcutol HP	NLC	Hot high-pressure homogenization	–	Sustained release up to 24h and an almost two-fold increase in drug concentration in the brain	[29]
Liver (hepatocellular carcinoma)	5-Fluorouracil	Glyceryl monostearate and oleic acid or Labrafac	NLC	Emulsification-solvent diffusion	galactosylation	Improved cytotoxic effect of the antitumor encapsulated drug and enhanced drug uptake by hepatic cells	[30]
Brain cancer	Curcumin	Tripalmitin and oleic acid	NLC	Hot high-pressure homogenization	–	Enhanced cytotoxicity and cellular uptake, in addition to 6.4 fold increase in plasma concentration and enhanced brain and tumor targeting	[31]

Continued

Table 1 Examples of parenteral delivered lipid-based nanoparticles for specific drug targeting—cont'd

Target	Encapsulated drug	Lipid matrix	Nanosystem	Production method	Ligand	Application	Ref.
Glioblastoma multiforme (brain cancer)	Temozolomide	Compritol ATO 888 and Cremophor ELP	NLC	Solvent diffusion	Arginine–glycine–aspartic acid peptide (RGD)	Higher cytotoxicity and three-fold enhanced tumor inhibition compared with the free drug solution	[32]
Brain targeting	Baicalein	Gelucires 48/9 and 62/5 and Vitamin E	NLC	Ultrasonication	–	Baicalein loaded into NLC showed increased plasma levels and prolonged half-life as compared with the free drug. Compared with the aqueous solution, NLC presented superior performance: 7.5 and 4.7 fold higher accumulation of baicalein in the cerebral cortex and brain stem, respectively, and two to three fold enhanced accumulation in other regions of the brain (hippocampus, striatum, thalamus, and olfactory tract)	[33]

Spleen (treatment of rheumatoid arthritis)	Actarit	Stearic acid	SLN	Solvent diffusion-evaporation	–	SLN loaded with actarit (antirheumatic drug) improved therapeutic efficacy and reduced side effects. Compared with drug solution, actarit-loaded SLN presented an enhanced targeting efficiency to the spleen (from 6.31% to 16.29%), while significantly lowering the renal distribution after intravenous administration to mice. Therefore injectable drug-loaded SLN enable passive targeting and a more efficient treatment of rheumatoid arthritis	[34]

Continued

Table 1 Examples of parenteral delivered lipid-based nanoparticles for specific drug targeting—cont'd

Target	Encapsulated drug	Lipid matrix	Nanosystem	Production method	Ligand	Application	Ref.
Glioblastoma multiforme (brain cancer)	Temozolomide and vincristine	Stearic acid (SLN); Compritol ATO 888 and Cremophor ELP (NLC)	SLN and NLC	Solvent displacement (SLN) and solvent diffusion (NLC)	–	NLC formulation outperformed SLN in delivering the drugs more efficiently into U87MG cells and achieving higher inhibition efficacy both in vitro and in vivo	[35]
Gastric cancer	5-Fluorouracil and cisplatin	Glyceryl monostearate and soybean oil	NLC	Emulsification solidification	Hyaluronic acid	Synergistic effect between the drugs when they were administered in ratios of 10:1 and 20:1 in vitro; reduction on tumor growth and reconstitution of the body weight of mice bearing BGC-823 xenografts	[36]
Lung cancer	Docetaxel and baicalein	Glyceryl monostearate	SLN	Emulsification	PEG, transferrin, and hydrazone	The combined drug system presented a strong synergistic effect, the best tumor inhibition ability, and the lowest systemic toxicity in A549 tumor-bearing mice	[37]

Breast cancer	β-lapachone and doxorubicin	Compritol ATO 888 and oleic acid	NLC	Ultra-sonication	PEG	Increased doxorubicin concentration in tumor tissues, when compared with the free doxorubicin, in MCF-7 ADR tumor bearing mice model	[38]
Glioma	Kaempferol	Egg-lecithin and medium-chain triglycerides	Nanoemulsion	High-pressure homogenization	Chitosan	The mucoadhesive nanoemulsion reduced C6 glioma cell viability through the induction of apoptosis. The drug concentration in brain tissue after nasal administration of the nanoemulsion containing chitosan was found to be significantly higher in comparison to the formulation without chitosan or drug solutions.	[39]

Continued

Table 1 Examples of parenteral delivered lipid-based nanoparticles for specific drug targeting—cont'd

Target	Encapsulated drug	Lipid matrix	Nanosystem	Production method	Ligand	Application	Ref.
Metastatic prostate cancer	Docetaxel	Soy phosphatidylcholine, cholesterol and DSPE–PEG2000	Liposome	Hydration of lipid film	anti–EGFR antibody	The developed immunoliposomes enhanced the selectivity of Docetaxel delivery to EGFR-positive prostate cancer cells.	[40]

EGFR, epidermal growth factor receptor.

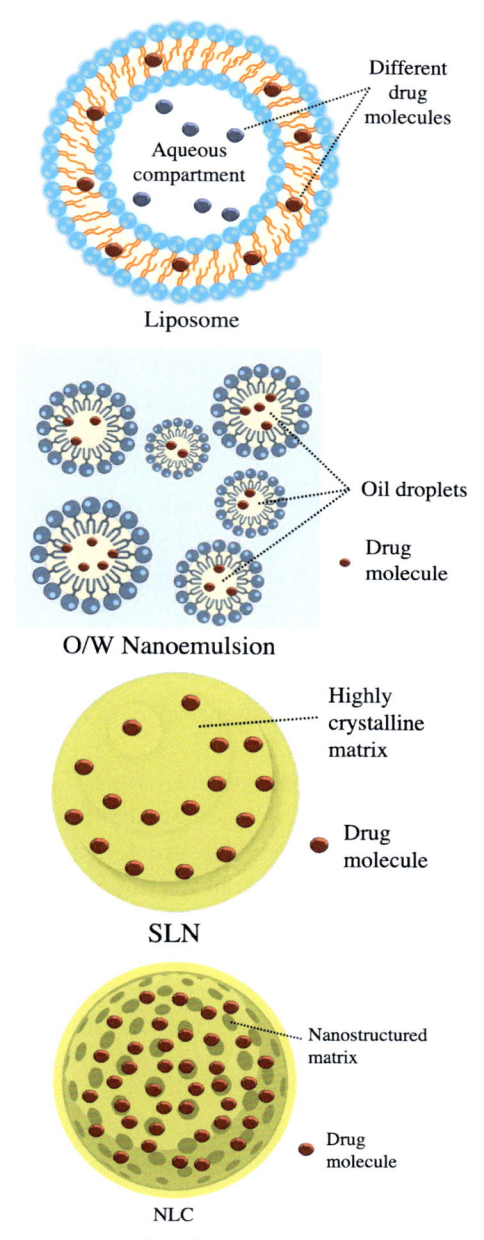

Fig. 3 Comparative structure of various lipid-based nanoparticles.

clearance can be minimized by PEGylation of the liposome surface, reducing unwanted phagocytic clearance.

Nanoemulsions are biphasic dispersions, generally oil dispersed in water, stabilized by one or more surfactant, with small dimension particles (around 200 nm or less). They are particularly suitable for the intravenous, oral, and ocular delivery of lipophilic compounds [5] and also represent a promising strategy to deliver drugs directly into the brain by using the intranasal route [55].

In contrast to the emulsions and liposomes that contain liquid lipid, a solid lipid matrix offers several advantages to the lipid-based nanosystem. SLN and NLC are systems that present great kinetic stability and rigid morphology, which enables a higher drug loading capacity and the modulation of drug release [20, 47, 56]. SLN were first designed to resolve some of the problems presented by parenteral nanoemulsions; later, the development of NLC brought solutions for some of the SLN problems related to higher crystallinity and lower drug loading [57]. Indeed, as NLC also contain a liquid lipid (as opposed to SLN matrix, which is composed of a solid lipid only), they show lower degree of crystallinity and higher drug loading [58, 59]. Together, SLN and NLC combine the advantages of liposomes, parenteral emulsions, and polymeric nanoparticles and overcome some drawbacks associated with these systems, such as drug leakage, membrane instability, and limited scaling production [53, 60]. The industrial scale production of lipid nanoparticles, including sterilization, is another point to be highlighted: a relatively low-cost production process, with the advantage of avoiding the use of organic solvents [3, 47].

Once injected intravenously, lipid nanoparticles are sufficiently small to achieve the microvascular system and prevent macrophage uptake when formulated with a hydrophilic coating. Surfactant plays a key role in the interaction with cells [61]. The coating with specific polyethylene glycol (PEG)-containing surfactants (e.g., polysorbate 80, P80 and poloxamer 188) increases the plasma half-life of lipid nanoparticles by reducing phagocytic uptake, improving bioavailability [62]. The reduction in phagocytic uptake by surface-coated hydrophilic polymers owes to the reduction of protein adsorption and thus suppression of opsonization in vivo [63]. This "stealth" system affords long blood circulation time [7]. Moreover, PEG coating is an interesting strategy for brain drug targeting due to its tropism by the BBB [53].

The reduced phagocytic uptake also allows the application of lipid nanoparticles for viral and nonviral gene delivery. Cationic SLN effectively bind genes directly via electrostatic interactions, with potential application for targeted gene therapy in the treatment of cancer. The charge of particles can thus be tuned through, favoring the encapsulation of oppositely charged molecules [25, 62].

The physicochemical characteristics of lipid nanoparticles should also be considered when addressing critical issues related to the development of suitable targeting formulations [47]. Among these characteristics, particle size is one of the most important parameters that governs the biodistribution and elimination of a colloidal drug delivery system, especially for the parenteral route of administration [64]. In the spleen the uptake of lipid nanoparticles is directly dependent on particle size. Regardless of whether nanoparticles are coated with a stealth component or not, particles bigger than 200 nm exhibit a faster rate of intravascular clearance via spleen and liver than those with a hydrodynamic diameter of less than 200 nm [22, 65]. Once on blood circulation, the incorporated drug is gradually released due to surface erosion (e.g., degradation by enzymes) or by diffusion from the nanoparticle matrix. The rate of release depends on the lipid composition, the particle size, the choice of surfactant, and the inner structure of the nanoparticle [46]. The mathematical modeling of drug release from nanoparticles has demonstrated their capacity to modify the release kinetics, following Higuchi, Korsmeyer-Peppas or the first-order kinetics [66–69], topical of modified release formulations.

These features underline the huge potential of lipid nanoparticles in different areas of drug delivery. Nanoemulsions have been used in medical practice for over five decades [70] and their importance as a drug targeting delivery system has been widely studied and reviewed by a number of authors around the world [55, 70–72]. Likewise, liposomes have been explored extensively aiming antitumor therapy. The possibility of integrating well-established targeting ligands into already approved liposome drug carriers and creating new potential alternatives to improve therapeutic delivery took liposomes to the various phases of clinical studies. Nowadays, there is a reasonable number of approved liposome-based therapeutic alternative for clinical use and several others at different stages of clinical trials [73]. They have been well documented in several reports, including from an innovative point of view [54, 73–76]. On the other hand, lipid nanoparticles (SLN and NLC) are "relatively young" compared with the liposomes and nanoemulsions. The studies related to these nanoparticles, mainly those regarding the drug targeting delivery system, although promising, are still restricted to research laboratories. Many companies and research groups are working on this theme. Better knowledge about the polymorphism of the lipids and the relation between lipid matrix and release profile and establishing relationships between physical chemistry and biofate of nanoparticles are some challenges that explain the longer time-to-market. In view of this, we dedicate the content of this chapter exclusively to studies regarding the drug targeting of lipid nanoparticles.

3.1 Tumor targeting

The development of lipid nanoparticles has received considerable attention in the field of oncology, as they can be highly selective for tumors, providing a local controlled release of the chemotherapeutic drug. SLN and NLC reduce drug toxicity by limiting systemic distribution and circulation time of the drug in the blood [22]. Furthermore, chemotherapeutics require a certain concentration to be effective in the site of action, and the clearance of these molecules by the RES can limit their bioavailability and activity [22]. The tumor microenvironment differs from healthy tissues in different aspects, such as the degree of oxygenation and perfusion, pH, and metabolic activity [77]. Because of their small size, nanoparticles take advantage of the leaky tumor microvasculature being retained, there by passive targeting and releasing the drug directly to neoplastic cells [77].

The higher vascular permeability coupled with the damaged lymphatic drainage in tumors provides a good opportunity for tumor targeting [3]. For passive tumor, targeting the nanoparticles should have long blood circulation time, not lose therapeutic activity while in circulation and keep the drug loaded until the target is achieved [3]. They should be prepared in pH-sensitive formulations to promote drug release in an acid environment and can be associated with antibodies that recognize tumor cells. Moreover, lipid nanoparticles may also have chemical modifications to prevent detection by the immune system or even to increase drug solubility [44].

3.2 Brain targeting

Considering the prevalence of CNS diseases and challenges related to potent drugs to pass the BBB, there is an urgent need for the development of a successful drug delivery system for the brain. One interesting application is the delivery of drugs to the CNS and, in particular, to the brain [14–17, 78, 79], being a promising system for the treatment of neurodegenerative disorders [53, 80]. Lipid nanoparticles can be surface-tailored for site-specific targeting to the BBB [15], they can escape from the RES, and bypass the liver [4]. There are limited studies related to active targeting of lipid nanoparticles coupling ligands to the particle surface. The main reason for that is related to the chemical structure of the lipid nanoparticles, with little flexibility for chemical modification as compared to polymeric nanoparticles. Thus the binders are attached to the surfactants or steric stabilizers. Different classes of drugs, such as antipsychotics, antiparkinsonian, antiischemic, and antibiotics, have been encapsulated in lipid nanoparticles with the purpose to modify the biodistribution and achieve brain targeting [81]. However, extensive studies are still required to establish the parenteral acceptability of the components for successful commercialization of lipid nanoparticle-based products.

4. Toxicity assessment of lipid nanoparticles

The expedition of the transition into clinical applications is a crucial issue that entails an urgent need for data relating to the safety profile of nanomedicines [82]. This section provides insights with respect to toxicological evaluations relevant for lipid nanoparticles for parenteral administrations. NLC are per se composed of biocompatible lipids; however, it should be noted that lipids might nonetheless exhibit toxicity when administered at high doses. Furthermore, other than the dose, the type of lipid used is another important experimental factor affecting toxicity. The positive charges on cationic lipids can promote binding of lipid nanoparticles to circulating blood cells, while the presence of unprotected surface negative charges on lipid molecules serve as binding sites for plasma opsonin, which favors their uptake by macrophages [83]. The most commonly used strategy to evaluate the toxicity of nanoparticles consists of in vitro assays with various cells lines. Till date, only a few publications have reported in vivo toxicity assessments [83]. In view of this the current section has been dedicated to in vitro and in vivo toxicity studies pertaining to lipid nanoparticles for parenteral delivery. Howbeit, there is paucity of publications dealing with toxicity of lipid nanoparticles formulated for parenteral drug delivery.

4.1 In vitro toxicity evaluation

Despite the proclamation that the lipids and surfactants used for the production of SLN are of GRAS status, the cytotoxicity of these raw materials is dependent on the concentration and cell model. The administration route also plays a decisive role in setting the limits of acceptance. In addition to the raw materials the cytotoxicity of SLN is dependent on the size and surface properties and on the concentration of the particles in contact with the cell. Any cytotoxic effect of particles can be ascribed to their adherence to the cell membrane, particle internalization and/or degradation of products in the cell culture medium or inside the cells [84]. With limited data available for support, detailed studies regarding the long-term effects of lipid nanoparticles, particularly charged ones, are still needed to reduce any possible threats.

4.1.1 Cytotoxicity

The interactions of lipid nanoparticles and their cytotoxicity profiles have been investigated against a range of cell lines. Current cytotoxicity assessments of lipid nanoparticles are particularly confined to cell viability measurement. Olbrich et al. investigated the influence of various surfactants on the cytotoxicity of Dynasan 114 (D114) SLN via MTT (3-(4,5-dimethylthiazol-2-yl)-2,5-diphenyl tetrazolium bromide) assay and cytokines levels assessment (such as interleukin [IL] 6, IL 12, and tumor necrosis factor-α [TNF-α]). The results demonstrated that surfactants such as sodium cholate, Poloxamer 188, Tween 80, Lipoid 575, Poloxamine 908, and sodium dodecyl sulfate were nontoxic to RAW 264.7 cells. With these surfactants, cytokine production was also retarded. However, with CPC (cetylpyridinium chloride) as stabilizing surfactant, SLN were shown to be cytotoxic in a concentration-dependent fashion. Similar results were derived when the results of the RAW 264.7 were compared with the data from toxicity evaluation on animals. This study also advocated the possibility of using RAW 264.7 cell lines for cytotoxicity evaluation of colloidal drug delivery systems, instead of laboratory animals, as source of macrophages [85]. In another study, Bondi et al. (2009) developed nimesulide-based SLN for intravenous administration. In addition to routine characterization, in vitro cytotoxicity studies were performed on SW-480 and HT-29 cell lines (human colorectal cancer cells). In this research the cells from chosen cell lines were incubated with blank SLN to ensure that cytotoxicity was due to drug itself and not by formulation excipients. Results showed no cytotoxicity for blank SLN after 72 h on both cell lines, even at the highest concentrations. In the presence of free nimesulide or nimesulide SLN the survival of SW-480 and HT-29 was found to decrease in a dose-dependent fashion. In addition, it was observed that nimesulide SLN showed antitumor activity comparable to free drug, ensuring that the presence of the nanocarriers did not reduce the drug activity [86]. Nimesulide has also been loaded into glyceryl behenate–SLN and their lack of cytotoxicity in Caco-2 cell lines has also been confirmed [87]. The cytotoxicity of several monoterpenes (known for their anticancer and antioxidant properties) was also reduced when loaded into SLN composed of glycerol monostearate and tested in several cell lines (e.g., HaCaT, A431, RAW 264.7, Caco-2, MCF-7 cell lines) [12, 67, 88, 89]. Dolatabadi et al. performed cytotoxicity evaluation of alendronate sodium SLN via MTT assay. The toxicity of alendronate sodium and its SLN was estimated by cell viability study using A549 cells. It was observed that at the highest concentration used, more than 90% of the cells remained viable. This group further performed genotoxicity studies on the prepared nanoformulation (please see ahead in this work, in the genotoxicity section) [90]. In the same year, Lopes et al. performed a comparative study of liposomes and lipid nanoparticles encapsulating oryzalin for improvement of its performance. For cytotoxic evaluation, THP-1 (human monocytic cell line) was chosen. The results were analyzed (quantitative measurement of cell damage after incubation with samples) for propidium iodide fluorescence by flow cytometry. It was revealed that free oryzalin exhibited cytotoxicity toward THP-1 cells (IC_{50} of 20 μM), whereas both oryzalin formulations evidenced no cytotoxicity effects up to the highest concentrations chosen. Hence, the optimized oryzalin nanoformulations demonstrated protective action to the selected mammalian cells [91]. In a recent study, Aparicio-Blanco et al. (2019) developed lipid nanoparticles enthused with cannabidiol (CBD) for brain targeting and performed both in vitro and in vivo evaluation. This study was devoted to in vitro cytotoxicity and cellular uptake assessment, using hCMEC/D3 cell lines (human brain endothelial cells). The fluorescence intensity of the treated cells (with fluorescent dyes DiO and DiD NLC) was evaluated using a flow cytometer and taking blank NLC as control. Results of the cytotoxicity study demonstrated no toxicity for any of the lipid nanoformulation, at any point of time. In this study, this nontoxic concentration of the prepared formulation was selected for subsequent in vitro experiments. Results of quantitative analysis by flow cytometer exhibited a time-dependent cellular uptake for all formulations with higher fluorescence intensity for 24 h. For both plain and CBD-loaded NLC, it was observed that smaller size particles showed higher cellular uptake by human cerebral endothelial cells at 4 and 24 h time period. Further, it was seen from the results that CBD-enthused NLC enhanced the blood–brain targeting property to a greater extent than plain NLC. This research group also evaluated the targeting abilities of NLC via laser scanning confocal microscopy and monolayer integrity of hCMEC/D3 cells. In addition, biodistribution study in healthy mice was also performed [92].

Also in 2019, Andrade et al. produced praziquantel SLN using supercritical fluid technique. In this study, in addition to physicochemical characterization, in vitro cytotoxicity was investigated using fibroblasts as the cellular model. The cytotoxicity profile of both blank and drug-loaded SLN was obtained as percent cell viability. For pure praziquantel the cell viability was more than 88%; however, blank SLN and praziquantel SLN reported 97.78% and 86.77% viability, respectively, thus confirming the biocompatibility and safety of the prepared nanoformulations against human fibroblast L2929 cells [84].

Mittal et al. formulated genistein NLC with the help of solvent evaporation and emulsification method [93]. The resultant nanoformulation was characterized appropriately for physicochemical characteristics. The formulations were further evaluated and pharmacokinetics and biodistribution assessments were performed in vivo using female Wistar rats. In this research, human ovary cancer PA-1 cell line was used for in vitro cytotoxicity evaluation via MTT assay. The results exhibited that genistein-loaded NLC had remarkably higher toxicity, which was found increasing with increase in concentration. However, the lipid components used for this formulation (without D-α-tocopherol PEG 1000 [TPGS]) did not show any cytotoxicity. The augmented anticancer potential of genistein NLC was attributed to the presence of TPGS. TPGS has been well established for its anticancer property as it inhibits efflux of drug moieties from the cancer cells. The author concluded that the enhanced cytotoxicity by TPGS will help in providing adjuvant therapy in the treatment of ovarian cancer [94].

In a recent study, SLN of 4-(N)-docosahexaenoyl 2′,2′-difluorodeoxycytidine were developed for improvement in solubility, stability, and antitumor efficacy. Cytotoxicity studies were performed against tumor cell line (B-16-F10, TC-1, and M-Wnt) and in vivo evaluation was performed in mice bearing B-16-F10 murine melanoma. Survival of tumor cell was determined after incubation with 4-(N)-docosahexaenoyl 2′, 2′-difluorodeoxycytidine (DHA-dFdc) SLN. It was observed that the drug-loaded SLN were more cytotoxic than the pure drug in M-Wnt and B-16-F10 cell lines. However, against TC-1 cell line, the cytotoxicity of drug-loaded SLN and pure drug was found to be similar. Blank SLN and the vehicle that used DMSO (dimethyl sulfoxide) did not exhibit significant cytotoxicity with chosen concentrations in any of the cell lines [95].

In another study, Zafar et al. optimized chitosan-grafted lipid nanoparticles for the coadministration of docetaxel and thymoquinone for the treatment of breast cancer. This group conducted endosome escape study, hemolytic toxicity assay, in vitro cell viability assay, and in vivo angiogenesis assay. In vitro cytotoxicity was evaluated using human breast cancer cell lines (MDA-MB-231 and MCF-7). The cytotoxic effects were found dependent upon drug concentration as well as incubation time [96].

Asthana et al. designed macrophage-targeted lipid polymer hybrid nanoparticles (LPNPs) (using cationic stearylamine) for improving the therapeutic efficacy and alleviating toxic effects of amphotericin-B. After appropriate characterization, safety of the formulation for parenteral delivery was evaluated using J774A.1 and attenuated erythrocytes. Further, in vitro and in vivo antileishmania efficacy for prepared nanoformulations was checked. In addition, biodistribution of amphotericin-B lipid polymeric nanoparticles was also assessed. The cytotoxicity evaluation was conducted on J774A.1 macrophages using colorimetric MTT assay. Herein, the cytotoxic potential was checked by measuring mitochondrial activity. It was demonstrated in this study that amphotericin-B LPNPs are remarkably less cytotoxic in comparison to free drug. This negligible toxicity to macrophages advocates their safe use [97].

The cytotoxicity of lipid nanoparticles is mainly attributed to the composition of the surfactants used for their stabilization in aqueous dispersions [98], which also govern the interaction of nanoparticles with the biological surroundings (e.g., cells). It has been reported that cationic nanoparticles have higher risk of cytotoxicity than anionic and nonsurface charged particles [99, 100].

4.1.2 Allergic reactions

Allergic reactions and immunological responses have not been investigated for lipid nanoparticles as drug carriers. Antigens entrapped or adsorbed on lipid nanoparticles are found to trigger a higher immunological response. Further, immune-based response can be affected by the fabrication technique or chosen excipients (lipids and additives). For example, using high-pressure homogenization (LAB 40 homogenizer made from V2A steel) may result in traces of nickel in the final product with a risk of allergic reactions [101]. Use of surfactant-like Cremophor EL was reported as allergic in paclitaxel emulsion, fabricated for parenteral use. Hence, extensive investigation with regards to immunological factors and ILs release exhibited nonsignificant IL concentrations (interferon [IFN], TNF, IL 12, and IL 6). Among these the absence of IL 12 advocated nontoxic profile of SLN as it is associated with severe toxicity in vivo [46, 102, 103]. Traversing through the literature pertaining to allergic reactions caused by parenteral delivery of lipid nanoparticles, it was inferred that the research in this sphere is still in infancy and that in-depth studies shall pave the way for deeper understanding of interactions occurring between the immune system and lipid nanoparticles.

4.1.3 Hemolytic toxicity

To evaluate the biocompatibility of lipid nanoparticles with red blood cells (RBCs), an in vitro hemolysis test has been used. This evaluation has been documented in limited research reports (to quantify the release of hemoglobin). Hemolysis study is generally performed using blood samples for the preparation of erythrocytes suspension. The acceptable limit of percentage hemolysis for various formulations is <5% [104]. Lopes et al.

evaluated the hemolytic potential of oryzalin liposomes and SLN. This study revealed that free oryzalin manifested hemolytic effect against RBCs; HC_{50}–425 μM. However, oryzalin nanoformulations demonstrated no hemolytic activity up to 500 μM concentration [91].

Recently, Nimtrakul et al. developed amphotericin-B–loaded NLC for parenteral delivery. This drug, in aggregation state, results in hematological toxicity. Therefore hemolysis assessment was performed using sheep blood. Fungizone (commercial amphotericin-B formulation) and free amphotericin-B are toxic even at low concentration (5 μg/mL) and both of these exhibited significant hemolysis in a dose-dependent fashion. The higher hemolysis reported with commercial formulation may be the result of sodium deoxycholate, which itself triggers hemolysis. Amphotericin-B NLC resulted in remarkably lower hemolysis, in comparison to free drug and commercial formulation, which may be attributed to lesser molecular aggregation of the drug in the formulation. Among the three fabricated NLC (using stearic, palmitic, and myristic acid), stearic and palmitic acid-based NLC showed less hemolysis at the highest drug concentration (20 μg/mL) used; however, myristic acid-based NLC resulted in high hemolytic potential (p<0.05). It is interesting to note that all the three blank NLC formulations exhibited negligible hemolysis at the selected dose, advocating that the blank lipid formulations were not responsible for hemolysis. Hemolysis obtained with myristic acid NLC can be correlated to the smallest size of this formulation. As a result, greater tendency to bind these with RBC's membrane is speculated. Hence, some aggregation of the drug in myristic acid NLC, locked on the surface of the nanoparticles, was likely to disrupt the membrane of RBCs [105].

In another study in the same year, Zafar and his research group performed hemolytic toxicity assay for blank chitosan-grafted lipid nanocapsules and compared them with docetaxel- and thymoquinone-loaded lipid formulations of similar nature. Double-distilled water, the positive control, demonstrated complete hemolysis, owing to hypotonic medium. However, blank formulation reduced the negligible hemolytic potential owing to isotonicity of the medium. The different concentrations of drug-loaded lipid formulations resulted in hemolysis between 2.5 ± 0.2% and 4.7 ± 0.2%. The authors suggested that a higher positive charge with a higher concentration of lipid formulations may be responsible for slightly higher hemolysis. The percentage hemolysis observed in this study was within the acceptable limits [96].

4.1.4 Genotoxicity

Genotoxicity is related to DNA damages on human cells which may be investigated via, for example, Comet and micronucleus assays to detect chromosomal aberrations, oxidative DNA damage, DNA strand breaks, mutations, DNA adducts, and micronucleus formation [51, 52]. 4′,6-Diamidino-2-phenylindole (DAPI) staining for nucleus shrinkage assessment in A549 cells treated with alendronate sodium and alendronate sodium-loaded

SLN was performed by Ezzati et al. Morphology of DAPI-stained cells exhibited no obvious fragmentation in the chromatin and DNA rings within the nucleus of cell's treated with alendronate sodium and alendronate sodium-loaded SLN and their morphology was also the same as untreated normal cells [90]. In a study by Roux et al., genotoxicity was investigated using Comet and micronucleus assay on human lymphocytes. As revealed from the result of the Comet assay, lipid nanoparticles showed no DNA damage at noncytotoxic concentration. Further, results of micronucleus assay did not disclose any micronucleated, binucleated or mononucleated cells [82]. In the following year, Bahadori and his collaborators evaluated the genotoxicity of lipid-based vesicles (liposomes) and micelles against L 929 healthy mouse fibroblast cells [106]. However, thorough literature search did not bring to light any published reports of genotoxicity studies conducted with lipid nanocarriers meant for parenteral delivery. Veritably, genotoxicity studies are warranted to complete the safety profile of these carriers for parenteral administration.

4.1.5 Toxicokinetics

Toxicokinetics deals with the uptake as well as the disposition of xenobiotics and its toxic response in the body, encompassing the assessment of absorption, degradation, metabolization, and elimination. It furnishes information regarding the biological and chemical basis for toxicological effects observed inside the body, relating them to the target dose, the drug on its concentration or concentration of its metabolites in the body compartments [107]. In purview of toxicokinetics of SLN, sole study has been published for epigallocatechin-3-gallate SLN meant for oral administration. As per the reported data the fabricated lipid nanoparticles did not exhibit any signs of acute or subchronic toxicity in rats when compared with free drug. Future research in the field of parenteral lipid nanoparticles should include this vital parameter [108]. Mitochondrial dysfunction as well as inefficiencies resulting in reactive oxygen species (ROS) production and nitric oxide quenching have been hypothesized as prime disruptors of normal microvascular reactions. These associated side effects caused by drug overdose may result in mitochondrial dysfunction. The controlled release of drug from lipid nanoparticles has been hypothesized as an approach to reduce the risk of mitochondrial dysfunction [81].

4.1.6 Tissue injury

Nanoparticles may induce injury and damage to various body tissues upon subcutaneous and intramuscular administration. Ma et al., in their research work, investigated hepatic and renal tissue injury (via oxidative stress) of magnetic Fe_3O_4 nanoparticles after intraperitoneal injection. This group detected the bio members (ROS, glutathione, malondialdehyde, DNA protein crosslinks, and 8-hydroxyl-2′-deoxyguanosine) to examine the level of oxidative stress and consequent damage incurred [109]. Literature pertaining to parenteral lipid nanoparticles lacks any report of investigations performed with similar objective.

4.1.7 Phototoxicity

Several drugs have the potential to induce phototoxicity that manifests in the skin. Phototoxic reactions are also triggered by incident sunlight. These effects have been reported for systemically delivered fluoroquinolone antibiotics [110]. The literature is, however, deficient in publications reporting the studies of this kind for parenteral lipid nanoparticles.

4.1.8 Macrophage uptake

The RES plays a significant role in clearing nanoformulations from systemic circulation. Further, it has been evidenced that nature and pattern of routine absorption on the surface of lipid nanocarriers determine their organ distribution. Therefore targeting of lipid nanostructure is governed by their physicochemical parameters. Depending on the surface properties of lipid nanoparticles, proteins adsorb on their surface after intravenous administration. As a result, these nanoparticles adhere to the cells possessing complementary receptors to the adsorbed proteins. For example, nanoparticles possessing adsorbed protein with an opsonic function, such as immunoglobulin IgG and complement factor C4γ, are cleared via RES cells, whereas the absence of opsonins and presence of dysopsonins like albumin IgA provides stealth features. Coating of nanoparticles with apolipoprotein E results in their selective brain targeting, as this protein plays a vital role in the transport of lipoprotein in the brain via LDL receptor. Therefore the apolipoprotein E–modified nanocarriers mimic the lipoprotein, leading to their uptake by the brain via endocytic pathway. Controlled fabrication of nanocarriers with selective surface properties results in appropriate absorption of the targeted protein, and subsequently, to site-specific intravenous administration [111].

Following intravenous delivery, nanoparticles are quickly removed from the blood circulation owing to their recognition by RES cells such as Kupffer cells present in liver, spleen, and bone marrow macrophages. Therefore Vonarbourg et al. evaluated PEG-coated lipid nanoparticles for their interaction with the immune system. PEG–lecithin shell is known to delay the recognition of lipid nanoparticles by the immune system affecting their macrophage uptake. Results were measured by means of complement activation crossed-immunoelectrophoresis. With an increase in the size of nanoparticles, enhanced macrophage uptake was demonstrated. Further, decreases in the flexibility of PEG chains can trigger longer contact period, sufficient for recognition by macrophages receptors, resulting in their quick removal from the bloodstream. It was reported that modifying the lipid nanoparticles with longer PEG chains may result in potent, intravenous, drug administration, accompanied by long circulation time [112].

In another study, lipid nanospheres for curcumin were formulated to administer the moiety into tissue macrophages via intravenous injection in rats. After a time period of 6 h postadministration the tissue sections were observed by confocal laser scanning microscopy, (CLSM) which illustrated that the nanoparticles were massively distributed in bone marrow and spleen macrophages [113].

Berger et al. produced targeted lipid-based nanoparticles for human immunodeficiency virus (HIV) treatment. Macrophages are important players in HIV infection as the HIV can survive for longer duration in these cells. In addition to other evaluations the prepared lipid nanoformulations were tested by CLSM. The results evidenced that the formulation is quickly internalized into THP-1 macrophages, being able to escape the lysosomes, allowing the delivery of small interfering RNAs. For evaluation purposes, THP-1 monocyte macrophages were used as in vitro cellular model [114].

In a study, calcitriol was encapsulated for macrophages targeting and treatment of inflammatory disorders. PEGylated drug-loaded lipid nanoparticles were designed to target macrophage-specific endocytic receptors (CD163). The specific uptake of fabricated lipid nanoformulations can be targeted to macrophages, as evidenced by flow cytometry, confocal microscopy of lipid nanoparticles in mice organs, in vivo [115].

4.2 In vivo toxicity evaluation

In vitro toxicity evaluation of nanoparticles gives a valuable insight into the adverse reactions associated with nanoparticulate drug delivery; nonetheless, in vivo studies bring us a step closer to their clinical benefit. The following section has been included with the view to throw some light on the in vivo toxicological evaluation of lipid nanoparticles for parenteral delivery.

4.2.1 Tissue distribution

Subsequent to parenteral delivery of lipid nanoparticles, they may be easily translocated throughout the various body parts via systemic circulation. Their distribution may be related to their size and surface properties such as lipophilicity, hydrophilicity, and polarity. It is also hypothesized that small-sized nanoparticles are taken up at a faster rate by the cells than their larger counterparts [116].

In vivo studies, in this regard, indicated that liver, lungs, and kidneys represent the major distribution sites as well as target organs for nanoparticles exposure. From this perspective the clearance behavior of nanomaterials in the above-mentioned organs is crucial for understanding their fate in vivo [117]. On injection of nonstealth SLN, their accumulation within the Kupffer cells of the liver was notably observed.

Yang et al. investigated the biodistribution of intravenously injected camptothecin (CA) SLN [27]. It was a detailed study in which body distribution of CA SLN in C57BL/6J mice was performed and concentration of drug in heart, liver, spleen, lungs, kidneys, and brain was measured, using reversed phase high performance liquid chromatography along with their mean resident time (MRT). CA-loaded SLN were fabricated using stearic acid (2.0% w/w), soy bean lecithin (1.5% w/w), and poloxamer 188 (0.5% w/w) using high-pressure homogenization. In addition to routine characterization the potential of CA SLN for brain targeting was evaluated. The results indicated that the observed sustained release behavior of the formulation was due to the protection conferred by the SLN matrix to inhibit the hydrolysis of CA lactone (active) form to carboxylate (inactive) form. In comparison to CA solution, CA SLN exhibited a higher uptake in RES organs, a prolonged MRT of drug in various tissues and its targeting to lungs, liver, spleen, heart, and brain after intravenous administration. The prepared formulations have shown effective brain targeting potential via crossing the BBB. However, in this study, no toxicity at selected drug concentration has been reported [27].

In another study, cationic stearylamine LPNPs, that combine characteristics of liposomes and polymeric nanoparticles, have been developed to improve the therapeutic efficacy and reduce toxic effects of amphotericin-B. After formulation and characterization, macrophage uptake study, in vitro evaluation of parasite growth inhibition, in vivo antileishmania activity and in vitro toxicity evaluation (hemolytic and cytotoxicity assay) was performed. In addition, in vivo biodistribution study and in vivo toxicity evaluation through biochemical markers was performed. For biodistribution study, male Wistar rats were chosen as animal model and plasma and tissues (spleen, liver, kidneys, heart, and lungs) were processed after intravenous administration of amphotericin-B poly(D,L-lactideco-glycolide polymeric nanoparticles, amphotericin-B LPNPs, and amphotericin-B. In vivo liver and spleen distribution data clearly demonstrated that higher in vitro phagocytosis leads to high accumulation in liver and spleen (in vivo). Further, a retardation in renal sequestration of amphotericin-B (in kidneys) was observed with drug-loaded LPNPs, in comparison with other tested formulations. Routine amphotericin therapy is found associated with nephrotoxicity; therefore the results of this study are clinically relevant. Evaluation of biomarkers (upregulation of IL-12 and IFN-γ in splenocytes, amplified levels of cytokines, TNF-α, and upregulation of NO secretion) advocated the T-cell–mediated immunomodulation, resulting in immunomodulatory targeting through stearylamine hybrid nanoparticles. This study also demonstrated the safety of amphotericin-B LPNPs supported by their minimal distribution in kidneys and reduced concentration of nephrotoxicity markers (blood urea nitrogen and creatinine) [97].

In a study performed with CBD lipid nanoparticles (discussed earlier), biodistribution of lipid nanoparticles was checked in healthy mice. The possible toxicity was ruled out in this study, as the dose taken for NLC was previously reported nontoxic for intravenous administration. The dose of CBD was also on the safe side in this study, in comparison to its tolerated dose in mice (up to 120 mg/kg) [92]. In another study, genistein-loaded NLC were evaluated for in vivo biodistribution in healthy female Wistar rats. After intravenous administration of the formulation the total amount of the drug accumulated after 48 h was computed, the area under the curve (AUC) was found less for heart, brain, lungs, and kidneys. With administration for pure drug, however, AUC values were found increased to a certain extent. When genistein nanoformulation was administered, extensive AUC enhancement was observed in liver and ovary, in comparison to pure drug solution. The higher tissue distribution of intravenously delivered nanoformulation was largely influenced by the nanosize of the particles. The nanoparticles administered to the systemic circulation were not directly absorbed there and were taken by RES as well as by ovarian tissue (estrogen receptors), resulting in their accumulation, thereby enabling passive targeting [94].

4.2.2 Immunotoxicity

Prior to parenteral application, lipid nanoparticles must be examined for any possible interactions with the immune system [50, 52]. Surface chemistry of the nanocarriers and any modification thereof dictates their immunogenic potential. It has been reported that nanoparticles are capable of stimulating as well as repressing immune responses [116]. There have been limited investigations in this context, concerning the parenteral administration of nanoparticles, particularly lipidic systems. As the evaluation of immunotoxicity depends upon the measurement of cytokine production, the in vivo studies may be directed toward the assessment of factors such as ILs, chemokines, macrophage inflammatory proteins, monocyte chemoattractant protein, granulocyte macrophage–colony stimulating factor and TNFs. Histological changes in cardinal immune organs (such as spleen) may be checked by hematoxylin and eosin staining. Studies performed with other types of nanoparticles suggest that nanoparticulate formulations could be the contributing factor toward the pathogenesis of proinflammatory conditions (such as allergy), particularly in lungs, via oxidative stress mechanism [116].

4.2.3 Organ-related toxicity

On account of their sub-micron size, SLN may be administered either by intravenous, intramuscular, or subcutaneous route, or specifically to the target organ [46]. Physiological interactions of nanoparticles result in pathological changes characterized by morphological alterations and functional impairment. With respect to in vivo toxicity assessment, selection of an appropriate animal model is imperative, and the need for corroboration of a standard, predictive model cannot be overemphasized. Mice and rats remain the most commonly used animals for nanotoxicological investigations. It is so because their genome is distinctly known for toxicity testing of organs such as liver, kidney, lungs, heart, brain, ovary/testis, and spleen. In recent studies, zebrafish (*Danio rerio*), *Caenorhabditis elegans* [118], and rabbit [119, 120] have also been used. In vivo assays may be categorized as: (i) those involving structural infiltration of the pivotal organs of the body and (ii) those targeting particular systems whose structural attributes tend to accumulate nanoparticles. Hepatic sinusoids and Kupffer cells are examples of the latter, which are the main structures responsible for hepatic metabolism and detoxification. Similarly, renal filtration membrane is also known to deposit nanoparticles. The establishment of a suitable animal model for toxicological evaluation of nanoparticulate systems must take into perspective the route of exposure and intended application [50, 52, 121].

Liver is the chief organ implicated in the metabolism and detoxification of administered xenobiotics. The data furnished by in vivo rodent models show that nanomaterials are liable to deposition in the liver, as upon intravenous injection, blood flowing at high rate delivers good concentrations of toxic substances to it, leading to high levels of exposure. Therefore hepatotoxicity evaluation of lipid nanoparticles assumes much importance. On the same lines, kidney and associated nephrotoxicity also call for attention of researchers. Kidney, being the chief site for the clearance of nanomaterials, faces the risk of damage of basement in the glomeruli [116]. Hepatotoxicological assessment encompasses the application of immunohistochemistry to track down fibrosis and inflammation of the liver. Analytic determination by serum enzymology detects hepatic dysfunction by quantitative measurement of aspartate aminotransferase, alanine aminotransferase, alkaline phosphatase, and c-glutamyl transferase. On the other hand, histopathological investigation can be performed to uncover degenerative changes in glomerulus, whereas immunohistochemistry can be used to confirm pathologic alterations caused by collagenous tubulointerstitial matrix and glomerulosclerosis. In this case, markers such as IFN-6, vimentin, and transforming growth factor- β_1 are detected. Kidney special dyeing and microscopic observation can also be used to ascertain pathologic changes. Measurement of kidney indices gives an insight into the functional aspect of the organ.

Weyhers et al. prepared SLN with two different lipid cores, such as compritol and cetyl palmitate, and characterized them for intravenous administration [122]. The resulting dispersions were subject to in vivo toxicity assessment in mice for their effect on liver and spleen, upon intravenous bolus injection, six times for 20 days period (high dose). In addition, multiple low-dose injections, as well as cetyl palmitate SLN dispersions, were also administered. Histological examination of hepatic and splenic tissues revealed that the results were dependent on the lipids composing the matrix and on the dose injected. High dose of Compritol–SLN formulation was found to be associated with the deposition of the lipid in liver, along with spleen, resulting in pathological changes. These were found to be partially reversible, within 6 weeks, subsequent to termination of dosing. On the other side, cetyl palmitate SLN and low-dose Compritol–SLN dispersions were found to be well tolerated.

In addition to liver and kidney, lungs have also been reported to accumulate considerable concentrations of nanoparticles [123]. The ensuing pneumocyte injury and/or inflammatory reactions need to be ascertained. While the former may be checked by biochemical analysis of bronchioalveolar lavage (BAL) fluid for lactate dehydrogenase, the latter can be tested by determining the number of BAL-recovered neutrophils. Furthermore, assessment of lipid peroxidation and glutathione synthesis accompanied by pathological and apoptotic investigation are also useful for pulmonary toxicity studies [124]. However, there are no published reports of pulmonary and renal toxicity studies performed with parenterally administered lipid nanoparticles.

The literature is barren with regard to information concerning in vivo toxicological investigations for parenteral lipid nanoparticulate formulations, on reproductive organs, heart, and brain. Considering the impact of intravenously injected nanoparticles on the cardiovascular system, evaluation of cardiotoxicity of lipid nanoparticles is of paramount importance. Hematological and serum analysis is quite useful in this respect. As observed clinically, phlebitis is the most common risk factor, which may be addressed by examining its typical symptoms in pathological sections. Hemolysis and thrombosis are also of concern, out of which thrombosis is particularly important. It is assayed using vascular thrombosis model in rats. Cardiac injury may be detected by analyzing serum biomarkers such as myoglobin, creatine kinase-MB, and troponin-T. Apart from these, cardiac calcium ion concentration, which is related to contractile function and DNA impairment, should also be measured. To account for oxidative stress, antioxidant enzymes such as glutathione peroxidase, superoxide dismutase, and catalase can also be measured. Radio-labeled lipid nanoparticles may be used to find out their cardiac uptake.

Reproductive toxicity study is a comparatively longer-term investigation that entails the measurement of relevant parameters for the male and female organism. Testicular tissue structure, sperm parameters, and serous testosterone levels are routinely examined in males. As for females, follicle stimulating hormone, luteinizing hormone, and estradiol serum levels need to be analyzed, in addition to functional evaluation of uterus, ovary, and vaginal tract. Histopathological examination may also be undertaken. As nanocarriers can penetrate placental–blood barrier, teratogenicity, reproductive index, and development of progeny must be given due emphasis.

To strengthen the viewpoint of safety of lipid nanoparticles, a holistic approach must be upheld that encompasses the investigation of all the parameters relevant to parenteral delivery. In the landscape of pharmaceutical research, formulation scientists focus on the development of novel and efficient drug delivery systems, whereas toxicologists direct their efforts toward the toxic effects of drugs and drug products on cells, tissues, organs, or organ systems. The two can, however, work in tandem and complement each other. Seemingly, there are a number of segments with regard to parenteral lipid nanoparticles that lack toxicological data (Table 2). Considering this, the said sections have been included in the chapter so that an outlook regarding their prospective investigation may be developed.

5. Conclusions and perspectives

The potential of lipid nanoparticles for the parenteral delivery and drug targeting has been successfully established. The involved challenges reveal the limitations still faced by the pharmaceutical industry in the development of new nanotechnologies for controlled and target drug delivery. In recent years the advances in this area are still limited to a few patent applications, mainly from the academy. Understanding the interactions between

Table 2 Toxicity assessment of lipid nanoparticles fabricated for parenteral delivery.

Drug	Type of lipid Nanocarriers	Toxicity assessment (In vitro and in vivo)	References
4-(N)-docosahexaenoyl 2′,2′-difluorodeoxycytidine	SLN	In vitro cytotoxicity	[95]
Amphotericin-B	NLC	In vitro (nephrotoxicity & hemolytic toxicity)	[105]
Cannabidiol	LNCs	In vitro and in vivo (cytotoxicity, cellular uptake, monolayer integrity and biodistribution study)	[92]
Nimesulide	SLN	In vitro (cytotoxicity and cell culture viability assay)	[86]
Genistein	NLC	In vitro and in vivo (cytotoxicity, biodistribution study)	[94]
Alendronate sodium	SLN	In vitro (cytotoxicity)	[90]
Docetaxel, Thymoquinone	LNCs	In vitro and in vivo (cytotoxicity, hemolytic toxicity, and cell viability study)	[96]
Oryzalin	SLN	In vitro (cytotoxicity and hemolytic toxicity)	[91]
Dynasan 114	SLN	In vitro cytotoxicity	[85]
Praziquantel	SLN	In vitro cytotoxicity	[84]
Amphotericin-B	Lipid–polymer hybrid nanoparticles (LPNPs)	In vitro and in vivo cytotoxicity and biodistribution study	[97]
Nucleic acid	Lipid nanoparticles	In vivo hepatotoxicity	[125]

lipid nanoparticles and their environment allows us to identify the characteristics that determine the fate of drug delivery devices during manufacture, storage, and after injection. This knowledge allows the design of highly specific delivery systems, for different classes of drugs and targets. Nearly 30 years after lipid nanoparticles creation, they are already present at several areas of application. It now remains to explore their promising potential as targeting mechanisms. Despite the encouraging results found so far, the development of lipid-based nanoparticles for parenteral use is still expected to grow exponentially as more information about in vivo conditions are available.

Acknowledgments

The authors would like to thank the financial support received from Portuguese Science and Technology Foundation (FCT/MCT) and from European Funds (PRODER/COMPETE) under the project references M-ERA-NET/0004/2015-PAIRED and UIDB/04469/2020, co-financed by FEDER, under the Partnership Agreement PT2020. The authors also acknowledge CAPES (Coordenação de Aperfeiçoamento de Pessoal de Nível Superior) for the financial support and for the fellowship of the first author (88887.368385/2019-00). This study was also supported by the Portuguese Foundation for Science and Technology (FCT) under the scope of the strategic funding of UIDB/04469/2020 unit and BioTecNorte operation (NORTE-01-0145-FEDER-000004) funded by the European Regional Development Fund under the scope of Norte2020 - Programa Operacional Regional do Norte.

References

[1]   M. Yokoyama, Drug targeting with nano-sized carrier systems, J. Artif. Organs 8 (2005) 77–84.

[2]   V.P. Torchilin, Drug targeting, Eur. J. Pharm. Sci. 11 (2000) S81–S91.

[3]   M.R. Mozafari, Nanocarrier Technologies, Springer Netherlands, Dordrecht, 2006.

[4]   S.G. Patel, M.D. Patel, A.J. Patel, M.B. Chougule, H. Choudhury, Solid lipid nanoparticles for targeted brain drug delivery, in: P. Kesharwani, U. Gupta (Eds.), Nanotechnology-Based Targeted Drug Delivery Systems for Brain Tumors, 2018, pp. 191–244.

[5]   N. Karra, S. Benita, The ligand nanoparticle conjugation approach for targeted cancer therapy, Curr. Drug Metab. 13 (2012) 22–41.

[6]   A.P. Singh, A. Biswas, A. Shukla, P. Maiti, Targeted therapy in chronic diseases using nanomaterial-based drug delivery vehicles, Signal Transduct. Target. Ther. 4 (2019) 33.

[7]   H.L. Wong, R. Bendayan, A.M. Rauth, Y. Li, X.Y. Wu, Chemotherapy with anticancer drugs encapsulated in solid lipid nanoparticles, Adv. Drug Deliv. Rev. 59 (2007) 491–504.

[8]   L. Gastaldi, L. Battaglia, E. Peira, D. Chirio, E. Muntoni, I. Solazzi, M. Gallarate, F. Dosio, Solid lipid nanoparticles as vehicles of drugs to the brain: current state of the art, Eur. J. Pharm. Biopharm.: Off. J. Arbeitsgemeinsch. Pharm. Verfahrenstech. 87 (2014) 433–444.

[9]   R. Langer, Drug delivery and targeting, Nature 392 (1998) 5–10.

[10] E.B. Souto, I. Baldim, W.P. Oliveira, R. Rao, N. Yadav, F.M. Gama, S. Mahant, SLN and NLC for Topical, Dermal and Transdermal Drug Delivery, Expert Opinion on Drug Delivery, Accepted, (2020).

[11] S. Mahant, R. Rao, E.B. Souto, S. Nanda, Analytical tools and evaluation strategies for nanostructured lipid carrier based topical delivery systems. Expert Opin. Drug Deliv. 17 (7) (2020) 963–992, https://doi.org/10.1080/17425247.2020.1772750.

[12] L.N. Andrade, M.S.S. Cavendish, S.P.M. Costa, R.G. Amaral, C.B. Corrêa, D.S. Oliveira, M. Morsink, E.H. Gokce, R.L.C. de Albuquerque Junior, E.B. Souto, P. Severino, Perillyl alcohol in solid lipid nanoparticles (SLN-PA): cytotoxicity and antitumor potential in sarcoma 180 mice model. Precis. Nanomed. 3 (5) (2020) 685–698, https://doi.org/10.33218/001c.

[13] T.N. Pashirova, I.V. Zueva, K.A. Petrov, V.M. Babaev, S.S. Lukashenko, I.Kh. Rizvanov, E.B. Souto, E.E. Nikolsky, L.Ya. Zakharova, P. Masson, O.G. Sinyashin, Nanoparticle-delivered 2-PAM for rat brain protection against paraoxon central toxicity. ACS Appl. Mater. Interfaces 9 (20) (2017) 16922–16932, https://doi.org/10.1021/acsami.7b04163.

[14] S. Jose, S.S. Anju, T.A. Cinu, N.A. Aleykutty, S. Thomas, E.B. Souto, In vivo pharmacokinetics and biodistribution of resveratrol-loaded solid lipid nanoparticles for brain delivery, Int. J. Pharm. 474 (2014) 6–13.

[15] M. Patel, E.B. Souto, K.K. Singh, Advances in brain drug targeting and delivery: limitations and challenges of solid lipid nanoparticles, Expert Opin. Drug Deliv. 10 (2013) 889–905.

[16] S. Martins, I. Tho, I. Reimold, G. Fricker, E. Souto, D. Ferreira, M. Brandl, Brain delivery of camptothecin by means of solid lipid nanoparticles: formulation design, in vitro and in vivo studies, Int. J. Pharm. 439 (2012) 49–62.

[17] S. Martins, S. Costa-Lima, T. Carneiro, A. Cordeiro-da-Silva, E.B. Souto, D.C. Ferreira, Solid lipid nanoparticles as intracellular drug transporters: an investigation of the uptake mechanism and pathway, Int. J. Pharm. 430 (2012) 216–227.

[18] E.B. Souto, R.H. Muller, Lipid nanoparticles: effect on bioavailability and pharmacokinetic changes, Handb. Exp. Pharmacol. (2010) 115–141.

[19] S. Doktorovova, R. Shegokar, P. Martins-Lopes, A.M. Silva, C.M. Lopes, R.H. Muller, E.B. Souto, Modified rose Bengal assay for surface hydrophobicity evaluation of cationic solid lipid nanoparticles (cSLN), Eur. J. Pharm. Sci. 45 (2012) 606–612.

[20] L. Battaglia, M. Gallarate, Lipid nanoparticles: state of the art, new preparation methods and challenges in drug delivery, Expert Opin. Drug Deliv. 9 (2012) 497–508.

[21] A.E. Nel, L. Mädler, D. Velegol, T. Xia, E.M.V. Hoek, P. Somasundaran, F. Klaessig, V. Castranova, M. Thompson, Understanding biophysicochemical interactions at the nano–bio interface, Nat. Mater. 8 (2009) 543–557.

[22] P. Yingchoncharoen, D.S. Kalinowski, D.R. Richardson, E.L. Barker, Lipid-based drug delivery systems in cancer therapy: what is available and what is yet to come, Pharmacol. Rev. 68 (2016) 701–787.

[23] R.A. Petros, J.M. DeSimone, Strategies in the design of nanoparticles for therapeutic applications, Nat. Rev. Drug Discov. 9 (2010) 615–627.

[24] S.M. Moghimi, A.C. Hunter, T.L. Andresen, Factors controlling nanoparticle pharmacokinetics: an integrated analysis and perspective, Annu. Rev. Pharmacol. Toxicol. 52 (2012) 481–503.

[25] N. Pedersen, S. Hansen, A.V. Heydenreich, H.G. Kristensen, H.S. Poulsen, Solid lipid nanoparticles can effectively bind DNA, streptavidin and biotinylated ligands, Eur. J. Pharm. Biopharm. 62 (2006) 155–162.

[26] A. Fundaro, R. Cavalli, A. Bargoni, D. Vighetto, G.P. Zara, M.R. Gasco, Non-stealth and stealth solid lipid nanoparticles (SLN) carrying doxorubicin: pharmacokinetics and tissue distribution after i.v. administration to rats, Pharmacol. Res. 42 (2000) 337–343.

[27] S.C. Yang, L.F. Lu, Y. Cai, J.B. Zhu, B.W. Liang, C.Z. Yang, Body distribution in mice of intravenously injected camptothecin solid lipid nanoparticles and targeting effect on brain, J. Control. Release 59 (1999) 299–307.

[28] W. Wang, F. Zhou, L. Ge, X. Liu, F. Kong, A promising targeted gene delivery system: folate-modified dexamethasone-conjugated solid lipid nanoparticles, Pharm. Biol. 52 (2014) 1039–1044.

[29] W.M. Lim, P.S. Rajinikanth, C. Mallikarjun, Y.B. Kang, Formulation and delivery of itraconazole to the brain using a nanolipid carrier system, Int. J. Nanomedicine 9 (2014) 2117–2126.

[30] J. Varshosaz, F. Hassanzadeh, H. Sadeghi, M. Khadem, Galactosylated nanostructured lipid carriers for delivery of 5-FU to hepatocellular carcinoma, J. Liposome Res. 22 (2012) 224–236.

[31] Y. Chen, L. Pan, M. Jiang, D. Li, L. Jin, Nanostructured lipid carriers enhance the bioavailability and brain cancer inhibitory efficacy of curcumin both in vitro and in vivo, Drug Deliv. 23 (2016) 1383–1392.

[32] S. Song, G. Mao, J. Du, X. Zhu, Novel RGD containing, temozolomide-loading nanostructured lipid carriers for glioblastoma multiforme chemotherapy, Drug Deliv. 23 (2016) 1404–1408.

[33] M.J. Tsai, P.C. Wu, Y.B. Huang, J.S. Chang, C.L. Lin, Y.H. Tsai, J.Y. Fang, Baicalein loaded in tocol nanostructured lipid carriers (tocol NLCs) for enhanced stability and brain targeting, Int. J. Pharm. 423 (2012) 461–470.

[34] J. Ye, Q. Wang, X. Zhou, N. Zhang, Injectable actarit-loaded solid lipid nanoparticles as passive targeting therapeutic agents for rheumatoid arthritis, Int. J. Pharm. 352 (2008) 273–279.

[35] M. Wu, Y. Fan, S. Lv, B. Xiao, M. Ye, X. Zhu, Vincristine and temozolomide combined chemotherapy for the treatment of glioma: a comparison of solid lipid nanoparticles and nanostructured lipid carriers for dual drugs delivery, Drug Deliv. 23 (2016) 2720–2725.

[36] C.Y. Qu, M. Zhou, Y.W. Chen, M.M. Chen, F. Shen, L.M. Xu, Engineering of lipid prodrug-based, hyaluronic acid-decorated nanostructured lipid carriers platform for 5-fluorouracil and cisplatin combination gastric cancer therapy, Int. J. Nanomedicine 10 (2015) 3911–3920.

[37] S. Li, L. Wang, N. Li, Y. Liu, H. Su, Combination lung cancer chemotherapy: design of a pH-sensitive transferrin-PEG-Hz-lipid conjugate for the co-delivery of docetaxel and baicalin, Biomed. Pharm.Biomed. Pharmacother. 95 (2017) 548–555.

[38] X. Li, X. Jia, H. Niu, Nanostructured lipid carriers co-delivering lapachone and doxorubicin for overcoming multidrug resistance in breast cancer therapy, Int. J. Nanomedicine 13 (2018) 4107–4119.

[39] M. Colombo, F. Figueiró, A. de Fraga Dias, H.F. Teixeira, A.M.O. Battastini, L.S. Koester, Kaemp-ferol-loaded mucoadhesive nanoemulsion for intranasal administration reduces glioma growth in vitro, Int. J. Pharm. 543 (2018) 214–223.

[40] J.O. Eloy, A. Ruiz, F.T. de Lima, R. Petrilli, G. Raspantini, K.A.B. Nogueira, E. Santos, C.S. de Oliveira, J.C. Borges, J.M. Marchetti, W.T. Al-Jamal, M. Chorilli, EGFR-targeted immunolipo-somes efficiently deliver docetaxel to prostate cancer cells, Colloids Surf. B Biointerfaces 194 (2020) 111185.

[41] L. Li, H. Wang, Z.Y. Ong, K. Xu, P.L.R. Ee, S. Zheng, J.L. Hedrick, Y.-Y. Yang, Polymer- and lipid-based nanoparticle therapeutics for the treatment of liver diseases, Nano Today 5 (2010) 296–312.

[42] J.D. Byrne, T. Betancourt, L. Brannon-Peppas, Active targeting schemes for nanoparticle systems in cancer therapeutics, Adv. Drug Deliv. Rev. 60 (2008) 1615–1626.

[43] N.T. Huynh, E. Roger, N. Lautram, J.P. Benoit, C. Passirani, The rise and rise of stealth nanocarriers for cancer therapy: passive versus active targeting, Nanomedicine 5 (2010) 1415–1433.

[44] B. Garcia-Pinel, C. Porras-Alcala, A. Ortega-Rodriguez, F. Sarabia, J. Prados, C. Melguizo, J. M. Lopez-Romero, Lipid-based nanoparticles: application and recent advances in cancer treatment, Nanomaterials 9 (4) (2019) 638–661.

[45] H. Shrestha, R. Bala, S. Arora, Lipid-based drug delivery systems, J. Pharm. 2014 (2014) 1–10.

[46] S.A. Wissing, O. Kayser, R.H. Müller, Solid lipid nanoparticles for parenteral drug delivery, Adv. Drug Deliv. Rev. 56 (2004) 1257–1272.

[47] P. Blasi, S. Giovagnoli, A. Schoubben, M. Ricci, C. Rossi, Solid lipid nanoparticles for targeted brain drug delivery, Adv. Drug Deliv. Rev. 59 (2007) 454–477.

[48] D.P. Gaspar, A.J. Almeida, Surface-functionalized lipid nanoparticles for site-specific drug delivery, in: Y.V. Pathak (Ed.), Surface Modification of Nanoparticles for Targeted Drug Delivery, Springer International Publishing, 2019, , pp. 73–98.

[49] V. Andonova, P. Peneva, Characterization methods for solid lipid nanoparticles (SLN) and nanostruc-tured lipid carriers (NLC), Curr. Pharm. Des. 23 (2018) 6630–6642.

[50] S. Doktorovova, A.B. Kovacevic, M.L. Garcia, E.B. Souto, Preclinical safety of solid lipid nanopar-ticles and nanostructured lipid carriers: current evidence from in vitro and in vivo evaluation, Eur. J. Pharm. Biopharm. 108 (2016) 235–252.

[51] S. Doktorovova, A.M. Silva, I. Gaivao, E.B. Souto, J.P. Teixeira, P. Martins-Lopes, Comet assay reveals no genotoxicity risk of cationic solid lipid nanoparticles, J. Appl. Toxicol. 34 (2014) 395–403.

[52] S. Doktorovova, E.B. Souto, A.M. Silva, Nanotoxicology applied to solid lipid nanoparticles and nanostructured lipid carriers—a systematic review of in vitro data, Eur. J. Pharm. Biopharm. 87 (2014) 1–18.

[53] P. Blasi, A. Schoubben, G.V. Romano, S. Giovagnoli, A. Di Michele, M. Ricci, Lipid nanoparticles for brain targeting II. Technological characterization, Colloids Surf. B Biointerfaces 110 (2013) 130–137.

[54] D. Bobo, K.J. Robinson, J. Islam, K.J. Thurecht, S.R. Corrie, Nanoparticle-based medicines: a review of FDA-approved materials and clinical trials to date, Pharm. Res. 33 (2016) 2373–2387.

[55] M.C. Bonferoni, S. Rossi, G. Sandri, F. Ferrari, E. Gavini, G. Rassu, P. Giunchedi, Nanoemulsions for "nose-to-brain" drug delivery, Pharmaceutics 11 (2019).

[56] Q. He, J. Liu, J. Liang, X. Liu, W. Li, Z. Liu, Z. Ding, D. Tuo, Towards improvements for pene-trating the blood-brain barrier-recent progress from a material and pharmaceutical perspective, Cell 7 (2018).

[57] H. Svilenov, C. Tzachev, Solid lipid nanoparticles—a promising drug delivery system, in: Nanomedicine, 2014, pp. 187–237.

[58] M. Cavendish, L. Nalone, T. Barbosa, R. Barbosa, S. Costa, R. Nunes, C.F. da Silva, M.V. Chaud, E. B. Souto, L. Hollanda, P. Severino, Study of pre-formulation and development of solid lipid nano-particles containing perillyl alcohol, J. Therm. Anal. Calorim. (2019) 1–8.

[59] P. Severino, S.C. Pinho, E.B. Souto, M.H. Santana, Polymorphism, crystallinity and hydrophilic-lipophilic balance of stearic acid and stearic acid-capric/caprylic triglyceride matrices for production of stable nanoparticles, Colloids Surf. B Biointerfaces 86 (2011) 125–130.

[60] Z.-r. Huang, S.-c. Hua, Y.-l. Yang, J.-y. Fang, Development and evaluation of lipid nanoparticles for camptothecin delivery: a comparison of solid lipid nanoparticles, nanostructured lipid carriers, and lipid emulsion, Acta Pharmacol. Sin. 29 (2008) 1094–1102.

[61] R. H. Muller, R. Shegokar, C. M. Keck, 20 years of lipid nanoparticles (SLN and NLC): present state of development and industrial applications, Curr. Drug Discov. Technol. 8 (2011) 207–227.

[62] M. Uner, G. Yener, Importance of solid lipid nanoparticles (SLN) in various administration routes and future perspectives, Int. J. Nanomedicine 2 (2007) 289–300.

[63] S.M. Moghimi, J. Szebeni, Stealth liposomes and long circulating nanoparticles: critical issues in pharmacokinetics, opsonization and protein-binding properties, Prog. Lipid Res. 42 (2003) 463–478.

[64] M. Wacker, Nanocarriers for intravenous injection—the long hard road to the market, Int. J. Pharm. 457 (2013) 50–62.

[65] S.M. Moghimi, H. Hedeman, I.S. Muir, L. Illum, S.S. Davis, An investigation of the filtration capacity and the fate of large filtered sterically-stabilized microspheres in rat spleen, Biochim. Biophys. Acta Gen. Subj. 1157 (1993) 233–240.

[66] A. Zielińska, N.R. Ferreira, A. Feliczak-Guzik, E.B. Souto, I. Nowak, Release kinetics and stability assessment of monoterpenes-loaded solid lipid nanoparticles (SLN), Pharm. Dev. Technol. (2020) revised.

[67] E.B. Souto, S.B. Souto, P. Severino, J. Dias-Ferreira, B.C. Naveros, A. Durazzo, M. Lucarini, A. G. Atanasov, S. El Mamouni, A. Santini, Croton argyrophyllus kunth essential oil—loaded SLN: Optimization and evaluation of antioxidant and antitumoral activities. Sustainability 12 (18) (2020) 7697–7709, https://doi.org/10.3390/su12187697.

[68] E.B. Souto, A. Zielinska, S.B. Souto, A. Durazzo, M. Lucarini, A. Santini, A.M. Silva, A.G. Atanasov, C. Marques, L.N. Andrade, P. Severino, (+)-Limonene 1,2-epoxide-loaded SLN: Evaluation of drug release, antioxidant activity and cytotoxicity in HaCaT cell line. Int. J. Mol. Sci. 21 (4) (2020) 1449–1460, https://doi.org/10.3390/ijms21041449.

[69] R. Vieira, P. Severino, L.A. Nalone, S.B. Souto, A.M. Silva, M. Lucarini, A. Durazzo, A. Santini, E. B. Souto, Sucupira oil-loaded nanostructured lipid carriers (NLC): lipid screening, factorial design, release profile and cytotoxicity. Molecules 25 (3) (2020) 685–707, https://doi.org/10.3390/molecules25030685.

[70] K. Hörmann, A. Zimmer, Drug delivery and drug targeting with parenteral lipid nanoemulsions—a review, J. Control. Release 223 (2016) 85–98.

[71] Y. Singh, J.G. Meher, K. Raval, F.A. Khan, M. Chaurasia, N.K. Jain, M.K. Chourasia, Nanoemulsion: concepts, development and applications in drug delivery, J. Control. Release 252 (2017) 28–49.

[72] Z. Karami, M.R. Saghatchi Zanjani, M. Hamidi, Nanoemulsions in CNS drug delivery: recent developments, impacts and challenges, Drug Discov. Today 24 (2019) 1104–1115.

[73] M.M. El-Hammadi, J.L. Arias, An update on liposomes in drug delivery: a patent review (2014-2018), Expert Opin. Ther. Pat. 29 (2019) 891–907.

[74] T.N. Pashirova, I.V. Zueva, K.A. Petrov, S.S. Lukashenko, I.R. Nizameev, N.V. Kulik, A. D. Voloshina, L. Almasy, M.K. Kadirov, P. Masson, E.B. Souto, L.Y. Zakharova, O.G. Sinyashin, Mixed cationic liposomes for brain delivery of drugs by the intranasal route: the acetylcholinesterase reactivator 2-PAM as encapsulated drug model, Colloids Surf. B Biointerfaces 171 (2018) 358–367.

[75] M.C. Teixeira, C. Carbone, E.B. Souto, Beyond liposomes: recent advances on lipid based nanostructures for poorly soluble/poorly permeable drug delivery, Prog. Lipid Res. 68 (2017) 1–11.

[76] B. Clares, A.C. Calpena, A. Parra, G. Abrego, H. Alvarado, J.F. Fangueiro, E.B. Souto, Nanoemulsions (NEs), liposomes (LPs) and solid lipid nanoparticles (SLNs) for retinyl palmitate: effect on skin permeation, Int. J. Pharm. 473 (2014) 591–598.

[77] F. Danhier, O. Feron, V. Preat, To exploit the tumor microenvironment: passive and active tumor targeting of nanocarriers for anti-cancer drug delivery, J. Control. Release 148 (2010) 135–146.

[78] S. Martins, I. Tho, D.C. Ferreira, E.B. Souto, M. Brandl, Physicochemical properties of lipid nanoparticles: effect of lipid and surfactant composition, Drug Dev. Ind. Pharm. 37 (2011) 815–824.

[79] T.N. Pashirova, A. Braiki, I.V. Zueva, K.A. Petrov, V.M. Babaev, E.A. Burilova, D.A. Samarkina, I. K. Rizvanov, E.B. Souto, L. Jean, P.Y. Renard, P. Masson, L.Y. Zakharova, O.G. Sinyashin, Combination delivery of two oxime-loaded lipid nanoparticles: time-dependent additive action for prolonged rat brain protection, J. Control. Release 290 (2018) 102–111.

[80] V.R. Sinha, S. Srivastava, H. Goel, V. Jindal, Solid lipid nanoparticles (SLN'S) – trends and implications in drug targeting, Int. J. Adv. Pharm. Sci. 1 (2010) 212–238.

[81] M.D. Joshi, R.H. Müller, Lipid nanoparticles for parenteral delivery of actives, Eur. J. Pharm. Biopharm: Off. J. Arbeitsgemeinsch. Pharm. Verfahrenstech. 71 (2009) 161–172.

[82] G. Le Roux, H. Moche, A. Nieto, J.P. Benoit, F. Nesslany, F. Lagarce, Cytotoxicity and genotoxicity of lipid nanocapsules, Toxicol. In Vitro 41 (2017) 189–199.

[83] E. Winter, C.D. Pizzol, C. Locatelli, T.B. Crezkynski-Pasa, Development and evaluation of lipid nanoparticles for drug delivery: study of toxicity in vitro and in vivo, J. Nanosci. Nanotechnol. 16 (2016) 1321–1330.

[84] L.N. Andrade, D.M. Oliveira, M.V. Chaud, T.F. Alves, M. Nery, C.F. da Silva, J.K. Gonsalves, R. S. Nunes, C.B. Corrêa, R.G. Amaral, Praziquantel-solid lipid nanoparticles produced by supercritical carbon dioxide extraction: physicochemical characterization, release profile, and cytotoxicity, Molecules 24 (2019) 3881.

[85] C. Olbrich, N. Schöler, K. Tabatt, O. Kayser, R.H. Müller, Cytotoxicity studies of Dynasan 114 solid lipid nanoparticles (SLN) on RAW 264.7 macrophages—impact of phagocytosis on viability and cytokine production, J. Pharm. Pharmacol. 56 (2004) 883–891.

[86] M.L. Bondi, A. Azzolina, E.F. Craparo, G. Capuano, N. Lampiasi, G. Giammona, M. Cervello, Solid lipid nanoparticles containing nimesulide: preparation, characterization and cytotoxicity studies, Curr. Nanosci. 5 (2009) 39.

[87] J.R. Campos, A.R. Fernandes, R. Sousa, J.F. Fangueiro, P. Boonme, M.L. Garcia, A.M. Silva, B. C. Naveros, E.B. Souto, Optimization of nimesulide-loaded solid lipid nanoparticles (SLN) by factorial design, release profile and cytotoxicity in human Colon adenocarcinoma cell line, Pharm. Dev. Technol. 24 (2019) 616–622.

[88] A. Zielinska, C. Martins-Gomes, N.R. Ferreira, A.M. Silva, I. Nowak, E.B. Souto, Anti-inflammatory and anti-cancer activity of citral: optimization of citral-loaded solid lipid nanoparticles (SLN) using experimental factorial design and LUMiSizer(R), Int. J. Pharm. 553 (2018) 428–440.

[89] E.B. Souto, S.B. Souto, A. Zielinska, A. Durazzo, M. Lucarini, A. Santini, O.K. Horbańczuk, A. G. Atanasov, C. Marques, L.N. Andrade, A.M. Silva, P. Severino, Perillaldehyde 1,2-epoxide loaded SLN-tailored mAb: production, physicochemical characterization and in vitro cytotoxicity profile in MCF-7 cell lines, Pharmaceutics 12 (2020) 161.

[90] J. Ezzati Nazhad Dolatabadi, H. Hamishehkar, M. Eskandani, H. Valizadeh, Formulation, characterization and cytotoxicity studies of alendronate sodium-loaded solid lipid nanoparticles, Colloids Surf. B Biointerfaces 117 (2014) 21–28.

[91] R. Lopes, M. Gaspar, J. Pereira, C. Eleutério, M. Carvalheiro, A. Almeida, M. Cruz, Liposomes versus lipid nanoparticles: comparative study of lipid-based systems as oryzalin carriers for the treatment of leishmaniasis, J. Biomed. Nanotechnol. 10 (2014) 3647–3657.

[92] J. Aparicio-Blanco, I.A. Romero, D.K. Male, K. Slowing, L. Garcia-Garcia, A.I. Torres-Suárez, Cannabidiol enhances the passage of lipid Nanocapsules across the blood–brain barrier both in vitro and in vivo, Mol. Pharm. 16 (2019) 1999–2010.

[93] P. Mittal, H. Vardhan, G. Ajmal, G.V. Bonde, R. Kapoor, A. Mittal, B. Mishra, Genistein-loaded nanostructured lipid carriers for intravenous administration: a quality by design based approach, Int. Res. J. Pharm. 20 (2019) 119–134.

[94] P. Mittal, H. Vrdhan, G. Ajmal, G. Bonde, R. Kapoor, B. Mishra, Formulation and characterization of Genistein-loaded nanostructured lipid carriers: pharmacokinetic, biodistribution and in vitro cytotoxicity studies, Curr. Drug Deliv. 16 (2019) 215–225.

[95] S.A. Valdes, R.F. Alzhrani, A. Rodriguez, D.S. Lansakara-P, S.G. Thakkar, Z. Cui, A solid lipid nanoparticle formulation of 4-(N)-docosahexaenoyl 2′,2′-difluorodeoxycytidine with increased solubility, stability, and antitumor activity, Int. J. Pharm. 570 (2019) 118609.

[96] S. Zafar, S. Akhter, I. Ahmad, Z. Hafeez, M.M.A. Rizvi, G.K. Jain, F.J. Ahmad, Improved chemotherapeutic efficacy against resistant human breast cancer cells with co-delivery of docetaxel and Thymoquinone by chitosan grafted lipid Nanocapsules: formulation optimization, in vitro and in vivo studies, Colloids Surf. B Biointerfaces (2019) 110603.

[97] S. Asthana, A.K. Jaiswal, P.K. Gupta, A. Dube, M.K. Chourasia, Th-1 biased immunomodulation and synergistic antileishmanial activity of stable cationic lipid-polymer hybrid nanoparticle: biodistribution and toxicity assessment of encapsulated amphotericin B, Eur. J. Pharm. Biopharm. 89 (2015) 62–73.

[98] L.Y. Zakharova, T.N. Pashirova, S. Doktorovova, A.R. Fernandes, E. Sanchez-Lopez, A.M. Silva, S. B. Souto, E.B. Souto, Cationic surfactants: self-assembly, structure-activity correlation and their biological applications, Int. J. Mol. Sci. 20 (2019).

[99] S. Doktorovova, D.L. Santos, I. Costa, T. Andreani, E.B. Souto, A.M. Silva, Cationic solid lipid nanoparticles interfere with the activity of antioxidant enzymes in hepatocellular carcinoma cells, Int. J. Pharm. 471 (2014) 18–27.

[100] S. Doktorovova, R. Shegokar, E. Rakovsky, E. Gonzalez-Mira, C.M. Lopes, A.M. Silva, P. Martins-Lopes, R.H. Muller, E.B. Souto, Cationic solid lipid nanoparticles (cSLN): structure, stability and DNA binding capacity correlation studies, Int. J. Pharm. 420 (2011) 341–349.

[101] K.P. Krause, O. Kayser, K. Mäder, R. Gust, R. Müller, Heavy metal contamination of nanosuspensions produced by high-pressure homogenisation, Int. J. Pharm. 196 (2000) 169–172.

[102] R. Müller, S. Maaben, H. Weyhers, W. Mehnert, Phagocytic uptake and cytotoxicity of solid lipid nanoparticles (SLN) sterically stabilized with poloxamine 908 and poloxamer 407, J. Drug Target. 4 (1996) 161–170.

[103] N. Schöler, E. Zimmermann, U. Katzfey, H. Hahn, R. Müller, O. Liesenfeld, Effect of solid lipid nanoparticles (SLN) on cytokine production and the viability of murine peritoneal macrophages, J. Microencapsul. 17 (2000) 639–650.

[104] W. Rao, H. Wang, J. Han, S. Zhao, J. Dumbleton, P. Agarwal, W. Zhang, G. Zhao, J. Yu, D. L. Zynger, Chitosan-decorated doxorubicin-encapsulated nanoparticle targets and eliminates tumor reinitiating cancer stem-like cells, ACS Nano 9 (2015) 5725–5740.

[105] P. Nimtrakul, W. Tiyaboonchai, S. Lamlertthon, Amphotericin B loaded nanostructured lipid carriers for parenteral delivery: characterization, antifungal and in vitro toxicity assessment, Curr. Drug Deliv. 16 (2019) 645–653.

[106] F. Bahadori, A. Kocyigit, H. Onyuksel, A. Dag, G. Topcu, Cytotoxic, apoptotic and genotoxic effects of lipid-based and polymeric nano micelles, an in vitro evaluation, Toxics 6 (2018) 7.

[107] C.D. Klaassen, M.O. Amdur, Casarett and Doull's Toxicology: The Basic Science of Poisons, McGraw-Hill, New York, 2013.

[108] N. Ramesh, A.k.A. Mandal, Pharmacokinetic, toxicokinetic, and bioavailability studies of epigallocatechin-3-gallate loaded solid lipid nanoparticle in rat model, Drug Dev. Ind. Pharm. (2019) 1–31.

[109] P. Ma, Q. Luo, J. Chen, Y. Gan, J. Du, S. Ding, Z. Xi, X. Yang, Intraperitoneal injection of magnetic Fe3O4-nanoparticle induces hepatic and renal tissue injury via oxidative stress in mice, Int. J. Nanomedicine 7 (2012) 4809.

[110] S.M. Boudon, G. Morandi, B. Prideaux, D. Staab, U. Junker, A. Odermatt, M. Stoeckli, D. Bauer, Evaluation of sparfloxacin distribution by mass spectrometry imaging in a phototoxicity model, J. Am. Soc. Mass Spectrom. 25 (2014) 1803–1809.

[111] M.D. Joshi, R.H. Muller, Lipid nanoparticles for parenteral delivery of actives, Eur. J. Pharm. Biopharm. 71 (2009) 161–172.

[112] A. Vonarbourg, C. Passirani, P. Saulnier, P. Simard, J. Leroux, J. Benoit, Evaluation of pegylated lipid nanocapsules versus complement system activation and macrophage uptake, J. Biomed. Mater. Res. A: Off. J. Soc. Biomater. Japan. Soc. Biomater. Aust. Soc. Biomater. Korean Soc. Biomater. 78 (2006) 620–628.

[113] K. Sou, S. Inenaga, S. Takeoka, E. Tsuchida, Loading of curcumin into macrophages using lipid-based nanoparticles, Int. J. Pharm. 352 (2008) 287–293.

[114] E. Berger, D. Breznan, S. Stals, V.J. Jasinghe, D. Gonçalves, D. Girard, S. Faucher, R. Vincent, A. R. Thierry, C. Lavigne, Cytotoxicity assessment, inflammatory properties, and cellular uptake of Neutraplex lipid-based nanoparticles in THP-1 monocyte-derived macrophages, Nano 4 (2017) 1849543517746259.

[115] A. Rafique, A. Etzerodt, J.H. Graversen, S.K. Moestrup, F. Dagnæs-Hansen, H.J. Møller, Targeted lipid nanoparticle delivery of calcitriol to human monocyte-derived macrophages in vitro and in vivo: investigation of the anti-inflammatory effects of calcitriol, Int. J. Nanomedicine 14 (2019) 2829.

[116] S.C. Sahu, A.W. Hayes, Toxicity of nanomaterials found in human environment: a literature review, Toxicol. Res. Appl. 1 (2017) 2397847317726352.

[117] B. Wang, X. He, Z. Zhang, Y. Zhao, W. Feng, Metabolism of nanomaterials in vivo: blood circulation and organ clearance, Acc. Chem. Res. 46 (2012) 761–769.

[118] M.T. Jacques, J.L. Oliveira, E.V. Campos, L.F. Fraceto, D.S. Avila, Safety assessment of nanopesticides using the roundworm Caenorhabditis elegans, Ecotoxicol. Environ. Saf. 139 (2017) 245–253.

[119] E. Gonzalez-Mira, M.A. Egea, M.L. Garcia, E.B. Souto, Design and ocular tolerance of flurbiprofen loaded ultrasound-engineered NLC, Colloids Surf. B Biointerfaces 81 (2010) 412–421.

[120] E. Gonzalez-Mira, S. Nikolic, M.L. Garcia, M.A. Egea, E.B. Souto, A.C. Calpena, Potential use of nanostructured lipid carriers for topical delivery of flurbiprofen, J. Pharm. Sci. 100 (2011) 242–251.

[121] E.B. Souto, J.R. Campos, S.B. Souto, M. Lucarini, A. Durazzo, A. Santini, Ocular cell lines and genotoxicity assessment. Int. J. Environ. Res. Public Health 17 (6) (2020) 2046–2063, https://doi.org/10.3390/ijerph17062046.

[122] H. Weyhers, S. Ehlers, H. Hahn, E B. Souto, R.H. Müller, Solid lipid nanoparticles (SLN)—effects of lipid composition on in vitro degradation and in vivo toxicity, Pharmazie 61 (2006) 539–544.

[123] L. Yang, H. Kuang, W. Zhang, Z.P. Aguilar, H. Wei, H. Xu, Comparisons of the biodistribution and toxicological examinations after repeated intravenous administration of silver and gold nanoparticles in mice, Sci. Rep. 7 (2017) 3303.

[124] H. Yang, Q.Y. Wu, M.Y. Li, C.S. Lao, Y.J. Zhang, Pulmonary toxicity in rats caused by exposure to Intratracheal instillation of SiO2 nanoparticles, Biomed. Environ. Sci. 30 (2017) 264–279.

[125] R. Kedmi, N. Ben-Arie, D. Peer, The systemic toxicity of positively charged lipid nanoparticles and the role of toll-like receptor 4 in immune activation, Biomaterials 31 (2010) 6867–6875.

CHAPTER 6

Cellular interactions of nanoparticles within the vasculature

Azziza Zaabalawi and May Azzawi
Cardiovascular Research Group, Department of Life Sciences, Faculty of Science and Engineering, Manchester Metropolitan University, Manchester, United Kingdom

1. Introduction

1.1 The application of nanoparticles as a drug delivery modality

The demand for specific targeted drug delivery and sensitive molecular imaging modalities in medical diagnostic and prognostic techniques has stimulated the application of nanotechnology as an alternative approach to conventional imaging and drug delivery systems [1–3]. Nanoparticles (NPs) present the foundation for this nanotechnology and have been exploited as promising novel tools for drug delivery in the treatment of a number of disease conditions, due to their easily modifiable properties such as size, shape and surface functionalization [2, 4]. As the different physical, chemical and biological properties of NPs depend on their chemical synthesis and material composition (including surface coating), this dictates their reactivity with biological media and cells/tissues [2, 5, 6]. This has laid the foundation for the surface modification of NPs by incorporating different functional groups to target surface molecules that are present on pathologically altered cells, ultimately enhancing site-specific drug delivery and improving treatment outcomes.

The application of NPs in the diagnosis and treatment of cardiovascular disease (CVD) has been gaining increasing attention in recent years [7]. CVD accounts for 3.9 million mortalities in Europe each year and exerts a significant economic burden of €210 billion annually as over 45 million individuals in Europe are living with this disease [8]. While there is an ever-increasing trend in the prevalence of CVD due to the persistent rise in obesity, smoking, and inactive lifestyles, treatment strategies remain challenging, compounded by the poor bioavailability and adverse off-target effects of traditional medications that limit their systemic administration [9, 10]. Nanotechnology offers a promising avenue to combat these problems by the rational design and development of NPs that allow targeted drug delivery directly to pathologically altered cells/diseased areas of the vasculature, overcoming biological barriers in the immune system and enhancing drug effectiveness. The vascular endothelium, located in the inner cellular lining of blood vessels, is an extremely differentiated single layer of endothelial cells that play a critical role in the regulation of cellular adhesion, vascular permeability, and vascular tone to ensure sufficient blood flow and perfusion in the body. Therefore any damage or dysfunction to the endothelium would have significant implications in the pathogenesis of a whole range of CVD including hypertension, atherosclerosis, stroke, ischemia, and inflammation. For this reason, the endothelial cell lining of the vasculature serves as a critical therapeutic target for treating the underlying cause of such diseases [11]. Furthermore, NPs can also be designed to deliver contrast agents to a targeted antigenic determinant of interest, enabling their successful use for noninvasive diagnostic imaging for the early detection of vascular inflammation which is an early marker for the formation of atherosclerotic plaques [12, 13]. This targeting approach can be achieved by functionalization of NPs via various surface modifications to target cells of the vessel wall, in particular, the lining endothelial cells toward therapeutic intervention in disease states. Herein, we will summarize our current understanding of the various uptake mechanisms of NPs in the first section, with specific focus on silica nanoparticles, liposomes, and nanostructured lipid nanocarriers. The final part discusses the various functionalization methods used to enhance site-specific drug delivery of these NPs, with a particular focus on endothelial cell-specific approaches.

2. Types of nanoparticles for drug delivery

Among the different types of inorganic NPs present, silicon dioxide; also referred to as silica nanoparticles (SiNPs) have displayed a greater potential for use in biomedical research due to their unique properties enabling their use for imaging diagnostics and targeted drug delivery [14]. SiNPs comprise two main types: mesoporous and rigid nanostructures. The latter has received great interest as an effective host material for the transport of molecules specifically enzymes within cells [15], whereas PEGylated mesoporous silicon nanoparticles (MSNs) have been extensively studied as a potential delivery method for the release of intravascular drugs into tissues in animal models of CVD [16]. The extensive development of MSNs as an excellent therapeutic delivery platform can be attributed to several factors including reduced cost and ease of production, their large surface area to volume ratio that allows for effective interaction with target cells, as well as their adjustable pore shape and sizes, as shown in Fig. 1A, that permit the loading and storage of pharmaceutical drugs and prevent its early release and breakdown before reaching target cells [14, 15]. Although SiNPs have been significantly exploited in the biomedical field for drug delivery and imaging diagnostics, some cytotoxic effects have been reported. To overcome any cytotoxicity, a number of strategies have been utilized, including surface functionalization using polymers and other types of materials, such as cerium oxide and lipid coating [17, 18]. Equally, it has initiated more research into the production of other efficient, robust and perhaps more biocompatible delivery systems.

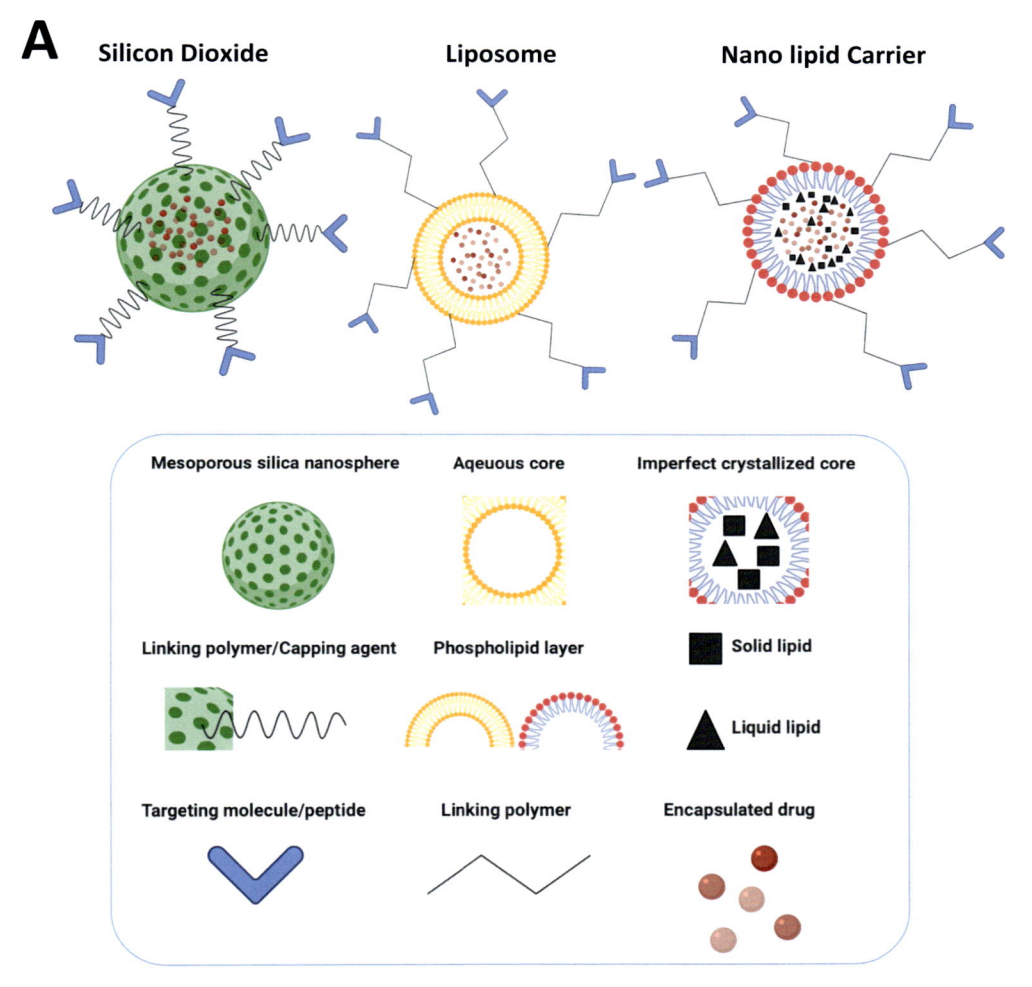

Fig. 1 Schematic diagram of nanoparticles explored for targeted drug delivery into vascular cells by endocytic mechanisms. Illustration of the various constituents of modified silicon, liposome, and nanolipid nanoparticles engineered for endothelial cell targeting and drug loading for the vascular delivery of therapeutic agents (A). Schematic representation of the major uptake pathways of nanoparticle endocytosis (B). Modification of nanoparticles via surface coating with targeting molecules to enhance their directed intracellular uptake by endocytosis (C).

(Continued)

Over the years, the application of nanosized phospholipid bilayer vesicles called liposomes for targeted drug delivery have gained much popularity and has become the most extensively studied nanocarrier [19]. This is due to their improved biocompatibility when compared with inorganic nanoparticles, effective encapsulation of a diverse range

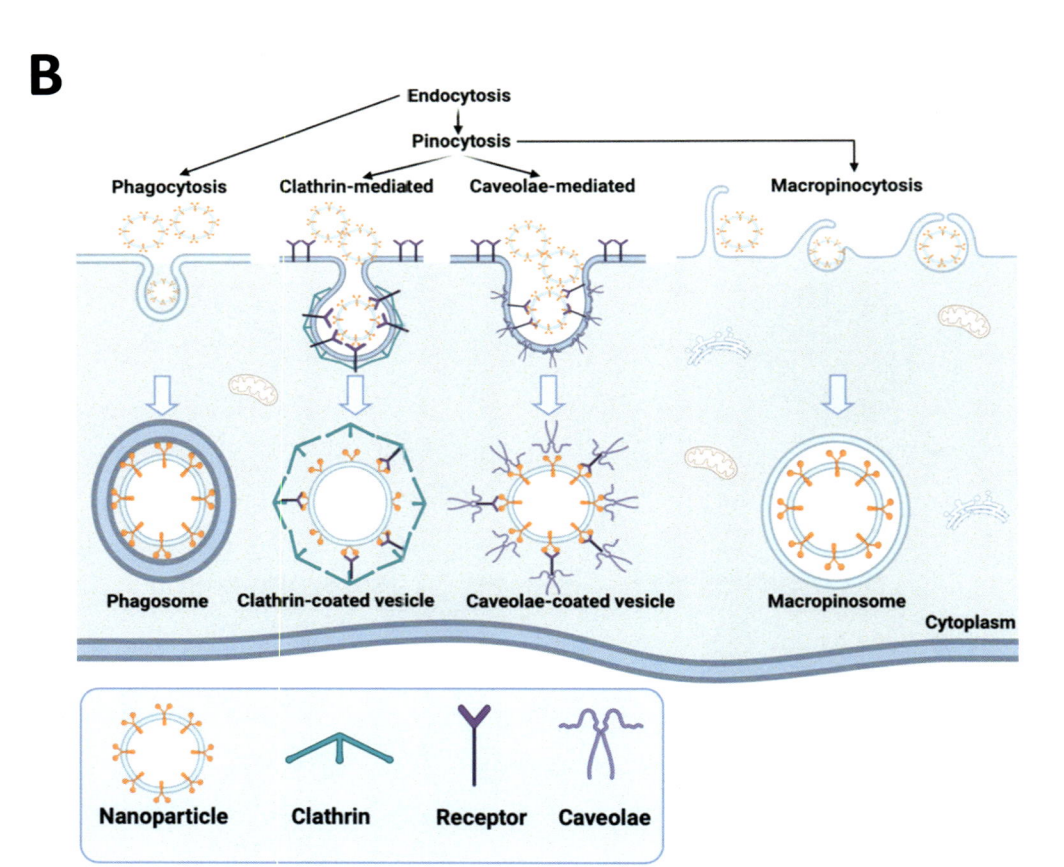

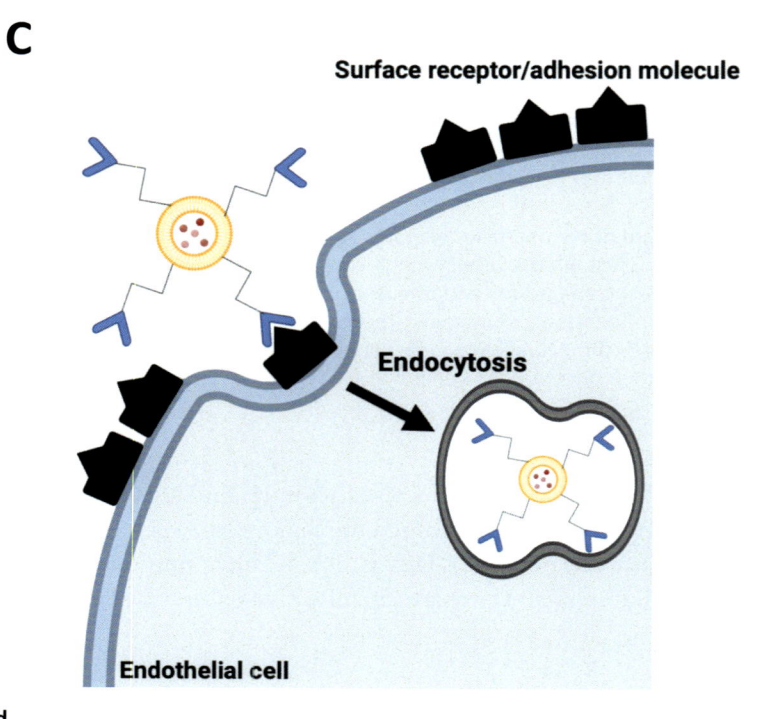

Fig. 1, cont'd

of drugs, lack of immunogenicity, and particularly because of their improved solubility and ability to cross complex physiological barriers such as the blood-brain-barrier [20, 21]. These spherical vesicles have a bilayer membrane made up of cholesterol, and an aqueous core which allows the entrapment of both hydrophobic and hydrophilic compounds, respectively, as demonstrated in Fig. 1A, giving liposomes its unique ability to encapsulate a wide range of drugs [19, 22]. Numerous liposome formulations have gained clinical approval and are being utilized for pain control, cancer treatment, and vaccination delivery [23]. Interestingly, Pfizer and Moderna's mRNA-based vaccine against the corona virus (SARS-CoV-2), delivered via nanolipid carriers, were the first vaccine candidates to be introduced into human clinical trials [24], further demonstrating their excellent biocompatibility and biodegradable properties that allow them to be successfully used as drug delivery platforms [25]. A number of ex vivo studies have demonstrated the efficacy of liposomes for targeted drug delivery to enhance vasodilator responses in acute hypertension models and to treat impaired uteroplacental perfusion in pregnancy [26, 27]. While useful, there have been several evident limitations of liposomal application as oral drug delivery systems due to their poor stability in the liquid state, short shelf-life, drug leakage/burst release, and difficulties in mass production [28]. For this reason, further research has been conducted to formulate more reliable delivery modalities such as nanostructured lipid carriers.

Nanostructured lipid carriers (NLCs) are emerging as potential lipid-based drug delivery systems developed to overcome the limitations associated with conventional carriers such as liposomes. The advantages of NLCs include higher drug loading capacity and reduced drug leakage; due to the presence of a blend of both solid and liquid lipids in their formulations to produce a partially crystallized matrix scattered within an aqueous phase of emulsifiers (imperfect crystal core) [29, 30] as shown in Fig. 1A, controlled drug release, low cytotoxicity, improved bioavailability and solubility as well as tissue-targeted delivery [31–33]. For instance, recent studies have documented improved bioavailability of weakly soluble antihypertensive drugs [34, 35] and vasoprotective BioCeuticals [36, 37] following their delivery by NLCs. Furthermore, in contrast to previous generation counterparts, NLCs can be intravenously or orally administered, enhancing patient stratification and ameliorating therapeutic outcomes [32, 34, 38]. The ability of NLCs to maintain drug stability while improving bioavailability makes it an appealing option for its development as site-specific delivery modalities in CVD.

3. Cellular uptake of nanoparticles

Understanding the mode of entry of NPs into target cells is crucial as it can have significant implications for their intracellular localization and cytotoxicity, to ultimately maximize their delivery efficiency and improve biocompatibility [39]. This section will focus particularly on the cellular uptake pathways utilized by SiNPs, liposomes, and NLCs within cells of the vessel wall. The plasma cell membrane of the cells then displays the first initial contact surface for NPs, where they can interact with the extracellular matrix to initiate entry into cells primarily by their engulfment into cell membrane invaginations and the formation of vesicles in a process called endocytosis. Endocytic mechanisms comprise two cellular uptake processes: (1) phagocytosis, which involves the engulfment of larger NPs and the formation of intracellular phagosomes, and (2) pinocytosis, a key route for the internalization of fluids and solutes containing smaller NPs of a few nanometers in size by the formation of small vesicles called pinosomes. Furthermore, pinocytosis can be subdivided into clathrin-mediated endocytosis, caveolin-mediated endocytosis, clathrin/caveolae-independent endocytosis, and macropinocytosis, depending on the cell type, as well as the shape, size, material composition and surface chemical modification of NPs involved in the process [40–42]. A description of each of these processes, with specific reference to endothelial cells of the vasculature is as follows.

3.1 Phagocytosis

Phagocytosis involves the engulfment of larger-sized NPs, ranging between 200 and 1500 nm, by phagocytes. This internalization pathway is triggered by the adsorption of opsonins, for example, antibodies or complement proteins, onto the surface of NPs in a process known as opsonization to allow their recognition and attachment to phagocytic cells by specific ligand-receptor interactions. This results in the opsonized NPs being enclosed and internalized via an internal compartment called a phagosome, as shown in Fig. 1B, with a diameter of up to 10 μm. The phagosomes finally undergo lysosomal fusion, resulting in the breakdown of the phagosome cargo by acidification and enzymatic degradation. For this reason, it is important to design NPs that bypass the lysosomal route to achieve the desired therapeutic effect.

Based on multiple experimental studies, NP internalization is a complex process that is governed by major factors including the size, shape, and surface chemical modification of NPs. For instance, larger-sized radiolabeled albumin NPs (>1 μm) and polystyrene NPs (>2 μm) exhibited maximal internalization by phagocytosis into human mononuclear cells and murine peritoneal macrophages, respectively [43, 44]. Additionally, research conducted by Agarwal et al. documented greater intracellular accumulation of disc-shaped NPs into mammalian endothelial cells, as compared with nanorods. This highlights the significance of NP geometry on their cellular uptake across biological barriers [45]. In addition to size and shape, surface chemical characteristics of NPs can also influence cellular uptake by phagocytosis. As the phagocytic process is greatly facilitated by opsonization of NPs, any attachment of shielding polymers on the surface of NPs can hinder this process and thus alter their cellular internalization by phagocytosis. For instance, functionalization of NPs with hydrophilic polyethylene glycol (PEG), commonly utilized as an approach for targeted drug delivery to improve encapsulated drug efficiency, significantly repels the adsorption of protein on the surface of NPs, and thus preventing opsonization. This can be attributed to the formation of a repulsive energy barrier by PEG between NPs, which defeats the force of attraction required for opsonization [46]. A study by McCright et al. established the vital role of PEG conformation in influencing NP transport across endothelial cells of the lymphatic system and into the vasculature. The conjugation of hydrophilic PEG to NP surfaces significantly improved their internalization into endothelial cells, and thus establishing PEG density as an important strategy in designing targeted NP formulations to improve therapeutic efficacy [47]. Interestingly, FDA-approved chemotherapy drug, Pegylated liposomal doxorubicin (Doxil) exhibited reduced phagocytic internalization as a result of an increased surface PEG density, and thus enhancing the half-life of the encapsulated drug. This ultimately boosts the pharmacokinetics of NPs. Contrariwise, functionalization of NPs with hydrophobic surfaces increase attraction to complement proteins to enhance opsonization and phagocytic uptake [41]. These results provide a fundamental understanding on the various factors that influence NP internalization, providing vital information for the production and design of improved NPs as well as predicting their toxicity [45].

3.2 Clathrin- and caveolae-mediated endocytosis

First, clathrin-mediated endocytosis is a receptor-dependent or -independent adsorptive uptake by which cells engulf nutrients and metabolites, including cholesterol and iron via low density lipoprotein and transferrin receptors, respectively. It is the major cellular uptake mechanism and occurs at regions of the plasma membrane with high clathrin density. During this uptake mechanism, ligand-receptor complexes are produced by the attachment of specific ligands in the extracellular fluid to surface receptors, followed by their movement to clathrin-rich regions of the plasma membrane. The ligand-receptor complexes are then engulfed via cell membrane invagination, through the formation of clathrin-coated vesicles, as demonstrated in Fig. 1B. Majority of silicon and lipid-based NPs utilize this uptake pathway [41]. Second, caveolin-mediated endocytosis is heavily involved in cell signaling pathways and regulating membrane proteins, and it relies on the formation of flask-shaped membrane invaginations called caveolae. Receptor activation results in the detachment of caveolae from the plasma membrane to form caveolae-coated vesicles, as illustrated in Fig. 1B. This entry route is particularly employed in nanomedical applications as the internalized vesicles can avoid lysosomes, thus protecting its contents from enzymatic degradation which is crucial in the enhancement of therapeutic delivery. Although clathrin and caveolae-mediated uptake both rely on receptor activation for vesicle engulfment, it is important to note that caveolae-mediated endocytosis involve smaller-sized coated vesicles and require a longer duration to occur, as compared with clathrin-mediated internalization [42].

Numerous studies have established cell type and size and charge-dependent differences in the uptake of SiNPs [39], with clathrin-mediated and caveolae-mediated endocytic pathways being the most prevalent mode of cellular internalization for SiNPs ranging between 50 and 200 nm in size [40, 42, 48, 49]. In endothelial cells, endocytic mechanisms have been documented to be the most common pathway associated with SiNPs cellular uptake [50]. Pharmacological inhibitors used to identify the major endocytic pathways involved in SiNPs uptake include chlorpromazine hydrochloride (CPZ) and methyl-β-cyclodextrin (MBCD), for the inhibition of clathrin-mediated and caveolae-mediated endocytosis, respectively [40]. Furthermore, surface charge has been documented to significantly affect the internalization of nanoparticles via clathrin- and caveolae-mediated endocytosis, as well as their intracellular fate [41]. To support this finding, studies have documented a greater role for clathrin-mediated endocytosis in the uptake of positively charged nanoparticles, whereas negatively charged nanoparticles were internalized via both clathrin- and caveolin-mediated endocytosis [51]. Additionally, uptake of SiNPs is reduced by size [52], with positively charged NPs exhibiting greater cellular internalization by clathrin-mediated endocytosis than negatively charged counterparts [39]. A study investigating the effects of physiological stretch conditions on the uptake of SiNPs (<100 nm) into human endothelial cells demonstrated reduced endocytic uptake of SiNPs under cyclic stretch conditions when compared with cells under static conditions. This indicates that in addition to shear stress, cyclic stretch may also impact uptake interactions between SiNPs and endothelial cells [53]. Furthermore, our research group have demonstrated the reduced uptake of mono-dispersed dye-encapsulated SiNPs (~97 nm) by endothelial cells in the vascular endothelium of rat mesenteric arteries under shear stress conditions [54].

Despite the significant research to develop lipid-based nanocarriers as promising drug delivery modalities, their cellular uptake mechanisms have not been extensively studied to allow a more detailed understanding on their cellular processing. Several pathways have been suggested for cellular internalization of lipid-based NPs depending on their size, charge and surface modification. While a large body of evidence suggests clathrin-mediated endocytosis to be the most prevalent liposomal mode of entry into cells [55], a study conducted by Yang et al. demonstrated that liposome-cell membrane fusion also plays a major role for the cellular internalization of liposome-based delivery systems. This is because, in addition to the high fluorescence intensity of dye-labeled liposomes detected in the cytosol of living cells confirming endocytic uptake, a strong signal remained at the plasma membrane suggesting that liposomes are initially taken up by endocytosis after which they quickly fuse with the endosomal membrane [56]. Another study has illustrated the role of caveolae-mediated endocytosis in the uptake of cationized dye-loaded liposomes (between 120 and 150 nm) into brain endothelial cells, which was shown to be concentrated in intracellular vesicles following intravenous administration [57]. Nonetheless, recent studies have also suggested the cellular uptake of positively charged liposomes via clathrin-mediated mechanisms for their successful application in gene therapy, due to strong interactions between cationic liposomes and negatively charged components of the endosomal membrane. This results in the effective release of encapsulated material from endosomal vesicles [55]. Similar to liposomes, recent studies have documented the reliance of cellular internalization of NLCs on energy, which is indicative of endocytosis. To further characterize the specific endocytic pathways involved, inhibitor studies with chlorpromazine and indomethacin were used to block clathrin-mediated and caveolin-mediated pathways, respectively. Results indicated caveolin-dependent endocytosis to be the major entry route for NLCs into endothelial cells, with less involvement of clathrin-mediated endocytosis. A large body of evidence confirms this finding as NLCs were found to be more effectively taken up in cells that expressed caveolin-1 [32, 58]. Furthermore, it is important to note that the kinetics, efficiency, and mode of NP uptake is also cell-type specific. For instance, while epithelial cells internalize NP mainly through caveolae-mediated endocytosis, human umbilical vein endothelial cells (HUVECs) uniquely utilize clathrin-mediated endocytic pathways for the uptake of NP with several shapes. The efficiency of NP uptake was much higher in HUVECs, as compared with epithelial cells [45].

3.3 Clathrin and caveolae-independent endocytosis

Clathrin/caveolae-independent endocytosis is common in cells that lack both clathrin and caveolae and occurs independently of receptors. During this process, cells utilize other cargos such as growth hormones and cellular fluids to initiate entry into cells, and thus are not reliant on coat proteins for vesicle formation [41, 42]. Folate-functionalized NPs have been commonly identified to be taken up into cells via this pathway [59].

3.4 Macropinocytosis

Macropinocytosis is a transient, actin-driven and dynamin-independent mechanism involving the rearrangement of the cell membrane's cytoskeleton. On activation of intracellular signaling pathways, outer large membrane circular ruffles are formed via protrusion of the plasma membrane. These large membrane extensions then fuse back with the cell membrane, engulfing high volumes of surrounding fluids into large vesicles called macropinosomes, as demonstrated in Fig. 1B. Macropinosomes have a large diameter of up to 10 µm, allowing the cellular internalization of larger-sized NPs that cannot be taken up by clathrin- or caveolae-dependent endocytosis [60, 61]. For this reason, NP delivery via micropinocytosis can overcome the size and charge limitations associated with the other types of pinocytosis [61]. There are many interesting approaches associated with the targeting of NP entry into site-specific cells via the uptake processes summarized in this section. These advancements enable the improved rational design of NPs to enhance their therapeutic delivery. The next section will discuss this in detail, with specific reference to pathologically altered endothelial cells of the vasculature.

4. Targeted drug delivery to vascular cells using nanoparticles

Drug-loaded NPs and delivery modalities targeted for the vasculature can be delivered via a number of routes. This includes direct delivery via the blood (subsequent to their injection intravascularly or intravenously), or indirectly, subsequent to their inhalation, or oral ingestion. The latter cases lead to the translocation of NPs across extracellular tissues and interstitial spaces to allow their direct contact with the outer adventitial layer of the vasculature (extravascular exposure) [62–65]. The effects of NP uptake and cellular internalization (actively or passively) strongly rely on the mode of administration [66]. Administering NPs via inhalation (pulmonary route) offers advantages such as rapid absorption of NPs owing to the large surface area of endothelial cells found in the highly vascularized pulmonary bed as it receives a high volume of cardiac output, making the lungs a significant focus in the treatment of vascular inflammation and oxidative stress-related conditions [67–70]. For this reason, NP administration by inhalation can be an effective drug delivery route into the vasculature. In relation to improving oral administration of NPs, approaches involving the formulation of NPs associated with peptide ligands to enhance targeted drug delivery into gastrointestinal tract (GIT) can overcome the limitations associated with susceptibility to enzymatic breakdown and poor drug stability owing to the acidic nature of the GIT [71].

Currently, most drugs used to treat CVD are associated with limited effectiveness, adverse off-target and systemic side effects, as well as complications with patient administration due to their lack of specificity for endothelial cells (no endothelial affinity) and thus inadequate cellular uptake of the administered dose [67, 72–74]. To overcome this problem, recent attempts to improve the performance of NPs for both therapeutic and diagnostic purposes have resulted in the development of multitargeting and multifunctional NPs for specific drug delivery to the diseased endothelial layer, via surface assimilation with affinity ligands of EC determinants [67, 75]. The plasma membrane of cells has received immense attention as a target for the development of therapeutic agents to treat CVD as it is heavily involved in a range of functions that are crucial for cellular health and vascular homeostasis [76]. For this reason, recent studies have focused on modifying the surface of NPs with adhesion molecules present on the surface of leucocytes at diseased sites, to mimic their interactions with the EC layer. This ultimately enhances the detection, active targeting (leukocyte migration inhibition), and drug delivery (surface anchoring of modified drug-encapsulated NPs) to effectively improve treatment for vascular inflammation and oxidative stress [67, 75, 77]. Functionalization of NPs involves surface chemical conjugation of ligands such as proteins, nucleic acid aptamers and peptides, due to their high specificity and stability, ease of surface assimilation with NPs, as well as direct accessibility to the vascular circulation. Following are some examples of the types of functionalization used for silica nanoparticles and liposomes, with some recent uses:

4.1 Functionalization with antibodies to adhesion molecules

To specifically target inflamed endothelial cells in the vasculature and enhance site-specific intracellular delivery of nanocarriers by endocytosis, NPs can be coated with antibodies to markers of inflammation. These include receptors of adhesion molecules such as vascular cell adhesion molecule-1 (VCAM-1), platelet and endothelial cell adhesion molecule 1 (PECAM-1), and intercellular adhesion molecule 1 (ICAM-1), as demonstrated in Fig. 1C, as they are overly expressed on the surface of pathologically affected endothelial cells which are key areas of reactive oxygen species (ROS) production. For instance, oxidative stress at the blood-brain barrier was significantly attenuated following the exposure to endothelial-targeting NPs conjugated with the endogenous antioxidant catalase bound to anti-ICAM-1 antibodies [78, 79]. These findings highlight the effective use of ICAM-1 as a suitable target for specific delivery of NPs to the activated endothelial layer at diseased regions [78]. However, it is important to note that individual antibody molecules bound to receptors do not undergo cellular internalization, and thus serving as an anchor for conjugated NPs inside blood vessels. Importantly, internalization via adhesion molecules has been shown to be dependent on the rate of blood flow [80]. Numerous modern nanocarriers, including liposomes, have incorporated a known ligand of VCAM-1 (VHP peptide) for their use as diagnostic agents in the vasculature and have been found to be effectively taken up into cells via clathrin-mediated endocytosis [81, 82]. Moreover, the expression of PECAM-1 and ICAM-1; actively involved in endothelial signaling, is exacerbated during pro-inflammatory states in CVD. For instance, lipid-based NPs have been extensively used for targeting drugs to the endothelium via adhesion molecules in several animal models of CVD, with studies showing >50% greater cellular internalization in inflamed regions as compared with nontargeted liposome-based formulations, demonstrating enhanced targeted drug delivery (reviewed in Ref. [67]).

4.2 Functionalization with peptides

NPs may also be modified with peptides such as the RGD (arginine–glycine–aspartic acid) peptide; the ligand for the adhesion molecule integrin that is overly expressed on the surface of pathologically altered endothelial cells. The high specificity of RGD and its great affinity for integrin has allowed its extensive application for targeted drug delivery, due to improved permeability rates and intracellular retention of drug-encapsulated RGD-modified NPs, by receptor-mediated endocytosis [83–85]. A large body of evidence demonstrated greater therapeutic efficacy of targeted NP delivery to the vascular endothelium when compared with conventional nontargeted approaches [67, 86, 87].

4.3 Functionalization with polymers

Polymer-based "stealth" technology which involves the linking of polymers such as polyethylene glycol (PEG) on the surfaces of NPs is a key strategy to enhance drug delivery by nanocarriers. In the case of delivering antioxidant enzymes, for example, PEG surface functionalization improved treatment efficacy in animal models of vascular oxidative stress due to the improvement in systemic bioavailability of PEGylated NPs, owing to increased aqueous solubility and protection against protein absorption to extend circulation time [67, 88, 89].

Targeted drug-loaded NPs can be further functionalized so that the drug release is triggered on exposure to certain environments only, using, for example, capping agents which act as stimuli-responsive gatekeepers. Capping agents block the outward diffusion of the entrapped drug through pore entrances of NPs until they are exposed to stimuli, initiating pore opening and triggering drug release [90]. For instance, FDA-approved peptide drug Protamine; clinically used to reverse heparin's anticoagulant effects, was used to cap mesoporous silica NPs, where exposure to enzyme activation results in breakdown of Protamine and subsequent drug release from the mesopores [91]. Moreover, several research groups have focused on the development of triggered drug release from NPs, whereby the NPs release their payload in response to an altered intracellular environment. This can include altered pH levels or elevated levels of reactive oxygen species for cells under oxidative stress conditions. The latter is a promising novel strategy to treat endothelial dysfunction. Although nontargeted antioxidant therapies have been documented to effectively suppress the harmful effects of ROS in CVD, their useful effects in human clinical trials have been disappointing [92–94]. A major reason for this may be the nonspecific elimination of ROS by these antioxidants, leading to unfavorable effects due to the disruption of essential ROS-mediated physiological and cellular signaling. Hence, targeting suppression directly to regions of ROS overproduction may have remarkable therapeutic potentials.

For diagnostic imaging purposes, NPs are required to exhibit high sensitivity to receptors, blood flow dependency, as well as generating signals that can be visualized in response to pathologically altered cellular functions. However, some limitations that could hinder the efficacy of targeted NPs include permeability and ability to cross the blood-brain barrier and ability to target relatively stable molecules. Examples of ROS-targeting NPs that hold promising potential for medical use include selenium-based (targets superoxide and hydrogen peroxide) and mesoporous silicon-based (targets superoxide) NPs [95, 96]. Although the potential use of SiNPs as a drug delivery modality has been previously discussed in this chapter, an innovative approach involving the attachment of drugs containing thiol groups and hydrophilic polymers to the inner and outer layer of SiNPs through the formation of disulfide bridges to enhance a redox-responsive release mechanism, has also been established [96]. Additionally, recent studies have documented the effective role of capping agents sealed on the surfaces of mesoporous SiNPs for stimulus responsive drug delivery and release, ultimately allowing for precise diagnosis and treatment due to their ability to respond to specific changes in the intracellular environment [97, 98]. Furthermore, recent findings have demonstrated the effectiveness of polylactic acid (PLA) coating of SiNPs in enhancing the release of encapsulated drugs across the blood-brain barrier via an increase in the breakdown of the polymer on NP surfaces by ROS. The improved drug release from these modified SiNPs under high oxidative stress environments demonstrates promising potential for the use of polymer coating in the treatment of a number of conditions, including cardiovascular and neurological conditions [99]. Generally the most critical aspects for the development of effective ROS-targeting delivery systems as theragnostic agents include biocompatibility and safety for human use, high stimuli sensitivity to specific redox changes and targeted antioxidant capacity [93]. As many CVDs and associated risk factors hypertension have an inflammatory component, more research is required to investigate the biocompatibility of this drug delivery approach, and whether targeting peptides to inflamed endothelial cells by modified NPs will exacerbate the inflammatory response, before it can be extensively translated into clinical applications [67, 100].

In conclusion, nanotechnology continues to make progress in the management of CVD, by the development of multitargeting and multifunctional NPs for their selective drug delivery to diseased regions, ultimately allowing for targeted imaging, diagnosis and effective treatment of CVD. Thus far, the successful use of cell adhesion molecules and surface assimilation of polymers for the formulation of endothelium-targeting NPs has exhibited favorable outcomes in animal studies, demonstrating promising potential for their therapeutic application in patients with various cardiovascular, inflammatory and metabolic diseases. Most importantly, continued effort to understand the multifaceted interactions of functionalized nanoparticles with biological systems is required in order to establish their safe and effective role as powerful tools for the diagnosis and treatment of disease in clinical settings.

References

[1] R. Santos-Oliveira, M. Albernaz, B. de Carvalho Patricio, Development of nano radiopharmaceuticals by labelling polymer nanoparticles with Tc-99m, World J. Nucl. Med. 12 (1) (2013) 24.

[2] M. Mauricio, S. Guerra-Ojeda, P. Marchio, S. Valles, M. Aldasoro, I. Escribano-Lopez, et al., Nanoparticles in medicine: a focus on vascular oxidative stress, Oxid. Med. Cell. Longev. 2018 (2018) 1–20.

[3] V. Prabhu, S. Uzzaman, V. Grace, C. Guruvayoorappan, Nanoparticles in drug delivery and cancer therapy: the giant rats tail, J. Cancer Ther. 02 (03) (2011) 325–334.

[4] Wahajuddin, S. Arora, Superparamagnetic iron oxide nanoparticles: magnetic nanoplatforms as drug carriers, Int. J. Nanomedicine 3445 (2012).

[5] A. Wiesenthal, L. Hunter, S. Wang, J. Wickliffe, M. Wilkerson, Nanoparticles: small and mighty, Int. J. Dermatol. 50 (3) (2011) 247–254.

[6] J. Suk, Q. Xu, N. Kim, J. Hanes, L. Ensign, PEGylation as a strategy for improving nanoparticle-based drug and gene delivery, Adv. Drug Deliv. Rev. 99 (2016) 28–51.

[7] R. Prajnamitra, H. Chen, C. Lin, L. Chen, P. Hsieh, Nanotechnology approaches in tackling cardiovascular diseases, Molecules 24 (10) (2019) 2017.

[8] A. Timmis, N. Townsend, C. Gale, A. Torbica, M. Lettino, S. Petersen, et al., European Society of Cardiology: cardiovascular disease statistics 2019, Eur. Heart J. 41 (1) (2019) 12–85.

[9] M. Chandarana, A. Curtis, C. Hoskins, The use of nanotechnology in cardiovascular disease, Appl. Nanosci. 8 (7) (2018) 1607–1619.

[10] A.M. Gorabi, N. Kiaie, Z. Reiner, F. Carbone, F. Montecucco, A. Sahebkar, The therapeutic potential of nanoparticles to reduce inflammation in atherosclerosis, Biomolecules 9 (9) (2019) 416.

[11] R. Kiseleva, P. Glassman, C. Greineder, E. Hood, V. Shuvaev, V. Muzykantov, Targeting therapeutics to endothelium: are we there yet? Drug Deliv. Transl. Res. 8 (4) (2017) 883–902.

[12] A. Flores, J. Ye, K. Jarr, N. Hosseini-Nassab, B. Smith, N. Leeper, Nanoparticle therapy for vascular diseases, Arterioscler. Thromb. Vasc. Biol. 39 (4) (2019) 635–646.

[13] R. Palekar, A. Jallouk, G. Lanza, H. Pan, S. Wickline, Molecular imaging of atherosclerosis with nanoparticle-based fluorinated MRI contrast agents, Nanomedicine 10 (11) (2015) 1817–1832.

[14] S. Jafari, H. Derakhshankhah, L. Alaei, A. Fattahi, B. Varnamkhasti, A. Saboury, Mesoporous silica nanoparticles for therapeutic/diagnostic applications, Biomed. Pharmacother. 109 (2019) 1100–1111.

[15] K. Lee, J. Lee, M. Kwak, Y. Cho, B. Hwang, M. Cho, et al., Two distinct cellular pathways leading to endothelial cell cytotoxicity by silica nanoparticle size, J. Nanobiotechnol. 17 (1) (2019).

[16] R. Pala, V. Anju, M. Dyavaiah, S. Busi, S. Nauli, Nanoparticle-mediated drug delivery for the treatment of cardiovascular diseases, Int. J. Nanomedicine 15 (2020) 3741–3769.

[17] A. Farooq, Restored endothelial dependent vasodilation in aortic vessels after uptake of ceria coated silica nanoparticles, ex vivo, J. Nanomed. Nanotechnol. 05 (02) (2014).

[18] M. Van Schooneveld, E. Vucic, R. Koole, Y. Zhou, J. Stocks, D. Cormode, et al., Improved biocompatibility and pharmacokinetics of silica nanoparticles by means of a lipid coating: a multimodality investigation, Nano Lett. 8 (8) (2008) 2517–2525.

[19] L. Sercombe, T. Veerati, F. Moheimani, S. Wu, A. Sood, S. Hua, Advances and challenges of liposome assisted drug delivery, Front. Pharmacol. 6 (2015).

[20] P. Deshpande, S. Biswas, V. Torchilin, Current trends in the use of liposomes for tumor targeting, Nanomedicine 8 (9) (2013) 1509–1528.

[21] R. Montesinos, Liposomal drug delivery to the central nervous system, Liposomes (2017) 213–242.

[22] B. Pattni, V. Chupin, V. Torchilin, New developments in liposomal drug delivery, Chem. Rev. 115 (19) (2015) 10938–10966.

[23] M. Li, C. Du, N. Guo, Y. Teng, X. Meng, H. Sun, et al., Composition design and medical application of liposomes, Eur. J. Med. Chem. 164 (2019) 640–653.

[24] M. Shin, S. Shukla, Y. Chung, V. Beiss, S. Chan, O. Ortega-Rivera, et al., COVID-19 vaccine development and a potential nanomaterial path forward, Nat. Nanotechnol. 15 (8) (2020) 646–655.

[25] V. Cardoso, B. Moreira, E. Comparetti, I. Sampaio, L. Ferreira, P. Lins, et al., Is nanotechnology helping in the fight against COVID-19? Front. Nanotechnol. 2 (2020).

[26] N. Cureton, I. Korotkova, B. Baker, S. Greenwood, M. Wareing, V. Kotamraju, et al., Selective targeting of a novel vasodilator to the uterine vasculature to treat impaired uteroplacental perfusion in pregnancy, Theranostics 7 (15) (2017) 3715–3731.

[27] A. Zaabalawi, C. Astley, L. Renshall, F. Beards, A. Lightfoot, H. Degens, et al., Tetramethoxystilbene-loaded liposomes restore reactive-oxygen-species-mediated attenuation of dilator responses in rat aortic vessels ex vivo, Molecules 24 (23) (2019) 4360.

[28] M. Lee, Liposomes for enhanced bioavailability of water-insoluble drugs: in vivo evidence and recent approaches, Pharmaceutics 12 (3) (2020) 264.

[29] P. Jaiswal, B. Gidwani, A. Vyas, Nanostructured lipid carriers and their current application in targeted drug delivery, Artif. Cells Nanomed. Biotechnol. 44 (1) (2014) 27–40.

[30] A. Khosa, S. Reddi, R. Saha, Nanostructured lipid carriers for site-specific drug delivery, Biomed. Pharmacother. 103 (2018) 598–613.

[31] C. Fang, S.A. Al-Suwayeh, J. Fang, Nanostructured lipid carriers (NLCs) for drug delivery and targeting, Recent Pat. Nanotechnol. 7 (1) (2012) 41–55.

[32] V. Piazzini, B. Lemmi, M. D'Ambrosio, L. Cinci, C. Luceri, A. Bilia, et al., Nanostructured lipid carriers as promising delivery systems for plant extracts: the case of silymarin, Appl. Sci. 8 (7) (2018) 1163.

[33] V. Salvi, P. Pawar, Nanostructured lipid carriers (NLC) system: a novel drug targeting carrier, J. Drug Deliv. Sci. Technol. 51 (2019) 255–267.

[34] M. Cirri, L. Maestrini, F. Maestrelli, N. Mennini, P. Mura, C. Ghelardini, et al., Design, characterization and in vivo evaluation of nanostructured lipid carriers (NLC) as a new drug delivery system for hydrochlorothiazide oral administration in pediatric therapy, Drug Deliv. 25 (1) (2018) 1910–1921.

[35] A. Khan, I. Abdulbaqi, R. Abou Assi, V. Murugaiyah, Y. Darwis, Lyophilized hybrid nanostructured lipid carriers to enhance the cellular uptake of verapamil: statistical optimization and in vitro evaluation, Nanoscale Res. Lett. 13 (1) (2018).

[36] K. Magyar, R. Halmosi, A. Palfi, G. Feher, L. Czopf, A. Fulop, et al., Cardio protection by resveratrol: a human clinical trial in patients with stable coronary artery disease, Clin. Hemorheol. Microcirc. 50 (3) (2012) 179–187.

[37] M. Theodotou, K. Fokianos, A. Mouzouridou, C. Konstantinou, A. Aristotelous, D. Prodromou, et al., The effect of resveratrol on hypertension: a clinical trial, Exp. Ther. Med. 13 (1) (2016) 295–301.

[38] N. Poonia, R. Kharb, V. Lather, D. Pandita, Nanostructured lipid carriers: versatile oral delivery vehicle, Future Sci. OA 2 (3) (2016), FSO135.

[39] J. Sun, Y. Liu, M. Ge, G. Zhou, W. Sun, D. Liu, et al., A distinct endocytic mechanism of functionalized-silica nanoparticles in breast cancer stem cells, Sci. Rep. 7 (1) (2017).

[40] J. Saikia, M. Yazdimamaghani, S. Hadipour Moghaddam, H. Ghandehari, Differential protein adsorption and cellular uptake of silica nanoparticles based on size and porosity, ACS Appl. Mater. Interfaces 8 (50) (2016) 34820–34832.

[41] S. Behzadi, V. Serpooshan, W. Tao, M. Hamaly, M. Alkawareek, E. Dreaden, et al., Cellular uptake of nanoparticles: journey inside the cell, Chem. Soc. Rev. 46 (14) (2017) 4218–4244.

[42] P. Foroozandeh, A. Aziz, Insight into cellular uptake and intracellular trafficking of nanoparticles, Nanoscale Res. Lett. 13 (1) (2018).

[43] V. Schäfer, H. von Briesen, R. Andreesen, A. Steffan, C. Royer, S. Tröster, et al., Phagocytosis of nanoparticles by human immunodeficiency virus (HIV)-infected macrophages: a possibility for antiviral drug targeting, Pharm. Res. 09 (4) (1992) 541–546.

[44] Y. Tabata, Y. Ikada, Effect of the size and surface charge of polymer microspheres on their phagocytosis by macrophage, Biomaterials 9 (4) (1988) 356–362.

[45] R. Agarwal, V. Singh, P. Jurney, L. Shi, S. Sreenivasan, K. Roy, Mammalian cells preferentially internalize hydrogel nanodiscs over nanorods and use shape-specific uptake mechanisms, Proc. Natl. Acad. Sci. 110 (43) (2013) 17247–17252.

[46] H. Gustafson, D. Holt-Casper, D. Grainger, H. Ghandehari, Nanoparticle uptake: the phagocyte problem, Nano Today 10 (4) (2015) 487–510.

[47] J. McCright, C. Skeen, J. Yarmovsky, Dense poly (ethylene glycol) coatings maximize nanoparticle transport across lymphatic endothelial cells, bioRxiv (2020). https://doi.org/10.1101/2020.08.01. 232249.

[48] I.-L. Hsiao, A.M. Gramatke, R. Joksimovic, M. Sokolowski, M. Gradzielski, A. Haase, Size and cell type dependent uptake of silica nanoparticles, J. Nanomed. Nanotechnol. 05 (06) (2014).

[49] C. Wu, Y. Wu, Y. Jin, P. Zhu, W. Shi, J. Li, et al., Endosomal/lysosomal location of organically modified silica nanoparticles following caveolae-mediated endocytosis, RSC Adv. 9 (24) (2019) 13855–13862.

[50] J. Duan, Y. Yu, Y. Yu, Y. Li, P. Huang, X. Zhou, et al., Silica nanoparticles enhance autophagic activity, disturb endothelial cell homeostasis and impair angiogenesis, Part. Fibre Toxicol. 11 (1) (2014).

[51] O. Harush-Frenkel, N. Debotton, S. Benita, Y. Altschuler, Targeting of nanoparticles to the clathrin-mediated endocytic pathway, Biochem. Biophys. Res. Commun. 353 (1) (2007) 26–32.

[52] K. Shapero, F. Fenaroli, I. Lynch, D. Cottell, A. Salvati, K. Dawson, Time and space resolved uptake study of silica nanoparticles by human cells, Mol. BioSyst. 7 (2) (2011) 371–378.

[53] C. Freese, D. Schreiner, L. Anspach, C. Bantz, M. Maskos, R. Unger, et al., In vitro investigation of silica nanoparticle uptake into human endothelial cells under physiological cyclic stretch, Part. Fibre Toxicol. 11 (1) (2014).

[54] A. Shukur, D. Whitehead, A. Seifalian, M. Azzawi, The influence of silica nanoparticles on small mesenteric arterial function, Nanomedicine 11 (16) (2016) 2131–2146.

[55] A. Alshehri, A. Grabowska, S. Stolnik, Pathways of cellular internalisation of liposomes delivered siRNA and effects on siRNA engagement with target mRNA and silencing in cancer cells, Sci. Rep. 8 (1) (2018).

[56] J. Yang, A. Bahreman, G. Daudey, J. Bussmann, R. Olsthoorn, A. Kros, Drug delivery via cell membrane fusion using lipopeptide modified liposomes, ACS Cent. Sci. 2 (9) (2016) 621–630.

[57] F. Helm, G. Fricker, Liposomal conjugates for drug delivery to the central nervous system, Pharmaceutics 7 (2) (2015) 27–42.

[58] M. Kardara, S. Hatziantoniou, A. Sfika, A. Vassiliou, E. Mourelatou, C. Magkou, et al., Caveolar uptake and endothelial-protective effects of nanostructured lipid carriers in acid aspiration murine acute lung injury, Pharm. Res. 30 (7) (2013) 1836–1847.

[59] Y. Lu, P. Low, Folate-mediated delivery of macromolecular anticancer therapeutic agents, Adv. Drug Deliv. Rev. 64 (2012) 342–352.

[60] L. Kou, J. Sun, Y. Zhai, Z. He, The endocytosis and intracellular fate of nanomedicines: implication for rational design, Asian J. Pharm. Sci. 8 (1) (2013) 1–10.

[61] A. Desai, M. Hunter, A. Kapustin, Using macropinocytosis for intracellular delivery of therapeutic nucleic acids to tumour cells, Philos. Trans. R. Soc. B: Biol. Sci. 374 (1765) (2018) 20180156.

[62] W. Arap, R. Pasqualini, M. Montalti, L. Petrizza, L. Prodi, E. Rampazzo, et al., Luminescent silica nanoparticles for cancer diagnosis, Curr. Med. Chem. 20 (17) (2013) 2195–2211.

[63] K. Wang, X. He, X. Yang, H. Shi, Functionalized silica nanoparticles: a platform for fluorescence imaging at the cell and small animal levels, Acc. Chem. Res. 46 (7) (2013) 1367–1376.

[64] J. Margolis, J. McDonald, R. Heuser, P. Klinke, R. Waksman, R. Virmani, et al., Systemic nanoparticle paclitaxel (nab–paclitaxel) for in-stent restenosis I (SNAPIST-I): a first-in-human safety and dose-finding study, Clin. Cardiol. 30 (4) (2007) 165–170.

[65] M. Miller, J. Raftis, J. Langrish, S. McLean, P. Samutrtai, S. Connell, et al., Correction to "inhaled nanoparticles accumulate at sites of vascular disease", ACS Nano 11 (10) (2017) 10623–10624.

[66] G. Oberdorster, Safety assessment for nanotechnology and nanomedicine: concepts of nanotoxicology, J. Intern. Med. 267 (1) (2010) 89–105.

[67] V. Muzykantov, Targeted drug delivery to endothelial adhesion molecules, ISRN Vasc. Med. 2013 (2013) 1–27.

[68] D. Chenthamara, S. Subramaniam, S. Ramakrishnan, S. Krishnaswamy, M. Essa, F. Lin, et al., Therapeutic efficacy of nanoparticles and routes of administration, Biomater. Res. 23 (1) (2019).

[69] R. Chen, L. Xu, Q. Fan, J. Li, J. Wang, L. Wu, et al., Hierarchical pulmonary target nanoparticles via inhaled administration for anticancer drug delivery, Drug Delivery 24 (1) (2017) 1191–1203.

[70] M. Miller, J. Raftis, J. Langrish, S. McLean, P. Samutrtai, S. Connell, et al., Inhaled nanoparticles accumulate at sites of vascular disease, ACS Nano 11 (5) (2017) 4542–4552.

[71] Y. Yun, Y. Cho, K. Park, Nanoparticles for oral delivery: targeted nanoparticles with peptidic ligands for oral protein delivery, Adv. Drug Deliv. Rev. 65 (6) (2013) 822–832.

[72] D. van der Laan, P. Elders, C. Boons, G. Nijpels, J. Krska, J. Hugtenburg, The impact of cardiovascular medication use on patients' daily lives: a cross-sectional study, Int. J. Clin. Pharm. 40 (2) (2018) 412–420.

[73] A. Olowofela, A. Isah, A profile of adverse effects of antihypertensive medicines in a tertiary care clinic in Nigeria, Annals Afr. Med. 16 (3) (2017) 114.

[74] E. Gebreyohannes, A. Bhagavathula, T. Abebe, Y. Tefera, T. Abegaz, Adverse effects and non-adherence to antihypertensive medications in University of Gondar Comprehensive Specialized Hospital, Clin. Hypertens. 25 (1) (2019).

[75] J. Yoo, C. Park, G. Yi, D. Lee, H. Koo, Active targeting strategies using biological ligands for nanoparticle drug delivery systems, Cancers 11 (5) (2019) 640.

[76] K. Burns, J. Delehanty, Targeting therapeutics to the plasma membrane: opportunities for nanoparticle-mediated delivery abound, Ther. Deliv. 8 (5) (2017) 235–237.

[77] N. Muhamad, T. Plengsuriyakarn, K. Na-Bangchang, Application of active targeting nanoparticle delivery system for chemotherapeutic drugs and traditional/herbal medicines in cancer therapy: a systematic review, Int. J. Nanomedicine 13 (2018) 3921–3935.

[78] E. Lutton, S. Farney, A. Andrews, V. Shuvaev, G. Chuang, V. Muzykantov, et al., Endothelial targeted strategies to combat oxidative stress: improving outcomes in traumatic brain injury, Front. Neurol. 10 (2019).

[79] E. Lutton, R. Razmpour, A. Andrews, L. Cannella, Y. Son, V. Shuvaev, et al., Acute administration of catalase targeted to ICAM-1 attenuates neuropathology in experimental traumatic brain injury, Sci. Rep. 7 (1) (2017).

[80] T. Bhowmick, E. Berk, X. Cui, V. Muzykantov, S. Muro, Effect of flow on endothelial endocytosis of nanocarriers targeted to ICAM-1, J. Control. Release 157 (3) (2012) 485–492.

[81] J. Kusunose, M. Gagnon, J. Seo, K. Ferrara, Quantitation of nanoparticle accumulation in flow using optimized microfluidic chambers, J. Drug Target. 22 (1) (2013) 48–56.

[82] M. Voinea, I. Manduteanu, E. Dragomir, M. Capraru, M. Simionescu, Immunoliposomes directed toward VCAM-1 interact specifically with activated endothelial cells—a potential tool for specific drug delivery, Pharm. Res. 22 (11) (2005) 1906–1917.

[83] F. Danhier, V. Pourcelle, J. Marchand-Brynaert, C. Jérôme, O. Feron, V. Préat, Targeting of tumor endothelium by RGD-grafted PLGA-nanoparticles, Methods Enzymol. (2012) 157–175.

[84] C. Yang, K. Bromma, D. Chithrani, Peptide mediated in vivo tumor targeting of nanoparticles through optimization in single and multilayer in vitro cell models, Cancers 10 (3) (2018) 84.

[85] Y. Sakurai, H. Akita, H. Harashima, Targeting tumor endothelial cells with nanoparticles, Int. J. Mol. Sci. 20 (23) (2019) 5819.

[86] P. Homem de Bittencourt, D. Lagranha, A. Maslinkiewicz, S. Senna, A. Tavares, L. Baldissera, et al., LipoCardium: endothelium-directed cyclopentenone prostaglandin-based liposome formulation that completely reverses atherosclerotic lesions, Atherosclerosis 193 (2) (2007) 245–258.

[87] P. Vader, B. Crielaard, S. van Dommelen, R. van der Meel, G. Storm, R. Schiffelers, Targeted delivery of small interfering RNA to angiogenic endothelial cells with liposome-polycation-DNA particles, J. Control. Release 160 (2) (2012) 211–216.

[88] S. Muro, V. Muzykantov, Targeting of antioxidant and anti-thrombotic drugs to endothelial cell adhesion molecules, Curr. Pharm. Des. 11 (18) (2005) 2383–2401.

[89] Z. Shen, A. Fisher, W. Liu, Y. Li, PEGylated "stealth" nanoparticles and liposomes, Eng. Biomater. Drug Deliv. Syst. (2018) 1–26.

[90] M. Regi, M. Colilla, I. Barba, M. Manzano, Mesoporous silica nanoparticles for drug delivery: current insights, Molecules 23 (1) (2018) 47.

[91] K. Radhakrishnan, S. Gupta, D. Gnanadhas, P. Ramamurthy, D. Chakravortty, A. Raichur, Protamine-capped mesoporous silica nanoparticles for biologically triggered drug release, Part. Part. Syst. Charact. 31 (4) (2013) 449–458.

[92] N. Cook, A randomized factorial trial of vitamins C and E and beta carotene in the secondary prevention of cardiovascular events in women, Arch. Intern. Med. 167 (15) (2007) 1610.

[93] K. Kim, D. Lee, C. Song, P. Kang, Reactive oxygen species-activated nanomaterials as theranostic agents, Nanomedicine 10 (17) (2015) 2709–2723.

[94] J. Himmelfarb, T. Ikizler, C. Ellis, P. Wu, A. Shintani, S. Dalal, et al., Provision of Antioxidant Therapy in Hemodialysis (PATH): a randomized clinical trial, J. Am. Soc. Nephrol. 25 (3) (2013) 623–633.

[95] N. Ma, Y. Li, H. Xu, Z. Wang, X. Zhang, Dual redox responsive assemblies formed from diselenide block copolymers, J. Am. Chem. Soc. 132 (2) (2010) 442–443.

[96] Q. Zhao, C. Wang, Y. Liu, J. Wang, Y. Gao, X. Zhang, et al., PEGylated mesoporous silica as a redox-responsive drug delivery system for loading thiol-containing drugs, Int. J. Pharm. 477 (1–2) (2014) 613–622.

[97] A. Bakhshian Nik, H. Zare, S. Razavi, H. Mohammadi, P. Torab Ahmadi, N. Yazdani, et al., Smart drug delivery: capping strategies for mesoporous silica nanoparticles, Microporous Mesoporous Mater. 299 (2020) 110115.

[98] Y. Song, Y. Li, Q. Xu, Z. Liu, Mesoporous silica nanoparticles for stimuli-responsive controlled drug delivery: advances, challenges, and outlook, Int. J. Nanomedicine 12 (2016) 87–110.

[99] Y. Shen, B. Cao, N. Snyder, K. Woeppel, J. Eles, X. Cui, ROS responsive resveratrol delivery from LDLR peptide conjugated PLA-coated mesoporous silica nanoparticles across the blood–brain barrier, J. Nanobiotechnol. 16 (1) (2018).

[100] E. Chung, Targeting and therapeutic peptides in nanomedicine for atherosclerosis, Exp. Biol. Med. 241 (9) (2016) 891–898.

CHAPTER 7

Pharmacokinetics and in vivo evaluation of nanoparticles

Largee Biswas[a],*, Asiya Mahtab[b],*, and Anita K. Verma[a]
[a]Nanobiotech Lab, Department of Zoology, Kirori Mal College, University of Delhi, Delhi, India
[b]Department of Pharmaceutics, School of Pharmaceutical Education and Research, Jamia Hamdard, New Delhi, India

1. Introduction

Over the past six decades, varied nanoparticulate systems, including liposomes, nanosomes, nanospheres, nanocarriers, and nanocapsules, are being used in drug delivery for targeting of ultrafine particles and macromolecules (e.g., RNA, DNA, and proteins) and therapeutic molecules [1–5]. The versatile nature and exceptional properties of nanoparticles make them promising candidates for a variety of applications in nanomedicine. Ensuring safe and effective biomedical applications of nanomaterials would require a complete understanding of the interaction between the material and its biological response in the body. Hypothetically, a drug-loaded nanocarrier protects the active molecules against degradation and inactivation of drug en route the desired site. This often results in degradation of the active agent necessary for optimal therapeutic effectiveness with negligible drug-induced side effects. For delivering chemotherapeutic molecules to solid tumors the notion of risk-to-benefit ratio is still crucial as a narrow therapeutic window exists for toxic molecules when compared to the steep dose-response curve (Fig. 1) [1].

Moreover, the morphology, including shape, size, and surface chemistry, can be manipulated to dictate the pharmacokinetics (PK) and biotargeting, depending on the nature of disease, developmental stage, as well as location of the disease. Biocompatibility should be comprehended with a focus on *biological milieu* in which the biomaterial has to function [6]. Mechanistic models for physiology-based PK (PBPK) studies are mathematical in nature because the data are collected and assembled based on human physiological and anatomical characteristics. The physicochemical parameters of the drug molecules undergo a complex process of absorption, distribution, metabolism, and excretion [7]. Therefore PBPK models are capable of predicting drug bioavailability, distribution, efficiency, and toxicity, once coupled with pharmacodynamic (PD) models that predict the pharmacological response based on drug exposure at the site of action [8].

---

* These authors contributed equally to this chapter.

Nanoparticle Therapeutics
https://doi.org/10.1016/B978-0-12-820757-4.00006-5

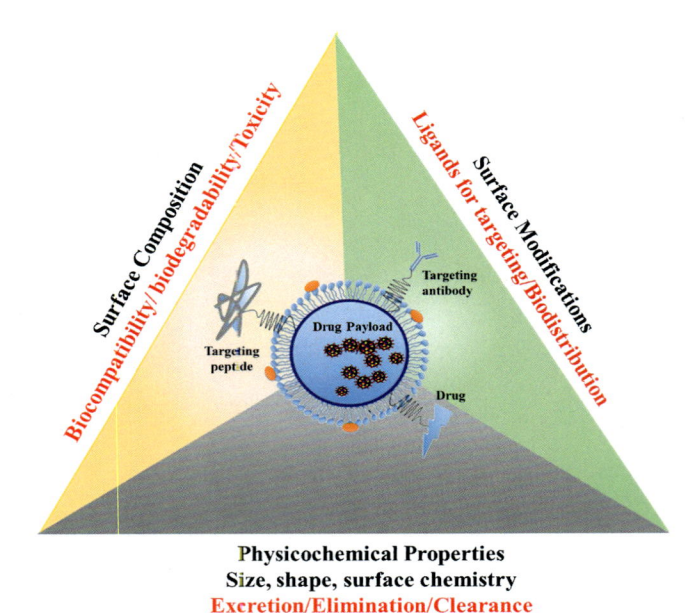

Fig. 1 Schematic representation of a multifunctional nanoparticle, its physicochemical properties such as size, shape, and surface chemistry can control the in vivo pharmacokinetics, pharmacodynamics, biodistribution, bioavailability, and toxicity.

For drug delivery a critical evaluation of nanoparticles' biocompatibility to minimize cytotoxicity is necessary, so factors affecting biocompatibility of nanoparticles are fundamental to safe delivery of drugs. The administered route of nanocarriers delivery in the body can be intravenous, intramuscular, subcutaneous or oral intake, all are bound to elicit differential immune responses [9]. This chapter comprehensively reviews the PK, PBPK, bioavailability, and PD for anticancer drug-loaded nanoparticles.

2. Pharmacokinetics

PK signifies drugs disposition in the body, primarily establishing drug distribution, absorption, metabolism, and excretion in a due course of time in the body [10]. It further helps in preclinical toxicology studies done in animals recognized as toxicokinetics, that is, the level of drug in plasma or tissues is more prognostic for human use than the actual dose to generalize the toxicity data [11]. Equated to drugs per se, nanomedicines can show novel issues of toxicity, as the PK functions of active pharmaceutical ingredient can be transformed upon encapsulation in nanoformulations [12].

2.1 Pharmacokinetic analysis

The first step in the determination of the PK data includes estimation of model-independent parameters such as maximum concentration (C_{max}), observed time of maximum concentration (T_{max}), area under the curve (AUC), half-life ($t_{1/2}$), clearance (Cl), and mean resident time (MRT), that is, average time that a drug molecule resides in the body and the steady state. These parameters well signify the data without the requirement of any complex mathematical model [13, 14]. Fig. 2 shows a representative PK profile, when measuring a drug in plasma over time after oral administration, which means that samples are collected before and postadministration of drug. It is pertinent to mention here that drug can be detected in either plasma which is blood minus the blood cells, but sera is blood without cells and clotting proteins. So, a PK profile would ideally be from samples collected as: predose, every 30 min postdose, every hour for the next few hours and then at longer time intervals every 12 h or 24 h. It is important to restrict the number of samples to be collected per animal, or as per the ethical norms.

Although Fig. 2A represents blood kinetics, this may not necessarily be an indicator of what happens in the organs or in tissues. But, at a particular time point, the plasma drug concentrations measure all the four PK processes that are happening in the body. Initial phase indicates absorption, which occurs at the highest rate, and the terminal phase exhibits excretion. When the plasma concentration reaches its peak, the highest point represents the amount of drug entering the plasma equals the amount of drug being eliminated from the blood plasma. These rates of absorption are dependent on the route of administration of drug. The oral drugs are required to be absorbed into the blood after they reach the gastrointestinal (GI) tract, whereas intravenously given drugs are injected directly in the systemic circulation. Fig. 2B illustrates the variation in the concentrations of drug in plasma after an oral or intravenous dose. Fig. 2C represents a generalized PK profile wherein we can observe and determine PK characteristics that indicate the drug exposure within the body along with the rate and amount of absorption from a single plasma concentration profile.

2.2 Bioavailability

Bioavailability specifies the percentage of drug absorbed into the systemic circulation. Fig. 2D showed that the entire intravenous dose is directly injected into the plasma; hence, it will be represented as 100% bioavailable. But the oral dose may not be entirely absorbed, probably because of insufficient time for absorption, or poor absorptive surfaces, thereby exhibiting reduced bioavailability when compared with an intravenous dose. Bioavailability measures the quantity of drug that is basically absorbed after a specified dose and denoted either as relative bioavailability or absolute bioavailability. Relative bioavailability corresponds to the rate and degree of absorption of drug, once compared to its nanoformulation or a product of the similar drug. It basically compares

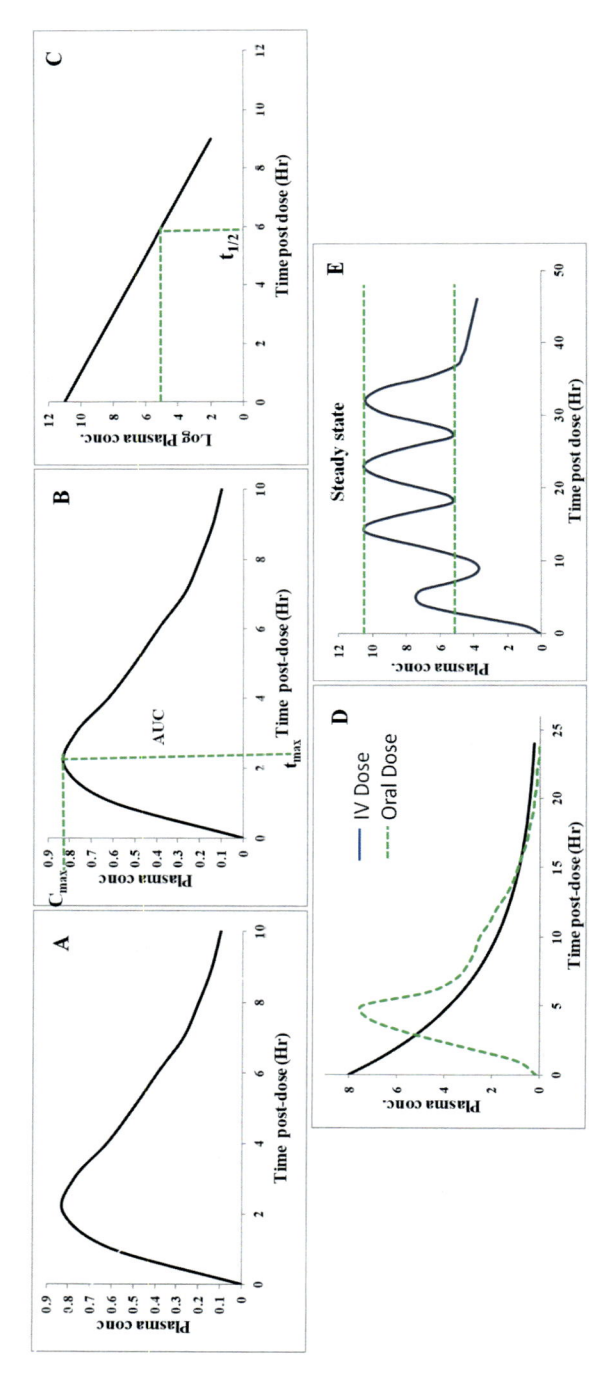

Fig. 2 (A): Illustration of a linear PK profile of a drug after oral administration. (B) (C) Measurements representing the elimination phase PK analysis using a semilogarithmic plot. Where, C_{max}—record of the maximum concentration, T_{max}—time taken to reach C_{max}, AUC—area under the curve, that measures the time of drug exposure and elimination half-life, t1/2—the time taken for the plasma concentration to fall by half its original value; (D) bioavailability of intravenous dose vs. oral dose; (E) steady-state representing the rate of absorption equal to the rate of elimination.

the difference in the PK profile of the drug per se with the release kinetics of drug from the nanoformulation. Absolute bioavailability evaluates an oral dose vs. an intravenous dose, where 100% drug bioavailability is expected after intravenous injection. The object of an absolute bioavailability study is to determine the amount of oral drug required to reach the drug concentration in systemic circulation that is a comparable concentration to the intravenous dose. It is evident that most of the oral drug is lost during the absorption.

2.3 Dose proportionality

Dose proportionality specifies the constant ratio among the administered dose and the detected PK profile, such that, if the dose is doubled, one may expect to observe a double value of C_{max}. T_{max}, AUC, and $t_{1/2}$, but interestingly, they remain constant. If the drug being tested exhibits dose proportionality, then one has to predict repeat-dose PK, from a single dose. That will help predict the dose adjustments, that should be extrapolated for application on subpopulations.

2.4 Steady state

A steady state is reached when the drug accumulates in the body after repeated drug administration over a period of time. From Fig. 2E one observes the increasing plasma concentrations when multiple drugs are administered. Steady state is reached once the absorption process is occurring at precisely the same rate as the elimination. The time interval to attain the steady state, assuming that the dosing intervals were kept constant, would be then dependent upon the drug half-life. Minimum doses required to attain the steady-state are two. In clinical practice, this would represent a higher number of doses being administered, although an initial loading dose is required to be given to reach the steady state rapidly.

For any in vivo evaluation, it is significant to evaluate the PK of a drug for phase I studies as the future of the engineered nanoparticle or drug would be an indicator for their clinical success. This includes data to be collected on whether to dose or fast, time interval of dose, use of associated medication and most importantly when to optimally collect the PK samples. It is not necessary that phase I studies be performed, occasionally, some phase I trials are done in parallel to phase II or III, especially for biliary or hepatic or renal studies.

2.5 Biocompatibility

Biocompatibility has formally been defined as "The ability of a material to perform with an appropriate host response in a specific situation" [15]. The three doctrines that are essential to fulfill this description are that: (i) a material must perform its requisite roles and not just exist in the tissue; (ii) the induced response must be appropriate for the envisioned application; and (iii) the premise of a reaction to that precise material and its appropriateness must be unlike to other [16]. Kohane and Langer explicated biocompatibility with respect to drug delivery and described biocompatibility as an "expression of the benignity of the relation between a material and its biological environment" [17].

2.6 Hepatic/renal impairment

Hepatic and renal impairment studies are conducted in parallel where subjects are categorized into groups based on levels of impairment: mild, moderate, or severe. Creatinine clearance is a marker for renal damage, and the creatinine clearance value is recorded to determine renal and hepatic impairment. The PK profiles of the healthy volunteers are matched with the profiles of impaired patients, and dose adjustment is advised based on significant statistical difference. A nanoparticle has to negotiate three layers of glomeruli: an endothelium having leaky fenestrae 70–90 nm, the basement membrane of glomeruli having 2–8 nm pores, and filtration slits on the outer epithelium 4–11 nm size range. Theoretically, any nanoparticle having a hydrodynamic diameter <6 nm can cross the capillary wall of glomeruli [18].

2.7 Mass balance analysis

Mass balance analysis helps regulate the routes of administration, the rates of excretion as well as the metabolic profile of the nanoparticles to be tested. These are performed by administering radiolabeled drug or the nanoparticle. Radiochemical labeling has been achieved using ^{64}Cu [19, 20], ^{65}Zn [21], ^{68}Ga [22], ^{109}Cd [23], ^{99}Tec [24], ^{18}F [25], ^{111}In [26], ^{141}Ce [21], ^{153}Sm [27], and ^{198}Au [28, 29]. The radioactive marking can keep track of the molecule and its degraded metabolites by assaying collected samples of blood, urine, and feces for radioactive label. It is easy to determine mass balance as we can correlate the administered routes and rates of elimination and then identify any circulatory and excretory metabolites.

Although a plethora of publications are available imparting such knowledge, restrictive technology is accessible to evaluate and monitor real-time measurements relating to nanoparticle features in circulation such as blood flow properties inside the vessels and flow rates at forks in the vessels and the capillaries, dynamics of plasma protein binding, peeling of nanoparticle components, and dynamic changes in shape and size. However, current advances in nanodevices and microfluidic devices allow understanding of some in vitro aspects of fluid dynamics to enable appropriate nanoparticle design in terms of performance in in vivo models [30–32]. Biological factors include anatomical, biochemical, physiological, and immunological barriers; nonetheless, it offers possibilities by diseased states for nanoparticles exploration and therapeutic exploitation based on available routes for delivery, expression of selective antigen or receptor [1, 3]. PK analysis gets further intricate when multiple doses are given with varying dosing schedules [1, 3], it has been observed that PK changes after the first dose for the patients who receive multiple injections, at regular intervals. Mostly, the researchers have estimated the individual factors, such as the impact of nanoparticle shape, size, zeta potential or surface charge, required for protein binding, for clearance kinetics, and the role of nanoparticle size and shape on extravasation from vessels. Few reports on the combinatorial approach explains the effect of both size and surface function on plasma protein–binding efficiency and their correlations with the circulating half-lives [1, 3, 31–37]. All these approaches are invaluable but restricted within structure-performance associations, a concerted strategy is obligatory to evaluate the dynamic interplay amid physicochemical and biological factors monitoring PK of nanoparticle (Fig. 2).

3. PK of drugs per se and encapsulated drugs in nanoparticles

Drug encapsulation in nanoparticles enhances drug delivery to the target tissue due to amplified drug permeability or absorption, thereby reducing the dose frequency with improved patient compliance as compared with free drug. Free drug and drug encapsulated in nanoparticles work contrarily when administered in living organisms that is evident in the variations on PK parameters. Drug encapsulated in nanoparticles leads to reduced drug exposure and is protected from the unfavorable condition in the GI tract, thus reducing both nonenzymatic and enzymatic degradations; that further leads to enhanced AUC [38]. Nanoparticles encapsulating drugs augment oral absorption by increased gastric residence time via mucosal entry of cell or tissue [39]. In the liver, before biliary excretion, lipophilic drugs undergo biotransformation to hydrophilic metabolites; therefore, encapsulation of drug in nanoparticles is essential as it moderates renal clearance due to increased size (renal clearance is <15 nm) [40]. Negatively charged glomeruli capillary wall allows positively charged nanoparticles to rapidly cross the filtration barrier, having a size of 6–8 nm, but negatively charged or neutral nanoparticles of same size are unable to pass the filtration barrier [18, 41]. Hence, drug encapsulation in nanoparticulate formulations possibly protects lipophilic drugs from the liver metabolizing enzymes [42]. A hydrophilic moiety when injected intravenously is rapidly eliminated from the blood by excretion, because negligible protein binds to drug; instead, a hydrophobic drug has significantly reduced renal clearance as there is higher binding of serum proteins. Drug encapsulation suggestively decreases the apparent clearance of drug from plasma, indicating prolonged circulation of drug, increased half-life, and the possibility of cumulative drug availability at the target tissues.

To calculate the mechanism and release kinetics of drug, the results of the in vitro drug release studies should be accessed with different kinetics equations like:

Zero order (% cumulative drug release vs. time), that is,

$$F(t) = kt \tag{1}$$

First-order (log% cumulative drug remaining vs. time), that is,

$$\text{In}[1 - M_t/M_\infty] = -kt \tag{2}$$

Higuchi matrix (% cumulative drug release vs. square root of time), that is,

$$M_t/M_\infty = kt^{1/2} \tag{3}$$

The data for release kinetics of drug should be analyzed by fitting it to Peppas equation to define a model which will constitute a best fit for the formulation, $M_t/M_\infty = kt^n$, where Mt is the amount of drug released at time 't' and 'M_∞' is the amount of drug released at ∞, Mt/M∞ is the part of the drug released at a time 't', the kinetic constant is represented by 'k', and 'n' is the diffusional exponent, a measure of the initial release of drug. r^2 values indicate the regression analysis and were calculated for the linear curves obtained from plots above [43].

Table 1 represents the examples of nanoparticle formulations with and without surface modifications following different release kinetic models.

Table 1 Release kinetic models of various nanoparticle formulations.

Drug	Surface modification	Nanoparticle	Model	Reference
Curcumin and resveratrol	Hyaluronic acid	Hyaluronic acid nanoparticle	Korsmeyer-Peppas model	Hussain et al. (2020)
Methotrexate	Calcium phosphate	Bioinspired calcium phosphate nanoparticles	Higuchi model	Pandey et al. (2019)
Lutein	Hyaluronic acid	Hyaluronic acid-coated PLGA nanoparticles	Zero-order kinetics	Chittasupho et al. (2019)
5-Fluorouracil	Silica	Polymeric mesoporous Silica nanoparticles	Korsmeyer-Peppas model	Moodley et al. (2019)
Betamethasone valerate	Hyaluronic acid	Hyaluronic acid coated chitosan nanoparticle	Higuchi model	Pandey et al. (2019)
Nimodipine	PEG	PEGylated nanoparticle	Zero order	Moreno et al. (2018)
Voriconazole	Chitosan	Chitosan-coated nanoparticles	Korsmeyer-Peppas kinetic model	Paul et al. (2018)
Doxorubicin	Oxidized-cellulose	Oxidized-cellulose nanoparticles	Korsmeyer-Peppas model	Bhagat et al. (2018)
Phytic acid	Chitosan	Chitosan iron oxide magnetic nanoparticles	Pseudo-second-order model	Barahuie et al. (2017)

Table 1 Release kinetic models of various nanoparticle formulations—cont'd

Drug	Surface modification	Nanoparticle	Model	Reference
Etoposide	Amino group	Silica nanoparticles	Weibull model	Saroj et al. (2017)
Chitosan	Chitosan-silver	Silver nanoparticles	Higuchi model	Verick et al. (2017)
Baicalin	PEG	PEGylated nanoparticles	Ritger-Peppas model	Zhang et al. (2016)
Diazepam	Poly (lactic-co-glycolic acid)	PLGA nanoparticles	Korsmeyer-Peppas model	Bohrey et al. (2016)
Paclitaxel	Gold	Gold nanoparticles	Korsmeyer-Peppas model	England (2015)
Doxorubicin	Chitosan	Chitosan nanoparticles	Korsmeyer-Peppas model	Verma et al. (2014)
Mitoxantrone dihydrochloride	Hyaluronic acid	Polymeric nanoparticles	Higuchi model	Zafar et al. (2014)
Cyclosporine A	–	Nanoparticles	Weibull model	Aksungur et al. (2011)
Paclitaxel	Pectin	Pectin nanoparticles	First order	Verma et al. (2011)
Tacrine	Chitosan	Chitosan nanoparticle	Korsmeyer-Peppas model	Wilson et al. (2010)
Cycloheximide	Gelatin	Cross-linked gelatin nanoparticle	First order	Verma et al. (2005)

4. Parameters influencing PK of nanoparticles

Innumerable parameters that influence the PK of nanoparticles, such as surface functionalization (PEGylation and surface charge), size, charge, shape, and biological barriers, are enumerated below. All these factors affect the half-life in blood and biodistribution of circulating nanoparticles by delaying opsonization, lowering the nonspecific uptake level, thereby augmenting the specific tissue accumulation.

4.1 Surface functionalization

Nanoparticles can be engineered to release the therapeutic cargo in a sustained, controlled, or a triggered manner. Surface functionalization of nanoparticles can be applied to upsurge residence time of the particle in the blood and decrease the nonspecific distribution by targeting the target tissues or by identifying detailed cell surface antigens and using targeting ligands such as small molecules: aptamer, peptide, antibody-antibody fragment. Surface functionalization facilitates conjugation, adsorption, or even grafting onto the nanoparticle surface by hydrophilic polymers such as like polyethylene glycol (PEG). This process offers steric stabilization and provide "stealth" characteristics like inhibition of protein absorption. Although surface functionalization improves PK, it may often be a major limiting factor too, as the proteins adsorption on the nanoparticle surface leads to opsonization, causing aggregation and rapid clearance of the particle from the bloodstream [44].

4.2 Impact of size and shape

Size is an essential element of PK and biodistribution as variability in the sizes of inter-endothelial pores lining the blood vessels exist. Nanoparticles having size below 6 nm are eradicated and rapidly excreted by the kidneys. But when degradable materials like polymers, lipids, or hydrogels are used, they cannot be eliminated via kidneys as they swell and their size exceeds 6 nm [45]. Size above 200 nm of the nanoparticles tend to accumulate in the liver, spleen, and are eliminated by the reticuloendothelial system (RES) cells [46, 47].

Biological parameters, such as hepatic filtration, tissue diffusion, vesicular extravasation, and kidney excretion, clearly indicate that particle size and shape dictate the biodistribution of long-circulating nanoparticles to achieve therapeutic potential. The uptake of nanoparticles by murine macrophages further established their impact on blood clearance kinetics. PEGylated nanoparticles having size 200 nm incubated for 2 h with serum protein displayed a substantial correlation among particle size and protein absorption [44, 48]. Smaller nanoparticles displayed slow blood clearance twice as compared with the larger nanoparticle formulations. Liu et al. have reported the biodistribution of radioisotope-labeled liposomes in the size range from 30 to 400 nm, injected intravenously in mice where <50 nm were in blood circulation, 60% of liposomes in 100–200 nm size range and only 20% of the liposomes in size >250 nm were in tissues after 4 h [46], accumulated in spleen, liver, blood, and tumor based on size. Shape of nanoparticles also governs the half-life of nanoparticles in blood. For instance, rod-shaped micelles have ten times extended circulation lifetime, when compared with the spherical micelles [49]. The cellular uptake depended upon the shape of the nanoparticles, such as elliptical nanoparticles are internalized faster than spherical shaped ones (Fig. 3) [50].

4.3 Effect of charge

It was observed that the half-life for neutral nanoparticles in blood is highest, whereas positively charged nanoparticles are eliminated rapidly from the blood and may cause complications such as platelet aggregation and even hemolysis [47, 51]. Drummond and colleagues have reported that the neutral and uncharged liposomes have a low clearance rate, when compared with negatively or positively charged liposomes. This could be attributed to decrease opsonization on the uncharged surface of liposome, which leads to reduction in mononuclear phagocytic system (MPS) uptake [52]. Levchenko et al. has

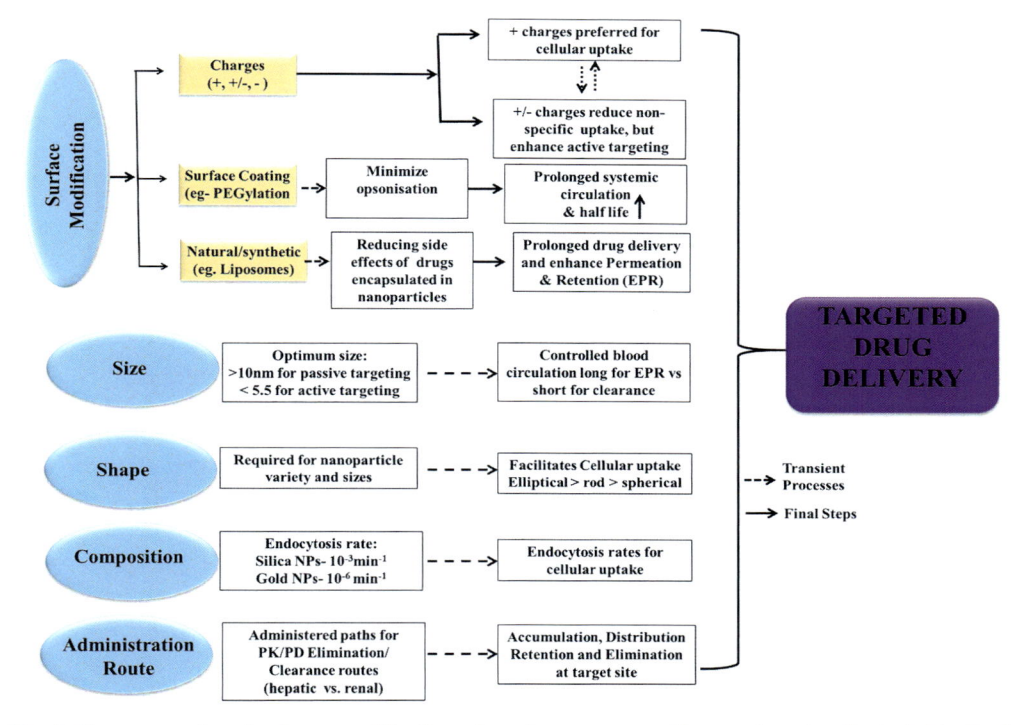

Fig. 3 Key parameters (surface modification, size, shape, composition, and route of administration) influencing the pharmacokinetics of nanoparticles.

further revealed that PEGylation or surface modification provides a negative charge on liposomes, thereby significantly reducing the liver uptake and prolonging the blood circulation [53].

4.4 Effect of nanoparticles rigidity

Degrees of flexibility of nanoparticle has been theorized to increase their binding efficiency to the cell surface [54]. Zwitterionic nanogels exhibiting variable degree of stiffness were synthesized by fine-tuning the densities of reactant excipients and their crosslinking. In vivo evaluation of nanogels revealed that the softer nanogels were able to cross the physiological barriers, predominantly the splenic filtration with ease, when compared with their stiffer counterparts, enabling enhanced circulation half-life and reduced accumulation in spleen [55].

4.5 Route of administration

The drug PK largely depends upon route of administration of the nanoparticle, as it modifies the pharmacological efficacy of the drug. as mentioned in Fig. 5. There are

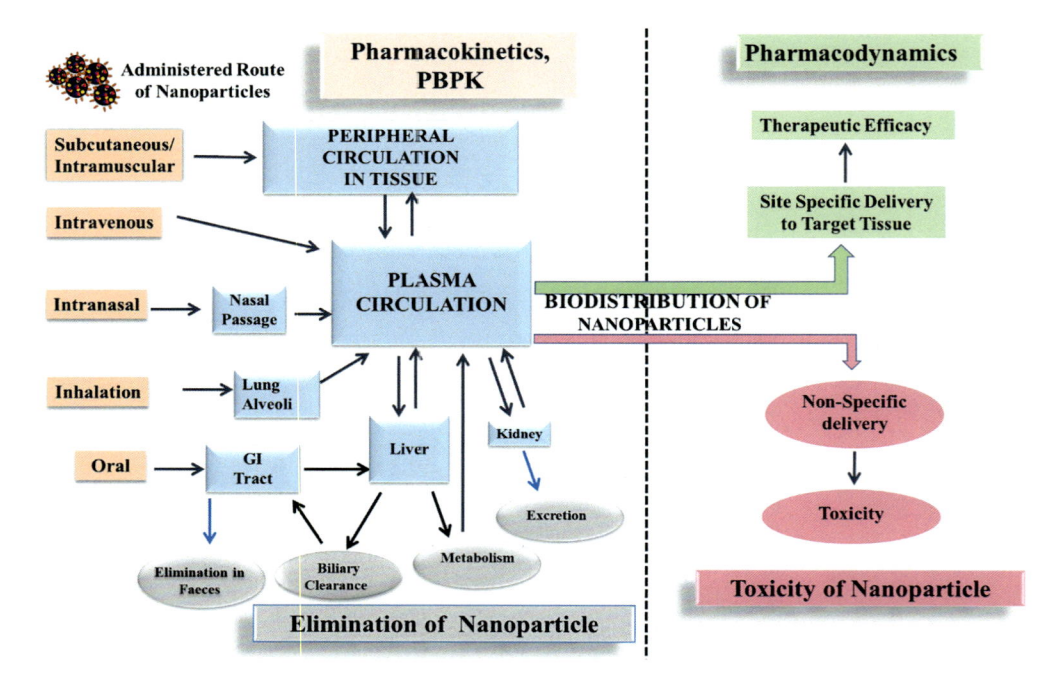

Fig. 4 Diagrammatic representation of nanoparticles following various routes of administration showing impact on PK, PD, elimination and toxicity of nanoparticles. The fate of multifunctional nanoparticle relies on their physicochemical parameters and administered route altering PK, PBPK, and PD.

various routes of nanoparticles administration such as oral, inhalation, intranasal, intravenous, intramuscular, or subcutaneous, which leads to the improved PK of nanoparticles that further enhanced the biodistribution kinetics. Huang et al. described the blood clearance, biodistribution, and tumor uptake of carbon dots and indicated that the renal clearance through three different administration routes was in the order of intravenous > intramuscular > subcutaneous injections for the blood clearance and urinary accumulation. It was further also observed that tumor uptake of carbon dots was higher by subcutaneous and intravenous routes, when compared with the intramuscular route (Fig. 4) [56, 57].

4.6 Effect of surface chemistry

Currently, surface chemistry has gained a lot of attention with regard to protein deposition [58–60], but a detailed material characterization is still lacking. Because the experiments are generally done with plasma or sera at low concentrations, the data do not correspond when applied to whole blood, thereby interpretation of data is misleading in terms of biological relevance. Reports on the comparative analysis of binding amine-functionalized, carboxyl-functionalized, and native polystyrene nanoparticles in the range of 50 and 100 nm of size plasma protein binding have suggested that there exists a difference in qualitative and quantitative protein binding efficiency, data on functional-group density, spacing or clustering was lacking [61]. This analysis had significant potential for triggering the blood enzymatic cascade, both clotting and the complement activation. It is fascinating to note that a gamut of plasma proteins in nature exist like acute-phase proteins, apolipoproteins, immunoglobulins, complement system proteins, clotting factors, and serum albumin, and with such vivacity, how they adhere themselves simultaneously on the surface of a 50-nm nanoparticle with varied geometric shapes and sizes is a puzzle. This definitely may be due to the surface heterogeneity of nanoparticles that cause a dynamic protein adsorption or a protein desorption process, whereby the binding affinity differs among specific clusters of proteins. We still lack the methodologies that could give precise individual binding efficiency, and generally, it is interpreted as protein-mediated aggregation of nanoparticles and reported as bulk effect.

5. Role of blood proteins in clearance kinetics

Because blood proteins can transport most of the drugs, the property of drug to bind with plasma proteins is crucial as it alters the pharmacological activity of a particular drug. Drug activity is generally affected by protein binding in two ways: one it may alter the active drug concentration at the site of action or it may alter the elimination rate of the drug [62, 63]. Drug elimination is another critical parameter known as *Clearance* that relates to the concentration of drug, that is, the rate of administration requisite to maintain a constant drug concentration in blood. Clearance indicates the speed of removal from the body and helps determine the range of distribution outside the plasma. The unbound clearance of a drug depends upon its binding within blood proteins. If unbound clearance is low, with respect to blood flow in the organs, then the extraction ratio followed by clearance will also be low and largely dependent on the plasma binding. But, if the drug extraction ratio is enhanced, then drug elimination becomes rate-limited by perfusion and clearance will be comparatively impervious to fluctuations in binding [64]. The probable impact of changed plasma binding on drug clearance can be comprehended by the extraction ratio of the drug by an organ. When the extraction ratio of a drug is around the upper limit of 1, the elimination becomes perfusion rate-limited and the clearance reaches the value similar to organ blood flow. Hence, all of the drug that passes through the organ, whether bound or unbound, has to be extracted, and basically, clearance becomes relatively independent to changes in binding within blood. In contrast, for drugs whose extraction ratio is usually low, clearance should be affected by plasma proteins [65]. The physicochemical features of nanoparticles, irrespective of their physical characteristics, are frequently altered the moment nanoparticles come in contact with blood and interact with plasma proteins, both qualitatively and quantitatively and correlate with the half-lives nanoparticle as well as biodistribution. Albumin is the ample plasma colloidal proteins present on almost every nanoparticles surface that had been tested, that is, there is either surface protein deposition or surface opsonization. Protein deposition facilitates recognition and subsequent plasma clearance of nanoparticles by the circulation of phagocytes, along with the tissue macrophages [66].

6. In vivo toxicological evaluations of nanoparticles

Indiscriminate use of nanoparticles will bring to fore the concerns regarding the usefulness as well as the possibly of adverse and unpredictable harm caused by them to humans on exposure. Their potential hazardous behavior cannot be correctly estimated in vitro; hence, in vivo evaluation is imperative. Toxicity of nanoparticles relates to its ability to unfavorably impact the normal physiological processes or indirectly or directly disturb the normal architecture of the organs and tissues of humans and animal models, such as mice, rats, and zebrafish.

Knowledge *priori* explicitly explains that toxicity is caused by the physicochemical parameters, such as nanoparticle shape, diameter, surface charge/chemistry, surface modification, stability. The precise fundamental mechanism is as yet indefinite, but literature extensively advocates that cytotoxicity should be related to metabolic insult induced by oxidative imbalance and proinflammatory responses and gene activation [67–69]. Additional factors like the administered route, dose, and level of tissue distribution are vital parameters to be considered for evaluating nano–cytotoxicity. Classically, evaluation of cellular toxicity uses increasing doses of nanoparticles with the aim of observing dose-dependent cellular or tissue toxicity. These dose-response simulations form the basis for determining the safe limits of nanoparticle concentrations for in vivo administration. Regardless of the hypothetical clarity, animal experiments and human trials have imparted information that is different and also highlighted the feasibility of comparing organ toxicity with a predetermined dose; doses are generally extrapolated based on in vitro concentrations for use in in vivo evaluation, which is again problematic. It may be subdivided in two aspects; one, it has yet not been determined how efficiently the administered dose of nanoparticle will reach the target tissue, and second, whether nanoparticles induce and exhibit any biochemical changes in vivo, that may have been overlooked in the isolated cell-based experiments?

Not only the dosing problem, but also the route of nanoparticle administration causes nanotoxicity that is fairly independent of the dose concentration, as absorption, biodistribution, metabolism, accumulation, and excretion of nanoparticles will differ based on the administered route. Nanoparticles can enter the body by oral, dermal penetration, inhalation and intravenous injection and then be distributed to the organ system (Fig. 5).

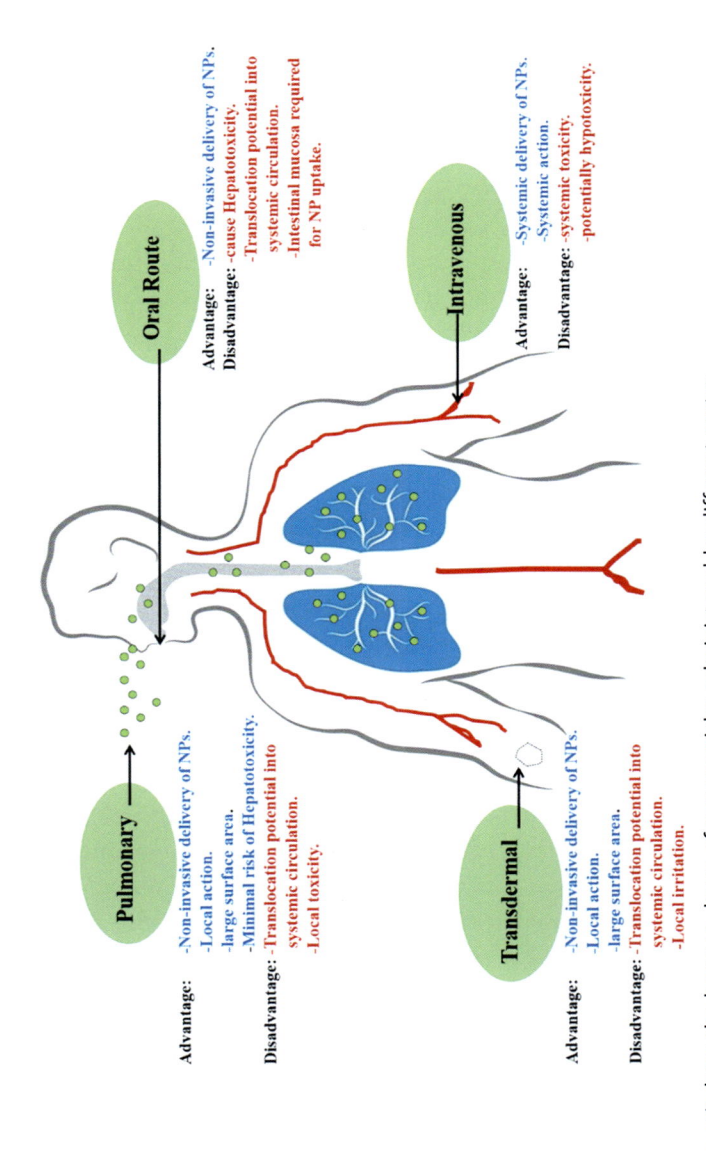

Fig. 5 Recapitulates the boon *vs.* bane of nanoparticles administered by different routes.

Pulmonary delivery has incredible potential, but is limited by systemic and local toxicity, which diminishes the enthusiasm [70]. Aggregation of nanoparticles leading to tissue inflammation has been assumed to be the fundamental mechanism [71–73]. Nanoformulations for topical applications in cosmetics and sunscreen use zinc and titanium dioxide by manipulating their ability to block UV rays. This penetrative capability of certain nanoparticles could be modulated for success in transdermal delivery. Hence, the mechanism of local and probably dermal or systemic toxicity essentially needs to be evaluated. On the other hand, intravenous and oral administrations of nanoparticles are rapidly taken up in systemic circulation, when compared with transdermal route. Once it reaches the circulation the nanoparticles are initially filtered, accumulated, and metabolized in the liver or distributed to various organs, including the spleen and brain. Although there exists a natural protection, that is, the blood-brain barrier against exogenous chemical insults, the possibility of nanoparticles to permeate the tight junctions makes the brain susceptible to particulate-mediated toxicity.

7. Physiology-based pharmacokinetics

The premise for a simple PBPK model is to compartmentalize the whole body as per individual organs that are then treated as building blocks and interlinked with one other by a network of capillaries of the circulatory system, which includes the lymphatic or drainage system [8]. The PBPK models envisage the drug release rates from the bulk flow of blood into each organ with respect to time, integrate the various PK factors responsible for unpredictability, and permit interspecies extrapolation to make animal studies relevant and acceptable [74].

8. Pharmacodynamics

PD deals with the effect of drug on the body with respect to exposure time and the concentration of the drug [75]. PK/PD models represent relevant and dynamic data throughout the drug development phases; advent of nanotechnology has once again raised this important issue for unknown biomaterials. PD models definitely offer relevant insights about dose optimization, evolving a regimen and reliable safety concerns and efficacy measures.

The evidence about the optimal dosage for treatment, their efficacy, and mathematical models can be useful and applied on population PD, where drug concentration validation at population level is essential. This modeling can be performed using the mixed effect model [76]. PK/PD modeling is severely limited by ethical issues, the huge costs to be incurred during the first clinical trials and reproducibility. Designing of a nanocarrier needs minimum side effect of the drugs, as drug per se has known toxicity. In addition, because the therapeutic indices of drugs are generally narrow, many preclinical studies have to be performed preceding application in humans. Therefore PK/PD model is a breakthrough to support with, throughout the phases efficiently. PD model is also represented by regimented clinical studies that will give significant indications regarding the drug effects, investigate the clinical endpoints and accept the surrogates, provide supporting evidence to establish the efficacy of the mode of action of the particular drug. Hence, PK/PD models are helpful for planning the pharmacological response associations and aids to differentiate the effects of the drug in various nonclinical species [77]. In case of anticancer activity evaluation, tumor size is considered as a reliable biomarker for PD data and PK models for time vs. drug release correlation.

9. Toxicity in cell-based targets and animal models

Owing to ultrasmall size and morphological semblance to physiological moieties such as proteins, nanoparticles have the ability to revolutionize biomedical diagnostics, imaging, and therapeutics facilitating biological functional processes and unwarrantedly causing toxicity. Also, based on the administered mode of nanoparticles along with their sites of deposition, severity in toxicity may vary. The therapeutic efficacy depends on the release kinetics, biodistribution, and cellular uptake but all this is limited by the size of nanoparticles. Owing to the small size of nanoparticles, they frequently evade the immune surveillance and circulate unrecognized by the macrophages and may then enter to macrophage cells via the membrane pores. Decrease in the size of particles increases the surface area of the particles; this enhanced surface area is then available for interaction of the nanoparticles with the cell surface to facilitate simple diffusion. As per the earlier investigations, clathrin-coated pits regulate the penetration of microspheres based on the shape and diameter of the nanoparticles [50, 78].

Because the size and shape of nanoparticles have substantial effect on cellular internalization, choice of nanoparticles with suitable shape and chemistry holds remarkable consequences on circulation time, absorption, biodistribution, and mean residence time of nanoparticles [79]. Elongated nanoparticles demonstrate higher efficacy as compared with spherical nanoparticles in adhering to the cells due to the curvature of spherical particles, which allows inadequate number of binding sites to interact with the receptors of target cells. Whereas multivalent interaction of elongated nanoparticles is correlated with its higher surface area interacting with the cell surface [80]. This results in extensive uptake of the elongated particles when compared with the spherical particles of the same size. The rod-shaped nanoparticles have revealed greater affinity toward the endothelial cells during in vitro/in vivo experiments [81, 82].

10. Aspects of particle duration in the circulation

Multifunctional nanoparticles are engineered for both systemic and targeted drug delivery. Moreover, allowing circulation of nanoparticles in blood for sufficiently prolonged period of time to ensure drug accumulation at the target site is a prodigious challenge.

Site-specific accumulation of nanoparticle is enhanced if it achieves prolonged systemic circulation allowing sufficient contact time to the nanoparticle to reach the target [83].

Nanoparticles can potentially ameliorate the issues of low solubility, poor stability and alter PK and biodistribution, without altering the drug properties. Because various types of drugs can be encapsulated and protected in nanoparticles from degradation by body, nanoparticles do act as a drug reservoir that releases the drugs in blood in a sustained and controlled manner [2]. Encapsulation of acyclovir improves its pharmacokinetic profiles and mitigates adverse effects triggered by recurrent dosage [84].

It is important for the nanoparticles to evade immune recognition so as to avoid capture by the macrophages that remove exogenous invaders. The RES contains circulating monocytes and macrophages in the liver, spleen, lung, and bone marrow. Once administered in blood, nanoparticles are rapidly opsonized and phagocytosed by macrophages, thus restricting their circulation times [85]. Rapid clearance of nanoparticles limits the success of nanotherapy because of its inability to accumulate at the target site owing to short half-life. Therefore surface modification decreases nanoparticle hydrophobicity as well as density of surface charge, both of which initially trigger opsonization. The most experimented agents for surface functionalization are carbohydrates, PEG, and PEG-derived polymer (poloxamers and poloxamines) [83]. Even though PEGylation does prolong half-life of nanoparticles, its efficacy is limited by its unstable and transient nature coupled with unsatisfactory particle-target interactions. Consequently, different methodologies are being experimented upon to enhance systemic circulation of nanoparticles, including functionalization with CD47, tweaking nanoparticle morphology and mechanical properties, and attaching to the red blood cells [83].

Although PEGylation was successful in inhibiting opsonization, conflicting evidence regarding PEG-triggered compliment activation was being reported [86]. PEGylation is known to reduce protein binding, the compliment system might still bind to the components eliciting an immune response. To further improvise the stealth properties, polysaccharides in combination with PEG have been designed. Sheng et al. suggested the dual coating of PEG and hydrophilic chitosan of polylactic acid nanoparticles, which significantly reduced the uptake by macrophages in in vitro and remarkably prolonged half-life in blood circulation, almost ~63.5 h [87]. Exceptionally long circulation times for nanoparticles might not essentially progress the benefit-to-risk ratio in some clinical outcomes. Correspondingly, brief half-lives also might have unfavorable effects due to fast clearance/elimination by macrophages of the RES. This causes Kupffer cell destruction, thereby cumulating the risk of septicaemia during macrophage deficiency as observed in cancer nanotherapeutics.

11. Enhanced permeation and retention and strategies

Nanoparticles injected intravenously can easily extravasate via leaky endothelium and get accumulated in tumor. Generally, escape of particle from the vasculature is restricted to the areas where there is open fenestration of capillaries; however, they can also extravasate in the areas where endothelium turns permeable leading to certain pathological progression as in the case of tumor growth or inflammation [88]. The endothelium in normal conditions is impermeable and does not permit nanoparticles to cross the endothelium. In solid tumors, the endothelial integrity is challenged by activation of proinflammatory cytokines and release of nitric oxide that increases the gap between endothelial cells [89]. Finally, the caveolae-mediated transcytosis in the endothelial cell facilitates transport of a molecule from blood into the tissue, which apparently is a promising strategy for overcoming the endothelial barrier. Rather than making enhanced permeation and retention (EPR) effect based on passive accumulation with limited capacity, the gold standard for nanotherapy, alternative targeting methodologies, such as active targeting with tissue-specific ligand to caveolae, that uses the active transport system is desired.

The leaky vasculature is of critical advantage while strategizing nanoparticulate cancer treatment. The blood vessels adjacent to the tumor being leaky show varied hyperpermeability and faulty drainage of lymphatic vessels as compared with the normal tissues. This leakiness of tumor neovasculatures causes penetration and retention of nanoparticles in tumor microenvironment and this phenomenon is called as EPR effect [88].

Once the nanoparticle reaches the advanced stage solid tumor, it has to negotiate the complex, heterogenous microenvironment. Hypoxic conditions prevail due to rapid growth and increased metabolic activity initiating the Warburg effect. Increased activity leads to anaerobic glycolytic pathway that causes acidification, metabolic reprogramming of tumor cell that modulates the microenvironment as per its requirement [90]. The interplay between surrounding cells and cancer cells, together with the immune and stromal cells (that are recruited as inflammatory response at tumor site), results in more modifications of the microenvironment cellular mechanisms, restructuring of the extracellular matrix (ECM) advances tumor cells and microenvironment constituents recurrently adapt to the environmental stimuli, impacting the overall tumor growth. Fig. 6 highlights the EPR effect and illustrates the current strategies used to target tumor microenvironment. The heterogeneity of cancer cells, hypoxia, and enhanced inflammation in the microenvironment cause modifications of the ECM proteins, ensuing increased stiffness and density, primarily because of collagen deposition called desmoplasia, matrix metalloproteins. Tumor cells divide rapidly and trigger angiogenesis that forms disordered neovessels that are irregularly branched. Stromal cells consisting of mesenchymal stromal cells, cancer-associated fibroblasts, and the immune system cells from lymphoid as well as myeloid lineage cells are involved in tumor development. A new strategy includes the exosomes, which are 30–100 nm vesicles synthesized by endosomes of both tumor and normal cells. These lipid bilayer endosomes encapsulate mRNA proteins and play an important paracrine or autocrine intercommunication within microenvironment, which impacts the expressed surface receptors, whether silenced or activated signaling pathways that will affect the therapeutic efficacy. These annotations highlight the significance of considerations while designing a multifunctional nanoparticle for achieving

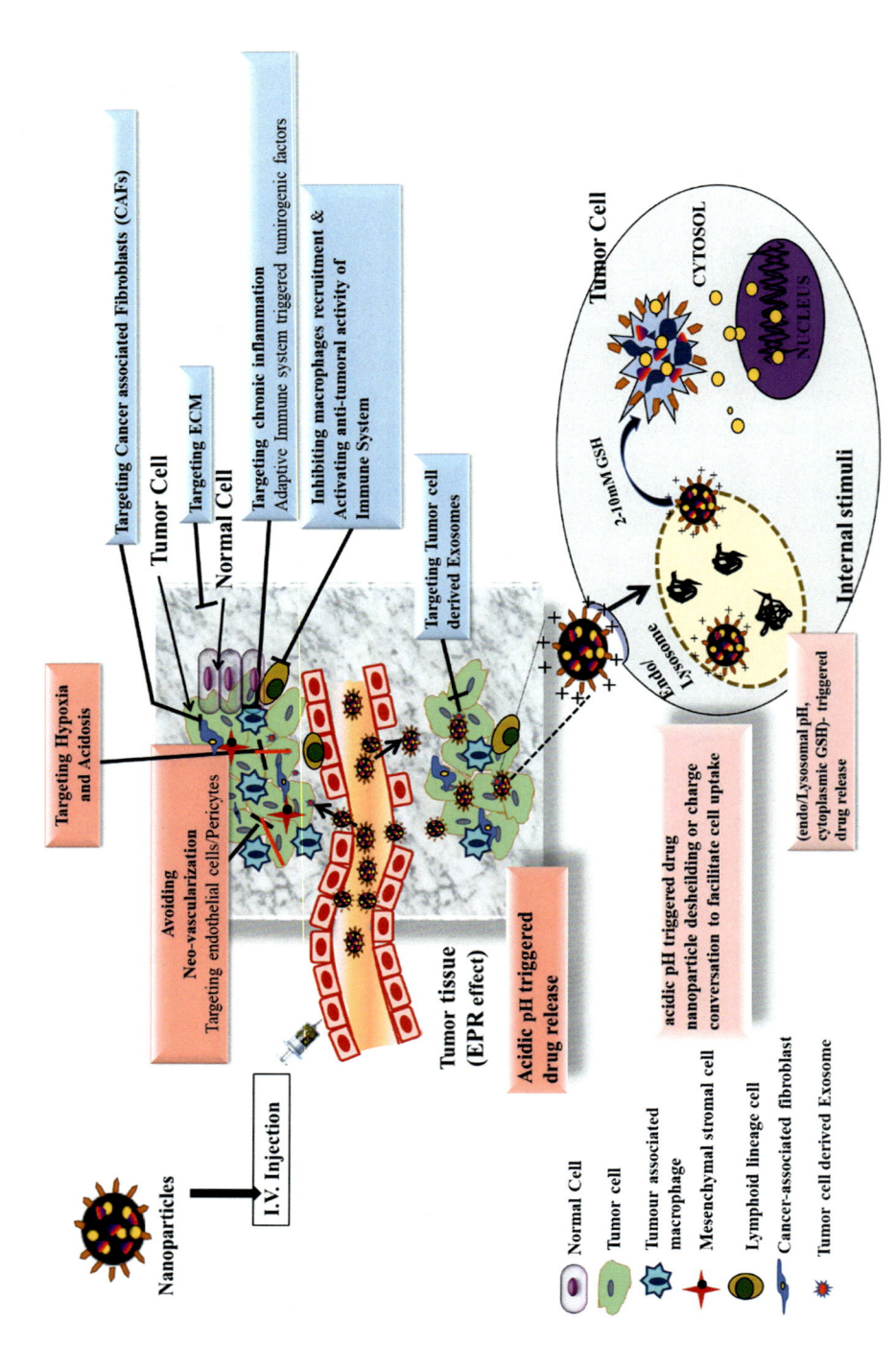

Fig. 6 Illustration depicting nanoparticle penetration and retention through leaky vasculature via enhanced EPR effect in tumor targeting by nanoparticles.

suitable half-lives for efficient nanotherapeutics in systemic circulation and respond as per the desired pathology. Stimuli-sensitive, magnetotherapy and photodynamic are still in preclinical phase. Although EPR is the key factor to substantiate efficacy of nanoparticles, the major dilemma of resistance or no response to therapy, each tumor representing a multifactorial disease and therefore each patient requires a different combination of therapeutics. In such a scenario, use of nanotechnology can simplify personalized medicine as different targeting molecules can easily be attached in a single nanomaterial.

12. Conclusion

Biomedical applications of nanoparticles have risen exponentially thereby increasing their experimental and clinical use for drug/gene delivery, cell-imaging and cell tracking. Physicochemical properties of nanoparticles, including size, charge, and surface chemistry, significantly influence the PK/PD and bioavailability in the tissues. Maneuvring these parameters (size $\sim 100\,nm$, ζ potential within $\pm 10\,mV$, and PEGylation) can effectively alter the rate of nanoparticle uptake by the RES, prolong blood circulation, and allow accumulation in desired target tissues (i.e., tumor). PK studies are valuable indicators that can be extrapolated from experiments being conducted in laboratory animals for drug research and development. Future of nanomedicine lies in biocompatible, nonimmunogenic, and effective nanoparticulate drug delivery systems. Emergence of novel smart, intelligent nanoformulations will facilitate advancement of existing delivery systems and neoformulations that will specifically deliver the therapeutic agents to target tissue and be stimuli-sensitive to perceive microenvironmental aberrations and respond accordingly. In addition, care should be taken to develop inexpensive but safe biomaterials to economize nanotechnology-based therapy for patient compliance and larger patient population benefit. It is important to validate the in vivo fate of the nanoparticles by analyzing blood, urine, or cerebrospinal fluid, with precision at the detection of isolating single cells, toxic molecules, or secreted exosomes. Only this will enhance the clinical benefits of theragnostic nanoparticles.

References

[1] S.M. Moghimi, A.C. Hunter, J. Murray, Long-circulating and target-specific nanoparticles: theory to practice, Pharmacol. Rev. 53 (2) (2001) 283–318.
[2] T.M. Allen, Drug delivery systems: entering the mainstream, Science 303 (5665) (2004) 1818–1822.
[3] S.M. Moghimi, A.C. Hunter, J.C. Murray, Nanomedicine: current status and future prospects, FASEB J. 19 (3) (2005) 311–330.
[4] D. Peer, J.M. Karp, S. Hong, O.C. Farokhzad, R. Margalit, R. Langer, Nanocarriers as an emerging platform for cancer therapy, Nat. Nanotech. 2 (12) (2007) 751–760.
[5] B.Y.S. Kim, J.T. Rutka, W.C.W. Chan, Current concepts: nanomedicine, N. Engl. J. Med. 363 (2010) 2434–2443.
[6] M.E. Davis, Z. Chen, D.M. Shin, Nanoparticle therapeutics: an emerging treatment modality for cancer, Nat. Rev. Drug Discov. 7 (9) (2008) 771–782.
[7] I. Nestorov, Whole body pharmacokinetic models, Clin. Pharmacokinet. 42 (10) (2003) 883–908.

[8] H. Jones, K. Rowland-Yeo, Basic concepts in physiologically based pharmacokinetic modeling in drug discovery and development, CPT: Pharmacomet. Syst. Pharmacol. 2 (8) (2013), e63.

[9] M.A. Dobrovolskaia, S.E. McNeil, Immunological properties of engineered nanomaterials, Nat. Nanotechnol. 2 (2007) 469–478.

[10] L. Yang, D.J. Watts, Particle surface characteristics may play an important role in phytotoxicity of alumina nanoparticles, Toxicol. Lett. 158 (2) (2005) 122–132.

[11] P.G. Welling, Differences between pharmacokinetics and Toxicokinetics, Toxicol. Pathol. 23 (2) (1995) 143–147.

[12] D. Yuan, H. He, Y. Wu, J. Fan, Y. Cao, Physiologically based pharmacokinetic modeling of nanoparticles, J. Pharm. Sci. 108 (1) (2019) 58–72.

[13] S. Mostafalou, H. Mohammadi, A. Ramazani, M. Abdollahi, Different biokinetics of nanomedicines linking to their toxicity; an overview, DARU J. Pharm. Sci. 21 (1) (2013) 14 (2008-2231-21–14).

[14] H. Twitchett, P. Grimsey, A peak at PK — an introduction to pharmacokinetics, Pharmaceut. Program. 5 (1–2) (2012) 42–49.

[15] L.G. Donaruma, Definitions in biomaterials, D. F. Williams, Ed., Elsevier, Amsterdam, 1987, 72 pp, J. Polym. Sci. B Polym. Lett. 26 (9) (1988) 414.

[16] D.F. Williams, The Williams Dictionary of Biomaterials, Liverpool University Press, 1999.

[17] D.S. Kohane, R. Langer, Biocompatibility and drug delivery systems, Chem. Sci. 1 (4) (2010) 441–446.

[18] J. Liu, M. Yu, C. Zhou, J. Zheng, Renal clearable inorganic nanoparticles: a new frontier of bionanotechnology, Mater. Today 16 (12) (2013) 477–486.

[19] M. Zhou, R. Zhang, M. Huang, W. Lu, S. Song, M.P. Melancon, et al., A chelator-free multifunctional [^{64}Cu]CuS nanoparticle platform for simultaneous micro-PET/CT imaging and photothermal ablation therapy, J. Am. Chem. Soc. 132 (43) (2010) 15351–15358.

[20] Y. Zhao, D. Sultan, L. Detering, S. Cho, G. Sun, R. Pierce, et al., Copper-64-alloyed gold nanoparticles for cancer imaging: improved radiolabel stability and diagnostic accuracy, Angew. Chem. Int. Ed. 53 (1) (2014) 156–159.

[21] L. Yang, G. Sundaresan, M. Sun, P. Jose, D. Hoffman, P.R. McDonagh, et al., Intrinsically radiolabeled multifunctional cerium oxide nanoparticles for in vivo studies, J. Mater. Chem. B 1 (10) (2013) 1421.

[22] J. Pellico, J. Ruiz-Cabello, M. Saiz-Alía, G. del Rosario, S. Caja, M. Montoya, et al., Fast synthesis and bioconjugation of ^{68}Ga core-doped extremely small iron oxide nanoparticles for PET/MR imaging: chelator-free ^{68}Ga-iron oxide nanoparticles, Contrast Media Mol. Imag. 11 (3) (2016) 203–210.

[23] M. Sun, D. Hoffman, G. Sundaresan, L. Yang, N. Lamichhane, Synthesis and characterization of intrinsically radio- labeled quantum dots for bimodal detection, Am. J. Nucl. Med. Mol. Imag. 2 (2) (2012) 122–135.

[24] V. Kumar, A. Leekha, A. Tyagi, A. Kaul, A.K. Mishra, A.K. Verma, Preparation and evaluation of biopolymeric nanoparticles as drug delivery system in effective treatment of rheumatoid arthritis, Pharm. Res. 34 (3) (2017) 654–667.

[25] S. Berke, A.-L. Kampmann, M. Wuest, J.J. Bailey, B. Glowacki, F. Wuest, et al., ^{18}F-radiolabeling and in vivo analysis of SiFA-derivatized polymeric core–shell nanoparticles, Bioconjug. Chem. 29 (1) (2018) 89–95.

[26] J. Zeng, B. Jia, R. Qiao, C. Wang, L. Jing, F. Wang, et al., In situ ^{111}In-doping for achieving biocompatible and non-leachable ^{111}In-labeled Fe$_3$O$_4$ nanoparticles, Chem. Commun. 50 (17) (2014) 2170.

[27] Y. Yang, Y. Sun, T. Cao, J. Peng, Y. Liu, Y. Wu, et al., Hydrothermal synthesis of NaLuF$_4$:^{153}Sm,Yb, Tm nanoparticles and their application in dual-modality upconversion luminescence and SPECT bioimaging, Biomaterials 34 (3) (2013) 774–783.

[28] K.C.L. Black, Y. Wang, H.P. Luehmann, X. Cai, W. Xing, B. Pang, et al., Radioactive ^{198}Au-doped nanostructures with different shapes for in vivo analyses of their biodistribution, tumor uptake, and intratumoral distribution, ACS Nano 8 (5) (2014) 4385–4394.

[29] Z. Wu, S. Yang, W. Wu, Shape control of inorganic nanoparticles from solution, Nanoscale 8 (3) (2016) 1237–1259.

[30] J.M. Rosano, N. Tousi, R.C. Scott, B. Krynska, V. Rizzo, B. Prabhakarpandian, et al., A physiologically realistic in vitro model of microvascular networks, Biomed. Microdevices 11 (5) (2009) 1051–1057.

[31] B. Prabhakarpandian, K. Pant, R.C. Scott, C.B. Patillo, D. Irimia, M.F. Kiani, et al., Synthetic microvascular networks for quantitative analysis of particle adhesion, Biomed. Microdevices 10 (4) (2008) 585–595.

[32] N. Doshi, B. Prabhakarpandian, A. Rea-Ramsey, K. Pant, S. Sundaram, S. Mitragotri, Flow and adhesion of drug carriers in blood vessels depend on their shape: a study using model synthetic microvascular networks, J. Control. Release 146 (2) (2010) 196–200.

[33] A. Ruggiero, C.H. Villa, E. Bander, D.A. Rey, M. Bergkvist, C.A. Batt, et al., Paradoxical glomerular filtration of carbon nanotubes, Proc. Natl. Acad. Sci. 107 (27) (2010) 12369–12374.

[34] S. Muro, C. Garnacho, J.A. Champion, J. Leferovich, C. Gajewski, E.H. Schuchman, et al., Control of endothelial targeting and intracellular delivery of therapeutic enzymes by modulating the size and shape of ICAM-1-targeted carriers, Mol. Ther. 16 (8) (2008) 1450–1458.

[35] P.P. Karmali, D. Simberg, Interactions of nanoparticles with plasma proteins: implication on clearance and toxicity of drug delivery systems, Expert Opin. Drug Deliv. 8 (3) (2011) 343–357.

[36] M.P. Monopoli, D. Walczyk, A. Campbell, G. Elia, I. Lynch, F. Baldelli Bombelli, et al., Physical—chemical Aspects of protein corona: relevance to *in vitro* and *in vivo* biological impacts of nanoparticles, J. Am. Chem. Soc. 133 (8) (2011) 2525–2534.

[37] I. Hamad, O. Al-Hanbali, A.C. Hunter, K.J. Rutt, T.L. Andresen, S.M. Moghimi, Distinct polymer architecture mediates switching of complement activation pathways at the nanosphere—serum interface: implications for stealth nanoparticle engineering, ACS Nano 4 (11) (2010) 6629–6638.

[38] R.S. Kadam, D.W.A. Bourne, U.B. Kompella, Nano-advantage in enhanced drug delivery with biodegradable nanoparticles: contribution of reduced clearance, Drug Metab. Dispos. 40 (7) (2012) 1380–1388.

[39] H. Takeuchi, H. Yamamoto, T. Niwa, T. Hino, Y. Kawashima, Enteral absorption of insulin in rats from mucoadhesive chitosan-coated liposomes, Pharm. Res. 13 (6) (1996) 896–901.

[40] H. Soo Choi, W. Liu, P. Misra, E. Tanaka, J.P. Zimmer, B. Itty Ipe, et al., Renal clearance of quantum dots, Nat. Biotechnol. 25 (10) (2007) 1165–1170.

[41] M. Ohlson, J. Sörensson, B. Haraldsson, A gel–membrane model of glomerular charge and size selectivity in series, Am. J. Physiol.–Renal Physiol. 280 (3) (2001) F396–F405.

[42] S.-D. Li, L. Huang, Pharmacokinetics and biodistribution of nanoparticles, Mol. Pharm. 5 (4) (2008) 496–504.

[43] K.S. Yadav, K. Chuttani, A.K. Mishra, K.K. Sawant, Long circulating nanoparticles of etoposide using PLGA-MPEG and PLGA–pluronic block copolymers: characterization, drug-release, blood-clearance, and biodistribution studies, Drug Dev. Res. 71 (4) (2010) 228–239.

[44] F. Alexis, E. Pridgen, L.K. Molnar, O.C. Farokhzad, Factors affecting the clearance and biodistribution of polymeric nanoparticles, Mol. Pharm. 5 (4) (2008) 505–515.

[45] M. Longmire, P.L. Choyke, H. Kobayashi, Clearance properties of nano-sized particles and molecules as imaging agents: considerations and caveats, Nanomedicine 3 (5) (2008) 703–717.

[46] B.S. Eliana, Patenting Nanomedicines: Legal Aspects, Intellectual Property and Grant, Springer, Heidelberg, 2012.

[47] N. Hoshyar, S. Gray, H. Han, G. Bao, The effect of nanoparticle size on *in vivo* pharmacokinetics and cellular interaction, Nanomedicine 11 (6) (2016) 673–692.

[48] C. Fang, B. Shi, Y.-Y. Pei, M.-H. Hong, J. Wu, H.-Z. Chen, In vivo tumor targeting of tumor necrosis factor-α-loaded stealth nanoparticles: effect of MePEG molecular weight and particle size, Eur. J. Pharm. Sci. 27 (1) (2006) 27–36.

[49] A. Albanese, P.S. Tang, W.C.W. Chan, The effect of nanoparticle size, shape, and surface chemistry on biological systems, Annu. Rev. Biomed. Eng. 14 (1) (2012) 1–16.

[50] S. Salatin, S. Maleki Dizaj, A. Yari Khosroushahi, Effect of the surface modification, size, and shape on cellular uptake of nanoparticles: cellular uptake of nanoparticles, Cell Biol. Int. 39 (8) (2015) 881–890.

[51] J.S. Petschauer, A.J. Madden, W.P. Kirschbrown, G. Song, W.C. Zamboni, The effects of nanoparticle drug loading on the pharmacokinetics of anticancer agents, Nanomedicine 10 (3) (2015) 447–463.

[52] D.C. Drummond, O. Meyer, K. Hong, D.B. Kirpotin, D. Papahadjopoulos, Optimizing liposomes for delivery of chemotherapeutic agents to solid tumors, Pharmacol. Rev. 51 (4) (1999) 691–744.

[53] T.S. Levchenko, R. Rammohan, A.N. Lukyanov, K.R. Whiteman, V.P. Torchilin, Liposome clearance in mice: the effect of a separate and combined presence of surface charge and polymer coating, Int. J. Pharm. 240 (1–2) (2002) 95–102

[54] S. Takeoka, Y. Teramura, Y. Okamura, E. Tsuchida, M. Handa, Y. Ikeda, Rolling properties of rGPIba-conjugated phospholipid vesicles with different membrane flexibilities on vWf surface under flow conditions, Biochem. Biophys. Res. Commun. 6 (2002).

[55] J. Zhang, P. Li, H. Guo, L. Liu, X. Liu, Pharmacokinetic-pharmacodynamic modeling of diclofenac in normal and Freund's complete adjuvant-induced arthritic rats, Acta Pharmacol. Sin. 33 (11) (2012) 1372–1378.

[56] L. Harivardhan Reddy, R.K. Sharma, K. Chuttani, A.K. Mishra, R.S.R. Murthy, Influence of administration route on tumor uptake and biodistribution of etoposide loaded solid lipid nanoparticles in Dalton's lymphoma tumor bearing mice, J. Control. Release 105 (3) (2005) 185–198.

[57] X. Huang, F. Zhang, L. Zhu, K.Y. Choi, N. Guo, J. Guo, et al., Effect of injection routes on the biodistribution, clearance, and tumor uptake of carbon dots, ACS Nano 7 (7) (2013) 5684–5693.

[58] I. Lynch, K.A. Dawson, Protein-nanoparticle interactions, Nano Today 3 (1–2) (2008) 40–47.

[59] T. Cedervall, I. Lynch, M. Foy, T. Berggård, S.C. Donnelly, G. Cagney, et al., Detailed identification of plasma proteins adsorbed on copolymer nanoparticles, Angew. Chem. Int. Ed. 46 (30) (2007) 5754–5756.

[60] C. Salvador-Morales, L. Zhang, R. Langer, O.C. Farokhzad, Immunocompatibility properties of lipid–polymer hybrid nanoparticles with heterogeneous surface functional groups, Biomaterials 30 (12) (2009) 2231–2240.

[61] M. Lundqvist, J. Stigler, G. Elia, I. Lynch, T. Cedervall, K.A. Dawson, Nanoparticle size and surface properties determine the protein corona with possible implications for biological impacts, Proc. Natl. Acad. Sci. 105 (38) (2008) 14265–14270.

[62] P. Keen, Effect of binding to plasma proteins on the distribution, activity and elimination of drugs, in: B.B. Brodie, J.R. Gillette, H.S. Ackerman (Eds.), Concepts in Biochemical Pharmacology [Internet], Springer Berlin Heidelberg, Berlin, Heidelberg, 1971, pp. 213–233, https://doi.org/10.1007/978-3-642-65052-9_10.)cited 21 April 2020).

[63] K.G. Gurevich, Effect of blood protein concentrations on drug-dosing regimes: practical guidance, Theor. Biol. Med. Model. 10 (1) (2013) 20.

[64] M. Baker, T. Parton, Kinetic determinants of hepatic clearance: plasma protein binding and hepatic uptake, Xenobiotica 37 (10–11) (2007) 1110–1134.

[65] M. Rowland, Protein binding and drug clearance, Clin. Pharmacokinet. 9 (Suppl. 1) (1984) 10–17.

[66] S.M. Moghimi, A.C. Hunter, T.L. Andresen, Factors controlling nanoparticle pharmacokinetics: an integrated analysis and perspective, Annu. Rev. Pharmacol. Toxicol. 52 (1) (2012) 481–503.

[67] S.M. Hussain, K.L. Hess, J.M. Gearhart, K.T. Geiss, J.J. Schlager, In vitro toxicity of nanoparticles in BRL 3A rat liver cells, Toxicol. in Vitro 19 (7) (2005) 975–983.

[68] S.J. Kang, B.M. Kim, Y.J. Lee, H.W. Chung, Titanium dioxide nanoparticles trigger p53-mediated damage response in peripheral blood lymphocytes, Environ. Mol. Mutagen. 49 (5) (2008) 399–405.

[69] T. Xia, M. Kovochich, J. Brant, M. Hotze, J. Sempf, T. Oberley, et al., Comparison of the abilities of ambient and manufactured nanoparticles to induce cellular toxicity according to an oxidative stress paradigm, Nano Lett. 6 (8) (2006) 1794–1807.

[70] Z. Liu, S. Tabakman, K. Welsher, H. Dai, Carbon nanotubes in biology and medicine: in vitro and in vivo detection, imaging and drug delivery, Nano Res. 2 (2) (2009) 85–120.

[71] G.M. Mutlu, G.R.S. Budinger, A.A. Green, D. Urich, S. Soberanes, S.E. Chiarella, et al., Biocompatible nanoscale dispersion of single-walled carbon nanotubes minimizes in vivo pulmonary toxicity, Nano Lett. 10 (5) (2010) 1664–1670.

[72] D.B. Warheit, Comparative pulmonary toxicity assessment of single-wall carbon nanotubes in rats, Toxicol. Sci. 77 (1) (2003) 117–125.

[73] J. Muller, F. Huaux, N. Moreau, P. Misson, J.-F. Heilier, M. Delos, et al., Respiratory toxicity of multi-wall carbon nanotubes, Toxicol. Appl. Pharmacol. 207 (3) (2005) 221–231.

[74] I. Nestorov, Whole-body physiologically based pharmacokinetic models, Expert Opin. Drug Metab. Toxicol. 3 (2) (2007) 235–249.

[75] P. Lees, F.M. Cunningham, J. Elliott, Principles of pharmacodynamics and their applications in veterinary pharmacology, J. Vet. Pharmacol. Ther. 27 (6) (2004) 397–414.

[76] J.H. Byun, D.-G. Han, H.-J. Cho, I.-S. Yoon, I.H. Jung, Recent advances in physiologically based pharmacokinetic and pharmacodynamic models for anticancer nanomedicines, Arch. Pharm. Res. 43 (1) (2020) 80–99.

[77] P. Macheras, A. Iliadis, Modeling in biopharmaceutics, pharmacokinetics and pharmacodynamics: homogeneous and heterogeneous Approaches [Internet], in: Interdisciplinary Applied Mathematics, vol. 30, Springer International Publishing, Cham, 2016, https://doi.org/10.1007/978-3-319-27598-7 (cited 28 April 2020).

[78] J. Choi, Q. Zhang, V. Reipa, N.S. Wang, M.E. Stratmeyer, V.M. Hitchins, et al., Comparison of cytotoxic and inflammatory responses of photoluminescent silicon nanoparticles with silicon micron-sized particles in RAW 264.7 macrophages, J. Appl. Toxicol. 29 (1) (2009) 52–60.

[79] N.P. Truong, M.R. Whittaker, C.W. Mak, T.P. Davis, The importance of nanoparticle shape in cancer drug delivery, Expert Opin. Drug Deliv. 12 (1) (2015) 129–142.

[80] R. Agarwal, V. Singh, P. Jurney, L. Shi, S.V. Sreenivasan, K. Roy, Mammalian cells preferentially internalize hydrogel nanodiscs over nanorods and use shape-specific uptake mechanisms, Proc. Natl. Acad. Sci. 110 (43) (2013) 17247–17252.

[81] S. Dasgupta, T. Auth, G. Gompper, Shape and orientation matter for the cellular uptake of nonspherical particles, Nano Lett. 14 (2) (2014) 687–693.

[82] P. Kolhar, A.C. Anselmo, V. Gupta, K. Pant, B. Prabhakarpandian, E. Ruoslahti, et al., Using shape effects to target antibody-coated nanoparticles to lung and brain endothelium, Proc. Natl. Acad. Sci. 110 (26) (2013) 10753–10758.

[83] J.-W. Yoo, E. Chambers, S. Mitragotri, Factors that control the circulation time of nanoparticles in blood: challenges, solutions and future prospects, CPD 16 (21) (2010) 2298–2307.

[84] A.O. Kamel, G.A.S. Awad, A.S. Geneidi, N.D. Mortada, Preparation of intravenous stealthy acyclovir nanoparticles with increased mean residence time, AAPS PharmSciTech 10 (4) (2009) 1427.

[85] D. Owensiii, N. Peppas, Opsonization, biodistribution, and pharmacokinetics of polymeric nanoparticles, Int. J. Pharm. 307 (1) (2006) 93–102.

[86] F. Meng, G.H.M. Engbers, A. Gessner, R.H. Müller, J. Feijen, Pegylated polystyrene particles as a model system for artificial cells: pegylated polystyrene particles, J. Biomed. Mater. Res. 70A (1) (2004) 97–106.

[87] Y. Sheng, C. Liu, Y. Yuan, X. Tao, F. Yang, X. Shan, et al., Long-circulating polymeric nanoparticles bearing a combinatorial coating of PEG and water-soluble chitosan, Biomaterials 30 (12) (2009) 2340–2348.

[88] M.S. Peter, S.S. Ulrike, Regional cancer therapy, Biomed. Sci. (2007).

[89] H.F. Galley, N.R. Webster, Physiology of the endothelium, Br. J. Anaesth. 93 (1) (2004) 105–113.

[90] C. Roma-Rodrigues, R. Mendes, P. Baptista, A. Fernandes, Targeting tumor microenvironment for cancer therapy, IJMS 20 (4) (2019) 840.

PART 2

Application of nanoparticles in drug delivery

CHAPTER 8

Biodegradable self-assembled nanocarriers as the drug delivery vehicles

Charu Misra[a], Rakesh Kumar Paul[a], Nagarani Thotakura, and Kaisar Raza
Department of Pharmacy, School of Chemical Sciences and Pharmacy, Central University of Rajasthan, Bandarsindri, Rajasthan, India

1. Introduction

As per the general understanding, the term "nanotechnology" was instigated in the scientific domain by N. Taniguchi in 1974 [1]. The term was used to describe the accuracy of the mechanisms associated with the material in the nanometric range [2]. The idea behind using various strategies employing nanotechnology was introduced by Feynman and further applied by E. Drexler for the first time in his book "*Vehicles of creation: the arrival of the nanotechnology era*," which was published in 1986 [3]. Numerous notable and remarkable applications have fascinated researchers in various domains to work on the upliftment of this novel technology. From the early 1990s, researchers have designed various approaches to enrich and widen the field of nanotechnology [4, 5].

Especially in medicine, nanotechnology for drug delivery is expected to transform the face of the pharmaceutical and biotechnology industries. In some instances, the pharmaceutical firms appear to be stagnating, but various blockbuster products will come off as magic using this novel approach in the long future. The use of the Hatch-Waxman Act by generic drug firms has also risen, allowing them to compete with the branded products, further deteriorating the future income of pharmaceutical companies [6]. Henceforth, nanotechnology is envisioned as an approach to revive the patents further.

The advancement of engineered nanomaterials can play a vital role in adding a newly proposed method to pharmaceutical industry reservoirs [7]. The use of nanotechnology has offered various benefits such as ease of poorly soluble drug delivery, early detection of the tumor stages and reduce toxicity associated with multidrug delivery [8]. It also provides site-specific delivery of various macromolecules and micromolecules, and thus increasing the bioavailability of the drug. This technique allows the co-delivery of various drugs, enhancing the therapeutic modality in combinational drug therapy [9, 10]. Besides

---

[a] These authors contributed equally to this work.

the advantages mentioned earlier, this also helps in transcytosis of drugs across tight epithelial, endothelial barriers and detecting various diseases through different biomarkers [11].

In addition, nanotechnology and nanocarrier-based drug delivery systems provide increased repairing potency and minimized unnecessary side effects associated with traditional dosage forms [12, 13]. Nanocarriers for the development of dosage forms have been widely researched, and Fig. 1 displays the structural visualization of a few nanocarriers employed in drug delivery.

Nanocarriers protect the drug from various biological and environmental factors, affecting the drug's efficacy [14]. They play a vital role in the controlled and sustained release of the drug at a diseased site [15]. In cancer, active targeting using ligand molecules helps to bind the nanocarriers with the over-expressed receptors, as shown in Fig. 2. Compared with conventional systems, targeted drug delivery systems and therapeutics are beneficial in several ways [16, 17]. Active targeting allows the nanocarriers to deliver

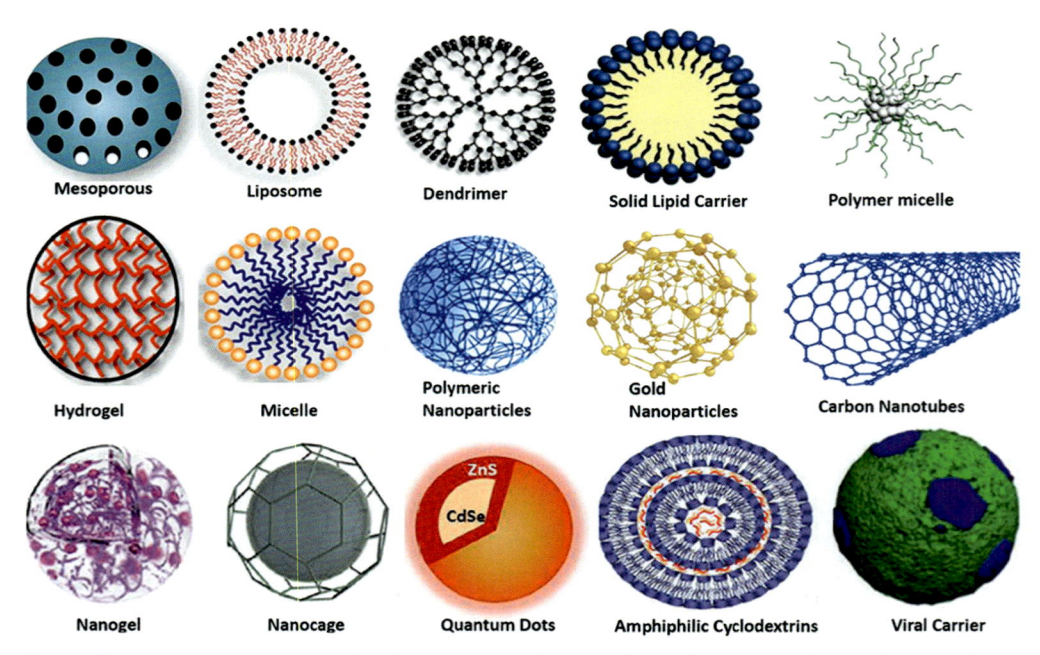

Fig. 1 Pictorial representation of various nanocarriers employed for drug delivery. *(Reprinted with permission from Creative Commons Attribution License (CC BY). A.A. Yaqoob, H. Ahmad, T. Parveen, A. Ahmad, M. Oves, I.M.I. Ismail, et al., Recent advances in metal decorated nanomaterials and their various biological applications: a review, Front. Chem. 8 (2020). https://doi.org/10.3389/fchem.2020.00341.)*

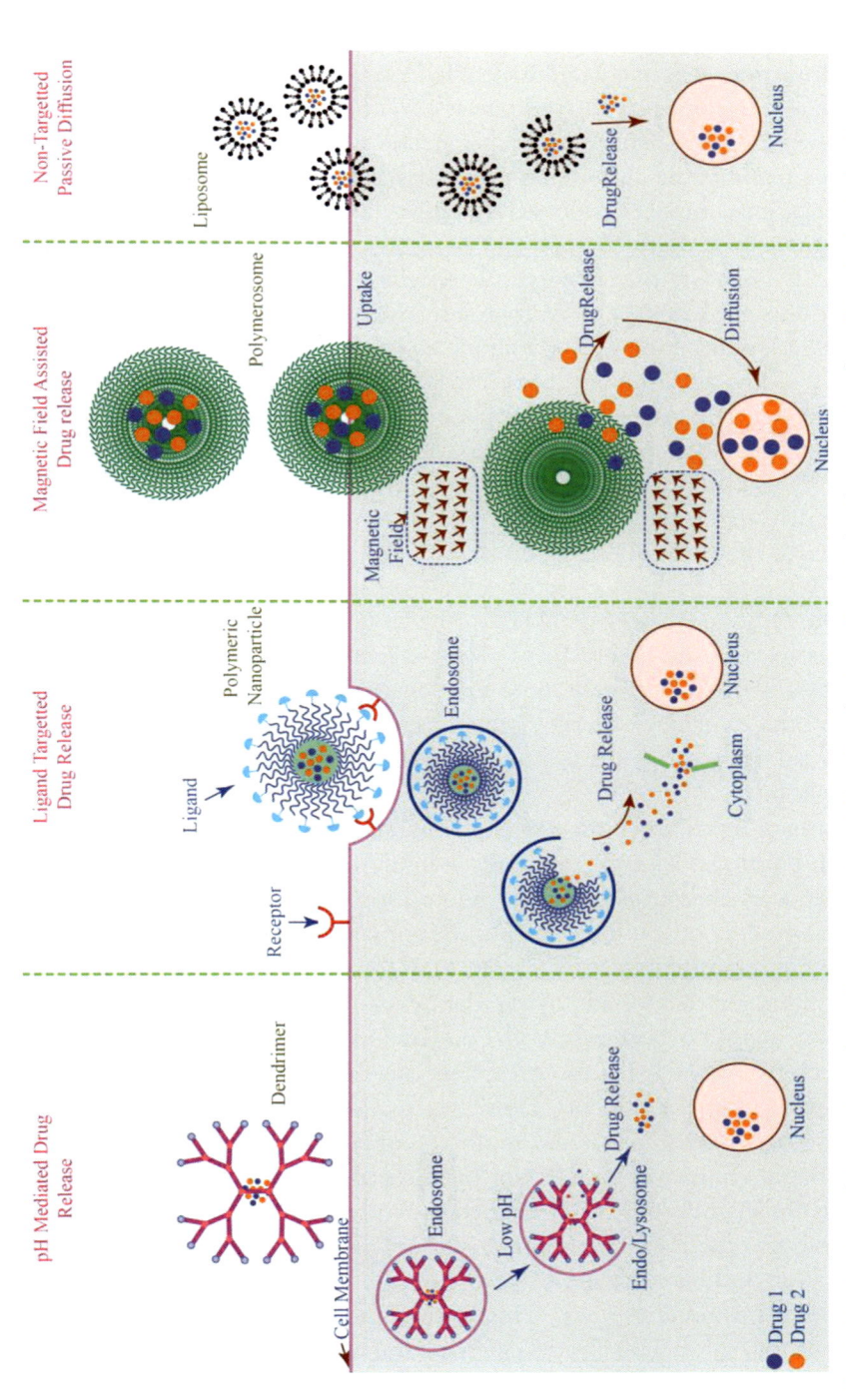

Fig. 2 Mechanism of drug release from the nanocarriers at the site of action. *(Reprinted with permission from R.S.S. Pushpalatha, Nanocarrier mediated combination drug delivery for chemotherapy—a review, J. Drug Delivery Sci. Technol. 39 (2017) 362–371.)*

the drugs to the target site of action without influencing other normal cells [18]. They can deliver the drug quite effectively to the specified location, providing improved treatment productivity, and decreased systemic toxicity [19].

Nanocarriers can also deliver other therapeutic molecules such as peptides, nucleic acids, proteins, or antibodies in addition to traditional drugs [20]. Targeted delivery is generally being examined for several diseased conditions like diabetes, inflammatory bowel disease, and brain-related disorders such as Parkinson's, Alzheimer's disease or other than cancer [21–23]. There are many benefits in designing the nanocarriers for drug delivery, including the improved stability of hydrophobic and hydrophilic drugs, enhanced pharmacokinetics and biodistribution of the drug molecule, enriched permeability, efficacy, and retention of the drug resulting in the specific targeting along with reduced toxicity [24].

Self-assembled nanocarriers have imparted their impact in the drug delivery process as these carriers noncovalently self-associate with the drug molecules to form a relatively more stable structure, which is more helpful in the drug delivery [25, 26]. These drug delivery systems show minimal impression on the activity of the drug and the ligand as they are generally bound to the carrier systems by intermolecular interactions. Owing to the high coupling efficiency, there will be no need for further purification to remove the undesirable by-products and reactants [26]. Biomimetic construction from basic building blocks to massive, complex structures is feasible through molecular assembly [27]. Accurate molecular identification mechanisms can be realized using the intrinsic specifics of biological molecules [28]. There is a structural resemblance between these drug carriers with the viral particles due to their virus-like architecture [29].

In the formation process of the self-assembled nanocarriers, two types of approaches are involved: positional assembly and self-assembly of the molecules [30]. In the self-assembled nanocarriers formed through the positional assembly method, particles (the atoms, or molecules, or clusters) are pushed into proximity and positioned one by one [31]. In the self-assembly method, the molecules or atoms are organized through hydrogen bonding or electrostatic or van der Waals interactions in such a way to form ordered nanostructures or patterns by a secondary force without external guidance [32]. The self-assembly method is a more convenient one due to the advantage of the inherent propensity among the components themselves toward unique, local interactions [33]. A key feature of self-assembly is the multiple cooperating or competing interactions to prepare hybrid nanocomposites based on the variety of materials and organization of these individual entities into well-defined structures [34]. Assembled nanostructured systems represent a significant advance in the field of nanotechnology. Essential biochemical identification and biocompatibility can contribute to the development of uniform self-assembled nanostructures [35]. This chapter discusses the evolution, chemistry involved in the formation, classification, preparation techniques, its applications in medicine, and prospects of the self-assembled nanocarrier systems.

2. Evolution of self-assembly

The word "self" implies "on its own (or) without any external force," and "assembly" implies "getting together." The term self-assembly is the spontaneous adsorption or arrangement of a particle in a chronological pattern leading to supra-molecular structure [36]. The concept was originated by an imaginative approach regarding the organized pattern of the solar system, including stars according to their size given by Greek and French philosophers Democritus and Descartes, respectively, in the initial 20th century [37]. Self-assembly has provided an effective way to create materials and assemble them into practical structures tailored for particular use in the context of nanofabrication by bottom-up synthetic chemistry. As a fundamental theory, self-assembly reveals that all sort of structures illustrated by atoms or polymer colloids will develop into a random structure, guided by a map of pressure exerted on several length scales, to a higher degree of structural complexies [38].

In 1935, Langmuir and Blodgett initiated the idea of self-assembly as a natural growth phenomenon [39]. Bigelow et al. studied the arrangement of monolayer packed platinum, but the term self-assembly was not explicitly stated to clarify the creation of well-ordered molecular monolayers [40]. Self-assembly worked in these structures. In 1991 Laibinis and Allara contacted gold surfaces with alkyl disulfides. They found that they formed close-packed monolayers, and for the first time, the term "self-assembled monolayers" was introduced for chemisorbed alkanethiols molecules [41].

3. The chemistry behind self-assembled systems

Self-assembly is defined by the term "spontaneity," which means the system is built from modular building modules in an ordered pattern. In these systems, the building blocks are self-bounded without the involvement of the external binding forces [42]. Self-assembling of the building blocks into well-organized structures determines the scale, form and surface property monitoring ability with a high degree of precision. Therefore a primary objective of self-assembly is to synthesize building blocks with defined dimensions and shapes and gain control over the electrostatic repulsions among them. Chemical control of surface properties enables self-assembled systems to dynamically assemble to create an integrated chemical, physical or biological framework with deliberate purpose and utility [43]. The dynamic simulation drives the major component of static self-assembly to form static equilibrium systems without external factors [44].

A dynamically self-assembling system can exist in the presence of emerging forces that can respond to its surrounding atmosphere by remaining at an energetic minimum, triggered by the introduction of energy into the system [45]. Once the energy stops flowing, the system disassembles itself. Any living organism is a perfect example of dynamic self-assembly. It reduces entropy by absorbing energy from the environment. This

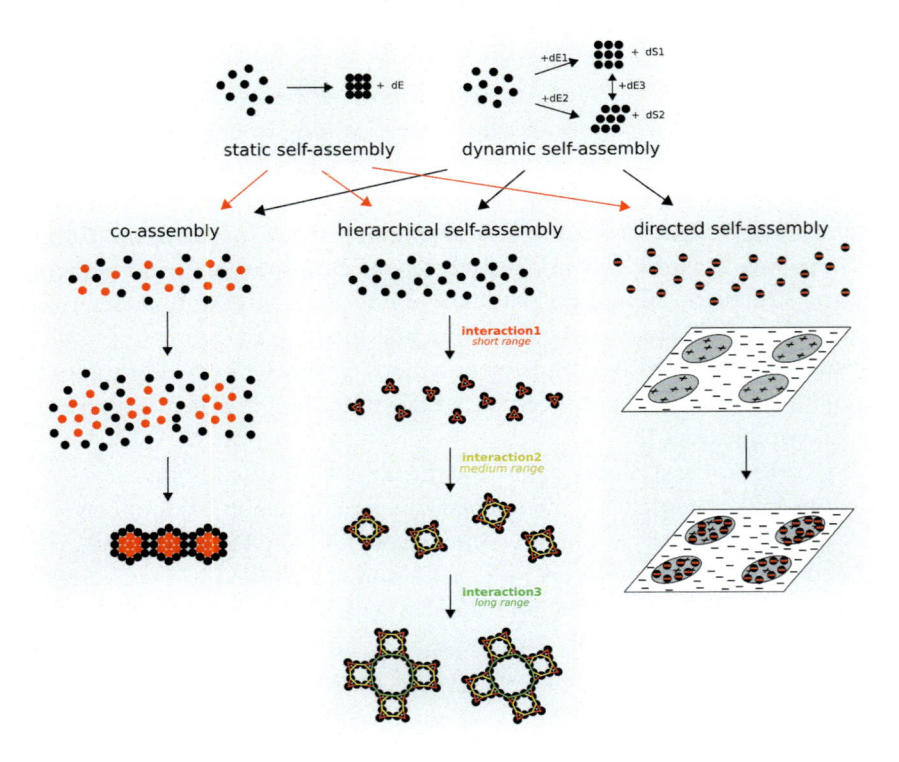

Fig. 3 Various types of interactions for the self-assemblies (dE states the rate of energy, dE1 and dE2 signify different rate of energy input, whereas dS1 and dS2 signify the energy output rate). *(Reprinted with permission from G.A. Ozin, K. Hou, B.V. Lotsch, L. Cademartiri, D.P. Puzzo, F. Scotognella, et al., Nanofabrication by self-assembly, Mater. Today 12 (2009) 12–23. https://doi.org/10.1016/S1369-7021(09)70156-7.)*

gradient in entropy between the organism and the environment can be maintained only as long as energy is driven from the environment into the organism in the form of food and heat. Once that flux ceases, the organism disassembles. Two main categories of self-assembly are called static, dynamic, and both are illustrated in Fig. 3.

3.1 Hydrogen bonds

Hydrogen bonds are the interactions between hydrogen atoms and a substantial array of biomolecular electronegative atoms [46]. This plays a significant role in the development of biological nanomaterials. The hydrogen bonds facilitate the growth of biomolecules in a single direction with a large order in the biomolecular self-assembly phase to form one-dimensional (1D) nanostructures [30]. Li et al. formulated self-assembled diphenylamine loaded microrods using hydrogen bonding. The formulation indicates the formation of the most stable bond between the peptide molecules and hexafluoroisopropanol, which is used as a solvent for the dissolution. The evaporation of hexafluoroisopropanol after dissolving of peptide in water leads to self-assembly resulting in the microfibers, microrods, or microtubes. Diphenylamines are the first molecules with self-assembled properties and can grow into nanotubes in the presence of water. They can further transform into microrods in the presence of hydrophilic bonds by the interaction among peptide residues [47].

Another researcher worked on the formation of nanowires with the help of diphenylamine on graphene surfaces. The peptide was dissolved in graphene solution and dropwise poured into the substrate to dry at 50°C. It was observed that graphene interacted with peptides through π-π interactions. Peptide nanowires were forming due to hydrogen bonding between graphene and peptide on the graphene surface [48].

Viruses also play an essential role in the formation of self-assembly. Lee et al. have formulated nanosheets using graphene oxide and the M13 virus. The self-assembly process was carried out using amino acids present in M13 viruses. These amino acids exhibit hydrogen and electrostatic bonding among the carboxyl group present at the edges of graphene oxide sheets [49]. Xi et al. fabricated a biosensor with a surface plasmon resonance technique and investigated the role of hydrogen bond interaction. They observed that the interaction of single-strand deoxyribonucleic acid and graphene oxide [50]. Dapeng et al. has explored the various molecular interactions between graphene and biomolecules such as DNA, carbohydrate, peptide, protein, enzyme, and virus [51].

3.2 Electrostatic interactions

Electrostatic interactions have a vital significance in the formation of various self-assembled nanostructures. They provide stable and robust interaction among various enzymes, peptides, and proteins. Wang et al. formulated nanofibers-based peptide and peptide nanofiber-based silver nanowires using graphene nanosheet [52]. The formulation was based on electrostatic interaction. It was observed that graphene nanosheets were positively charged, having a zeta potential of +35.87 mV. Fabrication of nanocomposite peptide nanofiber-based silver nanowires having negative charge was self-assembled through a synthetic process. Strong interactions were observed between the molecules. Yang et al. prepared a nanoformulation to treat cancer using chitosan and poly(lactic-*co*-glycolic acid) (PLGA) as a polymer. The researcher formulated paclitaxel-loaded chitosan-modified PLGA nanoparticles. It was observed that the uptake efficacy of the cancer cells was enhanced at lower pH due to strong electrostatic interaction between the two polymers [53]. Denise et al., in their study, demonstrated the significance of surface electrostatic bonding in preventing the toxicity of nanoparticles. The research group formulated zinc oxide metallic nanoparticles having a negative charge [54]. The formulation was modified using polyacrylic acid to maintain the electrostatic interactions. The study revealed that the surface modification enhanced the reactive oxygen species, which extended the apoptosis in the cancer cells. Tan et al. used a self-assembled method for the formulation of nanocomposites using PLGA-based folic acid-modified chitosan. The effective anti-cancer effect was observed through electrostatic interaction between the polymer and drug [55].

3.3 Hydrophobic and hydrophilic interactions

Various studies have been reported for self-assembled interactions between the molecules via hydrophobic and hydrophilic interactions. Xu et al. observed the self-assembled nature of small amphiphilic peptides. These peptides, via hydrophobic interaction, first grow into nanofibers through nucleation and then extend their network in the form of nanosheets or mixed nanofibers [56]. Zhao et al. observed hydrophobic interaction in preparation of self-assembled ultrathin gelatin and tannin oxide composed membrane. The hydrophilicity of the surface was enhanced as compared with gelatin alone. It was also observed that the membrane stability was enhanced as compared with the pristine one. Separation of the membrane was done using an ethanol solution [57].

Cai et al. formulated a complex structure using poly(N–isopropyl acrylamide) and bovine serum albumin. It was observed that the hydrophobic interaction of these two complexes played a vital role in the self-assembly process. The approach suggested the formation of protein and polymer conjugate with enhance in the efficacy. It also protected protein by preventing it from getting damage [58]. Zou et al., in their research, used porphyrin for the formation of peptide-linked self-assembled nanodots. The self-assembled nature of porphyrin helped in forming aggregated structures in light and heat, resulting in the fabrication of nanodots. The excessive growth in the nanodots by porphyrin was prevented by peptide moiety. Apart from it, the structural stability was improved due to hydrophilic interaction [59].

3.4 π-π interactions

Research evidence has proclaimed that π-π interactions between various self-assembled molecules have fetched beneficial results in structural organization, which helped to deliver the drugs for communicable and noncommunicable diseases. Jian et al. synthesized 2-carboxyterthiophene self-assembled magnetic nanoparticles. The uniformity of aggregated particles was maintained by the π-π interactions. Thiophene was responsible for the formation of a self-assembled nanoparticle [60]. Wang et al. developed peptide molecules, which further self-assembled to form peptide nanofibers (PNFs) [61]. As a two-dimensional scaffold, the graphene oxide (GO) nanosheet was then modified with the prepared PNFs to create a new form of GO-PNF nanohybrid, which was further used as a model for biomimetic mineralization. It was observed that nanohybrid facilitated the growth of hydroxyapatite (HA) crystals. In short-term incubation, the produced PNFs facilitated the formation of HA nanocrystals along the axis of PNFs. The GO nanosheet mediates the formation after long-term mineralization of the HA microspheres. For the more bioinspired synthesis of cotton-flower-like platinum nanoparticles (PtNPs), Li et al. developed composites employing the π–π interactions [62]. The single-stranded DNA molecules (ssDNA) were bound to reduced graphene oxide (rGO), and the bioinspired synthesis formed PtNPs on the surface of ssDNA–rGO. The material provided by ssDNA–rGO–PtNPs showed high catalytic activity for methanol oxidation and carbon monoxide tolerance.

4. Classifications of self-assembled systems

Self-assembled systems can be classified into several types based on molecular structures such as size and shape. The classification of self-assembled systems has been depicted in Fig. 4.

4.1 Polymeric self-assembled system

Polymeric self-assembled nanostructures can be classified based on the type of polymer employed, like natural or synthetic polymers. These polymeric self-assembled systems are applicable in drug delivery to improve bioactivity, control the drug delivery process, and improve drug bioavailability [63]. The increased concentration of amphiphilic polymers results in the formation of polymeric micelles [64]. These polymeric micelles, which are also amphiphiles, are designed using both hydrophobic and hydrophilic molecules. In general, the core is composed of hydrophobic molecules, whereas the exterior portion is hydrophilic. These micelles are used to encapsulate the poorly water–soluble drugs

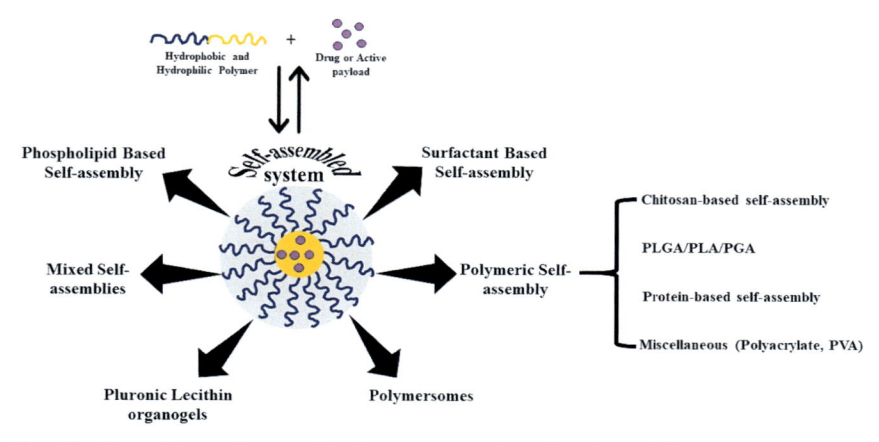

Fig. 4 Classification of the self-assembled systems employed in drug delivery.

[65]. Researchers have also shown that some polymeric micelles can respond to different stimuli such as pH, temperature, and ionic strength. An example of pH-sensitive polymeric micellar systems includes folate-poly(ethylene glycol)-poly(aspartate hydrazone adriamycin). The study explained the conjugation and the controlled release mechanism of anti-cancer drug adriamycin with polymeric micelles at different pH environments such as in the endosomes (pH 5–6) and lysosomes (pH 4–5) [66].

Similarly, in another study, the two-block copolymer has been prepared for controlled release of doxorubicin to the MCF-7 cells using poly(L-histidine)-b-poly(ethylene glycol) and poly(L-lactic acid)-b-PEG-b-poly(L-histidine)-biotin. The dissociation of these polymeric micelles was based on the pH-dependent mechanism [67]. In another study, thermo-sensitive self-assembled polymeric micelles were also prepared by loading honokiol as a payload in poly(ε-caprolactone)-poly(ethylene glycol)-poly(ε-caprolactone) copolymer for cancer chemotherapy [68].

4.1.1 Chitosan-based self-assembly

Chitosan is widely used in the pharmaceutical field because of its biochemical activity, biodegradability, and low toxicity [69]. Chitosan comprises of positive charge and can bind with the negatively charged groups of the cell membranes. It also has mucoadhesive properties, which are desired for the uptake of nanoparticles [70]. Chitosan-based polymeric micelles have been prepared using N-octyl-N-(2-carboxyl-cyclohexamethenyl) for the delivery of paclitaxel. The pH-sensitivity study revealed that the micelles were highly sensitive at pH 5.5 and physiological pH 7.4, and the structures were relatively stable. This indicates that chitosan can form micelles in the aqueous solution by a self-assembly mechanism [69]. Xu et al. have also demonstrated the formation of doxorubicin-loaded polymeric micelles by N-succinyl-N-octyl chitosan [71]. Chenguang et al. conjugated the adriamycin with linoleic acid-modified carboxymethyl-chitosan. The studies showed that adriamycin was released from chitosan by forming the self-assembly mechanism for about 3 days [72]. Some of the chitosan-based self-assembled nanocarriers used in cancer are being listed in Table 1.

4.1.2 PLGA/PLA/PGA-based self-assembled systems

PLGA is also known as "Smart biodegradable polymer" due to the stimuli sensitive behavior [78]. It is one of the biodegradable and biocompatible copolymers that decomposes and eliminates H_2O and CO_2 products in the body through the Krebs cycle. PLGA is composed of lactic acid (LA) and glycolic acid (GA), conjugated together by the copolymerization process. Poly(glycolic acid) (PGA) is a hydrophilic polymer, while poly(lactic acid) (PLA) is hydrophobic [79]. The block copolymers have numerous applications in tissue engineering, drug discovery, and biomedical fields. Highly hydrophilic and low permeable drugs can be delivered by PLGA self-assembled nanoparticles [78]. Mohana et al. have prepared the transdermally administered self-assembled nanoparticles

Table 1 List of some drugs encapsulated in chitosan-based self-assembled nanocarriers.

Drug	Nanocarriers	Disease	References
Tamoxifen	Stearic acid-based polymeric micelles	Breast cancer	[73]
Paclitaxel	Steric acid-grafted chitosan-based self-assembled micelles	Cancer	[74]
Tamoxifen	PLGA polymeric nanocarriers	Cancer	[75]
Tamoxifen	Palmitic acid-based polymeric micelles	Breast cancer	[76]
Quercetin	Chitosan-based nanoparticles	Breast cancer	[77]

by conjugating PLGA with glucosamine (GlcN) to deliver glucosamine. There was the sustained release of glucosamine from self-assembled nanoparticles for about 48 h [80].

Self-assembled micelles of PEG–PLGA copolymer formed a core shell composed of a hydrophobic PLGA core and a hydrophilic PEG shell [81]. The formation of PLGA–PEG amphiphilic block copolymer has been employed to deliver insulin self-assembled Nanomicelles [82]. The triblock copolymeric micelles of PLGA-PEG-PLGA have also demonstrated the formation of self-assembly by loading the curcumin. The self-assembly of a triblock copolymer (PLGA-PEG-PLGA) can be explained by the incorporation of a hydrophilic segment (PEG) into the hydrophobic chain of PLGA, which facilitated the release of the drug [83].

4.1.3 Miscellaneous (polyacrylate, PVA, polymers) based self-assembled systems

Studies related to polyacrylate-based dendritic polymers have shown a successive approach for the construction of self-assembled systems. These assemblies are formed via the atom-transfer radical polymerization [84]. Liu et al. prepared polyacrylate-based self-assembled nanoparticles to produce E7 protein to eradicate tumor cells [85]. Chung et al. have reported the formation of self-assembled nanoparticle carriers using cyclodextrin and adamantane polyacrylates for tumor-targeted drug delivery [86]. Polyvinyl alcohol (PVA) is a biocompatible and hydrophilic polymer used to apply drug delivery and tissue engineering [87]. Uttam et al. prepared the borax-mediated self-assembly of PVA and chitosan in which an anti-cancer drug, doxorubicin, is incorporated into the multi-thin film [88].

4.2 Protein-based self-assembled systems

It is one of the attractive self-assembly that aggregates itself into a polymeric-like architecture. These supramolecular protein structures build up by self-assembled molecules by forming a noncovalent bond. The mechanism by which protein-based supramolecular structures grow is isodesmic, cooperative and ring-chain competition. Several macrocycles and bio-inspired polymeric-like architectures such as polypeptides and DNAs associate together and form one-, two-, and three-dimensional structures [89]. High selectivity and molecular recognition characteristics are required to be considered while synthesizing the protein-based supramolecular systems. Several architectures such as branched, cross-linked, and diverse structures have shown the formation of supramolecular structures [89]. Matthew et al. have also explained the self-assembled, thermally responsive elastin-like polypeptides (ELPs) for the release of doxorubicin. This acid-labile hydrazone promotes the release of doxorubicin at pH 5 from the lysosomes [90].

4.3 Surfactant-based self-assembled systems

Surfactant-based self-assembled systems are used widely in pharmaceutical formulations. The surfactants are amphiphilic molecules that form micelles at the critical micelles concentration (CMC). Self-assembled amphiphiles depend on the molecule's charge and can be classified into anionic, cationic, zwitterionic, or nonionic [91]. The prominent interactions in these self-assembled amphiphilic systems are van der Waals forces, electrostatic interaction, hydrogen bonding, and hydrophobic effects. The two major driving forces that occur during the formation of these self-assembled supramolecular structures are hydrogen bonding and the hydrophobic effects [92]. Nagarajan et al. explained the formation of aggregates of surfactant using the Tanford concept. In their paper, the proposed idea of Israelachvili, Mitchell, and Ninham is discussed and explained how molecules were packed during the formation of aggregates and the subsequent thermodynamic properties [93]. Some of the surfactants used for forming these self-assembled supramolecular structures were Polysorbate 80 (Tween-80), Cremophor EL, Brij-35 [63]. For an amphiphilic compound, the most challenging task is the development of supramolecular assembly, as they can change their properties when they come in contact with the aqueous medium. It is observed that cationic surfactants cause toxicity, so numerous approaches have been employed to overcome these limitations [94]. Another class of surfactants is carbamates have an imidazolium moiety. The CMC value of these carbamate surfactants is lower than that of the cationic surfactants [95]. Ruslan et al. have demonstrated the mixed supramolecular formation of anionic surfactant sodium dodecyl sulfate (SDS) and methylated amino calix [4] resorcinol (ACR). The formation of supramolecular assembly occurred in the acidic medium, where amine groups of ACRs were protonated [96].

4.4 Phospholipid-based self-assembled system

Lipids, especially phospholipids and monoglycerides, have drawn particular interest in drug delivery. These lipids are thermodynamically stable and ideal materials to form self-assembled supramolecular structures. Phospholipids-based self-assembled liposome-like systems are most commonly used for the delivery of the drugs. These liposome-like structures can entrap the hydrophobic and hydrophilic drugs with low toxicity and biocompatible nature [97]. Darshan et al. demonstrated the self-assembled phytosomal soft nanoparticles using mangiferin complexed phospholipid (Phospholipon 90H). These self-assembled phytosomal soft nanoparticles were evaluated for the antioxidant properties using an in vivo carbon tetrachloride (CCl_4)-intoxicated albino rat model [98].

Similarly, May et al. has also prepared a self-assembled system of phospholipid-based phytosomal nanocarriers to improve oral bioavailability by using celastrol (a natural drug). The self-assembled phytosomes were designed using Soy phosphatidylcholine (SPC) (Lipoid S 100). The scheme followed by the team is depicted in Fig. 5 [99]. Another research group has also demonstrated the formation of self-assembled monodispersed microparticles using berberine as a natural drug and SPC as phospholipid. These monodispersed self-assembled supramolecular structures were evaluated for oral bioavailability and anti-diabetic efficacy [100]. Yang et al. has prepared the supramolecular systems using water-soluble drug-phospholipid complex. The self-assembled structure of the mitomycin C-SPC complex has been loaded into the PEG-lipid-PLA hybrid nanoparticles with folate functionalization for targeted drug delivery and dual-controlled release of the drug [101].

4.5 Mixed self-assemblies

Self-assembled mixed micelles are composed of two or more different block copolymers. These self-assembled micelles improve physical stability and enhance the loading of drugs into the polymeric micelles. The use of amphiphilic block copolymers aids in forming core-shell type self-assembled micelles to protect therapeutic cargos from the external medium. Block copolymers can form weak hydrophobic interaction by molecular recognition, such as H-bonding, ionic interactions, and cross-linking, they form self-assembled mixed micelles. The formation of the self-assembled supramolecular structure of mixed micelles has been depicted in Fig. 6 [102]. Pluronic copolymers having different hydrophobicity/hydrophilicity ratios have shown the formation of self-assembled mixed micelles. Valery et al. used the combination of two polypropylene oxide block copolymers (Pluronics) with doxorubicin-loaded into it for the high efficacy against both drug-resistant and drug-sensitive tumors. These self-assembled micelles were formed by hydrophilic block (Pluronic F127) and hydrophobic block (Pluronic L61) and were thermodynamically stable [103]. The preparation of mixed micelles with a multifunctional core and shell is revealed by the use of graft copolymer poly(N-isopropyl acrylamide-co-methacrylic acid)-g-poly(D,L-lactide) along with two diblock copolymers, poly(ethylene glycol)-b-poly(D,L-lactide), and poly (2-ethyl-2-oxazoline)-b-poly(D,L-lactide). Doxorubicin, an anti-cancer drug, has been encapsulated into the copolymer to form a mixed self-assembled supramolecular structure [104].

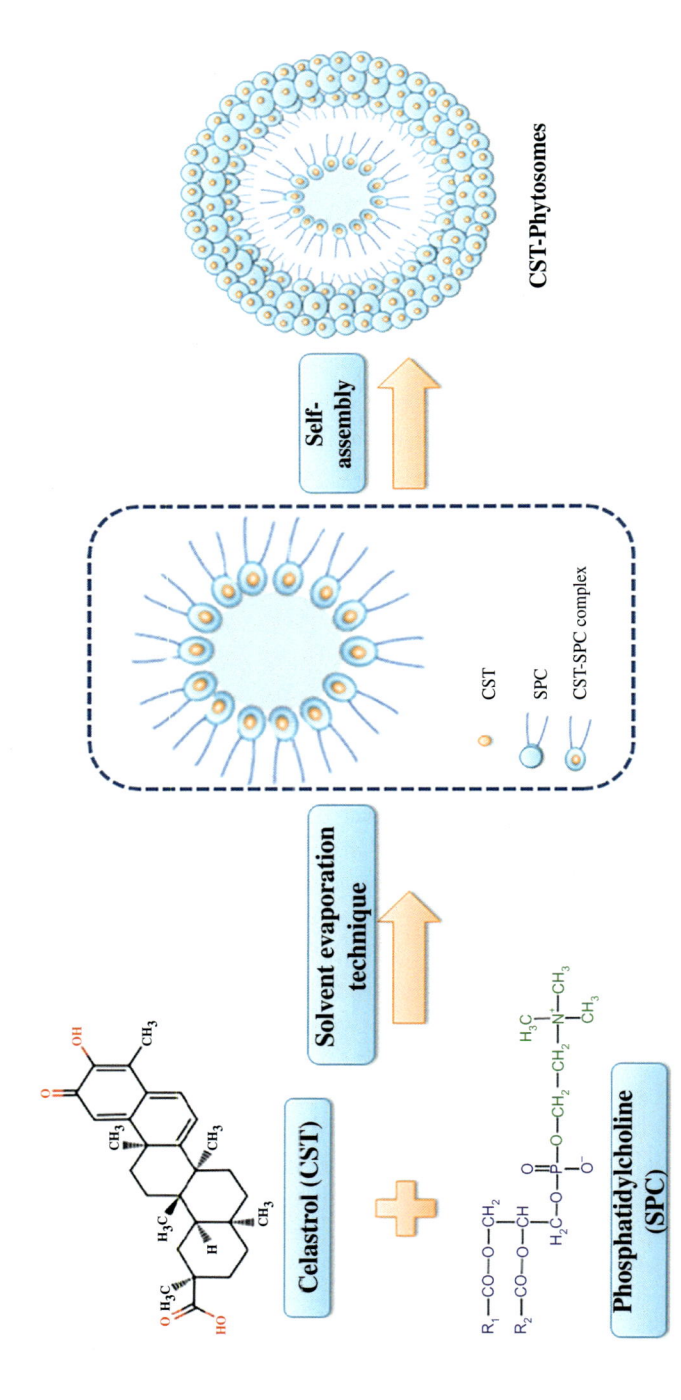

Fig. 5 Schematic representation illustrating the formation of self-assembled CST phytosomes by applying celastrol and phosphatidylcholine. (Reprinted with permission from M.S. Freag, W.M. Saleh, O.Y. Abdallah, Self-assembled phospholipid-based phytosomal nanocarriers as promising platforms for improving oral bioavailability of the anticancer celastrol, Int. J. Pharm. 535 (2018) 18–26. https://doi.org/10.1016/j.ijpharm.2017.10.053.)

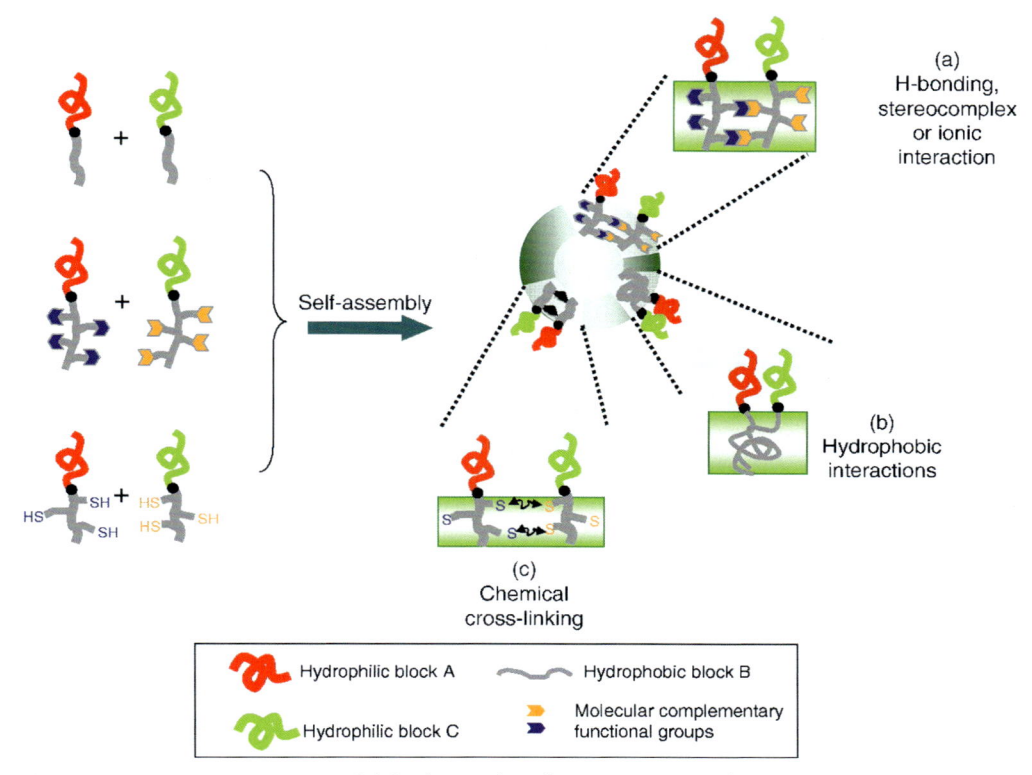

Fig. 6 Various core interactions (A) hydrogen bonding, stereo complexation, or ionic interaction; (B) hydrophobic interactions; (C) chemical cross-linking (disulfide bond) for the formation of mixed micelles. *(Reprinted with permission from A.B. Ebrahim Attia, Z.Y. Ong, J.L. Hedrick, P.P. Lee, P.L.R. Ee, P.T. Hammond, et al., Mixed micelles self-assembled from block copolymers for drug delivery, Curr. Opin. Colloid Interface Sci. 16 (2011) 182–194. https://doi.org/10.1016/j.cocis.2010.10.003.)*

4.6 Pluronic lecithin organogels

Organogels are small organic molecules that self-assemble into three-dimensional networks and convert into the gel from a liquid. There are various kinds of organogel bases; among these, pluronic lecithin organogels (PLO) consists of lecithin [105]. Ahmed et al. have demonstrated self-assembled organogels based on the pluronic and lecithin for the sustained release of a nonsteroidal anti-inflammatory drug, Etodolac [106]. Similarly, pluronic lecithin self-assembled organogels were prepared using pluronic F127, lecithin, flurbiprofen, isopropyl palmitate, water, ascorbic acid, and potassium sorbate to release the nonsteroidal anti-inflammatory drug, Flurbiprofen [107]. Researchers from other groups have demonstrated sumatriptan for transdermal delivery by supramolecular structures [108]. Apart from these, a combination approach of docetaxel and cisplatin has been delivered by pluronic lecithin organogel at tumor sites [109]. Bhatia et al. prepared lecithin organogel for tamoxifen. The formulation was highly stable and safe for oral delivery [110].

4.7 Polymersomes

Polymersomes consist of amphiphilic block copolymers, which aggregate to form a self-assembled supramolecular structure [111]. These polymersomes are hollow spherical structures with bilayered membranes. The bilayer membrane is composed of hydrophilic polymers in the internal and external surfaces, and in between, hydrophobic polymers are incorporated [112]. In the construction of polymersomes, stimuli-sensitive polymers were used, which depends on various stimuli. Thermosensitive polymer like poly(-N-isopropyl acrylamide) is mostly used as a hydrophobic building block. Owing to the lower critical solution temperature (LCST) and sharp transition behavior, this thermosensitive polymer was used. Apart from temperature-dependent polymers, pH soluble polymers can be used to construct a self-assembled system [112]. Fariyal et al. have prepared the biodegradable self-assembled polymersomes loaded with doxorubicin and paclitaxel. These prepared self-assembled polymersomes revealed a significant reduction in the tumor by inhibiting cell growth. The diblock copolymer, PEG-PLA, and PEG-butadiene were used to incorporate doxorubicin and paclitaxel [113]. Prajakta et al. have synthesized a hypoxia-responsive, amphiphilic diblock copolymer to treat hypoxic pancreatic cancer cells. These polymersomes were prepared by conjugating the poly(lactic acid) with poly(ethylene glycol) in which gemcitabine and erlotinib were encapsulated [114].

5. Preparation of self-assembled nanocarriers

5.1 Preparation of self-assembled micelles

Generally, there are three methods involved in preparing self-assembled micelles, i.e., direct dissolution, dialysis method, and dry down or an evaporation method. In the direct dissolution method, drug and copolymers are directly dissolved in buffer or aqueous solution. For loading the drug into the nanomicellar form, external forces like heating, sonicating or simple stirring are provided [115]. Cores are formed by the dehydration process, and thus resulting in forming blocks that initiate the micelle formation. This technique has usually been used to formulate polyion complex micelles for mildly hydrophobic polymers such as poloxamers.

The dialysis method is one of the frequently used methods for the preparation of micelles. In this method, drug and copolymer are dissolved in an organic solvent such as dichloromethane, acetone, acetonitrile, and dimethyl sulfoxide. The formation of micelles gets stimulated by the addition of water to the drug-copolymer mixture. To remove the organic solvent, the micelles are then dialyzed against water for longer time spans. In addition, the selection of the solvent significantly affects the micelles physical and drug encapsulation properties [116].

Dry down or evaporation method can also be used to prepare micelles where the copolymer and drug are dissolved in a miscible solvent. The mixture is stirred till it forms a thin layer. The thin layer of film is slowly reconstituted with lukewarm water or buffer, resulting in micelles. The sample is then sonicated and lyophilized [117].

5.2 Preparation of self-assembled nanoparticles

Nanoparticles can be prepared using various techniques such as the solvent injection method, the double emulsion solvent evaporation method [117], the single emulsion method, and the salting-out method [118], and nanoprecipitation. In the solvent injection method, the active constituent is dissolved in an organic solvent such as ethanol, dichloromethane-containing polymer and then injected dropwise into the aqueous solution with continuous stirring at room temperature. The particles are collected by centrifugation method [119].

Other than this, double emulsion method can also be used for the preparation of nanoparticles. In this method, drug and polymer are dissolved in organic solvents. Further, the drug solution is dissolved dropwise into the primary emulsion at continuous stirring with the homogenization process. The solution is then poured into the secondary emulsion. After 3–4 h, particles are collected through the centrifugation process [120].

6. Biomedical applications of self-assembled system

Self-assembled supramolecular structures are formed spontaneously by the arrangement of molecules. Most of the biological nanostructures are assembled themselves by constructing cell membranes, the helical structure of DNA and polypeptide chain folding. Self-assembled supramolecular structures are formed by ligand-receptor interactions [121]. The application of small molecule self-assembled supramolecular structures is a valuable strategy for delivering drugs and biomedical applications.

6.1 Cancer management

Most of the cancers are treated with the use of chemotherapy. Most chemotherapeutic drugs have less water solubility, highly toxic to the normal tissues, and off-target effects. Many hydrophobic drugs show less solubility and bioavailability during administration. Apart from the solubility and bioavailability, many drugs have stability issues, have a short half-life, and abnormalities in biodistribution [121]. This creates a challenging task for delivering the active payloads to the target site. These problems can be minimized by forming a self-assembly system. The list of anti-cancer drugs which are encapsulated in the self-assembled systems is listed in Table 2.

The anti-cancer drugs were encapsulated in the polymeric self-assembled nanocarriers to achieve the optimum therapeutic effect and controlled drug release [132]. The amphiphilic block copolymer has much attracted in the drug delivery [133]. Several self-assembled functional copolymers tagged with targeting ligands were used for the

Table 2 List of a few drugs encapsulated in the self-assembled systems.

Drug used	Nanocarriers	Application(s)	References
Camptothecin	Nanofibers	For the management of cancer	[122]
Carboplatin	Chitosan nanoparticles	For the treatment of breast cancer	[123]
Cisplatin	Chitosan nanoparticles	For the management of tumor	[124]
Docetaxel	PLGA-based polymeric micelles	For the management of breast cancer	[125]
Doxorubicin	Polymeric micelle	For cancer therapy	[126]
Doxorubicin, Paclitaxel	Nanoparticles	For the management of breast cancer	[127]
Ellagic acid	Chitosan nanoparticles	For the treatment of oral cancer	[128]
Indomethacin	Heparin nanoparticles	For the management of human nasopharyngeal carcinoma	[129]
Methotrexate	PLGA-based polymeric micelles	For the management of breast cancer	[130]
Tamoxifen	Stearic acid-based polymeric micelles	For the management of breast cancer	[73]
Tamoxifen	PLGA polymeric nanocarriers	For the management of breast cancer	[75]
Tamoxifen	Palmitic acid-based polymeric micelles	For the management of breast cancer	[76]
Tamoxifen	Phospholipid-based mixed micelles	For the treatment of anti-cancer	[131]

site-specific delivery of drugs at the cancerous sites. The different types of block copolymers used to construct the self-assembled supramolecular structures in the various cancers were listed in Table 3 [132].

6.2 Self-assembled peptides in the vaccine delivery

Antigen plays a crucial role in the immune response. Therefore, the development of an effective vaccine by applying synthetic antigens such as peptide epitopes is essential. Several proteins such as b-nerve growth factor (b-NGF), bone morphogenetic protein-2 (BMP-2), and vascular endothelial growth factors were used for bone mineralization, axon sprouting, and to promote angiogenesis [134]. Researchers have prepared the self-assembled peptides for the vaccination purpose by using peptides like β-sheet or α-helical coiled. Self-assembly of peptides was extensively used due to alternating hydrophobic and hydrophilic amino acids, further converted into extended fibrillar nanostructures [135]. These self-assembled peptides were more beneficial compared with other antigen carrier systems [136]. Boato et al. prepared a lipo-peptide virus-like particle synthetically. These peptide monomers turned into micelles which showed high antigen-specific antibodies in vaccinated rabbits. The studies claimed that the lipid tail could

Table 3 List of block copolymers involved in forming self-assembled supramolecular structures to release anti-cancer drugs [132].

Drug	Block copolymers	Application(s)
Cisplatin	PEG-*b*-poly(L-glutamic acid)	Pancreatic cancer
Docetaxel	PLGA-dextran	For management of cancers
Doxorubicin	PEG-*b*-poly(propylene oxide)-PEG	Advanced adenocarcinoma
Doxorubicin	PEG-poly aspartate modified with 4-phenyl-butanol	Various solid tumors
Epirubicin	PEG-*b*-poly(aspartate-hydrazone)	Various solid tumors
Oxaliplatin	PEG-*b*-poly(L-glutamic acid)	Advanced solid tumor or lymphoma
Paclitaxel	PEG-*b*-poly(α,β-aspartic acid)	For the management of gastric cancer/breast cancer
SN-38	PEG-*b*-poly(L-glutamic acid)	For the management of breast cancer, small-cell lung cancer

be coupled to the N-terminus of the self-assembling peptide [137]. Some of the supramolecular peptide-based vaccines were listed in Table 4 [138].

The designing of a peptide-based vaccine requires an antigen, an adjuvant, and a delivery vehicle. For the induction of humoral immunity, epitopes are necessary on the peptides. These peptides were used as an antigen and can induce humoral or cellular immune responses to protect from pathogens. The immune stimulation by the

Table 4 A few examples of self-assembled supramolecular peptide-based vaccines.

Peptide	Features	Applications
RADA16	β-Sheet-rich nanofibrous hydrogel	For the EG7-OVA tumor model
Fmoc-KCRGDK	Micellar hydrogel	Postoperative immunotherapy in the 4T1 tumor model
OVA$_{253-266}$ peptide	Cylindrical micelles	For peptide tumor antigen in EG7-OVA tumor model
K$_2$(SL)$_6$K$_2$	Nanofibrous hydrogel	The delivery system in an MOC$_2$-E$_6$E$_7$ tumor model
Nap-GFFpY-OMe	Nanofibrous hydrogel	Vaccine adjuvant for EG7-OVA tumor model
Nap-GFFY-NMe	Nanofibrous hydrogel	Strong cellular and humoral immune response for HIV
GE11 (EGFR ligand)	Self-assembling peptide nanovesicle	Targeted toward EGFR expressing cancer
Cholesterol-aK-ChaVAaWTLKAa-LEEKKGNYVVTDH	Lipopeptide micelles with self-adjuvant properties	Cellular and humoral immune response in B16-EGFRvIII tumor model

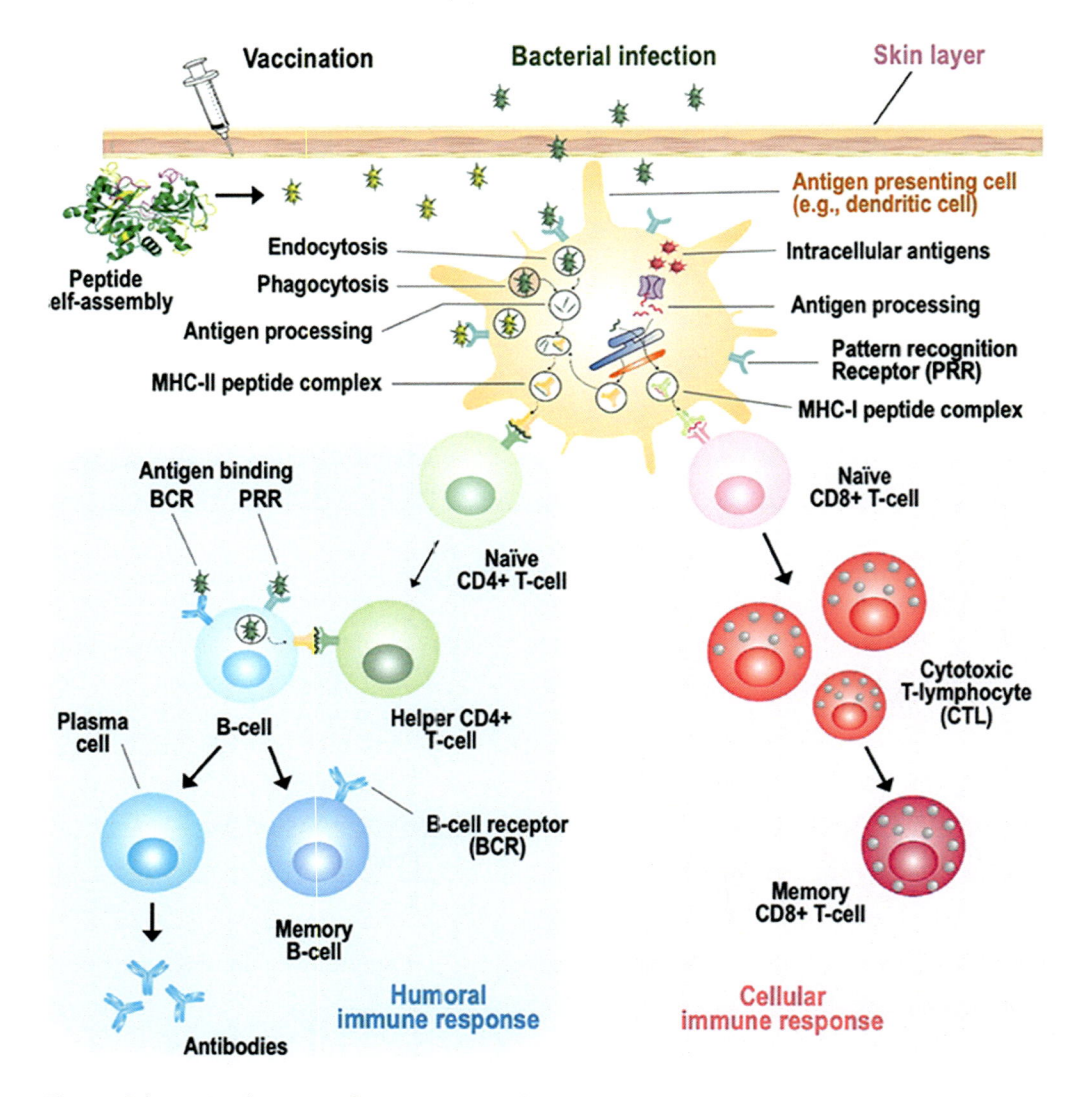

Fig. 7 Schematic diagram of immune stimulation by self-assembled vaccines. *(Reprinted with permission Creative Commons Attribution License (CC BY) from T. Abudula, K. Bhatt, L.J. Eggermont, N. O'Hare, A. Memic, S.A. Bencherif, Supramolecular self-assembled peptide-based vaccines: current state and future perspectives, Front. Chem. 8 (2020) 1–11. https://doi.org/10.3389/fchem.2020.598160.)*

self-assembled vaccines is illustrated in Fig. 7. The antigenic epitopes are an essential consideration while designing the vaccines [138]. Wang et al. prepared a peptide consisting of Fmoc-KCRGDK-based hydrogel in which BRD4 inhibitor, indocyanine green and autologous tumor cells were incorporated. Using laser irradiation, tumor–associated antigens are released from the cancer vaccine, which significantly inhibits the tumor relapse [139]. Another example of the self-assembled peptide is the nanofibrous RADA16 hydrogel. RADA16 peptide consists of alternating hydrophilic and hydrophobic amino acids, which are assembled to form a hydrogel. This peptide-based hydrogel vaccine comprises anti-PD-1 antibodies, DCs, and tumor antigens. The findings showed that vaccines stimulate and enhance T-cell immunity [140]. Tian et al. prepared a supramolecular hydrogel-based nanovector to create strong immune responses against HIV. This supramolecular nanovector is composed of NapGFFY-NMe (naphthyl acetic acid-modified tetrapeptide GFFY with C-terminal methyl amide group), the encapsulation of DNA sequence of the glycoprotein of HIV. The strong cellular and humoral immune responses were due to DNA protection from degradation, and thus enhancing the DNA transfection and gene expression [141].

6.3 Orthopedics applications

Human bone is composed of organic matrix and inorganic calcium phosphate. The matrix of the bone is called a type I collagen. Several techniques are established to mimic the formation of natural collagen [142]. The self-assembled peptides have gained interest in the field of orthopedic. The approaches are listed below.

The phosphorylation of serine prepared self-assembled amphiphilic nanofibers peptide. Mineralization of calcium phosphate (Hydroxyapatite) into the peptide nanofiber indicates hydroxyapatite crystals within the collagen fibrils [143]. Mata and his co-workers have also demonstrated the preparation of molecularly designed peptide amphiphile (PA), which further self-assembled to form nanofibers. The studies found that within 4 weeks, the bone was formed, and these hydrogels could be the potential candidate for the replacement of autografts or allografts [144]. Apart from these peptides, another most crucial process for the regeneration of the bone is by using growth factor, bone morphogenetic protein-2 (BMP-2). This biomimetic composite BMP-2 was incorporated into PLGA-(PEG-ASP)n copolymer to form bone [145]. Some examples were also found that BMP-2 enhances bone regeneration by using poly(ethylene glycol). The hydrogel is formed by applying integrin-binding RGD ligand, and for degradation of hydrogel matrix metalloproteinase (MMP) is required [146]. This indicates that bone regeneration and bone tissue engineering are vital tools for constructing the bones.

6.4 Wound healing

The tissue-engineering process for the skin was prepared for the treatment of wound healing. Loo et al. prepared two self-assembled hydrogels of ultrashort peptide for the recovery of burn wounds. These ultrashort peptides hydrogels showed faster-wound healing properties compared with Mepitel and have more extended stability at room temperature. The studies claimed that wound healing was promoted by the regeneration of the epithelial and dermal tissues without exogenous growth factors and achieved 86.2% and 92.9% wound closure in 14 days [147]. The wound healing properties by self-assembled peptides were depicted in Fig. 8.

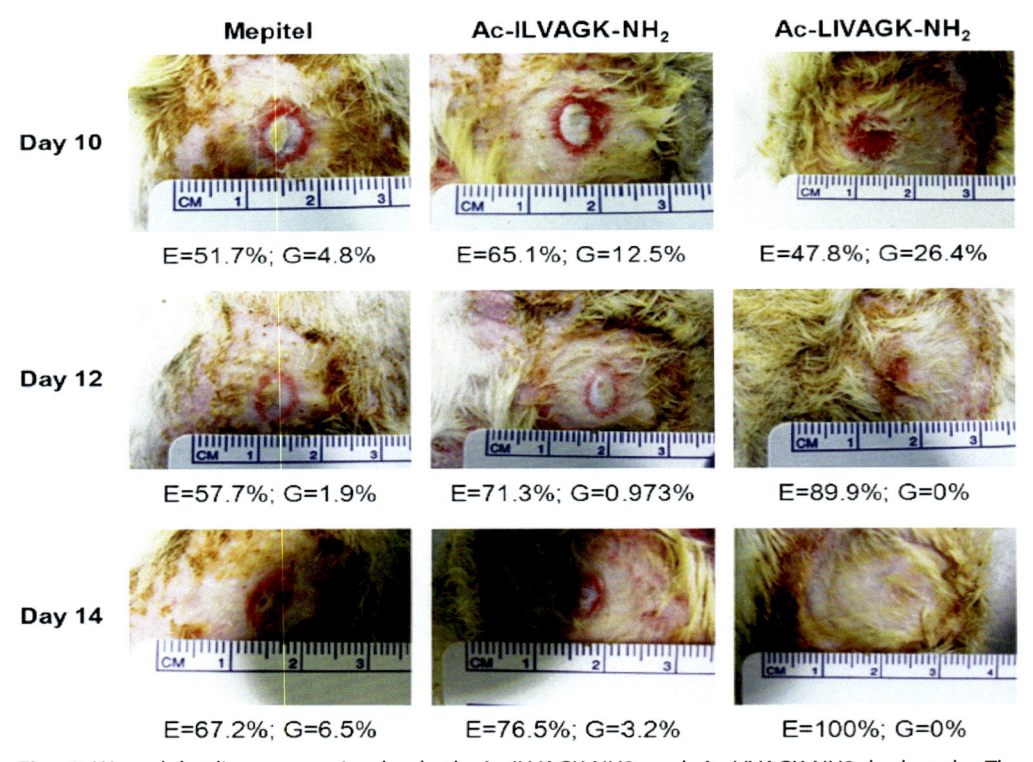

Fig. 8 Wound healing properties by both Ac-ILVAGK-NH2 and Ac-LIVAGK-NH2 hydrogels. The regenerated epidermis layer is healed completely by Ac-LIVAGK-NH2 hydrogel at 14 days. *(Reprinted with permission Y. Loo, Y.C. Wong, E.Z. Cai, C.H. Ang, A. Raju, A. Lakshmanan, et al., Ultrashort peptide nanofibrous hydrogels for the acceleration of healing of burn wounds, Biomaterials 35 (2014) 4805–4814. https://doi.org/10.1016/j.biomaterials.2014.02.047.)*

In other studies, the wound healing properties were observed for the infections. The self-assembled hydrogel releases the silver nanoparticles and prevents aggregation of these nanoparticles. These hydrogels release the nanoparticle for about 14 days and inhibit bacterial growth (both gram-positive and gram-negative bacteria). The biocompatibility studies of self-assembled hydrogels on HDFa cells indicate no effect on cell viability and no cytotoxic effect [148]. So, these self-assembled peptides could help in the wound healing process.

6.5 Pancreatic islet transplantation

Pancreatic islet cells were the most important cells for the secretion of insulin, which reduce the blood glucose level in diabetes. Glucagon-like peptide 1 (GLP-1) are the peptides that are responsible for the secretion of insulin. Researchers have prepared the self-assembled amphiphilic mimetic peptide (glucagon-like peptide 1) for the enhancement of insulin. This self-assembled system is converted into nanofibers and other forms of a macroscopic gel. The prepared nanofibers were tested in rat insulinoma (RINm5f) cells which stimulate the secretion of insulin. Due to the self-assembled system, the β-cell viability functions and the proliferation of encapsulated β-cells were significantly improved to transplant the pancreatic islet cells [149].

6.6 Tissue engineering for neurons

The nerves of the peripheral can regenerate after the injury. The use of synthetic nerve conduits is essential for the regeneration of the nerves. The development of neurons was revealed by a research group in which type I collagen and PA were used. The self-assembly of PA and collagen consist of IKVAV and YIGSR laminin epitopes. Neuronal subtypes of cerebellar cortex such as Granule cells (GC) and Purkinje cells (PC) showed a change in epitope concentration. This provides a versatile development of the neurons and the use of bioscaffolds [150]. In another study, a self-assembled peptide (sapeptide) scaffold is used for the growth of neurite. The formation of the synapse was by assembling the ionic self-complementary β-sheet oligopeptides. These self-assembled oligopeptides lead to the formation of a hydrogel. This shows the neuronal cell attachment and the neurite outgrowth, which could be helpful in tissue engineering [151].

Apart from peripheral neural repairing, spinal cord repairing is the most challenging task because axons do not regenerate [152]. Strategies were made for the recovery of spinal cord injury by the use of self-assembled nanofibers. These self-assembled nanofibers were prepared by injecting neural stem/progenitor cells (NPCs) together with QL6 peptide. The studies claimed that the engrafted NPCs preserve the motor neurons and attenuate perilesional inflammation, which resembles hydrogels' applicability in neuronal diseases [153].

6.7 Cartilage tissue engineering

Cartilage tissue engineering is required nowadays for the treatment of cartilage defects. Restoration of damaged or lost cartilages creates a new opportunity in the field of tissue engineering. Various tissue-engineered techniques were developed for the restoration process. Kisiday and his co-workers demonstrate the self-assembling peptide KLD-12 hydrogel as a 3D scaffold. These hydrogels encapsulate the chondrocytes and secrete proteoglycans and collagen type II, indicating stable chondrocyte phenotype [154]. The supramolecular design of self-assembling nanofibers was also prepared for the regeneration of the cartilage. The regeneration process was done by the transforming growth factor β-1 (TGFβ-1). The chondrogenic differentiation of human mesenchymal stem cells was also revealed by the in vivo studies [155]. These synthetic bioactive strategies have shown the regeneration process of cartilages.

6.8 Vascular tissue engineering

Cardiovascular tissue engineering methods are gaining importance nowadays in the western world. Several synthetic vascular grafts are present in the market, but these vascular tissue engineering methods are highlighted in the current scenario due to the grafts' failure. Adinarayana and co-workers developed self-assembled nanomatrices using peptides amphiphiles to evaluate the effects on adhesion and the spreading of endothelial cells and smooth muscle cells (SMCs). The studies claimed that the PA findings could replace the synthetic graft techniques [156]. In another study, tissue-engineered blood vessels were constructed for the replacement of damaged arteries. The developed PA nanofibers can form gels, and SMCs were encapsulated. This reveals that the artificial blood vessels can be regenerated [157].

6.9 Tissue engineering for liver

Tissue engineering for liver diseases is gaining importance due to the limited organ donors. Semino et al. demonstrated the differentiation properties of a putative rat liver progenitor cell line (Lig-8) self-assembled peptide scaffold. Expression of transcription factor C/EBPa, hepatocyte maturation, cytochrome P450s CYP1A1, CYP1A2, CYP2E1 and the upregulation of albumin activities are revealed by differentiated progeny cells [158]. In another study, self-assembling peptide nanofiber gels were prepared to investigate the mechanism of fibronectin-derived adhesion ligands and different modes of epidermal growth factor (EGF). This study's results claimed that the phenotypic and signaling responses were observed during the survival process to the differentiation of the cell functioning [159]. These studies proved to be promising for the regeneration of the liver tissues.

7. Future prospects

Exploration of self-assembled supramolecular structures has created a successful story in the field of therapeutics. These self-assembled systems are constructed by different types of materials, which may have biocompatibility and toxicity issues. Small molecule-based self-assembled systems may be a suitable carrier for therapeutic delivery of the payloads. Besides this, the stimuli-sensitive approaches have also proved to be promising for the delivery of cargoes. Research in self-assembled mediated cancer therapy has widened the application, but there is a need to discover self-assembled approaches for other diseases. Thus it is expected that the next-generation self-assembled supramolecular systems could be more promising to improve human healthcare and minimize the problems associated with biocompatibility and toxicity; meanwhile, a more significant number of candidates have received federal approval and reached the clinic for human benefits.

References

[1] S. Bayda, M. Adeel, T. Tuccinardi, M. Cordani, F. Rizzolio, The history of nanoscience and nanotechnology: from chemical-physical applications to nanomedicine, Molecules 25 (2020) 112, https://doi.org/10.3390/molecules25010112.

[2] J. Jeevanandam, A. Barhoum, Y.S. Chan, A. Dufresne, M.K. Danquah, Review on nanoparticles and nanostructured materials: history, sources, toxicity and regulations, Beilstein J. Nanotechnol. 9 (2018) 1050–1074, https://doi.org/10.3762/bjnano.9.98.

[3] D. Mehta, S. Guvva, M. Patil, Future impact of nanotechnology on medicine and dentistry, J. Indian Soc. Periodontol. 12 (2008) 34, https://doi.org/10.4103/0972-124x.44088.

[4] M. Mishra, P. Kumar, J.S. Rajawat, R. Malik, G. Sharma, A. Modgil, Nanotechnology: revolutionizing the science of drug delivery, Curr. Pharm. Des. 24 (2018) 5086–50107, https://doi.org/10.2174/1381612825666190206222415.

[5] K. Raza, Nanotechnology-based drug delivery products: need, design, pharmacokinetics and regulations, Curr. Pharm. Des. 24 (2019) 5085, https://doi.org/10.2174/1381612824431903280859117.

[6] J. Shi, A.R. Votruba, O.C. Farokhzad, R. Langer, Nanotechnology in drug delivery and tissue engineering: from discovery to applications, Nano Lett. 10 (2010) 3223–3230, https://doi.org/10.1021/nl102184c.

[7] G. Boehm, L. Yao, L. Han, Q. Zheng, Development of the generic drug industry in the US after the Hatch–Waxman Act of 1984, Acta Pharm. Sin. B 3 (2013) 297–311, https://doi.org/10.1016/j.apsb.2013.07.004.

[8] D. Chenthamara, S. Subramaniam, S.G. Ramakrishnan, S. Krishnaswamy, M.M. Essa, F.H. Lin, et al., Therapeutic efficacy of nanoparticles and routes of administration, Biomater. Res. 23 (2019), https://doi.org/10.1186/s40824-019-0166-x.

[9] A. Dadwal, A. Baldi, N.R. Kumar, Nanoparticles as carriers for drug delivery in cancer, Artif. Cells Nanomed. Biotechnol. 46 (2018) 295–305, https://doi.org/10.1080/21691401.2018.1457039.

[10] P. Kumar, K. Raza, L. Kaushik, R. Malik, S. Arora, O.P. Katare, Role of colloidal drug delivery carriers in taxane-mediated chemotherapy: a review, Curr. Pharm. Des. 22 (2016) 5127–5143, https://doi.org/10.2174/1381612822666160524.

[11] Y. Zhang, M. Li, X. Gao, Y. Chen, T. Liu, Nanotechnology in cancer diagnosis: progress, challenges and opportunities, J. Hematol. Oncol. 12 (2019) 137, https://doi.org/10.1186/s13045-019-0833-3.

[12] M. Chamundeeswari, J. Jeslin, M.L. Verma, Nanocarriers for drug delivery applications, Environ. Chem. Lett. 17 (2019) 849–865, https://doi.org/10.1007/s10311-018-00841-1.

[13] O. Katare, K. Raza, B. Singh, S. Dogra, Novel drug delivery systems in topical treatment of psoriasis: rigors and vigors, Indian J. Dermatol. Venereol. Leprol. 76 (2010) 612–621, https://doi.org/10.4103/0378-6323.72451.

[14] S. Arpicco, L. Battaglia, P. Brusa, R. Cavalli, D. Chirio, F. Dosio, et al., Recent studies on the delivery of hydrophilic drugs in nanoparticulate systems, J. Drug Delivery Sci. Technol. 32 (2016) 298–312, https://doi.org/10.1016/j.jddst.2015.09.004.

[15] S. Bamrungsap, Z. Zhao, T. Chen, L. Wang, C. Li, T. Fu, et al., Nanotechnology in therapeutics: a focus on nanoparticles as a drug delivery system, Nanomedicine 7 (2012) 1253–1271, https://doi.org/10.2217/nnm.12.87.

[16] M.L. Bondì, E.F. Craparo, G. Giammona, M. Cervello, A. Azzolina, P. Diana, et al., Nanostructured lipid carriers-containing anticancer compounds: preparation, characterization, and cytotoxicity studies, Drug Deliv. 14 (2007) 61–67, https://doi.org/10.1080/10717540600739914.

[17] K. Raza, M. Kumar, P. Kumar, R. Malik, G. Sharma, M. Kaur, et al., Topical delivery of aceclofenac: challenges and promises of novel drug delivery systems, Biomed. Res. Int. 2014 (2014), https://doi.org/10.1155/2014/406731.

[18] B. Mukherjee, B. Satapathy, L. Mondal, N. Dey, R. Maji, Potentials and challenges of active targeting at the tumor cells by engineered polymeric nanoparticles, Curr. Pharm. Biotechnol. 14 (2014) 1250–1263, https://doi.org/10.2174/1389201015666140608143235.

[19] Y. Zhong, F. Meng, C. Deng, Z. Zhong, Ligand-directed active tumor-targeting polymeric nanoparticles for cancer chemotherapy, Biomacromolecules 15 (2014) 1955–1969, https://doi.org/10.1021/bm5003009.

[20] H. Hillaireau, P. Couvreur, Nanocarriers' entry into the cell: relevance to drug delivery, Cell. Mol. Life Sci. 66 (2009) 2873–2896, https://doi.org/10.1007/s00018-009-0053-z.

[21] A.Z. Wang, F. Gu, L. Zhang, J.M. Chan, A. Radovic-Moreno, M.R. Shaikh, et al., Biofunctionalized targeted nanoparticles for therapeutic applications, Expert Opin. Biol. Ther. 8 (2008) 1063–1070, https://doi.org/10.1517/14712598.8.8.1063.

[22] A. Trompetero, A. Gordillo, M. Carolina Del Pilar, V.M. Cristina, R.H. Bustos Cruz, Alzheimer's disease and Parkinson's disease: a review of current treatment adopting a nanotechnology approach, Curr. Pharm. Des. 24 (2018) 22–45, https://doi.org/10.2174/1381612823666170828133059.

[23] M. Nazıroğlu, S. Muhamad, L. Pecze, Nanoparticles as potential clinical therapeutic agents in Alzheimer's disease: focus on selenium nanoparticles, Expert. Rev. Clin. Pharmacol. 10 (2017) 773–782, https://doi.org/10.1080/17512433.2017.1324781. Taylor Fr.

[24] T. Sun, Y.S. Zhang, B. Pang, D.C. Hyun, M. Yang, Y. Xia, Engineered nanoparticles for drug delivery in cancer therapy, Angew. Chem. Int. Ed. 53 (2014) 12320–12364, https://doi.org/10.1002/anie.201403036.

[25] J. Li, C. Fan, H. Pei, J. Shi, Q. Huang, Smart drug delivery nanocarriers with self-assembled DNA nanostructures, Adv. Mater. 25 (2013) 4386–4396, https://doi.org/10.1002/adma.201300875.

[26] W. Cui, J. Li, G. Decher, Self-assembled smart nanocarriers for targeted drug delivery, Adv. Mater. 28 (2016) 1302–1311, https://doi.org/10.1002/adma.201502479.

[27] N.A. Peppas, J.H. Ward, Biomimetic materials and micropatterned structures using iniferters, Adv. Drug Deliv. Rev. 56 (2004) 1587–1597, https://doi.org/10.1016/j.addr.2003.10.046.

[28] R.S. Tu, M. Tirrell, Bottom-up design of biomimetic assemblies, Adv. Drug Deliv. Rev. 56 (2004) 1537–1563, https://doi.org/10.1016/j.addr.2003.10.047.

[29] H.E. Van Kan-Davelaar, J.C.M. Van Hest, J.J.L.M. Cornelissen, M.S.T. Koay, Using viruses as nanomedicines, Br. J. Pharmacol. 171 (2014) 4001–4009, https://doi.org/10.1111/bph.12662.

[30] X. Zhang, C. Gong, O.U. Akakuru, Z. Su, A. Wu, G. Wei, The design and biomedical applications of self-assembled two-dimensional organic biomaterials, Chem. Soc. Rev. 48 (2019) 5564–5595, https://doi.org/10.1039/c8cs01003j.

[31] Y. Chen, A.A. Orr, K. Tao, Z. Wang, A. Ruggiero, L.J.W. Shimon, et al., High-efficiency fluorescence through bioinspired supramolecular self-assembly, Am. Chem. Soc. 14 (2020) 2798–2807, https://doi.org/10.1021/acsnano.9b10024.

[32] S.I. Stupp, L.C. Palmer, Supramolecular chemistry and self-assembly in organic materials design, Chem. Mater. 26 (2014) 507–518, https://doi.org/10.1021/cm403028b.

[33] N. Stephanopoulos, J.H. Ortony, S.I. Stupp, Self-assembly for the synthesis of functional biomaterials, Acta Mater. 61 (2013) 912–930, https://doi.org/10.1016/j.actamat.2012.10.046.

[34] E. Piccinini, D. Pallarola, F. Battaglini, O. Azzaroni, Self-limited self-assembly of nanoparticles into supraparticles: towards supramolecular colloidal materials by design, Mol. Syst. Des. Eng. 1 (2016) 155–162, https://doi.org/10.1039/c6me00016a.

[35] W. Zhang, S. Mo, M. Liu, L. Liu, L. Yu, C. Wang, Rationally designed protein building blocks for programmable hierarchical architectures, Front. Chem. 8 (2020) 587975, https://doi.org/10.3389/fchem.2020.587975.

[36] D. Philp, J. Fraser Stoddart, Self-assembly in natural and unnatural systems, Angew. Chem. Int. Ed. Engl. 35 (1996) 1154–1196, https://doi.org/10.1002/anie.199611541.

[37] G.A. Ozin, K. Hou, B.V. Lotsch, L. Cademartiri, D.P. Puzzo, F. Scotognella, et al., Nanofabrication by self-assembly, Mater. Today 12 (2009) 12–23, https://doi.org/10.1016/S1369-7021(09)70156-7.

[38] J.N. Israelachvili, D.J. Mitchell, B.W. Ninham, Theory of self-assembly of lipid bilayers and vesicles, Biomembranes 470 (1977) 185–201, https://doi.org/10.1016/0005-2736(77)90099-2.

[39] D.J. Wales, J.A. Kitchen, Surface-based molecular self-assembly: Langmuir-Blodgett films of amphiphilic Ln(III) complexes, Chem. Cent. J. 10 (2016) 72, https://doi.org/10.1186/s13065-016-0224-6.

[40] W. Bigelow, D. Pickett, W.A. Zisman, Oleophobic monolayers: I. films adsorbed from solution in non-polar liquids, J. Colloid Sci. 1 (1946) 513–538.

[41] P.E. Laibinis, G.M. Whitesides, D.L. Allara, Y.-T. Tao, A.N. Parikh, R.G. Nuzzo, Comparison of the structures and wetting properties of self-assembled monolayers of n-alkanethiols on the coinage metal surfaces, Cu, Ag, Au1, J. Am. Chem. Soc. 113 (1991) 7152–7167.

[42] B. Olenyuk, J.A. Whiteford, A. Fechtenkötter, P.J. Stang, Self-assembly of nanoscale cuboctahedra by coordination chemistry, Nature 398 (1999) 796–799, https://doi.org/10.1038/19740.

[43] M.T. Pope, A. Müller, Introduction to polyoxometalate chemistry: from topology via self-assembly to applications, in: Polyoxometalate Chemistry From Topology via Self-Assembly to Applications, Kluwer Academic Publishers, 2006, pp. 1–6, https://doi.org/10.1007/0-306-47625-8_1.

[44] B.A. Grzybowski, K. Fitzner, J. Paczesny, S. Granick, From dynamic self-assembly to networked chemical systems, Chem. Soc. Rev. 46 (2017) 5647–5678, https://doi.org/10.1039/c7cs00089h.

[45] C. Zhou, P. Yue, J.J. Feng, Dynamic simulation of droplet interaction and self-assembly in a nematic liquid crystal, Langmuir 24 (2008) 3099–3110, https://doi.org/10.1021/la703312f.

[46] J. Fan, X. Xu, W. Yu, Z. Wei, D. Zhang, Hydrogen-bond-driven supramolecular self-assembly of diacetylene derivatives for topochemical polymerization in solution, Polym. Chem. 11 (2020) 1947–1953, https://doi.org/10.1039/c9py01745c.

[47] Q. Li, Y. Jia, L. Dai, Y. Yang, J. Li, Controlled rod nanostructured assembly of diphenylalanine and their optical waveguide properties, Am. Chem. Soc. 9 (2015) 2689–2695, https://doi.org/10.1021/acsnano.5b00623.

[48] P. Li, X. Chen, W. Yang, Graphene-induced self-assembly of peptides into macroscopic-scale organized nanowire arrays for electrochemical NADH sensing, Langmuir 29 (2013) 8629–8635, https://doi.org/10.1021/la401881a.

[49] P. Passaretti, Y. Sun, T.R. Dafforn, P.G. Oppenheimer, Determination and characterisation of the surface charge properties of the bacteriophage M13 to assist bio-nanoengineering, RSC Adv. 10 (2020) 25385–25392, https://doi.org/10.1039/d0ra04086j.

[50] T. Xue, X. Cui, W. Guan, Q. Wang, C. Liu, H. Wang, et al., Surface plasmon resonance technique for directly probing the interaction of DNA and graphene oxide and ultra-sensitive biosensing, Biosens. Bioelectron. 58 (2014) 374–379, https://doi.org/10.1016/j.bios.2014.03.002.

[51] D. Li, W. Zhang, X. Yu, Z. Wang, Z. Su, G. Wei, When biomolecules meet graphene: from molecular level interactions to material design and applications, Nanoscale 8 (2016) 19491–19509, https://doi.org/10.1039/c6nr07249f.

[52] J. Wang, X. Zhao, J. Li, X. Kuang, Y. Fan, G. Wei, et al., Electrostatic assembly of peptide nanofiber-biomimetic silver nanowires onto graphene for electrochemical sensors, ACS Macro Lett. 3 (2014) 529–533, https://doi.org/10.1021/mz500213w.

[53] R. Yang, W.S. Shim, F. De Cui, G. Cheng, X. Han, Q.R. Jin, et al., Enhanced electrostatic interaction between chitosan-modified PLGA nanoparticle and tumor, Int. J. Pharm. 371 (2009) 142–147, https://doi.org/10.1016/j.ijpharm.2008.12.007.

[54] D. Wingett, P. Louka, C.B. Anders, J. Zhang, A. Punnoose, A role of ZnO nanoparticle electrostatic properties in cancer cell cytotoxicity, Nanotechnol. Sci. Appl. 9 (2016) 29–45, https://doi.org/10.2147/NSA.S99747.

[55] Y.L. Tan, C.G. Liu, Preparation and characterization of self-assembled nanoparticles based on folic acid modified carboxymethyl chitosan, J. Mater. Sci. Mater. Med. 22 (2011) 1213–1220, https://doi.org/10.1007/s10856-011-4302-y.

[56] H. Xu, J. Wang, S. Han, J. Wang, D. Yu, H. Zhang, et al., Hydrophobic-region-induced transitions in self-assembled peptide nanostructures, Langmuir 25 (2009) 4115–4123, https://doi.org/10.1021/la802499n.

[57] J. Zhao, F. Pan, P. Li, C. Zhao, Z. Jiang, P. Zhang, et al., Fabrication of ultrathin membrane via layer-by-layer self-assembly driven by hydrophobic interaction towards high separation performance, ACS Appl. Mater. Interfaces 5 (2013) 13275–13283, https://doi.org/10.1021/am404268z.

[58] Y. Cai, F. Liu, X. Ma, X. Yang, H. Zhao, Hydrophobic interaction-induced coassembly of homopolymers and proteins, Langmuir 35 (2019) 10958–10964, https://doi.org/10.1021/acs.langmuir.9b01749.

[59] Q. Zou, M. Abbas, L. Zhao, S. Li, G. Shen, X. Yan, Biological photothermal nanodots based on self-assembly of peptide-porphyrin conjugates for antitumor therapy, J. Am. Chem. Soc. 139 (2017) 1921–1927, https://doi.org/10.1021/jacs.6b11382.

[60] J. Jin, T. Iyoda, C. Cao, Y. Song, L. Jiang, et al., Self-assembly of uniform spherical aggregates of magnetic nanoparticles through π–π interactions, Angew. Chem. 113 (2001) 2193–2196.

[61] L. Wang, Y. Zhang, A. Wu, G. Wei, Designed graphene-peptide nanocomposites for biosensor applications: a review, Anal. Chim. Acta 985 (2017) 24–40, https://doi.org/10.1016/j.aca.2017.06.054.

[62] M. Li, Y. Pan, X. Guo, Y. Liang, Y. Wu, Y. Wen, et al., Pt/single-stranded DNA/graphene nanocomposite with improved catalytic activity and CO tolerance, J. Mater. Chem. A 3 (2015) 10353–10359, https://doi.org/10.1039/c5ta00891c.

[63] G. Verma, P.A. Hassan, Self assembled materials: design strategies and drug delivery perspectives, Phys. Chem. Chem. Phys. 15 (2013) 17016–17028, https://doi.org/10.1039/c3cp51207j.

[64] B. Felice, M.P. Prabhakaran, A.P. Rodríguez, S. Ramakrishna, Drug delivery vehicles on a nano-engineering perspective, Mater. Sci. Eng. C 41 (2014) 178–195, https://doi.org/10.1016/j.msec.2014.04.049.

[65] D. Bobo, K.J. Robinson, J. Islam, K.J. Thurecht, S.R. Corrie, Nanoparticle-based medicines: a review of FDA-approved materials and clinical trials to date, Pharm. Res. 33 (2016) 2373–2387, https://doi.org/10.1007/s11095-016-1958-5.

[66] Y. Bae, W.D. Jang, N. Nishiyama, S. Fukushima, K. Kataoka, Multifunctional polymeric micelles with folate-mediated cancer cell targeting and pH-triggered drug releasing properties for active intracellular drug delivery, Mol. Biosyst. 1 (2005) 242–250, https://doi.org/10.1039/b500266d.

[67] E.S. Lee, K. Na, Y.M. Bae, Super pH-sensitive multifunctional polymeric micelle, Nano Lett. 5 (2005) 325–329, https://doi.org/10.1021/nl0479987.

[68] X.W. Wei, C.Y. Gong, S. Shi, S.Z. Fu, K. Men, S. Zeng, et al., Self-assembled honokiol-loaded micelles based on poly(ε-caprolactone)-poly(ethylene glycol)-poly(ε-caprolactone) copolymer, Int. J. Pharm. 369 (2009) 170–175, https://doi.org/10.1016/j.ijpharm.2008.10.027.

[69] J. Liu, H. Li, X. Jiang, C. Zhang, Q. Ping, Novel pH-sensitive chitosan-derived micelles loaded with paclitaxel, Carbohydr. Polym. 82 (2010) 432–439, https://doi.org/10.1016/j.carbpol.2010.04.084.

[70] N. Duceppe, M. Tabrizian, Advances in using chitosan-based nanoparticles for in vitro and in vivo drug and gene delivery, Expert Opin. Drug Deliv. 7 (2010) 1191–1207, https://doi.org/10.1517/17425247.2010.514604.

[71] X. Xiangyang, L. Ling, Z. Jianping, L. Shiyue, Y. Jie, Y. Xiaojin, et al., Preparation and characterization of N-succinyl-N′-octyl chitosan micelles as doxorubicin carriers for effective anti-tumor activity, Colloids Surf. B Biointerfaces 55 (2007) 222–228, https://doi.org/10.1016/j.colsurfb.2006.12.006.

[72] C. Liu, W. Fan, X. Chen, C. Liu, X. Meng, H.J. Park, Self-assembled nanoparticles based on linoleic-acid modified carboxymethyl-chitosan as carrier of adriamycin (ADR), Curr. Appl. Phys. 7 (2007) 125–129, https://doi.org/10.1016/j.cap.2006.11.031.

[73] N. Thotakura, M. Dadarwal, P. Kumar, G. Sharma, S.K. Guru, S. Bhushan, et al., Chitosan-stearic acid based polymeric micelles for the effective delivery of tamoxifen: cytotoxic and pharmacokinetic evaluation, AAPS PharmSciTech 18 (2017) 759–768, https://doi.org/10.1208/s12249-016-0563-6.

[74] F.Q. Hu, G.F. Ren, H. Yuan, Y.Z. Du, S. Zeng, Shell cross-linked stearic acid grafted chitosan oligosaccharide self-aggregated micelles for controlled release of paclitaxel, Colloids Surf. B Biointerfaces 50 (2006) 97–103, https://doi.org/10.1016/j.colsurfb.2006.04.009.

[75] C.K. Thakur, N. Thotakura, R. Kumar, P. Kumar, B. Singh, D. Chitkara, et al., Chitosan-modified PLGA polymeric nanocarriers with better delivery potential for tamoxifen, Int. J. Biol. Macromol. 93 (2016) 381–389, https://doi.org/10.1016/j.ijbiomac.2016.08.080.

[76] N. Thotakura, M. Dadarwal, R. Kumar, B. Singh, G. Sharma, P. Kumar, et al., Chitosan–palmitic acid based polymeric micelles as promising carrier for circumventing pharmacokinetic and drug delivery concerns of tamoxifen, Int. J. Biol. Macromol. 102 (2017) 1220–1225, https://doi.org/10.1016/j.ijbiomac.2017.05.016.

[77] R.O. de Pedro, S. Pereira, F.M. Goycoolea, C.C. Schmitt, M.G. Neumann, Self-aggregated nanoparticles of N-dodecyl,N′-glycidyl(chitosan) as pH-responsive drug delivery systems for quercetin, J. Appl. Polym. Sci. 135 (2018) 1–12, https://doi.org/10.1002/app.45678.

[78] D.N. Kapoor, A. Bhatia, R. Kaur, R. Sharma, G. Kaur, S. Dhawan, PLGA: a unique polymer for drug delivery, Ther. Deliv. 6 (2015) 41–58, https://doi.org/10.4155/tde.14.91.

[79] S. Rezvantalab, M.M. Keshavarz, Microfluidic assisted synthesis of PLGA drug delivery systems, RSC Adv. 9 (2019) 2055–2072, https://doi.org/10.1039/C8RA08972H.

[80] M. Marimuthu, D. Bennet, S. Kim, Self-assembled nanoparticles of PLGA-conjugated glucosamine as a sustained transdermal drug delivery vehicle, Polym. J. 45 (2013) 202–209, https://doi.org/10.1038/pj.2012.103.

[81] K. Zhang, X. Tang, J. Zhang, W. Lu, X. Lin, Y. Zhang, et al., PEG-PLGA copolymers: their structure and structure-influenced drug delivery applications, J. Control. Release 183 (2014) 77–86, https://doi.org/10.1016/j.jconrel.2014.03.026.

[82] M. Ashjari, S. Khoee, A.R. Mahdavian, R. Rahmatolahzadeh, Self-assembled nanomicelles using PLGA-PEG amphiphilic block copolymer for insulin delivery: a physicochemical investigation and determination of CMC values, J. Mater. Sci. Mater. Med. 23 (2012) 943–953, https://doi.org/10.1007/s10856-012-4562-1.

[83] Z. Song, R. Feng, M. Sun, C. Guo, Y. Gao, L. Li, et al., Curcumin-loaded PLGA-PEG-PLGA triblock copolymeric micelles: preparation, pharmacokinetics and distribution in vivo, J. Colloid Interface Sci. 354 (2011) 116–123, https://doi.org/10.1016/j.jcis.2010.10.024.

[84] G. Zhao, S. Chandrudu, M. Skwarczynski, I. Toth, The application of self-assembled nanostructures in peptide-based subunit vaccine development, Eur. Polym. J. 93 (2017) 670–681, https://doi.org/10.1016/j.eurpolymj.2017.02.014.

[85] T.-Y. Liu, W.M. Hussein, Z. Jia, Z.M. Ziora, N.A.J. McMillan, M.J. Monteiro, et al., Self-adjuvanting polymer–peptide conjugates as therapeutic vaccine candidates against cervical cancer, Biomacromolecules 14 (2013) 2798–2806, https://doi.org/10.1021/bm400626w.

[86] C.Y. Ang, S.Y. Tan, X. Wang, Q. Zhang, M. Khan, L. Bai, et al., Supramolecular nanoparticle carriers self-assembled from cyclodextrin- and adamantane-functionalized polyacrylates for tumor-targeted drug delivery, J. Mater. Chem. B 2 (2014) 1879–1890, https://doi.org/10.1039/c3tb21325k.

[87] H.F. Alharbi, M. Luqman, K.A. Khalil, Y.A. Elnakady, O.H. Abd-Elkader, A.M. Rady, et al., Fabrication of core-shell structured nanofibers of poly (lactic acid) and poly (vinyl alcohol) by coaxial electrospinning for tissue engineering, Eur. Polym. J. 98 (2018) 483–491, https://doi.org/10.1016/j.eurpolymj.2017.11.052.

[88] U. Manna, S. Patil, Borax mediated layer-by-layer self-assembly of neutral poly(vinyl alcohol) and chitosan, J. Phys. Chem. B 113 (2009) 9137–9142, https://doi.org/10.1021/jp9025333.

[89] Q. Luo, Z. Dong, C. Hou, J. Liu, Protein-based supramolecular polymers: progress and prospect, Chem. Commun. 50 (2014) 9997, https://doi.org/10.1039/C4CC03143A.

[90] M.R. Dreher, D. Raucher, N. Balu, O.M. Colvin, S.M. Ludeman, A. Chilkoti, Evaluation of an elastin-like polypeptide-doxorubicin conjugate for cancer therapy, J. Control. Release 91 (2003) 31–43, https://doi.org/10.1016/S0168-3659(03)00216-5.

[91] G.H. Sagar, M.A. Arunagirinathan, J.R. Bellare, Self-assembled surfactant nano-structures important in drug delivery: a review, Indian J. Exp. Biol. 45 (2007) 133–159.

[92] D. Lombardo, M.A. Kiselev, S. Magazù, P. Calandra, Amphiphiles self-assembly: basic concepts and future perspectives of supramolecular approaches, Adv. Condens. Matter Phys. 2015 (2015), https://doi.org/10.1155/2015/151683.

[93] R. Nagarajan, Molecular packing parameter and surfactant self-assembly: the neglected role of the surfactant tail, Langmuir 18 (2002) 31–38, https://doi.org/10.1021/la010831y.

[94] R. Kashapov, G. Gaynanova, D. Gabdrakhmanov, D. Kuznetsov, R. Pavlov, K. Petrov, et al., Self-assembly of amphiphilic compounds as a versatile tool for construction of nanoscale drug carriers, Int. J. Mol. Sci. 21 (2020) 1–47, https://doi.org/10.3390/ijms21186961.

[95] A.B. Mirgorodskaya, R.A. Kushnazarova, S.S. Lukashenko, L.Y. Zakharova, Self-assembly of mixed systems based on nonionic and carbamate-bearing cationic surfactants as a tool for fabrication of biocompatible nanocontainers, J. Mol. Liq. 292 (2019) 111407, https://doi.org/10.1016/j.molliq.2019.111407.

[96] R.R. Kashapov, S.V. Kharlamov, Y.S. Razuvayeva, A.Y. Ziganshina, I.R. Nizameev, M.K. Kadirov, et al., Supramolecular assemblies involving calix[4]resorcinol and surfactant with pH-induced morphology transition for drug encapsulation, J. Mol. Liq. 261 (2018) 218–224, https://doi.org/10.1016/j.molliq.2018.04.018.

[97] J.D. Du, W.K. Fong, S. Salentinig, S.M. Caliph, A. Hawley, B.J. Boyd, Phospholipid-based self-assembled mesophase systems for light-activated drug delivery, Phys. Chem. Chem. Phys. 17 (2015) 14021–14027, https://doi.org/10.1039/c5cp01229e.

[98] D.R. Telange, N.K. Sohail, A.T. Hemke, P.S. Kharkar, A.M. Pethe, Phospholipid complex-loaded self-assembled phytosomal soft nanoparticles: evidence of enhanced solubility, dissolution rate, ex vivo permeability, oral bioavailability, and antioxidant potential of mangiferin, Drug Deliv. Transl. Res. (2020), https://doi.org/10.1007/s13346-020-00822-4.

[99] M.S. Freag, W.M. Saleh, O.Y. Abdallah, Self-assembled phospholipid-based phytosomal nanocarriers as promising platforms for improving oral bioavailability of the anticancer celastrol, Int. J. Pharm. 535 (2018) 18–26, https://doi.org/10.1016/j.ijpharm.2017.10.053.

[100] F. Yu, Y. Li, Q. Chen, Y. He, H. Wang, L. Yang, et al., Monodisperse microparticles loaded with the self-assembled berberine-phospholipid complex-based phytosomes for improving oral bioavailability and enhancing hypoglycemic efficiency, Eur. J. Pharm. Biopharm. 103 (2016) 136–148, https://doi.org/10.1016/j.ejpb.2016.03.019.

[101] Y. Li, H. Wu, X. Yang, M. Jia, Y. Li, Y. Huang, et al., Mitomycin C-soybean phosphatidylcholine complex-loaded self-assembled PEG-lipid-PLA hybrid nanoparticles for targeted drug delivery and dual-controlled drug release, Mol. Pharm. 11 (2014) 2915–2927, https://doi.org/10.1021/mp500254j.

[102] A.B. Ebrahim Attia, Z.Y. Ong, J.L. Hedrick, P.P. Lee, P.L.R. Ee, P.T. Hammond, et al., Mixed micelles self-assembled from block copolymers for drug delivery, Curr. Opin. Colloid Interface Sci. 16 (2011) 182–194, https://doi.org/10.1016/j.cocis.2010.10.003.

[103] V. Alakhov, E. Klinski, S. Li, G. Pietrzynski, A. Venne, E. Batrakova, et al., Block copolymer-based formulation of doxorubicin. From cell screen to clinical trials, Colloids Surf. B Biointerfaces 16 (1999) 113–134, https://doi.org/10.1016/S0927-7765(99)00064-8.

[104] C.L. Lo, K.M. Lin, C.K. Huang, G.H. Hsiue, Self-assembly of a micelle structure from graft and diblock copolymers: an example of overcoming the limitations of polyions in drug delivery, Adv. Funct. Mater. 16 (2006) 2309–2316, https://doi.org/10.1002/adfm.200500627.

[105] V. Belgamwar, M. Pandey, D. Chauk, S. Surana, Pluronic lecithin organogel, Asian J. Pharm. 2 (2008) 134, https://doi.org/10.4103/0973-8398.43295.

[106] A.M. Mohammed, W. Faisal, K.I. Saleh, S.K. Osman, Self-assembling organogels based on pluronic and lecithin for sustained release of Etodolac: in vitro and in vivo correlation, Curr. Drug Deliv. 14 (2016) 926–934, https://doi.org/10.2174/1567201813666160902151514.

[107] M. Pandey, V. Belgamwar, S. Gattani, S. Surana, A. Tekade, Pluronic lecithin organogel as a topical drug delivery system, Drug Deliv. 17 (2010) 38–47, https://doi.org/10.3109/10717540903508961.

[108] V. Agrawal, V. Gupta, S. Ramteke, P. Trivedi, Preparation and evaluation of tubular micelles of pluronic lecithin organogel for transdermal delivery of sumatriptan, AAPS PharmSciTech 11 (2010) 1718–1725, https://doi.org/10.1208/s12249-010-9540-7.

[109] C.E. Chang, C.M. Hsieh, L.C. Chen, C.Y. Su, D.Z. Liu, H.J. Jhan, et al., Novel application of pluronic lecithin organogels (PLOs) for local delivery of synergistic combination of docetaxel and cisplatin to improve therapeutic efficacy against ovarian cancer, Drug Deliv. 25 (2018) 632–643, https://doi.org/10.1080/10717544.2018.1440444.

[110] A. Bhatia, B. Singh, K. Raza, S. Wadhwa, O.P. Katare, Tamoxifen-loaded lecithin organogel (LO) for topical application: development, optimization and characterization, Int. J. Pharm. 444 (2013) 47–59, https://doi.org/10.1016/j.ijpharm.2013.01.029.

[111] A. Albisa, L. Espanol, M. Prieto, V. Sebastian, Polymeric nanomaterials as nanomembrane entities for biomolecule and drug delivery, Curr. Pharm. Des. 23 (2016) 263–280, https://doi.org/10.2174/1381612822666161010111741.

[112] J.S. Lee, J. Feijen, Polymersomes for drug delivery: design, formation and characterization, J. Control. Release 161 (2012) 473–483, https://doi.org/10.1016/j.jconrel.2011.10.005.

[113] F. Ahmed, R.I. Pakunlu, A. Brannan, F. Bates, T. Minko, D.E. Discher, Biodegradable polymersomes loaded with both paclitaxel and doxorubicin permeate and shrink tumors, inducing apoptosis in proportion to accumulated drug, J. Control. Release 116 (2006) 150–158, https://doi.org/10.1016/j.jconrel.2006.07.012.

[114] P. Kulkarni, M.K. Haldar, S. You, Y. Choi, S. Mallik, Hypoxia-responsive polymersomes for drug delivery to hypoxic pancreatic cancer cells, Biomacromolecules 17 (2016) 2507–2513, https://doi.org/10.1021/acs.biomac.6b00350.

[115] F. Bian, L. Jia, W. Yu, M. Liu, Self-assembled micelles of N-phthaloylchitosan-g-polyvinylpyrrolidone for drug delivery, Carbohydr. Polym. 76 (2009) 454–459, https://doi.org/10.1016/j.carbpol.2008.11.008.

[116] P. Opanasopit, T. Ngawhirunpat, A. Chaidedgumjorn, T. Rojanarata, A. Apirakaramwong, S. Phongying, et al., Incorporation of camptothecin into N-phthaloyl chitosan-g-mPEG self-assembly micellar system, Eur. J. Pharm. Biopharm. 64 (2006) 269–276, https://doi.org/10.1016/j.ejpb.2006.06.001.

[117] R. Trivedi, U.B. Kompella, Nanomicellar formulations for sustained drug delivery: strategies and underlying principles, Nanomedicine 5 (2010) 485–505, https://doi.org/10.2217/nnm.10.10.

[118] E. Allémann, J.C. Leroux, R. Gurny, E. Doelker, In vitro extended-release properties of drug-loaded poly(DL-lactic acid) nanoparticles produced by a salting-out procedure, Pharm. Res. 10 (1993) 1732–1737, https://doi.org/10.1023/A:1018970030327.

[119] B.A. Yegin, J.P. Benoît, A. Lamprecht, Paclitaxel-loaded lipid nanoparticles prepared by solvent injection or ultrasound emulsification, Drug Dev. Ind. Pharm. 32 (2006) 1089–1094, https://doi.org/10.1080/03639040600683501.

[120] I.D. Rosca, F. Watari, M. Uo, Microparticle formation and its mechanism in single and double emulsion solvent evaporation, J. Control. Release 99 (2004) 271–280, https://doi.org/10.1016/j.jconrel.2004.07.007.

[121] S. Yadav, A.K. Sharma, P. Kumar, Nanoscale self-assembly for therapeutic delivery, Front. Bioeng. Biotechnol. 8 (2020) 1–24, https://doi.org/10.3389/fbioe.2020.00127.

[122] S. Soukasene, D.J. Toft, T.J. Moyer, H. Lu, H.K. Lee, S.M. Standley, et al., Antitumor activity of peptide amphiphile nanofiber-encapsulated camptothecin, ACS Nano 5 (2011) 9113–9121, https://doi.org/10.1021/nn203343z.

[123] M.A. Khan, M. Zafaryab, S.H. Mehdi, J. Quadri, M.M.A. Rizvi, Characterization and carboplatin loaded chitosan nanoparticles for the chemotherapy against breast cancer in vitro studies, Int. J. Biol. Macromol. 97 (2017) 115–122, https://doi.org/10.1016/j.ijbiomac.2016.12.090.

[124] J.H. Kim, Y.S. Kim, K. Park, S. Lee, H.Y. Nam, K.H. Min, et al., Antitumor efficacy of cisplatin-loaded glycol chitosan nanoparticles in tumor-bearing mice, J. Control. Release 127 (2008) 41–49, https://doi.org/10.1016/j.jconrel.2007.12.014.

[125] K. Raza, N. Kumar, C. Misra, L. Kaushik, S.K. Guru, P. Kumar, et al., Dextran-PLGA-loaded docetaxel micelles with enhanced cytotoxicity and better pharmacokinetic profile, Int. J. Biol. Macromol. 88 (2016) 206–212, https://doi.org/10.1016/j.ijbiomac.2016.03.064.

[126] M. Hrubý, Č. Koňák, K. Ulbrich, Polymeric micellar pH-sensitive drug delivery system for doxorubicin, J. Control. Release 103 (2005) 137–148, https://doi.org/10.1016/j.jconrel.2004.11.017.

[127] E. Jabbari, X. Yang, S. Moeinzadeh, X. He, Drug release kinetics, cell uptake, and tumor toxicity of hybrid VVVVVVKK peptide-assembled polylactide nanoparticles, Eur. J. Pharm. Biopharm. 84 (2013) 49–62, https://doi.org/10.1016/j.ejpb.2012.12.012.

[128] V. Arulmozhi, K. Pandian, S. Mirunalini, Ellagic acid encapsulated chitosan nanoparticles for drug delivery system in human oral cancer cell line (KB), Colloids Surf. B Biointerfaces 110 (2013) 313–320, https://doi.org/10.1016/j.colsurfb.2013.03.039.

[129] N.N. Li, B.N. Zheng, J.T. Lin, L.M. Zhang, New heparin-indomethacin conjugate with an ester linkage: synthesis, self aggregation and drug delivery behavior, Mater. Sci. Eng. C 34 (2014) 229–235, https://doi.org/10.1016/j.msec.2013.09.024.

[130] K.R. Madhwi, P. Kumar, B. Singh, G. Sharma, O.P. Katare, et al., In vivo pharmacokinetic studies and intracellular delivery of methotrexate by means of glycine-tethered PLGA-based polymeric micelles, Int. J. Pharm. 519 (2017) 138–144, https://doi.org/10.1016/j.ijpharm.2017.01.021.

[131] P. Kumar, R. Kumar, B. Singh, R. Malik, G. Sharma, D. Chitkara, et al., Biocompatible phospholipid-based mixed micelles for tamoxifen delivery: promising evidences from in-vitro anticancer activity and dermatokinetic studies, AAPS PharmSciTech 18 (2017) 2037–2044, https://doi.org/10.1208/s12249-016-0681-1.

[132] D. Dutta, W. Ke, L. Xi, W. Yin, M. Zhou, Z. Ge, Block copolymer prodrugs: synthesis, self-assembly, and applications for cancer therapy, Nanomed. Nanotechnol. 12 (2020) 1–19, https://doi.org/10.1002/wnan.1585.

[133] V. Delplace, P. Couvreur, J. Nicolas, Recent trends in the design of anticancer polymer prodrug nanocarriers, Polym. Chem. 5 (2014) 1529–1544, https://doi.org/10.1039/c3py01384g.

[134] A.N. Moore, T.L. Lopez Silva, N.C. Carrejo, C.A. Origel Marmolejo, I.C. Li, J.D. Hartgerink, Nanofibrous peptide hydrogel elicits angiogenesis and neurogenesis without drugs, proteins, or cells, Biomaterials 161 (2018) 154–163, https://doi.org/10.1016/j.biomaterials.2018.01.033.

[135] M.J. Sis, M.J. Webber, Drug delivery with designed peptide assemblies, Trends Pharmacol. Sci. 40 (2019) 747–762, https://doi.org/10.1016/j.tips.2019.08.003.

[136] M. Rad-Malekshahi, L. Lempsink, M. Amidi, W.E. Hennink, E. Mastrobattista, Biomedical applications of self-assembling peptides, Bioconjug. Chem. 27 (2016) 3–18, https://doi.org/10.1021/acs.bioconjchem.5b00487.

[137] F. Boato, R.M. Thomas, A. Ghasparian, A. Freund-Renard, K. Moehle, J.A. Robinson, Synthetic virus-like particles from self-assembling coiled-coil lipopeptides and their use in antigen display to the immune system, Angew. Chem. Int. Ed. 46 (2007) 9015–9018, https://doi.org/10.1002/anie.200702805.

[138] T. Abudula, K. Bhatt, L.J. Eggermont, N. O'Hare, A. Memic, S.A. Bencherif, Supramolecular self-assembled peptide-based vaccines: current state and future perspectives, Front. Chem. 8 (2020) 1–11, https://doi.org/10.3389/fchem.2020.598160.

[139] T. Wang, D. Wang, H. Yu, B. Feng, F. Zhou, H. Zhang, et al., A cancer vaccine-mediated postoperative immunotherapy for recurrent and metastatic tumors, Nat. Commun. 9 (2018) 1–12, https://doi.org/10.1038/s41467-018-03915-4.

[140] P. Yang, H. Song, Y. Qin, P. Huang, C. Zhang, D. Kong, et al., Engineering dendritic-cell-based vaccines and PD-1 blockade in self-assembled peptide nanofibrous hydrogel to amplify antitumor T-cell immunity, Nano Lett. 18 (2018) 4377–4385, https://doi.org/10.1021/acs.nanolett.8b01406.

[141] Y. Tian, H. Wang, Y. Liu, L. Mao, W. Chen, Z. Zhu, et al., A peptide-based nanofibrous hydrogel as a promising DNA nanovector for optimizing the efficacy of HIV vaccine, Nano Lett. 14 (2014) 1439–1445, https://doi.org/10.1021/nl404560v.

[142] L.E.R. O'Leary, J.A. Fallas, E.L. Bakota, M.K. Kang, J.D. Hartgerink, Multi-hierarchical self-assembly of a collagen mimetic peptide from triple helix to nanofibre and hydrogel, Nat. Chem. 3 (2011) 821–828, https://doi.org/10.1038/nchem.1123.

[143] J.D. Hartgerink, E. Beniash, S.I. Stupp, Self-assembly and mineralization of peptide-amphiphile nanofibers, Science (80-) 294 (2001) 1684–1688, https://doi.org/10.1126/science.1063187.

[144] A. Mata, Y. Geng, K.J. Henrikson, C. Aparicio, S.R. Stock, R.L. Satcher, et al., Bone regeneration mediated by biomimetic mineralization of a nanofiber matrix, Biomaterials 31 (2010) 6004–6012, https://doi.org/10.1016/j.biomaterials.2010.04.013.

[145] Z.Y. Lin, Z.X. Duan, X.D. Guo, J.F. Li, H.W. Lu, Q.X. Zheng, et al., Bone induction by biomimetic PLGA-(PEG-ASP)n copolymer loaded with a novel synthetic BMP-2-related peptide in vitro and in vivo, J. Control. Release 144 (2010) 190–195, https://doi.org/10.1016/j.jconrel.2010.02.016.

[146] M.P. Lutolf, F.E. Weber, H.G. Schmoekel, J.C. Schense, T. Kohler, R. Müller, et al., Repair of bone defects using synthetic mimetics of collagenous extracellular matrices, Nat. Biotechnol. 21 (2003) 513–518, https://doi.org/10.1038/nbt818.

[147] Y. Loo, Y.C. Wong, E.Z. Cai, C.H. Ang, A. Raju, A. Lakshmanan, et al., Ultrashort peptide nanofibrous hydrogels for the acceleration of healing of burn wounds, Biomaterials 35 (2014) 4805–4814, https://doi.org/10.1016/j.biomaterials.2014.02.047.

[148] M.R. Reithofer, A. Lakshmanan, A.T.K. Ping, J.M. Chin, C.A.E. Hauser, In situ synthesis of size-controlled, stable silver nanoparticles within ultrashort peptide hydrogels and their anti-bacterial properties, Biomaterials 35 (2014) 7535–7542, https://doi.org/10.1016/j.biomaterials.2014.04.102.

[149] S. Khan, S. Sur, C.J. Newcomb, E.A. Appelt, S.I. Stupp, Self-assembling glucagon-like peptide 1-mimetic peptide amphiphiles for enhanced activity and proliferation of insulin-secreting cells, Acta Biomater. 8 (2012) 1685–1692, https://doi.org/10.1016/j.actbio.2012.01.036.

[150] S. Sur, E.T. Pashuck, M.O. Guler, M. Ito, S.I. Stupp, T. Launey, A hybrid nanofiber matrix to control the survival and maturation of brain neurons, Biomaterials 33 (2012) 545–555, https://doi.org/10.1016/j.biomaterials.2011.09.093.

[151] S. Zhang, Extensive neurite outgrowth and active synapse formation on self-assembling peptide matrix scaffolds, in: Second Smith Nephew Int. Symp.—Tissue Eng. 2000 Adv. Tissue Eng. Biomater. Cell Signal, 2000, p. 12.

[152] C.E. Schmidt, J.B. Leach, Neural tissue engineering: strategies for repair and regeneration, Annu. Rev. Biomed. Eng. 5 (2003) 293–347, https://doi.org/10.1146/annurev.bioeng.5.011303.120731.

[153] M. Iwasaki, J.T. Wilcox, Y. Nishimura, K. Zweckberger, H. Suzuki, J. Wang, et al., Synergistic effects of self-assembling peptide and neural stem/progenitor cells to promote tissue repair and forelimb functional recovery in cervical spinal cord injury, Biomaterials 35 (2014) 2617–2629, https://doi.org/10.1016/j.biomaterials.2013.12.019.

[154] J. Kisiday, M. Jin, B. Kurz, H. Hung, C. Semino, S. Zhang, et al., Self-assembling peptide hydrogel fosters chondrocyte extracellular matrix production and cell division: implications for cartilage tissue repair, Proc. Natl. Acad. Sci. U. S. A. 99 (2002) 9996–10001, https://doi.org/10.1073/pnas.142309999.

[155] R.N. Shah, N.A. Shah, M.M.D.R. Lim, C. Hsieh, G. Nuber, S.I. Stupp, Supramolecular design of self-assembling nanofibers for cartilage regeneration, Proc. Natl. Acad. Sci. U. S. A. 107 (2010) 3293–3298, https://doi.org/10.1073/pnas.0906501107.

[156] A. Andukuri, W.P. Minor, M. Kushwaha, J.M. Anderson, H.W. Jun, Effect of endothelium mimicking self-assembled nanomatrices on cell adhesion and spreading of human endothelial cells and smooth muscle cells, Nanomed. Nanotechnol. Biol. Med. 6 (2010) 289–297, https://doi.org/10.1016/j.nano.2009.09.004.

[157] M.T. McClendon, S.I. Stupp, Tubular hydrogels of circumferentially aligned nanofibers to encapsulate and orient vascular cells, Biomaterials 33 (2012) 5713–5722, https://doi.org/10.1016/j.biomaterials.2012.04.040.

[158] C.E. Semino, J.R. Merok, G.G. Crane, G. Panagiotakos, S. Zhang, Functional differentiation of hepatocyte-like spheroid structures from putative liver progenitor cells in three-dimensional peptide scaffolds, Differentiation 71 (2003) 262–270, https://doi.org/10.1046/j.1432-0436.2003.7104503.x.

[159] G. Mehta, C.M. Williams, L. Alvarez, M. Lesniewski, R.D. Kamm, L.G. Griffith, Synergistic effects of tethered growth factors and adhesion ligands on DNA synthesis and function of primary hepatocytes cultured on soft synthetic hydrogels, Biomaterials 31 (2010) 4657–4671, https://doi.org/10.1016/j.biomaterials.2010.01.138.

CHAPTER 9

Albumin nanoparticles—A versatile and a safe platform for drug delivery applications

Tamara Zwain[a,g], Neetika Taneja[b], Suha Zwayen[a,c], Aditi Shidhaye[d], Aparana Palshetkar[e,f], and Kamalinder K. Singh[a,g,h]

[a]School of Pharmacy and Biomedical Sciences, Faculty of Clinical and Biomedical Sciences, University of Central Lancashire, Preston, United Kingdom
[b]Sun Pharmaceutical Industries Ltd, Vadodara, Gujarat, India
[c]Precision Nanosystems Inc., Vancouver, BC, Canada
[d]Colgate-Palmolive (India) Limited, Colgate Research Centre, Mumbai, India
[e]C U Shah College of Pharmacy, SNDT Women's University, Mumbai, India
[f]VES College of Pharmacy, Mumbai, India
[g]UCLan Research Centre for Smart Materials, University of Central Lancashire, Preston, United Kingdom
[h]UCLan Research Centre for Translational Biosciences & Behaviour, University of Central Lancashire, Preston, United Kingdom

1. Introduction

Nanotechnology has been extensively researched and long been used to overcome drug delivery system downsides, including obstacles for encapsulation of therapeutics and/or imaging compounds for targeted delivery [1, 2]. Albumin is one of the biomolecules used for targeted delivery and as carrier system that has captured the interest of many researchers and scientists due to its selective delivery capabilities, nontoxicity, and non-immunogenicity [3]. The cardinal intrinsical features of albumin are its high stability and high solubility in water as well as in dilute salt solutions [4]. Albumin has an extraordinary ability to bind to lipophilic and other molecules due to its structural advantages [5]. This unique property of albumin has been exploited in the design of nanoparticles. Thus, albumin, a multifunctional protein and nature's own drug delivery vehicle, is an ideal carrier for drug delivery applications because of its abundance, high binding capacity, improved bioavailability, biodegradability, and high safety profile [6].

This chapter focuses on design and synthesis of various types of nanostructures based on albumin and their drug-loading modalities, modification of surface of nanoparticles to enable targeting properties, and detailed discussion on their application for treatment of various therapeutic conditions.

2. Human serum albumin (HSA) structure and binding sites

Out of the various common sources of albumin like bovine serum albumin (BSA) and egg, HSA is the most extensively used for drug delivery applications because of its

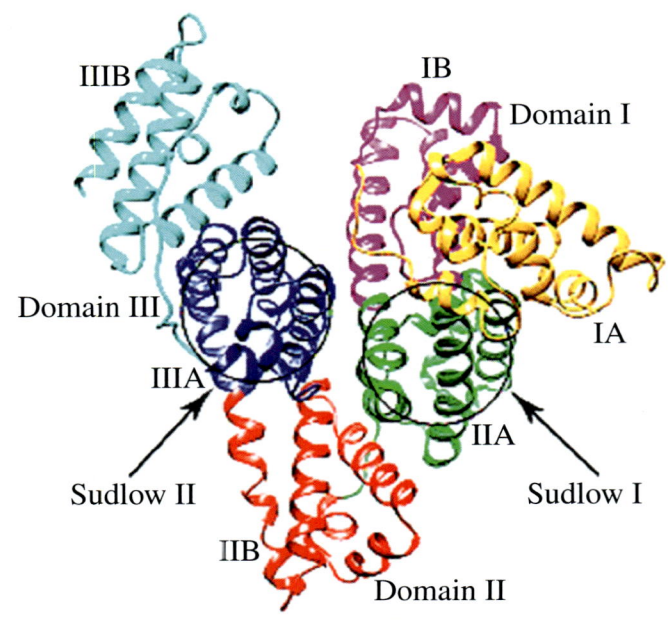

Fig. 1 Human serum albumin (HSA) structure with the representation of domains, subdomains, and Sudlow's binding sites I and II. *(From M. Mondal, P. Lakshmi T, R. Krishna, N. Sakthivel, Molecular interaction between human serum albumin (HSA) and phloroglucinol derivative that shows selective anti-proliferative potential, J. Lumin. 192 (2017) 990–998. https://doi.org/10.1016/j.jlumin.2017.08.007. Copyrights 0022-2313/© 2017 Elsevier B.V. All rights reserved.)*

endogenous nature, biocompatibility, and nontoxic nature. HSA is a plasma protein that is produced by the liver in humans and functions as a transporter of fatty acids in the plasma and interstitial fluid. Albumin plays a vital role in balancing fluid in the blood and contributes to the colloidal osmotic pressure with high multifunctional capacity for binding to ligands [7]. HSA is a water-soluble protein and constitutes 35–50 g/L of human blood plasma. Albumin molecule is heart-shaped (Fig. 1) with a molecular weight in the range of 65–70 kDa, stable at 4–9 pH range, and can endure heating up to temperature of 60°C for ten hours [8]. HSA is a single polypeptide consisting of five hundred eighty five amino acids with low tryptophan and high cysteine content [9]. The secondary structure of albumin contains of 67% α-helix and has seventeen disulfide bridges togather with six turns. The ellipsoidal shape of HSA contains three homologous domains (I, II, III) which are cross-linked by disulfide bonds (Fig. 1) [5]. All these three domains have a pair of subdomains called "A" and "B" [8].

Models to understand binding regions of HSA for various drugs has been proposed. Sudlow's classification proposes two major binding sites as shown in Fig. 1 viz. subdomain IIA and IIIA [5, 8]. The site IIA also known as the 'warfarin binding site' is a prearranged, flexible, large, multi-cell cavity and binds to heterocyclic, negatively charged bulky compounds such as azidothymidine, warfarin, etc. Site IIIA is named the 'indole and benzodiazepine binding site' and is topographically similar to the IIA site and binds to compounds such as, propofol, ibuprofen, diazepam [8]. Subdomain IB has a binding pocket which acts as a binding site drugs such as bilirubin, naproxen, warfarin, etc. Further third binding region subdomain IB has also been proposed for HSA molecule [5, 8].

Interestingly, HSA binds not only to low molecular weight compounds but also to large molecules like proteins and peptides. Around 35 known proteins were found to be associated to HSA including angiotensinogen, hemoglobin, apolipoproteins, and transferrin and prothrombin) [10]. Because of the drug binding sites available on the HSA molecule, drug loading on to albumin particles is facilitated and high drug loading can be easily achieved [5]. Therefore, the prevalence of binding sites on HSA have a significant effect on the fabrication of NPs, drug loading, drug entrapment efficiency, pharmacokinetics, and their therapeutic efficacy.

3. Albumin nanocarriers—Synthesis, preparation methods, and lyophilization

Albumin is notably useful for fabrication of NPs as it is stable and can be readily engineered with attachment of ligands via free cysteine in position C34 or its binding pockets. Different methods have been employed for the synthesis of HSA NPs. Basis of synthesis of the albumin NPs is highly dependent on multiple factors like the type of drug (hydrophobic or hydrophilic) to be encapsulated in the NPs, desired particle size, targeting or intended delivery site, etc. Some of the commonly used methods for the synthesis of albumin NPs are discussed below:

3.1 Desolvation/coacervation

Desolvation technique is one of methods of preparing the albumin NPs that is been most commonly used for preparation of albumin NPs. The method is also known as coacervation and is represented schematically in Fig. 2. It is robust and reproducible for a laboratory set-up and involves slow and continuous dropwise addition of desolating agents for albumin like ethanol, acetone, and/or methanol until the solution becomes turbid. The NPs are formed by phase separation of the albumin from the solution with the addition of any of these non-solvents for albumin. It has been observed that a combination of ethanol and acetone result in more spherical NPs than either of the solvent used alone [11]. The formed NPs are then crosslinked either by heating the solution or by using chemicals reagents like glutaraldehyde. The particle size of the NPs prepared using desolvation mainly depend upon, the concentration and rate at which the desolvating agent

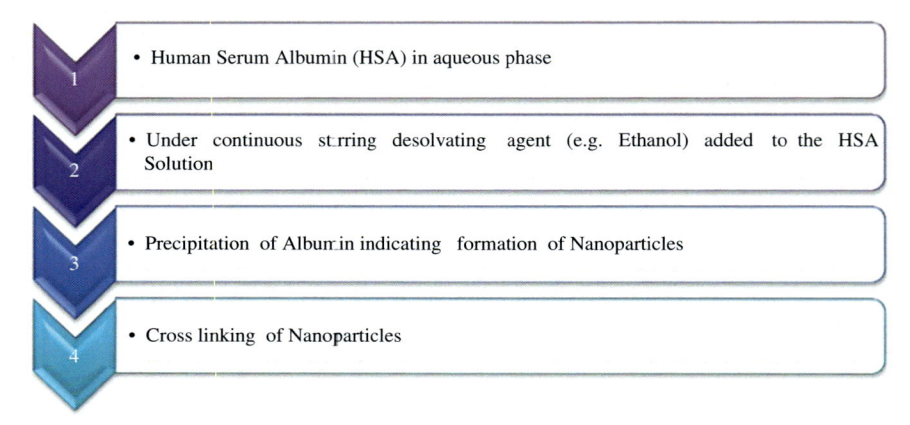

1
- Human Serum Albumin (HSA) in aqueous phase

2
- Under continuous stirring desolvating agent (e.g. Ethanol) added to the HSA Solution

3
- Precipitation of Albumin indicating formation of Nanoparticles

4
- Cross linking of Nanoparticles

Fig. 2 Schematic representation of desolvation method for preparation of human serum albumin (HSA) nanoparticles.

is added. The size, however, will not be affected with either the nature or the amount of crosslinking agent added to the dispersion [12].

Modified desolvation method involves monitoring of pH and ionic balance in addition to the albumin concentration. It has been observed that pH values from 7 and above help in controlling the size of the NPs to a greater extent because of decreased coagulation of the albumin at the basic pH which is far from the isoelectric point of albumin, which is 4.9, thus, resulting in dispersions with more uniform particle size [11]. The particle size can also be controlled by the addition of ionic salts like sodium chloride (NaCl). Larger particles are formed when the content of NaCl is higher. This is due to the enhanced shielding of the surface charges on the albumin molecules that reduces the net charge of the NPs [13].

3.2 Emulsification/microencapsulation

Emulsification is another widely used technique for the fabrication of protein NPs (Fig. 3). It has the advantage of being one of the fastest and scalable methods [14]. The NPs prepared by emulsification technique in an inert oil can be stabilized either by thermal or chemical treatment. Albumin NPs prepared by emulsification method utilize homogenization of albumin dropped into cottonseed oil at high speed then thermally stabilize it by heating in range of 175–180°C for period of ten minutes. Once the mixture is cooled, it is diluted by adding ethyl ether to reduce viscosity and allow easy centrifugation to separate the NPs. This method can be particularly applied for the drugs that are heat resistant. For heat-sensitive drugs, the NPs can be prepared by chemical stabilization where, HSA dissolved in water is emulsified in a fixed oil at temperature of 25°C

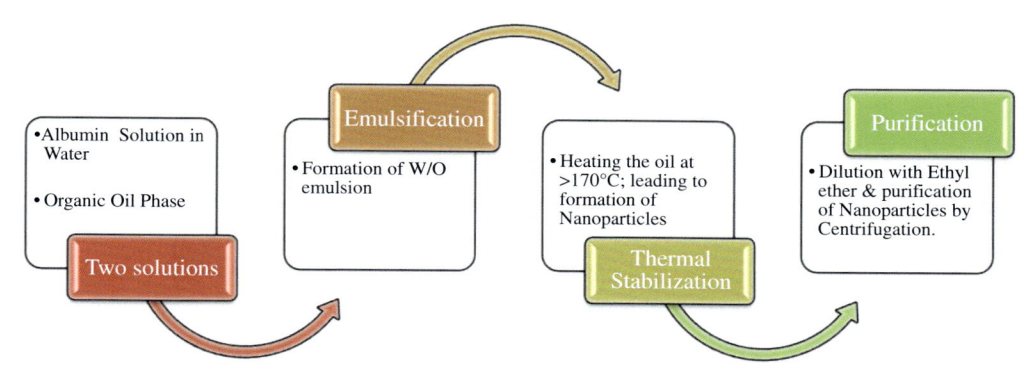

Fig. 3 Schematic diagram for the formation of the nanoparticle by emulsification method.

followed by denaturation by suspending it in an organic solvent containing the cross-linking agent like formaldehyde or glutaraldehyde [15].

In another study, a very poorly soluble drug 10-hydroxycamptothecin was combined with albumin to form NPs by reformative emulsion–heat stabilization technique. In this method, the drug and albumin were dissolved water and emulsified in the castor oil under high-speed homogenization. This emulsion was further added in a controlled manner to castor oil maintained at $140\pm5°C$. The formed NPs were cooled and washed several times with ethyl ether to remove the castor oil content [16].

3.3 Nanoparticle albumin-bound technology (Nab-technology)

HSA is very well known for its affinity toward the hydrophobic molecule like fatty acids, hormones and fat-soluble vitamins. Hydrophobic drug molecules like paclitaxel and docetaxel have been widely explored with this Nab-Technology for delivery of these agents to treat cancer. Landmark approval of Abraxane (nab-paclitaxel; paclitaxel-albumin NPs) used Nab-Technology resulting in NPs with an average particle size of 130 nm [17].

The NPs are synthesized by mixing the drug with HSA in an aqueous solvent, and the dispersion is then homogenized under high pressure to obtain NPs complexes with average size in the range of 100–200 nm [15]. These NPs were formed without any covalent bonds and proved to be stable in its galenic formulation. Hence, they dissolve rapidly when administered intravenously forming water-soluble drug-albumin complexes. In another study, albumin NPs compared two versions, i.e., cross-linked NPs and not cross-linked NPs both loaded with an anticancer drug paclitaxel; and it was observed that there was not much difference in the properties and composition of the two formulations. However, cross-linking stabilized the drug in the matrix, which resulted in delayed disintegration of the NPs while the non-cross-linked NPs disintegrated readily and resulted in initial higher drug-plasma concentration [18].

3.4 Self-assembly

Albumin is an excellent carrier for hydrophobic drugs which reversibly bind to this protein. It is very much possible that hydrophobic drug and albumin self-assemble to form NPs in aqueous solution. Such a process could prove to be simpler as compared to Nab technology. There have been reports of self-assembled BSA–Dextran conjugates by natural and non-toxic Maillard reaction. Optimization of the pH and heating temperature of the conjugate showed effective drug loading of hydrophilic drug ibuprofen in the NPs [19].

The hydrophilicity of the albumin has been altered by conjugating the HSA with cholesterol to make it more hydrophobic. The amino group of albumin reacted with the cholesterol through hydroxyl group through a coupling agent. NPs synthesized by this method have shown an improved drug-loading capacity and cellular uptake [20]. Another method reported novel octyl modified albumin micelles when the free amino groups of the albumin reacted with octaldehyde. These micelles have shown a higher percentage of drug loading (33.1%) with around 90% entrapment efficiency [21]. Curcumin-loaded albumin NPs have also been prepared by self-assembly [22]. It was observed that, in addition to pH and temperature, the ionic strength of the buffer used for preparing the albumin solution was also important and the increase in the ionic strength would lead to higher particle size of the NPs [22].

3.5 Nano-spray drying

Spray drying is a continuous process to convert liquids to dry powders in the presence of the hot air as a drying medium. This avenue of the spray drying is now being explored to prepare NPs as well [23]. Albumin NPs have been prepared using Nano Spray Dryer equipment where the surfactant mixed aqueous solution of albumin is passed through a vibrating mesh to form low particle size NPs, that were collected with the help of an electrostatic particle collector fitted [24].

3.6 Thermal gelation

Gelation property of albumin at high temperature has also been explored to produce NPs. It is a controlled process where monitoring the temperature of the albumin solution is of critical importance. At high temperature, albumin, being a protein, denaturizes to aggregates. Increase in the temperature during the process leads to the unfolding of the various bonds [25]. It has also been observed that the changes in the pH of the solution impact the thermal gelation of albumin [26].

3.7 Lyophilisation of albumin nanoparticles

The albumin NPs obtained from either of the above methods must be suitable for administration as a safe pharmaceutical dosage form. To impart better stability to such a product, it must be free from any contamination and water. The final product should also be easy to store. One of the ways to fulfill all the above requirements for the NPs is lyophilization [27].

Lyophilization involves removal of water where the water in the samples is frozen then samples go through a sublimation process (primary drying), which allows the removal of water from the sample, followed by desorption (secondary drying). The technique is used in preparation of albumin NPs incorporating thermolabile or water sensitive drugs to improve their stability. The freeze-dried albumin NPs are obtained as a free-flowing powder which are easy to store and re-disperse in the desired medium for administration. Lyophilization is carried out under low temperature and pressure; therefore, the stability of the drug and duration of the drying cycle are the two important factors to be considered. The selection of right cryoprotectants which aid in providing protection to the structure of NPs during the freeze-drying is important. Some of the excipients which have found to be useful as cryoprotectant include sugars like trehalose, sucrose, mannitol polymers, surfactants, amino acids, and some buffering agents or inorganic salts have found be useful as cryoprotectants for stabilizing the NPs (Fig. 4) [28]. The concentration of these excipients must be optimised to prevent aggregation of the final product [29].

4. Types of albumin nanocarriers

Albumin based nanocarrier may have varied shapes, structures and particle size and the various carriers include albumin NPs, nanocapsules, microspheres, microbubbles and

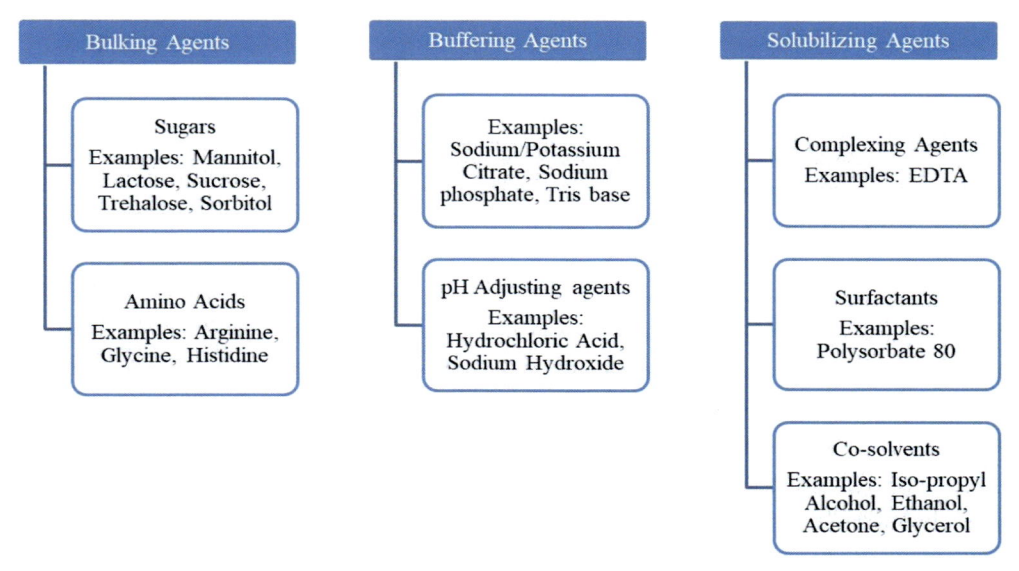

Fig. 4 Examples of cryoprotectants used in lyophilization.

core-shell albumin-coated nanostructures which could impact their physicochemical, drug release and pharmacokinetic properties.

4.1 Albumin microspheres

The first biodegradable and biocompatible microspheres based on albumin were reported in the 1970s [30]. These colloidal forms of albumin in the particle size range of 100 μm have potential to be carriers of drugs for targeting specific localization within the body or local application. The factors like size of particles, their stability influence the rate of metabolism of albumin microspheres at the target site. The rate of release of the bioactive from the microspheres can be modified by the particle size, extent of cross-linking, the position and concentration of drug incorporated in the microspheres. Thakkar et al. prepared celecoxib loaded BSA microspheres by emulsification chemical crosslinking method showing the sustained release of drug from microspheres in ∼6 days. In vivo, pharmacokinetic studies showed that albumin microspheres loaded with celecoxib remained in circulation in blood for a longer period of time as compared to native drug solution [31]. In addition, ketorolac tromethamine loaded albumin microspheres were formulated by emulsion crosslinking method for intramuscular administration which were found useful for once a day therapy [32]. Another study reported propranolol hydrochloride loaded bioadhesive albumin microspheres prepared by emulsion heat stabilization technique and depicted 70% adhesion on the rat jejunum mucosal surface. These microspheres revealed high burst release during the initial time periods which was followed by a more gradual terminal release [33].

4.2 Albumin nanoparticles

Albumin is widely utilized in NPs preparation because of its distinct advantages of being biodegradable, ease of preparation, and size in the nanometer range. Being biodegradable means the drug release from albumin NPs can be achieved naturally by protease digestion. Albumin-bound paclitaxel prepared by Nab technology is a successful example of the albumin NPs in clinics since 2005 when it was first approved by the FDA for metastatic breast cancer. Albumin-bound paclitaxel was found to be safer and patients could tolerate a higher dose with this nanoformulation. Hence, the dose of paclitaxel albumin NPs could be increased from 175 mg m^{-2} to 260 mg m^{-2} [34]. Besides, the patients showed a better response rate and longer time for tumor progression without increasing the drug toxicity [35]. Both water soluble and water insoluble drugs have been loaded into albumin NPs using various techniques to maximize their drug loading.

4.2.1 Albumin nanoparticles with loading of water soluble drugs

For water soluble hydrophilic drugs, the drugs can either be adsorbed on the surface of NPs or incorporated into the main matrix of NPs. There are three main methods to load the drug into albumin NPs [15]:

(a) Incubation method: The drug is incubated with the already formed albumin NPs.

(b) Incorporation method: The drug is added to the solution of albumin which then cross-linked to stabilize the NPs.

(c) The drug is added to the solution of cross linking agent like glutaraldehyde and then cross-linking carried out with this solution to form NPs.

The incubation method resulted in high loading of drug with maximum drug loading achieved within the first hour of incubation, followed by a plateau thereafter. The incorporation method has been utilized when a high amount of drug loading is required, while the last method though similar to the incubation method but requires a longer period of incubation.

Tacrine NPs have used coacervation method for its preparation followed by incubation to load the drug. It was observed that tacrine had a great affinity for albumin-based NPs and with sulfobutylether derivative of βCD highest amount of drug was entrapped. The presence of a negative charge on sulfobutylether βCD increased the density of the charge on the NPs surface, thus enhancing the absorption of the positively charged tacrine [36]. Similarly, TRAIL–DOX–albumin NPs were prepared by the incorporation method whereby albumin was conjugated with Dox and octyl aldehyde and adsorbed with apoptotic TRAIL protein to give TRAIL/Dox albumin NPs. A wide range of water-soluble drugs with varied therapeutic applications have been formulated as albumin NPs. Some of the examples include; 5 sodium ferulate, 5-fluorouracil and gabapentin [37–39].

4.2.2 Albumin nanoparticles with loading of water soluble drugs

Various methods have been used to incorporate poorly water soluble drugs into the albumin NPs. The two main methods are encapsulation and entrapment. Drugs may be loaded on to NPs either by covalent bonding or adsorption of the charged drug by electrostatic attraction [15]. Nab-technology and self-assembly preparation methods also accomplish a good quality drug loading into the albumin NPs.

Paclitaxel has low water solubility, but the entrapment of the drug into BSA NPs resulted in a markedly increased its solubility in water. The developed NPs elicited good stability, with high drug loading (27.2%) and very good entrapment efficiency (95.3%) [40]. Zu et al investigated vinblastine sulfate loaded BSA NPs and successfully reported high entrapment efficiency (84.83%) and superior drug loading (42.37%) [41]. Few examples of the delivery of some insoluble and poorly water-soluble drugs with albumin NPs includes; methotrexate [42], cisplatin [43], ruthenium-based anticancer drugs [44], diclofenac and tacrolimus [45].

4.3 Core-shell structure

The core-shell type NPs can be defined as comprising a inner core (inner material) and an outer shell. This type of NPs combines the functional properties of both core and coat and can be modified either by replacing the materials used for fabricating or the ratio of core to shell. Albumin has been utilized as a shell to coat the liposomes to improve its functionality. A study successfully used HSA coated liposomes to deliver antisense oligo-deoxyribonucleotides (ODN) and significantly increase the chemosensitivity of doxorubicin [46].   Another study prepared surface-modified cationic liposomes with BSA, where the electrostatic interaction between cationic liposomes and anionic albumin molecules was used for encapsulating doxorubicin. The NPs resulted in high loading efficiency of doxorubicin (83.8% $\pm$ 3%), with zeta potential (-15 ± 0.3 mV) and mean particle diameter of 107.8 ± 3 nm [47].

4.4 Albumin microbubbles

Ultrasound-activated microbubbles containing inert gases in the core and polymer as shell have been employed for a wide range of diagnostic and biomedical applications including cancer and cardiovascular therapy. Such structures have shown to prolong the circulation time, enable higher drug loading, improve stability and enable targeting at molecular level [48]. To improve longevity of microbubbles, high molecular weight inert gas having low solubility and diffusivity were incorporated into the core such as perfluorocarbons and sulfur hexafluoride [49]. The shell is made of rigid polymer with no active oscillation. Therefore, during sonification, the polymeric shell fractures allowing the encapsulated gas to escape. The fragmentation of shell coupled with expulsion under high acoustic pressure efficiently propels the drugs for entry into cells.

Curcumin loaded BSA microbubbles using perfluorocarbon gas were prepared, and the in vitro ultrasound triggered drug release study showed that the release of drug from albumin microbubbles was significantly increased after sonication. Additionally, albumin microbubbles evidently enhanced imaging, which indicates that the microbubbles possess good acoustic properties. Intravenous administration of albumin microbubbles in tumor-bearing mice when exposed to ultrasound resulted in an enhanced drug accumulation in tumor tissues and a significant increase in tumor growth inhibition rate (57.1%) compared with curcumin microbubbles (28.8%) as well as compared to curcumin-loaded albumin NPs (26.2%) [50]. Another study, where NPs with particle size ($\approx$150nm) were prepared with poly (lactic–co–glycolic acid) and covalently linked them to albumin-shelled microbubbles formulation and labeled with the fluorescent compound. When administered through intravenous injection NPs successfully delivery to skeletal muscle in an experimental model of peripheral arterial disease [51].   Furthermore, in another study, albumin microbubbles (PAMB) were coated with polyethylenimine (PEI) by sonication. PEI, PEI + albumin, PAMB and Lipofectamine 2000 were individually used to transfect 293T, CHO, and COS cells. The in vitro data demonstrated the ability of albumin microbubbles to bind plasmid DNA through the surface potential elevation and the albumin microbubbles modified by PEI had high efficiency for gene transfection and low cytotoxicity making it a useful method for nonviral gene delivery [52].

5. Surface modification and targeting of albumin nanoparticles

Albumin is widely known for its exceptional ligand binding capacity due to its functional groups (amino and carboxylic groups) that are present on the surface of the NPs. Easy modification of reactive amino and carboxyl groups can confer many functionalities to albumin NPs such as enhanced stability, improved circulation, sustain release and targeted delivery. The ligands are either covalently conjugated to the surface of albumin NPs or non-covalent bonding modifying the surface characteristics of albumin NPs. The efficiency of the albumin NPs can be improved by using different linkers and spacers. Also, surface coating or electrostatic adsorption techniques can also be utilized in enhancing drug targeting [12]. The selectivity of albumin NPs may be increased by active targeting of NPs to surface receptors or other membrane proteins overexpressed on target cells (Fig. 5).

In the albumin ligand combination, albumin is the carrier for loading the drug whereas the ligand serve to modify the pharmacokinetic parameters (e.g., surfactants), enhance the stability (e.g., poly-L-lysine), prolong the circulation time (e.g., PEG), slow the release of drug (e.g., cationic polymers) or act as a targeting agent (e.g., folate, thermosensitive polymers, transferrin, apolipoproteins and monoclonal antibodies) as detailed below.

5.1 Surfactants

Surface modification of HSA bound drug NPs with non-ionic surfactants like polysorbate 80 is reported to effectively reduce toxicity of doxorubicin. Pereverzeva et al. demonstrated the lower cardiotoxicity of doxorubicin NPs when surface coated with

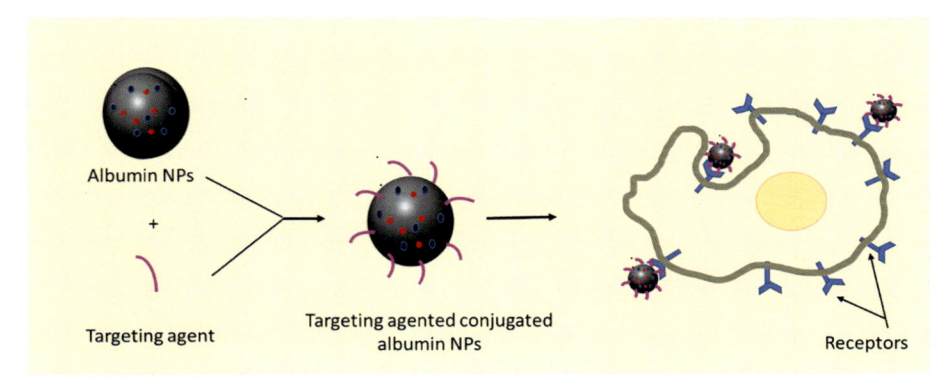

Fig. 5 Schematic representation of active cellular targeting, which can be achieved by the functionalization of the surface of nanoparticles with the targeting agents to promote cell-specific recognition and binding.

polysorbate 80 compared to uncoated NPs, thereby increasing their circulation time, bioavailability and reduced clearance of doxorubicin [53].

5.2 Polyethylene glycol (PEG)

PEGylation is the chemical modification of surface of proteins or nanoparticles with PEG. Molecular weight of PEG regulates the plasma protein binding and circulation characteristics of the PEGylated systems. Methoxy PEG (mPEG)–succinimidyl propionate have been used to PEGylate the BSA conjugated 5-flurouracil NPs which resulted in 50-fold prolongation of their circulation half-life as compared to non-PEGylated NPs with slower drug release and higher drug accumulation in tumors due to enhanced permeability [54] demonstrated by lower loading of Rose Bengal in HSA–mPEG NPs due to less number of drug protein-binding sites being available as compared to uncoated albumin NPs. It has also been shown that presence of the PEG on the surface of the NPs served as barrier and prevented enzymatic digestion of NP retarding the drug release from surface-modified albumin NPs [55]. PEG-modified polyethylenimine (PEI)-coated BSA NPs loaded with bone morphogenetic protein-2 (BMP2) were utilized for local regeneration of the bone and to stimulate formation of bone post implantation in the rat model and also displayed reduction in the cytotoxicity as compared to PEI and uncoated albumin NPs [56].

5.3 Folate

Folic acid, a low-molecular weight vitamin (441 Da), has been identified as a tumor marker in cancer cells, especially epithelial ovarian carcinoma cells, and has limited presence in most normal tissues. fFolic acid as a targeting agent has numerous advantages including its stability, cheap availability, nonimmunogenicity as compared to monoclonal antibodies, Folate conjugated to albumin is selectively internalized by receptor-mediated endocytosis in the cytoplasm [57]. Due to the encapsulation of anticancer drugs, they promote uptake and protect the drugs from degradation. Some studies explained the Carbodiimide (EDC: 1-ethyl-3-(3-dimethylaminopropyl)) coupling technique in which carboxylic group of folic acid was covalently conjugated with the amino groups on the surface of albumin NPs [58]. Paclitaxel-loaded BSA NPs covalently conjugated with folic acid observed effective targeting to human prostate cancer cells as compared to the uncoated NPs. Conjugation with folate enhanced cellular uptake and significantly increased water solubility of paclitaxel [40]. Folate-conjugated doxorubicin-loaded BSA NPs formulated using desolvation method (particle size 150-200 nm) were taken up by HeLa cells after 2 h incubation, whereas HeLa cells failed to incorporate the unconjugated NPs even after 4 h incubation. It was also seen that the cells treated with folate-conjugated doxorubicin-loaded albumin NPs showed increased viability in AoSMC compared to HeLa cells which do not express FRα, as compared to FRα overexpressing HeLa cells [58]. There are many reports of anticancer drugs loaded into folate-conjugated albumin NPs, including doxorubicin [58], cisplatin [59], vinblastine sulfate [41], and mitoxantrone [60], which have shown higher binding and improved uptake to cancer cells.

5.4 Apolipoproteins

Lipoproteins are naturally occurring and biodegradable and have the ability to transport the drug across the blood-brain barrier (BBB) into the brain. It has been reported that ApoE (Apo recognized by endothelial cells) covalently bound to HSA NPs and strongly enhanced drug transport protecting the drug from degradation and uptake into the brain to reach neurons These Apo-modified albumin NPs mimic endogenously circulating lipoproteins and can bind to lipoprotein receptors at the BBB followed by transcytosis through the endothelium, leading to accumulation into the brain [61]. The ApoE-modified albumin NPs prepared by incubation of thiolated ApoE solution with albumin NPs showed presence in the brain capillary endothelial cells after intravenous injection into mice, whereas uncoated NPs were not detected [62]. Presence of low-density lipoprotein (LDL) when ApoE-HSA NPs were incubated with bEnd3 cells showed an increase in binding affinity of ApoE to the LDL receptor and increased cellular uptake. It was suggested that LDL induced a conformational change in the ApoE structure, which could lead to an increase in the binding affinity [61]. Another study demonstrated a significant antinociceptive effect of loperamide ApoE-modified albumin NPs after 15 min of intravenous injection in mice which lasted more than an hour , whereas uncoated loperamide did not show any effect [61].

5.5 Monoclonal antibodies

Monoclonal antibodies (mAbs) are an interesting group of ligands that have applications for tumor-targeting therapy. Surface overexpression of the epidermal growth factor receptor (EGFR) was observed in many malignancies like ovarian, breast, colorectal, nonsmall cell lung, head, neck, and prostate cancers, as well as in glioma. The human epidermal factor receptor 2 (HER2) serves as a tumor-targeting marker in the treatment of patients with metastatic breast cancer. The surface of HSA NPs was modified with a humanized anti-HER2-specific antibody, trastuzumab (Herceptin), through avidin-biotin complex between the biotin-binding protein (NeutrAvidin) attached to the NPs and the biotinylated antibody [63]. The trastuzumab-conjugated HSA NPs demonstrated an effective internalization via receptor-mediated endocytosis in time- and dose-dependent manner in HER2-overexpressing cells (cell lines BT474, MCF7, and SK-BR-3) [63]. Trastuzumab-modified albumin NPs loaded with antisense oligonucleotides (ASOs) showed cell-specific targeting in HER2-overexpressing breast cancer cells and exhibited a significant downregulation of the Plk1 (Polo-like kinase 1) mRNA and

Plk1 protein [64]. The HSA NPs modified with cetuximab, a humanized IgG1 mAb, depict a promising carrier system for EGFR-expressing colon carcinoma cells [65], where a specific accumulation of drug could be shown in EGFR-expressing colon carcinoma cells with no nonspecific uptake into normal cells.

5.6 Cationic polymers

Coating of NPs with cationic polymers is mainly used for hindering the release of drug from the NPs. They protect protein NPs from enzymatic degradation and also eliminate the use of hazardous cross-linkers required for stabilization of albumin NPs. Cationic polymers such as polyethylenimine (PEI) is used for surface coating of anionic BSA loaded with bone morphogenetic protein 2 (BMP2) NPs, which control the rate of drug release, and the release rate is dependent on the concentrations of PEI used for coating. The PEI surface modification shifted the charge on the particles from negative to neutral or slightly positive, which may reduce the plasma protein adsorption on particle surfaces and thus facilitate the in vivo application of NPs. However, the osteoinductive activity of BMP-2 encapsulated in NPs was not readily achieved in a rat ectopic model, and this undesirable result was attributed to the toxic effect of the PEI on locally present cells [66]. In another study, doxorubicin-loaded HSA NPs surface-modified with PEI showed higher cell transfection and enhanced cell penetration of the particles, improving the therapeutic index of doxorubicin against MCF-7 breast cancer cells [67]. Poly-L-lysine (PLL), another cationic polymer, used for coating albumin NPs was found to enhance proteolytic resistance producing more stable NPs. 0.9 kDa PLL encapsulation of short interfering ribonucleic acid (siRNA)-loaded BSA NPs improved the aqueous solution stability and a continual release of Fluorescein isothiocyanate (FITC) loaded on the albumin NPs was seen in the release medium [68].

5.7 Thermosensitive polymers

Thermoresponsive polymer, poly(N-isopropylacrylamide-coacrylamide)-*block*-polyallylamine (PNIPAMAAm-b-PAA), was conjugated to carboxylic group of albumin through carbodiimide (EDC) coupling reaction to develop nanospheres of anticancer drug adriamycin [69]. With the increase in conjugation amounts or molecular weight of PNIPAM-AAm-AA, the release of the drug was decreased because of the existence of a hydrophilic steric barrier. Moreover, the release rate of the drug from conjugated albumin nanospheres above the cloud-point temperature of PNIPAM-AAm-AA became faster due to shrinkage of the hairy thermosensitive polymer. When the human hepatocellular carcinoma HepG2 cells were treated with nanospheres conjugated with PNIPAM-AAm-AA above the cloud-point temperature of PNIPAM-AAm-AA, it led to increased targeting to cancer cells [69].

5.8 Peptides and proteins

Expression of high levels of αvβ3 integrin (a membrane receptor for extracellular matrix ligands such as vitronectin and fibronectin) in cancer cells from various entities is reported in the literature [70]. The αvβ3 integrin has shown a high binding affinity to the cyclic arginine–glycine–aspartic acid (RGD) peptide ligand. An RGD peptide–anchored sterically stabilized BSA nanospheres (RGD-SN) bearing 5-fluorouracil showed that RGD-SN were significantly effective in the prevention of lung metastasis, angiogenesis, and ineffective regression of tumors compared with free fluorouracil [71]. For targeted delivery of doxorubicin to cells expressing the αvβ3 integrin, pegylated HSA nano micelles surface conjugated with cyclic RGD peptides resulted in improved uptake and prolonged retention of doxorubicin when incubated with human melanoma cells (M21+ cells) expressing αvβ3 integrin. Once accumulated in cells, covalently linked and noncovalently absorbed doxorubicin was readily released from the NPs by a combination of reversal of disulfide bonds by intracellular thiols and by proteolytic processes in endosomes and lysosomes [72].

5.9 Other compounds

A new approach for cancer therapy called sonodynamic therapy (SDT) was used whereby efficient sonodynamic activity of BSA conjugated to a water-soluble phthalocyanine ZnPcC4 (ZnPcC4-BSA) against HepG2 human hepatocarcinoma cells was caused by the production of intracellular reactive oxygen species. SDT is the synergistic effect of sonosensitizers excited by ultrasound [73]. The deliverability of albumin-based nanostructures was improved when attached with biomolecule such as alginate to improve the swelling behavior of NPs [74].

6. Biomedical applications of albumin nanocarriers

6.1 Albumin nanoparticles in targeting cancer

One of the major obstacles associated with cancer treatment is the lack of specificity [75]. NPs as delivery systems are being extensively investigated as a drug delivery strategy to overcome the specificity issue originated from conventional chemotherapy. Along with the other unique characteristics of albumin to carry a wide range of drugs, long systemic circulation, and natural degradation, albumin is highly uptaken and metabolized by rapidly growing and nutrient-starved cancer cells. Moreover, the pathophysiology of tumor tissue characterized by angiogenesis, leaky vasculature, and impaired lymphatic drainage enhances the uptake of albumin-based NPs in solid tumors. With the traditional passive targeting via enhanced permeation effect (EPR) that arises from abnormalities of the tumor, albumin also has active targeting potential. Albumin is taken up by cells via caveolin-1–mediated transcytosis when it binds to a 60-kDa glycoprotein (gp60) receptor present on the cancer cells. The enhanced uptake of albumin-based NPs is also mediated by secreted protein, acidic and rich in cysteine (SPARC), an extracellular matrix glyco-protein that is overexpressed in various cancers [17].

Abraxane (nab-paclitaxel), HSA NPs of paclitaxel, is one of the most clinically suc-cessful examples of nanotherapeutics based on the albumin-bound (nab) technology for cancer therapy [76]. Paclitaxel is a poorly water-soluble drug, and the conventional pac-litaxel formulation contains cremophor EL as solubilizer and ethanol as a solvent in order to enhance the solubility of paclitaxel, which leads to prolonged systemic exposure and toxicity. Nab technology is a drug delivery platform that delivers albumin-bound hydro-phobic drugs to tumors without using solubilizers or toxic solvents. The technology yields NPs with a particle size of 130 nm. It is used as a colloidal suspension derived from the lyophilized formulation of HSA-bound paclitaxel diluted in saline solution (0.9% NaCl). This formulation proved to have improved pharmacokinetic and biodistribution properties of drug with enhanced tumor efficacy and safety profile in preclinical and clin-ical studies [77]. The enhanced antitumor activity of Abraxane was also found to be due to increased transendothelial gp60-mediated transport and increased intratumoral accu-mulation as a result of the SPARC–albumin interaction.

The enhanced uptake of HSA NPs has been substantiated by many researchers, thus confirming the potential of HSA-based NPs as promising platform for the targeted deliv-ery of antineoplastic agents by enhancing their distribution and accumulation at tumor site with reduced toxicity. Doxorubicin, an anthracyclin derivative, is a DNA intercalatar entrapped in albumin nanocarrier which has been investigated for improved anticancer potential [78]. Albumin NPs of various other drugs including cabazitaxel and irinotecan have also been prepared and studied for anticancer activity [79, 80]. Similarly, a pro-longed blood circulation up to 96 h and tumor accumulation could be achieved with self-assembled HSA NPs of Paclitaxel [81].

In addition to albumin's own targeting ability to cancer cells, specific interactions between the ligands on the surface of nanocarriers and receptors expressed on the tumor cells may facilitate albumin NPs internalization by triggering receptor-mediated endocy-tosis [82]. RGD (arginine-glycine-aspartate) [83], hyaluronic acid [84], trastuzumab [85], folate [86], and tumor necrosis factor-related apoptosis-inducing ligand (TRAIL), when appended to albumin NPs, have shown more specificity toward various tumors, with reduced toxicity.

6.2 Albumin nanoparticles in malaria chemotherapy

The emergence of drug resistance and nonspecific drug targeting of conventional anti-malarial therapies resulted in the need for high-dose administration and subsequent intol-erable side effects leading to patient noncompliance and toxicity [87]. Though nanomedicine has been used more widely for delivery of anticancer drugs, more recently, nanocarrier-based drug delivery has acknowledged particular attention in the field of malaria chemotherapy. It holds promising treatment revolution being less toxic, more potent, and smart as in specific targeted antimalarial chemotherapy. Targeting drugs, especially to their site of action, would indeed enable the optimal concentration of the drug in the malaria parasite-infected erythrocyte and liver tissues [88].

Generally, antimalarial drugs need to bypass various barriers such as the host red blood cell membrane (RBCM), the parasitophorous vascular membrane (PVM), the parasite plasma membrane (PPM), tubulovesicular membrane (TVM) and, possibly a further organelle membrane such as the food vacuole membrane (FVM) or the mitochondrial membrane (MM) in an infected red blood cell, depending on the site of action of the drug [89, 90]. Mature erythrocytes do not show endocytosis, while *Plasmodium falciparum* parasitized RBCs (pRBCs) undergo morphological changes, the formation of metabolic window (MW), a specific entry and exit site for metabolites [91–93]. Solutes having low molecular weight are transported into the tubulovesicular network (TVN) by specific regions of the membrane formed between TVM and RBCM, and then will be taken up by the parasite [93, 94]. TVM is a part of the secretory pathways involved in parasite protein transport across the infected erythrocyte [92, 93, 95]. It has been hypothesized that solutes and macromolecules move between the intracellular parasite and the external milieu successively across the PVM and PPM by parasitophorous duct (PD); it is a tubular membranous structure that extends between PVM and RBCM and allows the parasites to have direct contact with the extracellular solution [96, 97].

Interestingly, native serum albumin has shown high specificity and affinity interaction with the parasitized red blood cells (pRBCs) [98]. In vitro malaria parasite cultures have exhibited uptake of labeled HSA and shown parasite membrane penetration through degradation in the parasite and membrane network [99]. HSA nanocarriers have been reported for selectively targeting pRBCs with a 50% reduction in dose of artemether—a first-line antimalarial drug when encapsulated in albumin nanocarrier and tested in *Plasmodium berghei*-infected mice. This dose reduction was also demonstrated in in vitro susceptibility studies in *P. falciparum* cultures, as demonstrated by lower IC_{50} values. These nanocarriers have potential to be used in cerebral malaria as they have been demonstrated safe for intravenous administration [100]. Another study demonstrated the usefulness of HSA nanocarrier in antimalarial therapy with improved mean survival and 97.5% reduction of parasitemia in "humanized" mice infected with *P. falciparum* with indolone-*N*-oxides albumin-bound NPs [101].

6.3 Albumin nanocarrier system—A possible next-generation for antiviral therapy

Viruses are submicroscopic molecular genetic parasites contained in a protein coat. They are considered to be biological entities since they possess ribonucleic acid (RNA) or deoxyribonucleic acid (DNA) genome and are able to adapt to host habits. Viruses lack many essential attributes for a living organism such as free energy capturing and storing, self-repairing, and replication; instead, they depend on the host cellular system for their replication [102]. The high rate of multiplication represents the viruses' living characteristics; the replication process occurs through the copying process carried out by the host cellular machinery [103]. Upon viral infection, the virus will inhibit host protein expression and attempt to elude the antiviral machinery of the host and direct the host cellular machinery in the late phase of infection to synthesize viral messenger RNA (mRNA) and proteins instead of normal cellular macromolecules; the viral mRNA is responsible for the virus replication and translation [104, 105].

Antiviral drugs can target the dependence of viruses on the host cell machinery for their replication [102], leading to limit the number of virus-specific metabolic functions without any damage to the host; and the specific functions of each virus, limiting the development of broad-spectrum antiviral therapy fighting against different viruses that cause similar symptoms, are several factors that hamper the development of antiviral drugs [103]. Developing new antiviral therapies involves changing the physicochemical and biopharmaceutical properties of antiviral molecules, which is a great challenge. Using new scientific strategies like nano-drug delivery systems, which are known for their physical, chemical, and biological advantages, can be utilized to enhance the delivery of existing antiviral therapies through biological barriers [103, 106]. NPs have biomimetic properties which result in intrinsic antiviral assets. The combination of several features in a stable and well-designed structure can offer a multifunctional nano-drug delivery system, which might prove beneficial to deliver an antiviral drug to the active site of action. Due to its surface modifications capabilities and targetability, albumin has been used for the delivery of interferon alpha2a (IFNα2a) to liver cells to combat chronic hepatitis B [107]. Also, it has been considered as a promising nanocarrier for the administration of ganciclovir owing to its controlled release properties; which might contribute to improving cytomegalovirus patient acceptance and compliance [108]. Furthermore, acyclovir-loaded albumin NPs were recommended as a potential ocular drug delivery system as they exerted permeation highly across human corneal epithelial (HCE-T) cell multilayers [109]. Albumin NPs have also been examined as a delivery system for antisense oligonucleotides; they can improve stability by protecting oligonucleotides from enzymatic degradation [110, 111]. Chemically modified albumin NPs exhibited antiherpetic activity [112]. Likewise, another study done using chemically modified albumin NPs shown the potential blocking entry of Ebola viruses [113].

The magnitude of developing new antiviral formulations to combat viral infections during epidemics and/or pandemic alert has been the driving force in antiviral research. From a scientific and ethical perspective, an antiviral nanomedicine should be safe, effective, and affordable, which might be achievable using nanomedicine approach.

6.4 Albumin nanoparticles for inflammatory conditions

Albumin is a versatile protein that accumulates in inflamed tissues to regulate the physiology and metabolism in inflamed tissues like in rheumatoid arthritis (RA), where patients frequently develop hypoalbuminemia, due to high HSA uptake at sites of inflammation. The metabolism of synovial cells is highly upregulated, and the HSA uptake is probably a relevant source covering their high demand for nitrogen and energy. The permeability of the blood-joint barrier for HSA in RA patients is noticeably increased [4]. A preclinical model of mice suffering from collagen-induced arthritis demonstrated accumulation of serum albumin in the mice's arthritic paws [114]. Since the antirheumatic drug methotrexate (MTX) bound to serum albumin has shown promising activity in the collagen-induced murine arthritis model, HSA may represent an attractive drug carrier to target drugs to inflamed joints of patients with RA [4]. Furthermore, the albumin-binding prodrug of methotrexate MTX-HSA accumulated in arthritic joints, and therefore, targeted MTX to these sites, resulting in higher concentrations of MTX at the site of inflammation, showing MTX-HSA is significantly more effective than MTX alone [115]. The potential of HSA-NPs on the delivery of tacrolimus (TAC) to enhance targetability and antiarthritic efficacy was assessed by in vitro studies of TAC-HSA-NPs using splenocytes excised from spleens of mice following induction of arthritis using collagen. The data clearly demonstrated the antiproliferative activity of TAC-HSA-NPs on activated T cells compared with nonactivated T cells. Furthermore, TAC-HSA-NPs displayed significantly more antiarthritic activity than TAC formulations including intravenously administered TAC solution or oral TAC suspension due to the targetability of HSA that facilitated the accumulation of TAC-HSA-NPs at inflamed arthritis sites [116]. Additionally, the use of albumin microcapsules (3–4 μm) containing cytokine inhibited drugs which include neutralizing antibodies to TNF and IL1, CNI-1493, antisense oligonucleotides to TNF and NF-kappaB, and the antioxidant catalase. Albumin microencapsulation produced high intracellular drug concentrations due to rapid uptake by phagocytic cells, including endothelial cells, without toxicity. Excellent inhibition of TNF and IL1, resulting in improved animal survival in a peritonitis model of septic shock and inflammation in an arthritis model, was observed [117]. In another study, HSA was used as carrier for targeting therapeutic agents to activated macrophages without disturbing normal cells and tissues using folic acid (FA)-functionalized-nanocapsules. The internalization of capsules is mediated through the folate receptor (FR), which is specifically expressed by activated macrophages that play a key role in RA [118]. In a new treatment strategy for delivering Celastrol (CLT) which displayed superior antiinflammatory activity, the CLT were loaded into HSA-NPs developed with Kolliphor HS 15 (CLT-loaded HSA-HS15-NPs). The CLT-based HSA-NPs demonstrated a higher accumulation in the inflamed joints which have significantly greater therapeutic effects against RA and concomitant pneumonia than free CLT. The HSA-based NPs exhibited significantly enhanced safety. Thus, it was considered a promising and valid therapeutic candidate for RA treatment [119]. In addition, co-delivery of prednisolone and curcumin in HSA-NPs that were studied in rats with adjuvant-induced arthritis model displayed an effective treatment of RA. Subsequently, a synergistic antiinflammation effect against macrophages was shown, and the accumulation of prednisolone and curcumin-based HSA-NPs at the inflamed joint sites was observed [120]. Therefore, HSA-NPs have a potential advantage as a therapeutic transporter for antiinflammatory drugs.

6.5 Albumin nanoparticles for brain delivery

For any drug intended for brain delivery, it is essential to bypass and penetrate the formidable blood-brain barrier (BBB), which regulates the passage of molecules from the bloodstream to the brain. The presence of BBB makes it impossible for more than 98% of potential new therapeutic molecules to cross the BBB [121]. The BBB comprises endothelial cells connected by tight junctions, in contact with the pericytes as well as astrocytes [122]. Also, few pinocytosis vesicles, and tight junctions, also known as zonula occludens, tightly regulate the movement of molecules through the paracellular pathway comprising an almost impermeable barrier for drugs administered through the peripheral circulation. The astrocytes give a further contribution to the BBB functions and play a vital multitude for brain homeostasis (maintenance of potassium ion levels, inactivation of neurotransmitters, and regulation and production of growth factors and cytokines) [123].

There are several pathways for transport across the BBB as schematically shown in Fig. 6. Briefly, many lipophilic molecules can be diffused through the BBB and reach the brain passively [124]. Molecular carriers present at the apical and basolateral side of the BBB selectively transport small molecules. Others, like transporting macromolecules like albumin and caveole, play an essential role in its transcytosis. The receptor-mediated transcytosis requires binding of a molecule to its specific receptor. However, adsorption-mediated transcytosis is nonselective, mediated by polycationic molecule binding to the negatively charged membrane [125]. Various HSA-NPs have been explored and fabricated for brain drug delivery, and their applications were investigated [15, 62]. The loperamide transport across the BBB was investigated when it was loaded into HSA-NPs and covalently surfaced modified with Transferrin (TfR) and transferrin-receptor-antibody-modified (TfR mAbs) (OX26 or R 17217). Loperamide-targeted HSA-NPs successfully crossed the BBB and induced antinociceptive analgesic effects in the brain, which was established in the tail-flick test [126]. In another study, the same authors used HSA-NPs with covalently attached insulin or an antiinsulin receptor monoclonal antibody to deliver loperamide across the BBB, which induced significant effects using a similar test [127].

Another formulation of HSA-NPs bound to HI 6 dimethanesulfonate and HI 6 dichloride monohydrate was developed and their effect was investigated for the

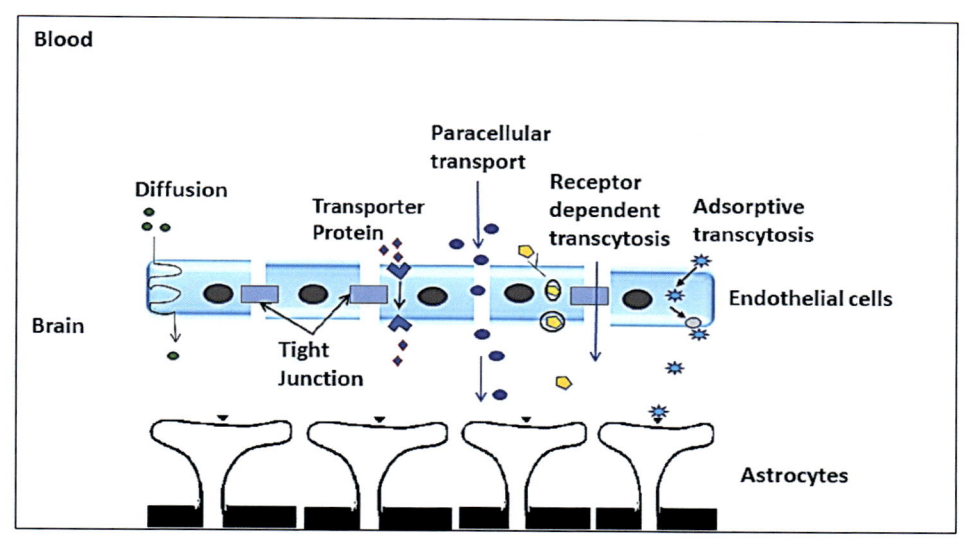

Fig. 6 Potential transport mechanisms across the blood-brain barrier, where diffusion and active transport are the main transport mechanisms.

treatment of poisoning by organophosphorus compounds. The data demonstrated HSA-NPs transported the bonded drugs through the BBB [128]. Additionally, Chitosan (CS)-coating on HSA-NPs for nasal delivery was investigated recently and was considered as a promising approach for nose-to-brain drug delivery [129]. A synthesized novel formulation for targeted glioma with paclitaxel (PTX)-loaded HSA-NPs with excellent properties in terms of drug–loading content, entrapment efficiency, spherical structure with desirable small particle size, and negative charge showed significant toxicity toward U87MG cells following the uptake by brain capillary endothelial cells and considered as an excellent carrier system for brain tumor delivery. The HSA-NPs were found to improve the drug accumulation in the targeted region after penetration of the BBB [130].

Another approach for brain delivery using HSA as a drug carrier has been investigated for treating Alzheimer's disease (AD). Tacrine was one of the first drug to receive FDA approval for treating AD in 1993, but suffered from drawbacks like low bioavailability due to hepatic first-pass effect, short elimination half-life, and incidence of side effects including hepatotoxicity which led to withdrawal of the drug. In order to improve delivery to brain [36], albumin-based NPs carrying β-cyclodextrins (and its hydrophilic derivatives) have been investigated to deliver tacrine via the intranasal route. It was noticeable that the presence of the different beta cyclodextrins in the polymeric network affected drug loading and could differently modulate NPs muco-adhesiveness and drug permeation behaviors [36]. In another study, rivastigmine tartrate (RT), a short–acting cholinesterase inhibitor used for AD, was loaded into HSA-NPs. The developed NPs were coated with polysorbate 80 to facilitate brain targeting via endocytosis; the data acquired from in vitro studies showed $55.59 \pm 3.80\%$ release of drug from RT-HSA-NPs in 12 h and demonstrated the suitability of HSA NPs as a potential carrier for providing sustained delivery of RT [131].

6.6 Albumin nanocarriers for pulmonary delivery

The pulmonary route has gained a lot of interest recently as a noninvasive administration for both systemic and local delivery of therapeutic agents, due to the high permeability and large absorptive surface area of lungs (approximately $70–140 \, m^2$ in adult humans having extremely thin absorptive mucosal membrane), and good blood supply HSA nanocarrier has a high capacity for drug loading that might help in reducing toxic effects of cytostatic agents and solve problems pertaining to multidrug resistance. The administration of drug-loaded HSA NPs intended for a pulmonary route for local delivery of a drug to the lungs and also for systemic drug delivery has shown excellent bioavailability on account of large absorption surface area, skipping of first-pass metabolism resulting in a rapid onset of action, resulting in better patient compliance [132, 133]. It was reported that tacrolimus (Tac), an immunosuppressant drug, was loaded into albumin NPs (Tac Alb-NPs) to give small particles with size <200 nm diameter with a negative charge and sustained release over 24 h after administration in lung tissues of mice. The Tac Alb-NPs' antifibrotic activity showed better effect through the inhalation route than Tac formulation given via intraperitoneal route [134]. In another study, TRAIL/Dox-HSA-NPs displayed synergistic cytotoxicity and apoptotic activity in H226 lung cancer cells. TRAIL/Dox-HSA-NPs were well-deposited in mouse lungs after administration by aerosolizer with a gradual release of the formulation over 3 days. It was reported that in BALB/c nu/nu mice bearing H226 cell-induced metastatic, notably tumors were smaller in size and lighter when treated with TRAIL/Dox-HSA-NPs rather than when TRAIL or Dox-HSA-NPs alone were administered, which could be beneficial to reduce Dox doses and minimize its side effects [135]. Additionally, albumin NPs can also be used as a carrier for treating lung diseases such as asthma and chronic obstructive pulmonary diseases. Albumin-NPs loaded with an antiasthmatic drug terbutaline sulfate were prepared by modified desolvation and evaluated through in vivo rat lung model. The data showed effective localization of the drug in the lungs with a trace of the drug in the blood samples of rat for up to 48 h which confirmed the presence of terbutaline sulfate-loaded albumin NPs in the lungs of the rat, which indicates the retention of drug in lungs without evident systemic delivery when administered via the pulmonary route [136]. Overall, albumin NPs have gained lot of popularity due to their significant advantages which opens opportunities for its application for local lung delivery for diseases like cancer, asthma, and infectious diseases. Albumin has a wide binding site to ligand/molecules that facilitated active targeting for diseases like cancer and targeted localization of drug, which proved to enhance its use as carrier system for treating pulmonary diseases [133].

6.7 Albumin nanoparticles for diagnosis

Albumin-based NPs have been investigated for their use for diagnostic purposes and bioimaging since they avoid capture by the reticuloendothelial system and reduce fluorescence interferences from untargeted organs during imaging. The diagnostic efficacy and safety of Albunex have been supported by various clinical studies [137]. Albunex is primarily used for myocardial contrast echocardiography, though the data demonstrated the ability of albumin in the diagnosis of various diseases. It is the first ultrasound contrast agent that is stable enough to show transpulmonary passage, which implies that the contrast agent can be injected intravenously and still give contrast in the left side of the heart. Albunex is a commercially prepared solution of air–filled albumin microspheres that have been proved to be effective contrast agents with no side effects on coronary blood flow, left ventricular function, and systemic hemodynamics prepared from sonicated 5% HSA. In another technique, the positron emission tomography (PET) imaging in an orthotopic lung cancer model was evaluated through the use of ultrasmall chelator-free radioactive [64Cu]Cu nanoclusters, which were developed using albumin as a scaffold. The model utilizing PET showed a preferable advantage of unlimited tissue penetration, high sensitivity, high temporal resolution, and deep penetration imaging of orthotopic lung cancer in vivo compared with near-infrared fluorescence imaging [138].

To further overcome the intrinsic drawbacks of optical imaging, a dual–modal nanoprobe was designed by modifying the albumin-templated Ag2S quantum dots (QDs) with gadolinium (Gd) by exploiting the Gd-diethylenetriaminepentaacetic acid (DTPAGd). The obtained nanoprobe combined both high sensitivity of optical imaging and high spatial resolution and unlimited penetration depth of Magnetic resonance imaging (MRI), which might benefit tiny tumor diagnosis [139]. Another multimodal nanoprobe that combined the complementary advantages of near-infrared (NIR) fluorescence, X-ray computed tomography (CT), and MRI utilized the high sensitivity of NIR fluorescence and the high spatial resolution and excellent tissue penetration depth of X–CT and MRI. The integration of multiple components in a single nanostructure helps obtain complete information ranging from noninvasive whole-body examinations to microscopic tissue examinations [140]. In a recent study, albumin–based Iohexol-NPs were developed; Iohexol is a commonly used second-generation nonionic iodinated contrast agent. The collected data showed an efficiently entrapped iohexol in IAP and P–IAP NPs. Additionally, in vivo CT imaging for the developed formulations manifested an enhancement in the anatomical structures of heart, liver, and kidneys along with enhanced residence time and produced stable and high opacification of the blood pool than the marketed Iopamidol [141]. Therefore, HSA protein was thought to be an eligible candidate for developing universal nanomedicine for targeted therapy and diagnostics and further studies in clinical conditions are required to improve the manufacturing quality and guarantee the clinical success of albumin-based nanomedicine.

7. Conclusion

It is quite clear that nanomedicines hold great potential to vanquish the problems associated with delivery of therapeutic drugs and imaging agents. Nanotechnology has gained much attention in recent years for producing nanosize systems with different compositions and biological properties which have been extensively investigated for drug delivery applications and other diagnostic aspects. Albumin-based nanoparticles are promising vehicles for multipurpose applications that show a great deal of potential in various in vitro and in vivo studies, demonstrating albumin nanoparticles have a bright future in the controlled delivery of therapeutic agents and ideal lead in the field of drug delivery due to their exceptional advantages and their exciting possibility to overcome the problems of drug resistance in targeting cells and to aid transport of drugs across various biological barriers. The challenges, however, remain and yet to be discovered; the precise characterization of molecular targets without affecting the normal cells/organs, reproducibility, and mass production are few of the many challenges and the future remains exciting and wide open to explore all the possibilities.

References

[1] L. Arms, D.W. Smith, J. Flynn, W. Palmer, A. Martin, A. Woldu, S. Hua, Advantages and limitations of current techniques for analyzing the biodistribution of nanoparticles, Front. Pharmacol. 9 (2018), https://doi.org/10.3389/fphar.2018.00802.

[2] M.R. Gwinn, V. Vallyathan, Nanoparticles: health effects—pros and cons, Environ. Health Perspect. 114 (2006) 1818–1825, https://doi.org/10.1289/ehp.8871.

[3] C. Yewale, D. Baradia, I. Vhora, A. Misra, Proteins: emerging carrier for delivery of cancer therapeutics, Expert Opin. Drug Deliv. (2013), https://doi.org/10.1517/17425247.2013.805200.

[4] F. Kratz, Albumin as a drug carrier: design of prodrugs, drug conjugates and nanoparticles, J. Control. Release 132 (2008) 171–183, https://doi.org/10.1016/j.jconrel.2008.05.010.

[5] M. Karimi, S. Bahrami, S.B. Ravari, P.S. Zangabad, H. Mirshekari, M. Bozorgomid, S. Shahreza, M. Sori, M.R. Hamblin, Albumin nanostructures as advanced drug delivery systems, Expert Opin. Drug Deliv. 13 (2016) 1609–1623, https://doi.org/10.1080/17425247.2016.1193149.

[6] D. Sleep, Albumin and its application in drug delivery, Expert Opin. Drug Deliv. 12 (2015) 793–812, https://doi.org/10.1517/17425247.2015.993313.

[7] K.A. Majorek, P.J. Porebski, A. Dayal, M.D. Zimmerman, K. Jablonska, A.J. Stewart, M. Chruszcz, W. Minor, Structural and immunologic characterization of bovine, horse, and rabbit serum albumins, Mol. Immunol. 52 (2012) 174–182, https://doi.org/10.1016/j.molimm.2012.05.011.

[8] F. Zsila, Subdomain IB is the third major drug binding region of human serum albumin: toward the three-sites model, Mol. Pharm. 10 (2013) 1668–1682, https://doi.org/10.1021/mp400027q.

[9] M. Mondal, P. Lakshmi T, R. Krishna, N. Sakthivel, Molecular interaction between human serum albumin (HSA) and phloroglucinol derivative that shows selective anti-proliferative potential, J. Lumin. 192 (2017) 990–998, https://doi.org/10.1016/j.jlumin.2017.08.007.

[10] M. Fasano, S. Curry, E. Terreno, M. Galliano, G. Fanali, P. Narciso, S. Notari, P. Ascenzi, The extraordinary ligand binding properties of human serum albumin, IUBMB Life 57 (2005) 787–796, https://doi.org/10.1080/15216540500404093.

[11] J.Y. Jun, H.H. Nguyen, S.Y.R. Paik, H.S. Chun, B.C. Kang, S. Ko, Preparation of size-controlled bovine serum albumin (BSA) nanoparticles by a modified desolvation method, Food Chem. 127 (2011) 1892–1898, https://doi.org/10.1016/j.foodchem.2011.02.040.

[12] C. Weber, J. Kreuter, K. Langer, Desolvation process and surface characteristics of HSA-nanoparticles, Int. J. Pharm. 196 (2000) 197–200, https://doi.org/10.1016/S0378-5173(99)00420-2.

[13] K. Langer, S. Balthasar, V. Vogel, N. Dinauer, H. von Briesen, D. Schubert, Optimization of the preparation process for human serum albumin (HSA) nanoparticles, Int. J. Pharm. 257 (2003) 169–180, https://doi.org/10.1016/S0378-5173(03)00134-0.

[14] C. Pinto Reis, R.J. Neufeld, A.J. Ribeiro, F. Veiga, Nanoencapsulation I. Methods for preparation of drug-loaded polymeric nanoparticles, Nanomed. Nanotechnol. Biol. Med. 2 (2006) 8–21, https://doi.org/10.1016/j.nano.2005.12.003.

[15] A.O. Elzoghby, W.M. Samy, N.A. Elgindy, Albumin-based nanoparticles as potential controlled release drug delivery systems, J. Control. Release 157 (2012) 168–182, https://doi.org/10.1016/j.jconrel.2011.07.031.

[16] L. Yang, F. Cui, D. Cun, A. Tao, K. Shi, W. Lin, Preparation, characterization and biodistribution of the lactone form of 10-hydroxycamptothecin (HCPT)-loaded bovine serum albumin (BSA) nanoparticles, Int. J. Pharm. 340 (2007) 163–172, https://doi.org/10.1016/j.ijpharm.2007.03.028.

[17] N. Desai, Nab technology: a drug delivery platform utilising endothelial gp60 receptor-based transport and tumour-derived SPARC for targeting, Drug Deliv. Rep. (2007) 37–41.

[18] C. Li, Y. Li, Y. Gao, N. Wei, X. Zhao, C. Wang, Y. Li, X. Xiu, J. Cui, Direct comparison of two albumin-based paclitaxel-loaded nanoparticle formulations: is the crosslinked version more advantageous? Int. J. Pharm. 468 (2014) 15–25, https://doi.org/10.1016/j.ijpharm.2014.04.010.

[19] J. Li, P. Yao, Self-assembly of ibuprofen and bovine serum albumin−dextran conjugates leading to effective loading of the drug, Langmuir 25 (2009) 6385–6391, https://doi.org/10.1021/la804288u.

[20] G. Battogtokh, J.H. Kang, Y.T. Ko, Long-circulating self-assembled cholesteryl albumin nanoparticles enhance tumor accumulation of hydrophobic anticancer drug, Eur. J. Pharm. Biopharm. 96 (2015) 96–105, https://doi.org/10.1016/j.ejpb.2015.07.013.

[21] J. Gong, M. Huo, J. Zhou, Y. Zhang, X. Peng, D. Yu, H. Zhang, J. Li, Synthesis, characterization, drug-loading capacity and safety of novel octyl modified serum albumin micelles, Int. J. Pharm. 376 (2009) 161–168, https://doi.org/10.1016/j.ijpharm.2009.04.033.

[22] M.S. Safavi, S.A. Shojaosadati, H.G. Yang, Y. Kim, E.J. Park, K.C. Lee, D.H. Na, Reducing agent-free synthesis of curcumin-loaded albumin nanoparticles by self-assembly at room temperature, Int. J. Pharm. 529 (2017) 303–309, https://doi.org/10.1016/j.ijpharm.2017.06.087.

[23] S.H. Lee, D. Heng, W.K. Ng, H.-K. Chan, R.B.H. Tan, Nano spray drying: a novel method for preparing protein nanoparticles for protein therapy, Int. J. Pharm. 403 (2011) 192–200, https://doi.org/10.1016/j.ijpharm.2010.10.012.

[24] C. Arpagaus, A novel laboratory-scale spray dryer to produce nanoparticles, Dry. Technol. (2012), https://doi.org/10.1080/07373937.2012.686949.

[25] R. Maiti, S. Panigrahi, Y. Tingjie, Bovine serum albumin nanoparticles constructing procedures on anticancer activities, Int. J. Adv. Res. Biol. Sci. (2018), https://doi.org/10.22192/ijarbs.

[26] J. Ong, J. Zhao, A.W. Justin, A.E. Markaki, Albumin-based hydrogels for regenerative engineering and cell transplantation, Biotechnol. Bioeng. 116 (2019) 3457–3468, https://doi.org/10.1002/bit.27167.

[27] J. Barley, Freeze Drying/Lyophilization Information: Basic Principles, SP Sci, 2009.

[28] A. Baheti, L. Kumar, A.K. Bansal, Excipients used in lyophilization of small molecules, J. Excipients Food Chem. 1 (2010) 41–54.

[29] A. Bansal, S. Lale, M. Goyal, Development of lyophilization cycle and effect of excipients on the stability of catalase during lyophilization, Int. J. Pharm. Investig. 1 (2011) 214, https://doi.org/10.4103/2230-973X.93007.

[30] P.A. Kramer, Albumin microspheres as vehicles for achieving specificity in drug delivery, J. Pharm. Sci. 63 (1974) 1646–1647, https://doi.org/10.1002/jps.2600631044.

[31] H. Thakkar, R.K. Sharma, A.K. Mishra, K. Chuttani, R.R. Murthy, Albumin microspheres as carriers for the antiarthritic drug celecoxib, AAPS PharmSciTech 6 (2005) E65–E73, https://doi.org/10.1208/pt060112.

[32] S.T. Mathew, S.G. Devi, K. Sandhya, K. Sandhya, Formulation and evaluation of ketorolac tromethamine-loaded albumin microspheres for potential intramuscular administration, AAPS PharmSciTech 8 (2007) E100–E108, https://doi.org/10.1208/pt0801014.

[33] S.A. Sajadi Tabassi, N. Razavi, Preparation and characterization of albumin microspheres encapsulated with propranolol HCl, Daru 11 (2003) 137–141.

[34] D.W. Nyman, K.J. Campbell, E. Hersh, K. Long, K. Richardson, V. Trieu, N. Desai, M.J. Hawkins, D.D. Von Hoff, Phase I and pharmacokinetics trial of ABI-007, a novel nanoparticle formulation of paclitaxel in patients with advanced nonhematologic malignancies, J. Clin. Oncol. 23 (2005) 7785–7793, https://doi.org/10.1200/JCO.2004.00.6148.

[35] M.J. Hawkins, P. Soon-Shiong, N. Desai, Protein nanoparticles as drug carriers in clinical medicine, Adv. Drug Deliv. Rev. 60 (2008) 876–885, https://doi.org/10.1016/j.addr.2007.08.044.

[36] B. Luppi, F. Bigucci, G. Corace, A. Delucca, T. Cerchiara, M. Sorrenti, L. Catenacci, A.M. Di Pietra, V. Zecchi, Albumin nanoparticles carrying cyclodextrins for nasal delivery of the anti-Alzheimer drug tacrine, Eur. J. Pharm. Sci. 44 (2011) 559–565, https://doi.org/10.1016/j.ejps.2011.10.002.

[37] B. Wilson, T.V. Ambika, R. Dharmesh Kumar Patel, J.L. Jenita, S.R.B. Priyadarshini, Nanoparticles based on albumin: preparation, characterization and the use for 5-flurouracil delivery, Int. J. Biol. Macromol. 51 (2012) 874–878, https://doi.org/10.1016/j.ijbiomac.2012.07.014.

[38] F.-Q. Li, H. Su, J. Wang, J.-Y. Liu, Q.-G. Zhu, Y.-B. Fei, Y.-H. Pan, J.-H. Hu, Preparation and characterization of sodium ferulate entrapped bovine serum albumin nanoparticles for liver targeting, Int. J. Pharm. 349 (2008) 274–282, https://doi.org/10.1016/j.ijpharm.2007.08.001.

[39] B. Wilson, Y. Lavanya, S.R.B. Priyadarshini, M. Ramasamy, J.L. Jenita, Albumin nanoparticles for the delivery of gabapentin: preparation, characterization and pharmacodynamic studies, Int. J. Pharm. 473 (2014) 73–79, https://doi.org/10.1016/j.ijpharm.2014.05.056.

[40] Zu, Preparation, characterization, and in vitro targeted delivery of folate-decorated paclitaxel-loaded bovine serum albumin nanoparticles, Int. J. Nanomedicine (2010) 669, https://doi.org/10.2147/IJN.S12918.

[41] Zu, Optimization of the preparation process of vinblastine sulfate (VBLS)-loaded folate-conjugated bovine serum albumin (BSA) nanoparticles for tumor-targeted drug delivery using response surface methodology (RSM), Int. J. Nanomedicine (2009) 321, https://doi.org/10.2147/IJN.S8501.

[42] A. Taheri, R. Dinarvand, F. Ahadi, M.R. Khorramizadeh, F. Atyabi, The in vivo antitumor activity of LHRH targeted methotrexate–human serum albumin nanoparticles in 4T1 tumor-bearing Balb/c mice, Int. J. Pharm. 431 (2012) 183–189, https://doi.org/10.1016/j.ijpharm.2012.04.033.

[43] Y.-R. Zheng, K. Suntharalingam, T.C. Johnstone, H. Yoo, W. Lin, J.G. Brooks, S.J. Lippard, Pt(IV) prodrugs designed to bind non-covalently to human serum albumin for drug delivery, J. Am. Chem. Soc. 136 (2014) 8790–8798, https://doi.org/10.1021/ja5038269.

[44] Y. Zhang, A. Ho, J. Yue, L. Kong, Z. Zhou, X. Wu, F. Yang, H. Liang, Structural basis and anticancer properties of ruthenium-based drug complexed with human serum albumin, Eur. J. Med. Chem. 86 (2014) 449–455, https://doi.org/10.1016/j.ejmech.2014.08.071.

[45] L. Zhao, Y. Zhou, Y. Gao, S. Ma, C. Zhang, J. Li, D. Wang, X. Li, C. Li, Y. Liu, X. Li, Bovine serum albumin nanoparticles for delivery of tacrolimus to reduce its kidney uptake and functional nephrotoxicity, Int. J. Pharm. 483 (2015) 180–187, https://doi.org/10.1016/j.ijpharm.2015.02.018.

[46] W. Weecharangsan, R.J. Lee, Growth inhibition and chemosensitization of human carcinoma cells by human serum albumin–coated liposomal antisense oligodeoxyribonucleotide against bcl-2, Drug Deliv. 19 (2012) 292–297, https://doi.org/10.3109/10717544.2012.714810.

[47] S.H. Jung, S.K. Kim, S.H. Jung, E.H. Kim, S.H. Cho, K.-S. Jeong, H. Seong, B.C. Shin, Increased stability in plasma and enhanced cellular uptake of thermally denatured albumin-coated liposomes, Colloids Surf. B Biointerfaces 76 (2010) 434–440, https://doi.org/10.1016/j.colsurfb.2009.12.002.

[48] K.H. Martin, P.A. Dayton, Current status and prospects for microbubbles in ultrasound theranostics, Wiley Interdiscip. Rev. Nanomed. Nanobiotechnol. 5 (2013) 329–345, https://doi.org/10.1002/wnan.1219.

[49] A.C. Anselmo, S. Mitragotri, Nanoparticles in the clinic, Bioeng. Transl. Med. 1 (2016) 10–29, https://doi.org/10.1002/btm2.10003.

[50] J. Wang, P. Li, R. Tian, W. Hu, Y. Zhang, P. Yuan, Y. Tang, Y. Jia, L. Zhang, A novel microbubble capable of ultrasound-triggered release of drug-loaded nanoparticles, J. Biomed. Nanotechnol. 12 (2016) 516–524, https://doi.org/10.1166/jbn.2016.2181.

[51] C.W. Burke, Y.-H.J. Hsiang, E. Alexander, A.L. Kilbanov, R.J. Price, Covalently linking poly(lactic-co–glycolic acid) nanoparticles to microbubbles before intravenous injection improves their ultrasound-targeted delivery to skeletal muscle, Small 7 (2011) 1227–1235, https://doi.org/10.1002/smll.201001934.

[52] S. Dang, R. Wang, M. Qin, Y. Zhang, Y. Gu, M. Wang, Q. Yang, X. Li, X. Zhang, A novel transfection method for eukaryotic cells using polyethylenimine coated albumin microbubbles, Plasmid 66 (2011) 19–25, https://doi.org/10.1016/j.plasmid.2011.03.003.

[53] E. Pereverzeva, I. Treschalin, D. Bodyagin, O. Maksimenko, K. Langer, S. Dreis, B. Asmussen, J. Kreuter, S. Gelperina, Influence of the formulation on the tolerance profile of nanoparticle-bound doxorubicin in healthy rats: focus on cardio- and testicular toxicity, Int. J. Pharm. 337 (2007) 346–356, https://doi.org/10.1016/j.ijpharm.2007.01.031.

[54] H. Kouchakzadeh, S.A. Shojaosadati, A. Maghsoudi, E. Vasheghani Farahani, Optimization of PEGylation conditions for BSA nanoparticles using response surface methodology, AAPS PharmSciTech 11 (2010) 1206–1211, https://doi.org/10.1208/s12249-010-9487-8.

[55] W. Lin, M.C. Garnett, S.S. Davis, E. Schacht, P. Ferruti, L. Illum, Preparation and characterisation of rose Bengal-loaded surface-modified albumin nanoparticles, J. Control. Release 71 (2001) 117–126, https://doi.org/10.1016/S0168-3659(01)00209-7.

[56] S. Zhang, C. Kucharski, M.R. Doschak, W. Sebald, H. Uludağ, Polyethylenimine–PEG coated albumin nanoparticles for BMP-2 delivery, Biomaterials 31 (2010) 952–963, https://doi.org/10.1016/j.biomaterials.2009.10.011.

[57] Y. Lu, P.S. Low, Folate-mediated delivery of macromolecular anticancer therapeutic agents, Adv. Drug Deliv. Rev. 64 (2012) 342–352, https://doi.org/10.1016/j.addr.2012.09.020.

[58] Z. Shen, Y. Li, K. Kohama, B. Oneill, J. Bi, Improved drug targeting of cancer cells by utilizing actively targetable folic acid-conjugated albumin nanospheres, Pharmacol. Res. 63 (2011) 51–58, https://doi.org/10.1016/j.phrs.2010.10.012.

[59] D. Chen, Q. Tang, W. Xue, J. Xiang, L. Zhang, X. Wang, The preparation and characterization of folate-conjugated human serum albumin magnetic cisplatin nanoparticles, J. Biomed. Res. 24 (2010) 26–32, https://doi.org/10.1016/S1674-8301(10)60005-X.

[60] L. Zhang, S. Hou, J. Zhang, W. Hu, C. Wang, Preparation, characterization, and in vivo evaluation of mitoxantrone-loaded, folate-conjugated albumin nanoparticles, Arch. Pharm. Res. 33 (2010) 1193–1198, https://doi.org/10.1007/s12272-010-0809-x.

[61] S. Wagner, A. Zensi, S.L. Wien, S.E. Tschickardt, W. Maier, T. Vogel, F. Worek, C.U. Pietrzik, J. Kreuter, H. von Briesen, Uptake mechanism of ApoE-modified nanoparticles on brain capillary endothelial cells as a blood-brain barrier model, PLoS One 7 (2012) e32568, https://doi.org/10.1371/journal.pone.0032568.

[62] A. Zensi, D. Begley, C. Pontikis, C. Legros, L. Mihoreanu, S. Wagner, C. Büchel, H. von Briesen, J. Kreuter, Albumin nanoparticles targeted with Apo E enter the CNS by transcytosis and are delivered to neurones, J. Control. Release 137 (2009) 78–86, https://doi.org/10.1016/j.jconrel.2009.03.002.

[63] H. Wartlick, K. Michaelis, S. Balthasar, K. Strebhardt, J. Kreuter, K. Langer, Highly specific HER2-mediated cellular uptake of antibody-modified nanoparticles in tumour cells, J. Drug Target. 12 (2004) 461–471, https://doi.org/10.1080/10611860400010697.

[64] I.M. Steinhauser, K. Langer, K.M. Strebhardt, B. Spänkuch, Effect of trastuzumab-modified antisense oligonucleotide-loaded human serum albumin nanoparticles prepared by heat denaturation, Biomaterials 29 (2008) 4022–4028, https://doi.org/10.1016/j.biomaterials.2008.07.001.

[65] K. Löw, M. Wacker, S. Wagner, K. Langer, H. von Briesen, Targeted human serum albumin nanoparticles for specific uptake in EGFR-expressing colon carcinoma cells, Nanomed. Nanotechnol. Biol. Med. 7 (2011) 454–463, https://doi.org/10.1016/j.nano.2010.12.003.

[66] S. Zhang, M.R. Doschak, H. Uludağ, Pharmacokinetics and bone formation by BMP-2 entrapped in polyethylenimine-coated albumin nanoparticles, Biomaterials 30 (2009) 5143–5155, https://doi.org/10.1016/j.biomaterials.2009.05.060.

[67] S. Abbasi, A. Paul, W. Shao, S. Prakash, Cationic albumin nanoparticles for enhanced drug delivery to treat breast cancer: preparation and in vitro assessment, J. Drug Deliv. 2012 (2012) 1–8, https://doi.org/10.1155/2012/686108.

[68] H.D. Singh, G. Wang, H. Uludağ, L.D. Unsworth, Poly-l-lysine-coated albumin nanoparticles: stability, mechanism for increasing in vitro enzymatic resilience, and siRNA release characteristics, Acta Biomater. 6 (2010) 4277–4284, https://doi.org/10.1016/j.actbio.2010.06.017.

[69] Z. Shen, W. Wei, Y. Zhao, G. Ma, T. Dobashi, Y. Maki, Z. Su, J. Wan, Thermosensitive polymer-conjugated albumin nanospheres as thermal targeting anti-cancer drug carrier, Eur. J. Pharm. Sci. 35 (2008) 271–282, https://doi.org/10.1016/j.ejps.2008.07.006.

[70] L. Bello, J. Zhang, D.C. Nikas, J.F. Strasser, R.M. Villani, D.A. Cheresh, R.S. Carroll, P.M. Black, Alpha(v)beta3 and alpha(v)beta5 integrin expression in meningiomas, Neurosurgery 47 (2000) 1185–1195.

[71] P.K. Dubey, D. Singodia, R.K. Verma, S.P. Vyas, RGD modified albumin nanospheres for tumour vasculature targeting, J. Pharm. Pharmacol. 63 (2011) 33–40, https://doi.org/10.1111/j.2042-7158.2010.01180.x.

[72] D. Chen, D. Chen, W. Xue, J. Xiang, Y. Gong, L. Zhang, C. Guo, Preparation and biodistribution of 188Re-labeled folate conjugated human serum albumin magnetic cisplatin nanoparticles (188Re-folate-CDDP/HSA MNPs) in vivo, Int. J. Nanomedicine (2011) 3077, https://doi.org/10.2147/IJN.S24322.

[73] H.-N. Xu, H.-J. Chen, B.-Y. Zheng, Y.-Q. Zheng, M.-R. Ke, J.-D. Huang, Preparation and sono-dynamic activities of water-soluble tetra-α-(3-carboxyphenoxyl) zinc(II) phthalocyanine and its bovine serum albumin conjugate, Ultrason. Sonochem. 22 (2015) 125–131, https://doi.org/10.1016/j.ultsonch.2014.05.019.

[74] A. Martínez, I. Iglesias, R. Lozano, J.M. Teijón, M.D. Blanco, Synthesis and characterization of thiolated alginate-albumin nanoparticles stabilized by disulfide bonds. Evaluation as drug delivery systems, Carbohydr. Polym. 83 (2011) 1311–1321, https://doi.org/10.1016/j.carbpol.2010.09.038.

[75] K. Banik, A.M. Ranaware, C. Harsha, T. Nitesh, S. Girisa, V. Deshpande, L. Fan, S.P. Nalawade, G. Sethi, A.B. Kunnumakkara, Piceatannol: a natural stilbene for the prevention and treatment of cancer, Pharmacol. Res. 153 (2020) 104635, https://doi.org/10.1016/j.phrs.2020.104635.

[76] S. Tomao, Albumin-bound formulation of paclitaxel (Abraxane® ABI-007) in the treatment of breast cancer, Int. J. Nanomedicine (2009) 99, https://doi.org/10.2147/IJN.S3061.

[77] M.R. Green, G.M. Manikhas, S. Orlov, B. Afanasyev, A.M. Makhson, P. Bhar, M.J. Hawkins, Abraxane®, a novel Cremophor®-free, albumin-bound particle form of paclitaxel for the treatment of advanced non-small-cell lung cancer, Ann. Oncol. 17 (2006) 1263–1268, https://doi.org/10.1093/annonc/mdl104.

[78] K. Kimura, K. Yamasaki, K. Nishi, K. Taguchi, M. Otagiri, Investigation of anti-tumor effect of doxorubicin-loaded human serum albumin nanoparticles prepared by a desolvation technique, Cancer Chemother. Pharmacol. 83 (2019) 1113–1120, https://doi.org/10.1007/s00280-019-03832-3.

[79] N. Taneja, K.K. Singh, Rational design of polysorbate 80 stabilized human serum albumin nanoparticles tailored for high drug loading and entrapment of irinotecan, Int. J. Pharm. 536 (2018) 82–94, https://doi.org/10.1016/j.ijpharm.2017.11.024.

[80] L. Teng, R. Lee, Y. Sun, G. Cai, J. Wang, M. Wang, J. Lu, Q. Meng, L. Teng, D. Wang, N. Qu, Cabazitaxel-loaded human serum albumin nanoparticles as a therapeutic agent against prostate cancer, Int. J. Nanomedicine 11 (2016) 3451–3459, https://doi.org/10.2147/IJN.S105420.

[81] J.E. Lee, M.G. Kim, Y.L. Jang, M.S. Lee, N.W. Kim, Y. Yin, J.H. Lee, S.Y. Lim, J.W. Park, J. Kim, D.-S. Lee, S.H. Kim, J.H. Jeong, Self-assembled PEGylated albumin nanoparticles (SPAN) as a platform for cancer chemotherapy and imaging, Drug Deliv. 25 (2018) 1570–1578, https://doi.org/10.1080/10717544.2018.1489430.

[82] H. Kouchakzadeh, M.S. Safavi, S.A. Shojaosadati, Efficient delivery of therapeutic agents by using targeted albumin nanoparticles, Adv. Protein Chem. Struct. Biol. (2015) 121–143, https://doi.org/10.1016/bs.apcsb.2014.11.002.

[83] T. Geng, X. Zhao, M. Ma, G. Zhu, L. Yin, Resveratrol-loaded albumin nanoparticles with prolonged blood circulation and improved biocompatibility for highly effective targeted pancreatic tumor therapy, Nanoscale Res. Lett. 12 (2017) 437, https://doi.org/10.1186/s11671-017-2206-6.

[84] Y. Shen, W. Li, HA/HSA co-modified erlotinib–albumin nanoparticles for lung cancer treatment, Drug Des. Devel. Ther. 12 (2018) 2285–2292, https://doi.org/10.2147/DDDT.S169734.

[85] A. Taheri, R. Dinarvand, F. Atyabi, M.H. Ghahremani, S.N. Ostad, Trastuzumab decorated methotrexate-human serum albumin conjugated nanoparticles for targeted delivery to HER2 positive tumor cells, Eur. J. Pharm. Sci. (2012), https://doi.org/10.1016/j.ejps.2012.06.016.

[86] B. Lian, M. Wu, Z. Feng, Y. Deng, C. Zhong, X. Zhao, Folate-conjugated human serum albumin-encapsulated resveratrol nanoparticles: preparation, characterization, bioavailability and targeting of liver tumors, Artif. Cells Nanomed. Biotechnol. 47 (2019) 154–165, https://doi.org/10.1080/21691401.2018.1548468.

[87] N.S. Santos-Magalhães, V.C.F. Mosqueira, Nanotechnology applied to the treatment of malaria, Adv. Drug Deliv. Rev. 62 (2010) 560–575, https://doi.org/10.1016/j.addr.2009.11.024.

[88] P. Urbán, J. Estelrich, A. Cortés, X. Fernàndez-Busquets, A nanovector with complete discrimination for targeted delivery to Plasmodium falciparum-infected versus non-infected red blood cells in vitro, J. Control. Release 151 (2011) 202–211, https://doi.org/10.1016/j.jconrel.2011.01.001.

[89] K. Kirk, Channels and transporters as drug targets in the Plasmodium-infected erythrocyte, Acta Trop. 89 (2004) 285–298, https://doi.org/10.1016/j.actatropica.2003.10.002.

[90] K. Kirk, Membrane transport in the malaria-infected erythrocyte, Physiol. Rev. 81 (2001) 495–537, https://doi.org/10.1152/physrev.2001.81.2.495.

[91] J.E. Bodammer, G.F. Bahr, The initiation of a "metabolic window" in the surface of host erythrocytes by Plasmodium berghei NYU 2, Lab. Investig. 28 (1973) 708–718.

[92] H. Wickert, Evidence for trafficking of PfEMP1 to the surface of -infected erythrocytes via a complex membrane network, Eur. J. Cell Biol. 82 (2003) 271–284, https://doi.org/10.1078/0171-9335-00319.

[93] N.P. Aditya, P.G. Vathsala, V. Vieira, R.S.R. Murthy, E.B. Souto, Advances in nanomedicines for malaria treatment, Adv. Colloid Interf. Sci. 201–202 (2013) 1–17, https://doi.org/10.1016/j.cis.2013.10.014.

[94] S.A. Lauer, A membrane network for nutrient import in red cells infected with the malaria parasite, Science (80-.) 276 (1997) 1122–1125, https://doi.org/10.1126/science.276.5315.1122.

[95] C. Bracho, I. Dunia, M. De La Rosa, E.-L. Benedetti, H. Perez, Traffic pathways of Plasmodium vivax antigens during intraerythrocytic parasite development, Parasitol. Res. 88 (2002) 253–258, https://doi.org/10.1007/s00436-001-0537-8.

[96] B. Pouvelle, R. Spiegel, L. Hsiao, R.J. Howard, R.L. Morris, A.P. Thomas, T.F. Taraschi, Direct access to serum macromolecules by intraerythrocytic malaria parasites, Nature 353 (1991) 73–75, https://doi.org/10.1038/353073a0.

[97] B. Pouvelle, J. Gysin, Presence of the parasitophorous duct in Plasmodium falciparum and P. vivax parasitized Saimiri monkey red blood cells, Parasitol. Today 13 (1997) 357–361, https://doi.org/10.1016/S0169-4758(97)01077-6.

[98] C. Duranton, V. Tanneur, C. Lang, V.B. Brand, S. Koka, R.S. Kasinathan, M. Dorsch, H.J. Hedrich, S. Baumeister, K. Lingelbach, F. Lang, S.M. Huber, A high specificity and affinity interaction with serum albumin stimulates an anion conductance in malaria-infected erythrocytes, Cell. Physiol. Biochem. 22 (2008) 395–404, https://doi.org/10.1159/000185483.

[99] A.E.L. Tahir, P. Malhotra, V.S. Chauhan, Uptake of proteins and degradation of human serum albumin by Plasmodium falciparum—infected human erythrocytes, Malar. J. (2003), https://doi.org/10.1186/1475-2875-2-1.

[100] A.A. Sidhaye, K.C. Bhuran, S. Zambare, M. Abubaker, N. Nirmalan, K.K. Singh, Bio-inspired artemether-loaded human serum albumin nanoparticles for effective control of malaria-infected erythrocytes, Nanomedicine (2016), https://doi.org/10.2217/nnm-2016-0235. nnm-2016-0235.

[101] N. Ibrahim, H. Ibrahim, A.M. Sabater, D. Mazier, A. Valentin, F. Nepveu, Artemisinin nanoformulation suitable for intravenous injection: preparation, characterization and antimalarial activities, Int. J. Pharm. 495 (2015) 671–679, https://doi.org/10.1016/j.ijpharm.2015.09.020.

[102] M. Tibayrenc, Genetics and Evolution of Infectious Disease, Elsevier, 2011, https://doi.org/10.1016/C2010-0-65658-6.

[103] F.-D. Cojocaru, D. Botezat, I. Gardikiotis, C.-M. Uritu, G. Dodi, L. Trandafir, C. Rezus, E. Rezus, B.-I. Tamba, C.-T. Mihai, Nanomaterials designed for antiviral drug delivery transport across biological barriers, Pharmaceutics 12 (2020) 171, https://doi.org/10.3390/pharmaceutics12020171.

[104] K. Herbert, A. Nag, A tale of two RNAs during viral infection: how viruses antagonize mRNAs and small non-coding RNAs in the host cell, Viruses 8 (2016) 154, https://doi.org/10.3390/v8060154.

[105] H. Lodish, A. Berk, S. Zipursky, et al., Viruses: structure, function and uses, in: Molecular Cell Biology, fourth ed., W.H. Freeman, New York, 2000.

[106] D. Lembo, R. Cavalli, Nanoparticulate delivery systems for antiviral drugs, Antivir. Chem. Chemother. 21 (2010) 53–70, https://doi.org/10.3851/IMP1684.

[107] Y. Ding, J. Lou, H. Chen, X. Li, M. Wu, C. Li, J. Liu, C. Liu, Q. Li, H. Zhang, J. Niu, Tolerability, pharmacokinetics and antiviral activity of rHSA/IFNα2a for the treatment of chronic hepatitis B infection, Br. J. Clin. Pharmacol. 83 (2017) 1056–1071, https://doi.org/10.1111/bcp.13184.

[108] M. Merodio, A. Arnedo, M.J. Renedo, J.M. Irache, Ganciclovir-loaded albumin nanoparticles: characterization and in vitro release properties, Eur. J. Pharm. Sci. 12 (2001) 251–259, https://doi.org/10.1016/S0928-0987(00)00169-X.

[109] P. Suwannoi, M. Chomnawang, N. Sarisuta, S. Reichl, C.C. Müller-Goymann, Development of acyclovir-loaded albumin nanoparticles and improvement of acyclovir permeation across human corneal epithelial T cells, J. Ocul. Pharmacol. Ther. 33 (2017) 743–752, https://doi.org/10.1089/jop.2017.0057.

[110] A. Arnedo, J.M. Irache, M. Merodio, M.S. Espuelas Millán, Albumin nanoparticles improved the stability, nuclear accumulation and anticytomegaloviral activity of a phosphodiester oligonucleotide, J. Control. Release 94 (2004) 217–227, https://doi.org/10.1016/j.jconrel.2003.10.009.

[111] F.-F. An, X.-H. Zhang, Strategies for preparing albumin-based nanoparticles for multifunctional bioimaging and drug delivery, Theranostics 7 (2017) 3667–3689, https://doi.org/10.7150/thno.19365.

[112] A. Oevermann, M. Engels, U. Thomas, A. Pellegrini, The antiviral activity of naturally occurring proteins and their peptide fragments after chemical modification, Antivir. Res. 59 (2003) 23–33, https://doi.org/10.1016/S0166-3542(03)00010-X.

[113] H. Li, F. Yu, S. Xia, Y. Yu, Q. Wang, M. Lv, Y. Wang, S. Jiang, L. Lu, Chemically modified human serum albumin potently blocks entry of ebola pseudoviruses and viruslike particles, Antimicrob. Agents Chemother. 61 (2017), https://doi.org/10.1128/AAC.02168-16.

[114] A. Wunder, U. Müller-Ladner, E.H.K. Stelzer, J. Funk, E. Neumann, G. Stehle, T. Pap, H. Sinn, S. Gay, C. Fiehn, Albumin-based drug delivery as novel therapeutic approach for rheumatoid arthritis, J. Immunol. 170 (2003) 4793–4801, https://doi.org/10.4049/jimmunol.170.9.4793.

[115] C. Fiehn, F. Kratz, G. Sass, U. Muller-Ladner, E. Neumann, Targeted drug delivery by in vivo coupling to endogenous albumin: an albumin-binding prodrug of methotrexate (MTX) is better than MTX in the treatment of murine collagen-induced arthritis, Ann. Rheum. Dis. 67 (2008) 1188–1191, https://doi.org/10.1136/ard.2007.086843.

[116] L.Q. Thao, H.J. Byeon, C. Lee, S. Lee, E.S. Lee, H.-G. Choi, E.-S. Park, Y.S. Youn, Pharmaceutical potential of tacrolimus-loaded albumin nanoparticles having targetability to rheumatoid arthritis tissues, Int. J. Pharm. 497 (2016) 268–276, https://doi.org/10.1016/j.ijpharm.2015.12.004.

[117] C.W. Oettinger, M.J. D'Souza, Microencapsulated drug delivery: a new approach to proinflammatory cytokine inhibition, J. Microencapsul. 29 (2012) 455–462, https://doi.org/10.3109/02652048.2012.658443.

[118] A. Rollett, T. Reiter, P. Nogueira, M. Cardinale, A. Loureiro, A. Gomes, A. Cavaco-Paulo, A. Moreira, A.M. Carmo, G.M. Guebitz, Folic acid-functionalized human serum albumin nanocapsules for targeted drug delivery to chronically activated macrophages, Int. J. Pharm. 427 (2012) 460–466, https://doi.org/10.1016/j.ijpharm.2012.02.028.

[119] T. Gong, P. Zhang, C. Deng, Y. Xiao, T. Gong, Z. Zhang, An effective and safe treatment strategy for rheumatoid arthritis based on human serum albumin and Kolliphor® HS 15, Nanomedicine 14 (2019) 2169–2187, https://doi.org/10.2217/nnm-2019-0110.

[120] F. Yan, H. Li, Z. Zhong, M. Zhou, Y. Lin, C. Tang, C. Li, Co-delivery of prednisolone and curcumin in human serum albumin nanoparticles for effective treatment of rheumatoid arthritis, Int. J. Nanomedicine 14 (2019) 9113–9125, https://doi.org/10.2147/IJN.S219413.

[121] W.M. Pardridge, Why is the global CNS pharmaceutical market so under-penetrated? Drug Discov. Today (2002), https://doi.org/10.1016/S1359-6446(01)02082-7.

[122] W.M. Pardridge, The blood-brain barrier: bottleneck in brain drug development, NeuroRx 2 (2005) 3–14, https://doi.org/10.1602/neurorx.2.1.3.

[123] P. Blasi, S. Giovagnoli, A. Schoubben, M. Ricci, C. Rossi, Solid lipid nanoparticles for targeted brain drug delivery☆, Adv. Drug Deliv. Rev. 59 (2007) 454–477, https://doi.org/10.1016/j.addr.2007.04.011.

[124] Y. Liu, R. Ran, J. Chen, Q. Kuang, J. Tang, L. Mei, Q. Zhang, H. Gao, Z. Zhang, Q. He, Paclitaxel loaded liposomes decorated with a multifunctional tandem peptide for glioma targeting, Biomaterials (2014), https://doi.org/10.1016/j.biomaterials.2014.02.031.

[125] N.J. Abbott, A.A.K. Patabendige, D.E.M. Dolman, S.R. Yusof, D.J. Begley, Structure and function of the blood–brain barrier, Neurobiol. Dis. 37 (2010) 13–25, https://doi.org/10.1016/j.nbd.2009.07.030.

[126] K. Ulbrich, T. Hekmatara, E. Herbert, J. Kreuter, Transferrin- and transferrin-receptor-antibody-modified nanoparticles enable drug delivery across the blood–brain barrier (BBB), Eur. J. Pharm. Biopharm. 71 (2009) 251–256, https://doi.org/10.1016/j.ejpb.2008.08.021.

[127] K. Ulbrich, M. Michaelis, F. Rothweiler, T. Knobloch, P. Sithisarn, J. Cinatl, J. Kreuter, Interaction of folate-conjugated human serum albumin (HSA) nanoparticles with tumour cells, Int. J. Pharm. 406 (2011) 128–134, https://doi.org/10.1016/j.ijpharm.2010.12.023.

[128] M. Dadparvar, S. Wagner, S. Wien, J. Kufleitner, F. Worek, H. von Briesen, J. Kreuter, HI 6 human serum albumin nanoparticles—development and transport over an in vitro blood–brain barrier model, Toxicol. Lett. 206 (2011) 60–66, https://doi.org/10.1016/j.toxlet.2011.06.027.

[129] V. Piazzini, E. Landucci, M. D'Ambrosio, L. Tiozzo Fasiolo, L. Cinci, G. Colombo, D.E. Pellegrini-Giampietro, A.R. Bilia, C. Luceri, M.C. Bergonzi, Chitosan coated human serum albumin nanoparticles: a promising strategy for nose-to-brain drug delivery, Int. J. Biol. Macromol. 129 (2019) 267–280, https://doi.org/10.1016/j.ijbiomac.2019.02.005.

[130] C. Ruan, L. Liu, Y. Lu, Y. Zhang, X. He, X. Chen, Y. Zhang, Q. Chen, Q. Guo, T. Sun, C. Jiang, Substance P-modified human serum albumin nanoparticles loaded with paclitaxel for targeted therapy of glioma, Acta Pharm. Sin. B 8 (2018) 85–96, https://doi.org/10.1016/j.apsb.2017.09.008.

[131] A. Avachat, Y. Oswal, K. Gujar, R. Shah, Preparation and characterization of rivastigmine loaded human serum albumin (HSA) nanoparticles, Curr. Drug Deliv. (2014), https://doi.org/10.2174/1567201811309990050.

[132] S. Dreis, F. Rothweiler, M. Michaelis, J. Cinatl, J. Kreuter, K. Langer, Preparation, characterisation and maintenance of drug efficacy of doxorubicin-loaded human serum albumin (HSA) nanoparticles, Int. J. Pharm. 341 (2007) 207–214, https://doi.org/10.1016/j.ijpharm.2007.03.036.

[133] M. Joshi, M. Nagarsenkar, B. Prabhakar, Albumin nanocarriers for pulmonary drug delivery: an attractive approach, J. Drug Deliv. Sci. Technol. 56 (2020) 101529, https://doi.org/10.1016/j.jddst.2020.101529.

[134] J. Seo, C. Lee, H.S. Hwang, B. Kim, L.Q. Thao, E.S. Lee, K.T. Oh, J.L. Lim, H.G. Choi, Y.S. Youn, Therapeutic advantage of inhaled tacrolimus-bound albumin nanoparticles in a bleomycin-induced pulmonary fibrosis mouse model, Pulm. Pharmacol. Ther. (2016), https://doi.org/10.1016/j.pupt.2016.01.001.

[135] S.H. Choi, H.J. Byeon, J.S. Choi, L. Thao, I. Kim, E.S. Lee, B.S. Shin, K.C. Lee, Y.S. Youn, Inhalable self-assembled albumin nanoparticles for treating drug-resistant lung cancer, J. Control. Release 197 (2015) 199–207, https://doi.org/10.1016/j.jconrel.2014.11.008.

[136] A. Woods, A. Patel, D. Spina, Y. Riffo-Vasquez, A. Babin-Morgan, R.T.M. de Rosales, K. Sunassee, S. Clark, H. Collins, K. Bruce, L.A. Dailey, B. Forbes, In vivo biocompatibility, clearance, and biodistribution of albumin vehicles for pulmonary drug delivery, J. Control. Release 210 (2015) 1–9, https://doi.org/10.1016/j.jconrel.2015.05.269.

[137] H.B. Levene, R. Moision, E. Villapando, J. Torgerson, J. Maniquis, R. Kleinhenz, R. Keen, J.L. Barnhart, Characterization of Albunex®, J. Acoust. Soc. Am. 87 (1990) S69–S70, https://doi.org/10.1121/1.2028331.

[138] F. Gao, P. Cai, W. Yang, J. Xue, L. Gao, R. Liu, Y. Wang, Y. Zhao, X. He, L. Zhao, G. Huang, F. Wu, Y. Zhao, Z. Chai, X. Gao, Ultrasmall [64 Cu]Cu nanoclusters for targeting orthotopic lung tumors using accurate positron emission tomography imaging, ACS Nano 9 (2015) 4976–4986, https://doi.org/10.1021/nn507130k.

[139] J. Zhang, G. Hao, C. Yao, J. Yu, J. Wang, W. Yang, C. Hu, B. Zhang, Albumin-mediated biomineralization of paramagnetic NIR Ag 2 S QDs for tiny tumor bimodal targeted imaging in vivo, ACS Appl. Mater. Interfaces 8 (2016) 16612–16621, https://doi.org/10.1021/acsami.6b04738.

[140] D.-H. Hu, Z.-H. Sheng, P.-F. Zhang, D.-Z. Yang, S.-H. Liu, P. Gong, D.-Y. Gao, S.-T. Fang, Y.-F. Ma, L.-T. Cai, Hybrid gold–gadolinium nanoclusters for tumor-targeted NIRF/CT/MRI triple-modal imaging in vivo, Nanoscale 5 (2013) 1624, https://doi.org/10.1039/c2nr33543c.

[141] T. Kale, K. Bendale, K.K. Singh, P. Chaudhari, Albumin based iohexol nanoparticles for computed tomography: an in vivo study, J. Biomed. Nanotechnol. (2019), https://doi.org/10.1166/jbn.2019.2690.

CHAPTER 10

Silver nanoparticles for biomedical applications

Vaikundamoorthy Ramalingam
Department of Animal Science, Jeonbuk National University, Jeonju, Republic of Korea
Centre for Natural Products & Traditional Knowledge, CSIR-Indian Institute of Chemical Technology, Hyderabad, India

1. Introduction

Nanotechnology is one of the influence field in the modern research which deals with the preparation, scheme, and the modification of nanoparticles properties including size (ranging from 1 to 100 nm) and various shapes [1]. The nanoparticles-based therapies have a wide array of treatments in medicine and biological sciences including but not limited to medical diagnosis, medicine, biosensing, health care, drug delivery, coating, medical devices, wound healing, the food industry, cosmetics, and environmental remediation [2, 3]. The researchers have focused on the preparation of nanoscale particles with different metals like copper [4], zinc [5], iron [6], titanium [7], palladium [8], gold [9], and silver [10]. These prepared metal nanoparticles with their low-dimensional nature has diverse applications in the fields of engineering and technology, though the preparation of silver nanoparticles (AgNPs) have excellent application in biomedical field [11]. Silver is a well-known soft, white, and shiny metal with exceptional physical properties such as high thermal and dielectrical activity and have been used for biomedical applications [12]. The silver-based nanoparticles have been used for several sectors including health and fitness, cleaning, food, house-hold apparatus, medicine, electronic devices, toys, etc. (Fig. 1). The silver-coated face creams are known for its excellent antimicrobial properties against microbial infections and wound healing [13]. There are many research reports that owing to their unique physical and chemical properties the nanosized AgNPs are deemed for biomedical applications than their macroscale form [14]. Hence, the evolving manifesto in nanotechnology is focusing on the preparation of AgNPs and development of Ag-tethered nanocomposite materials with emphasis on better treatment strategies in biomedical field.

2. Properties

The excellent biomedical applications of AgNPs are associated with their precise interactions with cells due to their chemical properties such as composition, surface coating, or

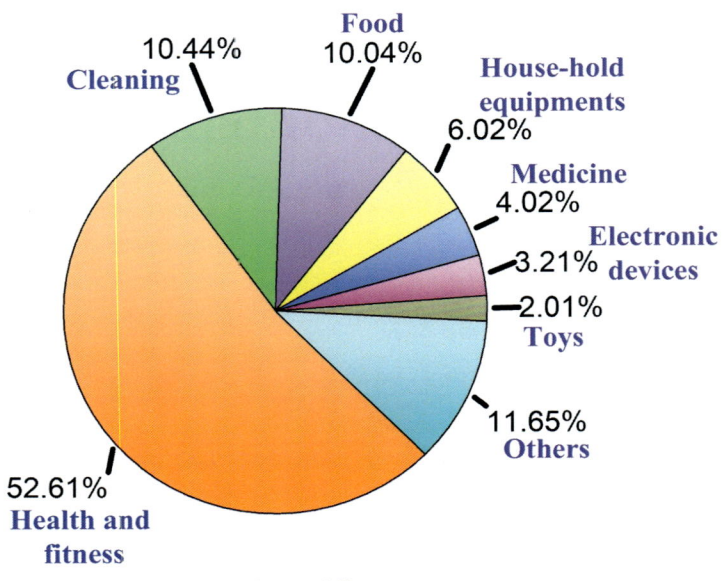

Fig. 1 Contribution of silver nanoparticles in different sectors.

the release of Ag^+ ions or the physical properties such as size and shape [15]. Moreover, the agglomeration and dissolution rate are also important controlling factors of biological interactions and activities [16]. Though, the larger surface area AgNPs could significantly affect the physiochemical properties and contribute to higher toxicity [17]. It is well documented that the surface charge of the AgNPs vary based on the surface coatings which are responsible for the communication with biological systems [18] and the dissolution rate of the AgNPs is dependent on the physicochemical properties and the surrounding media [19]. The surface plasmon resonance (SPR) of the AgNPs determines the optical properties which is resultant of the interaction with light, and thus causes the conversion of photon energy into thermal energy [20]. This property of AgNPs is key for exploiting in bio-diagnostic and bioimaging applications, in which the radioactive SPR is important, whereas in therapeutic applications the nonradioactive SPR is important [21]. Hence, there properties of AgNPs could enhance the effect of materials in the biomedical applications.

3. Synthesis of silver nanoparticles

There are several methods that have been established for the preparation of AgNPs and they can be classified as (1) physical methods, (2) chemical methods, and (3) biological methods. The physical synthesis of the AgNPs can be done through physical vapor condensation [36], arc-discharge [37], energy ball milling method [38], and direct current

(DC) magnetic sputtering methods [39]. Though, the physical-mediated synthesis has advantage of narrow size and shape distribution, the usage of high energy consumption is a major drawback for using these methods [40]. In case of chemical synthesis, it subclassifies into chemical reduction [41], electrochemical techniques [42], irradiation-assisted chemical methods [43], and spray pyrolysis [44]. Though the chemical methods are used to prepare narrow size and shape AgNPs, the use of toxic chemicals is in high contrast to physical methods [45]. For biological synthesis of AgNPs, the nontoxic molecules such as proteins, carbohydrates, antioxidants produced by living organisms are being used as the replacement of the toxic reducing and stabilizing agents [45, 46]. There are various biological organisms including microorganisms (Table 1) (bacteria, yeast, and fungi) and plants (Table 2) that have been used for the preparation of AgNPs and are well documented. The probable mechanism of biological preparation was either enzymatic (NADH reductase) or nonenzymatic reduction and is considered as environment friendly, cheap, and nontoxic [56].

4. Characterization of the nanoparticles

The synthesized AgNPs would be characterized using various spectroscopic techniques such as UV-visible spectroscopy, X-ray diffraction (XRD), dynamic light scattering (DLS), Fourier transform infrared (FTIR), and X-ray photoelectron spectroscopy

Table 1 Microbial synthesis of silver nanoparticles and its size, shape, and biomedical applications.

Organism	Size (nm)	Shape	Application	References
Lactobacillus brevis	45	Spherical	Antioxidant activity	[22]
Pseudomonas aeruginosa	13–76	Spherical	Antimicrobial, antibiofilm activity	[23]
Trichoderma viride	5–40	Spherical	Antimicrobial activity	[24]
Acinetobacter spp.	8–12	Spherical	Antibacterial activity	[25]
Gluconobacter roseus	10	Spherical	Antiplatelet activity	[26]
Klebsiella pneumoniae	15–37	Spherical	Antibacterial activity	[27]
Stenotrophomonas maltophilia	93	Spherical	Cytotoxic activity	[28]
Bacillus licheniformis	18–63	Spherical	Antibiofilm activity	[29]
Bacillus methylotrophicus	10–30	Spherical	Antimicrobial activity	[30]
Bacillus thuringiensis	45–140	Spherical	Larvicidal activity	[31]
Rhodococcus sp.	30	Spherical	Anticatalytic activity	[32]
Macrophomina phaseolina	16–20	Spherical	Antimicrobial activity	[33]
Humicola sp.	5–25	Spherical	Cytotoxic activity	[34]
Phoma glomerate	60–80	Spherical	Antimicrobial activity	[35]

Table 2 Phytosynthesis of silver nanoparticles and its size, shape, and biomedical applications.

Organism	Size (nm)	Shape	Application	References
Alpinia nigra	6–8	Spherical	Antimicrobial activity	[47]
Gelidium corneum	20–40	Spherical	Antibiofilm activity	[48]
Teucrium polium	70–100	Spherical	Anticancer activity against gastric cancer	[49]
Tamarindus indica	20–30	Spherical	Anticancer activity against breast cancer	[50]
Nauclea latifolia	10	Irregular	Antioxidant activity	[51]
Curcuma longa	102	Spherical	Bioimaging	[52]
Annona muricate	80–87	Spherical	Cell cycle arrest in lung cancer	[53]
Pandanus odorifer	10–50	Spherical	Antibiofilm activity	[54]
Rhododendron ponticum	10–20	Spherical	Antibiofilm activity	[55]

(XPS), and various microscopic techniques including scanning electron microscopy (SEM), transmission electron microscopy (TEM), and atomic force microscopy (AFM). Initially, the prepared AgNPs were confirmed by the appearance of brown color using naked eye, however, their preparation is further confirmed using various sophisticated analysis [57]. The UV–vis spectroscopy is used to measure the SPR of the AgNPs which arises due to the movement of electrons in the conduction and valence band and the absorption peak observed around 420–450 nm for AgNPs [58]. The XRD analysis is used to identify the crystal nature of the AgNPs based on the formation of diffraction patterns on the reflection of X-ray lights [59]. The working principle of X-ray diffraction is Bragg's law and each material has a unique diffraction pattern which can identified by comparing the peaks with the reference database in the Joint Committee on Powder Diffraction Standards (JCPDS) library [60]. The DLS analysis can probe the size distribution of nanosized particles ranging from submicron to 3 nm in the solution or suspension based on the interaction of light with the particles [61]. The FTIR analysis is commonly used to find out the molecules influencing the synthesis of AgNPs and also to confirm the functional molecules that covalently bind on the surface of the AgNPs and interactions occurring between the enzyme and substrate during the catalytic process [62].

The XPS analysis is designed to quantitatively estimate the surface chemicals associated on the AgNPs in a parts per thousand range [63]. The spectra are obtained by exposing the material with X-rays and concurrently estimating the kinetic energy and number of electrons that leave from nanosized material and analyzed [64]. The structural morphology such as size, shape, and nature of surface area is observed using electron microscopy. By using the SEM analysis, particle size, size distributions, particle shape, and the surface area of the AgNPs can be measured at micron and nanoscale levels [65]. Moreover, the SEM operating with energy-dispersive X-ray spectroscopy (EDX) can be used to quantify the elements presents in the silver nanomaterials [66]. As well, the elemental composition is observed as images using SAM nanoprobe in the SEM operating EDX analysis [67]. However, the drawback in the SEM analysis is not able to observe the internal structure of the particles, though the SEM analysis provides the purity and the degree of particle aggregation [68]. TEM and high-resolution TEM (HR–TEM) analysis is an important characterization technique which provides more insights about the material size, morphology of AgNPs and lattice fringes [69]. Unlike the SEM, the TEM could afford excellent spatial resolution images and also the additional analytical data and the quality of the images is completely based on the preparation of the samples on the grid [70]. Finally, the AFM is frequently used to examine the dispersion and aggregation of the nanomaterials along with size, shape and there are three different modes being used to acquire the data [71]. The drawbacks in AFM analysis are overestimation of the dimension, and selecting the operating mode are crucial factor in sample analysis [72].

5. Biomedical applications of silver nanoparticles

5.1 Antimicrobial activity

There are several commercial antibiotics available and were golden bullet to inhibit the growth of pathogenic bacteria and to control the infectious disease [73]. However, the prompt and the development of multidrug resistant (MDR) microbial pathogens has quizzed the future use of antibiotics [74]. The recent emergence of nanotechnology in the medicine attracts the researchers to prepare the novel nanomaterials to fight against the MDR pathogens [75]. Amongst the various nanomaterials, the AgNPs are renowned for their antimicrobial properties and utilized for decades in the biomedical field for antimicrobial applications [76]. Previously it has been reported that the AgNPs treatment influence the reactive oxygen species (ROS) homeostasis and morphological changes in the bacteria and is based on the chemical composition, concentration, size, and shape of the AgNPs [77]. The possible mechanism for potent antimicrobial activity is based on the attachment and penetration ability of AgNPs on the surface of the bacterial cell wall and disturbing the cell membrane and respiratory functions [78]. Also, the antimicrobial activity of AgNPs depend on the size of material, like the smaller AgNPs showed more cellular interaction and penetration ability than the larger AgNPs due to their larger surface area [79]. Together the AgNPs are well known for the excellent antimicrobial activity and substantial improvements have been accomplished on the elucidation of antimicrobial mechanism of AgNPs, however, the exact mechanism of action is still not completely known.

5.2 Antimalarial activity

Malaria is one of the most dangerous disease influenced by microscopic parasites called Plasmodium that are transmitted to human beings by mosquitoes [80]. Mosquitoes act as parasite vectors to transmit the malarial parasite to humans and the use of traditional and conventional methods to eradicate such vectors have few drawbacks including the development of resistance [81]. The modern antimalarial drugs are classified into quinolines, antifolates, artemisinin derivatives and hydroxynaphthaquinones, and each antimalarial drug has specific target (either passive or active targeting) of action (Fig. 2). Since the commercially available drugs, which are currently used for treatment of malaria, are not preferably suitable for the elimination of malaria campaigns and researchers are

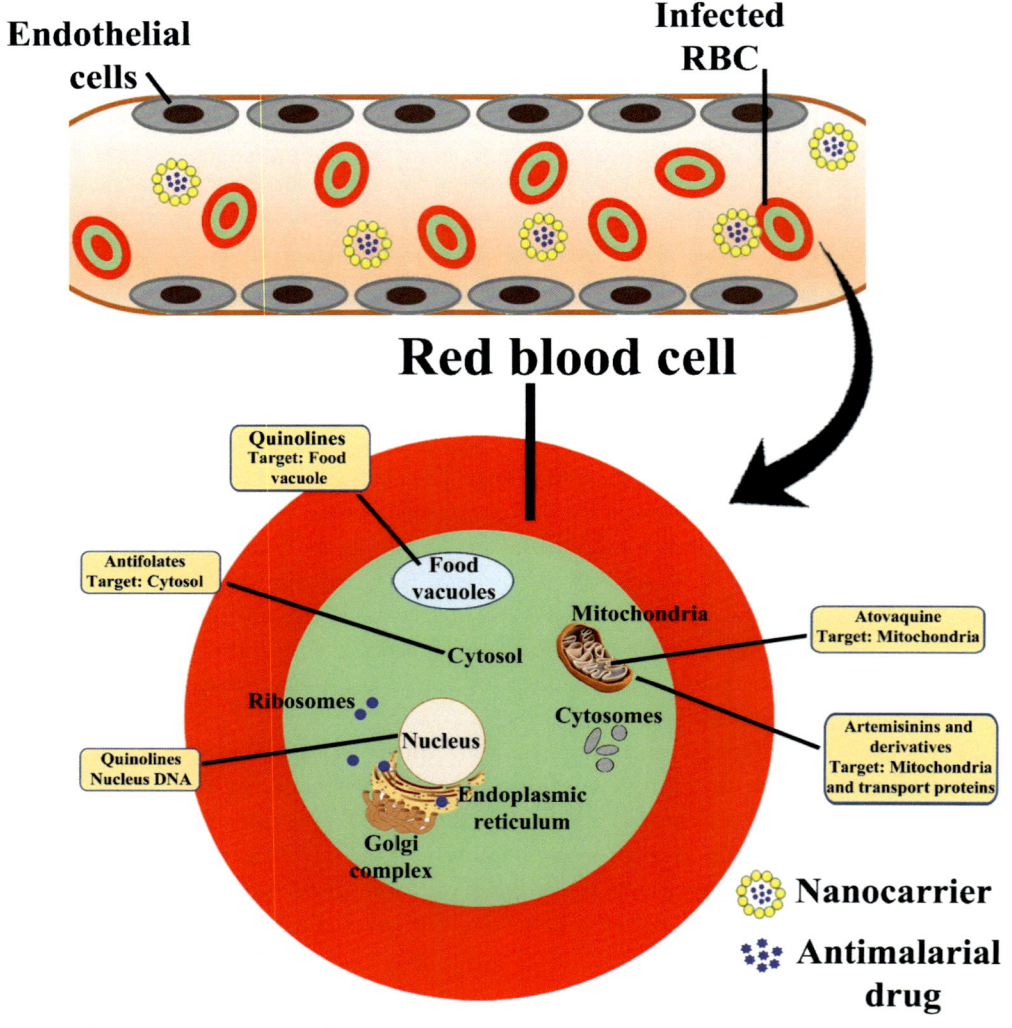

Fig. 2 Schematic representation showing the mechanism of silver nanoparticles mediated antimalarial drug delivery in plasmodium infected red blood cells (RBCs).

focusing on the development of novel drugs. Owing to the vast biomedical applications of nanotechnology, the various nanoparticles can be used for the treatment of malaria. AgNPs have been reported for potent activity against malarial parasite and could be one of the future solutions for the control of malaria. However, the AgNPs have been used for the delivery of malarial drugs into the specific target, instead of the direct treatment. The AgNP-encapsulated antimalarial drugs is injected intravenously into the blood stream, it binds with the infected red blood cells (RBCs) and release the drug at the surface of the infected RBCs [82]. Further, the drugs interact with the surface molecules and penetrates into the infected RBCs which leads to the disruption of RBCs by interfering with trophozoites in the infected RBCs [83].

5.3 Antibiofilm activity

Generally, the biofilms are formed by various pathogens which result into the development of resistance against the antimicrobial drugs [84]. The recent molecular analysis using pyrosequencing reported that the diversity of microbial communities are involved to build a structural and active complex environment where the microbial biofilms are the rule [85]. The extracellular polymeric matrix (EPM) in the cell wall of the bacteria acts as an architecture for the biofilm and have strong ability to impede the activity of antimicrobial drugs on bacteria which are surrounded and sealed beneath of the biofilm [86]. The different complicated stages involved in the formation of biofilm from the initial attachment to biotic or abiotic surfaces, development to dispersion pose a serious threat to public health [87]. Recent and past studies explored the alternative approaches to inhibit the formation of biofilm by different agents isolated from plants, animals, and microbes. However, the emergence of new MDR strains and up to 80% of bacterial infections including leprosy, pneumonia, plague, sepsis, tuberculosis, and inner ear infections, along with many hospital-acquired infections from tubes and ports are caused by biofilms [7]. To combat against biofilm formation, the research is focused on development of alternative strategies to prevent the pathogenesis by inhibiting biofilm formation at the initial stage and also eradicating the preformed mature biofilm by several active agents. Among these agents, the nanoparticles have received more attention due to their properties such as biodegradability, biocompatibility, nonallergenic, and nontoxicity [88]. Recent research has focused on employing AgNPs as effective agents to inhibit biofilm formation and attenuate virulence properties by various pathogenic bacteria. Silver is known to produce an antibacterial effect by acting on multiple targets starting from interaction with the sulfhydryl groups of proteins and DNA, alter the hydrogen bonding/respiratory chain, unwind DNA, and interfere with cell-wall synthesis/cell division [89]. AgNPs are known to further destabilize the bacterial membrane and increase permeability, leading to leakage of cell constituent (Fig. 3). However, the direct effect of AgNPs on quenching the quorum sensing signaling of pathogenic biofilm formation

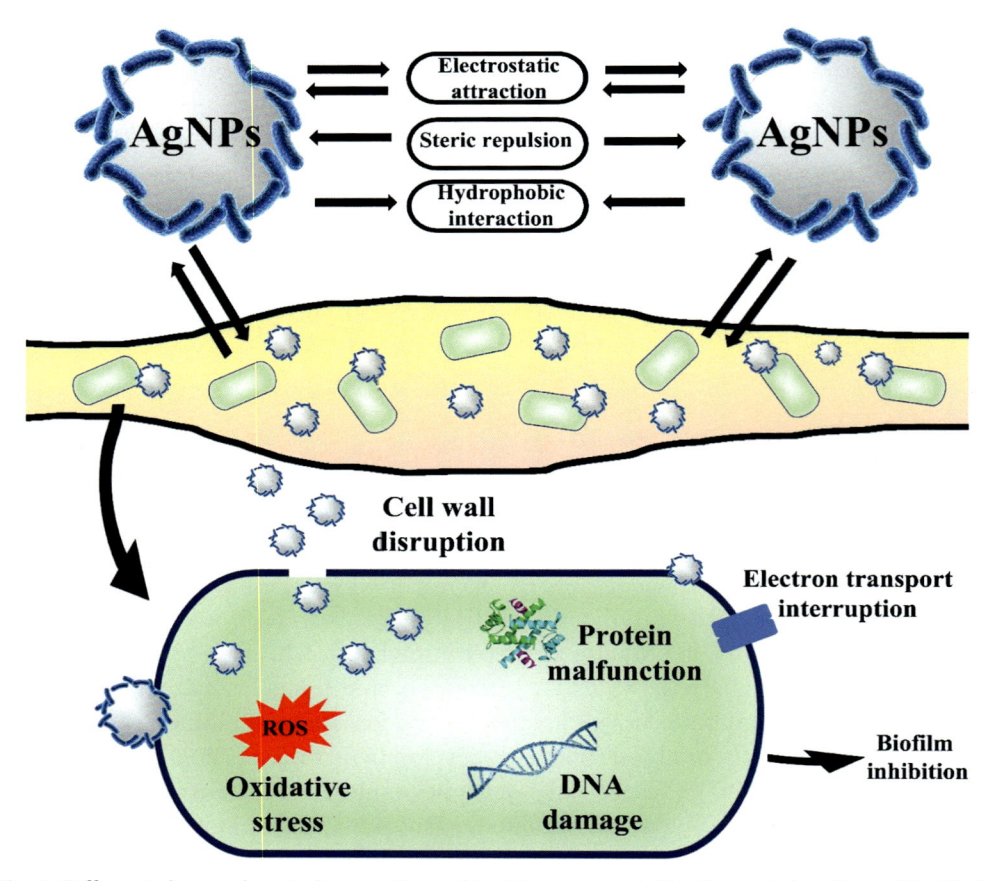

Fig. 3 Different physicochemical interactions of AgNPs on bacterial biofilm and the effect of AgNPs in inhibition of biofilm formation.

is still not clearly understood and also in vivo trials of the novel nanomaterials must be tested.

5.4 Antioxidant activity

The generation of ROS in multiple cell components is normal phenomenon, and it plays a key role in various signaling pathways [90]. However, lower amount of variation in ROS homeostasis is caused by the signaling molecules, whereas the excess ROS generation leads to oxidative stress which results in damage of the proteins, lipids, and DNA leading to malignancy, apoptosis, autophagy, etc. [91]. On other hand, the physiological role of ROS generation in cellular signaling also have been reported and it regulates the tissue microenvironment and tissue regeneration [92]. The ROS has significant effects on the activation of signaling pathways such as Jun N-terminal kinase (JNK) and p38 mitogen–activated protein kinases (MAPK), which are key regulators of cellular proliferation and differentiation [93]. For many decades, the plant-mediated phytochemical constituents were used to balance the redox imbalance and decrease the oxidative stress and play a key role in cellular growth [94]. However, the recent emergence of nanotechnology in the medicine draws the attention of the researchers to use the nanomaterials as potential antioxidant agent against various disease [95]. AgNPs represent one of the most promising frontiers in the research as improved antioxidants which demonstrate the intrinsic redox activity [96] and often are associated with radical blocking and superoxide dismutase-like and catalase-like activities [97]. However, the synergism and the antagonism in antioxidant activity of AgNPs are not clearly elucidated and the effect of AgNPs in molecular and cellular functions in in vivo system should be elucidated.

5.5 Anticancer activity

Cancer is a disease of abnormal cell division or growth which may occur in almost every part of the human body [98] and the common human cancers are breast cancer, liver cancer, lung cancer, oral cavity cancer, prostate cancer, cervical cancer, etc. Currently, cancer is one of the most important reasons of deaths worldwide with a projected 7.6 million people death each year and accounting for 13% of all deaths, and the cancer-related death is estimated to increase to 13.1 million by 2030 [99]. Detection, therapy, and diagnosis of cancer is mostly difficult which is responsible for the low survival rate of patients and only about 16% of cases are diagnosed before malignancy and most of them are detected in locally advanced or metastatic stages [100]. However, the traditional cancer therapy methods such as surgery, chemotherapy, and radiotherapy have experienced some significant development for the treatment of cancer, but due to their limitation in efficacy and unpredicted side effects regularly fail to accomplish the comprehensive treatment of the cancer [101]. Moreover, the emergence of MDR as a result of repeated drug administration over a chemotherapy treatment course, as well as insensitivity of hypoxic cancer cells to ionizing radiation are the leading causes of treatment failure following chemotherapy [102]. Hence, the recent research focuses on the nanoparticle-based therapy for the treatment of cancer because of the unique properties of the nanomaterials. Among the various nanoparticles, the AgNPs have been utilized for the treatment of cancer for its excellent SPR, unique size, shape, and increased cellular uptake [103]. The efficiency of AgNPs for cancer treatment is based on the continuous generation of ROS in the mitochondria during the treatment [104]. The excess ROS generation in the mitochondria leads to the oxidative stress which disrupts the mitochondrial membrane potential and thus causing the release of cytochrome C into the cytosol [105]. The increased apoptotic proteins in cytosol from mitochondria increased the expression of p53 and mitochondria-mediated activation of caspases, a final executioner of apoptosis [106] (Fig. 4). Although technology has developed toward new efficient and precise preparation methods,

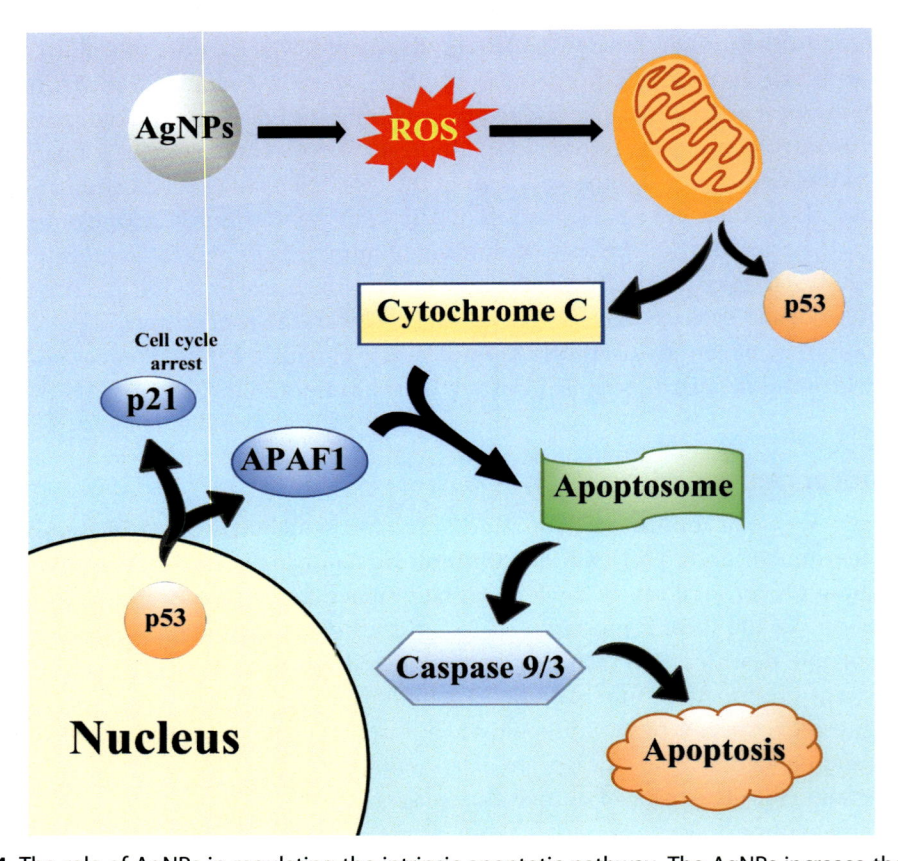

Fig. 4 The role of AgNPs in regulating the intrinsic apoptotic pathway. The AgNPs increase the ROS generation in mitochondria leading to the release of cytochrome C, which binds to Apaf-1 which is upregulated by the p53 in nucleus upon response to oxidative stress. The Apaf-1 triggers the formation of a heptameric apoptosome, which assembles with procaspase-9 to form the holo-apoptosome. The Apaf-1 apoptosome catalyzes the cleavage and activation of procaspase-9, which triggers the caspase cascade, and the activation of caspase-3 leads to eventual apoptosis.

characterization of the materials and the strategies to improve the efficiency of the AgNPs, there are still many drawbacks to overcome, and finally, reach the clinics.

5.6 Other applications

Nanoparticle-based therapies have a broad range of applications including wound healing, SERS, and photocatalytic degradation of carcinogens [107]. The treatment methods use the AgNPs to heal the wound, the AgNPs are preferably used as a nanocarrier to deliver the active compounds to the wound site [108]. Moreover, the potent antimicrobial properties of AgNPs efficiently inhibit the growth of pathogens in the wound which is a key for utilizing the AgNPs for the treatment of both acute and chronic wounds [109]. Raman spectroscopy, amplified by surface-enhanced Raman scattering (SERS) showed AgNPs can provide an in vivo imaging modality due to its high molecular specificity, high sensitivity, and negligible autofluorescence [110]. Most of the SERS substrates with different structures are simple and efficient with high enhancement activity and the SERS substrates are silver and gold, and they both have good SERS enhancement activity [111], and Ag has a better effect on enhancement than the gold. The key factor for the role of AgNPs in SERS applications is to improve the combining enhancement and reaction process which adequately increase the efficiency to observe and examine the chemical reactions [112]. The addition of AgNPs with photocatalysts increases the photocatalytic efficiency by inhibiting the photogeneration of electrons through trapping mechanisms and increases the visible light absorption by excellent SPR enhancement of AgNPs [113]. Through this photocatalytic disinfection process, the AgNPs are good candidates for disinfection and bactericidal activity of the catalyst can act in performance to afford the highly effective microbial inactivation under irradiation.

Though the AgNPs have attracted remarkable attention due to their powerful biomedical applications, their interaction with the environment and toxicity in live terrestrial or aquatic organisms is still a matter of intense debate [114]. The severe effect of free silver ions in the industrial wastes are renowned to be responsible for the occurrence of discoloration of eyes and skin and other related side effects on renal, hepatic, gastrointestinal, respiratory adverse effects [115]. However, the degree of toxic effect of AgNPs is based on the size and concentration of materials used for the treatment [116]. Previously it has been well documented that the half maximal effective concentration could increase the ROS generation, DNA damage, oxidative stress, enzyme expression, and reproductive or growth inhibition in different vertebrates and invertebrates [117].

6. Conclusion and future prospective

The AgNPs have earned more attention due to their unique properties and demonstrated applications in physics, chemistry, biology, and medicine. The AgNPs proven excellent antimicrobial activity against both gram-positive and gram-negative pathogens, however, these activities are functional under the acidic condition. Though, the clinical application of AgNPs is limited as the body pH is close to the neutral. Thus, modification in the preparation of AgNPs should be done in such a way so that it can function in neutral pH. As the AgNPs demonstrated their excellent antimalarial activity, further studies are needed to prove the properties of AgNPs at molecular and biochemical levels. The AgNPs have excellent antibiofilm activity by inhibiting the EPS production and inhibit the formation in the substrate, even though the role of AgNPs in targeting the QS signaling in biofilm-forming pathogenic bacteria is not clearly understood. Hence, quenching of QS signaling circuits by AgNPs must be also investigated. The synergism and the antagonism in antioxidant activity of AgNPs should be investigated further and the effect of AgNPs in molecular and cellular functions in in-vivo system should be elucidated. However, the biocompatibility and selectivity in tumor targeting remains most important challenge for AgNPs. The question of poor-to-moderate biocompatibility is dominant in all the systems and new preparation methods are required to overcome the rigorous principles in clinical settings. Overall, the future research on the health and risk assessment of AgNPs will provide the sound knowledge to prepare the AgNPs for clinical and commercial settings.

References

[1] L.K. Foong, M.M. Foroughi, A.F. Mirhosseini, M. Safaei, S. Jahani, M. Mostafavi, N. Ebrahimpoor, M. Sharifi, R.S. Varma, M. Khatami, Applications of nano-materials in diverse dentistry regimes, RSC Adv. 10 (2020) 15430–15460.

[2] R. Madannejad, N. Shoaie, F. Jahanpeyma, M.H. Darvishi, M. Azimzadeh, H. Javadi, Toxicity of carbon-based nanomaterials: reviewing recent reports in medical and biological systems, Chem. Biol. Interact. 307 (2019) 206–222.

[3] H.S. Rahman, H.H. Othman, N.I. Hammadi, S.K. Yeap, K.M. Amin, N. Abdul Samad, N.B. Alitheen, Novel drug delivery systems for loading of natural plant extracts and their biomedical applications, Int. J. Nanomedicine 15 (2020) 2439–2483.

[4] N.K. Ojha, G.V. Zyryanov, A. Majee, V.N. Charushin, O.N. Chupakhin, S. Santra, Copper nanoparticles as inexpensive and efficient catalyst: a valuable contribution in organic synthesis, Coord. Chem. Rev. 353 (2017) 1–57.

[5] V. Ramalingam, I. Hwang, Zinc oxide nanoparticles promoting the formation of myogenic differentiation into myotubes in mouse myoblast C2C12 cells, J. Ind. Eng. Chem. 83 (2020) 315–322.

[6] V. Ramalingam, M. Harshavardhan, S. Dinesh Kumar, S. Malathi devi, Wet chemical mediated hematite α-Fe_2O_3 nanoparticles synthesis: preparation, characterization and anticancer activity against human metastatic ovarian cancer, J. Alloys Compd. 834 (2020), 155118.

[7] V. Ramalingam, S. Sundaramahalingam, R. Rajaram, Size-dependent antimycobacterial activity of titanium oxide nanoparticles against *Mycobacterium tuberculosis*, J. Mater. Chem. B 7 (2019) 4338–4346.

[8] V. Ramalingam, S. Raja, M. Harshavardhan, In situ one-step synthesis of polymer-functionalized palladium nanoparticles: an efficient anticancer agent against breast cancer, Dalton Trans. 49 (2020) 3510–3518.

[9] V. Ramalingam, Multifunctionality of gold nanoparticles: plausible and convincing properties, Adv. Colloid Interf. Sci. 271 (2019) 101989.

[10] M. Marchioni, P.-H. Jouneau, M. Chevallet, I. Michaud-Soret, A. Deniaud, Silver nanoparticle fate in mammals: bridging in vitro and in vivo studies, Coord. Chem. Rev. 364 (2018) 118–136.

[11] A.C. Burdusel, O. Gherasim, A.M. Grumezescu, L. Mogoanta, A. Ficai, E. Andronescu, Biomedical applications of silver nanoparticles: an up-to-date overview, Nano 8 (2018) 681.

[12] L. Mo, Z. Guo, L. Yang, Q. Zhang, Y. Fang, Z. Xin, Z. Chen, K. Hu, L. Han, L. Li, Silver nanoparticles based ink with moderate sintering in flexible and printed electronics, Int. J. Mol. Sci. 20 (2019) 2124.

[13] M.M. Mihai, M.B. Dima, B. Dima, A.M. Holban, Nanomaterials for wound healing and infection control, Materials 12 (2019) 2176.

[14] B. Calderon-Jimenez, M.E. Johnson, A.R. Montoro Bustos, K.E. Murphy, M.R. Winchester, J.R. Vega Baudrit, Silver nanoparticles: technological advances, societal impacts, and metrological challenges, Front. Chem. 5 (2017) 6.

[15] M. Akter, M.T. Sikder, M.M. Rahman, A. Ullah, K.F.B. Hossain, S. Banik, T. Hosokawa, T. Saito, M. Kurasaki, A systematic review on silver nanoparticles-induced cytotoxicity: physicochemical properties and perspectives, J. Adv. Res. 9 (2018) 1–16.

[16] M.N. Martin, A.J. Allen, R.I. MacCuspie, V.A. Hackley, Dissolution, agglomerate morphology, and stability limits of protein-coated silver nanoparticles, Langmuir 30 (2014) 11442–11452.

[17] J. Zhang, W. Guo, Q. Li, Z. Wang, S. Liu, The effects and the potential mechanism of environmental transformation of metal nanoparticles on their toxicity in organisms, Environ. Sci. Nano 5 (2018) 2482–2499.

[18] H.M. Fahmy, A.M. Mosleh, A.A. Elghany, E. Shams-Eldin, E.S. Abu Serea, S.A. Ali, A.E. Shalan, Coated silver nanoparticles: synthesis, cytotoxicity, and optical properties, RSC Adv. 9 (2019) 20118–20136.

[19] W. Lee, E. Kim, H.-J. Cho, T. Kang, B. Kim, M. Kim, Y. Kim, N. Song, J.-S. Lee, J. Jeong, The relationship between dissolution behavior and the toxicity of silver nanoparticles on zebrafish embryos in different ionic environments, Nano 8 (2018) 652.

[20] L.A. Austin, M.A. Mackey, E.C. Dreaden, M.A. El-Sayed, The optical, photothermal, and facile surface chemical properties of gold and silver nanoparticles in biodiagnostics, therapy, and drug delivery, Arch. Toxicol. 88 (2014) 1391–1417.

[21] A. Loiseau, V. Asila, G. Boitel-Aullen, M. Lam, M. Salmain, S. Boujday, Silver-based plasmonic nanoparticles for and their use in biosensing, Biosensors 9 (2019) 78.

[22] M.S. Riaz Rajoka, H.M. Mehwish, H. Zhang, M. Ashraf, H. Fang, X. Zeng, Y. Wu, M. Khurshid, L. Zhao, Z. He, Antibacterial and antioxidant activity of exopolysaccharide mediated silver nanoparticle synthesized by *Lactobacillus brevis* isolated from Chinese koumiss, Colloids Surf. B: Biointerfaces 186 (2020) 110734.

[23] V. Ramalingam, R. Rajaram, C. PremKumar, P. Santhanam, P. Dhinesh, S. Vinothkumar, K. Kaleshkumar, Biosynthesis of silver nanoparticles from deep sea bacterium *Pseudomonas aeruginosa* JQ989348 for antimicrobial, antibiofilm, and cytotoxic activity, J. Basic Microbiol. 54 (2014) 928–936.

[24] A.M. Fayaz, K. Balaji, M. Girilal, R. Yadav, P.T. Kalaichelvan, R. Venketesan, Biogenic synthesis of silver nanoparticles and their synergistic effect with antibiotics: a study against gram-positive and gram-negative bacteria, Nanomedicine 6 (2010) 103–109.

[25] B.A. Chopade, R. Singh, P. Wagh, S. Wadhwani, S. Gaidhani, A. Kumbhar, J. Bellare, Synthesis, optimization, and characterization of silver nanoparticles from *Acinetobacter calcoaceticus* and their enhanced antibacterial activity when combined with antibiotics, Int. J. Nanomedicine 8 (2013) 4277–4290.

[26] R.N. Krishnaraj, S. Berchmans, In vitro antiplatelet activity of silver nanoparticles synthesized using the microorganism *Gluconobacter roseus*: an AFM-based study, RSC Adv. 3 (2013) 8953.

[27] D. Kalpana, Y.S. Lee, Synthesis and characterization of bactericidal silver nanoparticles using cultural filtrate of simulated microgravity grown *Klebsiella pneumoniae*, Enzym. Microb. Technol. 52 (2013) 151–156.

[28] V. Bansal, M. Oves, M.S. Khan, A. Zaidi, A.S. Ahmed, F. Ahmed, E. Ahmad, A. Sherwani, M. Owais, A. Azam, Antibacterial and cytotoxic efficacy of extracellular silver nanoparticles biofabricated from chromium reducing novel OS4 strain of *Stenotrophomonas maltophilia*, PLoS ONE 8 (2013), e59140.

[29] S. Shanthi, B. David Jayaseelan, P. Velusamy, S. Vijayakumar, C.T. Chih, B. Vaseeharan, Biosynthesis of silver nanoparticles using a probiotic *Bacillus licheniformis* Dahb1 and their antibiofilm activity and toxicity effects in *Ceriodaphnia cornuta*, Microb. Pathog. 93 (2016) 70–77.

[30] C. Wang, Y.J. Kim, P. Singh, R. Mathiyalagan, Y. Jin, D.C. Yang, Green synthesis of silver nanoparticles by *Bacillus methylotrophicus*, and their antimicrobial activity, Artif. Cells Nanomed. Biotechnol. 44 (2015) 1–6.

[31] A. Najitha Banu, C. Balasubramanian, P.V. Moorthi, Biosynthesis of silver nanoparticles using *Bacillus thuringiensis* against dengue vector, *Aedes aegypti* (Diptera: Culicidae), Parasitol. Res. 113 (2013) 311–316.

[32] S.V. Otari, R.M. Patil, N.H. Nadaf, S.J. Ghosh, S.H. Pawar, Green synthesis of silver nanoparticles by microorganism using organic pollutant: its antimicrobial and catalytic application, Environ. Sci. Pollut. Res. 21 (2013) 1503–1513.

[33] S. Chowdhury, A. Basu, S. Kundu, Green synthesis of protein capped silver nanoparticles from phytopathogenic fungus *Macrophomina phaseolina* (Tassi) Goid with antimicrobial properties against multidrug-resistant bacteria, Nanoscale Res. Lett. 9 (2014) 365.

[34] A. Syed, S. Saraswati, G.C. Kuncu, A. Ahmad, Biological synthesis of silver nanoparticles using the fungus *Humicola* sp. and evaluation of their cytoxicity using normal and cancer cell lines, Spectrochim. Acta A Mol. Biomol. Spectrosc. 114 (2013) 144–147.

[35] S.S. Birla, V.V. Tiwari, A.K. Gade, A.P. Ingle, A.P. Yadav, M.K. Rai, Fabrication of silver nanoparticles by *Phoma glomerata* and its combined effect against *Escherichia coli*, *Pseudomonas aeruginosa* and *Staphylococcus aureus*, Lett. Appl. Microbiol. 48 (2009) 173–179.

[36] J. Harra, J. Mäkitalo, R. Siikanen, M. Virkki, G. Genty, T. Kobayashi, M. Kauranen, J.M. Mäkelä, Size-controlled aerosol synthesis of silver nanoparticles for plasmonic materials, J. Nanopart. Res. 14 (2012) 870.

[37] H. Zhang, G. Zou, L. Liu, H. Tong, Y. Li, H. Bai, A. Wu, Synthesis of silver nanoparticles using large-area arc discharge and its application in electronic packaging, J. Mater. Sci. 52 (2016) 3375–3387.

[38] G.R. Khayati, K. Janghorban, Preparation of nanostructure silver powders by mechanical decomposing and mechanochemical reduction of silver oxide, Trans. Nonferrous Metals Soc. China 23 (2013) 1520–1524.

[39] P. Asanithi, S. Chaiyakun, P. Limsuwan, Growth of silver nanoparticles by DC magnetron sputtering, J. Nanomater. 2012 (2012) 1–8.

[40] Z. Zhang, W. Shen, J. Xue, Y. Liu, Y. Liu, P. Yan, J. Liu, J. Tang, Recent advances in synthetic methods and applications of silver nanostructures, Nanoscale Res. Lett. 13 (2018) 54.

[41] C. Quintero-Quiroz, N. Acevedo, J. Zapata-Giraldo, L.E. Botero, J. Quintero, D. Zárate-Triviño, J. Saldarriaga, V.Z. Pérez, Optimization of silver nanoparticle synthesis by chemical reduction and evaluation of its antimicrobial and toxic activity, Biomater. Res. 23 (2019) 27.

[42] R.A. Khaydarov, R.R. Khaydarov, O. Gapurova, Y. Estrin, T. Scheper, Electrochemical method for the synthesis of silver nanoparticles, J. Nanopart. Res. 11 (2008) 1193–1200.

[43] A.C. Dhayagude, A. Das, S.S. Joshi, S. Kapoor, γ-Radiation induced synthesis of silver nanoparticles in aqueous poly (N-vinylpyrrolidone) solution, Colloids Surf. A Physicochem. Eng. Asp. 556 (2018) 148–156.

[44] K.C. Pingali, D.A. Rockstraw, S. Deng, Silver nanoparticles from ultrasonic spray pyrolysis of aqueous silver nitrate, Aerosol Sci. Technol. 39 (2005) 1010–1014.

[45] A. Roy, O. Bulut, S. Some, A.K. Mandal, M.D. Yilmaz, Green synthesis of silver nanoparticles: biomolecule-nanoparticle organizations targeting antimicrobial activity, RSC Adv. 9 (2019) 2673–2702.

[46] K.S. Siddiqi, A. Husen, R.A.K. Rao, A review on biosynthesis of silver nanoparticles and their biocidal properties, J. Nanobiotechnol. 16 (2018) 14.

[47] D. Baruah, R.N.S. Yadav, A. Yadav, A.M. Das, *Alpinia nigra* fruits mediated synthesis of silver nanoparticles and their antimicrobial and photocatalytic activities, J. Photochem. Photobiol. B Biol. 201 (2019) 111649.

[48] B. Yılmaz Öztürk, B. Yenice Gürsu, İ. Dağ, Antibiofilm and antimicrobial activities of green synthesized silver nanoparticles using marine red algae *Gelidium corneum*, Process Biochem. 89 (2020) 208–219.

[49] S.F. Hashemi, N. Tasharrofi, M.M. Saber, Green synthesis of silver nanoparticles using *Teucrium polium* leaf extract and assessment of their antitumor effects against MNK45 human gastric cancer cell line, J. Mol. Struct. 1208 (2020) 127889.

[50] A.C. Gomathi, S.R. Xavier Rajarathinam, A. Mohammed Sadiq, S. Rajeshkumar, Anticancer activity of silver nanoparticles synthesized using aqueous fruit shell extract of *Tamarindus indica* on MCF-7 human breast cancer cell line, J. Drug Delivery Sci. Technol. 55 (2020) 101376.

[51] M.A. Odeniyi, V.C. Okumah, B.C. Adebayo-Tayo, O.A. Odeniyi, Green synthesis and cream formulations of silver nanoparticles of *Nauclea latifolia* (African peach) fruit extracts and evaluation of antimicrobial and antioxidant activities, Sustain. Chem. Pharm. 15 (2020) 100197.

[52] R. Sankar, P.K.S.M. Rahman, K. Varunkumar, C. Anusha, A. Kalaiarasi, K.S. Shivashangari, V. Ravikumar, Facile synthesis of *Curcuma longa* tuber powder engineered metal nanoparticles for bioimaging applications, J. Mol. Struct. 1129 (2017) 8–16.

[53] S. Meenakshisundaram, V. Krishnamoorthy, Y. Jagadeesan, R. Vilwanathan, A. Balaiah, *Annona muricata* assisted biogenic synthesis of silver nanoparticles regulates cell cycle arrest in NSCLC cell lines, Bioorg. Chem. 95 (2020) 103451.

[54] A. Hussain, M.F. Alajmi, M.A. Khan, S.A. Pervez, F. Ahmed, S. Amir, F.M. Husain, M.S. Khan, G.-M. Shaik, I. Hassan, R.A. Khan, M.T. Rehman, Biosynthesized silver nanoparticle (AgNP) from *Pandanus odorifer* leaf extract exhibits anti-metastasis and anti-biofilm potentials, Front. Microbiol. 10 (2019) 8.

[55] K. Nesrin, C. Yusuf, K. Ahmet, S.B. Ali, N.A. Muhammad, S. Suna, S. Fatih, Biogenic silver nanoparticles synthesized from *Rhododendron ponticum* and their antibacterial, antibiofilm and cytotoxic activities, J. Pharm. Biomed. Anal. 179 (2020) 112993.

[56] M. Shah, D. Fawcett, S. Sharma, S. Tripathy, G. Poinern, Green synthesis of metallic nanoparticles via biological entities, Materials 8 (2015) 7278–7308.

[57] G. Sathishkumar, C. Gobinath, K. Karpagam, V. Hemamalini, K. Premkumar, S. Sivaramakrishnan, Phyto-synthesis of silver nanoscale particles using *Morinda citrifolia* L. and its inhibitory activity against human pathogens, Colloids Surf. B: Biointerfaces 95 (2012) 235–240.

[58] G. Sharma, J.-S. Nam, A. Sharma, S.-S. Lee, Antimicrobial potential of silver nanoparticles synthesized using medicinal herb *Coptidis rhizome*, Molecules 23 (2018) 2268.

[59] T. Roshmi, P. Jishma, E.K. Radhakrishnan, Photocatalytic and antibacterial effects of silver nanoparticles fabricated by *Bacillus subtilis* SJ 15, Inorg. Nano-Met. Chem. 47 (2016) 901–908.

[60] B. Ingham, X-ray scattering characterisation of nanoparticles, Crystallogr. Rev. 21 (2015) 229–303.

[61] R. Xu, Light scattering: a review of particle characterization applications, Particuology 18 (2015) 11–21.

[62] C. Baudot, C.M. Tan, J.C. Kong, FTIR spectroscopy as a tool for nano-material characterization, Infrared Phys. Technol. 53 (2010) 434–438.

[63] S. Mourdikoudis, R.M. Pallares, N.T.K. Thanh, Characterization techniques for nanoparticles: comparison and complementarity upon studying nanoparticle properties, Nanoscale 10 (2018) 12871–12934.

[64] W.H. Doh, V. Papaefthimiou, T. Dintzer, V. Dupuis, S. Zafeiratos, Synchrotron radiation X-ray photoelectron spectroscopy as a tool to resolve the dimensions of spherical core/shell nanoparticles, J. Phys. Chem. C 118 (2014) 26621–26628.

[65] S. Agnihotri, S. Mukherji, S. Mukherji, Size-controlled silver nanoparticles synthesized over the range 5–100 nm using the same protocol and their antibacterial efficacy, RSC Adv. 4 (2014) 3974–3983.

[66] M.S.I. Khan, S.W. Oh, Y.J. Kim, Power of scanning electron microscopy and energy dispersive X-ray analysis in rapid microbial detection and identification at the single cell level, Sci. Rep. 10 (2020) 2368.

[67] S. Rades, V.-D. Hodoroaba, T. Salge, T. Wirth, M.P. Lobera, R.H. Labrador, K. Natte, T. Behnke, T. Gross, W.E.S. Unger, High-resolution imaging with SEM/T-SEM, EDX and SAM as a combined methodical approach for morphological and elemental analyses of single engineered nanoparticles, RSC Adv. 4 (2014) 49577–49587.

[68] P.-C. Lin, S. Lin, P.C. Wang, R. Sridhar, Techniques for physicochemical characterization of nanomaterials, Biotechnol. Adv. 32 (2014) 711–726.

[69] W. Wan, J. Su, X.D. Zou, T. Willhammar, Transmission electron microscopy as an important tool for characterization of zeolite structures, Inorg. Chem. Front. 5 (2018) 2836–2855.

[70] A. Haase, H.F. Arlinghaus, J. Tentschert, H. Jungnickel, P. Graf, A. Mantion, F. Draude, S. Galla, J. Plendl, M.E. Goetz, A. Masic, W. Meier, A.F. Thünemann, A. Taubert, A. Luch, Application of laser postionization secondary neutral mass spectrometry/time-of-flight secondary ion mass spectrometry in nanotoxicology: visualization of nanosilver in human macrophages and cellular responses, ACS Nano 5 (2011) 3059–3068.

[71] M. Naghdi, S. Metahni, Y. Ouarda, S.K. Brar, R.K. Das, M. Cledon, Instrumental approach toward understanding nano-pollutants, Nanotechnol. Environ. Eng. 2 (2017) 3.

[72] F.S. Ruggeri, T. Šneideris, M. Vendruscolo, T.P.J. Knowles, Atomic force microscopy for single molecule characterisation of protein aggregation, Arch. Biochem. Biophys. 664 (2019) 134–148.

[73] R.I. Aminov, A brief history of the antibiotic era: lessons learned and challenges for the future, Front. Microbiol. 1 (2010) 134.

[74] H. Nikaido, Multidrug resistance in bacteria, Annu. Rev. Biochem. 78 (2009) 119–146.

[75] L. Wang, C. Hu, L. Shao, The antimicrobial activity of nanoparticles: present situation and prospects for the future, Int. J. Nanomedicine 12 (2017) 1227–1249.

[76] M. Azharuddin, G.H. Zhu, D. Das, E. Ozgur, L. Uzun, A.P.F. Turner, H.K. Patra, A repertoire of biomedical applications of noble metal nanoparticles, Chem. Commun. 55 (2019) 6964–6996.

[77] M.K. Rai, S.D. Deshmukh, A.P. Ingle, A.K. Gade, Silver nanoparticles: the powerful nanoweapon against multidrug-resistant bacteria, J. Appl. Microbiol. 112 (2012) 841–852.

[78] T.C. Dakal, A. Kumar, R.S. Majumdar, V. Yadav, Mechanistic basis of antimicrobial actions of silver nanoparticles, Front. Microbiol. 7 (2016) 1831.

[79] R.A. Hamouda, M.H. Hussein, R.A. Abo-Elmagd, S.S. Bawazir, Synthesis and biological characterization of silver nanoparticles derived from the cyanobacterium *Oscillatoria limnetica*, Sci. Rep. 9 (2019) 13071.

[80] T.R. Schleicher, J. Yang, M. Freudzon, A. Rembisz, S. Craft, M. Hamilton, M. Graham, G. Mlambo, A.K. Tripathi, Y. Li, P. Cresswell, P. Sinnis, G. Dimopoulos, E. Fikrig, A mosquito salivary gland protein partially inhibits *Plasmodium sporozoite* cell traversal and transmission, Nat. Commun. 9 (2018) 2908.

[81] S. Huijben, K.P. Paaijmans, Putting evolution in elimination: winning our ongoing battle with evolving malaria mosquitoes and parasites, Evol. Appl. 11 (2018) 415–430.

[82] M. Rai, A.P. Ingle, P. Paralikar, I. Gupta, S. Medici, C.A. Santos, Recent advances in use of silver nanoparticles as antimalarial agents, Int. J. Pharm. 526 (2017) 254–270.

[83] P. Urbán, J. Estelrich, A. Cortés, X. Fernàndez-Busquets, A nanovector with complete discrimination for targeted delivery to *Plasmodium falciparum*-infected versus non-infected red blood cells in vitro, J. Control. Release 151 (2011) 202–211.

[84] R. Vaikundamoorthy, R. Rajendran, A. Selvaraju, K. Moorthy, S. Perumal, Development of thermostable amylase enzyme from *Bacillus cereus* for potential antibiofilm activity, Bioorg. Chem. 77 (2018) 494–506.

[85] H. Dang, C.R. Lovell, Microbial surface colonization and biofilm development in marine environments, Microbiol. Mol. Biol. Rev. 80 (2015) 91–138.

[86] D. Sharma, L. Misba, A.U. Khan, Antibiotics versus biofilm: an emerging battleground in microbial communities, Antimicrob. Resist. Infect. Control 8 (2019) 76.

[87] S. Galié, C. García-Gutiérrez, E.M. Miguélez, C.J. Villar, F. Lombó, Biofilms in the food industry: health aspects and control methods, Front. Microbiol. 9 (2018) 898.

[88] A. Regiel-Futyra, M. Kus-Liśkiewicz, V. Sebastian, S. Irusta, M. Arruebo, A. Kyzioł, G. Stochel, Development of noncytotoxic silver–chitosan nanocomposites for efficient control of biofilm forming microbes, RSC Adv. 7 (2017) 52398–52413.

[89] A. Kishen, A. Shrestha, Nanoparticles for endodontic disinfection, Clin. Dent. Rev. 2 (2018) 11.

[90] M. Redza-Dutordoir, D.A. Averill-Bates, Activation of apoptosis signalling pathways by reactive oxygen species, Biochim. Biophys Acta (BBA) 1863 (2016) 2977–2992.

[91] G.-Y. Liou, P. Storz, Reactive oxygen species in cancer, Free Radic. Res. 44 (2010) 479–496.

[92] Y. Yao, H. Zhang, Z. Wang, J. Ding, S. Wang, B. Huang, S. Ke, C. Gao, Reactive oxygen species (ROS)-responsive biomaterials mediate tissue microenvironments and tissue regeneration, J. Mater. Chem. B 7 (2019) 5019–5037.

[93] E.K. Kim, E.-J. Choi, Pathological roles of MAPK signaling pathways in human diseases, Biochim. Biophys. Acta (BBA)—Mol. Basis Dis. 1802 (2010) 396–405.

[94] K. Das, A. Roychoudhury, Reactive oxygen species (ROS) and response of antioxidants as ROS-scavengers during environmental stress in plants, Front. Environ. Sci. 2 (2014) 53.

[95] J.K. Patra, G. Das, L.F. Fraceto, E.V.R. Campos, M.d.P. Rodriguez-Torres, L.S. Acosta-Torres, L.A. Diaz-Torres, R. Grillo, M.K. Swamy, S. Sharma, S. Habtemariam, H.-S. Shin, Nano based drug delivery systems: recent developments and future prospects, J. Nanobiotechnol. 16 (2018) 71.

[96] N. Kanipandian, S. Kannan, R. Ramesh, P. Subramanian, R. Thirumurugan, Characterization, antioxidant and cytotoxicity evaluation of green synthesized silver nanoparticles using *Cleistanthus collinus* extract as surface modifier, Mater. Res. Bull. 49 (2014) 494–502.

[97] L. Valgimigli, A. Baschieri, R. Amorati, Antioxidant activity of nanomaterials, J. Mater. Chem. B 6 (2018) 2036–2051.

[98] K. Varunkumar, C. Anusha, T. Saranya, V. Ramalingam, S. Raja, V. Ravikumar, Avicennia marina engineered nanoparticles induce apoptosis in adenocarcinoma lung cancer cell line through p53 mediated signaling pathways, Process Biochem. 94 (2020) 349–358.

[99] F. Bray, J. Ferlay, I. Soerjomataram, R.L. Siegel, L.A. Torre, A. Jemal, Global cancer statistics 2018: GLOBOCAN estimates of incidence and mortality worldwide for 36 cancers in 185 countries, CA Cancer J. Clin. 68 (2018) 394–424.

[100] N. Harbeck, F. Penault-Llorca, J. Cortes, M. Gnant, N. Houssami, P. Poortmans, K. Ruddy, J. Tsang, F. Cardoso, Breast cancer, Nat. Rev. Dis. Primers. 5 (2019) 66.

[101] S. Kruger, M. Ilmer, S. Kobold, B.L. Cadilha, S. Endres, S. Ormanns, G. Schuebbe, B.W. Renz, J.G. D'Haese, H. Schloesser, V. Heinemann, M. Subklewe, S. Boeck, J. Werner, M. von Bergwelt-Baildon, Advances in cancer immunotherapy 2019—latest trends, J. Exp. Clin. Cancer Res. 38 (2019) 268.

[102] C. Wigerup, S. Påhlman, D. Bexell, Therapeutic targeting of hypoxia and hypoxia-inducible factors in cancer, Pharmacol. Ther. 164 (2016) 152–169.

[103] K. Sarkar, S.L. Banerjee, P.P. Kundu, G. Madras, K. Chatterjee, Biofunctionalized surface-modified silver nanoparticles for gene delivery, J. Mater. Chem. B 3 (2015) 5266–5276.

[104] B.P.A. George, N. Kumar, H. Abrahamse, S.S. Ray, Apoptotic efficacy of multifaceted biosynthesized silver nanoparticles on human adenocarcinoma cells, Sci. Rep. 8 (2018) 14368.

[105] R. Vaikundamoorthy, R. Sundaramoorthy, V. Krishnamoorthy, R. Vilwanathan, R. Rajendran, Marine steroid derived from *Acropora formosa* enhances mitochondrial-mediated apoptosis in non-small cell lung cancer cells, Tumor Biol. 37 (2016) 10517–10531.

[106] M. Jeyaraj, M. Rajesh, R. Arun, D. MubarakAli, G. Sathishkumar, G. Sivanandhan, G.K. Dev, M. Manickavasagam, K. Premkumar, N. Thajuddin, A. Ganapathi, An investigation on the cytotoxicity and caspase-mediated apoptotic effect of biologically synthesized silver nanoparticles using *Podophyllum hexandrum* on human cervical carcinoma cells, Colloids Surf. B: Biointerfaces 102 (2013) 708–717.

[107] D. Sharma, C.M. Hussain, Smart nanomaterials in pharmaceutical analysis, Arab. J. Chem. 13 (2020) 3319–3343.

[108] F. Paladini, M. Pollini, Antimicrobial silver nanoparticles for wound healing application: progress and future trends, Materials 12 (2019) 2540.

[109] A. Gupta, S.M. Briffa, S. Swingler, H. Gibson, V. Kannappan, G. Adamus, M. Kowalczuk, C. Martin, I. Radecka, Synthesis of silver nanoparticles using curcumin-cyclodextrins loaded into bacterial cellulose-based hydrogels for wound dressing applications, Biomacromolecules 21 (2020) 1802–1811.

[110] Y. Li, Q. Wei, F. Ma, X. Li, F. Liu, M. Zhou, Surface-enhanced Raman nanoparticles for tumor theranostics applications, Acta Pharm. Sin. B 8 (2018) 349–359.

[111] M. Rycenga, P.H.C. Camargo, W. Li, C.H. Moran, Y. Xia, Understanding the SERS effects of single silver nanoparticles and their dimers, one at a time, J. Phys. Chem. Lett. 1 (2010) 696–703.

[112] W. Wei, Y. Du, L. Zhang, Y. Yang, Y. Gao, Improving SERS hot spots for on-site pesticide detection by combining silver nanoparticles with nanowires, J. Mater. Chem. C 6 (2018) 8793–8803.

[113] Z. Wei, L. Feng, J. Zhi-Ming, S. Xiao-Bo, Y. Peng-Hui, W. Xue-Ren, S. Cheng, G. Zhan-Qi, L. Liang-Sheng, Efficient plasmonic photocatalytic activity on silver-nanoparticle-decorated $AgVO_3$ nanoribbons, J. Mater. Chem. A 2 (2014) 13226–13231.

[114] C.A. Dos Santos, M.M. Seckler, A.P. Ingle, I. Gupta, S. Galdiero, M. Galdiero, A. Gade, M. Rai, Silver nanoparticles: therapeutical uses, toxicity, and safety issues, J. Pharm. Sci. 103 (2014) 1931–1944.

[115] N.R. Panyala, E.M. Peña-Méndez, J. Havel, Silver or silver nanoparticles: a hazardous threat to the environment and human health? J. Appl. Biomed. 6 (2008) 117–129.

[116] M.C. Stensberg, Q. Wei, E.S. McLamore, D.M. Porterfield, A. Wei, M.S. Sepúlveda, Toxicological studies on silver nanoparticles: challenges and opportunities in assessment, monitoring and imaging, Nanomedicine 6 (2011) 879–898.

[117] B.H. Mao, Z.Y. Chen, Y.J. Wang, S.J. Yan, Silver nanoparticles have lethal and sublethal adverse effects on development and longevity by inducing ROS-mediated stress responses, Sci. Rep. 8 (2018) 2445.

CHAPTER 11

Synthesis of silica nanoparticles for biological applications

Ali Shukur[a], Asima Farooq[b], Debra Whitehead[b], and May Azzawi[a]
[a]Cardiovascular Research Group, Department of Life Sciences, Faculty of Science and Engineering, Manchester Metropolitan University, Manchester, United Kingdom
[b]Department of Natural Sciences, Faculty of Science and Engineering, Manchester Metropolitan University, Manchester, United Kingdom

1. Introduction

The requirement for an accurate delineation, sensitive visualization, specific targeting [1], and monitoring in medical diagnostic and prognostic procedures has promoted the application of nanotechnology as an alternative strategy to conventional histological, imaging, and drug-delivery systems [2]. Nanobiotechnology and the use of nanomaterials have displayed a platform for intervention in medicine in aiding new pathways to diagnostic and therapeutic options [3–5]. Nanoparticles (NPs) display the basis for this nanotechnology and behave as whole units in terms of their transport and properties [6]. Because of their small size, NPs exhibit various characteristics that make them attractive tools for biomedical applications. Among the different NPs present, silica nanoparticles (SiNPs) are of special interest because of their size, unique properties, and stability under flow conditions enabling their use for cell tracking, imaging diagnostics [7–13], drug delivery [14,15], and medical therapeutics [16–18]. There are a number of strategies being used to improve their biocompatibility, such as coating with polymers, lipids, and antibodies [19]. As such, detailed characterization of nanomaterials is essential to enable their safe use in imaging diagnostics [20] and therapeutics [21]. This chapter focuses on the potential use of SiNPs as diagnostic tools and therapeutic intervention modalities and provides an overview of the methods used to fabricate and characterize them. The chapter also highlights the strategies used to maximize the biocompatibility of SiNPs to improve their applicability and bioavailability, minimize their off-target localization and effects, and promote their guided delivery.

2. Overview of biomedical applications of silica nanoparticles

The biomedical application of nanomaterials depends on their material composition [22], shape [23], size [24,25], concentration [26], cellular permeability, stability, biodegradability, bioactivity [27], and toxicity [28–30]. SiNPs have been shown to be highly reactive and can interact efficiently with biological materials because of their large surface area to volume ratio [29]. SiNPs can be incorporated into various nanomaterials and nano-composites for use in biomedical research, including tissue engineering scaffolds [31,32]. Furthermore, SiNPs can be tagged by thousands of luminescent molecules making them attractive candidates for imaging diagnostics [33].

SiNPs of 20–200-nm size were previously used in the targeting of vessels in ischemia for diagnostic and therapeutic purposes [34]. SiNPs-based ultrasmall inorganic hybrid NPs, "C dots" (Cornell dots), were used for the first time in patients with metastatic melanoma as part of a clinical study assessing the safety, pharmacokinetics, clearance properties, and radiation dosimetry of iodine-labeled and cRGDY peptide-modified positron emission tomography (PET) C dots (^{124}I-cRGDY-PEG-C dots) after intravenous administration [35]. These hybrid PET-optical imaging agents were suggested to be biocompatible and well-tolerated as they exhibited in vivo stability and were capable of being preferentially taken up and localized within disease sites and were excreted over a 2-week period [35]. Such studies may form a platform for the clinical translation or use of silica-based NPs as a delivery system for imaging probes replacing the invasive surgical procedures in human cancer lesion detection, staging, diagnosis, and management. Other silica-based NPs that had undergone clinical trials include the Auroshell NPs used for the treatment of lung and prostate cancer, with promising results [5]. These 120-nm SiNPs included a 1–3-nm gold core with PEG surface functionalization and were designed for precise thermal ablation of solid tumors after stimulation with near infrared energy source. Another study by Kharlamov et al., 2015, suggested the usefulness of silica-gold NPs in causing a regression of coronary atherosclerosis and the potential they may have in lowering the risk of cardiovascular death [36].

Dysregulation of microRNAs is associated with a number of diseases, including cancer and may display a marker for a particular disease and/or its progression. A previous study by Li et al. had developed fluorescent dye-doped SiNPs (fabricated by the reverse microemulsion method) for target-cell-specific delivery and intracellular microRNA imaging [37]. Surfactant molecules were coimmobilized on the surface of SiNPs to aid the guided delivery of the latter agents specifically to human breast cancer cells in vitro by adhering to cell surface molecules with high affinity [37]. Thus, the highly sensitive luminescent nonviral vector transfection nanoagents may display a platform for nano-based diagnostic research in a clinical setting. There are other studies that highlighted the potential of SiNPs in cancer treatment as well as diagnosis. A study by Badr et al. showed that SiNPs loaded with snake venom (*Walterinnesia aegyptia* venom; WEV) were able to enhance and robustly sensitize the human breast cancer cells isolated from cancer biopsies to cellular growth arrest and death over WEV alone ex vivo [38]. The route of cancer cell apoptosis was suggested via the increase in the levels of free radicals (including reactive oxygen species [ROS], hydroperoxide, and nitric oxide [NO]) and caspases activities (including caspase-3, caspase-8, and caspase-9) [38].

The phenomenon of drug-loading and slow drug release via the aid of SiNPs loaded with chemical therapeutics may have a pivotal role in effects observed on enhanced toxin and/or drug action in vivo. This can be explained in part by the ability of NPs to be endocytosed and/or phagocytosed by cells, resulting in the internalization of the encapsulated drug [39]. Because of the therapeutic effects of small interfering RNA (siRNAs) and the obstacles accompanying their specific targeting, studies have developed methods using SiNPs as candidates for siRNA delivery in vivo. A study by Chen et al. used magnetic mesoporous SiNPs (MSNs) for lung cancer treatment [40]. The siRNA designed by the Chen et al. group was loaded into the mesopores of SiNPs, followed by polyethylenimine (PEI) capping, PEGylation, and fusogenic peptide KALA modification. The resultant SiNPs exhibited a prolonged half-life in the bloodstream, enhanced cell membrane translocation, were capable of escaping endosomes, and possessed favorable tissue biocompatibility and biosafety [40]. Together, the latter properties of SiNPs enabled the in vivo imaging of target tissues and the specific suppression of tumor growth and metastasis in subdermal and orthotopic lung cancer models when loaded with vascular endothelial growth factor (VEGF) siRNA [40]. It has also been shown that polycationic SiNPs decorated with PEI were capable of delivering siRNA for the purposes of gene silencing and hence disease prevention [41]. PEI is a positively charged branched polymer of 25 kDa in size of biocompatible nature [41]. PEI was attached to the SiNPs via a non-covalent inorganic linker cerium (III) cations (Ce^{3+}) between both polyNH$_2$–SiO$_2$ NP surface and the polycationic PEI polymer [41]. Thus the use of PEI-modified SiNPs for siRNA delivery may have a therapeutic potential overcoming the obstacles of inefficient delivery and/or the off-target effect of sole siRNA.

Various studies have shown that SiNPs can be used in aiding guided stem cell delivery and/or implantation to injured sites that require stem cell therapy, thereby overcoming cells death and the consequences of ischemia, inflammation, immune response, misinjection, and/or implantation into fibrotic tissue [42]. In the study by Jokerst et al., multimodal, fluorescent silica-based NPs (300 nm) were used for cell sorting, real-time ultrasound-guided cell delivery and implantation at high resolution, and long-term monitoring by magnetic resonance imaging (MRI) [42]. The study demonstrated that SiNPs increased the ultrasound and MRI contrast of labeled human mesenchymal stem cells (hMSCs) 700% and 200% versus unlabeled cells, respectively, and allowed cell imaging to be performed in animal models for 13 days after implantation [42]. The SiNPs used were biocompatible with no effects observed on hMSC cell metabolic activity, proliferation, or pluripotency [42]. This may be in part due to the in vivo aggregation of SiNPs into larger silica frameworks that amplify the ultrasound backscatter. This guided cell delivery and multimodal optical/ultrasound/MRI intracardiac cell tracking may act as a platform for improved cell therapy in the clinic by minimizing misdelivery or implantation into fibrotic tissues [42]. In their mesoporous forms; MSNs loaded with sodium nitroprusside (SNP) were able to cause relaxation of aortic vessels that were sustained over the 3-hour period ex vivo [43]. Titania (Ti) coating improved MSN biocompatibility (relaxation to ACh and endothelial cell uptake) and enabled a more sustained drug-release profiles at 0.08% relaxation/min [44]. This utility of MSNs may allow them to display a platform for using NPs as drug-delivery vehicles for various therapeutic applications. In addition to silica-based NPs, other nanomaterials have also shown interesting applications in the treatment of long-term life conditions. For example; fortification of milk with nano-calcium citrate was demonstrated to increase the bioavailability of calcium and bone microarchitecture stiffness and strength in ovariectomized and ovariectomized-osteoporosis rats [45] in an attempt to prevent osteoporosis [46]. Furthermore, magnetite- and folate-based NPs were endorsed as a single dose treatment for iron deficiency anemia [47]. Further modifications to magnetite NPs, including the capping with vitamin C (20 nm), improved their efficiency for treatment of iron deficiency [48]. The latter NPs stimulated erythropoiesis, where the myeloid-erythroid ratio decreases from 1.3 to 0.3, without any apparent toxicity [48]. As well as drug-delivery components, NPs have recently been applied to restore the function of critical organs and tissues ex vivo. Ceria-coated SiNPs were shown to improve endothelial-dependent dilatory effect of rat aortic vessels compared with SiNPs alone [49]. Restoration of vascular function may be attributed to the anti-ROS properties of ceria [50], thus promoting the use of the latter as a potential decorating material for therapeutic NPs. Unpublished observations suggest that ceria-coated SiNPs effects may be mediated via an alteration to specific signal transduction and/or oxidative pathways.

Surface-functionalized, nanoconjugated, and carrier SiNPs have been promoted as nano-biotags and photodynamic agents for labeling and ablating tumors and pathogens [51,52]. A study Renner et al. demonstrated the ability of mesoporous silica (SiO_2) shell of silica-gold NPs to serve as a reservoir for anticancer drugs such as doxorubicin and quercetin for sustained drug release [53]. Their covalent attachment to surface-bound ligands of synthetic estrogen molecules facilitated their uptake by breast cancer cells, in vivo [53]. Red fluorescence-emitting silica-coated core-shell upconverting NPs modified with polyethylene glycol (PEG5k)-folic acid and tetrakis(4-carboxyphenyl) porphyrin (TCPP) (UCNPs@SiO_2-NH_2@FA/PEG/TCPP; 50 nm) have been demonstrated to suppress the viability of cervical cancer HeLa cells following cellular uptake [52]. The chemoselective recognition between NPs and cell membranes of tumor tissues may therefore display a new promising drug-delivery platform for selective cancer targeting and suppression. In addition, a number of studies have evaluated the role of SiNPs as antimicrobial agents [54]. Gonzalo-Juan et al. proposed the use of silver-silica-based glassy bioactive (ABG) nanocomposites as antibacterial agents against *Escherichia coli* [55]. A study by Mapukata et al. suggested the use of electrospun silicon dioxide nanofibers (SiO_2 NFs) for antibacterial treatment [56]. The antibacterial activity of the latter NPs on methicillin-resistant *Staphylococcus aureus* [56] may be attributed to silver NPs [57] that were used to decorate SiO_2 NFs. The emergence of functionalized SiNPs as alternative delivery system for DNA/RNA vaccines [58] demonstrates the effectiveness of the latter NPs in providing preventative interventions in tackling the spread of infectious diseases [59] such as the coronavirus (COVID-19) pandemic.

3. Silica nanoparticle synthesis

There is a growing interest in the fabrication of SiNPs for various clinical applications because of their biocompatible nature. Colloidal SiNPs can be synthesized using the Stöber sol–gel method [60] and the water-in-oil (W/O) reverse microemulsion method as performed by Chen et al. [61]. Stöber sol–gel method uses tetraethyl orthosilicate (TEOS), aqueous ammonia solution ($NH_3 \cdot H_2O$), and water in absolute ethanol to generate the SiNPs; hence, the silica matrix can be produced by the controlled hydrolysis of TEOS in water nanodroplets with the initiation of ammonia [61]. SiNPs can also be fabricated to form ordered pore structures (MSNs) with a greater surface area in contrast to nonporous (amorphous) SiNPs; hence, the capability to accommodate various molecular species for potential system drug delivery [62].

Silicon dioxide represents the basic molecular structure of silica-based materials. It has a characteristic crystalline structure, where the atoms or molecules are arranged in an ordered array in a liquid or solid phase. The sol–gel method is used to synthesize SiNPs with nonporous amorphous (nonordered) or crystalline (regular lattice) structure. The term sol–gel refers to the fabrication of particles in the form of a gel or monodispersed solution and was first described by Stöber et al. in 1968 [60]. It uses the principle of precipitation and encompasses the use of inorganic precursors such as metal halides, metal salts, and alkoxides ($M(OR)_x$), all mixed in solution to form the initial particles in a single synthetic step. The initial synthesis of SiNPs involves the hydrolysis of the inorganic precursor followed by condensation to synthesize the silicon oxide (Fig. 1). The colloidal silica generated as a result of the sol–gel method facilitates the production of siloxane polymer from

$$\equiv Si-OR + H_2O \Longleftrightarrow \equiv Si-OH + ROH \qquad (1)$$

$$\equiv Si-OR + \equiv Si-OH \Longleftrightarrow \equiv Si-O-Si\equiv + ROH \qquad (2)$$

$$\equiv Si-OH + \equiv Si-OH \Longleftrightarrow \equiv Si-O-Si\equiv + H_2O \qquad (3)$$

$$Si(OR)_4 + 2H_2O \xrightarrow{OH^-} SiO_2 + 4ROH \qquad (4)$$

Fig. 1 The reaction scheme for the fabrication of a silica nanoparticle; where (1) is the initial hydrolysis reaction step, (2) is the alcohol condensation reaction step, and (3) is the water condensation reaction step. The overall reaction is given in step (4), where R denotes C_2H_5.

organosilica compounds. The sol–gel synthetic route for SiNPs allows particle size control by altering the temperature, pH, dispersants, and precursor concentrations [63]. Hence, the reaction rate and growth rate of SiNPs can be manipulated by the use of for example acids or bases to affect the hydrolysis or condensation rate. Using low pH results in fast hydrolysis and in turn slow condensation favoring the formation of a gel. This phenomenon is reversed with high pH conditions leading into slow hydrolysis and fast condensation favoring a monodispersed solution [64]. The change in the silica precursor and solvent concentrations can also affect the size and monodispersity of SiNPs. Stöber et al. were able to fabricate SiNPs at different size ranges (50 nm to 2 μm) using the latter phenomenon [60]. Furthermore, the change in temperature conditions was shown to affect the nucleation of SiNPs generated from hydrolysis of a silica precursor, known as tetraethyl orthosilicate (TEOS), in an alcohol solution and catalyzed by ammonia [65]. The hydrolysis reaction is followed by a condensation of the alkoxide silica ($Si(OR)_4$) occurring by alcohol or water molecules (Fig. 2) to fabricate monodispersed silica spheres [65].

A number of attempts have been carried out to produce smaller sizes of monodispersive SiNPs (below 100 nm) with a smooth surface since the Stöber et al. method was accomplished [66,67]. The demand for such NPs is of huge interest for biomedical

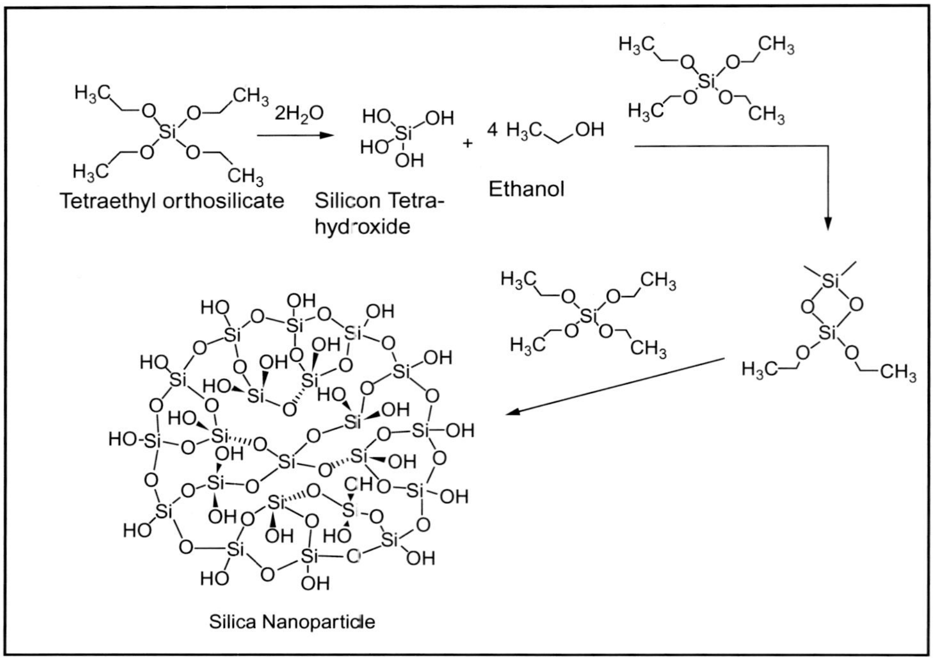

Fig. 2 Schematic illustration of the Stöber sol-gel reaction for the production of silica spheres. The initial reaction is the hydrolysis of tetraethyl orthosilicate (TOES) followed by the condensation of the alkoxide silica ($Si(OR)_4$) to fabricate monodispersed silica spheres.

applications. These attempts combine different factors that affect the reaction conditions such as changing the temperature as well as altering reagent concentrations (such as TEOS, NH_3, and ethanol) to fabricate SiNPs at the desired sizes. Stöber et al. used the phenomenon of reactant concentration (e.g., tetraethyl silicate concentration at $0.28\,mol\,dm^{-3}$) and certain physical conditions (such as the occurrence of the reaction at room temperature) as the determinant factor in fabricating monodispersed SiNPs [60]. Stöber et al. represented the relationship between particle diameter and reactant (water and ammonia) concentration graphically [60]. However, using the Stöber et al. method to obtain larger particles ($1-2\,\mu m$) tends to yield a poor size distribution, whereas small particles (approx. 30 nm) tend to lack a smooth surface. Bogush et al. had generated a model (growth-only) for synthesizing silica in an effort to enhance the monodispersity of SiNPs [66]. The "growth-only" model relies initially on SiNPs nucleation where the starting materials fully react to form small core SiNPs known as "seeds," followed by a diffusion-limited growth where TEOS is added in a molar ratio of 1:2 to water to coat the small SiNPs seeds [68,69]. This latter model causes the gradual increase in particle size or seeding growth, preventing the aggregation between particles of the same size as experienced with the "aggregation-only" model [68]. The relationship between reactant concentration and particle size is expressed through an equation, which was derived through the fabrication of over 100 samples at 25°C [66]. The particle diameter size generated depends on the total volume of TEOS added during the seeded growth [66] and derived as follows:

$$d = A[H_2O]^2 exp\left(-B[H_2O]^{1/2}\right)$$

where d is the average particle diameter in nm and the reactant concentrations with the units of $mol\,dm^{-3}$. The term A, specified in the above equation was corrected by Razink et al. [70] as follows:

$$A = [TEOS]^{1/2}\left(82 + 151[NH_3] + 1200[NH_3]^2 - 366[NH_3]^3\right)$$

Furthermore, the B in the former equation was expressed as follows:

$$B = 1.05 + 0.523[NH_3] - 0.128[NH_3]^2$$

Achieving smooth monodispersive SiNPs at a small size (10–100 nm) is difficult, as the particles tend to be granular in appearance because of the presence of ultrafine particles surrounding the surface of SiNPs.

The microemulsion method is another route for the synthesis of SiNPs. Microemulsions are a type of colloidal system where there are two phases, an oily substance and a water-based substance. The two are not miscible, but with the aid of a surfactant, they may become miscible. The oil and water mixture of microemulsion are known to be thermodynamically stabilized when a surfactant is present. A surfactant molecule has a long chain with a hydrophobic tail and a hydrophilic head; its structure allows it to dissolve in both water and in organic solvents. Surfactant molecules form micelles or reverse micelles, which have spherical structure. There is an interfacial tension between the water and the oil that is lowered by the surfactant because the surfactants' hydrophilic head will adsorb in the water and the hydrophobic tail will adsorb in the oil. Above certain surfactant concentrations, micelles are formed in the water phase where the tails are in the core and the heads are facing outward. Reverse micelles are formed in the oil phase with the tails facing outward and the heads in the core, as illustrated in Fig. 3.

The difference between a microemulsion and emulsion is the transparency, the thermodynamic stability, and the heterogeneity. Because of the microemulsion being optically isotropic, it is transparent. Also, unlike microemulsion, emulsions are thermodynamically unstable. The characteristics of the microemulsion medium have caused researchers to use this technique to synthesize various materials, such as metals, metal halides, chalcogenides, carbonates, organic polymers, and SiNPs [71]. When this technique is used to form NPs, the reactants maybe contained in separate aqueous microemulsion. The two are mixed and because of Brownian motion, they react to form small NPs. To fabricate SiNPs via microemulsion, the preparation involves alkoxide hydrolysis of a silica precursor, such as TEOS, in water-in-oil microemulsion [72]. The reversed micelles consist of water nanodroplets in an organic medium to be used as nanoreactors for the fabrication of NPs (Fig. 4). The particle size may be adjusted by altering the ratio of water to oil [73]. The water-to-surfactant molar ratio (R) and ammonia concentration have been shown to have an effect on the particle size [72]. The early stage of the reaction is a first order with respect to TEOS concentration, and as R increases, the first-order kinetics decrease. Initially, the reaction entails a hydrolysis of TEOS in alcohol, such as ethanol, which is catalyzed by ammonium hydroxide base in water solution. This follows a first-order kinetics reaction with respect to ammonia and TEOS concentrations [72]. Kay et al. concluded that independently of

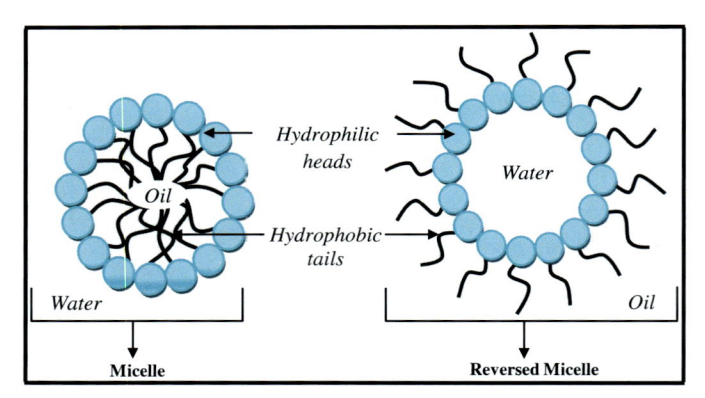

Fig. 3 Schematic illustration of the micelle and reversed micelle, illustrating the conformation of the hydrophilic heads in relation to the hydrophobic tails in oil and water, respectively.

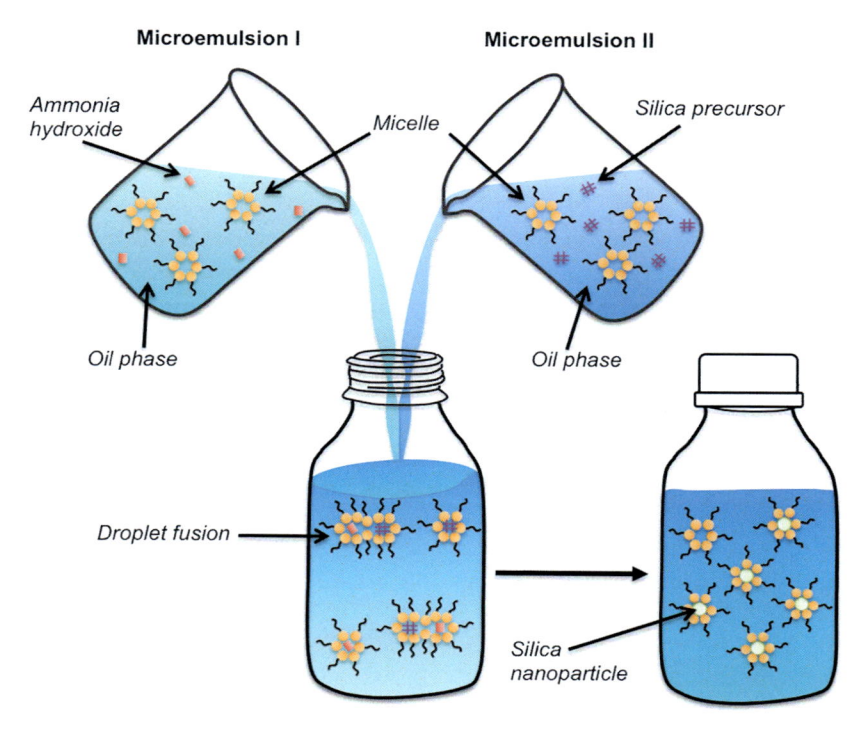

Fig. 4 An illustration of the silica nanoparticle formation from a microemulsion reaction.

the catalyst used for TEOS hydrolysis, acid, or base, the reaction will still remain first order with respect to the water concentration (1:2 molar ratio of TEOS:H_2O) [74]. The water concentration has been found to play an important role in the base-catalyzed hydrolysis rate [69]. The type of solvent mixed with TEOS also has an effect on the rate of hydrolysis [75]. If the solvent has long alkyl chains, then the rate of hydrolysis decreases because of steric stabilization. The rate of hydrolysis also decreases if the solvent has a tendency to form hydrogen bonds with water [69].

Porous materials that are predominately synthesized via a sol-gel precipitation method which contains surfactant molecules that behave as a template for the growth of material around them. Porous materials may be macroporous if the pore sizes are greater than 50 nm. If the pore sizes are between 2 and 50 nm, the material is mesoporous, whereas pore sizes smaller than 2 nm are known as microporous. In 1992, Beck et al. discovered that these mesoporous materials consisted of ordered porous structures [76]. This allows MSNs to be used in sensing and drug-delivery applications. However, the precipitations of these porous NPs alone do not make them a sufficient vehicle for drug-delivery systems. For these particles to be effective in these systems the internal channel must be surfactant free to produce a porous network [77]. The main focus of investigation for MSNs is silica-based, this is because they are inexpensive, chemically inert, thermally stable, and seem to have no toxicological side effects [78]. The synthesis of mesoporous silica materials of various morphologies has been reported: the two main families are M41S and SBA. Mobil Research and Development Corporation in the 1990s was the first to report the synthesis of the MCM-41 (Fig. 5), which is the most common of the M41S family and is produced with the aid of a cationic surfactant [78]. The extensive use of these MSNs in various applications, such as catalytic sieves, is because of the regular arrangement of pores within the NPs, thermal stability, and large surface area [79].

The discovery of this series has caused researchers to alter the structure directing agent (surfactant) to see how it affects the pore channels, such as the use of anionic surfactant like quaternary ammonium salts or nonionic polyethylene oxide surfactant [80]. The SBA-15 MNP has uniform hexagonal pores that range between 5 and 15 nm. The SBA-15 are synthesized with the aid of copolymer Pluronic 123 ($EO_{20}PO_{70}EO_{20}$) acting as the template under acidic conditions in the presence of silica precursor such as TEOS [81]. The pore sizes for this MCM-41 are hexagonal and normally smaller than 5 nm. These synthetic reactions commonly require the aid of an organic template (such as exadecyltrimethyammonium bromide). Depending on the molecules that are placed into the

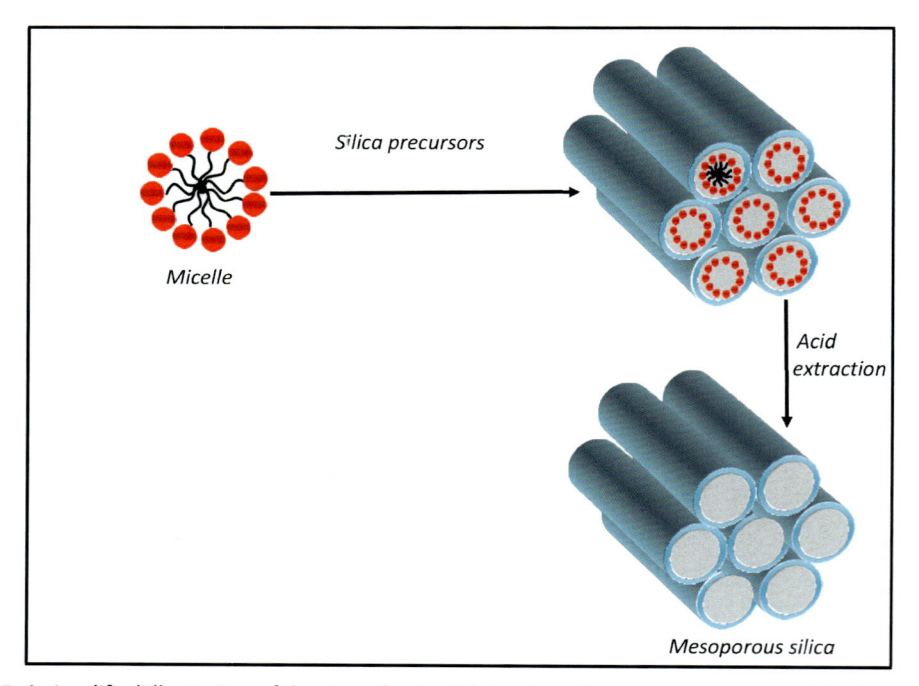

Fig. 5 A simplified illustration of the steps for the fabrication of mesoporous nanoparticles (MCM-41).

pores, for its diffusivity application, the pore diameter may be tuned. Nandiyanto et al. fabricated spherical silica MNP with pore sizes between 4 and 15 nm by changing the styrene (template) and hydrophobic molecule concentration [78].

3.1 Silica nanoparticle stability

Particles in the size range between 1 and 1000 nm are referred to as being colloidal, where they have a characteristic lyophobic or solvent hating behavior when dispersed in a continuous medium [82]. This minimizes the interaction between the particles and the continuous dispersed phase [82]. The laws of free Gibbs energy and entropy change accompany random "Brownian" motion of NPs within a solution according to their kinetic energy and particle collision [83]. The stability of SiNPs is characterized by their ability to remain dispersed in suspension and have a reduced tendency to aggregate. Small-size SiNPs disperse in a better fashion than larger ones, which have a tendency to settle quickly [84] in a colloidal suspension according to gravity and viscous drag force laws. The increased ratio of electrostatic repulsive forces over the attractive forces (Van der Waals forces) that exist between the NPs [85] helps to prevent the particles from aggregating. There are different types of aggregation, including conservation, gelation, coagulation, and flocculation (Fig. 6). In addition, the pH and ionic strength of the solution containing the NPs can affect the stability of the latter NPs [86]. There is a demand for stable SiNPs in biological media to fulfill their role [87]. Particles can be linked into branches forming a three-dimensional network leading to increased viscosity followed by gelation, which is characterized by the formation of a solidified network that retains the liquid. Particles can also form bridges together that link them leading into flocculation or become as closely packed clumps with each other leading into coagulation [84,88]. The absorption of materials onto the particle surface leads to reduced hydrophilicity and more concentrated liquid phase, which is immiscible in the aqueous phase termed "Conservation."

The stabilization of NPs can lead to more monodispersed NPs and can be achieved sterically, by creating a surface charge on the NP surfaces or the use of capping agents. Capping agents have the limitation of changing the properties of the intact NPs surfaces. The overall surface charge of the SiNP, for instance, is related to the carboxylic groups surrounding the surface. This charge determines NP reactivity and interaction with other NPs and the surrounding tissue, cells, and/or molecules. Hence, SiNP interaction with the surrounding tissue, cells, and/or molecules is responsible for minimizing aggregation [89], dispersive stability, the viscosity, flocculation, and film-forming abilities of NPs. NPs are prone to coating with ions and/or molecules when left in an electrolyte solution, leading to changes in their surface charge, stability, and hence, properties. The coating leads to the formation of what is termed "the electrochemical double layer (EDL)," which is described as the interface between two phases, whether it is solid/liquid,

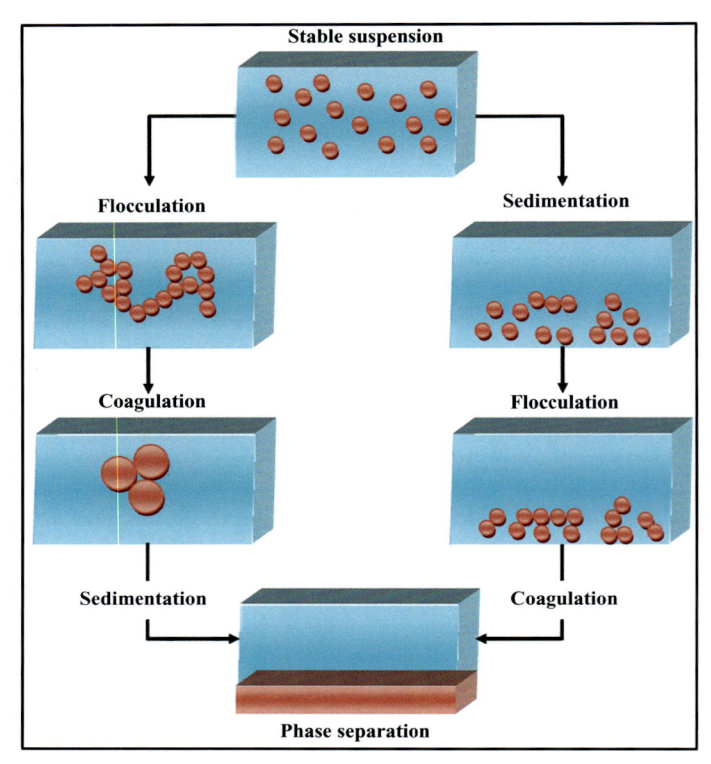

Fig. 6 A schematic illustration showing stable colloidal nanoparticles undergoing the flocculation and coagulation process of aggregation.

liquid/liquid, or liquid/gas, which possesses properties in the bulk phase. The EDL is regarded as a phase boundary having different distribution of electrical charges, therefore producing an electrical double layer, which has a potential that is different to both phases. The Stern model provided a graphical model for the charge distribution of particles in the EDL (Fig. 7). When NPs are placed in a solution that contains electrolytes of anion or cation origin, a monolayer of the electrolyte is adsorbed onto the particle surface in a nonspecific and/or specific manner forming an inner Helmholtz plane (IHP). This causes the NPs to become positively charged with a higher tendency to interact with other NP surface(s), whereas the counterions become dehydrated in the inner layer. The extent of the overall potential (ψ_i) of the NP surface depends on the strength of interaction between NPs and ions with the possibility of formation of a second layer of hydrated rigidly bound counterions referred to as the outer Helmholtz plane (OHP). Together, both the IHP and OHP with a particular thickness (δ) form the Stern layer. The control of the kinetics of the crystal growth of SiNPs and NPs aggregation using the sol-gel precipitation method with some alterations can lead to the generation of an organized array

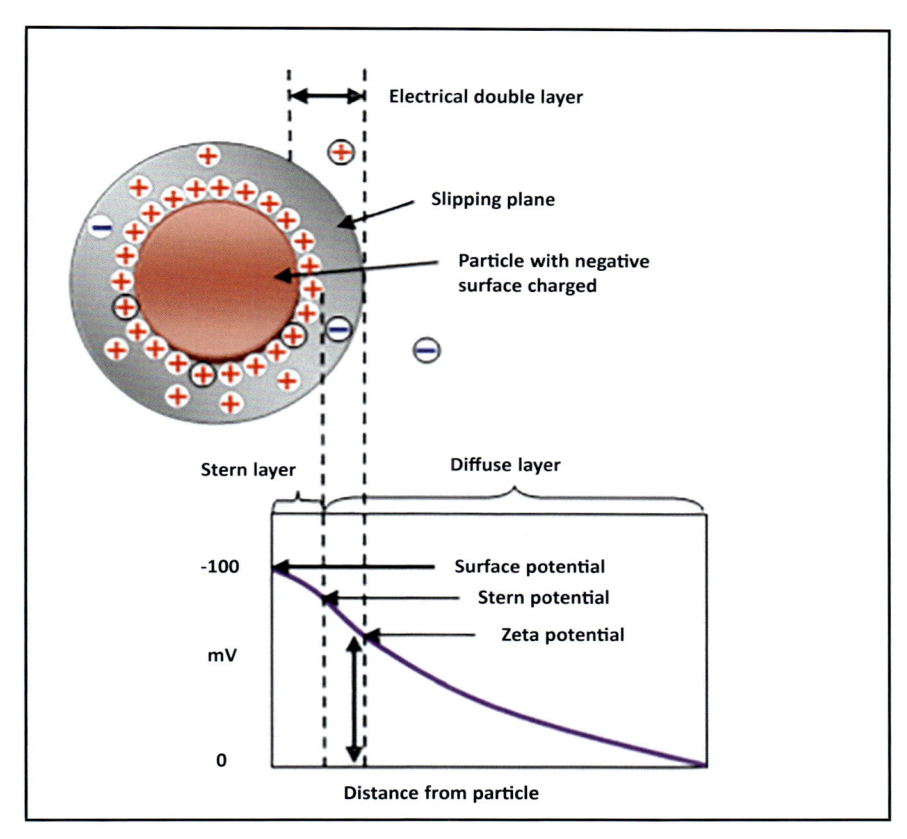

Fig. 7 Scheme of the electrical double layer surrounding a nanoparticle in solution. The negatively charged particle is surrounded by firmly attached positively charged ions forming the Stern layer. Zeta potential gives an indication of the colloidal systems stability.

of small and monodispersed SiNPs with surface stability [67]. The latter can be achieved by incorporating dye molecules [90] and coating NPs with inorganic groups to overcome the polydistribution [91,92] and surface instability.

The Debye Landau Verwey Overbeek (DLVO) theory is one theory that is used to describe the stability of lyophobic colloids, such as the SiNPs. Other non–DLVO theories are also proposed. These will be described in the section below. The DLVO theory was developed by Derjaguin and Landau in 1941 to describe the stability of colloidal solutions, which was dependent on Van der Waals attraction and electrostatic repulsion. The DLVO, which contributes to interface science, was further confirmed by Verwey and Overbeek to determine the interaction between charged surfaces in solution. The nanoscale provides a higher surface area per unit mass compared with bulk materials. When NP size is reduced, the high surface area generated possesses high surface energy, which can cause the NPs to react with each other, resulting in aggregation [89].

The DLVO theory is translated as the sum of attractive and repulsive forces because of double layer of counterions. This is where the attractive forces (V_{Total}) are the Van der Waals attraction (V_A) and the repulsive forces are original electrostatic forces (V_R) [93,94]. Particles with their charged surfaces in suspension are considered as electroneutral entities because of their ability to form a surrounding diffuse layer of oppositely charged ions. When particles come into close contact, a diffuse layer overlap is created forming an osmotic pressure between the particle surfaces. Hence, particle separation in suspension is achieved through the presence of the electrostatic repulsive forces that results from such overlap. The degree of dispersion within the lyophobic colloidal solution depends on the maximum repulsive energy that can be created between particle surfaces (the total potential energy curves for a lyophobic colloid V (1) and for a colloid with a maximal repulsive energy V_R (1), respectively; Fig. 8).

The addition of electrolytes into the colloidal solution (the total potential energy curve V (2); Fig. 8) favors Van der Waals attraction between particle surfaces as opposed to the repulsive energy that exists (V_R (2); Fig. 8). There is a correlation between particle surface potential and the electrolyte concentration within a colloidal suspension. When

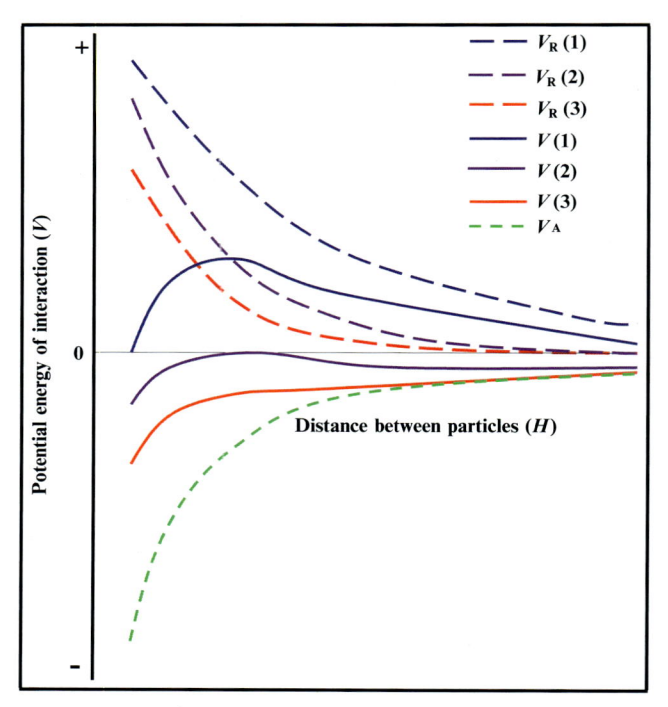

Fig. 8 A graphic representation of the potential energy of interaction between particles and their distance. The total energy of interaction (V (1), V (2), and V (3)) between two particles, acquired from the sum of attraction curve (V_A) and repulsion curves (V_R (1), V_R (2), and V_R (3)).

the electrolyte concentration is increased, particle surface potential is decreased, leading to the compression of the diffused layer and the reduction in the repulsive forces. The latter helps the particles to overcome the electrostatic energy barrier, thus leading to aggregation (Fig. 8). The minimal amount of reduction in repulsive force required to remove the energy barrier is referred to as the critical coagulation concentration of electrolytes. Higher concentrations of electrolytes lead to the further reduction in the energy barrier (the total potential energy curve V (3); Fig. 8) and thus rapid aggregation of NPs [95].

The DLVO theory used to describe the stability of particles within numerous colloidal systems [96–100] does not account for the additional interactions within silica solutions [101,102]. These include NP surface interactions with the other surfaces, the solvents as well as ionic species. Because of the dielectric-continuum nature of the solvent, additional repulsion (V_{Solv}) may be produced by the interaction between the highly structured hydration layer surrounding the particle surface-to-solvent and solvent-to-solvent. The existence of steric repulsion (V_{Ster}) between surfaces containing grafted polymers can also cause them to come close to each other. The DLVO theory differs from the non-DLVO theory in that it considers point charges, defined as the dimensions of the charge or charged ions to be infinitely interacting with particle surfaces at equal concentrations and different electrolyte types in an ion-specific manner. The degree of point charges depends on the distance between an ion and a NP surface, solvent, or another ion, but does not provide accurate estimation of any interaction between the ions to NP surface, solvent, or another ion. According to the DLVO theory, large SiNPs in an alkali solution (pH 2) will have no charge and be in the least stable region in contrast to previous studies, which suggest NP steric stabilization due to the hydrogen bonding between the water molecules and the hydroxyl groups attached to the particle surface forming an interactive monolayer [84,103,104]. The stability and dispersion of NPs therefore depends on the surface properties, surface interactions, and the surrounding environment.

Ion-specific adsorption influences the stability of the dispersion of colloidal NPs. The dispersion of NPs is affected by the ions surrounding its surface. When ions are in close contact with the particle surface, they are able to indiscriminately interact with the surface with a great screening efficiency leading to NPs' dispersion and destabilization even when lower concentrations of electrolyte are present [105]. According to the Stern model, the absorbability of ions to the SiNPs surfaces is dependent on ion charge and size. Thus, highly charged ions and small ions have greater absorbability, whereas large-sized ions and less charged ions have lower absorbability [106,107]. Ions can be categorized depending on how they interact with the solvent [108,109]. The presence of structure-maker water-forming ions such as Li^+ and Na^+ ions can cause the solvent to become structured on the particle surface. In contrast, the presence of structure-breaker ions can cause the solvent to become less structured on the particle surface compared with the bulk phase. Hence, the ions can promote or destroy water structures on the particle surface

[107,110]. Oxides with a high isoelectric point (IEP), such as aluminum, are referred to as structure-maker surfaces and thus the structure-maker ions will adsorb on their surface, whereas structure-breaker surfaces such as silica have low IEP [110,111]. Depending on the charge fluctuations and strength of the ion-to-ion correlation, ion-to-ion interactions within the double layer may lead to the buildup of an attractive electrostatic pressure between two NP surfaces [105,112]. The latter interactions are believed to be stronger than the Van der Waals attractive forces [112,113]. The double layer is generated when two approaching surfaces containing high ion density interact, causing a repulsion to occur. However, depending on the ionic density of either surface, when the ions are adsorbed close to one surface, lower ion density is generated at the mid-plane, thereby causing the attractive force to predominate.

Surfactant agents such as polymers, biological, or organic molecules that encapsulate or surround (dope, coat, or cap) NPs can add surface stability and prevent NP aggregation and sedimentation while maintaining dispersed NPs. For instance, because of their large surface to volume ratio, gold nanoparticles (AuNPs) have a tendency to aggregate [114]. To overcome this, AuNPs can be coated with organic polymer composites. Kobayashi et al. confirmed that larger SiNPs follow the DLVO theory [103]. At high pH where there is a low ionic strength, the rate constant decreases and there is a slow tendency for the large SiNPs to aggregate, whereas at low pH where there is a high ionic strength, the aggregation becomes fast. In contrast, small SiNPs below 80 nm in size become well dispersed at low and high pH values with less tendency to aggregate. Hence, Kobayashi et al. proposed the hairy model theory considering small SiNPs as stable NPs due to the presence of additional repulsive forces as postulating hairy layers with polysilicic acid chains surrounding the particle surface [103].

3.2 Fluorescent nanoprobes (mechanisms for nanoparticle probe loading)

The doping of SiNPs with luminescent dyes such as rhodamine promotes their use as agents for bioimaging [115] and biosensing [7] because of their fluorescent capabilities [116]. The fluorescent signal detected from dye-doped SiNPs may be used in the labeling and tracking of cellular organelles and biological molecules in a targeted and/or untargeted manner [117]. Core SiNPs can be doped with fluorescent and luminescent dyes such as tris (2,2′-bipyridyl)ruthenium (II) chloride hexahydrate (Rubpy) [118], fluorescein isothiocyanate (FITC), or rhodamine B isothiocyanate (RBITC) as fluorescent signal elements either by steric hindrance or by covalent bonding [61]. The fluorescent signal detected from dye-doped SiNPs may vary in intensity depending on the type of dye used, its bonding to other conjugates, and its corresponding stability; hence, SiNPs can be used in the targeted recognition (labeling and tracking) of cellular organelles and biological markers in a targeted and/or untargeted manner [117].

The dye-doped fluorescent SiNPs can be further stabilized using photostable linker compounds that are used to recognize, for example, cancer cells in vitro [61]. The study by Chen et al. doped SiNPs of 60 ± 5 nm in size with FITC or RBITC conjugated with dextran (RBITC-dextran) [61]. These NPs were further conjugated (with a secondary antibody [goat antirabbit IgG]) or functionalized with rabbit antiepithelial cell adhesion molecule (EpCAM) antibody on the surface of the polyethylene glycol (PEG)-terminated modified FITC-doped or RBITC-dextran-doped SiNPs by covalent binding to the PEG linkers using the cyanogen bromide method [61]. The functionalized rabbit anti-EpCAM NPs, used for the targeting of the human breast cancer SK-Br-3 cell surface tumor marker, as compared with fluorescent dye labeled IgG SiNPs, displayed better stability of fluorescence as well as photostability under continuous irradiation [61]. Another study by Wu et al. suggested the use of Rubpy dye-encapsulated SiNPs for the sensitive detection of interleukin (IL)-6 (IL-6), on a microarray format, thus promoting the latter use as labels with high fluorescent intensity, photostability, and biocompatibility for the clinical detection of IL-6-related diseases in vivo [118]. The 50-nm sized Rubpy-doped SiNPs were fabricated using a simple one-step microemulsion synthesis and the nonionic surfactant system entailing Igepal CA520 $((C_8H_{17})-C_6H_4-O-(CH_2-CH_2-O)_5H)$ and n-heptane [118]. The CA520 and heptane system was suggested to have a good solubility for both the surfactant and water and has a large, stable single-phase microemulsion region in the phase diagram [118]. Hence, the encapsulation of dye molecules within a silica matrix or the doping of SiNPs with dyes may add stability to SiNPs, enhance their optical properties, and thus display a platform for the specific targeting of cytokines or proteins to aid the clinical diagnosis of biomarkers.

4. Shelf-life assessment of nanoparticles

There are numerous medical applications of NPs. However, it is important to assess the suitability of different chemical and biological media on NPs before we consider packaging NPs for medical applications. We have evaluated the shelf-life of our RBITC dye-encapsulated monodispersed SiNPs in various media (ethanol, water, and physiological salt solution [PSS]) over a 15-month period. SiNPs were selected as the nanomaterial of choice because of their ability to decompose into silicic acid which is a relatively harmless by-product, in comparison to nanomaterials that are not metabolized such as AuNPs and carbon nanotubes [119]. This property makes silica an attractive metal of choice for NP fabrication and application. The SiNP size and morphologies of the different sizes were determined by dynamic light scattering readings, transmission electron micrographs, scanning electron micrographs, and Zeta potential readings using the Malvern Zetasizer and electron microscopy. The DLVO theory [120,121] describes colloidal dispersions stability.

The theory describes that the total interaction energy between two particles is given by the sum of the attraction due to Van der Waals interactions and the electrostatic repulsion originating from the overlap of the charged diffuse layers. The Zeta potential provides a charge for the surface potential and can be used to indicate particle stability.

The RBITC dye-encapsulated monodispersed SiNPs samples were characterized over a 15-month period in different media (ethanol, water, and PSS). The granular appearance of the SiNPs in water, may be caused by the silica seeds surrounding the individual SiNPs. In contrast, the SiNPs maintained their smooth surface characteristics in ethanol over time. The polydispersity index (PDI) of the SiNP samples did not significantly change over the storage period, with PDI values scoring below 0.3 nm, confirming the monodispersity of the samples. SiNP in water became aggregated and rough in its outer surfaces with time. During prolonged storage of the SiNPs in water suspension, silicic acid release [119] may have contributed to their aggregation with time. The latter may also be responsible for raising the pH of the storage medium, which can contribute further to the SiNP changes in characteristics. The breakage of the Van der Waal forces keeping the nanospheres in shape and the distortion of the lattice structures of these NPs may be one cause. The other possible explanation for the variation in the morphology of SiNPs samples in water over time may be due to the dissolution of these NPs in water with time. The dissolution of the SiNPs may give rise to silica seeds and enhance the aggregation of the NPs as well as the responsibility of the crunchy-looking surface due to the surrounding tiny seeds. Dissolution is a process where a solution is formed after a solute is added to a solvent. The dissolution mechanism of SiNPs initially undergoes the dissociation of the silanol group ($\equiv$SiO—H $\leftrightharpoons$ SiO$^-$ + H$^+$). Then, the silica (Si^{4+}) atom undergoes a nucleophilic attack by water molecules to form a silica intermediate (Si$_2$O(OH)$_7^-$). This is followed by the breaking of the siloxane bond $\equiv$Si—O—Si$\equiv$. Thus distorting and breaking the nanosphere structure. The dissolution of the silica may have been initiated by the NPs surface area, concentration, and/or water media over time. The large surface area due to the relatively small size of our SiNPs in water sample, may have contributed to the enhanced dissolution of the NPs due to the greater accessibility to water molecules, thereby reducing the Zeta potential. This phenomenon was previously demonstrated in MSNs [122,123]. Other factors such as temperature, pH, silicate concentration, and metal oxide formation can affect the dynamic equilibrium between silicate dissolution and decomposition [124].

Our results demonstrated that the Zeta potentials of the SiNPs samples were unaffected by storage in water, suggesting that water maintains SiNP stability. The Zeta potential measurements of SiNPs in ethanol were significantly decreased, suggesting that ethanol may not maintain the stability of silica samples overtime. The neutralization effect of PSS on the SiNPs surfaces may have decreased its negativity of the Zeta potential, which is expected because of a decrease in the extension of the diffuse repulsive layer caused by counterion screening. The existence of the surface charge on SiNPs may be due to the deprotonation of silanol groups projecting from the SiNPs surfaces. An increase in cation concentration causes an exchange of the hydrogens on the silanol groups ($SiOH \leftrightharpoons SiO^- + H^+$) [120]. The other possibility is that the PSS may compete for SiNPs with water molecules and lead to reduced hydration layer formation. However, PSS over the course of 15 months improved the stability of SiNPs. Interestingly, incubating the SiNPs in blood resulted in an increase in the size of SiNPs and a greater decrease in Zeta potential (hence increased stability) after 8 hours of incubation. This may be due to corona formation and serum protein adsorption onto their surface [125].

The calculations used for working out the number of NPs per milliliter, required in an experiment are described below:

Eq. (1): Calculates the volume of a sphere, where v is the volume of sphere, π is 3.14, and r is the radius of the particle in cm. This equation was used to calculate the number of particles per milliliter.

$$V = (4/3)\pi r^3 \tag{1}$$

Eq. (2): Calculates the mass of a nanoparticles sphere, where M is mass of a particles sphere, v is volume of a sphere, and d is the density of silica, which is 1.9.

$$m = V \times D \tag{2}$$

Eq. (3): Calculates the number of particles within a given solution, where V is the volume of a sphere at a given diameter; m is the mass of dry NP product in 1 mL of SiNPs suspension.

$$N = m/V \tag{3}$$

For example, the volume of a sphere for a 98-nm sized SiNPs was calculated as follows:

$$4/3 \times \pi \times \left(\text{half of nanoparticle size } (49\,\text{nm}) \times 10^{-7}\right)^3 \times 1.9 \,(\text{silica density})$$

To calculate NP/mL:

$$\frac{\text{Dry mass } (0.0044\,\text{g})}{\text{Volume of sphere } (9.36 \times 10^{-16})}$$

Therefore the NP/mL $= 4.70 \times 10^{12}$.

5. The influence of silica nanoparticles on biological systems

SiNPs, whether used as a delivery system and/or in guiding routes for therapeutic strategies, are a platform for future therapy for a wide range of diseases. However, the drawbacks to such potential of SiNPs may be due to the lack of strategies to tackle SiNPs' ultimate fate and/or the indiscriminate interference with other tissues or biological processing leading to injury and disease. SiNPs have numerous applications and high demand, SiNPs role(s), their effects, and their biocompatibility in living systems have been less studied. Various attempts had been carried out to investigate the role and effects of SiNPs on whole animals and cell models, and to assess their biosafety to justify their administration in vivo for diagnostic purposes and therapeutic interventions. In vivo studies have shown the ability of SiNPs to enter the blood stream, leading to their systemic uptake where they can be detected within organs that are distal to the exposure site [126,127]. It was suggested that the biodistribution of SiNPs was dependent on their surface functionalization [128]. SiNPs may exhibit cytotoxic behavior depending on their shape, aspect ratio, the surface area per unit mass, and interactions with target molecules [129].

A study performed by Thomassen et al. using SiNPs (2–335 nm in size) showed that their toxicity on a human endothelial cell line was related to the degree of NP contact with the cell surface [130]. SiNPs incubation with human lung epithelial cells [129] and myocardial cells [131] resulted in their uptake. SiNPs uptake was accompanied by the generation of ROS [132] and the subsequent oxidative stress damage to the different cells. A previous study by Kasper et al. suggested the harmful effects of small-sized SiNPs (30 and 70 nm) on reducing lung epithelial cell viability, triggering NP-induced inflammatory responses [133] and illustrated the dependence of SiNP size on mediating the cellular effects. Incubating cultured human lung epithelial cells with a concentration up to 0.5 mg/mL of luminescent SiNPs (equivalent to 5×10^{11} NP/mL; 50 ± 3 nm in diameter) caused a drastic reduction in cell survival in contrast to a concentration of 0.1 mg/mL, which did not show a significant reduction in cell survival rates [134]. It was previously reported that 50 nm but not 500 nm sized amorphous SiNPs were shown to induce human umbilical vein endothelial cells' (HUVECs) cytotoxicity in a size-dependent fashion after 24 hours of incubation [135]. The latter model indicates the importance of higher-surface-area-to-volume ratio in favoring biochemical interactions. The influence of SiNPs was also tested under flow conditions in vitro [136]. The study by Freese et al. indicated that HUVECs under physiological cyclic stretch culture conditions did not enhance the uptake and/or the cytotoxicity of amorphous SiNPs when compared with those internalized by the static culture conditions [136].

The lining of endothelial cell layer displays the first initial contact surface for NPs when intravenously injected for diagnostic and/or therapeutic purposes. In light of the limited awareness regarding nanomaterial fate in biological systems, the blood vessels may be a victim to such innovative technology [137]. Hence, an appreciation of the role and fate of NPs in blood vessels [138–144] is essential to draw conclusions and set up decisions on the applicability and possible incorporation of such technology in the clinic. The fate of NPs may be determined by their deposition, biodistribution, bioaccumulation, and/or incomplete breakdown in the vasculature, as well as to being prone to

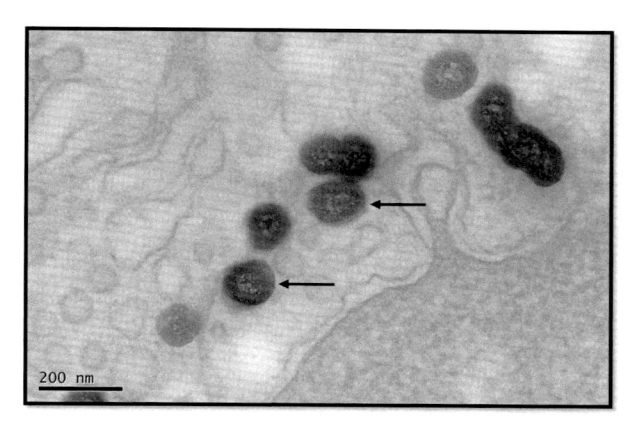

Fig. 9 Transmission electron micrograph (scale bar = 200 nm) illustrating the internalization of silica nanoparticles (SiNPs) into the vascular endothelium of a mesenteric artery after incubation, ex vivo. The SiNPs are represented by *black spherical structures* inside endothelial cells *(arrows)*.

immunological attack and/or clearance due to prolonged exposure. These foreign materials may thus lead to chronic macrophage activation and in turn to chronic inflammation [145]. For instance; a study conducted by Corbalan et al. suggested the capability of amorphous SiNPs of causing endothelial dysfunction in HUVECs [146], thus promoting disruption of vascular homeostasis [147]. These effects were mediated via a NO/peroxynitrite imbalance associated with an increased peroxynitrite ($ONOO^-$) production leading to a low [NO]/[$ONOO^-$] ratio and platelet aggregation within the blood stream via adenosine diphosphate and matrix metalloproteinase 2-dependent mechanisms, respectively [147]. Du et al. suggested that intratracheal-instilled SiNPs could pass through the alveolar–capillary barrier into systemic circulation with their concentration in the heart and serum being dependent on the particles size and dosage [148]. It may be speculated that the effects of NPs on blood vessels may vary in large [149] and small [150] blood vessels, based on the elaboration of endothelial-derived mediators and hence NPs may have different consequences on each type of vascular bed as well as segments within the vascular tree [151]. Studies investigating the long-term biodistribution of SiNPs in vivo suggested their clearance from the brain, muscle, liver, spleen, and adipose tissue, and the distribution and accumulation within the reticuloendothelial system, including the lungs and kidneys for at least 8 weeks [152].

Findings from our laboratory have demonstrated a size [148] and dosage dependent influence of SiNPs on arterial function [150,153]. These results will help identify strategies for their safe clinical administration in the future. They also highlight the importance of assessing the biosafety of nanoparticles for use in imaging diagnostics and medical interventions in order to minimise their toxicological influence on vessel contractility and function (Fig. 9).

6. Strategies to improve the biocompatibility of nanoparticles in vivo

The utility of nanoscale materials as foreign candidates in imaging diagnostics and medical therapeutics has provided the grounds for intensive biocompatibility studies as an opportunity to create innovative approaches to ensure their safety [154]. The large surface area, high surface reactivity, strong adsorbing ability [155], and the different modalities of nanomaterial interactions within biological systems, including direct skin contact, inhalation [156], and intravenous injection, pose a biosafety concern when applied in vivo. Previous studies suggested the toxic behavior of selected NPs [157]. Their toxicity was dependent on a number of factors, including nanomaterial composition, synthetic approach, agglomeration, size [158], surface coating [159], and functionalization, production contaminants, surface area, surface charge and chemistry, crystal structure, shape, concentration, and biopersistence. NPs are small with a high penetrability into tissues [160], cells, and organelles [161]. They display a large surface area of exposure for nonspecific interactivity with different biological molecules [162,163]. This includes the interaction and/or adsorption of biomolecules such as proteins and lipids onto NP surfaces in biological environments [164], leading to the formation of a "Corona" [165] and/or disturbance of their stability and/or coating material. This phenomenon may lead to the reorganization of the NP surface, altered surface chemistry, and conformational changes leading to varied interaction [165].

Thus it is vital to appreciate and assess the health risks associated with the application of nanomaterials to benefit from this technology, ensure the safe administration and/or targeting in vivo, and minimize their toxicity, accumulation, retention, and degradation in undesired sites. To fulfill this motion, a number of alternative approaches can be adopted to improve the biocompatibility of NPs; in particular, SiNPs to better monitor and understand their behavior in vitro as well as in vivo in acute and long-term settings. These strategies include the choice of nanomaterial, size, surface modifications [166], charge, dosage, and shelf-life assessment. The choice and dosage of nanomaterial has a huge impact on the stability and biocompatibility of nanostructures. For instance, despite their optical properties [167], semiconductor quantum dots (QDs) are known for their associated in vivo toxicity [168] that is induced by the heavy metals used and driven by the ability of their materials to become oxidized [169]. However, the use of surface coatings [170] and/or the incorporation of graphene to their core-shell architecture [171] makes them biocompatible and less prone to metal release [172], photobleaching, and biodegradation [173]. In an effort to protect QDs, modified siloxanes with varying functional groups can be incorporated to QDs structures via conjugation, providing an added layer of protection and long-term stability [174]. This added layer protects the QD against degradation and photobleaching, whereas decreasing the toxicity to the surrounding environment by preventing or minimizing the release of core materials. This is particularly important as QDs can be encapsulated in a silica shell to allow the NPs to become water soluble [175]. QDs coated with polymerized silica shells have greater stability prior to conjugation, because the extensive polymer network provides additional stability [176].

PEG is common shell-attached polymer used in crosslinking NPs to enhance their solubility [177], and hence, biocompatibility and prolonged blood circulation time [178]. Previous in vivo studies suggest that the coating of some NPs such as SiNPs with PEG lipids increases their blood circulation lifetime and prevents SiNPs from aggregation, and hence, the clogging of capillaries in the lung and liver of SiNP-injected mice [19]. Furthermore, a number of studies demonstrated the ability of PEG-modified silica in adding a surface stability to QDs and thereby eliminating their associated cytotoxicity [170]. PEG-QDs demonstrated extremely low levels of toxicity, at high concentrations (up to $30\,\mu M$), but also diminished uptake by cells as compared with other surface coatings [172]. Antibody surface-functionalized PEGylated (Ln^{3+}-doped $LaVO_4$-based) nanocapsules were previously used in fluorescence cellular imaging [179]. The latter nanocapsules displayed enhanced biocompatibility and cellular uptake through macropinocytosis without compromising the fluorescent signal detection compared with nonfunctionalized ones [180]. PEG was also previously reported to be more biocompatible in contrast to polyvinylpyrrolidone (PVP) polymers [181]. Mohamed et al. demonstrated that at a therapeutic dosage, mercapto PEG (mPEG)-modified AuNPs had less inhibitory effects on endothelial cell proliferation (through extracellular-signal-regulated kinase 1/2), viability, and ACh-induced cell and vascular functions than PVP-modified AuNPs [181], thus reiterating PEG biocompatibility.

A number of studies suggested the addition of large amphiphilic polymers and naturally derived biomolecules (such as bovine serum albumin), capping agents (e.g., silica shells [172] or that of a magnetic nature), coating and/or conjugation of NPs with antibodies to: (1) Form a physical barrier to the core shielding the environment from the toxic elements of the NP core (e.g., metals, fluorescent and/or luminescent dyes); (2) Defer NP entry into tissues and/or redirect NPs from critical tissues to renal clearance and thus help in avoiding retention; (3) Enable them to bypass off-target organs and tissues; (4) Prevent them from being retained in organs and minimize temporal or permanent damage to end organs, such as the vasculature; (5) Enhance NPs' specificity to target the desired site within the body. Various studies revealed that coating of SiNPs with micelles [182], peptides, polysaccharides, or polymers of biocompatible nature such as dextran [118], chitosan [183], PEI [184], and PEG [185,186] may stabilize and safely process SiNPs in vivo. The conjugation of SiNPs with peptides or proteins such as antibodies [187] may minimize their nonspecific off-target interaction and enhance their selective targeting in vivo. A study by Ma et al. suggested the improvement of SiNPs compatibility, using HUVECs cell line, following the incorporation of the liposomal architecture into cerasomes made from SiNPs core structures [188]. The latter bilayered nanostructures displayed potential in cancer theranostics [189].

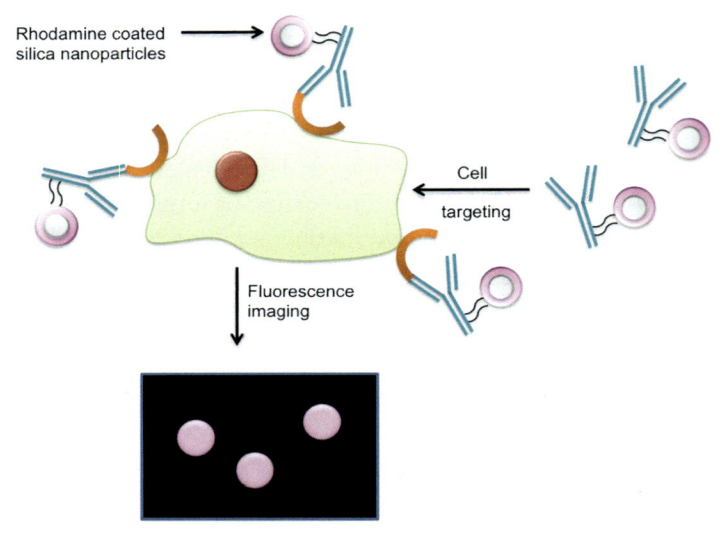

Fig. 10 A schematic illustration of a cell line imaging with high sensitivity and specificity silica nanoparticles (SiNPs). This is achieved via the use of cell-targeted dye-coated SiNP-antibody conjugates.

Preliminary studies performed by Gao et al. lead to the development of 15-nm-sized QDs linked to human prostate-specific membrane antibodies to enable the imaging of induced tumor sites in a murine model [190]. The later strategy was adapted and developed later to minimize the off-target localization of NPs in vivo (Fig. 10). A study by Mohajeri et al. described a method for NP-antibody conjugation and labeling (DNA-dot@EGFR antibody), which can facilitate fluorescent nanoimaging and thus serves as a toxic-free, swift, and efficient method in cancer cell detection [191]. The method described by Mohajeri et al. used DNA as the prime nanomaterial, a more feasible probe that was shown to reduce the turnaround time for cancer diagnosis compared with biomarker detection [191]. The latter strategy may have a potential in targeted gene therapy as well as diagnosis and hence new leeway to cancer treatment. The coupling of natural molecules with magnetic NPs has also been effective in differential diagnosis and guided treatment. Iron oxide–based nanocomposites have been demonstrated as reliable contrast agents in MRI diagnostics when they are fine-tuned and stabilized with precision-engineered polymers (e.g., poly(2-acrylamido-2-methylpropane sodium sulfonate) [P(AMPS)] polymers) [192].

The selectivity of nanomaterials and targeted drug delivery exhibits an exceptional area of concern within nanomedicine. This is due to the nondiscriminatory behavior of "bare or naked" NPs in having various off-target effects when they are administered, metabolized, engulfed, and/or cleared from the renal system. Bare NPs, for example, because of the effects of misfolding, decapping, and/or dissolution can be hydrated, cross-linked, and/or accumulated in target and/or surrounding tissues in vivo. Unpublished observations suggest that SiNPs for instance can dissolute over time in water media suspensions. When SiNPs are administered, they may dissolute over time, and hence, this needs to be taken into account when loading drugs with, for example, MSNs to minimize such effects. In an effort to implement a safe strategy for targeted drug delivery, a number of studies have modified NPs that can encapsulate various tissue-specific targeted drug carriers to enhance their biocompatibility and on-target delivery [193]. However, the latter strategy does not guarantee the possible deposition and bioaccumulation of NPs in their targeted tissues, delivery of the desired drug, and/or dissolution over time. Inorganic core or shell of nanostructured magnetic iron oxide [194], such as magnetite (Fe_3O_4) [195], and maghemite (γ-Fe_2O_3) [196] may be useful candidates in this regard. Iron oxide-based NPs may be prone to clearance from the body, perhaps through iron processing in the spleen; hence, NPs' attachment, incorporation, and decoration with iron oxide may thus bypass NPs' effects in vivo, enhance their chemical stability, and permit their renal clearance. Such superparamagnetic NPs (iron oxide–NPs hybrids) may also have a dual function or property for use as diagnostic probes in MRI [34].

In light of the applications, toxicity assessment, biodistribution, and biocompatibility studies; careful selection, design, and considerable attention to fine-tune, characterize, and optimize NPs must be considered before they can be suited for bedside administration [21]. The latter may aid in our understanding of nanomaterials' effects, role (interaction and interference) [197], impact, fate (deposition, dissolution, and/or renal clearance) as well as the downstream mechanism(s) that accompany NP exposure and/or uptake at the physiological, cellular, and molecular levels. In addition to the evaluation of their stability, porosity, dynamic behavior, and toxicity, NPs' shelf life may need to be assessed to better understand NP degradability over time if they were retained within our bodies and thus more informed decisions regarding clinical use. Hence, careful measures are inevitable to bypass the negative influences of nanomaterials, promote, and standardize [198,199] them as therapeutic agents and drug-delivery vehicles in biomedicine. Nanomaterial and polymer choice may play a phenomenal part in the adaptation of better nanoprobes to improve their biocompatibility. While optimizing NPs' dosage, chemical composition, and/or surface chemistry can aid in their specific targeting to localized tissues within the body.

7. Conclusion

Although SiNPs exhibit exceptional physicochemical properties driven by their chemical structure, composition, and morphology, these features can be enhanced by altering the surface topologies (coating and polymerization) to enable increased stability and fine tune their interactivity with biological systems. SiNPs are now endorsed as tracking agents in intravenous injections and drug loading for diagnostic and therapeutic purposes. Dye-encapsulated SiNPs are one of the most popular NPs recently being explored for medical intervention. Here, we discussed a number of methods used to fabricate and characterize them for different purposes. We also highlighted a number of studies that have attempted to understand NPs' behavior and assess the consequences of their biochemical interactions. These studies have provided a platform for improving their biocompatibility by optimizing their dosage, chemical composition, and/or surface chemistry and their specific targeting in vivo. Thus in-depth understanding of their nanochemistry and the strategies to improve their nanobiosafety can form the basis for the next generation of multifunctional nanoplatforms, which will revolutionize the way we diagnose and treat life-threatening illnesses.

References

[1] C. Seliger, R. Jurgons, F. Wiekhorst, D. Eberbeck, L. Trahms, H. Iro, et al., In vitro investigation of the behaviour of magnetic particles by a circulating artery model, J. Magn. Magn. Mater. 311 (2007) 358–362, https://doi.org/10.1016/j.jmmm.2006.10.1205.

[2] B. Patricio, G. Weissmuller, R. Santos-Oliveira, Development of the 153-sm-edtmp nanoradiopharmaceutical, J. Label. Compd. Radiopharm. 54 (2011).

[3] J.S. Hrkach, M.T. Peracchia, A. Bcmb, N. Lotan, R. Langer, Nanotechnology for biomaterials engineering: structural characterization of amphiphilic polymeric nanoparticles by 1H NMR spectroscopy, Biomaterials 18 (1997) 27–30, https://doi.org/10.1016/s0142-9612(96)00077-4.

[4] J.F. Pires, B.F. de Carvalho Patricio, M. de Souza Albernaz, G.D. Mendonca, B.F. Coelho, G. Weissmuller, et al., Preparation of biodegradable poly(L-lactide) (PLA) nanoparticles containing DMSA (dimercaptosuccinic acid) as novel radiopharmaceutical, Adv. Sci. Lett. 10 (2012) 143–145, https://doi.org/10.1166/asl.2012.2147.

[5] J.M. Stern, V.V. Kibanov Solomonov, E. Sazykina, J.A. Schwartz, S.C. Gad, G.P. Goodrich, Initial evaluation of the safety of nanoshell-directed photothermal therapy in the treatment of prostate disease, Int. J. Toxicol. 35 (2016) 38–46, https://doi.org/10.1177/1091581815600170.

[6] V. Prabhu, S. Uzzaman, V.M.B. Grace, C. Guruvayoorappan, Nanoparticles in drug delivery and cancer therapy: the giant rats tail, J. Cancer Ther. 02 (2011) 325–334, https://doi.org/10.4236/jct.2011.23045.

[7] M. Barshan-Tashnizi, S. Ahmadian, K. Niknam, S.-F. Torabi, S.-O. Ranaei-Siadat, Covalent immobilization of Drosophila acetylcholinesterase for biosensor applications, Biotechnol. Appl. Biochem. 52 (2009) 257, https://doi.org/10.1042/ba20080005.

[8] H. Choi, I.-W. Chen, Surface-modified silica colloid for diagnostic imaging, J. Colloid Interface Sci. 258 (2003) 435–437, https://doi.org/10.1016/s0021-9797(02)00130-3.

[9] S. Giri, B.G. Trewyn, M.P. Stellmaker, V.S.-Y. Lin, Stimuli-responsive controlled-release delivery system based on mesoporous silica nanorods capped with magnetic nanoparticles, Angew. Chem. Int. Ed. 44 (2005) 5038–5044, https://doi.org/10.1002/anie.200501819.

[10] J.-K. Hsiao, C.-P. Tsai, T.-H. Chung, Y. Hung, M. Yao, H.-M. Liu, et al., Mesoporous silica nanoparticles as a delivery system of gadolinium for effective human stem cell tracking, Small 4 (2008) 1445–1452, https://doi.org/10.1002/smll.200701316.

[11] M.S. Motwani, Y. Rafiei, A. Tzifa, A.M. Seifalian, In situ endothelialization of intravascular stents from progenitor stem cells coated with nanocomposite and functionalized biomolecules, Biotechnol. Appl. Biochem. 58 (2011) 2–13, https://doi.org/10.1002/bab.10.

[12] I. Sokolov, S. Naik, Novel fluorescent silica nanoparticles: towards ultrabright silica nanoparticles, Small 4 (2008) 934–939, https://doi.org/10.1002/smll.200700236.

[13] C.-P. Tsai, Y. Hung, Y.-H. Chou, D.-M. Huang, J.-K. Hsiao, C. Chang, et al., High-contrast paramagnetic fluorescent mesoporous silica nanorods as a multifunctional cell-imaging probe, Small 4 (2008) 186–191, https://doi.org/10.1002/smll.200700457.

[14] M. Christofidou-Solomidou, S. Kennel, A. Scherpereel, R. Wiewrodt, C.C. Solomides, G.G. Pietra, et al., Vascular immunotargeting of glucose oxidase to the endothelial antigens induces distinct forms of oxidant acute lung injury: targeting to thrombomodulin, but not to PECAM-1, causes pulmonary thrombosis and neutrophil transmigration, Am. J. Pathol. 160 (2002) 1155–1169, https://doi.org/10.1016/S0002-9440(10)64935-8.

[15] I.I. Slowing, C.-W. Wu, J.L. Vivero-Escoto, V.S.-Y. Lin, Mesoporous silica nanoparticles for reducing hemolytic activity towards mammalian red blood cells, Small 5 (2009) 57–62, https://doi.org/10.1002/smll.200800926.

[16] J. Lu, M. Liong, J.I. Zink, F. Tamanoi, Mesoporous silica nanoparticles as a delivery system for hydrophobic anticancer drugs, Small 3 (2007) 1341–1346, https://doi.org/10.1002/smll.200700005.

[17] H. Zhou, X. Wu, J. Wei, X. Lu, S. Zhang, J. Shi, et al., Stimulated osteoblastic proliferation by mesoporous silica xerogel with high specific surface area, J. Mater. Sci. Mater. Med. 22 (2011) 731–739, https://doi.org/10.1007/s10856-011-4239-1.

[18] J. Margolis, J. McDonald, R. Heuser, P. Klinke, R. Waksman, R. Virmani, et al., Systemic nanoparticle paclitaxel (nab-paclitaxel) for in-stent restenosis I (SNAPIST-I): a first-in-human safety and dose-finding study, Clin. Cardiol. 30 (2007) 165–170, https://doi.org/10.1002/clc.20066.

[19] M.M. van Schooneveld, E. Vucic, R. Koole, Y. Zhou, J. Stocks, D.P. Cormode, et al., Improved biocompatibility and pharmacokinetics of silica nanoparticles by means of a lipid coating: a multimodality investigation, Nano Lett. 8 (2008) 2517–2525, https://doi.org/10.1021/nl801596a.

[20] W. Arap, R. Pasqualini, M. Montalti, L. Petrizza, L. Prodi, E. Rampazzo, et al., Luminescent silica nanoparticles for cancer diagnosis, Curr. Med. Chem. 20 (2013) 2195–2211, https://doi.org/10.2174/0929867311320170005.

[21] H.S. Leong, K.S. Butler, C.J. Brinker, M. Azzawi, S. Conlan, C. Dufés, et al., On the issue of transparency and reproducibility in nanomedicine, Nat. Nanotechnol. 14 (2019) 629–635, https://doi.org/10.1038/s41565-019-0496-9.

[22] N.R. Yacobi, H.C. Phuleria, L. Demaio, C.H. Liang, C.-A. Peng, C. Sioutas, et al., Nanoparticle effects on rat alveolar epithelial cell monolayer barrier properties, Toxicol. In Vitro 21 (2007) 1373–1381, https://doi.org/10.1016/j.tiv.2007.04.003.

[23] X. Gong, Y. Yang, L. Zhang, C. Zou, P. Cai, G. Chen, et al., Controlled synthesis of Pt nanoparticles via seeding growth and their shape-dependent catalytic activity, J. Colloid Interface Sci. 352 (2010) 379–385, https://doi.org/10.1016/j.jcis.2010.08.069.

[24] L. Sun, L.C. Chow, Preparation and properties of nano-sized calcium fluoride for dental applications, Dent. Mater. 24 (2008) 111–116, https://doi.org/10.1016/j.dental.2007.03.003.

[25] V. Manolova, A. Flace, M. Bauer, K. Schwarz, P. Saudan, M.F. Bachmann, Nanoparticles target distinct dendritic cell populations according to their size, Eur. J. Immunol. 38 (2008) 1404–1413, https://doi.org/10.1002/eji.200737984.

[26] J. Davda, V. Labhasetwar, Characterization of nanoparticle uptake by endothelial cells, Int. J. Pharm. 233 (2002) 51–59, https://doi.org/10.1016/s0378-5173(01)00923-1.

[27] J.S. Suh, J.Y. Lee, Y.S. Choi, F. Yu, V. Yang, S.J. Lee, et al., Efficient labeling of mesenchymal stem cells using cell permeable magnetic nanoparticles, Biochem. Biophys. Res. Commun. 379 (2009) 669–675, https://doi.org/10.1016/j.bbrc.2008.12.041.

[28] D.B. Warheit, Comparative pulmonary toxicity assessment of single-wall carbon nanotubes in rats, Toxicol. Sci. 77 (2003) 117–125, https://doi.org/10.1093/toxsci/kfg228.

[29] L.J. Mortensen, G. Oberdörster, A.P. Pentland, L.A. Delouise, In vivo skin penetration of quantum dot nanoparticles in the murine model: the effect of UVR, Nano Lett. 8 (2008) 2779–2787, https://doi.org/10.1021/nl801323y.

[30] G. Oberdörster, E. Oberdörster, J. Oberdörster, Nanotoxicology: an emerging discipline evolving from studies of ultrafine particles, Environ. Health Perspect. 113 (2005) 823–839, https://doi.org/10.1289/ehp.7339.

[31] X. Wang, H.C. Schröder, M. Wiens, U. Schloßmacher, W.E.G. Müller, Biosilica, Adv. Mar. Biol. (2012) 231–271, https://doi.org/10.1016/b978-0-12-394283-8.00005-9.

[32] M.B.C. de Matos, A.P. Piedade, C. Alvarez-Lorenzo, A. Concheiro, M.E.M. Braga, H.C. de Sousa, Dexamethasone-loaded poly(ε-caprolactone)/silica nanoparticles composites prepared by supercritical CO_2 foaming/mixing and deposition, Int. J. Pharm. 456 (2013) 269–281, https://doi.org/10.1016/j.ijpharm.2013.08.042.

[33] D. Babaei, The in vitro investigation on the effect of infrared waves combined with silica-gold nanoparticle on the breast cancerous cells, J. Nanomed. Nanotechnol. 08 (2017) 7439, https://doi.org/10.4172/2157-7439-c1-056.

[34] J. Kim, L. Cao, D. Shvartsman, E.A. Silva, D.J. Mooney, Targeted delivery of nanoparticles to ischemic muscle for imaging and therapeutic angiogenesis, Nano Lett. 11 (2011) 694–700, https://doi.org/10.1021/nl103812a.

[35] E. Phillips, O. Penate-Medina, P.B. Zanzonico, R.D. Carvajal, P. Mohan, Y. Ye, et al., Clinical translation of an ultrasmall inorganic optical-PET imaging nanoparticle probe, Sci. Transl. Med. 6 (2014) 260ra149, https://doi.org/10.1126/scitranslmed.3009524.

[36] A.N. Kharlamov, A.E. Tyurnina, V.S. Veselova, O.P. Kovtun, V.Y. Shur, J.L. Gabinsky, Silica–gold nanoparticles for atheroprotective management of plaques: results of the NANOM-FIM trial, Nanoscale 7 (2015) 8003–8015, https://doi.org/10.1039/c5nr01050k.

[37] H. Li, Y. Mu, S. Qian, J. Lu, Y. Wan, G. Fu, et al., Synthesis of fluorescent dye-doped silica nanoparticles for target-cell-specific delivery and intracellular microRNA imaging, Analyst 140 (2015) 567–573, https://doi.org/10.1039/c4an01706d.

[38] G. Badr, D. Sayed, D. Maximous, A.O. Mohamed, M. Gul, Increased susceptibility to apoptosis and growth arrest of human breast cancer cells treated by a snake venom-loaded silica nanoparticles, Cell. Physiol. Biochem. 34 (2014) 1640–1651, https://doi.org/10.1159/000366366.

[39] G. Barratt, Colloidal drug carriers: achievements and perspectives, Cell. Mol. Life Sci. 60 (2003) 21–37, https://doi.org/10.1007/s000180300002.

[40] Y. Chen, H. Gu, D.S.-Z. Zhang, F. Li, T. Liu, W. Xia, Highly effective inhibition of lung cancer growth and metastasis by systemic delivery of siRNA via multimodal mesoporous silica-based nanocarrier, Biomaterials 35 (2014) 10058–10069, https://doi.org/10.1016/j.biomaterials.2014.09.003.

[41] Y. Kapilov-Buchman, E. Lellouche, S. Michaeli, J.-P. Lellouche, Unique surface modification of silica nanoparticles with polyethylenimine (PEI) for siRNA delivery using cerium cation coordination chemistry, Bioconjug. Chem. 26 (2015) 880–889, https://doi.org/10.1021/acs.bioconjchem.5b00100.

[42] J.V. Jokerst, C. Khademi, S.S. Gambhir, Intracellular aggregation of multimodal silica nanoparticles for ultrasound-guided stem cell implantation, Sci. Transl. Med. 5 (2013) 177ra35, https://doi.org/10.1126/scitranslmed.3005228.

[43] A. Farooq, L. Tosheva, M. Azzawi, D. Whitehead, Real-time observation of aortic vessel dilation through delivery of sodium nitroprusside via slow release mesoporous nanoparticles, J. Colloid Interface Sci. 478 (2016) 127–135, https://doi.org/10.1016/j.jcis.2016.06.004.

[44] A. Farooq, A. Shukur, C. Astley, L. Tosheva, P. Kelly, D. Whitehead, et al., Titania coating of mesoporous silica nanoparticles for improved biocompatibility and drug release within blood vessels, Acta Biomater. 76 (2018), https://doi.org/10.1016/j.actbio.2018.06.024.

[45] A. Erfanian, H. Mirhosseini, B. Rasti, M. Hair-Bejo, S.B. Mustafa, M.Y.A. Manap, Absorption and bioavailability of nano-size reduced calcium citrate fortified milk powder in ovariectomized and ovariectomized-osteoporosis rats, J. Agric. Food Chem. 63 (2015) 5795–5804, https://doi.org/10.1021/acs.jafc.5b01468.

[46] A. Shukla, N. Dasgupta, S. Ranjan, S. Singh, R. Chidambram, Nanotechnology towards prevention of anaemia and osteoporosis: from concept to market, Biotechnol. Biotechnol. Equip. 31 (2017) 863–879, https://doi.org/10.1080/13102818.2017.1335615.

[47] H.H. Elsayed, A.S.A.M. Al-Sherbini, E.E. Abd-Elhady, K.A.E.A. Ahmed, Treatment of anemia progression via magnetite and folate nanoparticles in vivo, ISRN Nanotechnol. 2014 (2014) 1–13, https://doi.org/10.1155/2014/287575.

[48] T.A. Salaheldin, Magnetite-vitamin C nanoparticles as a potent route for treatment of iron deficiency anemia, J. Nanomed. Nanotechnol. 8 (2017) 6, https://doi.org/10.4172/2157-7439-C1-056.

[49] A. Farooq, Restored endothelial dependent vasodilation in aortic vessels after uptake of ceria coated silica nanoparticles, ex vivo, J. Nanomed. Nanotechnol. 05 (2014), https://doi.org/10.4172/2157-7439.1000195.

[50] V.C. Minarchick, P.A. Stapleton, E.M. Sabolsky, T.R. Nurkiewicz, Cerium dioxide nanoparticle exposure improves microvascular dysfunction and reduces oxidative stress in spontaneously hypertensive rats, Front. Physiol. 6 (2015) 339, https://doi.org/10.3389/fphys.2015.00339.

[51] S.R. Paliwal, R. Kenwat, S. Maiti, R. Paliwal, Nanotheranostics for cancer therapy and detection: state of the art, Curr. Pharm. Des. (2020), https://doi.org/10.2174/1381612826666201116120422.

[52] K. Lim, H.K. Kim, X.T. Le, N.T. Nguyen, E.S. Lee, K.T. Oh, et al., Highly red light-emitting erbium- and lutetium-doped core-shell upconverting nanoparticles surface-modified with PEG-folic acid/TCPP for suppressing cervical cancer HeLa cells, Pharmaceutics 12 (2020), https://doi.org/10.3390/pharmaceutics12111102.

[53] A.M. Renner, S. Ilyas, H.A. Schlößer, A. Szymura, S. Roitsch, K. Wennhold, et al., Receptor-mediated in vivo targeting of breast cancer cells with 17α-ethynylestradiol-conjugated silica-coated gold nanoparticles, Langmuir (2020), https://doi.org/10.1021/acs.langmuir.0c02820.

[54] M. Colilla, M. Vallet-Regí, Targeted stimuli-responsive mesoporous silica nanoparticles for bacterial infection treatment, Int. J. Mol. Sci. 21 (2020), https://doi.org/10.3390/ijms21228605.

[55] I. Gonzalo-Juan, F. Xie, M. Becker, D.U. Tulyaganov, E. Ionescu, S. Lauterbach, et al., Synthesis of silver modified bioactive glassy materials with antibacterial properties via facile and low-temperature route, Materials (Basel, Switzerland) 13 (2020), https://doi.org/10.3390/ma13225115.

[56] S. Mapukata, J. Britton, O.L. Osifeko, T. Nyokong, The improved antibacterial efficiency of a zinc phthalocyanine when embedded on silver nanoparticle modified silica nanofibers, Photodiagn. Photodyn. Ther. (2020), https://doi.org/10.1016/j.pdpdt.2020.102100, 102100.

[57] H.-L. Su, C.-C. Chou, D.-J. Hung, S.-H. Lin, I.-C. Pao, J.-H. Lin, et al., The disruption of bacterial membrane integrity through ROS generation induced by nanohybrids of silver and clay, Biomaterials 30 (2009) 5979–5987, https://doi.org/10.1016/j.biomaterials.2009.07.030.

[58] N. Theobald, Emerging vaccine delivery systems for COVID-19: functionalised silica nanoparticles offer a potentially safe and effective alternative delivery system for DNA/RNA vaccines and may be useful in the hunt for a COVID-19 vaccine, Drug Discov. Today 25 (2020) 1556–1558, https://doi.org/10.1016/j.drudis.2020.06.020.

[59] M. Colilla, M. Vallet-Regí, Targeted stimuli-responsive mesoporous silica nanoparticles for bacterial infection treatment, Int. J. Mol. Sci. 21 (22) (2020), https://doi.org/10.3390/ijms21228605.

[60] W. Stöber, A. Fink, E. Bohn, Controlled growth of monodisperse silica spheres in the micron size range, J. Colloid Interface Sci. 26 (1968) 62–69, https://doi.org/10.1016/0021-9797(68)90272-5.

[61] M.-Y. Chen, Z.-Z. Chen, L.-L. Wu, H.-W. Tang, D.-W. Pang, Goat anti-rabbit IgG conjugated fluorescent dye-doped silica nanoparticles for human breast carcinoma cell recognition, Analyst 138 (2013) 7411, https://doi.org/10.1039/c3an01654d.

[62] J.L. Vivero-Escoto, I.I. Slowing, B.G. Trewyn, V.S.-Y. Lin, Mesoporous silica nanoparticles for intracellular controlled drug delivery, Small 6 (2010) 1952–1967, https://doi.org/10.1002/smll.200901789.

[63] H. Giesche, Synthesis of monodispersed silica powders I. Particle properties and reaction kinetics, J. Eur. Ceram. Soc. 14 (1994) 189–204, https://doi.org/10.1016/0955-2219(94)90087-6.

[64] G. Øye, W.R. Glomm, T. Vrålstad, S. Volden, H. Magnusson, M. Stöcker, et al., Synthesis, functionalisation and characterisation of mesoporous materials and sol–gel glasses for applications in catalysis, adsorption and photonics, Adv. Colloid Interf. Sci. 123–126 (2006) 17–32, https://doi.org/10.1016/j.cis.2006.05.010.

[65] C.-L. Chang, H.S. Fogler, Controlled formation of silica particles from tetraethyl orthosilicate in nonionic water-in-oil microemulsions, Langmuir 13 (1997) 3295–3307, https://doi.org/10.1021/la961062z.

[66] G.H. Bogush, M.A. Tracy, C.F. Zukoski, Preparation of monodisperse silica particles: control of size and mass fraction, J. Non-Cryst. Solids 104 (1988) 95–106, https://doi.org/10.1016/0022-3093(88)90187-1.

[67] H. Giesche, Synthesis of monodispersed silica powders II. Controlled growth reaction and continuous production process, J. Eur. Ceram. Soc. 14 (1994) 205–214, https://doi.org/10.1016/0955-2219(94)90088-4.

[68] G.H. Bogush, C.F. Zukoski, Studies of the kinetics of the precipitation of uniform silica particles through the hydrolysis and condensation of silicon alkoxides, J. Colloid Interface Sci. 142 (1991) 1–18, https://doi.org/10.1016/0021-9797(91)90029-8.

[69] C.H. Byers, M.T. Harris, D.F. Williams, Controlled microcrystalline growth studies by dynamic laser-light-scattering methods, Ind. Eng. Chem. Res. 26 (1987) 1916–1923, https://doi.org/10.1021/ie00069a033.

[70] J.J. Razink, N.E. Schlotter, Correction to "Preparation of monodisperse silica particles: control of size and mass fraction" by G.H. Bogush, M.A. Tracy and C.F. Zukoski IV, Journal of Non-Crystalline Solids 104 (1988) 95–106, J. Non-Cryst. Solids 353 (2007) 2932–2933, https://doi.org/10.1016/j.jnoncrysol.2007.06.067.

[71] F.J. Arriagada, K. Osseo-Asare, Synthesis of nanosize silica in a nonionic water-in-oil microemulsion: effects of the water/surfactant molar ratio and ammonia concentration, J. Colloid Interface Sci. 211 (1999) 210–220, https://doi.org/10.1006/jcis.1998.5985.

[72] K. Osseo-Asare, F.J. Arriagada, Growth kinetics of nanosize silica in a nonionic water-in-oil microemulsion: a reverse micellar pseudophase reaction model, J. Colloid Interface Sci. 218 (1999) 68–76, https://doi.org/10.1006/jcis.1999.6232.

[73] F. Gao, L. Wang, L. Tang, C. Zhu, A novel nano-sensor based on rhodamine-β-isothiocyanate − doped silica nanoparticle for pH measurement, Microchim. Acta 152 (2005) 131–135, https://doi.org/10.1007/s00604-005-0418-4.

[74] B.D. Kay, R.A. Assink, Sol-gel kinetics, J. Non-Cryst. Solids 104 (1988) 112–122, https://doi.org/10.1016/0022-3093(88)90189-5.

[75] M. Chatterjee, D. Ganguli, Alkoxy-derived monodisperse silica microspheres: the role of solvents in synthesis, Trans. Indian Ceram. Soc. 45 (1986) 95–99, https://doi.org/10.1080/0371750x.1986.10822798.

[76] J.S. Beck, J.C. Vartuli, W.J. Roth, M.E. Leonowicz, C.T. Kresge, K.D. Schmitt, et al., A new family of mesoporous molecular sieves prepared with liquid crystal templates, J. Am. Chem. Soc. 114 (1992) 10834–10843, https://doi.org/10.1021/ja00053a020.

[77] J. Kecht, T. Bein, Oxidative removal of template molecules and organic functionalities in mesoporous silica nanoparticles by H_2O_2 treatment, Microporous Mesoporous Mater. 116 (2008) 123–130, https://doi.org/10.1016/j.micromeso.2008.03.027.

[78] A.B.D. Nandiyanto, S.-G. Kim, F. Iskandar, K. Okuyama, Synthesis of spherical mesoporous silica nanoparticles with nanometer-size controllable pores and outer diameters, Microporous Mesoporous Mater. 120 (2009) 447–453, https://doi.org/10.1016/j.micromeso.2008.12.019.

[79] Y. Kong, S.-Y. Jiang, J. Wang, S. Wang, Q. Yan, Y. Lu, Synthesis and characterization of Cu–Ti–MCM41, Microporous Mesoporous Mater. 86 (2005) 191–197, https://doi.org/10.1016/j.micromeso.2005.07.006.

[80] I. Argatov, A. Davies, L. Dyson, R. Dyson, G. Lang, L. Mayaud, et al., How Do Manufactured Nanoparticles Enter Cells? Packing Nanoparticles Into Vesicles, 2011, pp. 1–33.

[81] J.P. Thielemann, F. Girgsdies, R. Schlögl, C. Hess, Pore structure and surface area of silica SBA-15: influence of washing and scale-up, Beilstein J. Nanotechnol. 2 (2011) 110–118, https://doi.org/10.3762/bjnano.2.13.

[82] M. Baalousha, F.V.D. Kammer, M. Motelica-Heino, H.S. Hilal, P. Le Coustumer, Size fractionation and characterization of natural colloids by flow-field flow fractionation coupled to multi-angle laser light scattering, J. Chromatogr. A 1104 (2006) 272–281, https://doi.org/10.1016/j.chroma.2005.11.095.

[83] P. Atkins, J. de Paula, The Second Law, Physical Chemistry, W.H. Freeman and Co., New York, 2006, pp. 76–116.

[84] A. Koohestanian, M. Hosseini, C.Z. Abbasian, The separation method for removing of colloidal particles from raw water, Am-Euras. J. Agric. Environ. Sci. 4 (2008).

[85] R.M. Pashley, M.E. Karaman, Van der Waals forces and colloid stability, Appl. Colloid Surf. Chem. (2005) 127–151, https://doi.org/10.1002/0470014709.ch7.

[86] K.S. Birdi, Colloidal systems, in: Surface and Colloid Chemistry, CRC Press, 2009, pp. 141–159.

[87] N. Akbar, T. Mohamed, D. Whitehead, M. Azzawi, Biocompatibility of amorphous silica nanoparticles: size and charge effect on vascular function, in vitro, Biotechnol. Appl. Biochem. 58 (2011) 353–362, https://doi.org/10.1002/bab.46.

[88] A. Yoshida, Silica nucleation, polymerization, and growth preparation of monodispersed sols, Adv. Chem. (1994) 51–66, https://doi.org/10.1021/ba-1994-0234.ch002.

[89] Y. Ju-Nam, J.R. Lead, Manufactured nanoparticles: an overview of their chemistry, interactions and potential environmental implications, Sci. Total Environ. 400 (2008) 396–414, https://doi.org/10.1016/j.scitotenv.2008.06.042.

[90] D.L. Marchisio, L. Rivautella, A.A. Barresi, Design and scale-up of chemical reactors for nanoparticle precipitation, AICHE J. 52 (2006) 1877–1887, https://doi.org/10.1002/aic.10786.

[91] H.-P. Boehm, The chemistry of silica. Solubility, Polymerization, Colloid and Surface Properties, and Biochemistry. Von R. K. Iler. John Wiley and Sons, Chichester 1979. XXIV, 886 S., geb. £ 39.50, Angew. Chem. 92 (1980) 328, https://doi.org/10.1002/ange.19800920433.

[92] N. Andersson, P.C.A. Alberius, J. Skov Pedersen, L. Bergström, Structural features and adsorption behaviour of mesoporous silica particles formed from droplets generated in a spraying chamber, Microporous Mesoporous Mater. 72 (2004) 175–183, https://doi.org/10.1016/j.micromeso.2004.04.019.

[93] A. Burns, H. Ow, U. Wiesner, Fluorescent core–shell silica nanoparticles: towards "Lab on a Particle" architectures for nanobiotechnology, Chem. Soc. Rev. 35 (2006) 1028–1042, https://doi.org/10.1039/b600562b.

[94] N.V. Churaev, The DLVO theory in Russian colloid science, Adv. Colloid Interf. Sci. 83 (1999) 19–32, https://doi.org/10.1016/s0001-8686(98)00067-0.

[95] D.J. Shaw, Colloid stability, in: Introduction to Colloid and Surface Chemistry, 1992, pp. 210–243, https://doi.org/10.1016/b978-0-08-050910-5.50012-8.

[96] B.W. Ninham, On progress in forces since the DLVO theory, Adv. Colloid Interf. Sci. 83 (1999) 1–17, https://doi.org/10.1016/s0001-8686(99)00008-1.

[97] K. Norrish, Low-angle X-ray diffraction studies of the swelling of montmorillonite and vermiculite, Clay Clay Miner. 10 (1961) 123–149, https://doi.org/10.1346/ccmn.1961.0100112.

[98] W.A.B. Donners, J.B. Rijnbout, A. Vrij, Light scattering from soap films, J. Colloid Interface Sci. 61 (1977) 249–260, https://doi.org/10.1016/0021-9797(77)90388-5.

[99] J.N. Israelachvili, G.E. Adams, Measurement of forces between two mica surfaces in aqueous electrolyte solutions in the range 0–100 nm, J. Chem. Soc. Faraday Trans. 1 Phys. Chem. Condens. Phases 74 (1978) 975, https://doi.org/10.1039/f19787400975.

[100] R.M. Pashley, P.M. McGuiggan, B.W. Ninham, J. Brady, D.F. Evans, Direct measurements of surface forces between bilayers of double-chained quaternary ammonium acetate and bromide surfactants, J. Phys. Chem. 90 (1986) 1637–1642, https://doi.org/10.1021/j100399a037.

[101] P.G. Hartley, I. Larson, P.J. Scales, Electrokinetic and direct force measurements between silica and mica surfaces in dilute electrolyte solutions, Langmuir 13 (1997) 2207–2214, https://doi.org/10.1021/la960997c.

[102] R.M. Pashley, DLVO and hydration forces between mica surfaces in Li+, Na+, K+, and Cs+ electrolyte solutions: a correlation of double-layer and hydration forces with surface cation exchange properties, J. Colloid Interface Sci. 83 (1981) 531–546, https://doi.org/10.1016/0021-9797(81)90348-9.

[103] M. Kobayashi, F. Juillerat, P. Galletto, P. Bowen, M. Borkovec, Aggregation and charging of colloidal silica particles: effect of particle size, Langmuir 21 (2005) 5761–5769, https://doi.org/10.1021/la046829z.

[104] L.T. Zhuravlev, The surface chemistry of amorphous silica. Zhuravlev model, Colloids Surf. A Physicochem. Eng. Asp. 173 (2000) 1–38, https://doi.org/10.1016/s0927-7757(00)00556-2.

[105] J. Lyklema, Lyotropic sequences in colloid stability revisited, Adv. Colloid Interf. Sci. 100–102 (2003) 1–12, https://doi.org/10.1016/s0001-8686(02)00075-1.

[106] D.C. Grahame, The role of the cation in the electrical double layer, J. Electrochem. Soc. 98 (1951) 343, https://doi.org/10.1149/1.2778217.

[107] L. Gierst, L. Vandenberghen, E. Nicolas, A. Fraboni, Ion pairing mechanisms in electrode processes, J. Electrochem. Soc. 113 (1966) 1025, https://doi.org/10.1149/1.2423746.

[108] Y. Marcus, Viscosity B–coefficients, structural entropies and heat capacities, and the effects of ions on the structure of water, J. Solut. Chem. 23 (1994) 831–848, https://doi.org/10.1007/bf00972677.

[109] Y. Marcus, Effect of ions on the structure of water: structure making and breaking, Chem. Rev. 109 (2009) 1346–1370, https://doi.org/10.1021/cr8003828.

[110] F. Dumont, J. Warlus, A. Watillon, Influence of the point of zero charge of titanium dioxide hydrosols on the ionic adsorption sequences, J. Colloid Interface Sci. 138 (1990) 543–554, https://doi.org/10.1016/0021-9797(90)90236-h.

[111] G.V. Franks, Zeta potentials and yield stresses of silica suspensions in concentrated monovalent electrolytes: isoelectric point shift and additional attraction, J. Colloid Interface Sci. 249 (2002) 44–51, https://doi.org/10.1006/jcis.2002.8250.

[112] R. Kjellander, Ion-ion correlations and effective charges in electrolyte and macroion systems, Ber. Bunsenges. Phys. Chem. 100 (1996) 894–904, https://doi.org/10.1002/bbpc.19961000635.

[113] C. Labbez, B. Jönsson, M. Skarba, M. Borkovec, Ion–ion correlation and charge reversal at titrating solid interfaces, Langmuir 25 (2009) 7209–7213, https://doi.org/10.1021/la900853e.

[114] I.-Y. Jeon, J.-B. Baek, Nanocomposites derived from polymers and inorganic nanoparticles, Materials (Basel) 3 (2010) 3654–3674, https://doi.org/10.3390/ma3063654.

[115] G. Canton, R. Riccò, F. Marinello, S. Carmignato, F. Enrichi, Modified Stöber synthesis of highly luminescent dye-doped silica nanoparticles, J. Nanopart. Res. 13 (2011) 4349–4356, https://doi.org/10.1007/s11051-011-0382-3.

[116] K. Wang, X. He, X. Yang, H. Shi, Functionalized silica nanoparticles: a platform for fluorescence imaging at the cell and small animal levels, Acc. Chem. Res. 46 (2013) 1367–1376, https://doi.org/10.1021/ar3001525.

[117] H. Shi, X. He, Y. Yuan, K. Wang, D. Liu, Nanoparticle-based biocompatible and long-life marker for lysosome labeling and tracking, Anal. Chem. 82 (2010) 2213–2220, https://doi.org/10.1021/ac902417s.

[118] H. Wu, Q. Huo, S. Varnum, J. Wang, G. Liu, Z. Nie, et al., Dye-doped silica nanoparticle labels/protein microarray for detection of protein biomarkers, Analyst 133 (2008) 1550–1555, https://doi.org/10.1039/b719810h.

[119] M. Diaconu, A. Tache, S. Eremia, F. Gatea, S.C. Litescu, G. Radu, Structural characterization of chitosan coated silicon nanoparticles -a FT-IR approach, UPB Sci. Bull. Ser. B Chem. Mater. Sci. 72 (2010) 115–122.

[120] B. Derjaguin, L. Landau, Theory of the stability of strongly charged lyophobic sols and of the adhesion of strongly charged particles in solutions of electrolytes, Prog. Surf. Sci. 43 (1993) 30–59, https://doi.org/10.1016/0079-6816(93)90013-l.

[121] E.J.W. Verwey, J.T. Overbeek, Theory of the stability of lyophobic colloids, J. Colloid Sci. 10 (1955) 224–225, https://doi.org/10.1016/0095-8522(55)90030-1.

[122] R. Finsy, On the critical radius in Ostwald ripening, Langmuir 20 (2004) 2975–2976, https://doi.org/10.1021/la035966d.

[123] X. Huang, N.P. Young, H.E. Townley, Characterization and comparison of mesoporous silica particles for optimized drug delivery, Nanomater. Nanotechnol. 4 (2014) 2, https://doi.org/10.5772/58290.

[124] J.D. Bass, D. Grosso, C. Boissiere, E. Belamie, T. Coradin, C. Sanchez, Stability of mesoporous oxide and mixed metal oxide materials under biologically relevant conditions, Chem. Mater. 19 (2007) 4349–4356, https://doi.org/10.1021/cm071305g.

[125] O. Gamucci, A. Bertero, M. Gagliardi, G. Bardi, Biomedical nanoparticles: overview of their surface immune-compatibility, Coatings 4 (2014) 139–159, https://doi.org/10.3390/coatings4010139.

[126] M. Cho, W.-S. Cho, M. Choi, S.J. Kim, B.S. Han, S.H. Kim, et al., The impact of size on tissue distribution and elimination by single intravenous injection of silica nanoparticles, Toxicol. Lett. 189 (2009) 177–183, https://doi.org/10.1016/j.toxlet.2009.04.017.

[127] S.J. So, I.S. Jang, C.S. Han, Effect of micro/nano silica particle feeding for mice, J. Nanosci. Nanotechnol. 8 (2008) 5367–5371, https://doi.org/10.1166/jnn.2008.1347.

[128] T. Yu, D. Hubbard, A. Ray, H. Ghandehari, In vivo biodistribution and pharmacokinetics of silica nanoparticles as a function of geometry, porosity and surface characteristics, J. Control. Release 163 (2012) 46–54, https://doi.org/10.1016/j.jconrel.2012.05.046.

[129] M.J. Akhtar, M. Ahamed, S. Kumar, H. Siddiqui, G. Patil, M. Ashquin, et al., Nanotoxicity of pure silica mediated through oxidant generation rather than glutathione depletion in human lung epithelial cells, Toxicology 276 (2010) 95–102, https://doi.org/10.1016/j.tox.2010.07.010.

[130] L.C.J. Thomassen, A. Aerts, V. Rabolli, D. Lison, L. Gonzalez, M. Kirsch-Volders, et al., Synthesis and characterization of stable monodisperse silica nanoparticle sols for in vitro cytotoxicity testing, Langmuir 26 (2010) 328–335, https://doi.org/10.1021/la902050k.

[131] Y. Ye, J. Liu, M. Chen, L. Sun, M. Lan, In vitro toxicity of silica nanoparticles in myocardial cells, Environ. Toxicol. Pharmacol. 29 (2010) 131–137, https://doi.org/10.1016/j.etap.2009.12.002.

[132] E.-J. Park, K. Park, Oxidative stress and pro-inflammatory responses induced by silica nanoparticles in vivo and in vitro, Toxicol. Lett. 184 (2009) 18–25, https://doi.org/10.1016/j.toxlet.2008.10.012.

[133] J. Kasper, M.I. Hermanns, C. Bantz, O. Koshkina, T. Lang, M. Maskos, et al., Interactions of silica nanoparticles with lung epithelial cells and the association to flotillins, Arch. Toxicol. 87 (2013) 1053–1065, https://doi.org/10.1007/s00204-012-0876-5.

[134] Y. Jin, S. Kannan, M. Wu, J.X. Zhao, Toxicity of luminescent silica nanoparticles to living cells, Chem. Res. Toxicol. 20 (2007) 1126–1133, https://doi.org/10.1021/tx7001959.

[135] A. Nemmar, S. Albarwani, S. Beegam, P. Yuvaraju, J. Yasin, S. Attoub, et al., Amorphous silica nanoparticles impair vascular homeostasis and induce systemic inflammation, Int. J. Nanomedicine 9 (2014) 2779–2789, https://doi.org/10.2147/IJN.S52818.

[136] C. Freese, D. Schreiner, L. Anspach, C. Bantz, M. Maskos, R.E. Unger, et al., In vitro investigation of silica nanoparticle uptake into human endothelial cells under physiological cyclic stretch, Part. Fibre Toxicol. 11 (2014) 68, https://doi.org/10.1186/s12989-014-0068-y.

[137] C. Guo, Y. Liu, Y. Li, Adverse effects of amorphous silica nanoparticles: focus on human cardiovascular health, J. Hazard. Mater. 406 (2020) 124626, https://doi.org/10.1016/j.jhazmat.2020.124626.

[138] A. Courtois, P. Andujar, Y. Ladeiro, I. Baudrimont, E. Delannoy, V. Leblais, et al., Impairment of NO-dependent relaxation in intralobar pulmonary arteries: comparison of urban particulate matter and manufactured nanoparticles, Environ. Health Perspect. 116 (2008) 1294–1299, https://doi.org/10.1289/ehp.11021.

[139] T.R. Nurkiewicz, D.W. Porter, M. Barger, V. Castranova, M.A. Boegehold, Particulate matter exposure impairs systemic microvascular endothelium-dependent dilation, Environ. Health Perspect. 112 (2004) 1299–1306, https://doi.org/10.1289/ehp.7001.

[140] T.R. Nurkiewicz, D.W. Porter, M. Barger, L. Millecchia, K.M.K. Rao, P.J. Marvar, et al., Systemic microvascular dysfunction and inflammation after pulmonary particulate matter exposure, Environ. Health Perspect. 114 (2006) 412–419, https://doi.org/10.1289/ehp.8413.

[141] T.R. Nurkiewicz, D.W. Porter, A.F. Hubbs, J.L. Cumpston, B.T. Chen, D.G. Frazer, et al., Nanoparticle inhalation augments particle-dependent systemic microvascular dysfunction, Part. Fibre Toxicol. 5 (2008) 1, https://doi.org/10.1186/1743-8977-5-1.

[142] A.J. LeBlanc, J.L. Cumpston, B.T. Chen, D. Frazer, V. Castranova, T.R. Nurkiewicz, Nanoparticle inhalation impairs endothelium-dependent vasodilation in subepicardial arterioles, J. Toxicol. Environ. Health A 72 (2009) 1576–1584, https://doi.org/10.1080/15287390903232467.

[143] P.A. Stapleton, V.C. Minarchick, A.M. Cumpston, W. McKinney, B.T. Chen, T.M. Sager, et al., Impairment of coronary arteriolar endothelium-dependent dilation after multi-walled carbon nanotube inhalation: a time-course study, Int. J. Mol. Sci. 13 (2012) 13781–13803, https://doi.org/10.3390/ijms131113781.

[144] A.J. LeBlanc, A.M. Moseley, B.T. Chen, D. Frazer, V. Castranova, T.R. Nurkiewicz, Nanoparticle inhalation impairs coronary microvascular reactivity via a local reactive oxygen species-dependent mechanism, Cardiovasc. Toxicol. 10 (2010) 27–36, https://doi.org/10.1007/s12012-009-9060-4.

[145] K. Donaldson, C.L. Tran, An introduction to the short-term toxicology of respirable industrial fibres, Mutat. Res. Mol. Mech. Mutagen 553 (2004) 5–9, https://doi.org/10.1016/j.mrfmmm.2004.06.011.

[146] J.J. Corbalan, C. Medina, A. Jacoby, T. Malinski, M.W. Radomski, Amorphous silica nanoparticles trigger nitric oxide/peroxynitrite imbalance in human endothelial cells: inflammatory and cytotoxic effects, Int. J. Nanomedicine 6 (2011) 2821–2835, https://doi.org/10.2147/IJN.S25071.

[147] J.J. Corbalan, C. Medina, A. Jacoby, T. Malinski, M.W. Radomski, Amorphous silica nanoparticles aggregate human platelets: potential implications for vascular homeostasis, Int. J. Nanomedicine 7 (2012) 631–639, https://doi.org/10.2147/IJN.S28293.

[148] Z. Du, D. Zhao, L. Jing, G. Cui, M. Jin, Y. Li, et al., Cardiovascular toxicity of different sizes amorphous silica nanoparticles in rats after intratracheal instillation, Cardiovasc. Toxicol. 13 (2013) 194–207, https://doi.org/10.1007/s12012-013-9198-y.

[149] A. Farooq, D. Whitehead, M. Azzawi, Attenuation of endothelial-dependent vasodilator responses, induced by dye-encapsulated silica nanoparticles, in aortic vessels, Nanomedicine 9 (2014) 413–425, https://doi.org/10.2217/nnm.12.213.

[150] A. Shukur, D. Whitehead, A. Seifalian, M. Azzawi, The influence of silica nanoparticles on small mesenteric arterial function, Nanomedicine 11 (2016) 2131–2146, https://doi.org/10.2217/nnm-2016-0124.

[151] V.C. Minarchick, P.A. Stapleton, D.W. Porter, M.G. Wolfarth, E. Çiftyürek, M. Barger, et al., Pulmonary cerium dioxide nanoparticle exposure differentially impairs coronary and mesenteric arteriolar reactivity, Cardiovasc. Toxicol. 13 (2013) 323–337, https://doi.org/10.1007/s12012-013-9213-3.

[152] M.A. Malfatti, H.A. Palko, E.A. Kuhn, K.W. Turteltaub, Determining the pharmacokinetics and long-term biodistribution of SiO_2 nanoparticles in vivo using accelerator mass spectrometry, Nano Lett. 12 (2012) 5532–5538, https://doi.org/10.1021/nl302412f.

[153] L. Feng, X. Yang, S. Liang, Q. Xu, M.R. Miller, J. Duan, Z. Sun, Silica nanoparticles trigger the vascular endothelial dysfunction and prethrombotic state via miR-451 directly regulating the IL6R signaling pathway, Part. Fibre Toxicol. 16 (1) (2019) 16, https://doi.org/10.1186/s12989-019-0300-x.

[154] G. Oberdörster, Safety assessment for nanotechnology and nanomedicine: concepts of nanotoxicology, J. Intern. Med. 267 (2010) 89–105, https://doi.org/10.1111/j.1365-2796.2009.02187.x.

[155] N. Elahi, M. Kamali, M.H. Baghersad, Recent biomedical applications of gold nanoparticles: a review, Talanta (2018), https://doi.org/10.1016/j.talanta.2018.02.088.

[156] T.L. Knuckles, J. Yi, D.G. Frazer, H.D. Leonard, B.T. Chen, V. Castranova, et al., Nanoparticle inhalation alters systemic arteriolar vasoreactivity through sympathetic and cyclooxygenase-mediated pathways, Nanotoxicology 6 (2012) 724–735, https://doi.org/10.3109/17435390.2011.606926.

[157] K.L. Aillon, Y. Xie, N. El-Gendy, C.J. Berkland, M.L. Forrest, Effects of nanomaterial physicochemical properties on in vivo toxicity, Adv. Drug Deliv. Rev. 61 (2009) 457–466, https://doi.org/10.1016/j.addr.2009.03.010.

[158] P. Jani, G.W. Halbert, J. Langridge, A.T. Florence, Nanoparticle uptake by the rat gastrointestinal mucosa: quantitation and particle size dependency, J. Pharm. Pharmacol. 42 (1990) 821–826, https://doi.org/10.1111/j.2042-7158.1990.tb07033.x.

[159] K. Pathakoti, H.-M. Hwang, H. Xu, Z.P. Aguilar, A. Wang, In vitro cytotoxicity of CdSe/ZnS quantum dots with different surface coatings to human keratinocytes HaCaT cells, J. Environ. Sci. 25 (2013) 163–171, https://doi.org/10.1016/S1001-0742(12)60015-1.

[160] A. Wiesenthal, L. Hunter, S. Wang, J. Wickliffe, M. Wilkerson, Nanoparticles: small and mighty, Int. J. Dermatol. 50 (2011) 247–254, https://doi.org/10.1111/j.1365-4632.2010.04815.x.

[161] A. Nemmar, Possible mechanisms of the cardiovascular effects of inhaled particles: systemic translocation and prothrombotic effects, Toxicol. Lett. 149 (2004) 243–253, https://doi.org/10.1016/j.toxlet.2003.12.061.

[162] J. Klein, Probing the interactions of proteins and nanoparticles, Proc. Natl. Acad. Sci. U. S. A. 104 (2007) 2029–2030, https://doi.org/10.1073/pnas.0611610104.

[163] T. Cedervall, I. Lynch, S. Lindman, T. Berggård, E. Thulin, H. Nilsson, et al., Understanding the nanoparticle-protein corona using methods to quantify exchange rates and affinities of proteins for nanoparticles, Proc. Natl. Acad. Sci. U. S. A. 104 (2007) 2050–2055, https://doi.org/10.1073/pnas.0608582104.

[164] M.S. Ehrenberg, A.E. Friedman, J.N. Finkelstein, G. Oberdörster, J.L. McGrath, The influence of protein adsorption on nanoparticle association with cultured endothelial cells, Biomaterials 30 (2009) 603–610, https://doi.org/10.1016/j.biomaterials.2008.09.050.

[165] M.P. Monopoli, D. Walczyk, A. Campbell, G. Elia, I. Lynch, F. Baldelli Bombelli, et al., Physical−chemical aspects of protein corona: relevance to in vitro and in vivo biological impacts of nanoparticles, J. Am. Chem. Soc. 133 (2011) 2525–2534, https://doi.org/10.1021/ja107583h.

[166] V. Labhasetwar, C. Song, W. Humphrey, R. Shebuski, R.J. Levy, Arterial uptake of biodegradable nanoparticles: effect of surface modifications, J. Pharm. Sci. 87 (1998) 1229–1234, https://doi.org/10.1021/js980021f.

[167] M.A. Walling, J.A. Novak, J.R.E. Shepard, Quantum dots for live cell and in vivo imaging, Int. J. Mol. Sci. 10 (2009) 441–491, https://doi.org/10.3390/ijms10020441.

[168] A.M. Derfus, W.C.W. Chan, S.N. Bhatia, Probing the cytotoxicity of semiconductor quantum dots, Nano Lett. 4 (2004) 11–18, https://doi.org/10.1021/nl0347334.

[169] N. Liu, M. Tang, Toxicity of different types of quantum dots to mammalian cells in vitro: an update review, J. Hazard. Mater. 399 (2020) 122606, https://doi.org/10.1016/j.jhazmat.2020.122606.

[170] J.P. Ryman-Rasmussen, J.E. Riviere, N.A. Monteiro-Riviere, Surface coatings determine cytotoxicity and irritation potential of quantum dot nanoparticles in epidermal keratinocytes, J. Invest. Dermatol. 127 (2007) 143–153, https://doi.org/10.1038/sj.jid.5700508.

[171] S. Chung, R.A. Revia, M. Zhang, Graphene quantum dots and their applications in bioimaging, biosensing, and therapy, Adv. Mater. (2019), https://doi.org/10.1002/adma.201904362, 1904362.

[172] C. Kirchner, T. Liedl, S. Kudera, T. Pellegrino, A. Muñoz Javier, H.E. Gaub, et al., Cytotoxicity of colloidal CdSe and CdSe/ZnS nanoparticles, Nano Lett. 5 (2005) 331–338, https://doi.org/10.1021/nl047996m.

[173] H. Sun, H. Ji, E. Ju, Y. Guan, J. Ren, X. Qu, Synthesis of fluorinated and nonfluorinated graphene quantum dots through a new top-down strategy for long-time cellular imaging, Chemistry 21 (2015) 3791–3797, https://doi.org/10.1002/chem.201406345.

[174] Y.H. Kim, S. Koh, H. Lee, S.-M. Kang, D.C. Lee, B.-S. Bae, Photo-patternable quantum dots/siloxane composite with long-term stability for quantum dot color filters, ACS Appl. Mater. Interfaces 12 (2020) 3961–3968, https://doi.org/10.1021/acsami.9b19586.

[175] Y. Xu, K. Meehan, L. Guido, G. Lu, C. Wyatt, N. Love, Synthesis and Characterization of Silica Coated CdSe/CdS Core/Shell Quantum Dots, 2005.

[176] X. Hu, X. Gao, Silica—polymer dual layer-encapsulated quantum dots with remarkable stability, ACS Nano 4 (2010) 6080–6086, https://doi.org/10.1021/nn1017044.

[177] J.D. Kingsley, H. Dou, J. Morehead, B. Rabinow, H.E. Gendelman, C.J. Destache, Nanotechnology: a focus on nanoparticles as a drug delivery system, J. NeuroImmune Pharmacol. 1 (2006) 340–350, https://doi.org/10.1007/s11481-006-9032-4.

[178] K. Knop, R. Hoogenboom, D. Fischer, U.S. Schubert, Poly(ethylene glycol) in drug delivery: pros and cons as well as potential alternatives, Angew. Chem. Int. Ed. 49 (2010) 6288–6308, https://doi.org/10.1002/anie.200902672.

[179] J. Jeyaraman, A. Shukla, S. Sivakumar, Targeted stealth polymer capsules encapsulating Ln^{3+}-doped $LaVO_4$ nanoparticles for bioimaging applications, ACS Biomater. Sci. Eng. 2 (2016) 1330–1340, https://doi.org/10.1021/acsbiomaterials.6b00252.

[180] C.R. Dhanya, J. Jeyaraman, P.A. Janeesh, A. Shukla, S. Sivakumar, A. Abraham, Bio-distribution and in vivo/in vitro toxicity profile of PEGylated polymer capsules encapsulating $LaVO_4:Tb^{3+}$ nanoparticles for bioimaging applications, RSC Adv. 6 (2016) 55125–55134, https://doi.org/10.1039/c6ra06719k.

[181] T. Mohamed, S. Matou-Nasri, A. Farooq, D. Whitehead, M. Azzawi, Polyvinylpyrrolidone-coated gold nanoparticles inhibit endothelial cell viability, proliferation, and ERK1/2 phosphorylation and reduce the magnitude of endothelial-independent dilator responses in isolated aortic vessels, Int. J. Nanomedicine 12 (2017) 8813–8830, https://doi.org/10.2147/IJN.S133093.

[182] Q. Huo, J. Liu, L.-Q. Wang, Y. Jiang, T.N. Lambert, E. Fang, A new class of silica cross-linked micellar core—shell nanoparticles, J. Am. Chem. Soc. 128 (2006) 6447–6453, https://doi.org/10.1021/ja060367p.

[183] L.-M. Zhao, L.-E. Shi, Z.-L. Zhang, J.-M. Chen, D.-D. Shi, J. Yang, et al., Preparation and application of chitosan nanoparticles and nanofibers, Braz. J. Chem. Eng. 28 (2011) 353–362.

[184] W.-T. He, Y.-N. Xue, N. Peng, W.-M. Liu, R.-X. Zhuo, S.-W. Huang, One-pot preparation of polyethylenimine-silica nanoparticles as serum-resistant gene delivery vectors: intracellular trafficking and transfection, J. Mater. Chem. 21 (2011) 10496–10503, https://doi.org/10.1039/C1JM11021G.

[185] J.S. Suk, Q. Xu, N. Kim, J. Hanes, L.M. Ensign, PEGylation as a strategy for improving nanoparticle-based drug and gene delivery, Adv. Drug Deliv. Rev. 99 (2016) 28–51, https://doi.org/10.1016/j.addr.2015.09.012.

[186] S.-A. Yang, S. Choi, S.M. Jeon, J. Yu, Silica nanoparticle stability in biological media revisited, Sci. Rep. 8 (2018) 185, https://doi.org/10.1038/s41598-017-18502-8.

[187] M. Bouchoucha, É. Béliveau, F. Kleitz, F. Calon, M.-A. Fortin, Antibody-conjugated mesoporous silica nanoparticles for brain microvessel endothelial cell targeting, J. Mater. Chem. B 5 (2017) 7721–7735, https://doi.org/10.1039/C7TB01385J.

[188] Y. Ma, Z. Dai, Y. Gao, Z. Cao, Z. Zha, X. Yue, et al., Liposomal architecture boosts biocompatibility of nanohybrid cerasomes, Nanotoxicology 5 (2011) 622–635, https://doi.org/10.3109/17435390.2010.546950.

[189] S. Hameed, P. Bhattarai, Z. Dai, Cerasomes and bicelles: hybrid bilayered nanostructures with silica-like surface in cancer theranostics, Front. Chem. 6 (2018) 127, https://doi.org/10.3389/fchem.2018.00127.

[190] X. Gao, Y. Cui, R.M. Levenson, L.W.K. Chung, S. Nie, In vivo cancer targeting and imaging with semiconductor quantum dots, Nat. Biotechnol. 22 (2004) 969–976, https://doi.org/10.1038/nbt994.

[191] N. Mohajeri, E. Mostafavi, N. Zarghami, The feasibility and usability of DNA-dot bioconjugation to antibody for targeted in vitro cancer cell fluorescence imaging, J. Photochem. Photobiol. B Biol. 209 (2020) 111944, https://doi.org/10.1016/j.jphotobiol.2020.111944.

[192] A.M. King, C. Bray, S.C.L. Hall, J.C. Bear, L.K. Bogart, S. Perrier, et al., Exploring precision polymers to fine-tune magnetic resonance imaging properties of iron oxide nanoparticles, J. Colloid Interface Sci. 579 (2020) 401–411, https://doi.org/10.1016/j.jcis.2020.06.036.

[193] H. Katsumi, S. Yamashita, M. Morishita, A. Yamamoto, Bone-targeted drug delivery systems and strategies for treatment of bone metastasis, Chem. Pharm. Bull. 68 (2020) 560–566, https://doi.org/10.1248/cpb.c20-00017.

[194] M.P. Zaytseva, A.G. Muradova, A.I. Sharapaev, E.V. Yurtov, I.S. Grebennikov, A.G. Savchenko, Fe₃O₄/SiO₂ core shell nanostructures: preparation and characterization, Russ. J. Inorg. Chem. 63 (2018) 1684–1688, https://doi.org/10.1134/S0036023618120239.

[195] M. Abbas, M.O. Abdel-Hamed, J. Chen, Efficient one-pot sonochemical synthesis of thickness-controlled silica-coated superparamagnetic iron oxide (Fe₃O₄/SiO₂) nanospheres, Appl. Phys. A Mater. Sci. Process. 123 (2017) 775, https://doi.org/10.1007/s00339-017-1397-0.

[196] M. Mahdavi, M. Ahmad, M.J. Haron, F. Namvar, B. Nadi, M. Rahman, et al., Synthesis, surface modification and characterisation of biocompatible magnetic iron oxide nanoparticles for biomedical applications, Molecules 18 (2013) 7533–7548, https://doi.org/10.3390/molecules18077533.

[197] A. Shukur, S.B. Rizvi, D. Whitehead, A. Seifalian, M. Azzawi, Altered sensitivity to nitric oxide donors, induced by intravascular infusion of quantum dots, in murine mesenteric arteries, Nanomed. Nanotechnol. Biol. Med. 9 (2013), https://doi.org/10.1016/j.nano.2012.10.004.

[198] M. Faria, M. Björnmalm, E.J. Crampin, F. Caruso, A few clarifications on MIRIBEL, Nat. Nanotechnol. 15 (1) (2020) 2–3, https://doi.org/10.1038/s41565-019-0612-x.

[199] H.F. Florindo, A. Madi, R. Satchi-Fainaro, Challenges in the implementation of MIRIBEL criteria on nanobiomed manuscripts, Nat. Nanotechnol. 14 (7) (2019) 627–628, https://doi.org/10.1038/s41565-019-0498-7.

CHAPTER 12

Advances in nanocrystals as drug delivery systems

Amanpreet Kaur[a], Prashantkumar Khodabhai Parmar[a], Sanika Jadhav[a], and Arvind Kumar Bansal
Department of Pharmaceutics, Solid State Pharmaceutics Lab, National Institute of Pharmaceutical Education and Research (NIPER), Mohali, Punjab, India

1. Introduction

Poor aqueous solubility poses a major challenge for optimal drug absorption and biological response. This is attributed to inadequate solvation of hydrophobic molecules by water, high crystal lattice energy, and insufficient flux across the epithelial membrane [1]. Variable drug absorption could be either solubility or dissolution rate-limited, depending on dose to solubility ratio and relative rate of drug dissolution and permeation. At high dose to solubility ratio, drug solubility may be low for the given dose and drug absorption becomes solubility-limited. However, at low dose to solubility ratio, drug solubility may be high for the dose, but slow dissolution may limit absorption [2, 3].

The challenges of solubility and/or dissolution rate-limited drug absorption can be overcome by plethora of techniques such as salt formation, metastable polymorphs, amorphization, cosolvents, lipidic systems, pH modifications, and size reduction to micro or nano size. However, there is no "one fits all" approach and suitable formulation is selected considering the physicochemical properties of the drug, route of administration, dose and dosage form, and in vivo performance. Nanocrystals are an enabling technology to eliminate variable drug absorption due to low solubility and dissolution rate [2, 3].

Nanocrystals are carrier-free pure drug crystals with a size in the nanometer range, i.e., between a few nanometers and 1000 nm (=1 μm). In contrast, micronized powders possess a size of 1–1000 μm. The particle size reduction should show improved product performance attributable to size [4].

Saturation solubility is a characteristic feature of crystalline thermodynamically most stable form. It is a constant that depends on the physicochemical properties of a drug, crystalline structure (lattice energy), dissolution medium, and temperature. However, below a critical size of 1000 nm, saturation solubility also becomes a function of particle size. Size reduction from microcrystals to nanocrystals (increase in particle curvature)

[a] All authors contributed equally.

marginally increases the apparent solubility due to increase in dissolution pressure. Ostwald–Freundlich equation, given in Eq. (1), explains this phenomenon where the saturation solubility of the drug is inversely proportional to the particle size [5]. In Eq. (1), C_S is the saturation solubility, $C\alpha$ is the solubility of large particles, σ is the interfacial tension of the substance, V is the molar volume of the particle material, R is the gas constant, T is the absolute temperature, ρ is the density of the solid, and r is the radius.

$$\log \frac{Cs}{C\alpha} = \frac{2\sigma V}{2.303 R T \rho r} \tag{1}$$

When the amount of solute dissolving exceeds the saturation solubility, a supersaturated solution is formed. This is a thermodynamically unstable state and the transient increase in drug concentration represents apparent solubility. This phenomenon of supersaturation achieved through nanocrystals has not been understood comprehensively. Its interplay with supersaturation-mediated crystallization and overall impact on solubility improvement need to be investigated more rigorously [6, 7].

Drug nanocrystals have higher dissolution rate than microcrystals due to large interfacial area, increase in concentration gradient, and decreased diffusion layer thickness [5]. This is explained by Noyes–Whitney equation depicted in Eq. (2), where dC/dt is the rate (concentration change as a function of time), D is the diffusion coefficient, S is the surface area, V is the dissolution volume, h is the diffusion layer thickness, C_S is the saturation concentration, and C_t is the concentration at time t.

$$\frac{dC}{dt} = \frac{DS(Cs - Ct)}{Vh} \tag{2}$$

The equation depicts that size reduction from 50 μm to 500 nm can increase the drug dissolution rate by 100 times due to increase in surface area.

Additionally, drug nanocrystals show increased adhesion to the mucus layer which might create high-localized concentration gradient, leading to faster drug absorption. It has been postulated that neat and stabilizer-coated nanocrystals demonstrate mucoadhesion due to electrostatic and van der Waals interactions or physical entanglement within the mucus membrane [5].

Drug nanocrystals may also be taken up as solid particles via endocytosis. Cellular uptake varies between cell types and their phagocytotic/endocytotic potential as well as nanocrystal properties such as size (prominent for <100 nm), shape, stabilizer type, and surface charge. Since nanocrystals are continuously dissolving and shrinking in size, endocytic uptake may be prominent at the end of their existence [5, 8, 9].

Drug absorption is a function of both solubility and permeability. Dissolution from nanocrystals is followed by permeation of the dissolved drug across the gastrointestinal wall [10]. The effect of drug's physicochemical properties on oral bioavailability should not be neglected [11]. Choice of drug nanocrystals as the delivery platform should be made based on sound principles of molecular and physicochemical properties of the drug.

Size reduction brings along challenge of crystal growth and aggregation. Nanocrystals are thermodynamically unstable due to high surface-free energy. The surface-free energy (ΔG or "Gibbs energy"), associated with area, is explained by Eq. (3).

$$\Delta G = \gamma SL \times \Delta A - T\Delta S \tag{3}$$

where, ΔA is the change in surface area, γSL is the interfacial tension between the solid and liquid interface, T is the absolute temperature, and ΔS is the change in entropy of the system. Thus, the nanocrystals tend to aggregate in order to minimize the surface energy of the system.

Additionally, Ostwald ripening can cause crystal growth in nanocrystals, where large particles grow at the expense of smaller particles. Drug concentration is high in the environment of smaller crystals due to increased apparent solubility. Consequently, the molecules in proximity of smaller crystals diffuse (high to low concentration) towards larger crystals resulting into recrystallization of dissolved molecules on the surface of larger particles [12, 13].

Surface-active agents (stabilizers) assist in stabilization of drug nanocrystals by increasing activation energy barrier to crystal growth. Ionic stabilizers such as sodium lauryl sulfate act by electrostatic repulsion, while nonionics (like polysorbates, sorbitan esters, and Pluronics) and polymers such as hydroxypropyl methylcellulose, polyvinyl pyrrolidone, and polyvinyl alcohols create steric barrier against aggregation. Moreover, a combination of electrostatic and steric stabilizer, also called 'electrosteric stabilization,' shows better effectiveness for stabilizing drug nanocrystals [14, 15]. Concentration of stabilizer has to be chosen wisely, as particle growth can occur due to incomplete covering of the nanocrystal surface, and on the other extreme overloading, can cause particle growth due to Ostwald ripening [12]. In addition to stabilization, surface-active agents may increase dissolution rate, interact with cell membranes, and modulate intestinal permeability. Stabilizers such as Vitamin E TPGS, poloxamers, and polysorbates can modify permeability due to their P-gp inhibitory effect [9, 10, 12].

Thus, increased apparent solubility/dissolution rate, supersaturation leading to enhanced permeation, manifests into pharmacokinetic advantages such as increased maximum plasma concentration (C_{max}), reduced time to maximum plasma concentration (T_{max}), enhanced area under the blood concentration-time curve (AUC), and reduced fasted/fed variability. This is especially important for drugs showing narrow absorption window in gastrointestinal tract such as aprepitant, marketed as Emend [16]. Drug absorption can be affected by food due to delayed gastric emptying, increased bile secretion, larger volume of the gastric fluid, increased gastric pH, and increased blood flow. Nanocrystals eliminate absorption differences due to variable dissolution rate in fasted and fed states [17, 18].

2. Formulation space

Nanocrystals have been used for improving solubility and dissolution rate of Biopharmaceutics Classification System (BCS) Class II and IV drugs. The BCS Class II drugs are subclassified into Class IIa and Class IIb as per Developability Classification System (DCS). Majorly, brick-dust molecules constitute Class IIa that shows dissolution rate-limited absorption, while Class IIb comprises grease ball molecules with solubility-limited absorption. Brick-dust molecules are insoluble in water as well as majority of organic solvents and lipids. This happens when the crystal energy is high (indicated by a high melting point) with limited capacity to dissociate from solid form. The grease ball molecules, on the other hand, may show solubility in lipids. Thus, grease ball molecules can also be formulated using lipid-based formulations. Brick-dust molecules are the candidates of choice for generation of nanocrystals.

In contrast to Class II drugs, Class IV drugs with poor permeability may also benefit from nanocrystals. This is because higher and faster drug dissolution can increase the concentration gradient owing to differences in dissolved *vs.* absorbed drug content [3, 19, 20].

3. Production of nanocrystals

Broadly, three approaches are used for production of drug nanocrystals: (1) top-down processes that use high-energy mechanical forces to comminute the drug microsuspension, (2) bottom-up processes where nucleation followed by crystal growth occurs from solution, and (3) a combination of both where the first step is precipitation followed by comminution. A common step in these processes is that the drug is either dispersed (top-down) or dissolved (bottom-up) in liquid media containing pre-dissolved stabilizer to prevent aggregation and crystal growth. The dispersion media used are usually water or a mixture of aqueous and nonaqueous solutions. This further implies that the final product obtained is a nanosuspension, where drug nanocrystals are dispersed in a liquid media [21–25].

The majority of the commercial products have been manufactured using top-down approaches because of robustness and ease of scalability. However, top-down process is energy-intensive and poses potential risk of physical instability due to high shear and contamination from the grinding media. On the other hand, the bottom-up processes need solvent removal and are difficult to control during large-scale manufacturing. In addition, the process itself is not universally applicable as most of the poorly water-soluble drugs are insoluble in aqueous as well as organic solvents [19, 24].

3.1 Miniaturization

In drug discovery and development, increase in number of BCS Class II/IV drugs has profoundly impacted the formulation strategies used in preclinical and clinical stages

[26]. Drug doses need to be optimized for low bioavailability drugs to obtain intended pharmacological response and establish clinical safety margins and toxicity [3]. To address these issues, solubilization in toxicology studies is carried out using high concentrations of surfactant (e.g., Tween 80 at 10%*w*/w), cosolvents (e.g., PEG 400), or lipid systems (e.g., Imwitor/Tween mixtures). Drug nanocrystals can be implemented as a formulation strategy from early screening studies to commercial manufacturing. The aqueous nature of the nanosuspension allows oral as well as parenteral administration at higher doses and fewer side effects in comparison to surfactant and lipid-based systems [16].

It is also desirable that the route of administration in preclinical stages matches that of the intended clinical use. During drug development, early clinical formulations are selected based on pharmacokinetic studies in dogs. These animal models are used to make biopharmaceutical decisions for formulations explored to improve drug bioavailability. The early identification of potential formulation approaches thus is critical for hastening product development [16, 26].

Compound availability is scarce in preformulation and testing stages; thus, the formulation must be able to accommodate milligrams of drugs for rapid and reliable screening. Systematic screening for optimal formulation composition and production parameters for nanosuspensions consumes a lot of time and resources [27]. Considering the limited amount of drug available, a simple, accessible, and cost-effective miniaturized 'top-down' approach for nanocrystal production has been developed [28]. It is possible to develop miniaturized batches using magnetic stirrer and only 0.5 mL (requiring approximately 50 mg or 5 mg of drug for 5% and 1% suspension, respectively) of suspension [29]. Van Eerdenbrugh et al. have also reported size reduction using pearls as milling media in 96-well plate with only 10 mg of drug stirred using orbital shaker [27]. Along with screening of formulations for preclinical applications, small-scale batches allow identification of type and concentration of stabilizer, assess comminution ability of crystals, and investigate physical stability, leading to acceleration of large-scale process development [26].

3.2 Optimization of process parameters using quality by design

Quality by Design (QbD) approach for production of nanocrystals is categorized into three stages: (i) selection of stabilizer and production method consistent with Quality Target Product Profile (QTPP), (ii) establishing Critical Quality Attributes (CQAs), and (iii) identification of design space using Design of Experiments (DoE). To achieve the desired CQA, Critical material (CMA) and Process attributes (CPA) need to be investigated. Regulatory agencies such as US FDA also emphasize on the need of QbD for development of pharmaceutical products right from early screening phase. Traditionally, the approach of studying "one factor at a time" consumed lot of time and resources; however, using QbD multiple factors/variables can be studied at a time. QbD helps to test for variable effects, study interactions, and relationship between process and material parameters and quality attributes for repeatable batch-to-batch performance [19]. It thus helps to develop a control strategy for process and quality control. The critical quality attributes indicating successful fabrication of nanocrystals are predominantly (i) crystalline form, (ii) size and polydispersity index (PDI), (iii) dissolution rate, (iv) degradation products, (iv) physical stability (aggregation), and (v) in vivo product performance [10].

Various examples for identification of CMAs and CPPs are reported in literature [30–32]. In a typical media milling process, drug concentration (vary between 2 and 30%w/w) in the milling chamber, weight ratio of drug to stabilizer (2:1 to 20:1), weight of milling pearls (10 to 50% of the weight per volume of the slurry), stirring speed, and time of milling process (30 min to days or weeks) are the CPPs that can significantly affect the quality of resulting nanosuspension. Identification and rank ordering of critical variables can establish quality of nanosuspension during early product development. Ghosh et al. optimized formulation and process parameters including agitation rate (rpm), size of milling media, and drug content and three different polymeric stabilizers (HPMC 3 cps, PVP K-30, and HPC-EXF) and Vitamin E TPGS as surface-active agent using 2^3 factorial design. The authors identified rpm and HPMC as the most significant process and material parameters influencing particle size [33].

Similarly, Singare et al. identified and optimized formulation (ratio of polymer to drug and ratio of surfactant to drug) and process variables (milling time and milling speed) using Box Behnken design at small and large scale. The study identified that high polymer concentration and high milling speed may affect manufacturing of nanocrystals at large scale [34]. Cerdeira et al. used a 2^3 factorial design to study the effects and interactions of miconazole %, concentration of sodium dodecyl sulfate (SDS), and HPC-LF on particle size distribution. It was realized that low contact angle between drug and dispersion medium indicates excellent dispersibility of the micronized drug particles in the medium. Likewise, authors concluded that in case of miconazole, SDS and HPC-LF provide a synergistic effect for particle size reduction and stabilization. SDS mediated wetting and facilitated adsorption of HPC-LF onto the miconazole particles, providing steric protection from agglomeration and crystal growth [35]. Jog et al. performed a comprehensive QbD approach with five DoE models at wet media milling and drying stages for identification and optimization of CQAs during production and downstream processing to obtain nanocrystal-based zileuton formulations [36].

Mahesh et al. have compared and evaluated effect of liquid anti-solvent precipitation and media milling on nanosuspension preparation using Box-Behnken design. Authors concluded that ratio of polymer to drug, milling time, and milling speed played significant role in controlling the zeta potential of nanosuspension prepared by media milling, while milling time and milling speed affected particle size distribution of nanosuspension. While in liquid anti-solvent precipitation, ratio of surfactant to drug and speed of mixing played a significant role in controlling the zeta potential of glipizide, ratio of polymer to drug and speed of mixing were considered to be the significant factors that affected the particle size distribution of nanosuspension [37].

Similarly in HPH and microfluidization, the mean crystal size depends primarily on homogenizer pressure, number of homogenization cycles, milling time, and temperature [38–40]. In a study reported by Cheng et al., Box-Behnken design was employed to optimize single value for four independent process variables, i.e., homogenization pressure, production temperature, puerarin amount, and PVP K30/puerarin ratio to prepare puerarin nanocrystals [41]. In an interesting study by Oktay et al., stabilizer type and concentration were optimized using 2^4 full factorial design. The results demonstrated that type and amount of stabilizer affected particle size and zeta potential and poloxamer was selected to impart optimal stability on storage [42].

3.3 Relation of crystal structure and nanonization

Drug properties that govern comminution kinetics and threshold of size reduction include ductility, elasticity, and brittleness. A material is considered to be brittle if crystal breaks without deformation on exposure to mechanical stress. On the other hand, a ductile material shows plastic deformation before size reduction. Therefore, a brittle material is likely to mill faster than a ductile one. Limited literature on comminution behavior of drug materials, especially for nanonization, is reported. Colombo et al. correlated impact of drug brittleness on milling behavior and subsequently its impact on crystallinity and solubility enhancement. Elastic modulus was used as the marker for brittleness of Tacrolimus (TAC; melting point 126°C; 12.4 GPa) and Dexamethasone (DEX; melting point 263°C; 16 GPa). The values for both the drugs were in between the two extremes of high (dicalcium phosphate, 31.3 GPa) and low elastic modulus (microcrystalline cellulose, 3.8 GPa). It was concluded that DEX was more brittle than TAC and the subsequent implications on milling include faster and shorter milling time (completed in <2 h) and high degree of crystallinity (plateau at 2 h). On the other hand, TAC started reducing in size after 2 h with reduction in degree of crystallinity (reduced to 74% in 5 h), larger mean particle size, and poor reproducibility. In terms of solubility, both the drugs showed increase in solubility due to decrease in degree of crystallinity and particle size [43].

In contrast, Malamatari et al. reported contrasting results for wet milling of low melting and ductile drug, Ibuprofen (IBU). Nanosuspensions were successfully prepared with HPMC and TPGS as stabilizers in 6 milling cycles in 3 h. Due to low melting point (75–78°C) of IBU, it was expected that drug may melt during the process and would show aggregation. However, cautious optimization of process such as milling cycle of 30 min was followed by a pause of 20 min to prevent heat-induced melting during nano comminution. The comminution kinetics of IBU were initially high and size continued to reduce on further milling but at a slower rate [44].

On the other hand, George et al. have suggested that enthalpy can be used to predict millability of drug for generation of nanocrystals. The drugs with higher melting enthalpies and hydrophobicity were found to have good millability, in contrast to hydrophilic drugs with low enthalpy. The stabilizer should be selected based on hydrophobicity of the drug. In addition, the authors suggested that the drugs with globular or plate-like structure have good millability as compared to rod-shaped particles [45].

In another study, Lestari et al. demonstrated the importance of selecting a suitable surface modifier to produce stable nanosuspensions with size <250 nm using wet ball milling. No correlation was found based on physicochemical properties of drugs such as molecular weight, log P value that represents lipophilicity, pK_a, and melting point, to classify them into good, bad, or medium millability. The particle size and stability of nanosuspension were more dependent on type and concentration of surface stabilizer employed. It was concluded that all drugs can be milled to nanocrystals using wet media milling. However, for the drugs with bad millability (as indicated by milling time and size obtained), sufficient efforts are needed for selection of stabilizer [15]. In view of the contrasting literature, more molecule-specific research is needed to understand comminution behavior and kinetics of drugs. However, it is clear that choice of right surfactant and drug-stabilizer interaction can guide formation of a stable nanosuspension.

3.4 Phase transformation during generation of nanocrystals

Mechanical energy induces strain in crystalline particles resulting in disruption of crystal lattice and particle fracturing. Under the stress, the ordered crystalline structure can lead to (i) transformation into a different polymorphic form and (ii) generation of regions of localized crystal defects, or (iii) disordered amorphous structures.

Polymorphism is known to affect drug solubility and dissolution rate. Typically, a thermodynamically most stable crystalline form is desirable to prevent the risk of solid-state transformations during storage and/or administration. Thus, the performance and stability of the solid form on nanonization need to be thoroughly investigated. Lai et al. formulated piroxicam nanocrystals using high-pressure homogenization. During processing, the starting material, i.e., Form I of Piroxicam, resulted into nanocrystals which were mixture of monohydrate and Form III. The solubility of Form I was lower than Form III. In this case, the increase in solubility was due to formation of metastable form and particle size reduction [46]. In some cases, polymorphic form can have more profound effect on solubility and dissolution rather than particle size reduction. For example, Lai et al. prepared nanosuspensions of Form 1 and 2 of Diclofenac using high-pressure homogenization. Form 1 was reported to have a higher solubility than Form 2. On processing, Form 2 partially transformed to Form 1, while there was no change in Form 1. Interestingly, in dissolution testing, the microsuspension of Form 1 performed better than the nanosuspension of Form 2. This indicated that dissolution is strongly affected by polymorphic form and selection of appropriate polymorph can impart higher dissolution and stability [47].

A more serious concern in nanocrystals generation is the purity of solid form. Nanonization can yield completely amorphous drugs, surface amorphization, or disordered crystals. In a study reported by Deng et al., wet milling of an undisclosed drug showed broadened X-ray pattern, which was confirmed as a disordered nanocrystalline structure but not amorphous form. The absence of a long-range crystal order was concluded to be due to smaller sized crystals and defects [48].

Generation of amorphous regions can contribute to solubility enhancement; however, uncontrolled crystallization on storage can create stability problems. Milling performed in an aqueous environment is expected to minimize the chances of surface amorphization since water functions as a plasticizer (raises molecular mobility) and aids in recrystallization. In a study reported by Sharma et al., the influence of wet milling using high-pressure homogenization on the solid state of indomethacin (IMC) and simvastatin was analyzed. A very low amount ($<1\%$) of surface amorphization was detected in milled IMC samples, with absence of bulk amorphization (as no change was detected in mDSC study). Cryogrinding of IMC in absence of water resulted in completely amorphous form, while presence of water (under same cryogrinding conditions) resulted in a solid state similar to that obtained in wet media milling (amorphization was limited to surface) [49]. Therefore, the water was recognized as the plasticizer to avoid the generation of bulk amorphous form.

4. Formulation of nanosuspension into oral solid dosage forms (OSDs)

4.1 Need for development of OSDs

Most of the top-down and bottom-up technologies employed for generation of nanocrystals produce a liquid dispersion of nanocrystals, i.e., nanosuspension. The nanosuspensions due to high surface energy and liquid medium pose numerous challenges (given in Fig. 1). Moreover, due to the aqueous nature of the medium, microbial growth may occur upon storage [50]. On the other hand, solid dosage forms exhibit better long-term physical and chemical stability. The nanosuspensions are thus subjected to downstream processing to improve stability and for administration in the form of OSDs such as tablet or capsule [51]. The first nanocrystal product Rapamune (Rapamycin) was an oral tablet, whereas the second product Emend (Aprepitant) was pellets filled in hard gelatin capsule [18].

4.2 Conversion of nanosuspension to OSDs

The nanosuspensions are solidified by two types of processes—(i) drying in the presence of excipient and (ii) granulation using nanosuspension as the granulating fluid. In the first

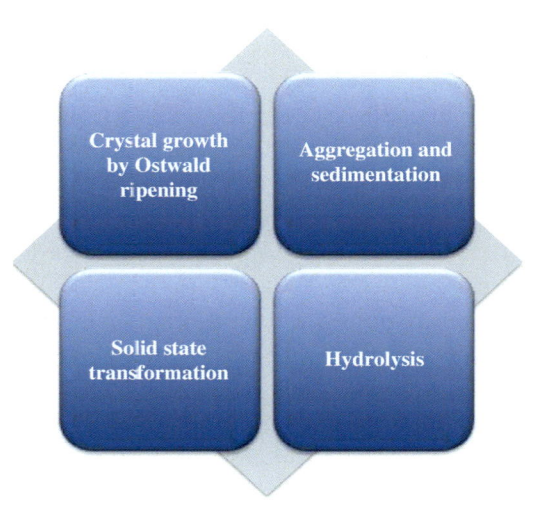

Fig. 1 Reasons for instability of nanosuspensions.

process, the dried nanosuspension is blended with additional excipients and processed into solid dosage forms by means of direct compression or capsule filling. Alternatively, in the second process, drying and shaping processes are combined in a single step wherein either nanosuspension is used as the granulating fluid or as layering dispersion in fluidized bed coating [51].

The drying process can result into reversible or irreversible aggregation of nanocrystals, upon reconstitution with water or gastric fluids [50]. The main challenge related to drying process lies in preserving re-dispersibility of the nanocrystals. Drying can also result in drug degradation, decreased dissolution kinetics, and additional thermal/freezing stresses due to heat or freezing conditions used during spray-drying and freeze-drying, respectively [51]. Hence, conversion of nanosuspensions to OSDs without compromising physical, chemical, and performance properties is imperative [52].

4.3 Techniques used for drying of nanosuspensions

The techniques used for solidification of nanosuspension are mainly spray-drying, freeze-drying, spray-freeze-drying, layering (coating) of nanosuspension, and granulation using nanosuspension. Although these techniques can be used in various applications, our discussion is limited to generation of oral solid dosage forms from nanosuspensions. The major highlights and challenges of these techniques are mentioned in Table 1.

4.3.1 Spray-drying

Spray-drying involves spraying of nanosuspension into a hot drying medium consisting of air or nitrogen. Fine droplets of nanosuspension are formed by atomization through a

Table 1 Advantages and limitations of various techniques used for drying of nanosuspensions [3, 50, 51].

	Spray-drying	Freeze-drying	Spray-freeze-drying	Coating and granulation
Advantages	• Easy to scale up • Fast and controlled drying • Cost efficient	• Useful for heat-sensitive drugs • Stable upon storage • High quality product • Industrial accessibility	• Good re-dispersibility • Useful for thermolabile compounds	• Lower cost • Simple downstream processing • Good flow properties
Disadvantages	• Not suitable for thermolabile compounds • Poor flow and need for further processing • Low yield	• Costly • Time consuming • Poor flow, low bulk density	• Need of sophisticated equipment • Low bulk density • Costly	• Long processing time • High pressure drops and energy consumption • High attrition potential

nozzle. The droplets are dried by solvent evaporation to form solid particles [53]. It is a unique drying process since it involves both particle formation and drying [54]. The spray-drying process encompasses the following four stages:

1. Atomization of nanosuspension into a spray: In this stage, fine droplets of nanosuspension having large surface area are generated to aid in faster drying. The optimal atomization pressure, feed rate, and nozzle diameter help in formation of desirable droplet size.

2. Spray-air contact: The droplets of nanosuspension and drying medium can come in contact with each other in two ways, i.e., cocurrent flow or countercurrent flow. In a cocurrent flow, the feed is atomized and sprayed through the drying chamber in the same direction as the flow of the heated drying medium. However, the atomized nanosuspension and heated drying medium come in opposite direction in countercurrent flow mode.

3. Drying of the spray: Evaporation of the solvent takes place from the droplet on coming in contact with the drying air. The drying process is rapid and occurs in milliseconds.

4. Separation of the dried product from the drying gas: Particle separation from the drying air and collection occurs in cyclone. The fines are usually retained in the filter bag scrubber.

4.3.1.1 Influence of formulation parameters on product properties

Spray-drying of nanosuspension alone can result into aggregation of nanocrystals, consequently decreasing the dissolution kinetics. Therefore, addition of matrix formers or carriers to the nanosuspension is necessary to reduce aggregation [55]. The carriers intersperse between drug nanocrystals and prevent their fusion. The matrix former should dissolve rapidly to release the nanocrystals into the medium. Typical matrix formers reported in the literature are sugars (sucrose, lactose), sugar alcohols (mannitol, sorbitol) and water-soluble polymers (Polyvinlyl Pyrollidine, polyvinyl alcohol, long chained Polyethylene Glycol), and microcrystalline cellulose [26, 56].

Several attributes related to matrix formers such as type, concentration, and ratio with respect to drug can affect product characteristics [57]. Vatanara et al. studied the influence of carrier type on size of spray-dried particles. The study showed that increase in mannitol concentration led to larger microparticles due to saturation of mannitol at the surface of the drying droplet. However, this phenomenon was absent in formulations prepared using combinations of sorbitol and microcrystalline cellulose (MCC) [58].

Alaei et al. investigated the effect of type and concentration of matrix former on yield and morphology of spray-dried products. The yield of powders was least for sorbitol due to hygroscopic nature; however, sorbitol as the carrier produced particles with smooth surfaces. Mannitol-based formulations showed compromised yield due to thermoplastic nature and particles were agglomerated with uneven surfaces. In contrast, microcrystalline cellulose containing formulations improved product yield due to nonhygroscopic nature and better thermodynamic behavior. In addition, MCC-based formulation had segregated particles with corrugated surfaces. Alaei et al. also showed that low nanocrystal to carrier ratio improved powders' flowability because higher concentration of carrier protected the nanocrystals from agglomeration [59]. Lee et al. also investigated the effect of HPC concentration on shape of spray-dried particles. Increase in HPC concentration from 2%w/w to 14%w/w resulted in spherical particles as higher amount of HPC could completely cover the surface of the nanocrystals [60].

Malamatari et al. demonstrated the impact of drug to mannitol ratio and inlet temperature on aggregation of nanocrystals. Solidification of fenofibrate nanosuspension showed that low drug to mannitol ratio and inlet temperature lower than the melting point of fenofibrate were optimal in order to minimize irreversible aggregation [50]. Van Eerdenbrugh et al. reported that a uniform dispersion of drug in InutecSP1 compared to Avicel and Aerosil resulted into better dissolution performance. This is because the polymeric surfactant concentrated on the interface between hydrophobic particles and water which led to improved wetting [56].

Chin et al. demonstrated the impact of spray-drying process on stabilizers used in the nanosuspension. An aqueous solution of polymeric dispersant, HPC, gelled during spray-drying. The restricted motion of drug nanocrystals resulted in entanglement of stabilizer chains on the surface of drug nanocrystals, which compromised the steric repulsive action of HPC. Hence, stabilizers with shorter chain length and lower molecular weight should be preferred [61].

Dolenc et al. also studied problems associated with stabilizers used in nanosuspensions during spray-drying. Nanosuspension stabilized with Tween 80 resulted in deposition of product on wall of the drying chamber unlike nanosuspensions stabilized with PVP/SDS [62]. Chaubal et al. evaluated the importance of charged surfactant during spray-drying. Itraconazole is a neutral molecule and required a charged surfactant (SDS) for electrostatic stabilization. Spray-dried powders in absence of a charged surfactant showed higher particle size and slow dissolution rate [57].

4.3.1.2 Influence of process parameters on product properties

Stability of drugs and particle shape are the leading parameters affected by inlet temperature variation. Higher inlet temperature promotes faster drying. However, high temperature might be unfavorable and can lead to formation of fused or donut-shaped particles [60]. Chaubal et al. studied the impact of inlet temperature on yield and moisture content of spray-dried powders. Spray-drying of Itraconazole particles at higher inlet temperature led to excessive collection of particles in the spray chamber, whereas lower temperature resulted in particles with high residual moisture content. Inlet temperature indirectly influenced the dissolution kinetics of the spray-dried particles. Formation of fused particles at high inlet temperature negatively impacted the re-dispersibility and dissolution rate of the nanocrystals [57].

Littringer et al. studied the effect of feed concentration on particle morphology and size. High feed concentration produced rough surfaces. On the other hand, lower feed concentrations along with higher rotation speeds resulted in smaller particles due to increase in breaking strength of the particles [63].

4.3.2 Freeze-drying

In this technique, nanosuspension is first frozen and then ice is sublimed under vacuum. Sublimation first occurs at the upper layer of the product, which gradually progresses to the core of the product. Freeze-drying generates a highly porous cake with low moisture content [64, 65]. Freeze-drying consists of three stages: (a) freezing, (b) primary drying, and (c) secondary drying. In the first step, around 80%–90% of water present in the nanosuspension is frozen. The frozen water is sublimated with the application of vacuum in the primary drying stage. A porous plug is formed wherein these pores correspond to the spaces that were previously occupied by the ice crystals. In secondary drying, the remaining liquid water is removed by maintaining temperature between 20°C and 40°C to achieve desired moisture content in the freeze-dried product [66–68].

4.3.2.1 Influence of formulation parameters on product properties

Freeze-drying may generate stress due to freezing and dehydration, which can induce fusion and irreversible aggregation of nanocrystals. Moreover, ice crystals may cause mechanical stress on the nanocrystals leading to their destabilization. Therefore, incorporation of cryoprotectant and lyoprotectant is essential to protect the nanosuspensions from freezing and drying stress, respectively. Sugars like trehalose, sucrose, and glucose and sugar alcohols such as mannitol and sorbitol are most commonly used cryoprotectants [68]. The cryoprotectant solution is freeze-concentrated to a stable glass with a high viscosity, which prevents nanocrystals fusion and aggregation [69].

Addition of steric stabilizer is essential along with cryoprotectant/lyoprotectant during freeze-drying. The stabilizer generates a dense steric layer on surface of the nanocrystals, which keeps them apart in the freeze-concentrated state. Steric stabilizer could also act as a cryoprotectant/lyoprotectant when the stabilizer is not fully adsorbed onto the surface of the drug particles. Hence, to avoid particle aggregation during freezing, an optimal concentration of both steric stabilizer and cryoprotectant must be present in the drug nanosuspension [70]. Cryoprotectant additionally ensures re-dispersion of freeze-dried product to nano size upon exposure to the dissolution medium. Freeze-drying of drugs with cryoprotectants such as danazol with sucrose and mannitol and loviride with sucrose and oridonin along with mannitol confirmed the role of cryoprotectants in re-dispersibility of the freeze-dried products [51].

Ma et al. investigated the effect of cryoprotectant concentration on agglomeration and re-dispersibility of freeze-dried ursodeoxycholic acid nanocrystals. A 400% concentration of sucrose (related to the weight of drug) significantly lowered irreversible aggregation of nanocrystals as compared to 100% or 200% sucrose. Sucrose and glucose showed minimal agglomeration and better re-dispersibility of freeze-dried nanocrystals of ursodeoxycholic acid, unlike mannitol and lactose [69]. Different concentration of cryoprotectant can lead to contrasting effects; thus the choice of cryoprotectants should be made judiciously. The lower concentration of sucrose acted as cryoprotectant and enhanced re-dispersibility of loviride. In contrast, higher amounts of sucrose resulted in noticeable agglomeration in Itraconazole formulation [56]. Gao et al. studied the effect of drug to cryoprotectant ratio on freeze-dried product properties. The lower drug/cryoprotectant ratio resulted in less agglomerated product with smaller particle size due to presence of higher amount of cryoprotectant [51].

Gao et al. also investigated the need of steric stabilizer during freeze-drying along with the cryoprotectant. Sugars alone in concentration of 25–250 $w/w\%$ were not effective in preventing aggregation of albendazole nanosuspensions. Addition of 12.5 $w/w\%$ HPMC or 2.5 $w/w\%$ Carbopol significantly prevented aggregation of the nanocrystals [51]. In another work, the significance of type of stabilizers used during freeze-drying was studied. Polymeric stabilizers were more effective as they adhered to the nanocrystals' surface and provided steric protection. The use of HPMC did not show any crystal growth, whereas incorporation of Cremophor RH40 or Tween 80 resulted in agglomeration of nanocrystals [69]. Furthermore, Chung et al. examined the effect of molecular weight of polymers. Higher molecular weight PEG prevented aggregation of drug nanocrystals more effectively [71].

Chin et al. demonstrated the impact of concentration of buffer salts and surfactants on drug solubilization. During freeze-drying, the concentration of buffer salts and surfactants changed, which led to pH variation and could affect re-dispersibility of the drug [61].

Drug properties strongly influence the amount of dispersants required for ensuring good re-dispersibility. A minimum carrageenan concentration of 0.5 w/w% was required for naproxen, while for itraconazole it was 3 w/w% [51]. Yue et al. studied the effect of cohesive energy and contact angle of drugs on stability of the freeze-dried product. Higher cohesive drugs were more stable during solidification as lower cohesive drugs easily underwent agglomeration. Also, hydrophobicity of drug played a key role in agglomeration [72].

4.3.2.2 Influence of process parameters on product properties

Although irreversible aggregation mainly occurs during drying and not in freezing stage, aggregation can be strongly affected by freezing rate. De Waard et al. investigated the impact of freezing rate on size and agglomeration of nanocrystals. Higher freezing rate resulted in smaller particles and prevented further agglomeration of the freeze-dried product. In contrast, coarser particles with pronounced agglomeration were obtained with lower freezing rate [73, 74]. Several freezing methods like liquid nitrogen freezing, loading vials onto precooled shelves, or ramped cooling on the shelves are employed to freeze nanosuspensions. These can cause different supercooling effects, which influence nanocrystals' stability. The highest supercooling was obtained with liquid nitrogen freezing and least for the precooled shelf method. Higher supercooling resulted in smaller ice crystals and larger ice-specific surface area. This might decrease the mechanical stress on nanocrystals, thus avoiding their aggregation [68].

Ma et al. evaluated the effect of different freezing temperatures on agglomeration of nanocrystals. Ursodeoxycholic acid nanocrystals were freeze-dried at conventional, i.e., $-20°C$, moderate $-80°C$, and rigorous $-196°C$ freezing temperatures. Re-dispersibility of nanocrystals frozen at conventional freezing temperature was better than those at moderate and rigorous temperature conditions with high freezing rate. With increase in freezing rate, water molecules excluded the nanocrystals, thus causing their aggregation [69].

4.3.3 Spray-drying vs. freeze-drying

Gubbala et al. explained the superiority of spray-dried powder with respect to freeze-dried product in terms of flow properties and drug release by virtue of formation of spherical particles. In contrast, freeze-drying resulted in product with poor flow properties due to formation of aggregates or large chunk of particles. The extent of drug release was higher for spray-dried product as compared to the lyophilized one [64]. The soluble stabilizers wrapped around the spherical spray-dried particle and dissolved rapidly to create hydrophilic microenvironment for the hydrophobic drug to dissolve. This mechanism facilitated in rapid dispersion of nanocrystals in the dissolution medium [75].

Chin et al. demonstrated that different drying technologies used for solidification of nanosuspension could give variable results with same excipients. MCC (Avicel PH 101) turned out to be superior to sucrose for stabilization of itraconazole nanosuspension during freeze-drying. Transition of sucrose from solution to glassy to crystalline state occurred and compromised the cryoprotectant effect. In contrast, MCC as a water-insoluble matrix former did not undergo these changes and led to better steric stabilization of the nanocrystals. However, in a different study involving spray-drying of bicalutamide nanosuspension, MCC was found inferior to sugars. The contradictory results could be explained due to differences in API and drying technology used in the two studies [61].

4.3.4 Spray-freeze-drying

Spray-freeze-drying (SFD) is an alternative technique to spray-drying for thermolabile particles. The first step involves atomization of nanosuspension into fine droplets. Then cold temperature provided by cryogen, i.e., liquid nitrogen, freezes the water molecules present in the tiny droplets. Extra liquid nitrogen is evaporated from the system and the frozen water is sublimed and water vapor is removed from the system. The sublimation of ice crystals present in the interstitial spaces leads to formation of porous structures as the frozen droplets retain their size and do not shrink upon sublimation. Thus, highly porous spray freeze-dried particles with good re-dispersibility are obtained. In this process, protection of nanocrystals from freezing and drying stresses is required. Cryoprotectants/lyoprotectants in nanosuspension before spray-freeze-drying are usually added [76].

4.3.4.1 Influence of formulation parameters on product properties

Niwa et al. investigated the effect of ratio of drug: dispersing agent (PVP) on microscopic structure of spray freeze-dried phenytoin particles. Particles with higher PVP loading showed rigid polymeric network with embedded drug nanocrystals, while agglomerated granules were formed at lesser PVP content. The particle size distribution of spray-freeze-dried particles differed with ratio of PVP: drug. In SFD, size of particles is affected by viscosity of the spray fluid. Increment in Phe: PVP ratio from 32:1 to 1:2 resulted in significant increase in viscosity of the spray mist. Thus, increase in PVP: drug ratio above 50% shifted D90 values of spray freeze-dried particles to higher side even if solid content in the spray suspension was kept constant.

In the same study, the impact of PVP content on dispersibility of spray freeze-dried particles was investigated. Higher PVP content not only produced smaller particles, but also facilitated wetting of hydrophobic drug crystals. Opposite effect was observed with higher amount of PVP (1:2 ratio of Phe:PVP), as this increased viscosity of the micro-environment of SFD particles. Also, morphology and size of spray freeze-dried particles were strongly influenced by solid content in the nanosuspension [77].

4.3.5 Layering of nanosuspension

This process is widely used to coat nanosuspensions onto the surface of multiparticulates like granules, pellets, sugar beads, and nonpareil seeds using a fluidized bed coater [60]. This involves solvent removal and simultaneous deposition of coating material onto non-active, spherical cores. First, particles to be coated are fluidized by air current. The nanosuspension is sprayed on the inert cores followed by spreading of droplets on surface, leading to flattening and adhesion of the droplet on the nonpareil core (mass transfer). Subsequently, solvent evaporation (heat transfer) leads to formation of a uniform layer on the surface of the core material. After the particle surface is wetted by nanosuspension, there is a competition between continuous layering of the nanosuspension and agglomeration of the wetted particles [78, 79].

Fluidized bed coating is much more efficient method for preparing solid formulations from bulk nanosuspensions compared to conventional spray-drying [33, 80]. Usually sugar spheres are used as core material since their smooth surface provides ideal base to build up successive layers of drug/excipient. This technology is already used for the commercial manufacturing of Rapamune tablets wherein the nanosuspension is coated onto inert core material [60].

4.3.5.1 Influence of formulation parameters on product properties

Several types of inert cores with divergent characteristics are used to coat nanosuspension. It is important to understand their impacts on coating process. Parmentier et al. studied the effect of bead size on size and dissolution profile of coated particles. Larger beads produced coated particles with higher D90 values, while coating on smaller beads resulted in comparatively smaller size particles. Therefore, smaller beads exhibited higher dissolution rate compared to larger beads, irrespective of bead type and coating polymer. The authors also examined the effect of bead type on dissolution kinetics. In case of sugar cores, disintegration of the beads enhanced dissolution of coated particles. However, drug release was dominated only by dissolution and erosion of the polymer matrix for water–insoluble cellulose beads. Thus, drug release from sugar beads was better than cellulose beads irrespective of the bead size. The authors also investigated the influence of type of coating polymer on dissolution performance of coated particles. HPMC VLV-coated beads displayed better release profile than Copovidone-coated beads irrespective of bead size and bead type. Higher hygroscopicity and production of larger coated particles compared to HPMC VLV led to slower release from Copovidone-coated beads. They also examined the effect of amount of nanosuspension layering on dissolution behavior of coated particles. The drug release of 50% layering level was higher and faster compared to 150% layering level, irrespective of bead size, bead type, and coating polymer [81].

Beads that are result of a layering process can easily be filled into capsules without further processing. The product obtained from this method has good flow ability, bulk density, and hygroscopicity unlike product obtained after spray-drying and freeze-drying [82].

Malamatari et al. demonstrated dual role of stabilizers during fluidized bed coating. Excipients such as TPGS and HPMC stabilized nanosuspension as well as acted as coating agents to adhere nanocrystals onto inert cores. Similarly, the triple role played by chitosan chloride was evaluated, i.e., as stabilizer for nanosuspension, binder in layering process, and mucoadhesive excipient [50]. The type of stabilizers used in nanosuspension should be selected judiciously if nanosuspension has to be processed using fluidized bed coating. Use of hydrophilic stabilizer such as HPMC in fenofibrate nanosuspension had positive impact on dissolution kinetics due to increased wettability of drug particles. Wang et al. evaluated the influence of surfactant addition before coating on dissolution rate. Increase in apparent saturation solubility and dissolution rate of fenofibrate was observed when surfactant such as SDS was added to the nanosuspension. They also demonstrated that adhesion preventer like talc decreased adhesion of particles and aggregation [80].

Parmentier et al. investigated the impact of surface hydrophobicity of drug molecule on dissolution performance of coated particles. Cinnarizine nanocrystals showed agglomeration after release from the coating and thus resulted in slower release compared to original nanosuspension. In contrast, no agglomeration and optimal dissolution profile was observed for naproxen. These differences were most likely caused by the difference of surface hydrophobicity between naproxen and cinnarizine [81].

4.3.6 Granulation using nanosuspension

Fluidized bed granulation is an alternative method to solidify nanosuspensions. During this process, the nanosuspension is used as a granulating fluid and is sprayed on to granulation substrate. Inert materials like lactose or microcrystalline cellulose are used for granulation. The process involves three stages: (a) wetting and nucleation, (b) consolidation and growth, and (c) breakage and attrition. Wetting of particles forms liquid and solid bonds between the particles, thus enabling growth of granules. These stages are governed by process parameters used during granulation, physicochemical properties of the nanosuspension, and the properties of the granulation substrate [78, 83, 84].

This method results in free-flowing granules that can be easily compressed into tablets or filled into capsules. Care must be taken to ensure that the ingredients and/or the processing does not cause aggregation of the dried drug nanocrystals [85]. The scale up of granulation in a fluid-bed system needs an understanding of excipient functionality, fluidization kinetics, and identification of critical process parameters [78].

4.3.6.1 Influence of formulation parameters on product properties

Bose et al. investigated the effect of type of granulation substrate on amount of fines and drug content in the final product. Lactose produced greater percentage of fines compared to mannitol. This was explained by higher brittleness of lactose compared to mannitol. Also, drug content of lactose-based batch (85.9%) was lower than mannitol-based batch (91.9%). Because of higher amount of fines, greater percentage of drug loss occurred during lactose-based granulation, subsequently lowering the product yield. In the same study, the impact of granulation substrate on dissolution performance of granules was studied. Mannitol-based granules showed better dissolution as compared to granules containing lactose wherein latter showed incomplete (70%) drug release. The authors also investigated the effect of granulation substrate on flow properties of granules. With 10% drug loading, the bulk density of lactose-based granules ($0.693\,g/cm^3$) was higher compared to mannitol-based granules ($0.411\,g/cm^3$). The authors also demonstrated the influence of drug loading on particle size of granules. Larger granules were obtained for a batch having 20% drug loading with D50 of 67.9 μm and D90 of 331.0 μm. In contrast, granulation with 10% drug loading showed D50 of 57.2 μm and D90 of 295.8 μm [83].

Beirowski et al. studied the effect of drug loading on dissolution profile. Increase in drug loading to 20% caused increase in surface hydrophobicity and harder agglomerates with slower dissolution rate [70].

4.3.6.2 Effect of process parameters on product characteristics

Figueroa et al. investigated the effect of spraying rate on dissolution profile of granules. Increase in spray rate resulted in faster growth of particles and agglomeration kinetics. This eventually decreased the dissolution rate of granules. In addition, higher atomization pressure produced smaller droplets with large surface area. These fine droplets were more uniformly distributed on granulating substrate and reduced likelihood of particle aggregation. Due to less particle aggregation at higher atomizing pressure, significant decrease in re-dispersed mean particle size was observed, at a constant spray rate. Thus, granules with better re-dispersibility could be obtained [86].

Flögel et al. investigated the effect of type of spray mode on various granule properties like particle size. The mean particle size of granules produced using bottom spray mode was smaller as compared to top spray. In case of bottom spray, a greater portion of sprayed nano-suspension is enclosed within the granule nuclei due to close contact between powder and nanosuspension. Thus, nanosuspension was not available for further growth of granules. Granules produced using top spray had bulk densities roughly 23% lower than those produced through bottom spray. Also, powders produced using top spray had BET surface area values that were 20% larger than those produced using bottom spray. Low bulk density and porous surfaces improved wicking of liquid into the granules, leading to enhanced dispersibility and disintegration. Hence, granules prepared using top spray demonstrated better re-dispersibility and faster release rates as compared to bottom spray mode [87].

4.4 Conversion of solidified nanosuspension into OSDs

The solidified drug nanosuspensions are finally converted into tablets, pellets, and capsules for oral administration. However, they should release nanocrystals after coming in contact with dissolution medium. This ensures conservation of the dissolution advantage of the nanocrystals. Rationale selection of excipients and optimizing process parameters can help in achieving this objective [51].

4.4.1 Factors associated with tableting process

Dolenc et al. investigated the difference in compaction force required to prepare tablets with nano and microparticles. This dissimilarity had remarkable impact on tensile strength, hardness, elastic recovery, and finally on dissolution performance. Hence, optimal compaction force should be selected for generation of tablets of nanoparticulate systems. The nanoparticles due to large surface area interact stronger with filler compared to microparticles. The stronger interparticulate interactions contribute to higher tensile strength. Much lower compaction force was needed for dried celecoxib nanosuspension-MCC blend compared to micronized celecoxib-MCC (refer Fig. 2). The lower forces also contributed to less deformation and less elastic behavior after removal of the compacting force [51, 62].

In a similar work, Lee explained the reason for variation in compaction force needed for both the systems. Homogeneous distribution of API and filler (HPC) was observed in nanoparticulate tablet as compared to microparticulate tablet. Lower mechanical performance of microparticulates was justified, as intraparticulate boundaries did not have significant amount of filler (HPC) and was more vulnerable to brittle deformations or fracture. In contrast, nanoparticulate system had fully entangled HPC chains which contributed optimal mechanical properties to primary nanoparticles [60].

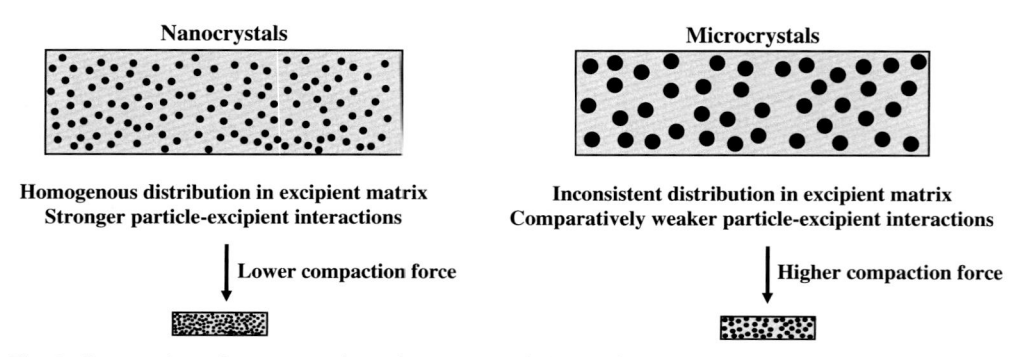

Fig. 2 Compaction of nanocrystals and microcrystals-based formulation.

4.4.2 Impact of excipients used during tableting

Gao et al. studied the effect of tableting procedure on particle aggregation. Drug nanoparticles have a large number of contact points for binding and tableting could potentially cause irreversible aggregation of particles. Additional matrix formers are required in tablet to prevent aggregation and enhance dissolution. Higher amount of matrix formers could significantly decrease the contact points among the drug nanocrystals and subsequently particle aggregation (refer Fig. 3). Higher amount of mannitol in the fenofibrate tablet resulted in enhanced dissolution. Also, the dissolution rate of ketoprofen matrix pellets was increased with increase in amount of starch derivatives [51].

Chin et al. examined the impact of tableting excipients on disintegration of tablet. Candesartan nanosuspension dried with mannitol was compressed with various excipients such as colloidal silicon dioxide, magnesium stearate, sodium starch glycolate, cornstarch, and crospovidone. Disintegration time varied more than twofold between the fastest disintegrating formulation and the slowest one. The formulation with crospovidone showed disintegration time of 10 min as against 20 min for formulation containing corn starch [61].

In another study, Tan et al. investigated influence of solubility of diluents used in the tablet formulation on rate and mechanism of tablet disintegration. Insoluble diluents such as Di–Calcium phosphate (DCP) did not interact with water due to repulsive forces and

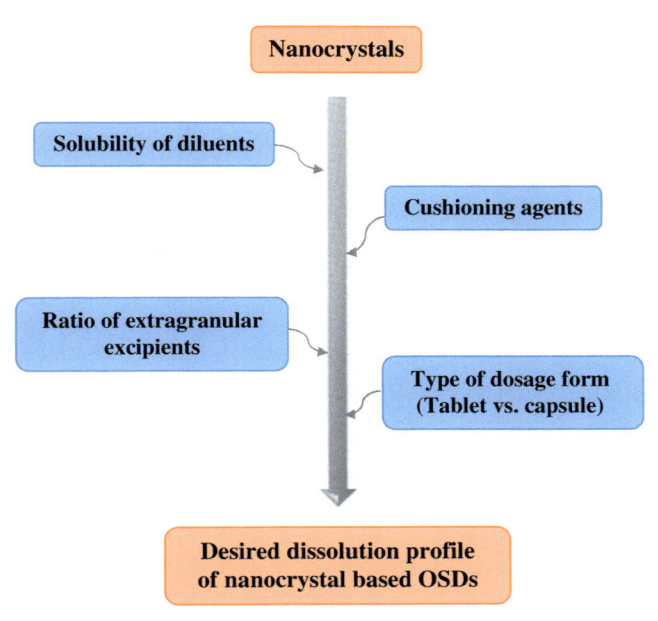

Fig. 3 Factors contributing to dissolution from nanocrystal-based OSDs.

facilitated rapid disintegration. On the other hand, water-soluble diluents increased disintegration time due to competition with other available soluble excipients in dissolution medium. Hence, the effect of disintegrant was more evident in an insoluble system than a soluble system, thus justifying the addition of DCP to improve disintegration time [88].

4.4.3 Effect of type of dosage form

Tan et al. demonstrated the differences in additional excipients required according to the type of solid dosage form. Powder formulations deformed plastically, producing strong compacts, while pellets of the same formulation exhibited plastic deformation and brittle fragmentation, resulting in compacts of lower tensile strength. More extra granular excipients were required for pellets as compared to powder formulations. Cushioning agents and binder excipients are added to improve compaction properties and confer adequate tensile strength, respectively [88].

Gao et al. studied the effect of type of solid dosage form on dissolution profile of nanocrystals. Changing tablet formulation to capsule having same composition markedly increased dissolution profile. This could be justified as tableting led to additional stresses and increased chances of aggregation, whereas only physical mixing with excipients was required in case of capsule [51].

5. In vitro in vivo correlation (IVIVC)

As defined by FDA, IVIVC is "a predictive mathematical model describing the relationship between an in vitro property of a dosage form and an in vivo response." In general, in vitro property is the rate or extent of drug dissolution or release, while in vivo response is the plasma drug concentration profile. Practically, the purpose of IVIVC is to use drug dissolution results from two or more products to predict similarity or dissimilarity of expected plasma drug concentration. Successful correlation can reduce the number of experiments during drug development, predict the effects of formulation changes on absorption of drug, and improve the final drug product quality.

The establishment of IVIVC is an essential step for Class II/IV drugs, since dissolution is the rate-limiting step in these cases. Conversion of these molecules into nanocrystals enhances their dissolution drastically, which should be reflected in vivo to obtain IVIVC. [89]. In addition, IVIVC can help optimize the process of production and drying of nanosuspensions. However, research reports on IVIVC of nanocrystal formulations are scarce and evolving.

Recently, attention has slowly shifted towards research in the field of IVIVC of nanocrystals. The early studies primarily demonstrated challenges in predicting in vivo behavior based on in vitro analysis. This is because not all aspects of behavior of nanocrystals in vivo are well-understood, thus making it tedious to achieve IVIVC. For example, Imono et al. investigated IVIVC of nanocrystal formulations of megestrol acetate and fenofibrate. In vitro, solubility advantage was 1.4-fold from nanocrystals, while permeation rate was threefold higher, in case of both the drugs. In vivo absorption in animals increased with decrease in particle size and a positive correlation was established between in vitro permeation rate (instead of dissolution alone) and in vivo maximum absorption rate. This study concluded that mechanism of enhanced drug absorption needs to be identified to reach IVIVC for nanocrystals formulations [90]. In another study, Sarnes et al. established that in superior in vitro dissolution of itraconazole nanocrystals over marketed amorphous solid dispersion, Sporanox was not replicated in vivo. The possible mechanisms reported for this anomalous behavior were faster gastric transit of drug nanocrystals from stomach to intestine, drug precipitation with increase in pH, and absence of precipitation inhibiting polymer [91]. This implies that in vitro methods that accurately predict the formulation effect on oral drug absorption are necessary in order to save time and resources during drug development. Kristin et al. demonstrated that instead of stand alone dissolution studies, dissolution/permeation experiments across Caco-2 cells were useful to predict in vivo performance of clarithromycin nanocrystals. A strong correlation of $AUC_{0-\infty}$ was obtained with the flux across Caco-2 cells, instead of excised rat intestinal sheets [92]. On similar lines, Warnken et al. have also established that membrane-permeation dissolution resembles the dissolution and absorption processes that occur in the gastrointestinal tract following oral administration [93].

Correlations may be influenced by methodology of in vitro tests as well as the approaches taken to correlate with the in vivo results. Jinno et al. highlighted the importance of selection of appropriate dissolution apparatus; media representing fasted and fed conditions and volume to establish IVIVC. They demonstrated that oral bioavailability of the wet-milled cilostazol in male beagle dogs was 13-fold higher than the hammer-milled commercial tablet in fasted condition. However, the same was not reflected initially in in vitro dissolution studies. To achieve IVIVC, significant efforts were dedicated for selection of appropriate biorelevant dissolution conditions. This indicated that, to achieve IVIVC, detailed dissolution studies in biorelevant media and absorption modeling can help in quantitative prediction of in vivo performance of nanocrystals [94]. However, the area has tremendous scope of growth and is currently receiving lot of focus in the field.

Several authors have reported use of "physiologically based absorption modeling (PBAM)" to support justification of the biowaiver. Authors have combined biorelevant solubility, intestinal permeability, and dissolution profiles to quantitatively explain the in vivo findings. Willmann et al. reported a combined PBPK model for gastrointestinal transit, absorption, and dissolution based on Noyes-Whitney for spherical particles. The model correctly predicted the influence of particle size on rate and extent of cilostazol absorption in beagle dogs under fasted and fed conditions [95]. Similarly, Shono et al. have used a model based on STELLA software to predict in vivo performance of micronized and nanonized aprepitant in fasted and fed states. The authors concluded that permeability was the rate-limiting step in oral absorption of nanonized aprepitant [96]. Juenemann et al. also demonstrated that biorelevant dissolution testing coupled with STELLA software enabled in vitro-in silico-in vivo correlation of both microsized and nanosized fenofibrate in fasted and fed state. The model indicated that in fasted state, absorption of fenofibrate from nanosized formulation is partly permeability-limited, while for the microsized formulation the dissolution of fenofibrate was the rate-determining step for drug absorption. While in the fed state, gastric emptying was the rate-determining step for absorption of both micro- and nanosized formulations of fenofibrate [97].

It can be concluded that selection of optimal dissolution conditions, mechanism of drug absorption, and limiting factors for differently sized formulations need to be identified to achieve IVIVC for nanonized formulation. PBAM is an important tool; however, a thorough understanding and expertise is needed to reach to scientifically sound conclusions. This can ultimately guide for selection of appropriate drug particle size for development of nanocrystals-based formulations.

6. Applications of nanocrystals in other routes of administration

6.1 Parenteral drug delivery using nanocrystals

Poor aqueous solubility remains a major obstacle for development of pharmaceutical formulations for clinical applications [52]. Besides, poorly water-soluble drugs need to be administered intravenously during preclinical phase for assessment of early pharmacokinetic (PK) profile. Several strategies are explored for the intravenous (IV) administration of poorly water-soluble drugs and most commonly include salt formation and use of cosolvents. As compared to above approaches, nanocrystals-based IV formulation offers high drug loading with minimal amount of stabilizers [98].

6.1.1 Formulation aspects

The IV nanosuspension can be generated by top–down approach, bottom–up approach, or combinations of both and must be sterile, isotonic, and safe for use. They can contain excipients like stabilizers, buffering agent, tonicity modifiers, antioxidants, and preservatives. Stabilizers are added to reduce interfacial free energy and stabilize the system. However, only a few stabilizers like poloxamer 188, poloxamer 407, Tween 20, Tween 80, lecithin, Solutol HS 15, vitamin E TPGS, proline, arginine, and sodium dodecyl sulfate (SDS) can be used in IV nanosuspensions. Tonicity modifiers are used to maintain isotonicity of formulation to reduce pain of injection. Sugars like mannitol, dextrose, and sucrose can be used as a tonicity modifier and salts should be avoided as they may change electric charge and affect the stability of nanosuspension. The major challenge of parenteral formulation is aggregation or agglomeration and crystals growth due to Ostwald ripening upon storage [99, 100].

6.1.2 Pharmacokinetic aspects

After IV administration of nanosuspension, poorly soluble drug encounters sink environment in the circulatory system. If the nanosize particles immediately dissolve in blood during circulation, the nanosuspension may exhibit PK and tissue distribution similar to a solution formulation. For example, the nanosuspension formulation of flurbiprofen showed similar PK and tissue distribution as solution formulation [101]. Additionally, the formulation was less irritant during administration. If the nanosized particles dissolve slowly after administration, they can be recognized as foreign materials by the immune system resulting in opsonization. Nanocrystals can be taken up by monocyte phagocytic system (MPS) and distributed to reticuloendothelial system (RES) organs (i.e., spleen, liver). Depot of drug particles in macrophages promotes sustained release and provide a longer duration of action. Besides, formation of phagolysosome takes place around drug particles in macrophage, where reduced pH of phagolysosome may allow dissolution of drug. The lipophilic drug molecule could travel through phagolysosomal membrane to cytoplasm and subsequently exit the cell by diffusing down the concentration gradient. Thus, depot effect results in modified PK profile with decreased maximum plasma concentration (C_{max}), increased area under the curve (AUC), and half-life ($t_{1/2}$) [20].

Few drug molecules show modified PK profile and higher tissue distribution through opsonization after IV administration. Itraconazole nanosuspension allowed higher tolerance dose and therapeutic efficacy than Sporanox in *Candida albicans*-induced immunosuppressed rat models. Besides kidney, higher accumulation of drug was found in lung and brain which are a potential site of opportunistic fungal infection [102]. The nanosuspension of itraconazole and Sporanox showed dose-dependent PK behavior, where nanosuspension sustained for a longer duration than Sporanox as indicated by increased $t_{1/2}$ and mean residence time (MRT). The nanosuspension exhibited maximum tolerable dose (MTD) of 80 mg/kg due to modified PK, resulting in the drop of C_{max} and higher AUC [103].

In case of enhanced permeation and retention effect (EPR), the protection of nanocrystals surface from recognition by RES and opsonin adsorption is required for prolonged circulation in the blood and subsequent accumulation in tumor tissue [20]. The surface of nanocrystals (<400 nm) can be modified with hydrophilic excipient (i.e., polyethylene glycol (PEG) with a molecular weight of 2000), which can extravasate from the blood through leaky endothelium and reach to solid tumors through EPR effect that facilitates the accumulation of the drug in tumor tissues [104]. This is particularly useful for the anticancer drugs for targeting to a specific site. The other factors such as physicochemical properties of drug, surface properties of nanocrystals, dose, infusion time, and the interaction between drug-plasma protein can influence PK profile and bio-distribution of drug nanocrystals [105].

The EPR effect has been shown by nanocrystals formulation of various anticancer drugs such as paclitaxel, camptothecin, deacetyl-mycoepoxydiene, oridonin, curcumin, and asulacrine [106]. The nanosuspension of paclitaxel showed higher efficacy for the reduction in median tumor burden in MV-522 human lung xenograft murine tumor model as compared to Taxol. Besides, paclitaxel nanosuspension was well-tolerated at 90 mg/kg dose in comparison to marketed product (30 mg/kg dose), which showed a 22% death rate. Thus, nanoformulation was able to significantly improve therapeutic efficacy and MTD up to 3-fold [107]. Angiotech Pharmaceutical Inc. has developed paclitaxel nanocrystals-based formulation (Paxeed) for IV administration free of Cremohor EL to reduce the incidence of hypersensitivity reactions [108]. Janseen Pharmaceutical Ltd. developed IV infusion of itraconazole nanocrystals (Sporanox IV) for the treatment of systemic fungal infection such as histoplasmosis [109].

Chen et al. prepared nanocrystals (NCs) of SN-38 (7-Ethyl-10-hydroxycamptothecin) like SN-38/NCs-A and SN-38/NCs-B with a mean particle size of 230 nm and 800 nm, respectively, for their antitumor effect. SN-38/NCs-A exhibited higher bioavailability and a significant reduction in tumor growth than SN-38/NCs-B. Moreover, tissue distribution study showed higher accumulation of SN-38/NCs-A in tumor tissues due to the EPR effect [110]. The nanosuspension of SN 30191 (investigational anticancer compound) showed a MTD of 10 mg/kg over solution of 2.5 mg/kg during dose-toxicity study in mice. Further, the compound showed rapid clearance from the blood and accumulation in liver, kidney, and heart as demonstrated by PK and tissue distribution studies [111].

Lipid-based nanosuspensions (LNS) were designed for drug which is insoluble in water and oil, improve drug loading, and increase therapeutic efficacy and reduction in toxicity. Wang et al. designed docetaxel (DTX)-LNS using soy lecithin by high-pressure homogenization method. The PK study of DTX-LNS exhibited rapid distribution phase, lower C_{max}, and higher $AUC_{0-\infty}$ and MRT by 1–31- and 1.92-fold, respectively, than Duopafei. The tissue distribution study of DTX-LNS showed higher drug level in liver, lung, and spleen than Duopafei due to uptake by MPS and distribution to RES organs. Overall, DTX-LNS showed higher therapeutic efficacy with reduced toxicity [112].

An antibacterial agent such as clofazimine nanosuspension showed higher concentration of drug in liver, spleen, and lung (more than MIC value for *Mycobacterium avium* strain) in mice and was as effective as control liposomal formulation in reducing microbial loads in above organs of *M. avium* strain-infected mice. This finding supported that nanocrystals effectively targeted the drug to principal host cells of mycobacterial infection such as tissue macrophage via RES sequestration [113].

The thermosensitive hydrogel can serve as a promising delivery system for in vivo sustained-release depot. Lin et al. prepared a thermosensitive hydrogel containing paclitaxel nanocrystals (PTX-NCs-Gel) using poloxamer 407 for sustained drug release, extended local retention, better efficacy, and lower toxicity. PTX-NCs-Gel improved localized retention in murine tumor up to 20 days and showed highest reduction in tumor volume as compared to PTX-NCs and Taxol [114].

6.1.3 Intramuscular injection

Janseen Pharmaceutical Ltd. developed paliperidone palmitate nanocrystals by wet media milling method and extended-release formulation (Invega Sustenna, ready-to-use pre-filled syringes) for intramuscular, once-monthly injection, for schizophrenia [115]. Alkermes prepared aripiprazole lauroxil nanosuspension by precipitation followed by media milling method and developed extended-release formulation (Aristada) for IM injection for the treatment of schizophrenia [116]. Otsuka Pharmaceutical Co., Ltd. developed aripiprazole nanosuspension-based extended-release formulation (Abilify Maintena) for once-monthly IM injection for the treatment of schizophrenia [117]. Eagle Pharmaceutical Ltd. developed dantrolene sodium IM infusion (Ryanodex) for the treatment of malignant hyperthermia [116]. Recently, Janseen Pharmaceutical Ltd. developed novel long-acting dual-drug injectable formulation comprising rilpivirine and cabotegravir nanosuspension and evaluated in phase 3 clinical trials. The formulation showed similar efficacy in maintaining viral suppression as compared to standard oral regime of three drugs (abacavir, dolutegravir, and lamivudine) and Triumeq by injection every 4 weeks up to 3-month study. Thus, dual-drug formulation provides option for management of HIV virus using 12 injections in a year. Besides, rilpivirine and cabotegravir are investigational nanosuspensions for IM injection for antiviral treatment, which are not yet approved by regulatory authorities [118].

6.2 Topical drug delivery using nanocrystals

Topical delivery of nanocrystals enhances saturation solubility and dissolution velocity of drug, which increases concentration gradient between the formulation and skin and subsequently leads to increased passive diffusion [119]. Nanocrystals because of their higher surface-free energy increase adhesiveness to the skin and adhered even after evaporation of the water phase of the formulation. Nanocrystals also facilitate the transport of drug through the hair follicle route, which can act as a depot for sustained release of the drug [120]. The nanocrystals for topical delivery were first investigated for cosmetic and subsequently for pharmaceuticals. Cosmetic products like Juvedical age-decoder cream of rutin by Juneva and Cellular serum platinum rare of hesperidin by La Prairie appeared in the market in 2007 [121].

6.2.1 Formulation aspects

The nanocrystals for topical application can be generated using methods similar to parenteral products. The nanosuspension generated using top-down approach can be directly incorporated into semisolid dosage forms. Nanosuspension generated using bottom-up or combination approaches is initially spray- or freeze-dried and subsequently incorporated into semisolid dosage forms. Semisolid dosage forms like gel, cream, lotion, and ointments are useful for topical delivery of nanocrystals. The nanocrystals-based topical formulations contain stabilizers, gelling agents, emulsifying agents, preservatives, antioxidants, and humectants. Stabilizers are added to prevent aggregation of nanosized particles by increasing the activation energy of the system. Nonionic stabilizers preferred to avoid skin irritation. Currently, stabilizers such as poloxamer grades, Plantacare grades, polysorbate grades, Inutec SP1, soy lecithin, vitamin E TPGS, hydroxypropyl methylcellulose (HPMC), polyvinylpyrrolidone (PVP), Carbopol 981, and SDS have been explored for topical nanocrystals [122].

6.2.2 Dermatokinetics

The application of topical nanocrystals-based formulations has been explored for antioxidants, antifungal, CNS-stimulant, antibacterial, antibiotic, and corticosteroid agents. Nanocrystals of antioxidants such as hesperetin, apigenin, lutein, rutin, and quercetin have been explored for increased antioxidant activity and improved topical delivery. Hesperetin nanocrystals generated using Plantacare 2000 and Inutec SP1 were stable during short-term stability [123]. The antioxidant capacity of apigenin nanocrystals was almost doubled as compared to a coarse powder [124]. The lutein nanocrystals showed 14-fold higher permeation through cellulose nitrate membrane as compared to coarse powder. However, pig ear skin did not allow lutein to permeate which resulted in localization of the lutein in the skin [125]. Nanocrystal-based gel (nanogel) of rutin had the highest antioxidant activity (90%) in comparison to commercial product and rutin drug powder. Quercetin-TPGS nanocrystals showed higher antioxidant activity as compared to quercetin powder and did not show cellular toxicity at higher concentration in an epithelial cell line (VERO cells) [126].

Nanogel of an antifungal agent such as clotrimazole was prepared to evaluate antifungal activity against *Candida albicans*. Nanogel provided 2.3-times higher accumulation in skin strata as compared to the marketed formulation and helped in inhibition of maximum zone by 2.9-fold [127]. Anti-inflammatory activity of diclofenac acid was evaluated in TPA-induced inflammatory mice model, where nanosuspension exhibited a higher reduction in myeloperoxidase activity as compared to coarse suspension and Voltaren Emulgel. Besides, dermatokinetic (DK) study of nanosuspension improved skin penetration by 5.3-folds as compared to Voltaren Emulgel [128].

Nanosuspension of an anti-acne agent such as tretinoin was prepared for the treatment of acne vulgaris. The nanosuspension favorably accumulates drug into different layers of the skin during DK study on pig ear skin. Besides, photodegradation studies revealed that nanosuspension improved photostability of a drug in comparison to methanol solution. Thus, nanosuspensions were able to significantly improve topical delivery and photostability of tretinoin [129]. The antibacterial activity of silver sulfadiazine nanocrystal-based hydrogel was improved against *Staphylococcus aureus*, *Escherichia coli,* and *Pseudomonas aeruginosa* than the positive control. The nanocrystal-based hydrogel showed a significant reduction of interleukin-6 and increment of vascular endothelial growth factor-A, transforming growth factor-β as compared to drug-based hydrogel in the wound and burn model, which indicates better healing and anti-inflammatory action [130].

Shen et al. prepared the ganoderma lucidum (GLT) nanogel to improve therapeutic efficacy in frostbite. The GLT nanogel showed 2.88-fold and 5.12-fold higher skin permeation and skin penetration, respectively, over GLT gel during DK study on rat skin. Besides, GLT nanogel showed higher percentage survival area of rat skin than GLT gel in the rat model for frostbite. Thus, nanogel was more effective and improved pathological changes in rat skin with frostbite [131].

Apart from nanocrystals of poorly soluble drugs for dermal application, nanocrystals of a sparingly soluble drug such as caffeine resulted in improved topical delivery. The nanocrystals were generated using low energy process (pearl milling) with low dielectric constant dispersion medium (ethanol-PG mixture) and Carbopol 981 as a stabilizer to avoid Ostwald ripening [121].

Nanocrystals-based topical formulations are useful for improved topical delivery of poorly as well as sparingly water-soluble drugs. At present, there is no commercial pharmaceutical product based on nanocrystals. Looking at the promising results of nanocrystals for topical delivery, it may activate pharmaceutical industry to bring nanocrystals-based topical formulation to the market in the near future.

6.3 Ophthalmic drug delivery using nanocrystals

The complex structure of eye, blood-retinal barrier, lacrimation, and nasolachyrymal drainage can limit the ocular bioavailability to less than 5%. Low ocular bioavailability is contributed by small volume of lacrimal fluids for drug dissolution, poor residence time in eye, and spilling of intilled dose by high turnover of tears and lacrimal fluids [132]. Recently, several researchers have explored the nanocrystals-based approach for poorly soluble drugs. The advantages of nanocrystals for ophthalmic delivery include improved ocular safety, prolonged retention of formulation in cul-de-sac, enhanced corneal penetration, enhanced ocular bioavailability, dual-drug release behavior in eye, and improved tolerability [133]. The increased saturation solubility and enhanced dissolution rate of nanocrystals lead to higher concentration gradient for absorption. Nanocrystals can be fabricated to exhibit immediate and sustained drug release profiles (dual-drug release) after instillation to eye.

6.3.1 Formulation aspects

The nanocrystals for ophthalmic application can be generated using previously reported methods. These drug nanocrystals can be formulated as eye drops. Small particle size of nanocrystals improves tolerability of ophthalmic formulation and patient compliance. Stabilizers are used during generation of nanocrystals to decrease the free energy of the system and to provide stabilization to the system. Nonionic stabilizers are more preferable in the formulation due to less irritating potential to eye. Poloxamer 188, poloxmer

407, polysorbate 80, vitamin E TPGS, HPMC, PVP, methyl cellulose, PVA, cetylpyridinium chloride (CPC), and BKC have been explored for ocular nanosuspensions. Additional parenterally excipients may be used such as adjusting tonicity and pH [133–135].

6.3.2 Ocular bioavailability

Tuomela et al. evaluated three nanocrystal formulations of brinzolamide in rat ocular hypertension model to evaluate their potential to decrease intraocular pressure (IOP) in glaucoma. Formulations I, II formulated at pH 7.2 showed similar IOP reduction as commercial product (Azopt). On the other hand, formulation III formulated at pH 4.5 exhibited highest IOP reduction due to acidic pH of the formulation. Later, increased solubility of drug was compared to neutral pH of formulation. Overall, particle size in the nanometer range and mucoadhesive property of HPMC (stabilizer) helped in efficient reduction of IOP by formulations (I-III) [136].

The sustained release of drug can be obtained by increasing viscosity of nanosuspension or inclusion into in situ gel system. Gupta et al. developed in situ gel forming forkolin nanosuspension (nanogel) using Noveon AA-1 polycarbophil/poloxamer 407 solution for the treatment of glaucoma. This aqueous composition undergoes gelation upon application to eye due to changes in pH and temperature, which provides sustained release and bioadhesive property. The nonogel improved precorneal residence time of drug up to 6 h over drug solution (0.5 h). The nanogel and conventional eye suspension decreased IOP by 31% and 18%, respectively, for 12 h and 4–6 h in dexamethasone-induced glaucomatous rabbits. Thus, nanogel improved therapeutic efficacy of the drug [137].

Ali et al. generated hydrocortisone nanosuspension (HCNS) for inflammatory disorders of the eye. HCNS due to their reduced particle size acts as a reservoir and immediately replenishes the absorbed drug and hence increased duration of action up to 8–9 h over drug solution (4–5 h). Besides, HCNS increased AUC value by twofold than drug solution indicating higher bioavailability. Thus, longer retention in ocular site and adhesion to corneal membrane may contribute to higher bioavailability and sustained release property [138]. Baba et al. prepared nanocrystals of fluorometholone using spray-drying for inflammatory disorders of the eye and applied nanoformulation to rabbit eyes for an evaluation of tissue permeability (intraocular migration) of drug. The nanoformulation increased tissue permeability by 7–8 times to cornea as compared to commercial eye drops [139].

The nanocrystals improved ocular safety due to less amount of surfactants present in the formulation as compared to conventional formulations. Kim et al. prepared cyclosporine-A-loaded nanosuspension (CsANS) with PVA as a stabilizer and compared it with commercial product (Restasis, ophthalmic emulsion) to evaluate irritation. The CsANS exhibited no change in tear flow rate and less irritation as compared to Restasis in Schirmer test. Thus, absence of solubilized drug, oil, and surfactant in nanosuspension caused less ocular irritation as compared to Restasis [140].

Pignatello et al. prepared ibuprofen sodium-loaded polymeric nanosuspension (IPNS) using Eudragit RS 100 for controlled delivery of drug. The IPNS showed higher and comparable mydriatic response to atropine and drug solution, respectively, on rabbit eye with paracentesis indicating valid therapeutic approach for the maintenance of mydriasis. The IPNS showed higher level of drug in aqueous humor than drug solution, indicating increased corneal residence time of IPNS [141]. Besides, IPNS significantly reduced polymorphonuclear leukocytes and protein level in aqueous humor and ocular inflammation as compared to drug solution [142]. Schopf et al. prepared novel mucus penetrating nanosuspension of loteprednol etabonate (LE-MPP, 0.4%) to improve ophthalmic delivery into tissues underlying the mucous barrier. The LE-MPP 0.4% increased C_{max} and ocular bioavailability (AUC) by approximately three- and twofold, respectively, in aqueous humor and ocular tissues than Lotemax 0.5% indicating higher absorption [143].

The cationic excipients present in the ophthalmic formulation can contribute to increased residence time of drug by interaction with negatively charged mucin layer. Romero et al. developed cationic nanocrystals formulation containing dexamethasone acetate nanocrystals using polymyxin B as positive charge drug and CPC and BKC as cationic stabilizers. The mucoadhesion of formulation showed reduction in zeta potential of cationic nanocrystals with increment of mucin (0.025–1 mg/mL) and subsequently reversed to negative value due to nanocrystals' surface covered by negatively charged mucin at higher concentration (1 mg/mL). Thus, cationic excipients are useful to increase residence time of drug [144].

Nagai et al. prepared ophthalmic formulation containing tranilast nanocrystals (TLNS) to evaluate corneal permeability, toxicity, and stability. The TLNS formulation showed better tolerability in human corneal epithelium cells over commercial formulation (Rizaben). The TLNS formulation showed 2.4-fold higher transcorneal penetration rate as compared to Rizaben during transcorneal penetration experiment. Besides, photodegradation studies of TLNS formulation showed higher photostabilty of drug as compared to solution of drug in dimethyl sulfoxide [145].

Thus, present research on nanocrystals showed promising drug delivery approach for improved ophthalmic delivery. Successful adoption of nanocrystals in the field of oral, parenteral, and topical may attract industries to bring more nanocrystals–based ophthalmic product in the near future. The performance benefits incurred by nanocrystals through various routes of administration are compiled in Fig. 4.

7. Functionalization of drug nanocrystals

Functionalization of the nanocrystals involves addition of a chemical functional group or biomolecule on their surface that enhances properties of nanocrystals for a specific application [146]. Various methods have been reported for the functionalization of nanoparticles which include suspension modification, surface modification, physical treatment

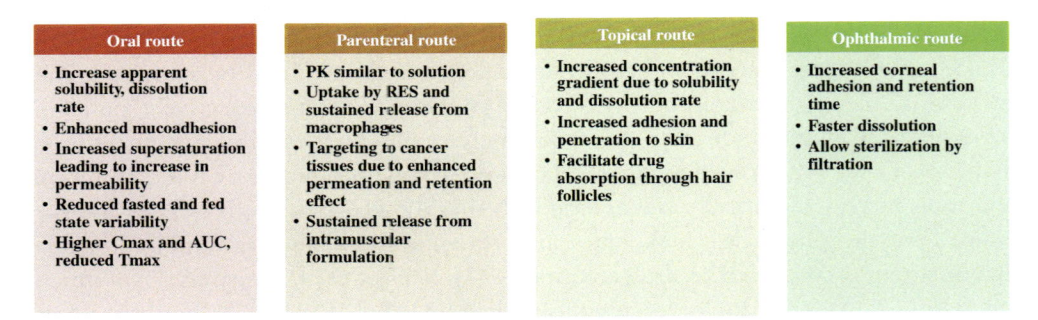

Fig. 4 Benefits of drug nanocrystals in different routes of administration.

such as molecular adsorption, hydrophobic interaction, ionic interaction, chemical modification such as surface oxidation, PEG chemistry, silane chemistry, and biological modification [147]. Various excipients such as chitosan, lecithin, dextran, albumin, cyclodextrin, hydrophobin, folic acid, PEG, stabilizers (poloxamer, SDS), oligonucleotides, peptides, and antibodies can be utilized for the functionalization [146]. The functionalization of nanocrystals is useful for targeting the drug to specific organ or tissue, increase therapeutic efficacy, and decrease toxicity of drug.

7.1 Targeting using functionalized nanocrystals

The functionalized nanocrystals can enhance specific targeting to brain, liver, and spleen which serve as an active and latent cellular reservoir during HIV infection, thus offering promising therapy of antiretroviral drugs for HIV/AIDS treatment. Shegokar et al. surface-modified nevirapine nanosuspension with serum albumin, PEG 1000, and dextran 60 to target active viral reservoirs site. Serum albumin and dextran-modified nanosuspension increased drug concentration after macrophage uptake in comparison to nanosuspension during cellular uptake study. On the other hand, PEGylated nanosuspension showed decreased cellular uptake into macrophages as compared to nanosuspension by avoiding opsonization and stealth property of nanoparticles. During tissue distribution study, serum albumin-modified nanosuspension showed a higher level of nevirapine with prolonged MRT and less clearance in the brain, which can be useful for the treatment of AIDS dementia. Besides, serum albumin-modified nanosuspension also showed the highest accumulation of the drug in spleen, lung, liver, and thymus as compared to dextran and PEG-modified nanosuspensions. Overall, surface-modified nanosuspensions affect the intracellular distribution of the drug, enhance therapeutic efficacy, and provide targeted drug delivery for the treatment of HIV [148].

It is known that the folate molecules have high affinity for the folate receptors (FR) which are overexpressed in human cancers. Thus, functionalized nanocrystals of an anticancer drug with folate can serve as a potential drug delivery system for selective tumor targeting to increase efficacy and decrease toxicity. Talekar et al. prepared folate-targeted nanosuspension (NS-FA) and nontargeted nanosuspension (NS) of PIK-75 to improve therapeutic efficacy in human ovarian cancer model. The drug uptake study of NS-FA showed 1.4-fold higher intracellular concentration of PIK-75 than NS after 6 h incubation possibly due to better internalization of nanocrystals via phagocytosis or endocytosis and adhesion to cells. This enhanced cytotoxicity of PIK-75 was attributed to increase in apoptosis. The PK study of NS-FA increased 3–4 fold C_{max} and AUC_{0-4} than NS. The NS and NS-FA achieved 5-fold and 10-fold higher tumor accumulation, respectively, as compared to PIK-75 suspension due to higher downregulation of pAkt Ser473 and pAkt Thr308 (serine/threonine-specific protein kinase). Overall, NS and NS-FA improved therapeutic efficacy of PIK-75 [149]. Liu et al. prepared nanocrystals and targeted nanocrystals of paclitaxel using poloxamer 407 (F-127) and folate-conjugated-F127, respectively. Nanocrystals with 10% folate-conjugated-F127 showed more cytotoxicity as compared to nanocrystals in FR-positive KB cells. Besides, nanocrystals with 5% folate-conjugated-F127 showed lower cytotoxicity than nanocrystals with 10% folate-conjugated-F127. This differential cytotoxicity effect was abolished in the presence of free folate as a competitor, indicating that drug uptake was associated with folate and receptor-mediated endocytosis [150].

The active molecule can be selectively targeted to specific site via its functionalization with constituent of targeted site. For example, surface functionalization of active molecule with carbonated hydroxyapatite (mineral phase of bone) can achieve controlled release of drug in biological environment of bone diseases. Palazzo et al. synthesized hydroxyapatite (HA) nanocrystals of plate shape (HAps) or needle shape (HAns) and investigated adsorption and desorption of cisplatin, alendronate, and di(ethylenediamineplatinum)medronate (DPM). Negatively charged alendronate and positively charged cisplatin showed higher adsorption as compared to neutral DPM towards negatively charged HA nanocrystals. Besides, the amount of drug released from nanocrystals was higher for DMP than cisplatin and alendronate, due to complete breakage of the platinum-medronate bond in DPM and strong electrostatic interaction in case of cisplatin and alendronate during desorption kinetics. Thus, HA nanocrystals can be modulated to produce HA-drug conjugates tailored for specific therapeutic applications like osteosarcoma with high efficacy and low toxicity [151].

The surface modification of nanocrystals with PEG can lead to longer blood circulation by reducing the tendency of protein adsorption and subsequent opsonization. This longer circulation time permits leaking of nanocrystals out of vasculature into tumor site through EPR effect. Wang et al. developed PEG-modified LNS of docetaxel (pLNS) and folate-modified LNS of docetaxel (fLNS). The fLNS showed higher cytotoxicity by 1.62-fold over pLNS in hepatocellular liver carcinoma B16 (FR +) and both showed similar cytotoxicity in HepG2 (FR-) cells due to fLNS had higher uptake in B16 cells. During the antitumor activity, fLNS exhibited higher tumor reduction rate of 92% as compared to pLNS (88%) and Duopafei (78%) by higher accumulation of drug in tumor due to folate modification. Besides, pLNS and fLNS significantly reduced C_{max} and improved $t_{1/2}$, MRT, and $AUC_{0-\infty}$ over Duopafei which prolonged duration of action [152]. Xiao et al. evaluated PEGylated azide-functionalized liposome containing ciprofloxacin nanocrystals to improve drug loading and achieve targeted drug delivery. This combination offers a promising delivery system for tumor-targeted drug delivery [153].

Overall, parenteral drug delivery using nanocrystals improves therapeutic efficacy and targeting of the drug, thus establishing potential of nanocrystals in parenteral drug delivery. The functionalization of nanocrystals is also useful for other applications such as controlled release of drug, improvement of oral bioavailability, enhanced uptake to specific site, and physical stability of formulation.

7.2 Functionalization for other applications

The release of drug can be controlled by functionalization of nanocrystals using suitable excipient. Castro et al. surface-modified the cellulose nanocrystals (CNC) with β-cyclodextrin (β-CD) using bridging agents such as succinic acid (SA) and fumaric acid (FA) to develop material which released antibacterial molecule (carvacrol) for a prolonged time. CNC-FA-β-CD increased reactivity and favored esterification as compared to CNC-SA-β-CD, and hence, was selected to evaluate the quantitative bacterial reduction in a dynamic medium. CNC-FA-β-CD showed antibacterial properties up to 48 h due to controlled release of carvacol as compared to CNC and CNC-FA until 36 h. This indicates that CNC-FA-β-CD maintained the necessary concentration of carvacrol for inhibition of microorganism and was more effective than CNC and CNC-FA [154].

The surfactants like SDS and poloxamer 188 (P188) are useful for the functionalization of nanosuspension and surface-modified nanosuspension can improve oral bioavailability and uptake of drug to specific diseased site. Shubar et al. prepared SDS-coated atovaquone nanosuspension (SDS-ANS) and P188-coated atovaquone nanosuspension (P188-ANS) to improve oral bioavailability and uptake into the brain. Besides, SDS-ANS at 50 mg/kg dose showed similar efficacy to Wellvone (100 mg/kg dose) and also enhanced passage through blood-brain barriers after oral administration. The possible explanation for improved therapeutic effect and brain uptake due to SDS may be opening of tight junction, enhanced uptake by paracellular pathway, receptor-mediated transport, or unknown receptor-binding process [155].

The functionalization of nanocrystals can improve the stability of naïve nanocrystals. Surface cationization of cellulose nanocrystals with epoxypropyltrimethylammonium chloride yielded cellulose nanocrystals with cationic hydroxypropyltrimethylammonium chloride (HPTMAC-CNC). Surface charge density of HPTMAC-CNC was decreased as compared to nonfunctionalized CNC. This indicated that functionalization introduced lesser cationic substituents than the original number of sulfate groups. Zeta potential also showed charge reversal from -39 to $+30$ mV after treatment. The aqueous suspension of HPTMAC-CNC is electrostatically stabilized by cationic trimethylammonium chloride groups. HPTMAC-CNC gelled in water before forming a liquid crystalline phase due to lower electrostatic repulsion and presence of quaternary ammonium substituent and counter-ion. Thus, functionalization led to stable aqueous suspensions of HPTMAC-CNC with unexpected gelling and shear thinning behavior [156].

8. Products in market

Nanocrystals were invented in the beginning of 1990s and products started to appear in market from 1998 onwards. In the beginning, all the marketed products were intended for oral administration and formulated as solid dosage forms (tablets, capsules) with exception of Megace that was formulated as nanosuspension.

Since then, nanocrystals have progressed as a formulation of choice for rescuing poorly water-soluble drugs. The products available in market establish the versatility and wide acceptance of nanocrystals for drug delivery through various routes. Tables 1 enlists the nanocrystals-based products available in market [51, 108, 157] (Table 2).

9. Conclusions and future prospects

Drugs with poor aqueous solubility were initially considered as risky candidates for drug development. The advent of nanocrystals created opportunity for novel technologies to emerge for their fabrication. Nanocrystals have progressed as a viable choice for improving dissolution rate of poorly water-soluble drugs. These are considered safe as they are dissolved and eliminated from body. In addition, nanocrystals have shown excellent in vivo performance for different administration routes. Significant research in the field of nanocrystals has already been carried out. However, there are aspects of size-dependent drug absorption that needs to be investigated individually for drugs. Successful IVIVC can be reached once the correct dissolution conditions and absorption mechanism are understood. The future research in the field of drug nanocrystals shall be directed towards more rationale selection of excipients, functionalization using targeting ligands, and mechanism-based performance evaluation. In future, more nanocrystal-based commercial products are expected for non-oral routes. In an attempt to hasten drug product development and reducing the developmental time, newer technologies that minimize downstream processing would be preferred.

Table 2 Compilation of nanocrystals-based products available in market.

Product/Company	Drug compound	Nano-sizing approach	Route	FDA approval
Gris-Peg/Novartis	Griseofulvin	Coprecipitation	Oral	1982
Verelan PM/Schwarz Pharmaceuticals	Verapamil	Media milling	Oral	1998
Azopt/Alcon	Brinzolamide	Media milling	Ocular	1998
Rapamune/Wyeth	Sirolimus	Media milling	Oral	2000
Focalin XR/Novartis	Dexmethyl phenidate HCl	Media milling	Oral	2001
Herbesser/Mitsubishi Tanabe Pharmaceuticals	Diltiazem	Media milling	Oral	2002
Avinza/King Pharmaceuticals	Morphine Sulphate	Media milling	Oral	2002
Ritalin LA/Novartis	Methyl phenidate HCl	Media milling	Oral	2002
Zanaflex/Acorda	Tizanidine	Media milling	Oral	2002
Emend/Merck	Aprepitant	Media milling	Oral	2003
Tricor/Abbott	Fenofibrate	Media milling	Oral	2004
Megace ES/Par Pharma	Megestrol acetate	Media milling	Oral	2005
TridlideTM/Skye Pharma	Fenofibrate	High-pressure homogenization	Oral	2005
Naprelan/Wyeth	Naproxen sodium	Media milling	Oral	2006
Theodur/Mitsubishi Tanabe Pharmaceuticals	Theophylline	Media milling	Oral	2008
Invega Sustenna/Johnson & Johnson	Paliperidone palmitate	High-pressure homogenization	Injection	2009
Ilevro/Novartis	Nepafenac	Media milling	Ocular	2012

References

[1] B.J. Boyd, et al., Successful oral delivery of poorly water-soluble drugs both depends on the intraluminal behavior of drugs and of appropriate advanced drug delivery systems, Eur. J. Pharm. Sci. 137 (2019) 104967.

[2] A. Narang, R.-K. Chang, M.A. Hussain, Pharmaceutical development and regulatory considerations for nanoparticles and nanoparticulate drug delivery systems, J. Pharm. Sci. 102 (11) (2013) 3867–3882.

[3] J.P. Möschwitzer, Drug nanocrystals in the commercial pharmaceutical development process, Int. J. Pharm. 453 (1) (2013) 142–156.

[4] US FDA, Drug Products, Including Biological Products, That Contain Nanomaterials—Guidance for Industry, US Department of Health and Human Services, Silver Spring, MD, 2017.

[5] J. Liu, et al., Mechanisms for oral absorption enhancement of drugs by nanocrystals, J. Drug Delivery Sci. Technol. 56 (2020) 101607.

[6] R. Mauludin, R.H. Müller, C.M. Keck, Kinetic solubility and dissolution velocity of rutin nanocrystals, Eur. J. Pharm. Sci. 36 (4–5) (2009) 502–510.

[7] M. Kakran, et al., Fabrication of quercetin nanocrystals: comparison of different methods, Eur. J. Pharm. Biopharm. 80 (1) (2012) 113–121.

[8] M. Guo, et al., Impacts of particle shapes on the oral delivery of drug nanocrystals: mucus permeation, transepithelial transport and bioavailability, J. Control. Release 307 (2019) 64–75.

[9] R.H. Müller, S. Gohla, C.M. Keck, State of the art of nanocrystals–special features, production, nano-toxicology aspects and intracellular delivery, Eur. J. Pharm. Biopharm. 78 (1) (2011) 1–9.

[10] L. Peltonen, C. Strachan, Understanding critical quality attributes for nanocrystals from preparation to delivery, Molecules 20 (12) (2015) 22286–22300.

[11] W. Li, et al., Influence of drug physicochemical properties on absorption of water insoluble drug nanosuspensions, Int. J. Pharm. 460 (1–2) (2014) 13–23.

[12] Y. Wang, et al., Stability of nanosuspensions in drug delivery, J. Control. Release 172 (3) (2013) 1126–1141.

[13] L. Wu, J. Zhang, W. Watanabe, Physical and chemical stability of drug nanoparticles, Adv. Drug Deliv. Rev. 63 (6) (2011) 456–469.

[14] L. Peltonen, J. Hirvonen, Pharmaceutical nanocrystals by nanomilling: critical process parameters, particle fracturing and stabilization methods, J. Pharm. Pharmacol. 62 (11) (2010) 1569–1579.

[15] M.L. Lestari, R.H. Müller, J.P. Möschwitzer, Systematic screening of different surface modifiers for the production of physically stable nanosuspensions, J. Pharm. Sci. 104 (3) (2015) 1128–1140.

[16] F. Kesisoglou, A. Mitra, Crystalline nanosuspensions as potential toxicology and clinical oral formulations for BCS II/IV compounds, AAPS J. 14 (4) (2012) 677–687.

[17] L. Gao, et al., Drug nanocrystals: in vivo performances, J. Control. Release 160 (3) (2012) 418–430.

[18] R. Shegokar, R.H. Müller, Nanocrystals: industrially feasible multifunctional formulation technology for poorly soluble actives, Int. J. Pharm. 399 (1–2) (2010) 129–139.

[19] L. Peltonen, J. Hirvonen, Drug nanocrystals–versatile option for formulation of poorly soluble materials, Int. J. Pharm. 537 (1–2) (2018) 73–83.

[20] B.E. Rabinow, Nanosuspensions in drug delivery, Nat. Rev. Drug Discov. 3 (9) (2004) 785–796.

[21] F. Fontana, et al., Production of pure drug nanocrystals and nano co-crystals by confinement methods, Adv. Drug Deliv. Rev. 131 (2018) 3–21.

[22] S.V. Jermain, C. Brough, R.O. Williams III, Amorphous solid dispersions and nanocrystal technologies for poorly water-soluble drug delivery–an update, Int. J. Pharm. 535 (1–2) (2018) 379–392.

[23] R. Al-Kassas, M. Bansal, J. Shaw, Nanosizing techniques for improving bioavailability of drugs, J. Control. Release 260 (2017) 202–212.

[24] J. Salazar, R.H. Müller, J.P. Möschwitzer, Combinative particle size reduction technologies for the production of drug nanocrystals, J. Pharm. 2014 (2014) 1–14.

[25] G. Shete, et al., Stabilizers used in nano-crystal based drug delivery systems, J. Excipients Food Chem. 5 (4) (2016) 941.

[26] B. Van Eerdenbrugh, G. Van den Mooter, P. Augustijns, Top-down production of drug nanocrystals: nanosuspension stabilization, miniaturization and transformation into solid products, Int. J. Pharm. 364 (1) (2008) 64–75.

[27] B. Van Eerdenbrugh, et al., Downscaling drug nanosuspension production: processing aspects and physicochemical characterization, AAPS PharmSciTech 10 (1) (2009) 44–53.

[28] M. Grau, O. Kayser, R. Müller, Nanosuspensions of poorly soluble drugs—reproducibility of small scale production, Int. J. Pharm. 196 (2) (2000) 155–159.

[29] G.B. Romero, C.M. Keck, R.H. Müller, Simple low-cost miniaturization approach for pharmaceutical nanocrystals production, Int. J. Pharm. 501 (1–2) (2016) 236–244.

[30] B.K. Ahuja, et al., Formulation, optimization and in vitro–in vivo evaluation of febuxostat nanosuspension, Int. J. Pharm. 478 (2) (2015) 540–552.

[31] S. Pattnaik, et al., Fabrication of aceclofenac nanocrystals for improved dissolution: process optimization and physicochemical characterization, J. Drug Delivery Sci. Technol. 29 (2015) 199–209.

[32] J. Zuo, et al., Design space approach in the development of esculetin nanocrystals by a small-scale wet-bead milling process, J. Drug Delivery Sci. Technol. 55 (2020) 101486.

[33] I. Ghosh, et al., Identification of critical process parameters and its interplay with nanosuspension formulation prepared by top down media milling technology–a QbD perspective, Pharm. Dev. Technol. 18 (3) (2013) 719–729.

[34] D.S. Singare, et al., Optimization of formulation and process variable of nanosuspension: an industrial perspective, Int. J. Pharm. 402 (1–2) (2010) 213–220.

[35] A.M. Cerdeira, M. Mazzotti, B. Gander, Miconazole nanosuspensions: influence of formulation variables on particle size reduction and physical stability, Int. J. Pharm. 396 (1–2) (2010) 210–218.

[36] R. Jog, D.J. Burgess, Comprehensive quality by design approach for stable nanocrystalline drug products, Int. J. Pharm. 564 (2019) 426–460.

[37] K.V. Mahesh, S.K. Singh, M. Gulati, A comparative study of top-down and bottom-up approaches for the preparation of nanosuspensions of glipizide, Powder Technol. 256 (2014) 436–449.

[38] S. Gora, et al., Nanosizing of valsartan by high pressure homogenization to produce dissolution enhanced nanosuspension: pharmacokinetics and pharmacodyanamic study, Drug Deliv. 23 (3) (2016) 930–940.

[39] J. Salazar, et al., Process optimization of a novel production method for nanosuspensions using design of experiments (DoE), Int. J. Pharm. 420 (2) (2011) 395–403.

[40] A. Karakucuk, N. Celebi, Z.S. Teksin, Preparation of ritonavir nanosuspensions by microfluidization using polymeric stabilizers: I. A Design of Experiment approach, Eur. J. Pharm. Sci. 95 (2016) 111–121.

[41] M. Cheng, et al., Fabrication of fine Puerarin Nanocrystals by box–Behnken design to enhance intestinal absorption, AAPS PharmSciTech 21 (3) (2020) 1–12.

[42] A.N. Oktay, et al., Screening of stabilizing agents to optimize flurbiprofen nanosuspensions using experimental design, J. Drug Delivery Sci. Technol. 57 (2020) 101690.

[43] M. Colombo, et al., Influence of drug brittleness, nanomilling time, and freeze-drying on the crystallinity of poorly water-soluble drugs and its implications for solubility enhancement, AAPS PharmSciTech 18 (7) (2017) 2437–2445.

[44] M. Malamatari, et al., Preparation of respirable nanoparticle agglomerates of the low melting and ductile drug ibuprofen: impact of formulation parameters, Powder Technol. 308 (2017) 123–134.

[45] M. George, I. Ghosh, Identifying the correlation between drug/stabilizer properties and critical quality attributes (CQAs) of nanosuspension formulation prepared by wet media milling technology, Eur. J. Pharm. Sci. 48 (1–2) (2013) 142–152.

[46] F. Lai, et al., Nanocrystals as tool to improve piroxicam dissolution rate in novel orally disintegrating tablets, Eur. J. Pharm. Biopharm. 79 (3) (2011) 552–558.

[47] F. Lai, et al., Diclofenac nanosuspensions: influence of preparation procedure and crystal form on drug dissolution behaviour, Int. J. Pharm. 373 (1–2) (2009) 124–132.

[48] Z. Deng, S. Xu, S. Li, Understanding a relaxation behavior in a nanoparticle suspension for drug delivery applications, Int. J. Pharm. 351 (1–2) (2008) 236–243.

[49] P. Sharma, W.A. Denny, S. Garg, Effect of wet milling process on the solid state of indomethacin and simvastatin, Int. J. Pharm. 380 (1–2) (2009) 40–48.

[50] M. Malamatari, et al., Solidification of nanosuspensions for the production of solid oral dosage forms and inhalable dry powders, Expert Opin. Drug Deliv. 13 (3) (2016) 435–450.

[51] L. Gao, et al., Application of drug nanocrystal technologies on oral drug delivery of poorly soluble drugs, Pharm. Res. 30 (2) (2013) 307–324.

[52] H. Chen, et al., Nanonization strategies for poorly water-soluble drugs, Drug Discov. Today 16 (7–8) (2011) 354–360.

[53] V.V. Pande, V.N. Abhale, Nanocrystal technology: a particle engineering formulation strategy for the poorly water soluble drugs, Int. J. Pharm. 453 (2013) 126–141.

[54] J. Broadhead, S. Edmond Rouan, C. Rhodes, The spray drying of pharmaceuticals, Drug Dev. Ind. Pharm. 18 (11 − 12) (1992) 1169–1206.

[55] B. Van Eerdenbrugh, et al., Drying of crystalline drug nanosuspensions—the importance of surface hydrophobicity on dissolution behavior upon redispersion, Eur. J. Pharm. Sci. 35 (1–2) (2008) 127–135.

[56] B. Van Eerdenbrugh, et al., Alternative matrix formers for nanosuspension solidification: dissolution performance and X-ray microanalysis as an evaluation tool for powder dispersion, Eur. J. Pharm. Sci. 35 (4) (2008) 344–353.

[57] M.V. Chaubal, C. Popescu, Conversion of nanosuspensions into dry powders by spray drying: a case study, Pharm. Res. 25 (10) (2008) 2302–2308.

[58] A. Vatanara, Spray drying of nanoparticles to form fast dissolving glipizide, Asian J. Pharm. Sci. 9 (3) (2015) 213–218.

[59] S. Alaei, E. Ghasemian, A. Vatanara, Spray drying of cefixime nanosuspension to form stabilized and fast dissolving powder, Powder Technol. 288 (2016) 241–248.

[60] J. Lee, Drug nano- and microparticles processed into solid dosage forms: physical properties, J. Pharm. Sci. 92 (10) (2003) 2057–2068.

[61] W.W.L. Chin, et al., A brief literature and patent review of nanosuspensions to a final drug product, J. Pharm. Sci. 103 (10) (2014) 2980–2999.

[62] A. Dolenc, et al., Advantages of celecoxib nanosuspension formulation and transformation into tablets, Int. J. Pharm. 376 (1–2) (2009) 204–212.

[63] E.M. Littringer, et al., Spray drying of mannitol as a drug carrier—the impact of process parameters on product properties, Drying Technol. 30 (1) (2012) 114–124.

[64] L. Gubbala, S. Arutla, V. Venkateshwarlu, Comparative evaluation of lyophilization, spray drying and spray granulation for converting quetiapine nanosuspension into dry powder, Int. J. Sci. Res. Methodol. 4 (1) (2016) 89–117.

[65] J.Y. Lim, et al., Process cycle development of freeze drying for therapeutic proteins with stability evaluation, J. Pharm. Investig. 46 (6) (2016) 519–536.

[66] S. Khairnar, et al., A review on freeze drying process of pharmaceuticals, Int. J. Res. Pharm. Sci. 4 (1) (2014).

[67] X.C. Tang, M.J. Pikal, Design of freeze-drying processes for pharmaceuticals: practical advice, Pharm. Res. 21 (2) (2004) 191–200.

[68] W. Abdelwahed, et al., Freeze-drying of nanoparticles: formulation, process and storage considerations, Adv. Drug Deliv. Rev. 58 (15) (2006) 1688–1713.

[69] Y.-Q. Ma, et al., Solidification drug nanosuspensions into nanocrystals by freeze-drying: a case study with ursodeoxycholic acid, Pharm. Dev. Technol. 21 (2) (2016) 180–188.

[70] J. Beirowski, et al., Freeze-drying of nanosuspensions, 1: freezing rate versus formulation design as critical factors to preserve the original particle size distribution, J. Pharm. Sci. 100 (5) (2011) 1958–1968.

[71] N.-O. Chung, M.K. Lee, J. Lee, Mechanism of freeze-drying drug nanosuspensions, Int. J. Pharm. 437 (1–2) (2012) 42–50.

[72] P.-F. Yue, et al., Study on formability of solid nanosuspensions during nanodispersion and solidification: I. novel role of stabilizer/drug property, Int. J. Pharm. 454 (1) (2013) 269–277.

[73] H. De Waard, W. Hinrichs, H. Frijlink, A novel bottom–up process to produce drug nanocrystals: controlled crystallization during freeze-drying, J. Control. Release 128 (2) (2008) 179–183.

[74] K. Lavanya, V. Senthil, V. Rathi, Pelletization technology: a quick review, Int. J. Pharm. Sci. Res. 2 (6) (2011) 1337.

[75] Y. Gao, S. Qian, J. Zhang, Physicochemical and pharmacokinetic characterization of a spray-dried cefpodoxime proxetil nanosuspension, Chem. Pharm. Bull. 58 (7) (2010) 912–917.

[76] W.S. Cheow, et al., Spray-freeze-drying production of thermally sensitive polymeric nanoparticle aggregates for inhaled drug delivery: effect of freeze-drying adjuvants, Int. J. Pharm. 404 (1–2) (2011) 289–300.

[77] T. Niwa, K. Danjo, Design of self-dispersible dry nanosuspension through wet milling and spray freeze-drying for poorly water-soluble drugs, Eur. J. Pharm. Sci. 50 (3–4) (2013) 272–281.

[78] R. Dixit, S. Puthli, Fluidization technologies: aerodynamic principles and process engineering, J. Pharm. Sci. 98 (11) (2009) 3933–3960.

[79] A. Perumalla, R. Manivannan, N.R. Rao, M. Radhakrishna, S. Devareddy, Formulation and evaluation of metaprolol succinate extended release pellets, Int. Res. J. Pharm. 3 (11) (2012) 96–99.

[80] P. Wang, et al., Improved dissolution rate and bioavailability of fenofibrate pellets prepared by wet-milled-drug layering, Drug Dev. Ind. Pharm. 38 (11) (2012) 1344–1353.

[81] J. Parmentier, et al., Downstream drug product processing of itraconazole nanosuspension: factors influencing drug particle size and dissolution from nanosuspension-layered beads, Int. J. Pharm. 524 (1–2) (2017) 443–453.

[82] P. Kayaert, M. Anné, G. Van den Mooter, Bead layering as a process to stabilize nanosuspensions: influence of drug hydrophobicity on nanocrystal reagglomeration following in-vitro release from sugar beads, J. Pharm. Pharmacol. 63 (11) (2011) 1446–1453.

[83] S. Bose, et al., Application of spray granulation for conversion of a nanosuspension into a dry powder form, Eur. J. Pharm. Sci. 47 (1) (2012) 35–43.

[84] E. Teunou, D. Poncelet, Batch and continuous fluid bed coating–review and state of the art, J. Food Eng. 53 (4) (2002) 325–340.

[85] E. Merisko-Liversidge, G.G. Liversidge, Nanosizing for oral and parenteral drug delivery: a perspective on formulating poorly-water soluble compounds using wet media milling technology, Adv. Drug Deliv. Rev. 63 (6) (2011) 427–440.

[86] C.E. Figueroa, S. Bose, Spray granulation: importance of process parameters on in vitro and in vivo behavior of dried nanosuspensions, Eur. J. Pharm. Biopharm. 85 (3) (2013) 1046–1055.

[87] S. Flögel, E.H. Chellai, Fluid bed granulation of lactose using bottom spray method, Eur. J. Pharm. Sci. 4 (1996) S185.

[88] E.H. Tan, et al., Downstream drug product processing of itraconazole nanosuspension: factors influencing tablet material properties and dissolution of compacted nanosuspension-layered sugar beads, Int. J. Pharm. 532 (1) (2017) 131–138.

[89] E. Gupta, et al., Review of global regulations concerning biowaivers for immediate release solid oral dosage forms, Eur. J. Pharm. Sci. 29 (3–4) (2006) 315–324.

[90] M. Imono, et al., The elucidation of key factors for oral absorption enhancement of nanocrystal formulations: in vitro–in vivo correlation of nanocrystals, Eur. J. Pharm. Biopharm. 146 (2020) 84–92.

[91] A. Sarnes, et al., Nanocrystal-based per-oral itraconazole delivery: superior in vitro dissolution enhancement versus Sporanox® is not realized in in vivo drug absorption, J. Control. Release 180 (2014) 109–116.

[92] F. Kristin, et al., Dissolution and dissolution/permeation experiments for predicting systemic exposure following oral administration of the BCS class II drug clarithromycin, Eur. J. Pharm. Sci. 101 (2017) 211–219.

[93] Z. Warnken, et al., In vitro–in vivo correlations of carbamazepine Nanodispersions for application in formulation development, J. Pharm. Sci. 107 (1) (2018) 453–465.

[94] W. Jiang, et al., The role of predictive biopharmaceutical modeling and simulation in drug development and regulatory evaluation, Int. J. Pharm. 418 (2) (2011) 151–160.

[95] S. Willmann, et al., Mechanism-based prediction of particle size-dependent dissolution and absorption: cilostazol pharmacokinetics in dogs, Eur. J. Pharm. Biopharm. 76 (1) (2010) 83–94.

[96] Y. Shono, et al., Forecasting in vivo oral absorption and food effect of micronized and nanosized aprepitant formulations in humans, Eur. J. Pharm. Biopharm. 76 (1) (2010) 95–104.

[97] D. Juenemann, et al., Biorelevant in vitro dissolution testing of products containing micronized or nanosized fenofibrate with a view to predicting plasma profiles, Eur. J. Pharm. Biopharm. 77 (2) (2011) 257–264.

[98] E.M. Merisko-Liversidge, G.G. Liversidge, Drug nanoparticles: formulating poorly water-soluble compounds, Toxicol. Pathol. 36 (1) (2008) 43–48.

[99] L. Lachman, H.A. Lieberman, J.L. Kanig, The Theory and Practice of Industrial Pharmacy, Lea & Febiger, 1986.

[100] D. Roethlisberger, et al., If euhydric and isotonic do not work, what are acceptable pH and osmolality for parenteral drug dosage forms? J. Pharm. Sci. 106 (2) (2017) 446–456.

[101] J. Wong, et al., Suspensions for intravenous (IV) injection: a review of development, preclinical and clinical aspects, Adv. Drug Deliv. Rev. 60 (8) (2008) 939–954.

[102] B. Rabinow, Pharmacokinetics of drugs administered in nanosuspension, Discov. Med. 5 (25) (2009) 74–79.

[103] B. Rabinow, et al., Itraconazole IV nanosuspension enhances efficacy through altered pharmacokinetics in the rat, Int. J. Pharm. 339 (1–2) (2007) 251–260.

[104] A.R. Fernandes, et al., Drug nanocrystals: Present, past and future, in: Applications of Nanocomposite Materials in Drug Delivery, Elsevier, 2018, , pp. 239–253.

[105] V. Patravale, A.A. Date, R. Kulkarni, Nanosuspensions: a promising drug delivery strategy, J. Pharm. Pharmacol. 56 (7) (2004) 827–840.

[106] B. Sun, Y. Yeo, Nanocrystals for the parenteral delivery of poorly water-soluble drugs, Curr. Opin. Solid State Mater. Sci. 16 (6) (2012) 295–301.

[107] E. Merisko-Liversidge, G.G. Liversidge, E.R. Cooper, Nanosizing: a formulation approach for poorly-water-soluble compounds, Eur. J. Pharm. Sci. 18 (2) (2003) 113–120.

[108] J.-U.A. Junghanns, R.H. Müller, Nanocrystal technology, drug delivery and clinical applications, Int. J. Nanomedicine 3 (3) (2008) 295.

[109] EMC, Sporanox I.V. 10 mg/ml Concentrate and Solvent for Solution for Infusion, Available from: https://www.medicines.org.uk/emc/product/5522/smpc, 2018.

[110] M. Chen, et al., In vitro and in vivo evaluation of SN-38 nanocrystals with different particle sizes, Int. J. Nanomedicine 12 (2017) 5487.

[111] P. Sharma, et al., Evaluation of a crystalline nanosuspension: polymorphism, process induced transformation and in vivo studies, Int. J. Pharm. 408 (1–2) (2011) 138–151.

[112] L. Wang, et al., Docetaxel-loaded-lipid-based-nanosuspensions (DTX-LNS): preparation, pharmacokinetics, tissue distribution and antitumor activity, Int. J. Pharm. 413 (1–2) (2011) 194–201.

[113] K. Peters, et al., Preparation of a clofazimine nanosuspension for intravenous use and evaluation of its therapeutic efficacy in murine Mycobacterium avium infection, J. Antimicrob. Chemother. 45 (1) (2000) 77–83.

[114] Z. Lin, et al., Novel thermo-sensitive hydrogel system with paclitaxel nanocrystals: high drug-loading, sustained drug release and extended local retention guaranteeing better efficacy and lower toxicity, J. Control. Release 174 (2014) 161–170.

[115] M. Malamatari, et al., Pharmaceutical nanocrystals: production by wet milling and applications, Drug Discov. Today 23 (3) (2018) 534–547.

[116] R. Lee, N. DiFranco, Nanomilling: A Key Option for Formulating Water-Insoluble APIs, Available from: https://lubrizolcdmo.com/blog/nanomilling-option-for-formulating-water-insoluble-apis/, 2019.

[117] H. Zhong, et al., A comprehensive map of FDA-approved pharmaceutical products, Pharmaceutics 10 (4) (2018) 263.

[118] Janssen, Janssen Announces Results of Two Phase 3 Studies Which Showed Long-Acting Injectable HIV Treatment Regimen of Rilpivirine and Cabotegravir Demonstrated Comparable Safety and Efficacy to Daily Oral HIV Therapy, Available from: https://www.janssen.com/janssen-announces-results-two-phase-3-studies-which-showed-long-acting-injectable-hiv-treatment, 2019.

[119] R.H. Müller, et al., Nanocrystals for passive dermal penetration enhancement, in: Percutaneous Penetration Enhancers Chemical Methods in Penetration Enhancement, Springer, 2016, , pp. 283–295.

[120] L. Vidlářová, et al., Nanocrystals for dermal penetration enhancement—effect of concentration and underlying mechanisms using curcumin as model, Eur. J. Pharm. Biopharm. 104 (2016) 216–225.

[121] X. Zhai, et al., Nanocrystals of medium soluble actives—novel concept for improved dermal delivery and production strategy, Int. J. Pharm. 470 (1–2) (2014) 141–150.

[122] V. Patel, O.P. Sharma, T. Mehta, Nanocrystal: a novel approach to overcome skin barriers for improved topical drug delivery, Expert Opin. Drug Deliv. 15 (4) (2018) 351–368.

[123] P.R. Mishra, et al., Production and characterization of Hesperetin nanosuspensions for dermal delivery, Int. J. Pharm. 371 (1–2) (2009) 182–189.

[124] L. Al Shaal, R. Shegokar, R.H. Müller, Production and characterization of antioxidant apigenin nanocrystals as a novel UV skin protective formulation, Int. J. Pharm. 420 (1) (2011) 133–140.

[125] K. Mitri, et al., Lutein nanocrystals as antioxidant formulation for oral and dermal delivery, Int. J. Pharm. 420 (1) (2011) 141–146.

[126] T. Hatahet, et al., Dermal quercetin smartCrystals®: formulation development, antioxidant activity and cellular safety, Eur. J. Pharm. Biopharm. 102 (2016) 51–63.

[127] V. Patel, O.P. Sharma, T.A. Mehta, Impact of process parameters on particle size involved in media milling technique used for preparing clotrimazole nanocrystals for the management of cutaneous candidiasis, AAPS PharmSciTech 20 (5) (2019) 175.

[128] R. Pireddu, et al., Novel nanosized formulations of two diclofenac acid polymorphs to improve topical bioavailability, Eur. J. Pharm. Sci. 77 (2015) 208–215.

[129] F. Lai, et al., Nanosuspension improves tretinoin photostability and delivery to the skin, Int. J. Pharm. 458 (1) (2013) 104–109.

[130] L. Gao, et al., Evaluation of genipin-crosslinked chitosan hydrogels as a potential carrier for silver sulfadiazine nanocrystals, Colloids Surf. B Biointerfaces 148 (2016) 343–353.

[131] C.-Y. Shen, et al., Nanogel for dermal application of the triterpenoids isolated from Ganoderma lucidum (GLT) for frostbite treatment, Drug Deliv. 23 (2) (2016) 610–618.

[132] R. Gaudana, et al., Recent perspectives in ocular drug delivery, Pharm. Res. 26 (5) (2009) 1197.

[133] O.P. Sharma, V. Patel, T. Mehta, Nanocrystal for ocular drug delivery: hope or hype, Drug Deliv. Transl. Res. 6 (4) (2016) 399–413.

[134] L. Allen, H.C. Ansel, Ansel's Pharmaceutical Dosage Forms and Drug Delivery Systems, Lippincott Williams & Wilkins, 2013.

[135] M. Reader, Influence of isotonic agents on the stability of Thimerosal in ophthalmic formulations, J. Pharm. Sci. 73 (6) (1984) 840–841.

[136] A. Tuomela, et al., Brinzolamide nanocrystal formulations for ophthalmic delivery: reduction of elevated intraocular pressure in vivo, Int. J. Pharm. 467 (1–2) (2014) 34–41.

[137] S. Gupta, M.K. Samanta, A.M. Raichur, Dual-drug delivery system based on in situ gel-forming nanosuspension of forskolin to enhance antiglaucoma efficacy, AAPS PharmSciTech 11 (1) (2010) 322–335.

[138] H.S. Ali, et al., Hydrocortisone nanosuspensions for ophthalmic delivery: a comparative study between microfluidic nanoprecipitation and wet milling, J. Control. Release 149 (2) (2011) 175–181.

[139] K. Baba, K. Nishida, N. Hashida, Method for Producing an Aqueous Dispersion of Drug Nanoparticles and Use Thereof, (2015)Google Patents.

[140] J.H. Kim, et al., Development of a novel ophthalmic ciclosporin A-loaded nanosuspension using top-down media milling methods, Pharmazie 66 (7) (2011) 491–495.

[141] R. Pignatello, et al., Eudragit RS100® nanosuspensions for the ophthalmic controlled delivery of ibuprofen, Eur. J. Pharm. Sci. 16 (1–2) (2002) 53–61.

[142] C. Bucolo, et al., Enhanced ocular anti-inflammatory activity of ibuprofen carried by an eudragit RS100® nanoparticle suspension, Ophthalmic Res. 34 (5) (2002) 319–323.

[143] L. Schopf, et al., Ocular pharmacokinetics of a novel loteprednol etabonate 0.4% ophthalmic formulation, Ophthalmol Therapy 3 (1–2) (2014) 63–72.

[144] G.B. Romero, et al., Development of cationic nanocrystals for ocular delivery, Eur. J. Pharm. Biopharm. 107 (2016) 215–222.

[145] N. Nagai, et al., Improved corneal toxicity and permeability of tranilast by the preparation of ophthalmic formulations containing its nanoparticles, J. Oleo Sci. 63 (2) (2014) 177–186.

[146] R. Thiruppathi, et al., Nanoparticle functionalization and its potentials for molecular imaging, Adv. Sci. 4 (3) (2017) 1600279.

[147] T. Xu, et al., Modification of nanostructured materials for biomedical applications, Mater. Sci. Eng. C 27 (3) (2007) 579–594.

[148] R. Shegokar, K.K. Singh, Surface modified nevirapine nanosuspensions for viral reservoir targeting: in vitro and in vivo evaluation, Int. J. Pharm. 421 (2) (2011) 341–352.

[149] M. Talekar, et al., Development of PIK-75 nanosuspension formulation with enhanced delivery efficiency and cytotoxicity for targeted anti-cancer therapy, Int. J. Pharm. 450 (1–2) (2013) 278–289.

[150] F. Liu, et al., Targeted cancer therapy with novel high drug-loading nanocrystals, J. Pharm. Sci. 99 (8) (2010) 3542–3551.

[151] B. Palazzo, et al., Biomimetic hydroxyapatite–drug nanocrystals as potential bone substitutes with antitumor drug delivery properties, Adv. Funct. Mater. 17 (13) (2007) 2180–2188.

[152] L. Wang, M. Li, N. Zhang, Folate-targeted docetaxel-lipid-based-nanosuspensions for active-targeted cancer therapy, Int. J. Nanomedicine 7 (2012) 3281.

[153] Y. Xiao, et al., PEGylation and surface functionalization of liposomes containing drug nanocrystals for cell-targeted delivery, Colloids Surf. B Biointerfaces 182 (2019) 110362.

[154] D. Castro, et al., Effect of different carboxylic acids in cyclodextrin functionalization of cellulose nanocrystals for prolonged release of carvacrol, Mater. Sci. Eng. C 69 (2016) 1018–1025.

[155] H.M. Shubar, et al., SDS-coated atovaquone nanosuspensions show improved therapeutic efficacy against experimental acquired and reactivated toxoplasmosis by improving passage of gastrointestinal and blood–brain barriers, J. Drug Target. 19 (2) (2011) 114–124.

[156] M. Hasani, et al., Cationic surface functionalization of cellulose nanocrystals, Soft Matter 4 (11) (2008) 2238–2244.

[157] M.C.C. Peters, et al., Advances in ophthalmic preparation: the role of drug nanocrystals and lipid-based nanosystems, J. Drug Target. 28 (3) (2020) 259–270.

CHAPTER 13

Superparamagnetic iron oxide nanoparticles (SPIONs) as therapeutic and diagnostic agents

Nisha Lamichhane, Maneea Eizadi Sharifabad, Ben Hodgson, Tim Mercer, and Tapas Sen
Nano-biomaterials Research Group, School of Natural Sciences, Faculty of Science & Technology, University of Central Lancashire, Preston, United Kingdom

1. Introduction to magnetism and magnetic materials

Magnetic materials are currently emerging materials due to their magnetic properties on the nanoscale that can be utilized across a range of applications such as bio-separations [1, 2], biomolecular testing, including therecent outbreak of Covid-19 [3], magnetically targeted drug delivery [4, 5], magnetic contrasting in magnetic resonance imaging (MRI) [6, 7], magnetic heating under an alternating magnetic field (AMF) for cancer therapy [8], economically viable industrial catalytic supports [9, 10], magnetically driven separation of toxic metal ions from polluted water [11, 12], etc.

Within their atomic structure, a magnetic field is created in these materials by the spin and orbital motion of unpaired electrons and their resultant magnetic moments. In solids, the orbital moment is often "quenched" to a significant degree as a consequence of the coulombic crystal field created by the surrounding atoms or ions in its crystal structure. Hence, the moment per atom or ion is, in the main, due to the unpaired electron spin [13]. How these moments align determines the classification and behavior of these materials as summarized in Table 1. Leaving aside diamagnetic materials (essentially nonmagnetic in zero field, with closed shells and *no* unpaired electrons), in all cases apart from some weakly paramagnetic materials, the atoms and their unpaired spins are close enough together for a quantum mechanical process known as Exchange Interaction to occur. For example, in a ferromagnetic material below its Curie temperature, this mechanism means it is energetically favorable for all the moments to align (lower net energy state) and so leads to a spontaneous magnetization throughout the material. Classically, it is thought of as an imaginary field intrinsic to the material (Weiss Molecular Field theory) that is tending to align the moments in a given direction and so can be considered as an equivalent aligning force. Above the Curie temperature, the energy of thermal agitation is greater than that of the aligning exchange interaction and the material becomes paramagnetic as the moments

Nanoparticle Therapeutics
https://doi.org/10.1016/B978-0-12-820757-4.00003-X

now flip randomly in the thermal field. In the cases of antiferromagnetic and ferrimagnetic materials, the exchange interaction is of a different form whereby the magnetic atoms or ions having unpaired electrons in the structure are separated by an interstitial nonmagnetic atom or ion between them. In this case the exchange interaction is then mediated via the nonmagnetic atom; an indirect process known as Super-exchange.

1.1 Superparamagnetism and SPIONs

Spontaneous magnetization occurs in both ferromagnetic and ferrimagnetic materials as illustrated schematically in Table 1. At first glance, this does not appear to be consistent with everyday experience of such materials in the bulk. For example, a nail or needle predominantly made up of iron is not normally "magnetized" unless some effort is made by, for example, rubbing it in one direction with a permanent bar magnet. The reason for this *apparent* lack of magnetization is readily explained by introducing the concept of domains. A domain is an area within the volume of the material where all the moments are aligned due to the spontaneous magnetization and is separated from other surrounding domains by a barrier called a domain wall. Neighboring domains also have their own spontaneous magnetization, but in different directions to each other and the domain of interest. In a fully magnetized material, there will only be a single domain and an associated magnetostatic energy (the energy due to the intrinsic magnetic field of the sample when in zero applied field). Hence, there is a tendency over time to reduce this higher energy state to a lower one by the formation of multiple domain structures that can be pictured as an assembly of small, but randomly orientated, bar magnets. In this manner, the overall stray field emanating from the total volume of the material is reduced by flux closure between opposite poles of neighboring domains and thus reduces the overall magnetostatic energy.

To understand how the main ferrimagnetic iron oxide of interest, magnetite (Fe_3O_4), becomes superparamagnetic, it is necessary to understand the effects of decreasing the volume of the material. It should be noted that this discussion is generally applicable to the other ferrimagnetically similar iron oxide, maghemite (γ-Fe_2O_3), that is often contained to some degree within an assembly of magnetite particles; in part due to it being the product of magnetite oxidation. If we consider a large particle in the context of this article, say with a diameter, d, of order microns (10^{-6} m), it is ferrimagnetic and multi-domain at room temperature as discussed above. By reducing the size into the submicron scale of 100s of nanometers (10^{-7} m), there is a point when the decreasing volume of the particle approaches the thickness of a domain wall. This critical size occurs at $d_c \sim 130$ nm in magnetite [14], at which point it is clear it can no longer support two domains separated by a domain wall that is of the same size order as the particle itself. The material is now single domain and, due to the associated spontaneous magnetization throughout its volume, will carry a permanent net magnetic

Table 1 Summary of various magnetic materials.

Type of magnetism	Characteristics	Direction of spins state
Ferromagnetic	Atoms hold parallel aligned magnetic moments due to exchange interaction between adjacent unpaired electrons. Ferromagnetic materials possess a permanent magnetic field.	Ferromagnetic
Paramagnetic	Atoms possess randomly angled magnetic moments due to thermal agitation. Under the influence of a magnetic field, the magnetic moments align to create a low magnetization with direction similar to that of the field.	Paramagnetic
Antiferromagnetic	Atoms hold antiparallel aligned magnetic moments. The magnetic fields counteract and the material behaves like a paramagnetic material above the Néel Temperature.	Antiferromagnetic
Ferrimagnetic	Atoms possess mixed parallel and antiparallel aligned magnetic moments of direction-dependent magnitude. Here the "up" moment is greater than the "down" moment and so results in an overall magnetization in the "up" direction (pointing to the right in the schematic). Hence, these materials behave like ferromagnetic materials but with lower saturation magnetizations.	Ferrimagnetic
Diamagnetic	Atoms have no net magnetic moment in zero field. In the presence of an externally applied field, a small negative magnetization is apparent, i.e., in the opposite direction to that of the applied field	No unpaired spins

moment in zero applied field. In terms of energy balance, it is the point when the contribution from the magnetostatic energy of a single domain, which varies with volume (proportional to d^3), reduces the total energy of a particle to a lower state than a system containing domain walls and their associated energy. For domain walls, this energy term varies with the domain wall area and hence with the particle's cross-sectional area (proportional to d^2).

In the single domain state, below the critical size, these particles tend to have a property described as uniaxial anisotropy. This means that each particle has a preferred single axis, the easy axis, along which the net magnetic moment of the spontaneous magnetization will relax back to when an applied field is removed. In this way, the moment can only have one of two opposing directions along this axis and can only be switched from one to the other of these two energy minima by application of a strong enough external field that rotates the moment to greater than $90°$ from its original easy axis direction. However, at room temperature, the effects of thermal agitation become increasingly important due to other size effects. As we further decrease d, and hence the particle volume size, V, there is a subsequent reduction in the energy barrier, $\triangle E = KV$, presented by the uniaxial anisotropy between the two minima directions. Here, K is the anisotropy energy density (ergs/cm^3 or J/m^3) that is *constant* for a material of a given particle shape and crystal structure and so is often termed the anisotropy constant. Hence, at a fixed temperature, T, there is a point where the reducing barrier $\triangle E(d)$ is $\leq$ the thermal energy, $k_B T$, agitating the particle moment (where k_B is Boltzmann's constant) and so it flips randomly between the two easy axis directions. As these fluctuations mean there is no overall net magnetic moment in zero applied field, it is behaving in the same way as a paramagnetic material, but with one significant difference. In this case, the application of an external field aligns the moments of a single domain ferrimagnetic material with a saturation magnetization, M_s, of 480 emu/cm^3 (480 kA/m) compared to a typical *paramagnetic* material with an M_s value of $\sim10^{-3}$ emu/cm^3 (1 A/m): a factor of 100s of thousands (10^5) times greater and hence the term *superparamagnetic* being an appropriate description of this phenomena.

From the previous discussion, it should be possible to determine the size, d_{spm}, (recall this is below the onset of single domains at size d_c,) at which superparamagnetism emerges at room temperature (RT). Taking a relatively simplistic approach, one imagines how the time taken to experimentally measure the magnetization curve of a sample of tightly packed nanoparticles (unable to physically move/rotate in the field) compares to the relaxation time of the moment back to its easy axis upon removal of the applied field. Assuming no interactions between the particles, this is given by the Néel relaxation time, τ_N,

$$\tau_N = \tau_0 \exp\left(\frac{KV}{k_B T}\right),$$

where τ_0 is considered as a constant (only shows a very weak dependency on temperature) and is usually assigned a value of 10^{-9} s, with all the other terms as before. As we expect the onset of superparamagnetic behavior to occur at the point when the energy barrier, $\triangle E = KV$, is equal to the thermal energy, $k_B T$, we could simply calculate the volume from $KV = k_B T$. However, as this effectively assumes the change is instantaneous, it is more usual to consider the realistic measuring time, τ_m, of ~100s of secs to ramp-down to a field of $H=0$ from saturation magnetization, M_s, as the means of setting the size limit for *measured* transition from stable to superparamagnetic behavior. By setting τ_m equal to τ_N in the above equation, and using the previously given value of τ_0, we find $KV/k_BT=25$. Rearranging, we see that.

$$V_{spm}=\frac{25k_BT}{K}.$$

Values of K for magnetite are usually assumed constant and given a value of $3.0\times10^{-5}\,\mathrm{ergs/cm^3}$ ($3.0\times10^{-6}\,\mathrm{J/m^3}$), at RT, we obtain the volume and subsequent particle size for a sphere of $d_{spm}\sim19\,\mathrm{nm}$. In reality, there is a size distribution in any assembly of SPIONs and the values of K also change significantly for relatively small particle shape deviations from that of an ideal sphere [15]. Nevertheless, SPIONs consisting predominantly of magnetite are found to be single domain at size d_c of $\sim130\,\mathrm{nm}$ and become increasingly superparamagnetic below d_{spm} sizes of $\sim50\,\mathrm{nm}$ [14, 16].

1.2 Standard magnetization curves

Magnetization, M, is defined as either the moment per unit volume, m/V, or the moment per unit mass, $m/\rho V$, where ρ is the material density. It has units of $\mathrm{emu/cm^3}$ (A/m) and $\mathrm{emu/g}$ ($\mathrm{A\,m^2/kg}$), respectively, where the cgs unit for m is known as the Electromagnetic Unit or emu. If a strong enough field, H, is applied, the moments of the material will tend towards complete alignment with the field direction and M will therefore tend to a saturation point, M_s. The cgs unit for H is the Oersted, abbreviated Oe, with the SI unit simply expressed as the base units, A/m.

In a standard M vs H magnetization curve such as Fig. 1B, the sample is initially saturated (M/Ms $\sim$ 1) in a strong positive field before the measurement begins. The field is

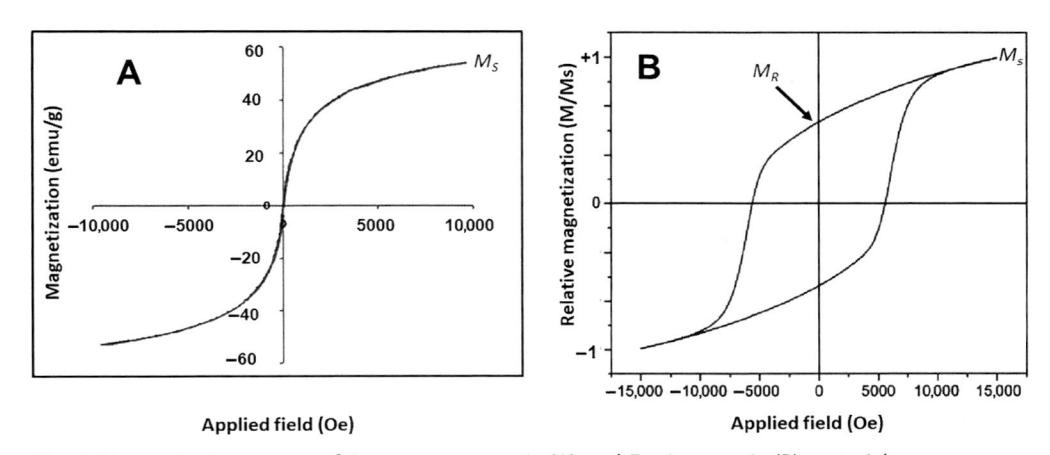

Fig. 1 Magnetization curves of Superparamagnetic (A) and Ferrimagnetic (B) materials.

then reduced in incremental steps towards zero with the resultant magnetization measured at each step. At $H=0$, the field direction is reversed and again increased up to the point of negative saturation. From $-M_s$, the process is repeated with a ramp-down of $-H$, reversal of field direction, and final ramp-up of H to the starting point. If we consider particles large enough to be ferrimagnetic (either multi-domain or single domain $> d_{spm}$), the M vs H curve results in the hysteresis loop of Fig. 1B. At zero applied field, the material retains a net magnetization known as the *remanence*, M_r, albeit at a lower value than M_s. The opposing field required to reduce M from remanence to zero is called the *coercive field* or *coercivity*, H_c. As the magnetization response "lags behind" the change in the applied field (hysteresis: the processes are irreversible), this loss of magnetic energy must go "somewhere," and it is dissipated as heat energy equal to the area of the loop.

A typical magnetization curve for a SPION is shown in Fig. 1A. As expected, there is zero or negligible remanence and coercivity, and it can be thought of as a closed or zero hysteresis loop. The observed experimental M_s values for SPIONs are reported [17] to be in the range of 30–60 emu/g, compared to 92 emu/g in bulk iron oxide Fe_3O_4. The tendency for M_s to reduce with particle size is not fully understood and may therefore be due to more than one mechanism. However, by comparison of experimental measurements with theory, it is generally accepted that the size of a particle determined from magnetic measurements is smaller than the physical size [18]. This is often attributed to effects at the surface resulting in a magnetically-dead outer layer of $\sim$1 nm thick [19] and is consistent with the dead layer volume becoming a larger fraction of the total volume with decreasing d.

1.3 Heating mechanism of SPIONs under an alternating magnetic field

When an assembly of nanoparticles is subjected to an alternating magnetic field (AMF), there is an increase in temperature due to the generation of heat by several processes known collectively as *magnetic hyperthermia*. These processes are classified by the underlying mechanism generating the heat and may be due to one or more of (i) susceptibility losses, (ii) hysteresis losses, (iii) viscous stirring, and (iv) induced eddy currents. The last of these is found to be insignificant for particles on the nanoscale, including the maximum value theoretically possible at the upper limit of field frequencies and strengths permitted for use on the human body of $\leq$1.2 MHz and $\leq$188 Oe (15 kA/m), respectively [20]. Furthermore, nanoparticles are often coated with biocompatible shells (e.g., silica, polymers, and lipids), which eliminate the setting up of current loops due to isolation or insulation.

It is important to note that in the case of magnetic hyperthermia pertinent to this article, we are considering suspensions of SPIONs in a carrier fluid (usually in water and sometimes referred to as ferrofluids) as this is how they are used in biomedical applications.

1.3.1 Susceptibility losses—Brownian and Néel rotations

In the earlier discussion on superparamagnetism, it was noted that there is a finite time, τ_N, required for the internal moments of an assembly of nanoparticles to relax back to a thermally agitated net moment of zero as given by the Néel equation earlier. It was assumed that the particles were physically constrained and so ensured that the only mechanism for relaxation, and its associated time as described by Néel, would be by rotation of the particle moments.

In the case of a magnetic fluid, the particles are no longer fully constrained, and another mechanism called Brownian rotation occurs along with that of Néel. Again, it is a consequence of the ambient thermal energy, but this time the agitation and subsequent rotation is of the particles themselves and results in a Brownian relaxation time, τ_B. It depends on the hydrodynamic properties of the fluid and particle and further details are given in Fig. 2. Upon removal of a magnetic field, the magnetization may relax back to zero by both processes in parallel and so yields an effective combined time of $\tau = \tau_N \tau_B / (\tau_N + \tau_B)$.

In both cases, the magnetization response will lag the change in applied field as a function of time and this energy loss is dissipated as heat. This M vs H response can be described in terms of complex susceptibility, $\chi = \chi' + i\chi''$, where susceptibility is generally defined as $\chi = M/H$ and under the AC field here as $\chi = dM/dH$. The out-of-phase χ'' component (due to the lag) gives rise to heat generation with power of

$$P_S = \pi f \chi'' H^2,$$

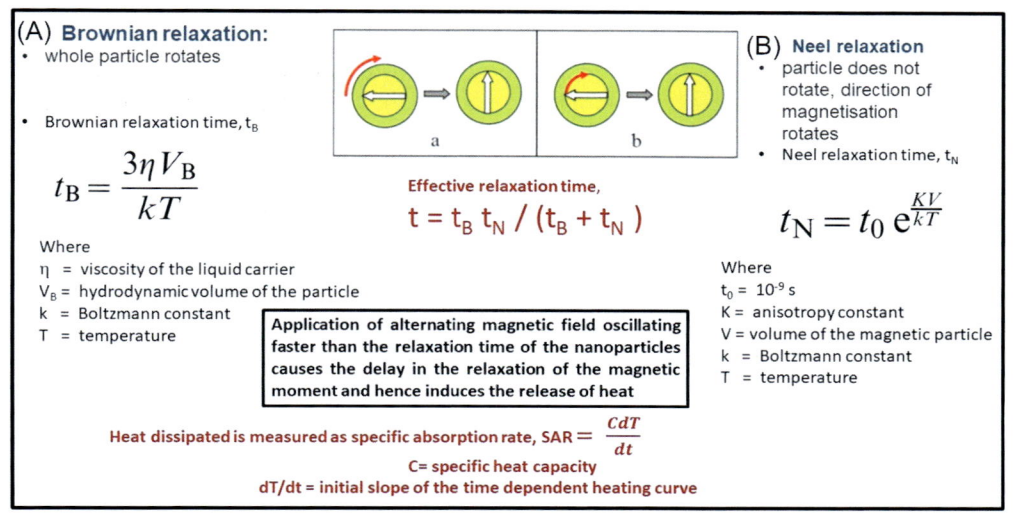

Fig. 2 Relaxation mechanism of superparamagnetic nanoparticles: Brownian (A) and Neel (B) and associated equations and parameters.

where f is the frequency of the applied field H. χ'' can be written in terms of the effective relaxation time, τ, due to both τ_N and τ_B, to give

$$\chi'' = \chi_0 \frac{2\pi f \tau}{1 + (2\pi f \tau)^2}$$

where χ_0 is the initial DC susceptibility obtained from the gradient through the origin of the DC magnetization curve such as that of Fig. 1A.

1.3.2 Hysteresis losses

An ideal magnetic fluid of SPIONs would consist of monodispersed and mono-sized particles of identical shape. In reality, there is a distribution of all of these parameters which can lead to a proportion having sufficient anisotropy (dominated by shape) and volume to be single domain and ferrimagnetic. Providing the applied field is of sufficient strength, it will reverse the magnetization along the easy axis and hysteresis will occur. The subsequent hysteresis loss is then calculated from the area enclosed by one cycle of the hysteresis loop multiplied by the permeability constant, μ_0, and the number of cycles per second, f, of the AMF to give a total power [20] of

$$P_H = \mu_0 f \oint H.dM.$$

1.3.3 Viscous stirring

In magnetic fluids with a wide enough size distribution, there is a point in the single domain hysteresis loss regime when the increasing coercivity with size is such that the particles can no longer be switched by the relatively low fields used in hyperthermia of about 200 Oe ($\sim$15 kA/m). At this point, hysteresis losses can no longer occur and the magnetic moments will remain fixed to one of their two easy axis directions and thereby cause physical movement in the alternating field in a stirring motion that causes viscous heating [21]. Determination of the friction in this highly complex (and likely turbulent) system is beyond the scope of any simple models and is little understood. It should also be noted that aggregates of smaller particles may also be subject to this process and that the possibility of further magnetic losses due to any lag in the response of the moment to the field cannot be ruled out.

In summary, suspensions of SPIONs are not ideal, having shape, size, and anisotropy distributions that often result in the target size $\lesssim$ 15 nm in the superparamagnetic range containing a significant enough fraction at larger sizes to include single domain ferrimagnetic particles. This leads to a number of possible heating mechanisms classified as susceptibility losses (superparamagnetic), hysteresis losses and frictional losses (both single domain and ferrimagnetic).

2. Synthesis and characterization

2.1 Synthesis of commonly used SPIONs of magnetite structure (Fe$_3$O$_4$)

Magnetite, Fe$_3$O$_4$, has an inverse-spinel structure. The unit cell consists of thirty-two oxygen atoms forming a face-centred-cubic (fcc) closed-packing structure, with iron cations located at octahedral and tetrahedral sites [22, 23]. The unit cell edge length is 0.839 nm. The crystal structure can be written as Fe^{2+}Fe$_2^{3+}$O$_4$ where the Fe^{2+} ions and half of the Fe^{3+} ions occupy octahedral sites and the second half of the Fe^{3+} ions occupy the tetrahedral sites; i.e., [Fe^{3+}]$_{Th}$[Fe^{3+} Fe^{2+}]$_{Oh}$O$_4$, where T$_h$ represents tetrahedral sites and O$_h$ represents octahedral sites. The characteristic black color of magnetite arises from inter-valence charge transfer between octahedral Fe^{2+} and Fe^{3+} ions within the crystal structure, perhaps due to its ability to absorb visible light of all wave lengths [23]. The structure is shown in Fig. 3 [24].

Sugimoto and Matijevic first reported the synthesis of uniform, spherical magnetite particles (SPIONs) of 30–1100 nm diameter using ferrous hydroxide gels in 1980 [25]. An example of this is the synthesis of magnetite nanoparticles via the oxidative hydrolysis of iron sulphate in alkaline media [25, 26]. Since then, many methods of synthesizing magnetite have been developed with the aim of controlling particle size, morphology,

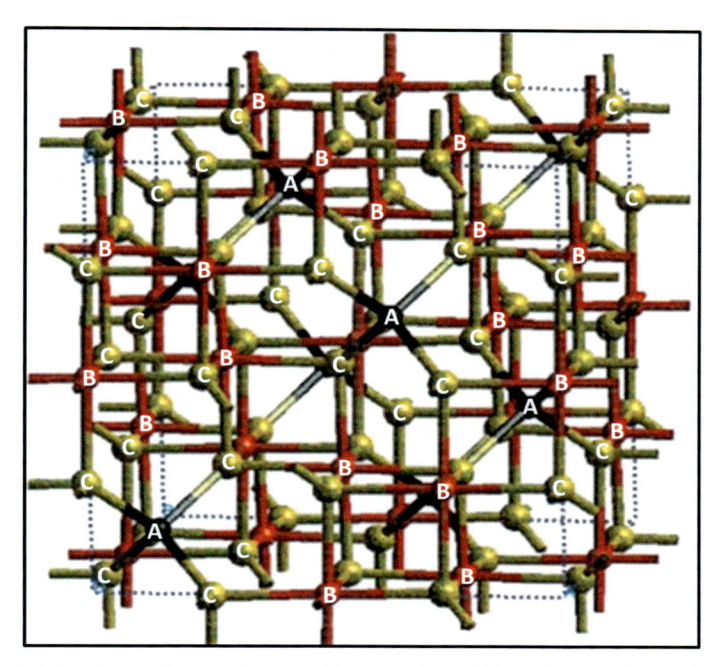

Fig. 3 The crystal structure of magnetite, Fe$_3$O$_4$, reproduced from Sorescu [24]. The black spheres (A) represent tetrahedral Fe^{3+} ions, red spheres (B) represent octahedral Fe^{2+} or Fe^{3+} ions, and the yellow spheres (C) correspond to oxygen anions.

stability, and dispersity. These factors have been found to be dependent on the reaction temperature, pH, ionic strength of the media, ferrous/ferric (Fe^{2+}/Fe^{3+}) chloride ratio, and the type of salts used [27].

Table 2 summarizes the most commonly used chemical methods of synthesis of SPIONs [28].

Table 2 Summary of the most common methods of SPIONs synthesis is outlined below [28].

Synthetic method	A summary of synthesis methods, their advantages and disadvantages
Coprecipitation	Coprecipitation is a simple and convenient synthesis method using aqueous iron salt solutions with the addition of a base under inert atmosphere at temperature ranging from room temperature up to 90°C. The resulting particles reported have polydispersity but with a high yield
Thermal decomposition	Thermal decomposition is performed under an inert atmosphere at high temperature from 240°C to 320°C. The particle size is monodispersed with a very narrow size range. The major disadvantage of this method is that the synthesized nanoparticles are poorly dispersed in water which limits the applications of the nanoparticles in nanomedicine. Therefore, coating is necessary for changing the surface property so that they can be dispersed in aqueous medium
Microemulsion	*The microemulsion method* uses water droplets as nano-reactors in an oil phase in the presence of surfactant molecules. Iron precursors precipitate as iron oxides in the water phase, inside the micelles core. The size of the synthesized nanoparticles is controlled by the size of micelles. The main disadvantage of this method is low yield. However, using a bi-continuous microemulsion method, a higher yield can be obtained
Hydrothermal	The hydrothermal procedure requires an autogenous pressure and a reasonable temperature. It is based on a general phase transfer followed by a separation of solid from the solution phase during the synthesis and results in nanoparticles with a very narrow size distribution
Sonochemical	This method uses high-intensity, high-energy ultrasonication via acoustic cavitation that produces localized heat with a temperature of about 5000 K. At this temperature, the formation and growth of nuclei occurs with the implosive collapse of bubbles. The method produces monodisperse nanoparticles in a variety of shapes. However, this method is not applicable to large-scale production.
Sol-gel combined with annealing	The sol-gel method includes a multistep complicated procedure, followed by annealing under high temperature of 200–400°C. The particles' size is controlled by the annealing temperature. Particles with a narrow size distribution may be prepared by this method.

Table 2 Summary of the most common methods of SPIONs synthesis is outlined below —cont'd

Synthetic method	A summary of synthesis methods, their advantages and disadvantages
Electrochemical deposition	Following this strategy, oxidation to metal ions in the solution occurs at the anode followed by further reduction of metal ions to metal at the cathode in the presence of stabilizers. The particle size is controlled by regulating the electrochemical oxidation, current density, or potential difference of the system
Microwave-assisted methods	The microwave-assisted method offers the advantages of rapid heating, high reaction rate, reduced reaction time, and increased product yield. This method reported the synthesis of uniform nanoparticles with a narrow size distribution at large scale
Biomimetic	Biomimetic production of magnetic nanoparticles by magnetotactic bacteria has been known for a long time. Magnetotactic bacteria contain intracellularly produced crystals of magnetite (Fe_3O_4). The morphology of the produced nanoparticles is species-specific. The mineralization processes are regulated to form uniform magnetic nanoparticles. The method lacks large-scale synthesis of monodisperse nanoparticles

2.2 Surface modification of SPIONs

2.2.1 Core-shell nanoparticles: SPIONs as core with various shell materials

There has been a significant progress in the synthesis of magnetic nanoparticles with different sizes, shapes, and properties, maintaining the stability of these particles without agglomeration and protection against oxidation or erosion by acidic or basic media. Stabilization and protection of the particles are closely linked with each other and are crucial requirements for almost all applications of magnetic nanoparticles. Therefore, it is important to develop efficient strategies to improve the chemical stability of magnetic nanoparticles. Of these, the coating of nanoparticles with a protective layer has attracted most interest. The coating layer may stabilize the nanoparticles and provide functional groups for the conjugation of targeting ligands and biomolecules. In the case of drug delivery application, the coating layer may be designed to limit nonspecific cell interactions, reduce the toxicity of the bare magnetite nanoparticles, and prolong circulation time. Furthermore, it can be tailored for drug loading and releasing behaviors at targeted sites.

These applied coating strategies can be categorized into two major groups of (i) organic shells, including surfactant and polymers [17, 29] or, (ii) inorganic shells, including silica [30] and precious metals [31]. Surface modification of nanoparticles is performed either during the synthesis process or in a post-synthesis process. The most commonly used materials for in situ surface modification of magnetite nanoparticles are surfactants such as oleic acid, lauric acid, alkane sulphonic, and alkane phosphonic acids [32]. Citric acid has been used commercially for the stabilization of SPIONs for

MRI contrast agent VSOP C184 [33]. Surfactant-mediated syntheses are mostly performed in organic solvents such as hexadecane, toluene, and n-hexane. The hydrophobic hydrocarbon chain of the surfactant forms a shell around nanoparticles rendering the nanoparticles less susceptible to aggregation. The post-synthesis modification of magnetite nanoparticles is mostly achieved using polymers, silica, liposomes, and chitosan [5, 17, 34, 35]. Table 3 summarizes some examples of materials which have been used to coat magnetite nanoparticles.

2.2.2 Surface functionalization of SPIONs and core-shell SPIONs

In order to facilitate core-shell SPIONs for attaching biomolecules such as single strand DNA, RNA, or antibodies, functionalization with specific surface functional groups is essential. Fig. 4 provides a schematic diagram of core-shell nanoparticles with

Table 3 Most common coating materials, applications, and advantages of core-shell SPIONs [28].

Coating materials	Applications	Advantages
Polylactic acid (PLA)	Drug delivery, MR imaging agent, cell labelling and magnetic cell separation, tissue repair, and hyperthermia	Biodegradable, biocompatible, and hemo-compatible
Polyvinyl alcohol (PVA)	Drug delivery and imaging contrast agent	Prevents aggregation and stabilizes nanoparticles in suspension
Polyvinyl pyrrolidine (PVP)	Drug delivery and MR imaging contrast agent	Enhances blood circulation time and stabilizes nanoparticles in suspension
Polyacrylic acid (PAA)	Target thrombolysis with recombinant tissue plasminogen activator, targeting drug resistance in mycobacteria	Increases stability and biocompatibility and helps in bio-adhesion
Polystyrene	MR cellular imaging and DNA hybridization	Stabilization of nanoparticles in suspension
Polymethyl methacrylate (PMMA)	DNA separation and amplification	Increases stability and can be applied in automated systems to achieve high-throughput detection of single nucleotide polymorph.
Dextran	Drug delivery, hyperthermia, and MRI contrast agent	Biocompatible, increased stability, and prolongs blood circulation time
Amorphous silica	Isolation and purification of DNA from soil samples, isolation of ultrapure plasmid DNA from bacterial cells	Eliminates the need for repeated centrifugation, vacuum filtration, or column separation

Table 3 Most common coating materials, applications, and advantages of core-shell SPIONs —cont'd

Coating materials	Applications	Advantages
Mesoporous silica	Controlled drug delivery, fluorescence imaging, mercury removal from industrial effluent, enzyme immobilization for biocatalysis, DNA extraction	High surface area with uniform pore size and high pore volume
Polyethylene glycol (PEG)	MRI contrast agent	Improves biocompatibility, blood circulation time, and internalization efficiency
Polyethylene glycol – Polyethylene imine (PEG-PEI)	DNA extraction, MRI contrast agent, drug delivery, and hyperthermia	Highly soluble and stable in water, biocompatible, hardly recognized by the macrophage system, and has prolonged circulation time
poly(ε-caprolactone)-polyethylene glycol (PEL-PEG)	Drug and gene delivery, hyperthermia, MRI contrast agent	Highly soluble and stable in water, biocompatible, hardly recognized by the macrophage system, and has prolonged circulation time
Poly-saccharide	MRI contrast agent, drug and gene delivery	Biocompatible, avoids elimination, provides high stability, and prevents agglomeration
Chitosan	Water treatment and extraction of trace pollutants, Tissue engineering, Hyperthermia	Biocompatible, Increases stability, hydrophilic
Lauric acid (LA)	Cellular tagging and MRI contrast agents, enzyme immobilization in the food industry	Biocompatible when used in low concentrations, stabilizes colloidal suspensions
Citric acid (CA)	Cellular tagging, magnetic hyperthermia	Increases stability and biocompatibility
Liposome	MRI contrast agent, drug delivery, and hyperthermia	Increases blood circulating time
Albumin	Magnetic tagging and magnetically targeted therapy	Biocompatible and not cytotoxic
Erythrocytes	MRI contrast agent and drug delivery	Avoids rapid clearance by the reticuloendothelial system (RES) and prolongs blood circulation
Gelatin	Genomic DNA extraction from bacterial cells, drug delivery	Hydrophilic, biocompatible, natural polymer. Improves drug loading and rapid DNA extraction

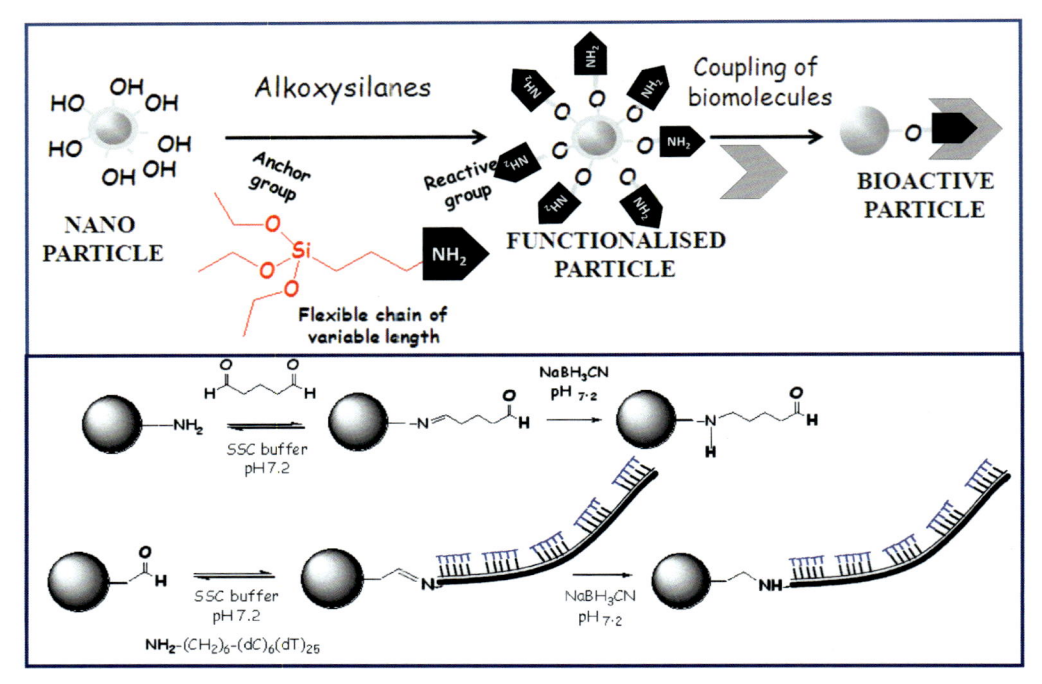

Fig. 4 Surface functionalization of core SPIONs or core-shell SPIONs with biomolecules to produce bioactive SPIONs for application in nanomedicine.

aminosilane for developing surface amino groups which can be easily linked with $-NH_2$ groups of DNA, proteins, and antibodies via a glutaraldehyde treatment using -CHO functional groups intermediate using imine bond ($N=C$) formation.

(a) Surface functionalization using aminosilanes:

Surface functionalization via aminosilanes involves using organosilanes, identified by the chemical formula $X-(CH_2)_n-SiR_n(OR')_{3-n}$. X represents the headgroup, $(CH_2)_n$, and is a flexible alkyl chain spacer group, with the $Si(OR')_n$ groups acting as anchor groups which attach to the silanol hydroxyl groups on the silica surface following hydrolysis of the alkoxy (OR') group. Aminosilanes are simply organosilanes possessing an amino group as the X functionality. Three commonly used aminosilanes for surface functionalization [(3-aminopropyl)-triethoxysilane (APTS), (3-aminopropyl)-diethoxymethylsilane (APDS), and (3-aminopropyl)-monoethoxydimethylsilane (APMS)] are shown in Fig. 5.

(b) Coupling biomolecules by conjugating with surface amine groups via glutaraldehyde:

The cross-linking of biomolecules (e.g., DNA, protein, antibodies, enzymes) typically happens via the amino groups of biomolecule residues, with aldehyde

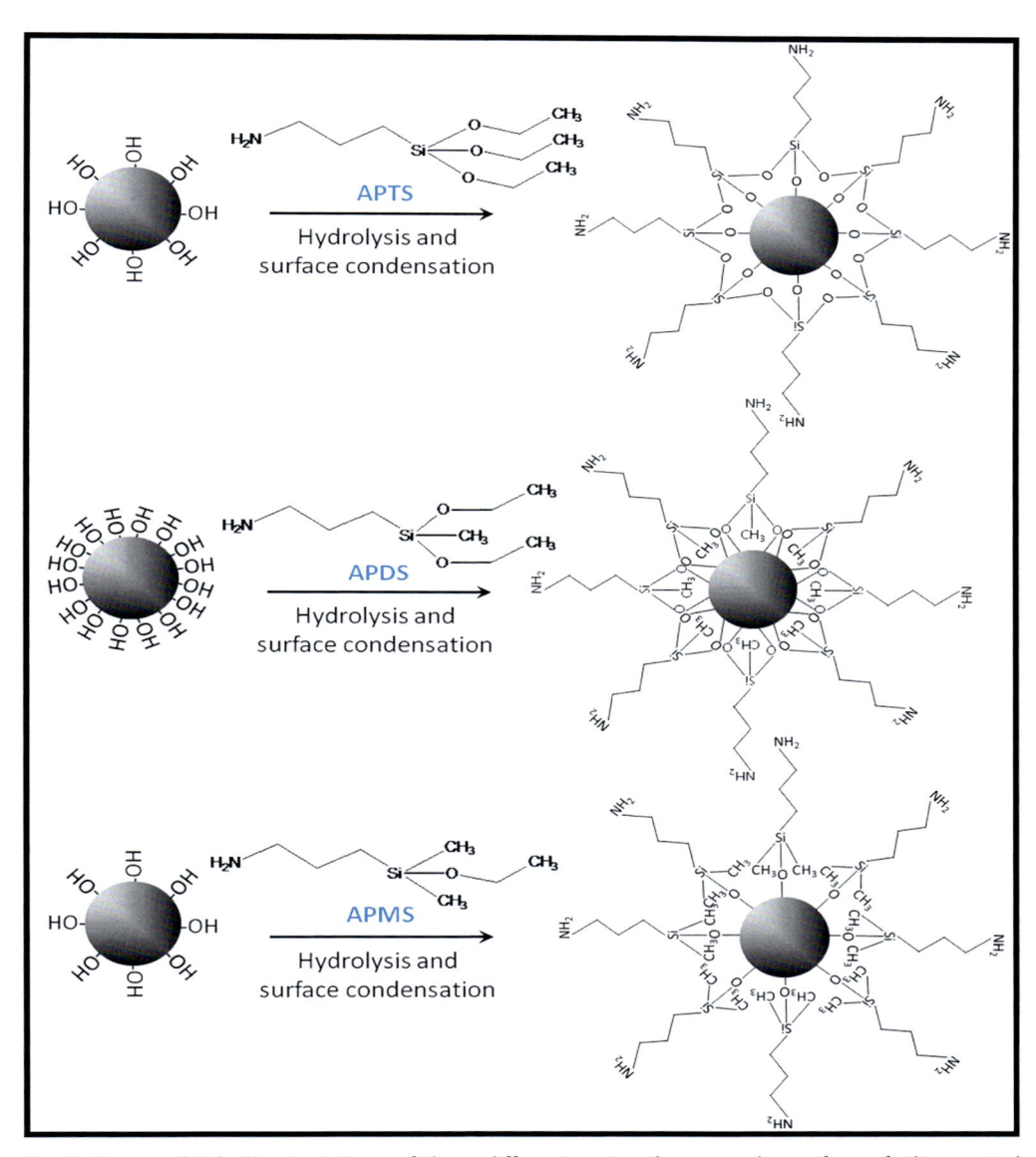

Fig. 5 Proposed "ideal" orientations of three different aminosilanes on the surface of silica-coated SPIONs before attaching to biomolecules.

groups on the surface of the SPIONs, involving an intermediate step of glutaraldehyde modification. Amino-functionalized SPIONs can be converte to aldehyde functional groups by treating with glutaraldehyde followed by covalent cross-linking with amino groups present in biomolecules (see Fig. 4).

(c) Coupling –COOH functional groups of core-shell SPIONs with –NH$_2$ groups of biomolecules via carbodiimide chemistry:

Carbodiimide reagents have also been used to couple surface carboxyl groups of core-shell SPIONs with amino groups of biomolecules, producing stable secondary amines/amides [36, 37]. It has been reported that it can be done by activating carboxyl groups in water, allowing them to react with amine groups [36–38].

3. Size, shape, surface charge of SPIONs, and their characterization

Uncoated magnetic nanoparticles tend to aggregate due to their ultrasmall sizes and magnetic properties. Coating the bare nanoparticles provides a range of stabilization strategies by changing various parameters such as coating composition, coating thickness, hydrophobicity, and surface charge that influence the characteristics of the SPIONs in different ways [5, 26, 39, 40]. For instance, coating SPIONs with diamagnetic materials results in decreased saturation magnetization [41–44]. Different methods of syntheses can also influence magnetization values depending on the surface properties such as hydrophilic or hydrophobic. For example, Lopez et al. [41] have reported a change in saturation magnetization for as-synthesized water-based magnetite nanoparticles from 0.067 emu/cm^3 to 0.335 emu/cm^3 by changing the coating using oleic acid kerosene-based magnetite nanoparticles. Voit et al. [43] have reported a decrease of saturation magnetization for polymer-coated magnetite nanoparticles similar to an earlier report by Sen and Bruce [44] for silica–coated magnetite nanoparticles.

Sen and coworkers reported [26] a systematic study for the synthesis and surface coating of a range of morphologies from spherical to rhombic by tuning reaction parameters. The stability of SPIONs in suspension is of paramount importance for applications in nanomedicine, as aggregation can restrict their diffusion through the cell membrane, and so the coating of SPIONs plays an essential role in improving this key parameter. Fig. 6 presents a range of SPIONs in suspension with or without coating by phospholipid vesicles [5]. Uncoated SPIONs are unstable in suspension and tend to aggregate, whereas coating can improve their stability for a period of up to several weeks. This is reflected in particle size data from dynamic light scattering experiments (Fig. 6H). Uncoated particle aggregates in the suspension are of size in the order of micrometer in diameter (DLS data) in comparison to the TEM data, where individual particulate sizes were measured to be of order 10s to 100s of nm. By coating with phospholipid, the resultant SPIONs have improved size and stability results in suspension. From DLS measurements, the average diameter after coating with phospholipids was consistent with the sizes directly measured by the TEM, being in the same 10s to 100s of nm range. The stability of nanoparticles in suspension is also directly linked with surface charge which can be correlated with the change in zeta potential values from high positive to high negative (Fig. 6I). Coating mesoporous silica with controlled shell thickness can improve the surface area of the

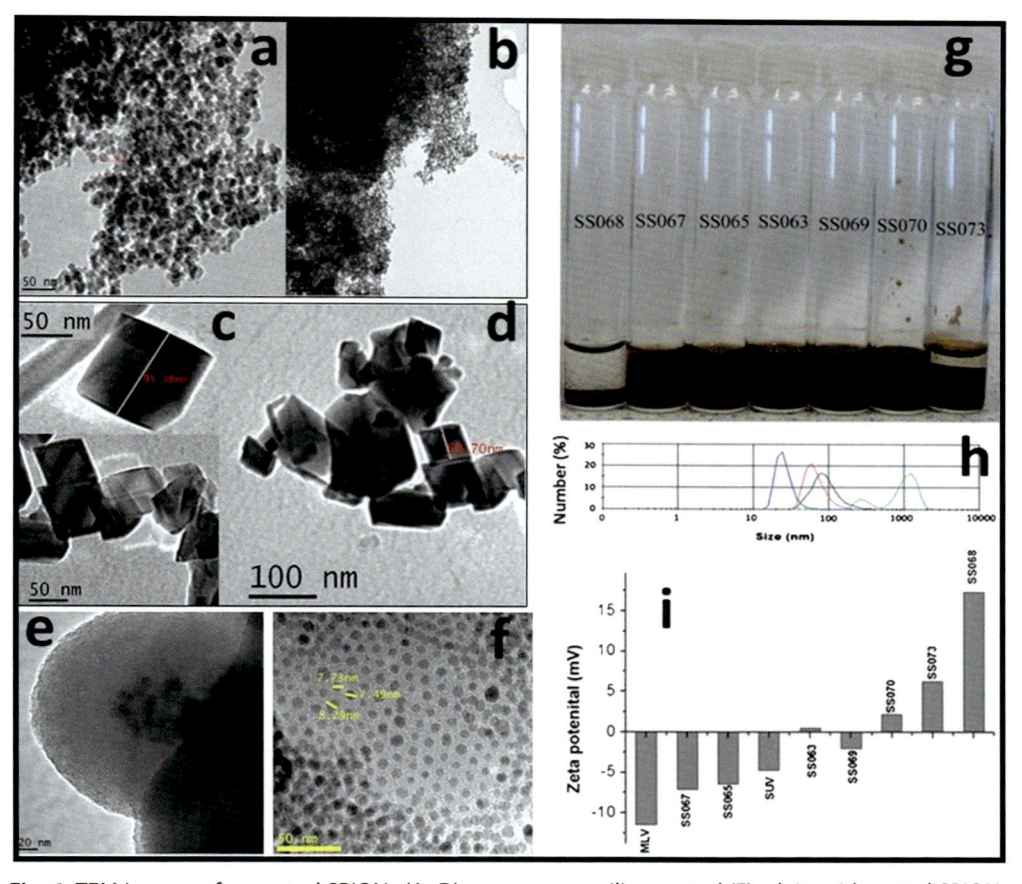

Fig. 6 TEM images of uncoated SPIONs (A–D), mesoporous silica-coated (E), oleic acid-coated SPIONs (F), bare and phospholipid-coated SPIONs as a photograph (G). Dynamic light scattering sizing and the zeta potentials of phospholipid-coated and uncoated SPIONs are shown in (H) and (I), respectively.

core-shell nanoparticles (Fig. 6E) due to internal porosity. In situ synthesis of magnetite in the presence of oleic acid produced nanoparticles (Fig. 6F) with ultrasmall sizes (diameter < 10 nm) and with high stability in organic solvent (hexane).

4. Applications of SPIONs as theranostic (therapeutic and diagnostics) agents in nanomedicine

SPIONs have shown great potential in nanomedicine. Current developments in the synthesis process of magnetic nanoparticles make it possible to develop SPIONs with controllable size and biocompatible coatings which enables them to interact with biological entities. Furthermore, SPIONs can be easily manipulated by means of an external magnetic field. These properties, with the intrinsic permeability of magnetic fields into human tissue, make the nanomaterials ideal for a wide range of biomedical applications involving the transport or immobilization of magnetic nanoparticles or magnetically tagged biological units, including delivery systems for therapeutic agents [20, 45]. Targeting allows localized delivery of the therapeutic agents, reduces the required dosage, and consequently has less toxic side effects. These therapeutic agents either consist of therapeutic drug-encapsulated nanoparticles or nanoparticles with their own therapeutic effects. Typically, these effects are toxic to healthy tissue as well, which makes it essential to optimize the therapeutic window with low enough toxicity to result in minimal damage to surrounding tissue, yet toxic enough to eradicate the intended target cells. These parameters should be considered when designing a therapeutic iron oxide nanoparticle, i.e., SPIONs by considering their hyperthermia heating properties applied to the killing of cancer cells. Magnetic nanoparticles respond to an AMF, resulting in thermal energy generation; hence, hyperthermia heating as described in Section 1.3. Furthermore, iron oxide nanoparticles may be used as a spin lattice relaxation time (T2) MRI contrast agent [6]. Iron oxide nanoparticles effectively lower the T2 contrast signal in MRI [40], giving significant differences in the image which allow the area of interest to be highlighted by the change in contrast. Adding an appropriate coating and incorporating functional ligands, such as fluorophores or radioactive ions, the resultant modified SPIONs can also be powerful for fluorescent and Positron Emission Tomography (PET) imaging [46].

4.1 Therapeutic applications

4.1.1 Magnetic drug targeting (MDT)

The main disadvantage of most chemotherapeutic drugs is their nonspecificity in drug action. An intravenously administered drug, via systemic distribution, results in harmful side effects such as bone marrow depression and reduced immunity, as the drug affects healthy cells as well as the targeted malignant cells. The use of magnetic carriers to target specific sites in the body was first proposed by Freeman et al. in the 1960s [47]. The main objectives of the magnetic drug delivery system are to target the specific site in order to reduce the systemic distribution and required dosage of the toxic drug and with it the unwanted side effects [4, 20]. Over the last two decades, much research has been carried out on developing and optimizing magnetic drug delivery systems with improved magnetic properties (size, charge, magnetization), increased nanoparticle concentration in targeted tissue (magnetically assisted targeting), and improving the circulation half-life and biodistribution of nanoparticles (size, charge, coating, and functionalization). [7, 19].

A magnetically targeted drug delivery system consists of a drug attached to biocompatible magnetic nanoparticles which could be inhaled or injected intravenously and be

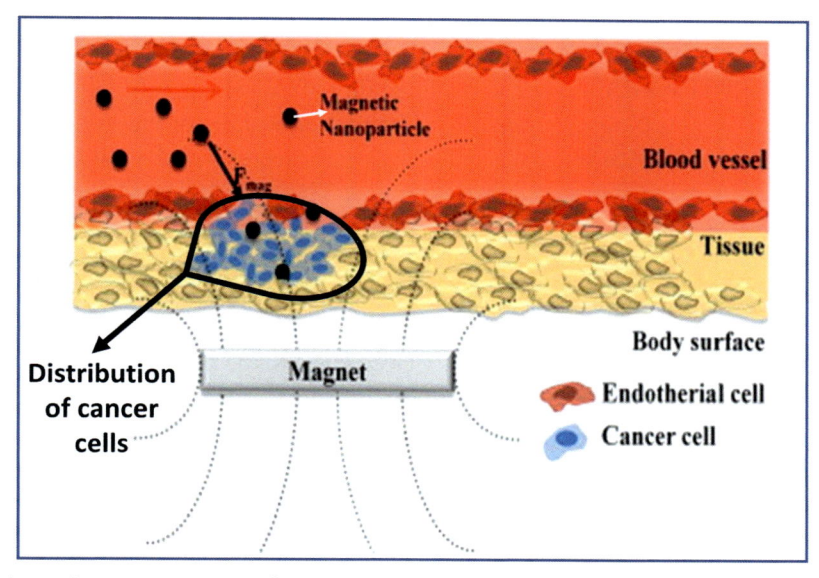

Fig. 7 Schematic representation of magnetic nanoparticle-based drug delivery system. *(Request permission from J.H. Park, G. Saravanakumar, K. Kim, I.C. Kwon, Targeted delivery of low molecular drugs using chitosan and its derivatives, Adv. Drug Deliv. Rev. 62 (1) (2010) 28–41.)*

guided to concentrate at the target site through an external magnetic field. Schematic representation of a magnetic-targeted drug delivery system is shown by Park et al. [48] in Fig. 7.

The physical concepts of magnetic targeting therapy are based on the magnetic force exerted on a magnetic nanoparticle under the influence of an external magnetic field gradient. The efficient magnetic targeting depends upon many physical parameters including the field strength, depth of the targeted tissue, magnetic moment of the nanoparticles, size and shape of the nanoparticles, blood flow rate, and nanoparticle concentration and viscosity [49,50].

Magnetic targeting studies on small animals have already been reported [51–53]. Alexiou et al. [51] investigated the treatment of squamous cell carcinoma in rabbits with mitoxantrone (MTX) bound to phosphate groups of magnetic particles such as Ferrofluids (FF) coated with starch derivatives (FF-MTX). The FF-MTX was injected either via the femoral artery or ear vein, while an external magnetic field was focused on the tumor. No negative side effects were observed and the intra artery FF-MTX administration produced a significant ($p < 0.05$), complete, and permanent remission of the squamous cell carcinoma in comparison with both the control group and the intravenously administered group. The accumulation of magnetic particles in the tumor site was visualized both histologically and by MRI [51]. The same group reported another study in 2006 which further confirmed the efficacy of magnetic targeting in concentrating the drug bonded to magnetic nanoparticles in the peritumoral region and inside the tumor [53].

Yanai et al. [54] have demonstrated the feasibility of using magnetically targeted delivery of stem cells in neurological tissue on the upper hemisphere of the rodent retina. Rat mesenchymal stem cells loaded with SPIONs were injected via the tail vein of the rat model with a disc magnet placed outside the eye. Their results demonstrated that cells were localized at the inner retina in a tightly confined area corresponding to the position of the magnet and stem cells [54].

Tietze et al. [55] have reported a study comparing the in vivo distribution of free and magnetic nanoparticles loaded with mitoxantrone (MTX). It was shown that while using magnetic targeting, 57.2% of the drug was accumulated in the tumor region compared to only 0.7% for systemic intravenous administration. Similar studies show the advantage of magnetically targeted chemotherapy over conventional chemotherapy [52]. However, their use is limited due to technical challenges that include a weakening of the magnetic field with distance and depth of the target.

4.1.2 Magnetic fluid hyperthermia (MFH)

The National Cancer Institute of the United States of America defines hyperthermia therapy as "A type of treatment in which body tissue is exposed to high temperatures to damage and kill cancer cells or to make cancer cells more sensitive to the effects of radiation and certain anticancer drugs" [56]. Hyperthermia therapy is commonly considered as a temperature rise between 41°C and 46°C, which alters the intercellular structures and leads to cellular degradation and induced apoptosis [57]. This type of hyperthermia is used in combination with chemotherapy or radiation for further efficiency [58]. Hyperthermia with a temperature rise above 46°C (thermal ablation) results in direct cell death by tissue necrosis or carbonization [57,59].

Different sources such as laser (see Section 4.1.3), ionizing radiation, ultrasound, and microwaves can be utilized for externally induced hyperthermia in cancerous tissue. Studies have shown that although many of these methods raise the tissue temperature, they are not able to specifically target heat to a cancer site and so result in harmful side effects within healthy tissues as well [56]. This observation has encouraged researchers to use magnetic-nanoparticle-induced hyperthermia.

Magnetically induced hyperthermia is based on the concept that placing magnetic particles (e.g., SPIONs) under the influence of an AMF with sufficient field strength and frequency induces heating which subsequently raises the surrounding tissue temperature. The concept of magnetically induced hyperthermia for cancer treatment was first proposed by Gilchrist et al. in 1957 [60]. They established that lymph nodes could be heated to kill lymphatic metastases using an external magnetic field after the administration of magnetic particles. Maghemite (γ-Fe_2O_3) nanoparticles were used in this experiment and showed a temperature rise of 14°C generated using 5 mg of γ-Fe_2O_3 per gram of tissue while exposed to a magnetic field of 200–240 Oe (16–19 kA/m) [60]. In 1979, Gordon et al. [61] reported inducing intracellular hyperthermia, by administering submicron magnetite nanoparticles intravenously to Sprague Dawley rats. They reported successful magnetic field-induced heating and destruction of the tumors occurred in their in vivo experiments.

4.1.3 Laser-induced hyperthermia with or without photosensitizer-conjugated SPIONs

Using magnetic nanoparticles alone can sometimes present limitations in heating efficiency and in monitoring their distribution in vivo [62]. This drawback can be solved by adding optical dyes or photosensitizers (PS), where the absorption of the laser light of specific wavelengths can help the transformation of emitted energies into heat [63]. The combination of magnetic nanoparticles and PS or optical dyes enhances the therapeutic system by dual modulation of both AMF and light-induced hyperthermia along with diagnosis but outside the scope of this book chapter.

It has been suggested that some of the drugs show improved efficiency under heat activation, demonstrating a synergistic effect of hyperthermia and chemotherapy over simple drug-induced monotherapy [64]. Fig. 8 is reproduced from published data on SPIONs which can be used for hyperthermia by simultaneous application of AMF and laser light without the presence of photosensitizer [65].

4.1.4 Chemodynamic therapy (CDT) by reactive oxygen species (ROS) due to the exposure of SPIONs and modified SPIONs

It is suggested that the cytotoxicity induced by magnetite nanoparticles is due to their physicochemical properties. Magnetite nanoparticles present a large surface area for the generation of free radicals in an acidic condition of the tumor sites in the presence of oxygen and water. Consequently, the reactive oxygen species (ROS), including superoxide anions, hydroxyl radicals, and nonradical hydrogen peroxide, are taken up by the cells where they can produce oxidative stress by activating transcription factors for proinflammatory mediators. Additionally, ROS react with cellular macromolecules and damage the cells by peroxidizing lipids, changing proteins, disrupting DNA, and interfering with signaling functions, and subsequently cause cell death [7]. The effect of SPIONs and core-shell SPIONs in cell viability on human HeLa cells is presented in Fig. 9.

4.2 Diagnostics applications
4.2.1 Magnetic resonance imaging (MRI)

MRI imaging is a noninvasive imaging modality based on the principle of Nuclear Magnetic Resonance (NMR). ^{1}H MRI is based on the precession of the magnetic moments of protons present in the hydrogen nuclei of cellular water in an applied magnetic field. To provide high-quality imaging, MRI contrast agents are utilized. The contrast agent

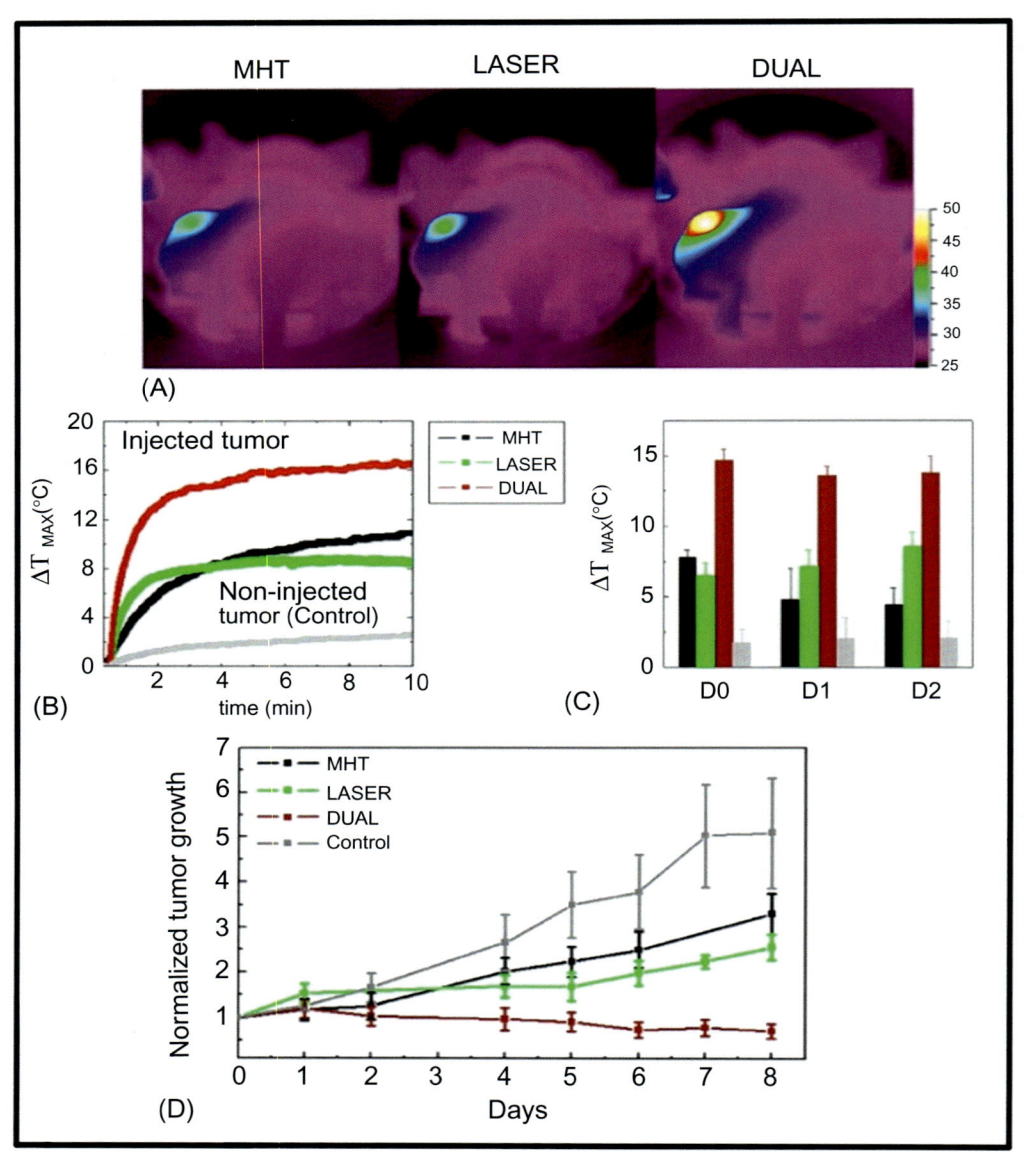

Fig. 8 In vivo heat therapy: (A) Thermal images obtained with an IR camera in mice. From left to right, (i) after intra-tumoral injection of nano cubes (50 µL at [Fe] = 250 mM) after 10 min application of magnetic hyperthermia (MHT, 110 kHz, 12 mT (120 G)), (ii) NIR-laser irradiation (LASER, 808 nm at 0.3 W/cm²), or (iii) DUAL (both effects). (B) Corresponding thermal elevation curves for all treatments and for the noninjected tumor in the DUAL condition. (C) Average final temperature increase obtained after 10 min (MHT, LASER, and DUAL) on day 0 (1 h after injection) and 1 and 2 days after injection and for noninjected tumors. (D) Average tumor growth (groups of six tumors each in noninjected mice submitted to no treatment (Control) and in nano cube-injected mice exposed to MHT, LASER, and DUAL during the 8 days following the 3 days of treatment). *(Reproduced by permission from A. Espinosa, R. Di Corato, J. Kolosnjaj-Tabi, P. Flaud, T. Pellegrino, C. Wilhelm, Duality of iron oxide nanoparticles in cancer therapy: amplification of heating efficiency by magnetic hyperthermia and photothermal bimodal treatment, ACS Nano 10 (2) (2016) 2436–2446.)*

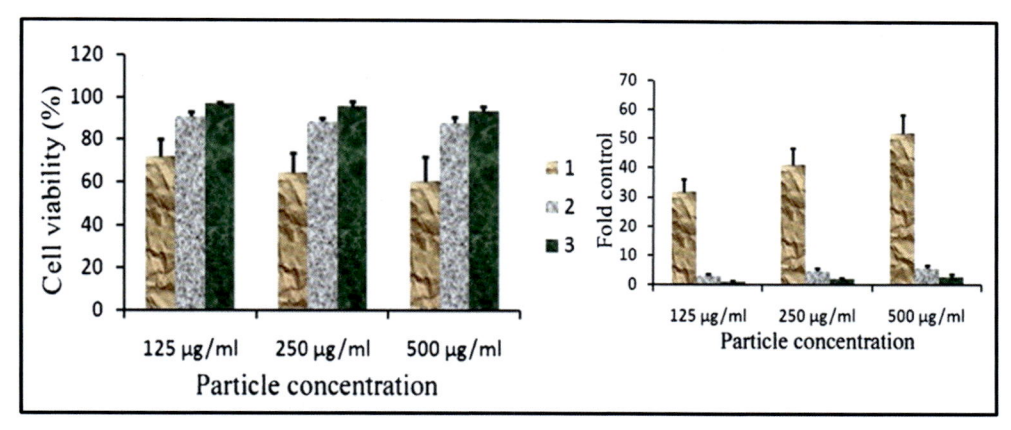

Fig. 9 Left: MTT assay values for HeLa cells incubated with various concentrations of uncoated and coated SPIONs with phospholipids; Right: Produced ROS in the HeLa cells after incubation with various concentrations (1: uncoated, 2 and 3: phospholipid-coated) [5].

shortens the relaxation time of the nuclear moments, measured after applying a Radio Frequency (RF) pulse. LaConte et al. [40] reported that the coating thickness of magnetite nanoparticles can significantly influence the NMR relaxation rates R2 (transverse) and R1 (longitudinal) and consequently the R2/R1 ratio of the MRI contrast agent. Using a polyethylene glycol (PEG)-modified phospholipid coating, they observed that an increase in coating thickness led to a decrease in R2 and an increase in R1. Duan et al. [39] have shown polymer coatings with higher hydrophilicity (PEI versus octadecene coating) yielded larger R2 values. They have suggested that these effects are the result of intrinsic surface spin disorders as well as rapid diffusion of water molecules between the bulk phase and adjacent layer surrounding around the particle surface [40, 42]. This demonstrates the importance of careful consideration in coating processes and the combinatorial effects of the coating on the final properties of a magnetic nanoparticle system. Fig. 10 shows the use of ultrasmall superparamagnetic iron oxide (USIOP) nanoparticles in detecting brain tumor and comparing the data with a conventional contrasting agent Gadolinium [66].

4.2.2 Magnetic particle imaging (MPI)

Magnetic particle imaging (MPI) is an emerging tracer imaging modality being extensively researched to detect SPIONs in cancerous tissues as well as inflammation. It provides 3D tomographic images of the iron oxide distribution within the tissues with several advantages over MRI [67]. Due to zero tissue background signal, the images produced are of high quality. In addition to imaging, MPI assists magnetic hyperthermia in cancer cells and minimizes exposure to normal cells. The high temporal and spatial resolution enables better quantification at any depth, with high sensitivity and fast imaging. This property allows researchers to plan and evaluate the amount of SPIONS required for

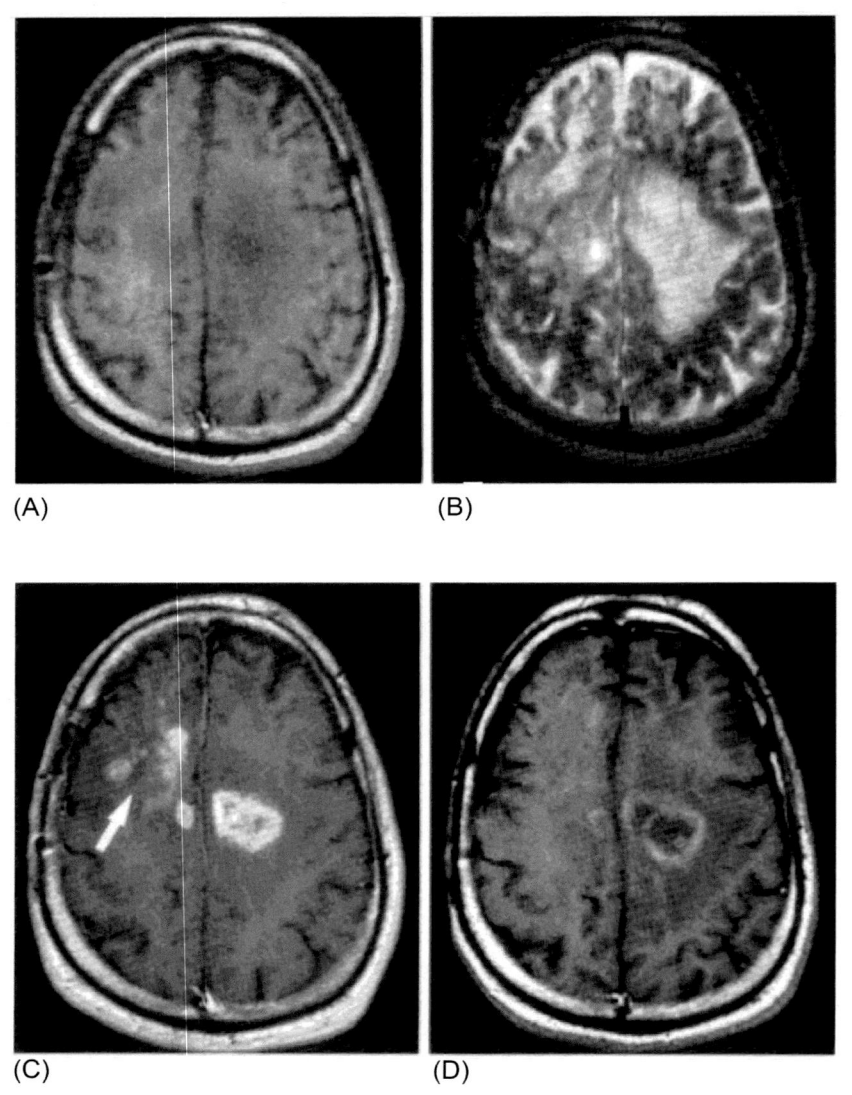

Fig. 10 Enhancement of a recurrent grade III/IV astrocytoma (patient G.P.) with a USPIO versus a Gd chelate. (A, B) Pre-contrast, T1-weighted SE and T2-weighted FSE images, respectively, which are somewhat degraded by motion but show vasogenic edema in the centra semiovale bilaterally, worse on the left. (C, D) T1-weighted SE images obtained, respectively, within minutes after intravenous administration of the Gd chelate and 24 h after administration of the USPIO. The dominant lesion(s) on the left show ring enhancement with both agents, but the degree of enhancement and thickness of the enhancing rind are less with the USPIO. Furthermore, multiple focal areas of enhancement with gadolinium (arrow, C) do not enhance with the USPIO. The pathological basis for the latter observation is unknown in this case. *(Reproduced with the permission form W.S. Enochs, G. Harsh, F. Hochberg, R. Weissleder, Improved delineation of human brain tumors on MR images using a long-circulating, superparamagnetic iron oxide agent, J. Magn. Reson. Imaging 9 (2) (1999) 228–232.)*

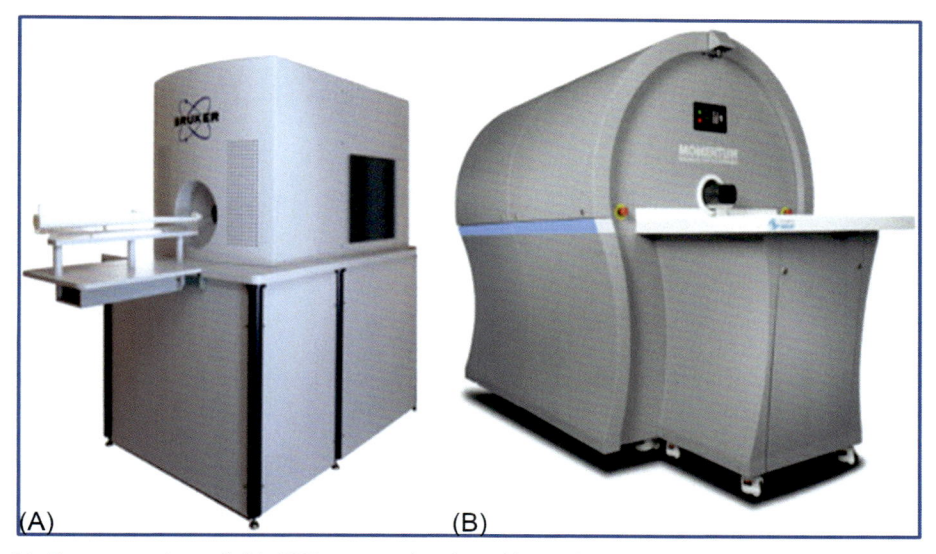

Fig. 11 Two currently available MPI systems developed by Bruker, Germany [68]. A: A Preclinical Bruker MPI designed for 3D high temporal resolution real time imaging application; B: Magnetic Insight Momentum MPI designed for 2D high sensitivity imaging.

subsequent hyperthermia treatment. However, the drawback of MPI over MRI is the lack of anatomical information provided on the images. To overcome this shortcoming, MPI is attached with other imaging modalities like MRI or computer tomography (CT). A commercial MPI instrument developed by Bruker, Germany, has been reported for this type of imaging by Bulte et al. [68] and is presented in Fig. 11.

In addition to imaging, MPI provides a platform for therapy using hyperthermia via magnetic heating. Fig. 12 provides a step-to-step guide on using MPI for localized hyperthermia as presented by Tay et al. [69].

5. SPIONs and surface-modified SPIONs in therapy and diagnosis for various diseases

In this section, an overview of a range of SPION formulations currently being studied for therapeutic and diagnostic purposes is given. Strategies to expand the use of SPIONs from imaging to targeted drug delivery are considered alongside a range of different health issues.

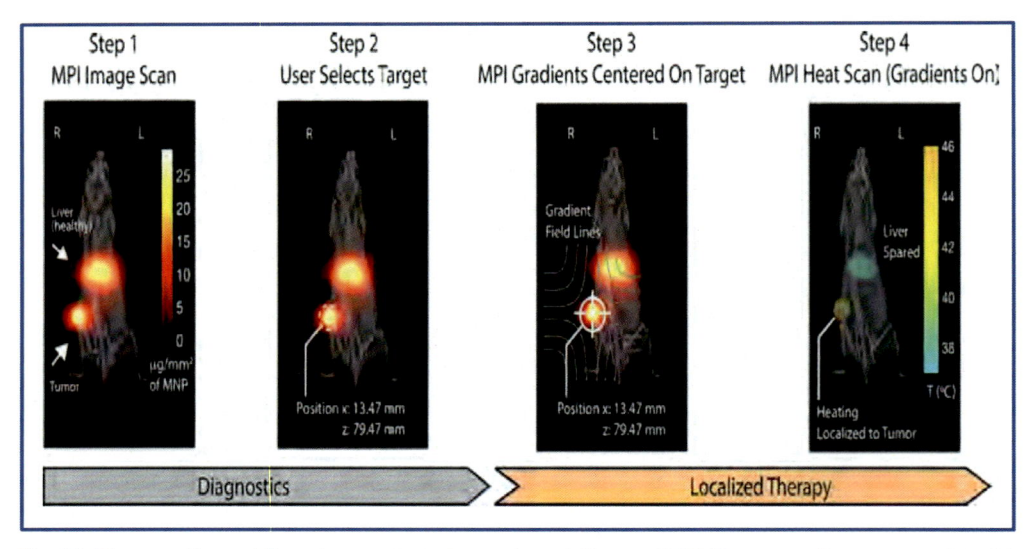

Fig. 12 Theranostic workflow demonstrated experimentally on a U87MG xenograft mouse model with superparamagnetic iron oxide nanoparticles (SPIONs) present in the liver and tumor.

Step 1: Magnetic particle imaging (MPI) scan at 20 kHz, 20 mT (200 G) enables clear visualization with high contrast of the SPION biodistribution in regions of pathology (tumor) and in healthy clearance organs (liver). Imaging parameters are such that the SPIONs do not heat.

Step 2: User selects a region, in this case the tumor, to localize the magnetic hyperthermia.

Step 3: MPI gradients are shifted to center the field-free region (FFR) on the target. This magnetically saturates SPIONs away from the FFR to prevent heating.

Step 4: Heat scan at 354 kHz, 13 mT (130 G) is performed while the MPI gradients are "on" and held in position. Heating is experimentally localized in the FFR (centered at the tumor) while minimizing collateral heat damage to the liver. Image courtesy: Tay et al. [69].

5.1 Cancer

Cancer treatment has greatly improved in the last few years, resulting in higher cure rates and reducing mortality rates of patients. The first Phase I clinical trial of magnetic drug targeting was conducted in 1996 with 14 unsuccessfully pretreated cancer patients [70]. In this study, Epirubicin chemically bounded to 100 nm biocompatible starch–coated iron oxide particles was used. The trial indicated that the drug was successfully guided to the target site using an external magnetic field for nearly half of the patients. A similar study was conducted in 2004 by Lemke et al. [71] with 11 patients, where magnetic targeting was monitored using MRI and which proved to be successful in magnetic drug targeting. The study showed that about half of the nanoparticles were successfully guided and concentrated at the target site. Fig. 13 shows a magnetic targeted drug delivery system presented by Shapiro et al. [72].

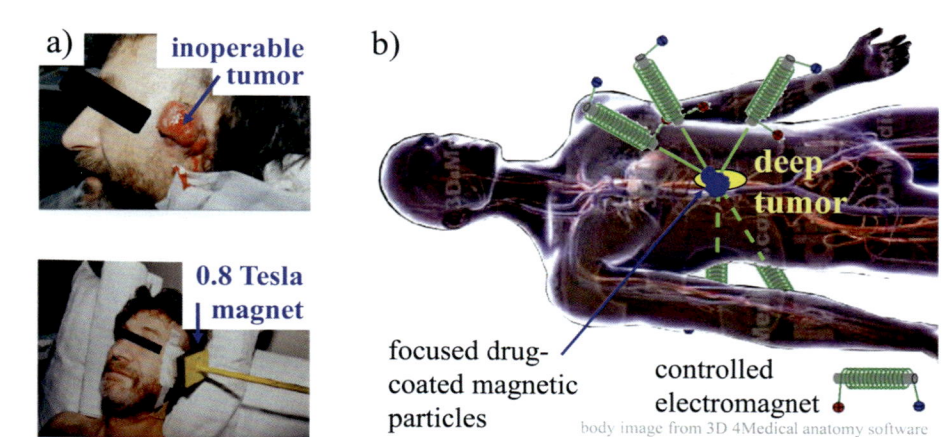

Fig. 13 (A) The first human trials in magnetic drug targeting (1). Epidoxorubicin-coated magnetic nanoparticles were administered systemically to advanced head and neck and breast cancer patients, and a single permanent magnet was held near inoperable but shallow tumors to concentrate the chemotherapy. (B) A goal in magnetic targeting is to use magnetic fields to focus therapy precisely to any desired target in the body, for example, to a deep tumor as illustrated. Currently, there are no magnetic systems that can achieve this kind of precise and deep focusing. *(Reproduced with the permission from B. Shapiro, S. Kulkarni, A. Nacev, S. Muro, P.Y. Stepanov, I.N. Weinberg, Open challenges in magnetic drug targeting, Wiley Interdiscip. Rev. Nanomed. Nanobiotechnol. 7 (3) (2015) 446–457.)*

The properties of superparamagnetism and magnetic hyperthermia allow SPIONs to be targeted at the site of cancer cells and their targeted treatment. Hilger [73] has provided a perspective for active and passive targeting of magnetic hyperthermia in in vivo studies. Johannsen et al. [74] have reported the first clinical study using SPIONs as therapeutic agents via hyperthermia treatment. Application of a magnetic field in a human trial was first recorded using a magnetic field applicator—MFH300F, manufactured by MagForce Nanotechnologies GmbH, Berlin, Germany (Fig. 14). The magnetic field frequency was 100 kHz with a variable field strength of up to 18 kA/m (226 Oe); however, it was kept at less than 5 kA/m (62 Oe) during the treatment. Hyperthermia sessions of 60 min were performed once a week for 6 weeks on 67-year-old patient with recurrent prostate cancer. Nanoparticle suspensions were injected transperineally into the prostate under transrectal ultrasound and fluoroscopy guidance. A CT scan of the prostate was used to plan the treatment. Invasive thermometry of the prostate was implemented in the first and last hyperthermia treatment. Additional CT-scans were carried out after the first and last treatment to investigate the magnetic nanoparticle distribution in the prostate and monitor their location by using temperature control probes. The results of this study indicated that hyperthermia treatment using magnetic nanoparticles was feasible and well-tolerated. The maximum and minimum intraprostatic temperatures measured were 48.5°C and 40.0°C during the 1st treatment and 42.5°C and 39.4°C during the 6th

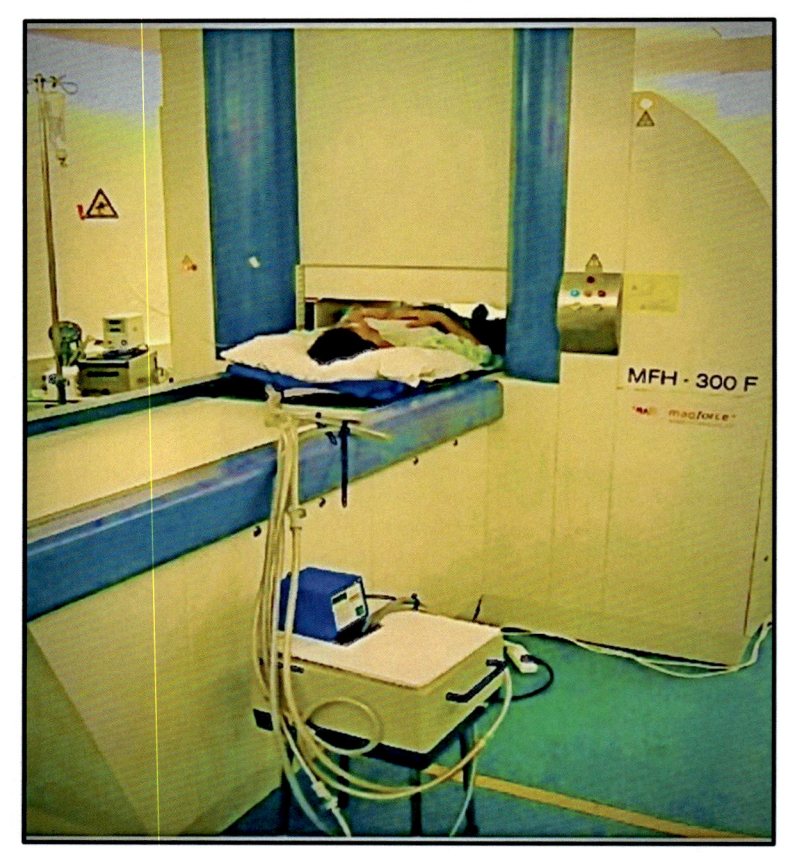

Fig. 14 AC magnetic field applicator (MFH300F, MagForce Nanotechnologies GmbH, Berlin). For cooling purposes, a closed loop of tubes with circulating cold water is placed around the patient's inner thigh, perineum, and the groin [74].

treatment, respectively [74]. These results encouraged a phase I clinical trial with 10 patients suffering from recurrence prostate cancer to investigate the feasibility and toxicity of magnetic induction hyperthermia [75]. The same hyperthermia treatment with intraprostatic injection of nanoparticles was used by the same group and a maximum temperature of 55°C was achieved in the prostates. Prostate-specific antigen (PSA) decreased in eight patients after treatment. No systemic toxicity was observed for 17.5 months (study period between 3 and 24 months) of the median follow-up and indication. The results indicated that magnetic fluid hyperthermia of recurring prostate cancer is feasible and well-tolerated and wouldn't cause significant side effects [75].

Wust et al. [76] have performed magnetic fluid hyperthermia (MFH), named earlier as Magnetic hyperthermia treatment, on 22 patients suffering from recurring tumor entities to evaluate the feasibility and efficacy of MFH in combination with irradiation and/or chemotherapy. Magnetic nanoparticles were implanted with three different methods of infiltration under CT fluoroscopy, transrectal ultrasound-guided implantation with X-ray fluoroscopy, and operative infiltration with visual control. The temperature distribution in the tumor region was based on direct temperature measurements. Magnetic fields with strengths of 3.0–6.0 kA/m (38 to 75 Oe), greater than 10.0 kA/m (126 Oe), and 7.5 kA/m (94 Oe) were used in the pelvis, head, and neck region, respectively. They have reported that the treatments were well-tolerated without significant side effects. However, they have reflected that an improvement in temperature distribution is required [76]. Maier-Hauff et al. [77] used magnetic hyperthermia treatment in combination with radiation therapy on 14 patients with recurring glioblastoma multiforme. The patients received 3D image-guided intra-tumoral injection of magnetic nanoparticles using MRI. The nanoparticle distribution after injection was studied using CT-scans, and the data compared to preoperative MRI data to calculate the expected heat distribution under the required magnetic field strength. A magnetic field with strength of 2.5–18 kA/m (31–226 Oe) and frequency of 100 kHz was used to induce particle heating. The results indicated that a median maximum temperature of 44.6°C (42.4–49.5°C) was reached in the tumor and the treatment was well-tolerated by patients with minor or no side effects [77]. Although there are still some challenges to overcome and optimize the effect of magnetic hyperthermia in cancer therapy, magnetic nanoparticle-based hyperthermia trials have passed the preclinical stages and received regulatory approval in 2010 as a new clinical therapy termed "thermotherapy" [78]. This approval follows successful completion of the conformity evaluation procedure of the NanoTherm magnetic fluid by Medcert GmbH and of its NanoActivator magnetic field applicator by Berlin Cert GmbH. The nanomaterials were aminosilane-coated iron oxide nanoparticles with an average diameter of around 15 nm [78]. Magnetic field parameters such as field strength and frequency are selected to be compliant with the approved protocols for magnetic fluid hyperthermia in Europe. For instance, for treatment of glioblastoma multiforme (MagForce, Berlin, Germany), magnetic field frequencies of order 100–200 kHz at around 20 mT (200 G) are typically chosen [79].

Moreover, cancer diagnosis through tumor imaging is highly effective and reduces the need for invasive tests, such as biopsy. Dual imaging (e.g., MRI and MPI) can provide structural and morphological information along with functional and molecular information which helps in identifying the disease and its precise location. Particularly, SPION-based nanocomposites have been extensively researched for MRI and MPI, for visualizing tumors and metastasis in liver, lymph nodes, and spleen [80–82]. Current investigations that visualize the temperature distribution in liver tumors for hyperthermia studies have also been reported using an MPI system [83]. A short overview of various core-shell SPIONs currently being researched in cancer treatment is presented in Table 4.

Table 4 SPIONs being researched for specific cancer therapeutics and diagnostics.

SPIONs	Coating	Drug/dyes	Size (nm) TEM/DLS	Magnetization (emu/g)	Types of cancer	Specificity	In vitro	In vivo	References
γ Fe₂O₃	Poly (ethylene glycol)-derivatized phosphine oxide (PO-PEG)	–	<4 (TEM)	10–50	Breast cancer	MRI	✓ MCF-7	–	[83]
	Phosphatidylcholine (PC) liposomes	–	90±20 (TEM)	34±4	Colon cancer	Passive targeting	✓ human colon carcinoma cell line T-84	–	[84]
	1,2-Diacyl-SN-glycero-3-phosphoethanolamine-N- [methoxy (polyethylene glycol)–2000] (DSPE-PEG2000)	–	180 (DLS)	–	Prostate cancer	Magnetic targeting	✓ human adenocarcinoma prostatic cell line PC3	–	[85]
Fe₃O₄	Poly (lactic-co-glycolic acid) (PLGA), AS1411 Aptamer (Apt)	Doxorubicin	173.8±8.3 (DLS)		Colon carcinoma	Targeted drug delivery and theranostics	✓ C26 colon carcinoma cell line	✓ (BALB/c mice)	[86]
	Poly (acrylic acid) (PAA), Pluronic F127-Folic acid	Nile red	9.6±2.6 (TEM), 246±11 (DLS)		Not specified	Targeted theranostics	✓ Hela cells	–	[87]
	Glyceryl monooleate (GMO); 2, 3 meso Dimercapto succinic acid (DMSA), human epidermal growth factor receptor 2 (HER2) antibody	Paclitaxel, Rapamycin	7 (TEM) 152 (DLS)	53.5±0.7	Breast cancer	Targeted dual therapy	✓ MCF-7	–	[88]
	Silica (SiO₂), polyethyleneimine (PEI), vascular endothelial growth	–	57 (TEM)	–	Breast cancer	Gene delivery and MRI as theranostics	✓ MCF-7	–	[89]

	factor small hairpin RNA (VEGF sh RNA)								
	Chitosan-silica	Sorafenib	>100	45	–	Drug delivery	–	–	[90]
	Calcium oxalate dehydrate (COD)	–	21 (TEM)	98–136	Not specified	Optical coherence tomography (OCT), theranostics	✓ Hela cells	–	[91]
	Ce6 nanocluster	Ce6	92 (DLS)	–	Breast cancer	Photothermal therapy, imaging	✓ 4T1, HUVEC	Athymic nude mice, 4T1 cancer cells	[92]
	Methoxy poly (ethyleneglycol)-*block*-poly (dopamine-ethylenediamine-2,3-dimethylmaleic anhydride)-L-glutamate [mPEG-b-P (Dopa-Ethy-DMMA) LG]	Ce6	100 ± 11 (DLS)	–	Liver cancer	Passive targeted bioimaging and PDT	✓ (HepG2 cell lines)	–	[93]
	Poly(ε-caprolactone-*co*-lactide)-b-poly (ethylene glyc*ol*)-b- poly(ε-caprolactone-co-lactide) (PCLA-PEG-PCLA)	PTX	212 (DLS)	–	Breast cancer	Tumor targeting and theranostics	✓ MCF-7	✓ (BALB/c mice)	[94]
	Triethanolamine (TEA), Diethylene glycol (DEG), Tetraethylene glycol (TTEG)	–	192 (DLS)	–	Liver cancer	Magnetic hyperthermia	✓ (HepG2 cell lines)	–	[95]
Iron oxide (not specified)	Polyethylene glycol- poly (β-benzyl-L-asparate) (PEG-PBLA)	Imidazole, Ce6	60 (TEM) 70 ± 5 (DLS)	30	Colorectal carcinoma	Passive targeted, pH-sensitive, MRI	✓ HCT116	✓ (BALB/c mice) homogenous HCT116 and heterogeneous CT26 tumors	[96]
	Polyethylene glycol- Folic acid (PEG-FA)	Dox	9 (TEM) 50 (DLS)	–	Breast cancer	Active targeting	✓ MCF-7	–	[97]
	Folic acid-Polyethylene glycol grafted polyethyleneimine-	–	120 (DLS)	–	HCC	Prognostic marker/	Hep3B	HCC xenograft	[98]

Continued

Table 4 SPIONs being researched for specific cancer therapeutics and diagnostics—cont'd

SPIONs	Coating	Drug/dyes	Size (nm) TEM/DLS	Magnetization (emu/g)	Types of cancer	Specificity	In vitro	In vivo	References
	siRNA plasmid-Transducin β-like protein 1-related protein (FA-PEG-g-PEI/psi RNA-TBLR1)					therapeutic targeting			
	1,2-Dipalmitoyl-sn-glycero-3 phosphoethanolamine-N-[methoxy(PEG)-2000 (PEG- 2 -PE)	Sorafenib	30–100 nm (TEM)	12–17	HCC	Magnetically targeted therapy	HepG2	–	[99]
	1,4-Diaminobenzene (14DAB), 4-aminobenzoic acid (4ABA), and 3,4-diaminobenzoic acid (34DABA), each combined with terephthalic acid (TA)	–	<20 (TEM) <200 (DLS)	55–71	Liver cancer	MFH and theranostics	HepG2	–	[100]

5.2 Anemia

SPIONs have been utilized as a source of iron supplement to individuals with iron deficiency. Ferumoxytol, consisting of dextran-coated SPIONs (Feraheme, AMAG Pharmaceuticals, Waltham, MA), is a clinically available intravenous SPION formulation approved by the FDA in 2009 to treat iron deficiency anemia to patients with chronic kidney diseases (CKD) [102]. The SPIONs are coated with carboxymethyl-dextran with hydrodynamic diameters of 17 to 31 nm. After intravenous injection, the iron content in the body increases, mainly in the liver and spleen, and over time is normalized in the body. It promotes erythrocyte production and relieves anemia. Later, in 2015, FDA changed the prescribing instructions and approved a Boxed Warning, strongly recommending the health professionals to consider the serious risks of this treatment on some patients as it can elicit potential fatal allergic reactions [103].

5.3 Inflammation

The identification of an active infection site is crucial for patients with a severe pathological condition such as atherosclerosis, since rupturing of the atherosclerosis plaque can lead to obstruction of the arteries and thereby induce myocardial infraction or stroke [104]. The infiltration of macrophages readily occurs in the infected tissues, making it an effective biomarker. As macrophage infiltration leads to increased vascular permeability, it assists in SPION uptake by macrophages. As SPIONs get phagocytosed by macrophages, they shorten the T2 and T2* relaxation times, providing contrast-enhanced MR images of inflammatory cells [105]. This imaging methodology is free of the ionizing radiation of other techniques, such as PET and CT scans, and is highly effective because of its excellent spatial resolution ($\sim$ 0.1–1 mm) combined with an ability to depict soft tissues and fluids [106].

Diagnosis and staging of multiple sclerosis (MS) is extensively done with conventional MRI. Using SPIONs over gadolinium-based contrast agents provides the advantage of imaging the macrophage infiltration [107]. Infiltration of myelin-containing phagocytes in MS is a hallmark of disease progression which can be detected by SPION-based contrast agents [108].

Interestingly, tissue and nerve inflammation in a pain condition was studied by Shen et al. [109] in animal and human models using Ferumoxytol-based MRI. In mice with an inflamed hind paw, after 72 h of Ferumoxytol injection; there was a marked signal loss on T2-weighted imaging. The mice with induced pain had a significantly higher change in T2 compared to the control mice, which would not have been feasible to image with conventional imaging system. Similar observation has been presented earlier in the context of brain in Fig. 10. In the case of human studies, the nerve root of the symptomatic neuronal tissue inflammation could be predicted [109]. This research is extremely valuable for individuals in pain conditions requiring epidural steroid injections in order to provide guidance along the treatment pathway.

5.4 Organ transplant (tissue engineering)

In recent years, a considerable amount of research has focussed on developing tissue and organ development techniques to overcome the shortage of organ donor and other serious immunological complications [110,111]. Tissue engineering through magnetic nanoparticles (MNPs) shows potential in developing a functional tissue substitute by means of a 3D porous scaffold. MNPs are used for manipulation of cellular functions under an appropriate external magnetic field in order to generate 3D tissues. The use of magnetic force-assisted tissue engineering is referred to as "Mag-TE" by Tripathi et al [112]. For this approach, MNPs are mainly used as a delivery vehicle for gene transfer (Magnetofection), labelling cells (Cell-patterning), and designing 3D tissues (Tissue-fabrication). An overview of the use of SPIONs for cell-directed tissue engineering and regenerative medicine is also provided [112]. The magnetically induced tissue-fabrication on skin, musculoskeletal, cardiac, and liver tissues is being investigated by a number of researchers [110,111].

To monitor the safety and fate of transplanted cells, Li et al. [113] utilized amino propyl triethoxysilane (APTS)-coated γ-Fe$_2$O$_3$ and found efficient labelling of human mesenchymal stromal cells (hMSCs) and monitored the in vivo MRI imaging of implanted cells. In this manner, the application of SPION-based platforms is proving to be useful in the field of tissue engineering.

5.5 Cardiovascular

Alam et al. [114] have summarized the use of ultrasmall SPIONs as a contrast agent in cardiovascular magnetic resonance (CMR), a noninvasive tool for assessment of cardiovascular pathology. Gadolinium-based contrast agents are the standard contrast agent for CMR. However, due to the limitation of diffusion of Gd within inflamed tissues, Gd enhancement is delayed in infarcted regions [115]. Additionally, Gd-based contrast agents possess the challenges of toxicity, deposition in neuronal cells [116], and the requirement of clinical diligence in order to avoid repeated doses and perform safe use [117]. SPION-based contrast agents possess less toxicity and can identify inflamed tissues, thereby making them possible candidates for CMR. Furthermore, quantification and characterization of macrophages using SPIONs assist in understanding the pathogenesis of cardiovascular disease and its prognosis and acts as a biomarker for theranostic purposes [114].

Cardiovascular magnetic resonance venography (CMRV) is a practical, accurate, and robust technique for high-resolution mapping of central thoracic, abdominal, and pelvic veins. Shahrouki et al. [118] studied the Ferumoxytol-enhanced CMRV (FE-CMRV) on 52 consecutive adult patients (47 years, IQR 32–61; 29 male) with renal impairment and suspected venous occlusion. They found that FE-CMRV compared to catheter venography showed 100% accuracy. They suggested that the FE-CMRV may play a pivotal role in the care of patients on whom conventional contrast agents (Gadolinium) may be contraindicated (cause renal failure) or ineffective.

6. Commercial SPIONs, toxicity, and regulatory aspects

Table 5 presents an updated list of various commercial products using SPIONs in nano-medicine. Ferumoxytol (Feraheme, AMAG Pharmaceuticals, Waltham, MA) is an FDA-approved IV iron supplement to treat iron deficiency anemia in adult patients with chronic kidney diseases (CKD). Ferumoxytol has a plasma half-life of 15 h. Spinowitz et al. [119] have reported on their Phase III trial that the overall treatment-related adverse event rate was similar to that of oral iron. This study was further supported by other clinical studies [120]. Ferumoxytol is being studied as "off-label" for vascular imaging, detection of stem cells transplantation, and cancer imaging purposes [121]. The toxicity of nanoparticles is one of the major areas of concern that needs to be addressed before they can be injected into human body. Many SPION-derived products such as Feridex, Resovist, and Lumirem have been discontinued in some countries owing to such concerns (see more information in Table 5). Therefore, biocompatibility and safety issues associated with SPIONs need to be dealt with for eventual use in approved applications.

One of the reasons for the cytotoxicity of SPIONs is their ability to create oxidative stress. After uptake, SPIONs release free iron ions which react with hydrogen peroxide and oxygen creating reactive hydroxyl radicals. These radicals are responsible for killing cancer cells by damaging DNA and causing oxidative stress. One research study in rats demonstrated dissimilarities in SPION toxicity vis-à-vis in vitro and in vivo studies [122]. Toxicity studies in vivo compared to responses in different cell lines suggest there is the possibility that individuals may be able to maintain homeostasis by storing excess iron in the body; considerably reducing the side effects of SPIONs. Therefore, researchers should work to modify their approach in in vitro cytotoxicity assessment for comparable results [123].

7. Conclusions and future perspectives

In this chapter, we have presented a comprehensive review of SPIONs by first introducing the physics of magnetic materials and the onset of superparamagnetism in nanosized single domain particles. This was followed and expanded upon with up-to-date synthesis protocols of core SPIONs, their magnetic properties, surface charge, stability in suspension, and the importance of developing core-shell structures. From various studies, it has been shown that bare or uncoated SPIONs tend to aggregate in suspension and hence the coating of SPIONs and preparation of various core-shell SPION-based composites can be a solution. In this context, we have presented up-to-date information on a range of shell materials with inorganic and organic compositions followed by the most commonly used functionalization techniques to conjugate with biomolecules for their application in nanomedicine. Coating with silica is safe due to its biocompatibility and coating with *meso*porous silica can dramatically improve the surface area due to internal porosity on

Table 5 Commercial SPIONs-based materials for therapeutic and diagnosis in cancer treatment.

Company name	Generic name	Brand name	Applications	Clinical trials	Clinically approved
AMAG Pharmaceuticals www. amagpharma.com	Ferumoxtran-10	Combidex (USA) Sinerem (EU)	lymph node and macrophage imaging	✓	
AMAG Pharmaceuticals	Ferumoxytol	Feraheme (USA) Rienso (EU)	Iron deficiency anemia (IDA), "off-label" imaging	✓	✓
Bayer healthcare www.Bayer.com	Ferucarbotran	Resovist (Japan) Cliavist (France)	Liver imaging, CNS imaging, cell labelling	✓	
Berlex Laboratories	Ferumoxide	Feridex IV (USA), Endorem (EU)	Liver imaging, CNS imaging, cell labelling	✓	
AMAG Pharmaceuticals	Ferumoxsil	Lumirem (USA) GastroMARK (EU)	Oral MRI contrast agent		✓
Imagion www. imagionbiosystems. com	–	PrecisionMRX	Cancer detection using MagSense™ technology	✓	
Endomag www.endomag. com	Magnecarbodex	Sienna$^+$ or MagTRACE™	Breast cancer lesion localization and tracer for sentinel node biopsies using Sentimag®		✓
Magforce www.magforce. com	Aminosilane-coated SPION	Nanotherm	Magnetic Hyperthermia		✓

the nm scale. Coating with organic polymer shells can be useful as well but can be toxic due to organic components. We have then introduced three possible therapeutic routes of core and core-shell SPIONs identified as (i) magnetic targeting, (ii) combination of hyperthermia–induced AMF and laser light in the NIR region; finally (iii) chemodynamic therapy (CDT) due to the formation of reactive oxygen species (ROS). Similarly, we have also introduced three possible diagnostic routes such as (i) Magnetic Resonance Imaging (MRI), (ii) recently developed Magnetic Particle Imaging (MPI) alongside (iii) thermal imaging. It has been observed that SPIONs can be a powerful agent for magnetic hyperthermia as well as light-induced hyperthermia. We have fully explored the therapeutic and diagnostic applications of SPIONs in various diseases and more extensively in Cancer, where we have described the potential of SPIONs as therapeutic agents. Finally, we have presented up-to-date details of commercially available SPIONs, including information pertaining to their clinical trials and possible disadvantages due to toxicity and subsequent regulatory approval.

SPIONs have been used in various applications, and more specifically as theranostic agents in nanomedicine, in a manner that takes advantage of their magnetic properties. However, biocompatibility and biodegradability of shell materials, stability in suspension, and deposition of SPIONs in certain organs in the body limit their full potential. Combining SPIONs with near infrared (NIR) photosensitizers, such as indocyanine green (ICG), is beginning to explore both diagnosis and therapeutic purposes as a dual-modality route. Finally, SPIONs do not only offer great potential as theranostic agents for diseases, but also provide powerful separation and detection mechanisms for microorganisms, bacteria, and viruses. In the current climate of the global pandemic due to COVID-19, surface-modified SPIONs can play an important role in the separation of SARS-CoV-2 RNA which can help immensely in detecting the virus using Reverse Transcription Polymerase Chain Reaction (RT-PCR) with negligible risk of false negative results.

References

[1] T. Sen, A. Sebastianelli, I.J. Bruce, Mesoporous silica-magnetite nanocomposite: fabrication and applications in magnetic bioseparations, J. Am. Chem. Soc. 128 (22) (2006) 7130–7131.

[2] A. Sebastianelli, T. Sen, I.J. Bruce, Extraction of DNA from soil using nanoparticles by magnetic bioseparation, Lett. Appl. Microbiol. 46 (4) (2008) 488–491.

[3] M. Mahmoudi, Emerging biomolecular testing to assess risk of mortality from COVID-19 infection. Mol. Pharm. (2020), https://doi.org/10.1021/acs.molpharmaceut.0c00371.

[4] M. Wahajuddin, S. Aurora, Superparamagnetic iron oxide nanoparticles: magnetic nanoplatforms as drug carriers, Int. J. Nanomedicine 7 (2012) 3445.

[5] T. Sen, S.J. Sheppard, T. Mercer, M. Eizadi-Sharifabad, M. Mahmoudi, A. Elhissi, Simple one-pot fabrication of ultra-stable core-shell superparamagnetic nanoparticles for potential application in drug delivery, RSC Adv. 2 (12) (2012) 5221–5228.

[6] D.L. Thorek, A.K. Chen, J. Czupryna, A. Tsourkas, Superparamagnetic iron oxide nanoparticle probes for molecular imaging, Ann. Biomed. Eng. 34 (1) (2006) 23–38.

[7] M. Mahmoudi, S. Sant, B. Wang, S. Laurent, T. Sen, Superparamagnetic iron oxide nanoparticles (SPIONs): development, surface modification and applications in chemotherapy, Adv. Drug Deliv. Rev. 63 (1–2) (2011) 24–46.

[8] R. Ivkov, S.J. Denardo, W. Daum, A.R. Foreman, R.C. Goldstein, V.S. Nemkov, G.L. Denardo, Application of high amplitude alternating magnetic fields for heat induction of nanoparticles localized in cancer, Clin. Cancer Res. 11 (19) (2005) 7093s–7103s.

[9] T. Sen, I.J. Bruce, T. Mercer, Fabrication of novel hierarchically ordered porous magnetic nanocomposites for bio-catalysis, Chem. Commun. 46 (36) (2010) 6807–6809.

[10] M. Eizadi-Sharifabad, B. Hodgson, M. Jellite, T. Mercer, T. Sen, Enzyme immobilised novel core-shell superparamagnetic nanocomposites for enantioselective formation of 4-(R)-hydroxycyclopent-2-en-1-(S)-acetate, Chem. Commun. 50 (2014) 11185–11188.

[11] A. Kaur, Synthesis and Charactersation of Multifunctional Porous Nanocomposites for the Removal of Pollutants from Water, Doctoral dissertationUniversity of Central Lancashire, 2019.

[12] M. Bhaumik, A. Maity, V.V. Srinivasu, M.S. Onyango, Enhanced removal of Cr (VI) from aqueous solution using polypyrrole/Fe$_3$O$_4$ magnetic nanocomposite, J. Hazard. Mater. 190 (1–3) (2011) 381–390.

[13] B.D. Cullity, C.D. Graham, Introduction to Magnetic Materials, second ed., IEEE Press, Wiley, 2009.

[14] G.F. Goya, T.S. Berquo, F.C. Fonseca, M.P. Morales, Static and dynamic magnetic properties of spherical magnetite nanoparticles. J. Appl. Phys. 94 (5) (2003) 3520–3528.

[15] G. Vallejo-Fernandez, K. O'grady, Effect of the distribution of anisotropy constants on hysteresis losses for magnetic hyperthermia applications, Appl. Phys. Lett. 103 (14) (2013) 142417.

[16] A.H. Lu, E.E. Salabas, F. Schüth, Magnetic nanoparticles: synthesis, protection, functionalization, and application, Angew. Chem. Int. Ed. 46 (8) (2007) 1222–1244.

[17] L.A. Harris, Polymer Stabilized Magnetite Nanoparticles and Poly (Propylene Oxide) Modified Styrene-Dimethacrylate Networks, Doctoral dissertationVirginia Tech, 2002.

[18] R. Chantrell, J. Popplewell, S. Charles, Measurements of particle size distribution parameters in ferrofluids, IEEE Trans. Magn. 14 (5) (1978) 975–977.

[19] A.K. Gupta, M. Gupta, Synthesis and surface engineering of iron oxide nanoparticles for biomedical applications, Biomaterials 26 (18) (2005) 3995–4021.

[20] Q.A. Pankhurst, J. Connolly, S.K. Jones, J.J. Dobson, Applications of magnetic nanoparticles in biomedicine, J. Phys. D Appl. Phys. 36 (13) (2003) R167.

[21] G. Vallejo-Fernandez, O. Whear, A.G. Roca, S. Hussain, J. Timmis, V. Patel, K. O'grady, Mechanisms of hyperthermia in magnetic nanoparticles, J. Phys. D Appl. Phys. 46 (31) (2013) 312001.

[22] D. Thapa, V.R. Palkar, M.B. Kurup, S.K. Malik, Properties of magnetite nanoparticles synthesized through a novel chemical route, Mater. Lett. 58 (21) (2004) 2692–2694.

[23] P. Roonasi, Adsorption and Surface Reaction Properties of Synthesized Magnetite Nanoparticles, Doctoral Dissertation, Luleå Tekniska Universitet, 2007.

[24] M. Sorescu, Phase transformations induced in magnetite by high energy ball milling, J. Mater. Sci. Lett. 17 (13) (1998) 1059–1061.

[25] T. Sugimoto, E. Matijević, Formation of uniform spherical magnetite particles by crystallization from ferrous hydroxide gels, J. Colloid Interface Sci. 74 (1) (1980) 227–243.

[26] I.J. Bruce, J. Taylor, M. Todd, M.J. Davies, E. Borioni, C. Sangregorio, T. Sen, Synthesis, characterisation and application of silica-magnetite nanocomposites, J. Magn. Magn. Mater. 284 (2004) 145–160.

[27] A.H. Lu, E.E. Salabas, F. Schüth, Magnetic nanoparticles: synthesis, protection, functionalization, and application, Angew. Chem. Int. Ed. 46 (8) (2007) 1222–1244.

[28] M.E. Sharifabad, Study of Porous Magnetic Nanocomposites for Bio-Catalysis and Drug Delivery, Doctoral dissertationUniversity of Central Lancashire, 2016.

[29] X. Liu, Y. Guan, Z. Ma, H. Liu, Surface modification and characterization of magnetic polymer nanospheres prepared by miniemulsion polymerization, Langmuir 20 (23) (2004) 10278–10282.

[30] A. Del Campo, T. Sen, J.P. Lellouche, I.J. Bruce, Multifunctional magnetite and silica-magnetite nanoparticles: synthesis, surface activation and applications in life sciences, J. Magn. Magn. Mater. 293 (1) (2005) 33–40.

[31] N.S. Sobal, M. Hilgendorff, H. Möhwald, M. Giersig, M. Spasova, T. Radetic, M. Farle, Synthesis and structure of colloidal bimetallic nanocrystals: the non-alloying system Ag/Co, Nano Lett. 2 (6) (2002) 621–624.

[32] J.B. Mamani, A.J. Costa-Filho, D.R. Cornejo, E.D. Vieira, L.F. Gamarra, Synthesis and characterization of magnetite nanoparticles coated with lauric acid, Mater Charact 81 (2013) 28–36.

[33] C. Boyer, M.R. Whittaker, V. Bulmus, J. Liu, T.P. Davis, The design and utility of polymer-stabilized iron-oxide nanoparticles for nanomedicine applications, NPG Asia Mater. 2 (1) (2010) 23–30.

[34] P. Pradhan, J. Giri, F. Rieken, C. Koch, O. Mykhaylyk, M. DÖblinger, R. Banerjee, D. Bahadur, C. Plank, Targeted temperature sensitive magnetic liposomes for thermo-chemotherapy, J. Control. Release 142 (1) (2010) 108–121.

[35] X. Zhang, H. Niu, Y. Pan, Y. Shi, Y. Cai, Chitosan-coated Octadecyl-functionalized magnetite nanoparticles: preparation and application in extraction of trace pollutants from environmental water samples, Anal. Chem. 82 (6) (2010) 2363–2371.

[36] N. Nakajima, Y. Ikada, Mechanism of amide formation by carbodiimide for bioconjugation in aqueous media, Bioconjug. Chem. 6 (1) (1995) 123–130.

[37] G.T. Hermanson, Zero-length crosslinkers, in: Bioconjugate Techniques, third ed., Academic Press, Elsevier, London, 2013, , pp. 259–273.

[38] D. Rother, T. Sen, D. East, I.J. Bruce, Silicon, silica and its surface patterning/activation with alkoxy- and amino-silanes for nanomedical applications, Nanomedicine 6 (2) (2011) 281–300.

[39] H. Duan, M. Kuang, X. Wang, Y.A. Wang, H. Mao, S. Nie, Reexamining the effects of particle size and surface chemistry on the magnetic properties of Iron oxide nanocrystals: new insights into spin disorder and proton relaxivity, J. Phys. Chem. C 112 (22) (2008) 8127–8131.

[40] L.E.W. Laconte, N. Nitin, O. Zurkiya, D. Caruntu, C.J. O'connor, X. Hu, G. Bao, Coating thickness of magnetic iron oxide nanoparticles affects R2 relaxivity, J. Magn. Reson. Imaging 26 (6) (2007) 1634–1641.

[41] J.A. Lopez, F. González, F.A. Bonilla, G. Zambrano, M.E. Gómez, Synthesis and characterization of Fe_3O_4 magnetic nanofluid, Revista Latinoamericana de Metalurgia y Materiales 30 (2010) 60–66.

[42] O. Veiseh, J.W. Gunn, M. Zhang, Design and fabrication of magnetic nanoparticles for targeted drug delivery and imaging, Adv. Drug Deliv. Rev. 62 (3) (2010) 284–304.

[43] W. Voit, D.K. Kim, W. Zapka, M. Muhammed, K.V. Rao, Magnetic behavior of coated superparamagnetic iron oxide nanoparticles in ferrofluids, MRS Online Proc. Libr. 676 (2001) Y7.8.1–Y7.8.6.

[44] T. Sen, I.J. Bruce, Mesoporous silica–magnetite nanocomposites: fabrication, characterisation and applications in biosciences, Microporous Mesoporous Mater. 120 (3) (2009) 246–251.

[45] S.C. Mcbain, H.H.P. Yiu, J. Dobson, Magnetic nanoparticles for gene and drug delivery, Int. J. Nanomedicine 3 (2) (2008) 169–180.

[46] S. Chen, Polymer-Coated Iron Oxide Nanoparticles for Medical Imaging, PhDMassachusetts Institute of Technology, 2010.

[47] M.W. Freeman, A. Arrott, J.H.L. Watson, Magnetism in medicine, J. Appl. Phys. 31 (5) (1960) S404–S405.

[48] J.H. Park, G. Saravanakumar, K. Kim, I.C. Kwon, Targeted delivery of low molecular drugs using chitosan and its derivatives, Adv. Drug Deliv. Rev. 62 (1) (2010) 28–41.

[49] R. Sensenig, Y. Sapir, C. Macdonald, S. Cohen, B. Polyak, Magnetic nanoparticle-based approaches to locally target therapy and enhance tissue regeneration in vivo, Nanomedicine 7 (9) (2012) 1425–1442.

[50] B. Chertok, A.E. David, V.C. Yang, Polyethyleneimine-modified iron oxide nanoparticles for brain tumor drug delivery using magnetic targeting and intra-carotid administration, Biomaterials 31 (24) (2010) 6317–6324.

[51] C. Alexiou, R. Jurgons, R.J. Schmid, C. Bergemann, J. Henke, W. Erhardt, E. Huenges, F. Parak, Magnetic drug targeting—biodistribution of the magnetic carrier and the chemotherapeutic agent mitoxantrone after locoregional cancer treatment, J. Drug Target. 11 (3) (2003) 139–149.

[52] J. Estelrich, E. Escribano, J. Queralt, M. Busquets, Iron oxide nanoparticles for magnetically-guided and magnetically-responsive drug delivery, Int. J. Mol. Sci. 16 (4) (2015) 8070–8101.

[53] R. Jurgons, C. Seliger, A. Hilpert, L. Trahms, S. Odenbach, C. Alexiou, Drug loaded magnetic nanoparticles for cancer therapy, J. Phys. Condens. Matter 18 (38) (2006) S2893–S2902.

[54] A. Yanai, U.O. Hafeli, A.L. Metcalfe, P. Soema, L. Addo, C.Y. Gregory-Evans, K. Po, X. Shan, O. L. Moritz, K. Gregory-Evans, Focused magnetic stem cell targeting to the retina using superparamagnetic iron oxide nanoparticles, Cell Transplant. 21 (6) (2012) 1137–1148.

[55] R. Tietze, S. Lyer, S. Durr, T. Struffert, T. Engelhorn, M. Schwarz, E. Eckert, T. Goen, S. Vasylyev, W. Peukert, F. Wiekhorst, L. Trahms, A. Dorfler, C. Alexiou, Efficient drug-delivery using magnetic nanoparticles—biodistribution and therapeutic effects in tumour bearing rabbits, Nanomedicine 9 (7) (2013) 961–971.

[56] M. Banobre-Lopez, A. Teijeiro, J. Rivas, Magnetic nanoparticle-based hyperthermia for cancer treatment, Rep. Pract. Oncol. Radiother. 18 (6) (2013) 397–400.

[57] A.E. Deatsch, B.A. Evans, Heating efficiency in magnetic nanoparticle hyperthermia, J. Magn. Magn. Mater. 354 (2014) 163–172.

[58] J. Overgaard, D.G. Gonzalez, M.C.C.H. Hulshof, G. Arcangeli, O. Dahl, O. Mella, S.M. Bentzen, Hyperthermia as an adjuvant to radiation therapy of recurrent or metastatic malignant melanoma. A multicentre randomized trial by the European Society for Hyperthermic Oncology, Int. J. Hyperthermia 12 (1) (1996) 3–20.

[59] C.S.S.R. Kumar, F. Mohammad, Magnetic nanomaterials for hyperthermia-based therapy and controlled drug delivery, Adv. Drug Deliv. Rev. 63 (9) (2011) 789–808.

[60] R.K. Gilchrist, R. Medal, W.D. Shorey, R.C. Hanselman, J.C. Parrott, C.B. Taylor, Selective inductive heating of lymph nodes, Ann. Surg. 146 (4) (1957) 596–606.

[61] R.T. Gordon, J.R. Hines, D. Gordon, Intracellular hyperthermia a biophysical approach to cancer treatment via intracellular temperature and biophysical alterations, Med. Hypotheses 5 (1) (1979) 83–102.

[62] S. Laurent, S. Dutz, U.O. Häfeli, M. Mahmoudi, Magnetic fluid hyperthermia: focus on superparamagnetic iron oxide nanoparticles, Adv. Colloid Interface Sci. 166 (1–2) (2011) 8–23.

[63] J. Estelrich, M.A. Busquets, Iron oxide nanoparticles in Photothermal therapy, Molecules 23 (7) (2018) 1567.

[64] C.M. Clavel, P. Nowak-Sliwinska, E. Păunescu, A.W. Griffioen, P.J. Dyson, In vivo evaluation of small-molecule thermoresponsive anticancer drugs potentiated by hyperthermia, Chem. Sci. 6 (5) (2015) 2795–2801.

[65] A. Espinosa, R. Di Corato, J. Kolosnjaj-Tabi, P. Flaud, T. Pellegrino, C. Wilhelm, Duality of iron oxide nanoparticles in cancer therapy: amplification of heating efficiency by magnetic hyperthermia and photothermal bimodal treatment, ACS Nano 10 (2) (2016) 2436–2446.

[66] W.S. Enochs, G. Harsh, F. Hochberg, R. Weissleder, Improved delineation of human brain tumors on MR images using a long-circulating, superparamagnetic iron oxide agent, J. Magn. Reson. Imaging 9 (2) (1999) 228–232.

[67] M.H. Pablico-Lansigan, S.F. Situ, A.C.S. Samia, Magnetic particle imaging: advancements and perspectives for real-time in vivo monitoring and image-guided therapy, Nanoscale 5 (10) (2013) 4040–4055.

[68] J.W. Bulte, Superparamagnetic iron oxides as MPI tracers: a primer and review of early applications, Adv. Drug Deliv. Rev. 138 (2019) 293–301.

[69] Z.W. Tay, P. Chandrasekharan, A. Chiu-Lam, D.W. Hensley, R. Dhavalikar, X.Y. Zhou, E.Y. Yu, P.W. Goodwill, B. Zheng, C. Rinaldi, S.M. Conolly, Magnetic particle imaging-guided heating in vivo using gradient fields for arbitrary localization of magnetic hyperthermia therapy, ACS Nano 12 (4) (2018) 3699–3713.

[70] A.S. Lubbe, C. Bergemann, H. Riess, F. Schriever, P. Reichardt, K. Possinger, M. Matthias, B. Dorken, F. Herrmann, R. Gurtler, P. Hohenberger, N. Haas, R. Sohr, B. Sander, A.J. Lemke, D. Ohlendorf, W. Huhnt, D. Huhn, Clinical experiences with magnetic drug targeting: a phase I study with 4'-epidoxorubicin in 14 patients with advanced solid tumors, Cancer Res. 56 (20) (1996) 4686–4693.

[71] A.J. Lemke, M.I. Senfft Von Pilsach, A. Lubbe, C. Bergemann, H. Riess, R. Felix, MRI after magnetic drug targeting in patients with advanced solid malignant tumors, Eur. Radiol. 14 (11) (2004) 1949–1955.

[72] B. Shapiro, S. Kulkarni, A. Nacev, S. Muro, P.Y. Stepanov, I.N. Weinberg, Open challenges in magnetic drug targeting, Wiley Interdiscip. Rev. Nanomed. Nanobiotechnol. 7 (3) (2015) 446–457.

[73] I. Hilger, In vivo applications of magnetic nanoparticle hyperthermia, Int. J. Hyperthermia 29 (8) (2013) 828–834.

[74] M. Johannsen, U. Gneveckow, L. Eckelt, A. Feussner, N. Waldöfner, R. Scholz, S. Deger, P. Wust, S.A. Loening, A. Jordan, Clinical hyperthermia of prostate cancer using magnetic nanoparticles: presentation of a new interstitial technique, Int. J. Hyperthermia 21 (7) (2005) 637–647.

[75] M. Johannsen, U. Gneveckow, K. Taymoorian, B. Thiesen, N. Waldofner, R. Scholz, K. Jung, A. Jordan, P. Wust, S.A. Loening, Morbidity and quality of life during thermotherapy using magnetic nanoparticles in locally recurrent prostate cancer: results of a prospective phase I trial, Int. J. Hyperthermia 23 (3) (2007) 315–323.

[76] P. Wust, U. Gneveckow, M. Johannsen, D. Bohmer, T. Henkel, F. Kahmann, J. Sehouli, R. Felix, J. Ricke, A. Jordan, Magnetic nanoparticles for interstitial thermotherapy—feasibility, tolerance and achieved temperatures, Int. J. Hyperthermia 22 (8) (2006) 673–685.

[77] K. Maier-Hauff, R. Rothe, R. Scholz, U. Gneveckow, P. Wust, B. Thiesen, A. Feussner, A. Von Deimling, N. Waldoefner, R. Felix, A. Jordan, Intracranial thermotherapy using magnetic nanoparticles combined with external beam radiotherapy: results of a feasibility study on patients with glioblastoma multiforme, J. Neurooncol 81 (1) (2007) 53–60.

[78] L. Asin, M.R. Ibarra, A. Tres, G.F. Goya, Controlled cell death by magnetic hyperthermia: effects of exposure time, field amplitude, and nanoparticle concentration, Pharm. Res. 29 (5) (2012) 1319–1327.

[79] C.A. Monnier, D. Burnand, B. Rothen-Rutishauser, M. Lattuada, A. Petri-Fink, Magnetoliposomes: opportunities and challenges, Eur. J. Nanomed. 6 (4) (2014) 201–215.

[80] F. Chen, J. Ward, P.J. Robinson, MR imaging of the liver and spleen: a comparison of the effects on signal intensity of two superparamagnetic iron oxide agents, Magn. Reson. Imaging 17 (4) (1999) 549–556.

[81] J.B. Weaver, J.R. Conejo-Garcia, S.N. Fiering, A.M. Rauwerdink, U.K. Scarlett, Dartmouth College, Magnetic particle imaging (MPI) system and method for use of iron-based nanoparticles in imaging and diagnosis, (2015)U.S. Patent 8,954,131.

[82] E.Y. Yu, M. Bishop, B. Zheng, R.M. Ferguson, A.P. Khandhar, S.J. Kemp, K.M. Krishnan, P. W. Goodwill, S.M. Conolly, Magnetic particle imaging: a novel in vivo imaging platform for cancer detection, Nano Lett. 17 (3) (2017) 1648–1654.

[83] J. Salamon, J. Dieckhoff, M.G. Kaul, C. Jung, G. Adam, M. Moddel, T. Knopp, S. Draack, F. Ludwig, H. Ittrich, Visualization of spatial and temporal temperature distributions with magnetic particle imaging for liver tumor ablation therapy, Sci. Rep. 10 (2020) 7480.

[84] B.H. Kim, N. Lee, H. Kim, K. An, Y.I. Park, Y. Choi, K. Shin, Y. Lee, S.G. Kwon, H.B. Na, J. G. Park, Large-scale synthesis of uniform and extremely small-sized iron oxide nanoparticles for high-resolution T 1 magnetic resonance imaging contrast agents, J. Am. Chem. Soc. 133 (32) (2011) 12624–12631.

[85] C. Lorente, L. Cabeza, B. Clares, R. Ortiz, L. Halbaut, Á.V. Delgado, G. Perazzoli, J. Prados, J. L. Arias, C. Melguizo, Formulation and in vitro evaluation of magnetoliposomes as a potential nanotool in colorectal cancer therapy, Colloids Surf. B Biointerfaces 171 (2018) 553–565.

[86] M.S. Martina, C. Wilhelm, S. Lesieur, The effect of magnetic targeting on the uptake of magnetic-fluid-loaded liposomes by human prostatic adenocarcinoma cells, Biomaterials 29 (30) (2008) 4137–4145.

[87] J. Mosafer, K. Abnous, M. Tafaghodi, A. Mokhtarzadeh, M. Ramezani, In vitro and in vivo evaluation of anti-nucleolin-targeted magnetic PLGA nanoparticles loaded with doxorubicin as a theranostic agent for enhanced targeted cancer imaging and therapy, Eur. J. Pharm. Biopharm. 113 (2017) 60–74.

[88] J.J. Lin, J.S. Chen, S.J. Huang, J.H. Ko, Y.M. Wang, T.L. Chen, L.F. Wang, Folic acid–Pluronic F127 magnetic nanoparticle clusters for combined targeting, diagnosis, and therapy applications, Biomaterials 30 (2009) 5114–5124.

[89] F. Dilnawaz, A. Singh, C. Mohanty, S.K. Sahoo, Dual drug loaded superparamagnetic iron oxide nanoparticles for targeted cancer therapy, Biomaterials 31 (13) (2010) 3694–3706.

[90] T.T. Li, X. Shen, Y. Chen, C.C. Zhang, J. Yan, H. Yang, C.H. Wu, H.J. Zeng, Y.Y. Liu, Polyetherimide-grafted $Fe_3O_4@SiO_2$ nanoparticles as theranostic agents for simultaneous VEGF siRNA delivery and magnetic resonance cell imaging, Int. J. Nanomedicine 10 (2015) 4279–4291.

[91] A. Heidarinasab, H.A. Panahi, M. Faramarzi, F. Farjadian, Synthesis of thermosensitive magnetic nanocarrier for controlled sorafenib delivery, Mater. Sci. Eng. C 67 (2016) 42–50.

[92] C.C. Huang, P.Y. Chang, C.L. Liu, J.P. Xu, S.P. Wu, W.C. Kuo, New insight on optical and magnetic Fe_3O_4 nanoclusters promising for near infrared theranostic applications, Nanoscale 7 (29) (2015) 12689–12697.

[93] A. Amirshaghaghi, L. Yan, J. Miller, Y. Daniel, J.M. Stein, T.M. Busch, Z. Cheng, A. Tsourkas, Chlorin e6-coated superparamagnetic Iron oxide nanoparticle (SPION) nanoclusters as a theranostic agent for dual-mode imaging and photodynamic therapy, Sci. Rep. 9 (1) (2019) 1–9.

[94] H.Y. Yang, M.S. Jang, Y. Li, Y. Fu, J.H. Lee, D.S. Lee, Hierarchical tumor acidity-responsive self-assembled magnetic nanotheranostics for bimodal bioimaging and photodynamic therapy, J. Control. Release 301 (2019) 157–165.

[95] J.F. Liao, X.W. Wei, B. Ran, J.R. Peng, Y. Qu, Z.Y. Qian, Polymer hybrid magnetic nanocapsules encapsulating IR820 and PTX for external magnetic field-guided tumor targeting and multifunctional theranostics, Nanoscale 9 (2017) 2479–2491.

[96] G. Kandasamy, S. Khan, J. Giri, S. Bose, N.S. Veerapu, D. Maity, One-pot synthesis of hydrophilic flower-shaped iron oxide nanoclusters (IONCs) based ferrofluids for magnetic fluid hyperthermia applications, J. Mol. Liq. 275 (2019) 699–712.

[97] D. Ling, W. Park, S.J. Park, Y. Lu, K.S. Kim, M.J. Hackett, B.H. Kim, H. Yim, Y.S. Jeon, K. Na, T. Hyeon, Multifunctional tumor pH-sensitive self-assembled nanoparticles for bimodal imaging and treatment of resistant heterogeneous Tumors, J. Am. Chem. Soc. 136 (2014) 5647–5655.

[98] K. Kaaki, K. Herve-Aubert, M. Chiper, A. Shkilnyy, M. Souce, R. Benoit, A. Paillard, P. Dubois, M. L. Saboungi, I. Chourpa, Magnetic nanocarriers of doxorubicin coated with poly (ethylene glycol) and folic acid: relation between coating structure, surface properties, colloidal stability, and cancer cell targeting, Langmuir 28 (2) (2012) 1496–1505.

[99] Y. Guo, J. Wang, L. Zhang, S. Shen, R. Guo, Y. Yang, W. Chen, Y. Wang, G. Chen, X. Shuai, Theranostical nanosystem-mediated identification of an oncogene and highly effective therapy in hepatocellular carcinoma, Hepatology 63 (4) (2016) 1240–1255.

[100] N. Depalo, R.M. Iacobazzi, G. Valente, I. Arduino, S. Villa, F. Canepa, V. Laquintana, E. Fanizza, M. Striccoli, A. Cutrignelli, A. Lopedota, Sorafenib delivery nanoplatform based on superparamagnetic iron oxide nanoparticles magnetically targets hepatocellular carcinoma, Nano Res. 10 (7) (2017) 2431–2448.

[101] G. Kandasamy, A. Sudame, T. Luthra, K. Saini, D. Maity, Functionalized hydrophilic superparamagnetic iron oxide nanoparticles for magnetic fluid hyperthermia application in liver cancer treatment, ACS Omega 3 (4) (2018) 3991–4005.

[102] M. Lu, M.H. Cohen, D. Rieves, R. Pazdur, FDA report: ferumoxytol for intravenous iron therapy in adult patients with chronic kidney disease, Am. J. Hematol. 85 (5) (2010) 315–319.

[103] FDA, FDA Drug Safety Communication: FDA Strengthens Warnings and Changes Prescribing Instructions to Decrease the Risk of Serious Allergic Reactions With Anemia Drug Feraheme (Ferumoxytol), [online] Available athttps://www.fda.gov/drugs/drug-safety-and-availability/fda-drug-safety-communication-fda-strengthens-warnings-and-changes-prescribing-instructions-decrease, 2015. Accessed 16 September 2020.

[104] F.K. Swirski, M. Nahrendorf, Imaging macrophage development and fate in atherosclerosis and myocardial infarction, Immunol. Cell Biol. 91 (4) (2013) 297–303.

[105] J.W. Bulte, D.L. Kraitchman, Iron oxide MR contrast agents for molecular and cellular imaging, NMR Biomed. 17 (7) (2004) 484–499.

[106] A. Neuwelt, N. Sidhu, C.A.A. Hu, G. Mlady, S.C. Eberhardt, L.O. Sillerud, Iron-based superparamagnetic nanoparticle contrast agents for MRI of infection and inflammation, Am. J. Roentgenol. 204 (3) (2015) W302–W313.

[107] T. Tourdias, S. Roggerone, M. Filippi, M. Kanagaki, M. Rovaris, D.H. Miller, K.G. Petry, B. Brochet, J.P. Pruvo, E.W. Radüe, V. Dousset, Assessment of disease activity in multiple sclerosis phenotypes with combined gadolinium-and superparamagnetic iron oxide–enhanced MR imaging, Radiology 264 (1) (2012) 225–233.

[108] M. Mahmoudi, M.A. Sahraian, M.A. Shokrgozar, S. Laurent, Superparamagnetic iron oxide nanoparticles: promises for diagnosis and treatment of multiple sclerosis, ACS Chem. Nerosci. 2 (3) (2011) 118–140.

[109] S. Shen, W. Ding, S. Ahmed, R. Hu, A. Opalacz, S. Roth, Z. You, G.R. Wotjkiewicz, G. Lim, L. Chen, J. Mao, J.W. Chen, Y. Zhang, Ultrasmall superparamagnetic Iron oxide imaging identifies tissue and nerve inflammation in pain conditions, Pain Med. 19 (4) (2018) 686–692.

[110] H. Yukawa, M. Watanabe, N. Kaji, Y. Okamoto, M. Tokeshi, Y. Miyamoto, H. Noguchi, Y. Baba, S. Hayashi, Monitoring transplanted adipose tissue-derived stem cells combined with heparin in the liver by fluorescence imaging using quantum dots, Biomaterials 33 (7) (2012) 2177–2186.

[111] J.D. Fisher, A.P. Acharya, S.R. Little, Micro and nanoparticle drug delivery systems for preventing allotransplant rejection, Clin. Immunol. 160 (1) (2015) 24–35.

[112] A. Tripathi, J.S. Melo, S.F. D'souza, Magnetic nanoparticles in tissue regeneration, in: Nanomaterials in Drug Delivery, Imaging, and Tissue Engineering, Scrivener Publishing, Danvar: USA, 2013, , pp. 443–492.

[113] X.X. Li, K.A. Li, J.B. Qin, K.C. Ye, X.R. Yang, W.M. Li, Q.S. Xie, M.E. Jiang, G.X. Zhang, X. W. Lu, In vivo MRI tracking of iron oxide nanoparticle-labeled human mesenchymal stem cells in limb ischemia, Int. J. Nanomedicine 8 (2013) 1063.

[114] S.R. Alam, C. Stirrat, J. Richards, S. Mirsadraee, S.I. Semple, G. Tse, P. Henriksen, D.E. Newby, Vascular and plaque imaging with ultra-small superparamagnetic particles of iron oxide, J. Cardiovasc. Magn. Reson. 17 (1) (2015) 83.

[115] J.A. Lima, R.M. Judd, A. Bazille, S.P. Schulman, E. Atalar, E.A. Zerhouni, Regional heterogeneity of human myocardial infarcts demonstrated by contrast-enhanced MRI: potential mechanisms, Circulation 92 (5) (1995) 1117–1125.

[116] S. Aime, P. Caravan, Biodistribution of gadolinium-based contrast agents, including gadolinium deposition, J. Magn. Reson. Imaging 30 (6) (2009) 1259–1267.

[117] A. Ranga, Y. Agarwal, K.J. Garg, Gadolinium based contrast agents in current practice: risks of accumulation and toxicity in patients with normal renal function, Indian J. Radiol. Imaging 27 (2) (2017) 141.

[118] P. Shahrouki, J.M. Moriarty, S.N. Khan, B. Bista, S.T. Kee, B.G. Derubertis, T. Yoshida, K. Nyugen, J.P. Finn, High resolution, 3-dimensional Ferumoxytol-enhanced cardiovascular magnetic resonance venography in central venous occlusion, J. Cardiovasc. Magn. Reson. 21 (2019) 17.

[119] B.S. Spinowitz, A.T. Kausz, J. Baptista, S.D. Noble, R. Sothinathan, M.V. Bernardo, L. Brenner, B. J. Pereira, Ferumoxytol for treating iron deficiency anaemia in CKD, J. Am. Soc. Nephrol. 19 (8) (2008) 1599–1605.

[120] J.M. Lewis, P.M. Jacobs, T.B. Frigo, Comparison of free iron in ferumoxytol with current iron therapeutics, J. Am. Soc. Nephrol. 14 (2003, November) 771A–772A Philadelphia, PA 19106-3621 USA: Lippincott Williams & Wilkins.

[121] H. Nejadnik, P. Pandit, O. Lenkov, A.P. Lahiji, K. Yerneni, H.E. Daldrup-Link, Ferumoxytol can be used for quantitative magnetic particle imaging of transplanted stem cells, Mol. Imaging Biol. 21 (3) (2019) 465–472.

[122] M. Mahmoudi, A. Simchi, M. Imani, M.A. Shokrgozar, A.S. Milani, U.O. HÄfeli, P. Stroeve, A new approach for the in vitro identification of the cytotoxicity of superparamagnetic iron oxide nanoparticles, Colloids Surf. B Biointerfaces 75 (1) (2010) 300–309.

[123] R. Weissleder, G. Elizondo, J. Wittenberg, A.S. Lee, L. Josephson, T.J. Brady, Ultra-small superparamagnetic iron oxide: an intravenous contrast agent for assessing lymph nodes with MR imaging, Radiology 175 (2) (1990) 494–498.

CHAPTER 14

Bio-inspired nanoparticles as drug delivery vectors

Mariacristina Gagliardi and Marco Cecchini
NEST, Scuola Normale Superiore and Istituto Nanoscienze-CNR, Pisa, Italy

1. Introduction

Conventional therapeutic practices are based on the administration of drugs with a proper pharmacological activity to treat a specific pathology. This approach suffers from some drawbacks related to the poor stability and biodistribution of the active molecules, their limited capabilities to cross specific physiological barriers, and their potential unwanted side-effects. Modern strategies to overcome limitations of the conventional drug administration involve the use of nanosized delivery systems.

Nanodelivery systems are commonly obtained as solid spherical dense (nanospheres, Fig. 1A) or hollow (nanocapsules, Fig. 1B) particles, also with "stretched" (nanorods) or "flatten" (nanodiscs) morphologies, as soft spherical (micelles, Fig. 1C) or stretchable (filomicelles) self-assembled structures, or highly branched (dendrimers) units.

Nanodelivery systems can effectively reduce a number of limitations that afflict the free administration of drugs. Enhanced stability and biodistribution, targeting capabilities, and prolonged circulation times can be achieved to limit the administered dosing and accumulation in healthy districts, and thus reducing related side-effects. With these prerogatives, nanodelivery systems also enable the use of drugs that are found to be effective but that do not find practical uses in their free form because of their drawbacks, also promoting the cost reduction for new drugs discovery.

In the following, we refer to nanosized delivery systems as NPs. NPs have a typical size ranging from 1 to 100 nm and are easily prepared with controlled dimensions and morphology [1]. As a single particle allocates several drug molecules, their high drug load capacity makes them good candidates for drug delivery applications [2].

NPs are a "homing system" that hides the active principles from the reticuloendothelial system (RES) immune response, and protects poorly stable active principles from chemical or enzymatic degradation maintaining the pharmacological activity intact. Behaving as homing systems, NPs increase the probability that the drug reaches the target site.

Additional benefits related to drug nanocarriers are their capabilities to be customized and functionalized, to modify their physicochemical properties, to redress their poor

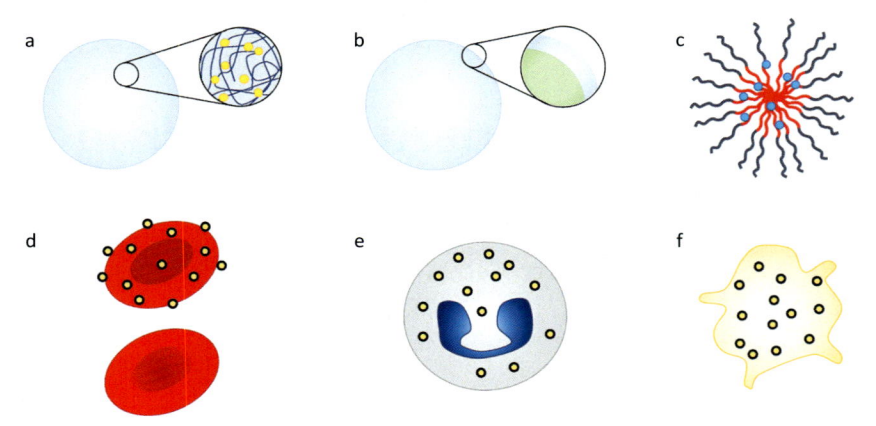

Fig. 1 (A) Dense nanoparticle containing a drug (hydrophilic or hydrophobic) dispersed in the matrix; (B) hollow nanocapsule containing a drug solution (hydrophilic); (C) polymer micelle composed of a hydrophilic shell and a hydrophobic core containing the solid drug (hydrophobic); (D) RBC-based drug carrier with nanoparticles decorating the RBC surface (top) or embedded in the RBC membrane (bottom); (E) whole WBC embedding drug-loaded nanoparticles; (F) drug-loaded nanoparticles cloaked with the platelet membrane.

bioavailability or to improve the targeting properties, limiting off-target drug deposition. Researchers are pushing hard in this direction making the most of biological systems. Recently, researches focused on novel cell-, virus- and bacteria-based therapies, and functionalities thereof, are significantly proliferating, indicating a significant interest in biomimetic systems.

2. Biomimicry lessons from natural systems

Biomimicry is a novel discipline that explores innovative materials and techniques to imitate natural elements, molecules, mechanisms and systems, to serve for our technological society [3].

In the past decades, significant advances in nanomedicine could improve several aspects of pharmacological treatments, in terms of therapeutic efficiency as well as for their safety concerns. However, there are some fundamental features, like long circulation time, precise targeting, and effective cell uptake, which cannot be optimized in a single synthetic NP [4] owing to its intrinsic nature of foreign material.

Synthetic drug delivery systems can implement cells, microorganisms, viruses, and materials extracted from them. Relevant biomimetic abilities are obtained in cell-based, cell ghost-based and cell-like-based therapies.

Although conventional cell therapies imply the administration only of viable cells, cell-based therapies embedding synthetic or semisynthetic systems can also benefit from the therapeutic effect of active agents. In cell-based therapies, a drug or a drug carrier is loaded within a living intact cell. Cells act as external envelope able to hide and transport the drug cargo, often in a specific body district. Whole cells as carriers present many advantages, but high costs and potential-associated risks are limiting their effective use.

To overcome limitations, some cells are currently under investigation in their ghost form. A ghost cell maintains its external proteolipid envelope intact but its biological activity is suppressed after a proper manipulation. This "empty phantom" is used to prepare core/shell NP systems in which the shell is the cell membrane and the core is the synthetic or semisynthetic NP. Cell membrane-coated NPs preserve the physicochemical properties of the nanosystem while acquiring complex cellular functions derived from biologically derived materials [5].

Apart from the whole cell membrane, synthetic NPs can be also functionalized with materials extracted from cells, like proteins or specific cell membrane components. Similarly to ghost systems, this strategy is a proven method to reduce the immune response thus prolonging the NP circulation time [6, 7].

Cell-related biomimetic modifications are mainly achieved by using erythrocytes (Fig. 1D), leukocytes (Fig. 1E) and stem cells (SCs). Erythrocytes are long-circulating cells and transferring their surface properties to synthetic systems can improve NPs circulation time. On the other hand, leukocytes and SCs possess unique recognition properties, thanks to their strong interactions with host body cells. Both cells can be exploited to add targeting properties on synthetic systems.

Virus capsids and their fragments are widely explored in the preparation of semisynthetic drug delivery systems. The high interest in such systems stems from their optimal targeting capabilities. Indeed, viruses are intrinsically targeted entities, and can strictly reach and strike specific districts.

Microorganism-based drug delivery therapies are obtained by using the bacterium envelope. Similar to the cell-based approach, bacteria can be used in drug delivery systems as living organisms or ghosts. Moreover, the surface functionalization of NPs with bacteria-derived proteins is also possible. Such applications suffer from some important safety concerns and thus they have limited application, as in cancer therapy. Bacteria have a strong capability to invade the host body. The biomimetic property that can be learnt from them is the biodistribution.

In the past decades, natural delivery systems have been extensively under investigation to overcome the poor bioavailability of synthetic carriers. Being composed of biological materials, cells and extracellular vesicles are safe and effective structures to carry drugs [8].

2.1 RBC-mediated drug delivery systems

Blood circulating cells, like erythrocytes or red blood cells (RBCs), not only possess intrinsic stealth properties but also can embed a variety of active molecules. These peculiar properties have justified the development of RBC-mediated drug delivery systems, and several studies report promising results.

RBCs are the most ubiquitous cells in the blood. They have a significant bloodstream circulation time (about 3 months), large internal volume (185–191 μm^3) that allows entrapping high drug amounts [9], and their plasma membrane surface is the most extended cellular surface accessible to blood [10]. Recent clinical trials have already demonstrated the excellent capability of RBCs to provide therapeutic effects [11], while a plethora of in vitro and in vivo studies focus on the improvement of drug loading [12] or surface properties [13, 14].

As their tailoring ability is well-established, natural RBCs are investigated in a wide range of usages, comprising imaging and diagnostic [15], and the selective "mechanical" deposition of synthetic NPs in organs [16]. Recently, modified RBCs were used to accelerate thrombus dissolution and enable thromboprophylaxis in patients with high risk of thrombosis [17]. However, their major potential is related to drug therapies, to refine the pharmacokinetic of active molecules.

By now, drug encapsulation in RBCs is a well-consolidated practice in preclinical studies [18, 19], with potential applications in the delivery of antiinflammatory drugs [20–22], enzymes [23, 24], antiviral agents [25], and in immunotherapy [26].

A notable application is related to the RBC-mediated antigen delivery, that is, hurdled, especially in autoimmune diseases, when the antigen is administered in its free form. Antigen-loaded RBCs overcome limitations related to the immune system clearance, inducing an effective and long-lasting immune tolerance [27].

On the other hand, the use of RBCs as drug vectors is severely limited in cancer treatment. Such limitation is due to the intrinsic functionality of RBCs to not extravasate, a characteristic that hampers their infiltration in cancer tissues. Only two applications for RBCs in cancer treatment were possible: (i) for organs belonging to the mononuclear phagocyte system (MPS) [28] and (ii) for the depletion of circulating cancer cells [29].

Drug can be loaded in the RBC core or can be attached/adsorbed on cell membrane, by means of ex vivo or in vivo (only surface attachment) procedures. After loading, RBCs can show a damaged physiology, such as the partial disruption of cell membrane or the modification of their mechanical properties.

2.2 WBC-mediated drug delivery systems

Leukocytes, or WBCs, belong to the immune system and have an important role in several local diseases, like infections and inflammatory processes.

WBC circulation time is shorter than that of RBCs (~20 days), and thus they are not indicated to promote biodistribution in NPs. On the other hand, WBCs are particularly attractive for localized drug delivery, thanks to their notable tissue penetration capabilities and their ability to cross some important physiological barriers. Such properties derive from specialized functions, based on cellular interactions that they exert in the body.

Although used in most cases to prepare biohybrid systems, WBCs possess a great potential in transport, migration toward inflammation sites, and adhesion to endothelial tissues [30]. Such properties make WBCs interesting candidates for targeted cancer therapies. Moreover, WBCs share some similarities to tumor cells in terms of migration in the body.

A few preclinical studies have demonstrated that WBCs can carry drugs-loaded NPs in poorly accessible zones of cancer lesions, like the necrotic hypoxic core of solid tumors. In necrotic tumor tissues, the lack of vasculature prevents the accumulation of nontargeted NPs. Conversely, chemoattractant gradients mostly recruit immune system cells from the peripheral blood. This mechanism can be exploited to obtain WBC-based Trojan Horses.

2.3 Stem cell-based therapies

SCs possess intrinsic targeting properties to be exploited in biomimetic drug delivery systems. Because of their inherent tropism toward tumors, SC-based therapies are successfully employed in cancer treatment [31], and in the control of metastatic processes [32]. SC-based delivery systems are particularly promising in the treatment of ovarian [33], brain [34–36], and breast [37] cancers. SCs are also a good vector for oncolytic virotherapy for cancer [38]. Oncolytic virotherapy is a novel approach exploiting the ability of a virus to promote cell lysis and to selectively destroy cancer cells. Preclinical studies are encouraging and clinical trials report that treatments are generally well tolerated. However, the clinical diffusion of oncolytic virotherapy is still limited due to the inability to target metastases. Efficacy of SC-based treatments is given by their high capability to protect oncolytic viral cargo from neutralizing antibodies and by their tumor homing ability [39]. Preclinical studies have shown that cancerous lesions can be treated with this therapy [40], which can also target metastases [41, 42].

2.4 Virus-based carriers

Virus-based carriers are prepared from capsids or envelope proteins extracted from living viruses self-assembled around a drug or a synthetic particle. Their safety is guaranteed by the complete absence of genetic material contained in natural viruses. Their functional properties can be improved with appropriate surface chemistry functionalization, for example, conjugating peptides and antibodies [43]. Moreover, virus-based vectors have the natural tropism of living parent viruses [44].

The production of virus-derived nanosized delivery systems is relatively easy and inexpensive. Home-made virus capsids have the same antigenicity of their parent viruses. Virus-based carriers find a wide application in the field of vaccine delivery [45]. The most diffused virus-based therapy is related to the human papillomavirus (HPV). HPV-like NPs have high similarities with the natural virion, and are good candidates for safe and highly immunogenic vaccines [46]. HPV-like NPs are used in clinical practice and FDA has already approved two commercial products: Gardasil [47], from Merck, and Cervarix [48], from GlaxoSmithKline.

Beside vaccines, virus-based NPs can also transport active principles like genes [49], proteins [50], and drugs [51]. Also, plant-derived virus-like NPs [52, 53] have some important advantages related to their biodegradability and biocompatibility and represent a valid alternative to obtain hybrid NP-based therapies.

2.5 Bacteria-based drug vectors

The majority of pathogens, as viruses and bacteria, possess an innate capability to invade the host body eluding the immune response, and to activate strong interactions with their target organs. Such capabilities derive from long pathogen evolution over time, and would also be desirable for an efficient drug carrier. Strategies inspired to pathogens in drug-delivery nanocarriers preparation are then advisable to improve their efficiency.

Some pathogens have unique bioavailability capabilities, being able to reach also organs that are difficult to penetrate, like the central nervous system. Molecular mechanisms that favor pathogen biodistribution are complex and, in most cases, multifactorial. The literature reports a plethora of studies explaining invasion underpinnings (e.g., see Refs. [54–58]), underlining multiple interactions between pathogens and receptors in the host body.

Bacteria-based drug nanovectors are currently obtained by genetic manipulation. Modern synthetic bioengineering techniques provide safe and efficient carriers for specific clinical needs.

Nonpathogenic anaerobic bacteria, like *Salmonella*, *Clostridium*, *Escherichia coli*, and *Listeria*, are currently investigated to treat some cancer types. This class of bacteria selectively accumulates in solid tumors and tumor-driven lymph nodes, and there can release chemotherapeutic drugs, also exhibiting an intrinsic antitumor activity. The hypoxic microenvironment in necrotic tissues can not only explain bacteria targeting ability [59], but also other microenvironment characteristics, such as the presence of chemoattracting compounds [60], aberrant neovasculature [61], multiple tumor-induced immunosuppressive mechanisms [62], and inhibition of immune cell-activating transduction [63], can favor their proliferation and colonization within tumors [64].

Recombinant living bacteria were obtained loading attenuated *Salmonella typhimurium* with plasmid-based small interfering RNAs and tested in vivo [65]. The system was not only able to produce and release in situ recombinant proteins at the target site, inhibiting the tumor growth, but also exerted a therapeutic effect on metastatic organs, reducing their overall number. An analogous strategy was successfully exploited to treat cervical cancer in mice [66], with the bacterial strain LH430. In this case, the therapeutic mechanism relies on the synergistic co-expression of plasmid-based E6-specific siRNA and the wild-type p53. The same approach was used to deliver the pEndo-Si-Stat3 protein from *S. typhimurium* attenuated bacteria, for the decrease of the Stat3 levels and thus the significant suppression of prostate cancer growth in in vivo tests [67]. The same bacterium in its living form was genetically modified and tested in vivo in mice for murine melanoma. Once again, the strain VPN20009 was used as vector to inhibit cancer growth and also related metastatic processes, also in mice with a compromised immune system [68]. For the same application, the VPN20009 vector loaded with the interleukin-21 expression plasmid was also evaluated with good results in terms of tumor growth control, inducing enhanced infiltrations of natural killer and T cells in the tumor area [69]. Several studies have demonstrated that engineered *Salmonella* is still a valid vector for conventional chemotherapeutics, like 5-fluorouracil [70] or cisplatin [71].

The particular tumor structure, with necrotic and hypoxic central regions and well perfused and oxygenated rims, makes the use of anaerobic bacteria particularly useful. From this point of view, interesting results were obtained by using the attenuated *Clostridium novyi-NT* bacterium. *C. novyi-NT* is an anaerobic bacterium that was found able to germinate exclusively in hypoxic cancerous regions after systemic administration [72]. For this characteristic, the bacterium was used to strength conventional chemotherapy, mainly hitting well-oxygenated tumor rims, in combination bacteriolytic therapy (COBALT) [73]. The co-administration of *C. novyi-NT* spores and conventional chemotherapeutic drugs, like dolastatin 10 and mitomycin C, produces an extensive hemorrhagic necrosis of tumors in few hours, providing prolonged antitumor effects [74]. Moreover, the doxorubicin level in tumor cells was significantly enhanced, up to five times, in the co-administration of Doxil and *C. novyi-NT* spores, inducing the complete tumor regression in 100% of tested mice [75]. Additional studies have demonstrated that the *C. novyi-NT* spores induce the complete tumor regression in animals without additional adjuvant therapies, and analogous results could be reasonably obtained in human patients [76].

Additional applications of living bacteria are related to vaccines [77] and antiinflammatory therapies [78]. Efficient drug vectors are also obtained from ghost nonliving bacteria that, after their manipulation, preserve the intrinsic ability to target different cells with their chemical surface properties. A very recent application confirmed the potential of bacteria-based drug carriers also for cerebral drug delivery [79].

Although relatively safe, bacteria-based drug vectors can maintain some degree of immunogenicity, and then concerns related to their administration in humans should be carefully considered before the clinical use.

3. Bio-responsive synthetic nanocarriers for drug delivery

In addition to natural systems, some synthetic or semisynthetic drug delivery systems are potential candidates for clinical applications. Synthetic carriers, composed of degradable polymers or inorganic materials, are constantly under investigation in preclinical and clinical studies and some of them result eligible to be marketed.

The "first generation" of synthetic drug carriers was conceived as "protection system" for drugs, able to preserve the activity of administered active principles, and to modulate its delivery kinetics. More recently, highly effective stealth and targeting properties based on elaborate surface chemistry techniques were also introduced. More complex NP systems are finally obtained by decorating their surface with natural molecules or synthetic bio-inspired units.

Polymer-based targeted NPs are particularly promising for therapies involving the administration of harmful drugs, like chemotherapeutics. Polymer NPs are engineered to optimize their physico-chemical properties, on the basis of the biology of the organ to be targeted and the characteristics of the loaded active molecule [80].

The majority of synthetic polymers used in modern NP preparation are degradable or biodegradable [81]. Engineered polymer nanocarriers can be tailored to show high biocompatibility and biodistribution, and good targeting properties. Excellent targeting abilities can be obtained by decorating NP surface with targeting ligands. This strategy shows a great potential also for the targeting of poorly accessible organs, like the brain [82]. Such structures are currently used to transport drugs, small molecules, proteins and genes, with applications in cancer therapy [83], brain-related disease treatments [84, 85], cardiovascular pathologies [86], and orthopedic applications [87].

Once the improved efficacy of natural systems was confirmed, synthetic systems have started to implement some biomimetic properties to boost their performances. A simple biomimetic property in polymer NPs is bio-responsiveness, or the capability to respond to biological signals or changes in the physiological environment [88]. Bio-responsive materials can adapt their behavior when exposed to pathological conditions, such as the change of pH, temperature, hypoxia or in redox conditions.

3.1 pH-responsive nanoparticles

pH-responsive NPs can improve the oral administration of active molecules, as reported by Guha et al. [89] This study reports the preparation and characterization of a core/shell nanosystem composed of mesoporous silica and covered with a pH-responsive polymer. Such core/shell NPs were tested for the oral administration of insulin and showed that the pH-responsiveness provided a preferential release at the systemic pH of 7.4.

3.2 Temperature-responsive nanoparticles

Temperature-responsive NPs can provide an on-demand drug delivery, triggered by a temperature change, for example, due to the altered physiology of pathological lesions [90].

Temperature change can also affect macroscopic particle properties, as demonstrated by He et al. [91]. Authors have reported the capability of fluorescent labeled NPs to change their shape on a thermal stimulus, with a significant diameter increase that enabled a fluorescence signal. The on-demand NP fluorescence can help in the image-guided drug release. Moreover, considering that size and shape govern the NP interactions with cells, the intravascular transport, barrier penetration and recognition properties, the tunability of NP size and shape in vivo is a significant improvement of their biomimicry [92].

3.3 Mechanical stimuli-responsive nanoparticles

Mechanical stimuli-responsive NPs can also tune drug delivery kinetics. To date, the control exerted by mechanical forces on drug delivery is well-accepted [93] and deformable NPs are good candidates for mechanically activated drug carriers. Along with the use as delivery systems, shape-adaptable NPs were also exploited as tool for the study of intracellular mechanical properties. In this field, Cheng and co-workers have reported significant results obtained with different deformed shapes in soft synthetic NPs incubated with different cell lines [94]. This study paves the way to more sophisticated drug delivery systems for a targeted intracellular delivery activated by mechanical stimuli.

3.4 Molecularly imprinted nanoparticles

Also, biomimetic targeting properties can be obtained by synthetic systems. The most useful and completely synthetic technology to do this is the molecular imprinting (MI) technique. Molecularly imprinted NPs have recognition properties able to selectively recognize homologous molecules, enabling a *lock and key* mechanism for molecular recognition. This promising technique is mostly used in the manufacture of synthetic antibodies to retain active molecules [95], for intracellular protein sequestration [96] or in intracellular imaging [97], but a few examples of drug delivery systems are also reported [98]. In the field of drug delivery from NPs, the major limitation of the MI technology is its application to degradable materials. A recent work [99] partially overcame this issue by limiting the imprinting process to a degradable shell polymerized around a nondegradable particle core. This approach partially reduces the amount of nondegradable material that should be administered in a clinical application, limiting the severe accumulation in the body, but larger efforts are needed to improve the overall biocompatibility of the proposed solution. More recently, a completely degradable bulk-imprinted NP system that combines, for the first time, degradability and recognition properties was proposed [100], demonstrating that functional properties are not significantly different than that obtained in nondegradable particles. Table 1 summarizes additional examples of bio-responsive synthetic nanocarriers used for drug delivery.

4. Biomimetic hybrid nanoparticulate systems

While the beneficial effects of natural cells are evident, safety drawbacks hinder their widespread use in clinical applications. Biomimetic hybrid NPs can be strictly controlled and thus can overcome such drawbacks. Hybrid systems take place by the combination of

Table 1 Biomimetic synthetic NPs.

NP composition	Main properties	Applications	References
pH-responsive systems			
Mesoporous silica covered with poly (methacrylic acid-co-vinyl triethoxylsilane)	Preferential delivery at pH 7.4	Oral administration of insulin	[89]
Poly(cyclohexane-1,4-diylacetone dimethylene ketal) and polyethylene glycol/PEGylated poly (lactic-co-glycolic acid) NPs functionalized with folic acid	Favored drug release at pH 5	Systemic treatment of rheumatoid arthritis (methotrexate administration)	[101]
Mesoporous silica NPs conjugated with hyaluronic acid	Zero premature release of the drug, conjugated to NPs via hydrazone bond, cleaved only at acidic pHs	Administration of doxorubicin	[102]
Hyaluronic acid NPs, functionalized with triphenylphosphonium	Mitochondrial-targeted delivery triggered in acidic environment	Administration of doxorubicin	[103]
Temperature-responsive systems			
PEGylated polyaspartamide derivative NPs	Highly efficient drug loading at 60°C	Systemic administration of paclitaxel	[104]
Polyamidoamine dendron lipids conjugated with alkyl chains	Improved drug release at 43°C	Systemic administration of doxorubicin	[105]
Gold NPs covered with elastin-like peptides	NPs aggregation induced by the temperature (40°C)	Photothermal cancer therapy	[106]
Gold NPs-grafted poly(dimethylacrylamide-co-acrylamide)/poly acrylic acid	Switchable release triggered at 50°C	Ofloxacin administration	[107]
Temperature and pH-responsive systems			
PEGylated poly (vinylcaprolactam)-based NPs	Drug delivery is enhanced at acidic pH and elevated temperature	Systemic administration of doxorubicin, preferentially for cancer treatment	[108]
Mesoporous silica NPs covered with poly(N-isopropylacrylamide-co-acrylic acid)	Delivered amount of drug doubled at 37°C and pH 2	Ibuprofene (model drug) delivery	[109]

Table 1 Biomimetic synthetic NPs—cont'd

NP composition	Main properties	Applications	References
Molecularly imprinted NPs			
(R)-Thalidomide-imprinted methacrylate-based NPs	Enantioselective controlled release	Potentially usable in cancer research	[110]
Rivastigmine-imprinted magnetic nanocomposite of graphene oxide/ Fe_3O_4 nanoparticles and acrylate monomers	Sustained release of rivastigmine	Alzheimer's disease treatment	[111]
Doxorubicin-imprinted mesoporous silica NPs	Drug loading during MIP synthesis, pH-controlled doxorubicin release	Potentially usable in cancer research	[112]
Paclitaxel-imprinted methacrylate NPs, functionalized with folic acid	High drug entrapment, sustained drug delivery	Targeted delivery in cancer cells	[113]

synthetic and natural materials, the latter making the most of their interactions with the physiological environment. The combination of the properties of synthetic materials, like the tailorability of physicochemical properties, and those of biological materials, like mimicry and targeting capabilities, endows final systems with unique performances.

In case of NPs with low or absent biomimetic capabilities, the host immune response after their administration can lead to a rapid clearance of the delivery system. It is the main limitation experienced with the majority of NP-based therapies. On the contrary, a good biomimetic hybrid system can be safely administered keeping low the immune response of the host body and limiting iatrogenic effects.

For all these reasons, the use of synthetic biomaterials conjugated to biologically derived materials, such as peptides and proteins is increasingly taking hold [114], confirming that hybrids show improved biocompatibility and enhanced functionalities.

To date, the functionalization of synthetic NPs with naturally derived materials is fairly consolidated but some safety concerns are still open. However, the practice to coat synthetic NPs with biological materials has a great potential and can be exploited to decrease the immunological response of the host body and to improve targeting properties.

4.1 Stealth nanoparticles with different functionalization strategies

NPs administered by intravenous injection are exposed to protein plasma components. Interactions between NPs and such components generate the so-called protein corona around particles. The protein corona modulates all biological responses, in particular those related to the human immune cells. Synthetic NPs are significantly subjected to the host immune response that affects their therapeutic efficacy. Strategies to limit immune response are strictly related to the NP surface properties, being directly connected to the corona formation. Large efforts in surface engineering of NPs have led to the minimization of interactions with plasma proteins, involving the use of specific moieties with antifouling properties. NPs with low affinity for plasma components are referred as "stealth," being able to elude their association with immune system cells and thus their premature capture and clearance off the bloodstream. As a rule of thumb, surface-engineered NPs with antifouling properties behave as stealthy NPs.

Stealth nanocarriers exhibit a significantly long circulation time in the blood flow. Stealthy properties are commonly obtained by increasing the hydrophilicity of the carrier surface with the functionalization or the conjugation of hydrophilic polymers. The most commonly used hydrophilic unit is the polyethylene glycol (PEG) [115, 116] (Fig. 2A) and its macromolecular derivatives [117]. Surface functionalization with PEG chains is called PEGylation. PEGylation is highly effective in the reduction of NP opsonization and, therefore, can limit the in vivo clearance by MPS. For these reasons, PEG functionalization has been introduced in several clinical products [118].

However, recent studies have underlined some safety concerns in the use of this polymer [119], pushing researchers to take different bioinspired alternatives into account. One of the most promising bioinspired approach to endow NPs with stealthy properties is to mimic the zwitterionic headgroups of phospholipids composing the red blood cell membrane, like phosphatidylcholine (PC) [120]. In vitro and in vivo tests have demonstrated that antifouling properties can be also achieved using zwitterionic synthetic polymers [121], like poly(carboxybetaine)s (PCB) [122, 123] (Fig. 2B), poly(sulfobetaine)s (PSB) [124–126] (Fig. 2C), poly(phosphorylcholine)s (PPC) [127–131] (Fig. 2D), and

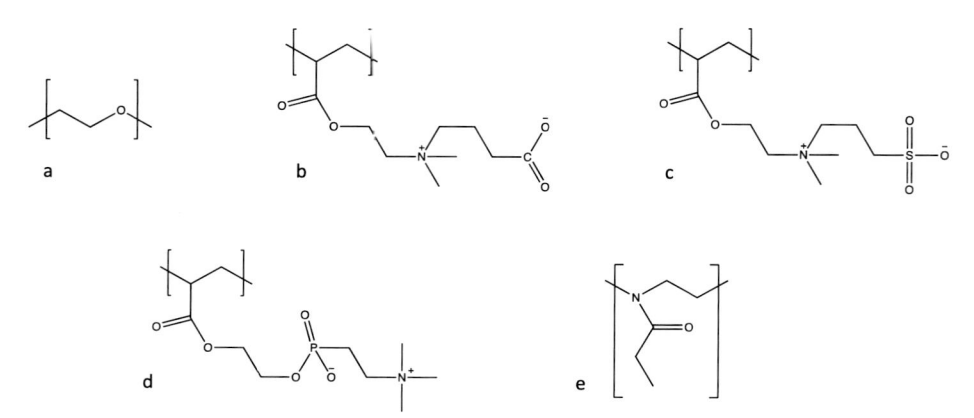

Fig. 2 Macromolecular structure of (A) polyethyleneglycol; (B) poly(carboxybetaine)s; (C) poly(sulfobetaine)s; (D) poly(phosphorylcholine); (E) poly(oxazoline)s.

zwitterionic polypeptides [132]. The functionalization with hydrophilic carbonates [133], polyoxazolines [134] (Fig. 2E), and some natural polymers [135, 136] are also valid alternatives to PEGylation. Table 2 summarizes additional literature related to stealth nanoparticles used for drug delivery.

4.2 RBC-cloaked or -functionalized nanoparticles

Taking inspiration from nature, the most valid alternative to chemical functionalization is the preparation of erythrocyte-like delivery systems. Some manufacturing techniques turned out to be helpful to reproduce the size, shape, and mechanical stiffness of RBCs in polymeric structures, like thin film stretching [147], electrospraying [148], and nonwetting templates (PRINT) technique [149, 150]. However, while RBC-shaped drug carriers are promising, the "synthetic" reproduction of RBC surface properties is still far away.

A very interesting approach to make particles "stealthier" with materials from biological derivation is the functionalization of the NP surface with membranes extracted from cells [151]. Final systems obtained are cell membrane-cloaked NPs. Cloaked NPs are embedded in intact cell membranes, forming a core/shell biohybrid structure. Copp et al. [152] have reported interesting results obtained with degradable PEGylated NPs, composed of poly(lactide-co-glycolide) (PLGA), enclosed within the intact RBC membrane. The system was tested for the selective depletion of disorder-causing antibodies. The rationale of this study was strong affinity between the RBC membrane surface and target antibodies. Since the absorption process is not specific, this approach is particularly valid when the disorder varies from patient to patient, as in antibody-mediated autoimmune diseases. Moreover, another study has demonstrated that RBC-coated PLGA NPs can have a more than doubled circulation time with respect to PEGylated PLGA NPs [153].

The literature reports a plethora of interesting results with RBC membrane-cloaked NPs as drug carrier, like vaccines [154], on-demand targeted drug delivery systems [155], or in cancer treatment to improve the pharmacokinetics of commercial drugs [156].

RBC-cloaked NPs can be further improved by functionalization of the cell membrane surface, to push performances of the hybrid carrier beyond those of natural cells, as successfully reported in Refs. [157, 158].

RBC-mediated delivery systems are currently considered very promising [159], thanks to the simple isolation and loading procedures, and their prolonged circulation. To date, functional capabilities of RBC-based drug delivery systems were extensively investigated ex vivo and in vivo as well [10].

As previously mentioned, RBCs can also serve as vectors for drug vectors. Anselmo et al. [16] have demonstrated how deposit polystyrene NPs to lung epithelium avoiding accumulation in spleen and liver. In a more recent work, this "RBC-hitchhiking" mechanism was majorly exploited to significantly boost drug delivery in a specific organ, after the accurate selection of the intravenous injection site [160]. This deposition mechanism

Table 2 Stealth NPs obtained with different functionalization strategies.

System	Main properties	Applications	References
PEGylated NPs			
Large unilamellar vesicles targeted with a PEGylated RICK peptide	NP cellular internalization and activity preserved with PEGylation (up to 20%), elimination time 34% longer than non-PEGylated NPs	Delivery of siRNA	[137]
PEGylated PLGA NPs	Higher stability and antinicotine IgG titer, optimal immunogenicity response at 20% PEGylation	Nicotine vaccine	[138]
PEGylated carbonate apatite NPs	Sixfold increased drug accumulation in tumor compared with free drug, higher plasma drug amount	Gemcitabine release for the treatment of breast cancer	[139]
PEGylated poly (ε-caprolactone)-based NPs covered with polydopamine	PEGylation reduces the local temperature reached in photothermal therapy	Delivery of siRNA and photothermal therapy in cancer	[140]
PEGylated chitosan micelles	Loading of hydrophobic drugs and micelle formation	Oral administration of camptothecin	[141]
Phosphatidylcholine-covered NPs			
Grafted iron oxide NPs	PC induces a lower complement activation than PEG	Stealth NPs	[142]
Redox-responsive liposomes	Improved biodistribution and antitumor activity	Delivery of paclitaxel for the treatment of solid tumors	[143]
Thioether phosphatidylcholine liposomes	ROS-responsive drug delivery, improved anticancer activity	Delivery of doxorubicin for the treatment of solid tumors	[144]
PLGA-based NPs	Improved drug bioavailability	Delivery of rutin for the treatment of Alzheimer's disease	[145]
Iron oxide NPs	Improved uptake and transcytosis across the blood-brain barrier (BBB) (in vitro)	Treatment of CNS disorders	[146]

Table 2 Stealth NPs obtained with different functionalization strategies—cont'd

System	Main properties	Applications	References
Poly(carboxybetaine)-coated NPs			
Methacrylated-based NPs	Low fouling properties, comparable with PEGylated NPs	Delivery of hydrophobic drugs for cancer therapy	[122]
Iron oxide NPs grafted with PCB brushes	Superior circulation time and anticancer activity	Photothermal therapy of solid tumors	[123]
Poly(sulfobetaine)-coated NPs			
Methacrylate-based NPs	Four times faster diffusion in the cytosol compared with PEGylated NPs	Intracellular imaging	[124]
Polyester-based NPs	Sustained release, high drug encapsulation (curcumin as model drug)	Passive drug targeting for cancer treatment	[125]
Zein NPs	Low macrophage activation in vitro, long circulation time in vivo	Systemic treatment of cancer	[126]
Poly(phosphorylcholine)-coated NPs			
Mesoporous silica NPs	Low fouling properties, control of protein corona formation	Stealth NPs	[127]
Methacrylate-based NPs	Protein-repellent surface	Delivery of 4-(*N*-(*S*-penicillaminylacetyl) amino) phenylarsonous acid for cancer treatment	[128]
Phosphorylcholine-based nanogel	Excellent protein adsorption resistance	Delivery of doxorubicin for cancer treatment	[129]
D-α-Tocopheryl polyethylene glycol succinate NPs	Prolonged in vivo half-life than PEGylated NPs, high tumor accumulation	Delivery of doxorubicin for liver cancer treatment	[130]
Bovine serum albumin NPs	Significantly prolonged drug half-life, high accumulation in tumor	Delivery of doxorubicin for cancer treatment	[131]

Continued

Table 2 Stealth NPs obtained with different functionalization strategies—cont'd

System	Main properties	Applications	References
Functionalization with zwitterionic polypeptides			
Zwitterionic polypeptide micelles	Drug half-life increased by 17-fold compared with free drug, and by 1.6-fold compared with the non-stealth NPs	Delivery of Abraxane in multiple tumor models	[132]
Polycarbonates			
Poly(amino acid)–based micelles	Non-fouling properties similar to PEG, enzymatically and hydrolytically degradable	Systemic drug delivery	[133]
Polyoxazolines			
Tobacco mosaic virus NPs	Reduced recognition and similar pharmacokinetic compared with PEGylated control	Nanocarrier for drug delivery and molecular imaging	[134]

occurs because particles attached to RBC surface can be immediately delivered to the first organ downstream the intravenous injection. RBC-hitchhiking vectors are particularly interesting in applications in which the organ to be treated is difficult to reach, as in the case of the brain. Brenner and co-workers have demonstrated that the intracarotid artery injection of RBC-hitchhiking vectors is more efficient, in terms of particle accumulation in the brain, than the strategy based on the decoration of particle surface with targeting unit ($\sim$10× higher). Remarkable results were also obtained in the inhibition of lung metastases growth, with or RBC-hitchhiking vectors providing not only an increased circulation time but also an effective redistribution of NPs in the body [161].

Another valid approach is the use of components of the RBC cell membrane to prepare core/shell NPs (Table 3). It allows transferring the complex biochemical structure of erythrocytes to NPs, as demonstrated in the study of Hu and co-workers [162]. The cited study reported that core/shell hybrid NPs had the same RBC surface density of CD47, the transmembrane protein that determines RBC circulation time in the bloodstream [165]. The camouflage provided by RBC membrane components effectively reduces the MPS activation in case of iron oxide NPs, providing a better in vivo biodistribution than that obtained for standard PEGylated NPs [163]; moreover, a significantly lowered nonspecific macrophage uptake in mice was obtained [164].

Table 3 RBC-cloaked or -functionalized nanoparticles.

System	Main properties	Applications	References
PLGA NPs	Target for pathological antibodies in an antibody-induced anemia disease model	Neutralization of anti-RBC polyclonal IgG, preservation of circulating RBCs	[152]
PLGA NPs	Superior circulation half-life compared with control particles, NP retention in the blood 72 h after injection	Systemic drug administration	[153]
PLGA NPs	Enhanced in vitro cell uptake, promoted retention in draining lymph nodes	Antitumor immunization against melanoma	[154]
Polyamide-based micelles	Limited drug self-leakage	Systemic delivery of doxorubicin	[155]
PLGA NPs	Sustained release, suppressed immunological response	Delivery of doxorubicin for the treatment of solid tumors	[156]
PLGA NPs	Prolonged circulation, reduction of inside-out membrane orientation on NPs, multifunctionality	Systemic treatment of cancer	[157]
Liposomes	40-fold increased uptake, suppressed end-organ toxicity	Augmented drug delivery in lung disease, stroke, and other diseases	[160]
Magnetic NPs, coated with polymers or uncoated	120-fold boosted drug delivery, significant targeting properties	Systemic delivery of doxorubicin for cancer treatment	[161]
PLGA NPs	Reduced susceptibility to macrophage uptake	Systemic drug delivery	[162]
Iron oxide NPs	Reduced reticuloendothelial system (RES) uptake, better biodistribution than PEGylated control NPs	Systemic administration of NPs	[163]
Iron oxide NPs	Prolonged blood circulation time, reduced RES capture, high tumor passive accumulation	Photothermal therapy in solid tumors	[164]

Carriers for active drug targeting need to accumulate in the pathological site through a passive mechanism, while active targeting is considered a complementary strategy to improve drug retention [166].

Naturally derived systems, based on pathogens or cells, possess interesting properties exploitable to improve drug therapies. Naturally derived vectors are intrinsically able to circulate in the blood flow and reach specific districts. Recent advances in nanomedicine demonstrated that safe carriers from living cells can be obtained, as well as the possibility to load sufficient amounts of drugs without compromising the overall properties of the native cell shell. Therapeutic approaches are mainly based on cell ghosts or cells devoid of their internal biological contents.

While synthetic NPs are well-characterized, and several consolidated strategies allow the modulation of their functional properties [167], performances of cell-based systems are not yet well recognized [168]. On the other hand, cell-based NPs are really promising in terms of targeting properties and thus are considerably under investigation.

4.3 WBC-cloaked nanoparticles

In a pioneering study, monocyte-based Trojan Horses were studied in photothermal treatment of human breast cancer cells using an in vitro model [169]. In this work, Au nanoshells were incubated with monocytes and phagocytized. Then, Au-loaded macrocytes infiltrated into the tumor tissue and differentiated in macrophages. Here macrophages released the nanoshells after photoablation induced by NIR irradiation, and free nanoshells were uptake by tumor cells. A further photoablation induced cell death. A more recent study has confirmed that photoablation was efficient also in vivo after the intratumor injection of gold nanorod-loaded macrocytes [170].

The same Trojan Horse mechanism can be exploited to treat brain metastases. While conventional drugs and a variety of NPs cannot penetrate the blood-brain barrier (BBB), macrophages and peripheral blood leukocytes are able to target brain metastases also in case of intact BBB [171]. The capability of particle-loaded macrophages to cross the BBB and accumulate in brain metastases was verified in mice bearing brain tumor xenografts. Macrophages loaded with the liposomal formulation of doxorubicin were injected in vivo in mice. Compared with the doxorubicin-loaded liposomes administration, the monocyte-based therapy demonstrated a major effectiveness in tumor reduction [172]. The penetration of monocytes and NPs across the BBB was further investigated for the treatment of depression. Liposomes with high affinity for integrin receptor of leukocytes were obtained by modification with a cyclic RGD peptide. Such liposomes developed interactions with monocytes in vivo, favoring the co-migration [173]. Similar to RBCs, leukocytes can be used as vector for drug vectors, and hitchhiking liposomes transported by WBCs are under investigation with promising preliminary in vitro results [174].

The cited delivery systems were all developed by loading, with drugs on NPs, living leukocytes that were used for their mimetic properties. Here the leukocytes provide a "biological interface" to their payload, being reasonable to attribute their intrinsic targeting properties to their surface. On this basis, the potential of a biomimetic camouflage by transferring the biological interface of leukocytes to synthetic systems was explored. It was recently demonstrated that nanoporous silica NPs embedded in a WBC membrane have enhanced permeability across tumor vasculature [175], thanks to the characteristic of

WBCs to invade inflamed tissues and extravasate across the endothelium. Nanoporous silicon NPs on which leukocyte membrane was transferred could interact with endothelial cell receptors and evade the immune response of the host. Moreover, they could selectively transport and release their drug cargo in a reconstructed inflamed endothelium [176]. He et al. have demonstrated that Janus NPs, half-coated with the WBC membrane, can have an improved uptake in cancer cells, also providing a significantly preferential recognition between different cancer cell types [177]. Moreover, Janus NPs with the same architecture were tested for photothermal cancer treatment [178]. Also Xuan et al. have recently reported the application of macrophages cell membrane-coated gold NPs for cancer photothermal therapy that could accumulate in cancer cells in vivo [179]. Owing to their envelope, these NPs showed specific recognition of tumor endothelium and a more efficient accumulation compared with RBC-coated NPs (Table 4). Similarly,

Table 4 WBC-cloaked NPs.

System	Main properties	Applications	References
Au nanoshells	Au-loaded macrocytes infiltrated into the tumor tissue and differentiated in macrophages	Photoablation of solid tumors	[169]
Au nanorods	Improved photothermal conversion than protein-coated nanorods	Photothermal therapy of solid tumors	[170]
Au nanoshells	Blood-brain barrier (BBB) penetration	Treatment of brain metastases	[172]
Liposomes	BBB penetration	Treatment of depression	[173]
Liposomes	High tumor accumulation, localized siRNA delivery	Treatment of solid tumors	[174]
Nanoporous silica NPs	Enhanced vascular permeability and accumulation in target tissues	Treatment of solid tumors	[175]
Nanoporous silica NPs	Transport of drugs across inflamed endothelium, reduced opsonization	Systemic administration of doxorubicin (drug model)	[176]
Polyelectrolyte-based Janus NPs	Discrimination of cancer cells from noncancerous cells	Efficient detection of cancer cells after administration	[177]
Polyelectrolyte-based Janus NPs	Better biocompatibility and target recognition	Photothermal therapy of solid tumors	[178]
Au nanoshells	Prolonged circulation time, active targeting ability by recognizing tumor endothelium	Photothermal therapy of solid tumors	[179]
Mesoporous silica NPs	Reduced RES clearance, prolonged circulation time, improved accumulation in target tissues	Cancer therapy	[180]

macrophage cell membranes were proposed to embed mesoporous silica NPs, providing an efficient NP platform for doxorubicin delivery [180].

The inventory of cell membranes used to cloak-targeted NPs have been extended by Fang and co-workers [181], who have developed an anticancer vaccine based on PLGA NPs coated with mouse melanoma cell membranes. The NPs showed the same adhesion capabilities of the source cells, with a significant cell-specific targeting activity. Li et al. have reported a biomimetic approach to limit metastatic processes by using platelet membrane-coated NPs [182, 183] (Fig. 1F). The rationale under this approach differs from those previously reported. The platelet targeting properties are attributed to their capability to be physically associated with blood-borne cancer cells in the blood flow, in which circulating cancer cells transit through. Works by Li et al. have demonstrated targeting properties of silica particles functionalized with membrane-derived vesicles from activated platelets, and their following conjugation with a tumor-specific apoptosis-inducing ligand cytokine (TRAIL). Polymer NPs coated with platelet membranes have been also tested for the co-delivery of antimyeloma drugs in the treatment and for the reduction of thrombus formation [184].

Other notable targeting properties were demonstrated by functionalizing synthetic NPs with bacteria-derived materials.

A biohybrid drug delivery system composed of the *Escherichia coli* bacteria envelop encapsulating elliptical disk-shaped polymeric microparticles was recently proposed [185]. It was shown that the chemotactic response of hybrid bacteria particles, presenting the tropism of the native bacteria, dominates the effect of body shape in extravascular transport.

5. Clinical trials

The urgent necessity of improved pharmacological therapies is increasingly pushing the translation from bench to bedside of drug-delivery NPs, and a growing number of clinical and preclinical studies are currently under investigation. In human, clinical trials can clarify particle/cell interactions and the actual performances of the therapy. According to clinicaltrials.gov, to date 18,057 clinical trials (under investigation or completed, all over the world) are testing some kinds of drug delivery systems, while 403 clinical trials involve the use of NPs, but only 27 studies are related to drug delivery systems based on NPs, 17 of them for the case of in cancer treatment. Liposomal formulations are the most investigated, with 122 clinical trials, 12 of them in the last 4 years [186].

Long-circulating NPs seem to be the most promising in clinical trials. The majority of them are PEGylated particles [118] while only one study focuses on non-PEGylated stealth NPs (Renagel, Sanofi) and is composed of poly(allylamine hydrochloride), for the treatment of chronic kidney diseases [187].

Among clinical trials involving hybrid systems, MesomiR (trial ID: NCT02369198) has concluded the Phase I trial [188]. This system is composed of a targeted bacterial minicells releasing miRNA, for the therapy of malignant pleural mesothelioma and non-small cell lung cancer. This medication exploits the EDV nanocells technology (EnGeneIC Ltd), already tested in vivo in preclinical studies for the delivery of chemotherapeutics [189], siRNAs [190], and miRNAs [191]. The same EDV nanocells technology is also involved in two additional clinical trials. The study ECREST (trial ID: NCT02687386) is in its Phase I and analyzes the effect of mitoxantrone delivery from EDVs for the treatment of recurrent or refractory solid or brain tumors expressing EGFR in children. The study CerebralEDV (trial ID: NCT02766699), also in Phase I, aims at evaluating the safety, tolerability and immunogenicity of doxorubicin-loaded EDVs administered to subjects with recurrent Glioblastoma Multiforme.

6. Conclusions

By now, biomimicry is the most performant strategy to enhance the functional properties of targeted drug delivery systems, improving the safety profile and the pharmacological efficiency of drug therapies. Targeting the pathological site, and thus limiting the off-target drug deposition, without increasing the administration dosing is one of the priorities in clinical practice and some encouraging results have shown that it can be achieved by using biomimetic NPs. Synthetic or semisynthetic biomimetic NPs enable significant benefits in terms of better drug biodistribution, longer circulation time, improved biocompatibility and accumulation in target organs.

Nowadays, the promising biomimetic approach, in which the combination of natural and synthetic materials provides a synergistic effect, seems to be a reliable way to tune the carrier properties. Preclinical and clinical studies have demonstrated the actual possibility to improve drug pharmacokinetics in this way. The combined knowledge of on material science and cell biology can contribute in the discovery of novel biomimetic strategies, and thus boosting the discovery of new and increasingly performing drug delivery systems.

References

[1] W. Wu, L. Luo, Y. Wang, Q. Wu, H.-B. Dai, J.-S. Li, et al., Endogenous pH-responsive nanoparticles with programmable size changes for targeted tumor therapy and imaging applications, Theranostics 8 (11) (2018) 3038–3058.
[2] J. Kreuter, Nanoparticles – a historical perspective, Int. J. Pharm. 331 (2007) 1–10.
[3] J.F.V. Vincent, O.A. Bogatyreva, N.R. Bogatyrev, A. Bowyer, A.-K. Pahl, Biomimetics: its practice and theory, J. R. Soc. Interface 3 (9) (2006) 471–482.
[4] Q. Sun, Z. Zhou, N. Qiu, Y. Shen, Rational design of cancer nanomedicine: nanoproperty integration and synchronization, Adv. Mater. 29 (14) (2017).

[5] Y. Huang, X. Gao, J. Chen, Leukocyte-derived biomimetic nanoparticulate drug delivery systems for cancer therapy, Acta Pharm. Sin. B 8 (1) (2018) 4–13.

[6] J. Simon, L.K. Müller, M. Kokkinopoulou, I. Lieberwirth, S. Morsbach, K. Landfester, et al., Exploiting the biomolecular corona: pre-coating of nanoparticles enables controlled cellular interactions, Nanoscale 10 (22) (2018) 10731–10739.

[7] J.Y. Oh, H.S. Kim, L. Palanikumar, E.M. Go, B. Jana, S.A. Park, et al., Cloaking nanoparticles with protein corona shield for targeted drug delivery, Nat. Commun. 9 (1) (2018) 4548.

[8] S. Tan, T. Wu, D. Zhang, Z. Zhang, Cell or cell membrane-based drug delivery systems, Theranostics 5 (8) (2015) 863–881.

[9] Y. Su, Z. Xie, G.B. Kim, C. Dong, J. Yang, Design strategies and applications of circulating cell-mediated drug delivery systems, ACS Biomater. Sci. Eng. 1 (4) (2015) 201–217.

[10] C.H. Villa, D.B. Cines, D.L. Siegel, V. Muzykantov, Erythrocytes as carriers for drug delivery in blood transfusion and beyond, Transfus. Med. Rev. 31 (1) (2017) 26–35.

[11] L. Chessa, V. Leuzzi, A. Plebani, A. Soresina, R. Micheli, D. D'Agnano, et al., Intra-erythrocyte infusion of dexamethasone reduces neurological symptoms in ataxia teleangiectasia patients: results of a phase 2 trial, Orphanet J. Rare Dis. 9 (2014) 5.

[12] H. He, J. Ye, Y. Wang, Q. Liu, H.S. Chung, Y.M. Kwon, et al., Cell-penetrating peptides mediated encapsulation of protein therapeutics into intact red blood cells and its application, J. Control. Release 176 (2014) 123–132.

[13] V.R. Muzykantov, Drug delivery by red blood cells: vascular carriers designed by mother nature, Expert Opin. Drug Deliv. 7 (4) (2010) 403–427.

[14] C.H. Villa, D.C. Pan, S. Zaitsev, D.B. Cines, D.L. Siegel, V.R. Muzykantov, Delivery of drugs bound to erythrocytes: new avenues for an old intravascular carrier, Ther. Deliv. 6 (7) (2015) 795–826.

[15] J. Shi, L. Kundrat, N. Pishesha, A. Bilate, C. Theile, T. Maruyama, et al., Engineered red blood cells as carriers for systemic delivery of a wide array of functional probes, Proc. Natl. Acad. Sci. U. S. A. 111 (28) (2014) 10131–10136.

[16] A.C. Anselmo, V. Gupta, B.J. Zern, D. Pan, M. Zakrewsky, V. Muzykantov, et al., Delivering nanoparticles to lungs while avoiding liver and spleen through adsorption on red blood cells, ACS Nano 7 (12) (2013) 11129–11137.

[17] C.H. Villa, V.R. Muzykantov, D.B. Cines, The emerging role for red blood cells in haemostasis: opportunity for intervention, ISBT Sci. Ser. 11 (S1) (2016) 158–164.

[18] G.M. Ihler, R.H. Glew, F.W. Schnure, Enzyme loading of erythrocytes, Proc. Natl. Acad. Sci. 70 (1973) 2663–2666.

[19] E. Ang, R. Glew, G. Ihler, Enzyme loading of nucleated chicken erythrocytes, Exp. Cell Res. 104 (1977) 430–434.

[20] X. Zhang, M. Qiu, P. Guo, Y. Lian, E. Xu, J. Su, Autologous red blood cell delivery of betamethasone phosphate sodium for long anti-inflammation, Pharmaceutics 10 (4) (2018) 286.

[21] E. Xu, X. Wu, X. Zhang, K. Zul, F. Raza, J. Su, et al., Study on the protection of dextran on erythrocytes during drug loading, Colloids Surf. B Biointerfaces 189 (2020) 110882.

[22] P. Dey, S. Banerjee, S. Mandal, P. Chattopadhyay, Design and evaluation of anti-fibrosis drug engineered resealed erythrocytes for targeted delivery, Drug Deliv. Transl. Res. 9 (5) (2019) 997–1007.

[23] V. Leuzzi, L. Rossi, C. Gabucci, F. Nardecchia, M. Magnani, Erythrocyte-mediated delivery of recombinant enzymes, J. Inherit. Metab. Dis. 39 (4) (2016) 519–530.

[24] T. Pascucci, L. Rossi, M. Colamartino, C. Gabucci, C. Carducci, A. Valzania, et al., A new therapy prevents intellectual disability in mouse with phenylketonuria, Mol. Genet. Metab. 124 (1) (2018) 39–49.

[25] L. Rossi, S. Serafini, L. Cappellacci, E. Balestra, G. Brandi, G.F. Schiavano, et al., Erythrocyte-mediated delivery of a new homodinucleotide active against human immunodeficiency virus and herpes simplex virus, J. Antimicrob. Chemother. 47 (6) (2001) 819–827.

[26] A. Banz, M. Cremel, A. Rembert, Y. Godfrin, In situ targeting of dendritic cells by antigen-loaded red blood cells: a novel approach to cancer immunotherapy, Vaccine 28 (17) (2010) 2965–2972.

[27] M. Cremel, N. Guérin, F. Horand, A. Banz, Y. Godfrin, Red blood cells as innovative antigen carrier to induce specific immune tolerance, Int. J. Pharm. 443 (1–2) (2013) 39–49.

[28] W.E. Lynch, G.P. Sartiano, A. Ghaffar, Erythrocytes as carriers of chemotherapeutic agents for targeting the reticuloendothelial system, Am. J. Hematol. 9 (3) (1980) 249–259.

[29] R. Mukthavaram, G. Shi, S. Kesari, D. Simberg, Targeting and depletion of circulating leukocytes and cancer cells by lipophilic antibody-modified erythrocytes, J. Control. Release 183 (2014) 146–153.

[30] M.J. Mitchell, M.R. King, Leukocytes as carriers for targeted cancer drug delivery, Expert Opin. Drug Deliv. 12 (3) (2015) 375–392.

[31] D.W. Stuckey, K. Shah, Stem cell-based therapies for cancer treatment: separating hope from hype, Nat. Rev. Cancer 14 (10) (2014) 683–691.

[32] T. Bagci-Onder, W. Du, J.-L. Figueiredo, J. Martinez-Quintanilla, K. Shah, Targeting breast to brain metastatic tumours with death receptor ligand expressing therapeutic stem cells, Brain 138 (6) (2015) 1710–1721.

[33] L. Li, D. Wang, J. Zhou, Y. Cheng, T. Liang, G. Zhang, Characteristics of human amniotic fluid mesenchymal stem cells and their tropism to human ovarian cancer, PLoS One 10 (4) (2015), e0123350.

[34] K.Y. Mapara, C.B. Stevenson, R.C. Thompson, M. Ehtesham, Stem cells as vehicles for the treatment of brain cancer, Neurosurg. Clin. N. Am. 18 (1) (2007) 71–80.

[35] J.S. Young, R.A. Morshed, J.W. Kim, I.V. Balyasnikova, A.U. Ahmed, M.S. Lesniak, Advances in stem cells, induced pluripotent stem cells, and engineered cells: delivery vehicles for anti–glioma therapy, Expert Opin. Drug Deliv. 11 (11) (2014) 1733–1746.

[36] K. Shah, Stem cell-based therapies for tumors in the brain: are we there yet? Neuro-Oncology 18 (8) (2016) 1066–1078.

[37] R. Chiotaki, H. Polioudaki, P.A. Theodoropoulos, Stem cell technology in breast cancer: current status and potential applications, Stem Cells Cloning 9 (2016) 17–29.

[38] K.J. Mahasa, L. de Pillis, R. Ouifki, A. Eladdadi, P. Maini, A.-R. Yoon, et al., Mesenchymal stem cells used as carrier cells of oncolytic adenovirus results in enhanced oncolytic virotherapy, Sci. Rep. 10 (1) (2020) 425.

[39] J. Kim, R.R. Hall, M.S. Lesniak, A.U. Ahmed, Stem cell-based cell carrier for targeted oncolytic virotherapy: translational opportunity and open questions, Viruses 7 (12) (2015) 6200–6217.

[40] A.U. Ahmed, B. Thaci, A.L. Tobias, B. Auffinger, L. Zhang, Y. Cheng, et al., A preclinical evaluation of neural stem cell-based cell carrier for targeted antiglioma oncolytic virotherapy, J. Natl. Cancer Inst. 105 (13) (2013) 968–977.

[41] R. Mooney, A.A. Majid, J. Batalla-Covello, D. Machado, X. Liu, J. Gonzaga, et al., Enhanced delivery of oncolytic adenovirus by neural stem cells for treatment of metastatic ovarian cancer, Mol. Ther. Oncolytics 12 (2019) 79–92.

[42] G.-T. Park, K.-C. Choi, Advanced new strategies for metastatic cancer treatment by therapeutic stem cells and oncolytic virotherapy, Oncotarget 7 (36) (2016) 58684–58695.

[43] E. Strable, M.G. Finn, Chemical modification of viruses and virus-like particles, Curr. Top. Microbiol. Immunol. 327 (2009) 1–21.

[44] Y. Seow, M.J. Wood, Biological gene delivery vehicles: beyond viral vectors, Mol. Ther. 17 (5) (2009) 767–777.

[45] K.L. Lee, R.M. Twyman, S. Fiering, N.F. Steinmetz, Virus-based nanoparticles as platform technologies for modern vaccines, Wiley Interdiscip. Rev. Nanomed. Nanobiotechnol. 8 (4) (2016) 554–578.

[46] S.M. Garland, M. Hernandez-Avila, C.M. Wheeler, G. Perez, D.M. Harper, S. Leodolter, et al., Quadrivalent vaccine against human papillomavirus to prevent anogenital diseases, N. Engl. J. Med. 356 (19) (2007) 1928–1943.

[47] D.M. Harper, S.L. Vierthaler, J.A. Santee, Review of Gardasil, J. Vaccines Vaccin. 1 (107) (2010) 1000107.

[48] A. Szarewski, HPV vaccine: Cervarix, Expert Opin. Biol. Ther. 10 (3) (2010) 477–487.

[49] M. Wang, T.-H. Tsou, L.-S. Chen, W.-C. Ou, P.-L. Chen, C.-F. Chang, et al., Inhibition of simian virus 40 large tumor antigen expression in human fetal glial cells by an antisense oligodeoxynucleotide delivered by the JC virus-like particle, Hum. Gene Ther. 15 (11) (2004) 1077–1090.

[50] L.K. Pattenden, A.P.J. Middelberg, M. Niebert, D.I. Lipin, Towards the preparative and large-scale precision manufacture of virus-like particles, Trends Biotechnol. 23 (10) (2005) 523–529.

[51] W. Wu, S.C. Hsiao, Z.M. Carrico, M.B. Francis, Genome-free viral capsids as multivalent carriers for taxol delivery, Angew. Chem. Int. Ed. Engl. 48 (50) (2009) 9493–9497.

[52] P.L. Chariou, K.L. Lee, A.M. Wen, N.M. Gulati, P.L. Stewart, N.F. Steinmetz, Detection and imaging of aggressive cancer cells using an epidermal growth factor receptor (EGFR)-targeted filamentous plant virus-based nanoparticle, Bioconjug. Chem. 26 (2) (2015) 262–269.

[53] A.S. Pitek, S.A. Jameson, F.A. Veliz, S. Shukla, N.F. Steinmetz, Serum albumin "camouflage" of plant virus based nanoparticles prevents their antibody recognition and enhances pharmacokinetics, Biomaterials 89 (2016) 89–97.

[54] A.W. Maresso, Bacterial invasion of the host cell, in: Bacterial Virulence, Springer, Cham, 2019, pp. 89–102.

[55] D. Ribet, P. Cossart, How bacterial pathogens colonize their hosts and invade deeper tissues, Microbes Infect. 17 (3) (2015) 173–183.

[56] P. Cossart, P.J. Sansonetti, Bacterial invasion: the paradigms of enteroinvasive pathogens, Science 304 (5668) (2004) 242–248.

[57] T. Van Acker, J. Tavernier, F. Peelman, The small GTPase Arf6: an overview of its mechanisms of action and of its role in host–pathogen interactions and innate immunity, Int. J. Mol. Sci. 20 (9) (2019) 2209.

[58] C.L. Libbing, A.R. McDevitt, R.-M.P. Azcueta, A. Ahila, M. Mulye, Lipid droplets: a significant but understudied contributor of host–bacterial interactions, Cells 8 (4) (2019) 354.

[59] I.Y.C. Lin, T.T.H. Van, P.M. Smooker, Live-attenuated bacterial vectors: tools for vaccine and therapeutic agent delivery, Vaccines (Basel) 3 (4) (2015) 940–972.

[60] R.W. Kasinskas, N.S. Forbes, Salmonella typhimurium lacking ribose chemoreceptors localize in tumor quiescence and induce apoptosis, Cancer Res. 67 (7) (2007) 3201–3209.

[61] Y.A. Yu, S. Shabahang, T.M. Timiryasova, Q. Zhang, R. Beltz, I. Gentschev, et al., Visualization of tumors and metastases in live animals with bacteria and vaccinia virus encoding light-emitting proteins, Nat. Biotechnol. 22 (3) (2004) 313–320.

[62] S.P. Kerkar, N.P. Restifo, Cellular constituents of immune escape within the tumor microenvironment, Cancer Res. 72 (13) (2012) 3125–3130.

[63] H.A. Schlößer, S. Theurich, A. Shimabukuro-Vornhagen, U. Holtick, D.L. Stippel, M. von Bergwelt-Baildon, Overcoming tumor-mediated immunosuppression, Immunotherapy 6 (9) (2014) 973–988.

[64] A.I. Bezio, B.E. Campbell, Vaccinations: Procedures, Types and Controversy, Nova Biomedical, 2012. 194 pp.

[65] L. Zhang, L. Gao, L. Zhao, B. Guo, K. Ji, Y. Tian, et al., Intratumoral delivery and suppression of prostate tumor growth by attenuated Salmonella enterica serovar typhimurium carrying plasmid-based small interfering RNAs, Cancer Res. 67 (12) (2007) 5859–5864.

[66] X. Li, Y. Li, J. Hu, B. Wang, L. Zhao, K. Ji, et al., Plasmid-based E6-specific siRNA and co-expression of wild-type p53 suppresses the growth of cervical cancer in vitro and in vivo, Cancer Lett. 335 (1) (2013) 242–250.

[67] X. Li, Y. Li, B. Wang, K. Ji, Z. Liang, B. Guo, et al., Delivery of the co-expression plasmid pEndo-Si-Stat3 by attenuated Salmonella serovar typhimurium for prostate cancer treatment, J. Cancer Res. Clin. Oncol. 139 (6) (2013) 971–980.

[68] X. Luo, Z. Li, S. Lin, T. Le, M. Ittensohn, D. Bermudes, et al., Antitumor effect of VNP20009, an attenuated Salmonella, in murine tumor models, Oncol. Res. 12 (11–12) (2001) 501–508.

[69] Y. Wang, J. Chen, B.O. Tang, X. Zhang, Z.-C. Hua, Systemic administration of attenuated Salmonella typhimurium in combination with interleukin-21 for cancer therapy, Mol. Clin. Oncol. 1 (3) (2013) 461–465.

[70] M.O. Din, T. Danino, A. Prindle, M. Skalak, J. Selimkhanov, K. Allen, et al., Synchronized cycles of bacterial lysis for in vivo delivery, Nature 536 (7614) (2016) 81–85.

[71] C.-H. Lee, C.-L. Wu, Y.-S. Tai, A.-L. Shiau, Systemic administration of attenuated Salmonella choleraesuis in combination with cisplatin for cancer therapy, Mol. Ther. 11 (5) (2005) 707–716.

[72] L.H. Dang, C. Bettegowda, D.L. Huso, K.W. Kinzler, B. Vogelstein, Combination bacteriolytic therapy for the treatment of experimental tumors, Proc. Natl. Acad. Sci. U. S. A. 98 (26) (2001) 15155–15160.

[73] C. Bettegowda, L.H. Dang, R. Abrams, D.L. Huso, L. Dillehay, I. Cheong, et al., Overcoming the hypoxic barrier to radiation therapy with anaerobic bacteria, Proc. Natl. Acad. Sci. U. S. A. 100 (25) (2003) 15083–15088.

[74] L.H. Dang, C. Bettegowda, N. Agrawal, I. Cheong, D. Huso, P. Frost, et al., Targeting vascular and avascular compartments of tumors with C. novyi-NT and anti-microtubule agents, Cancer Biol Ther. 3 (3) (2004) 326–337.

[75] I. Cheong, X. Huang, C. Bettegowda, L.A. Diaz Jr., K.W. Kinzler, S. Zhou, et al., A bacterial protein enhances the release and efficacy of liposomal cancer drugs, Science 314 (5803) (2006) 1308–1311.

[76] N. Agrawal, C. Bettegowda, I. Cheong, J.-F. Geschwind, C.G. Drake, E.L. Hipkiss, et al., Bacteriolytic therapy can generate a potent immune response against experimental tumors, Proc. Natl. Acad. Sci. U. S. A. 101 (42) (2004) 15172–15177.

[77] M.I. Husseiny, J. Rawson, A. Kaye, I. Nair, I. Todorov, M. Hensel, et al., An oral vaccine for type 1 diabetes based on live attenuated Salmonella, Vaccine 32 (20) (2014) 2300–2307.

[78] C. Lautenschläger, C. Schmidt, D. Fischer, A. Stallmach, Drug delivery strategies in the therapy of inflammatory bowel disease, Adv. Drug Deliv. Rev. 71 (2014) 58–76.

[79] J.R. Whittle, J.D. Lickliter, H.K. Gan, A.M. Scott, J. Simes, B.J. Solomon, et al., First in human nanotechnology doxorubicin delivery system to target epidermal growth factor receptors in recurrent glioblastoma, J. Clin. Neurosci. 22 (12) (2015) 1889–1894.

[80] S. Biffi, R. Voltan, B. Bortot, G. Zauli, P. Secchiero, Actively targeted nanocarriers for drug delivery to cancer cells, Expert Opin. Drug Deliv. 16 (5) (2019) 481–496.

[81] M. Gagliardi, Novel biodegradable nanocarriers for enhanced drug delivery, Ther. Deliv. 7 (12) (2016) 809–826.

[82] A. Del Grosso, M. Galliani, L. Angella, M. Santi, I. Tonazzini, G. Parlanti, et al., Brain-targeted enzyme-loaded nanoparticles: A breach through the blood-brain barrier for enzyme replacement therapy in Krabbe disease, Sci. Adv. 5 (11) (2019) eaax7462.

[83] B. Mishra, S. Chaurasia, Design of novel chemotherapeutic delivery systems for colon cancer therapy based on oral polymeric nanoparticles, Ther. Deliv. 8 (1) (2017) 29–47.

[84] M. Gagliardi, C. Borri, Polymer nanoparticles as smart carriers for the enhanced release of therapeutic agents to the CNS, Curr. Pharm. Des. 23 (3) (2017) 393–410.

[85] M. Gagliardi, G. Bardi, A. Bifone, Polymeric nanocarriers for controlled and enhanced delivery of therapeutic agents to the CNS, Ther. Deliv. 3 (7) (2012) 875–887.

[86] J.O. Morales, S. Sepulveda-Rivas, F. Oyarzun-Ampuero, S. Lavandero, M.J. Kogan, Novel nanostructured polymeric carriers to enable drug delivery for cardiovascular diseases, Curr. Pharm. Des. 21 (29) (2015) 4276–4284.

[87] J.E. Pullan, A.T. Pullan, V.B. Taylor, B.D. Brooks, D. Ewert, A.E. Brooks, Energy-triggered drug release from polymer nanoparticles for orthopedic applications, Ther. Deliv. 8 (1) (2017) 5–14.

[88] Y. Lu, A.A. Aimetti, R. Langer, Z. Gu, Bioresponsive materials, Nat. Rev. Mater. 2 (1) (2016) 16075.

[89] A. Guha, N. Biswas, K. Bhattacharjee, N. Sahoo, K. Kuotsu, pH responsive cylindrical MSN for oral delivery of insulin-design, fabrication and evaluation, Drug Deliv. 23 (9) (2016) 3552–3561.

[90] M. Karimi, P. Sahandi Zangabad, A. Ghasemi, M. Amiri, M. Bahrami, H. Malekzad, et al., Temperature-responsive smart nanocarriers for delivery of therapeutic agents: applications and recent advances, ACS Appl. Mater. Interfaces 8 (33) (2016) 21107–21133.

[91] S. He, G. Tourkakis, O. Berezin, N. Gerasimchuk, H. Zhang, H. Zhou, et al., Temperature-dependent shape-responsive fluorescent nanospheres for image-guided drug delivery, J. Mater. Chem. 4 (14) (2016) 3028–3035.

[92] V. Kozlovskaya, B. Xue, E. Kharlampieva, Shape-adaptable polymeric particles for controlled delivery, Macromolecules 49 (22) (2016) 8373–8386.

[93] Y. Zhang, J. Yu, H.N. Bomba, Y. Zhu, Z. Gu, Mechanical force-triggered drug delivery, Chem. Rev. 116 (19) (2016) 12536–12563.

[94] X. Chen, J. Cui, H. Sun, M. Müllner, Y. Yan, K.F. Noi, et al., Analysing intracellular deformation of polymer capsules using structured illumination microscopy, Nanoscale 8 (23) (2016) 11924–11931.

[95] J.K. Awino, Y. Zhao, Polymeric nanoparticle receptors as synthetic antibodies for nonsteroidal anti-inflammatory drugs (NSAIDs), ACS Biomater. Sci. Eng. 1 (6) (2015) 425–430.

[96] Y. Liu, S. Fang, J. Zhai, M. Zhao, Construction of antibody-like nanoparticles for selective protein sequestration in living cells, Nanoscale 7 (16) (2015) 7162–7167.

[97] S. Kunath, M. Panagiotopoulou, J. Maximilien, N. Marchyk, J. Sänger, K. Haupt, Cell and tissue imaging with molecularly imprinted polymers as plastic antibody mimics, Adv. Healthc. Mater. 4 (9) (2015) 1322–1326.

[98] M. Gagliardi, B. Mazzolai, Molecularly imprinted polymeric micro- and nano-particles for the targeted delivery of active molecules, Future Med. Chem. 7 (2) (2015) 123–138.

[99] H.R. Culver, S.D. Steichen, N.A. Peppas, A closer look at the impact of molecular imprinting on adsorption capacity and selectivity for protein templates, Biomacromolecules 17 (12) (2016) 4045–4053.

[100] M. Gagliardi, A. Bertero, A. Bifone, Molecularly imprinted biodegradable nanoparticles, Sci. Rep. 7 (2017) 40046.

[101] J. Zhao, M. Zhao, C. Yu, X. Zhang, J. Liu, X. Cheng, et al., Multifunctional folate receptor-targeting and pH-responsive nanocarriers loaded with methotrexate for treatment of rheumatoid arthritis, Int. J. Nanomedicine 12 (2017) 6735–6746.

[102] J.-T. Lin, J.-K. Du, Y.-Q. Yang, L. Li, D.-W. Zhang, C.-L. Liang, et al., pH and redox dual stimulate-responsive nanocarriers based on hyaluronic acid coated mesoporous silica for targeted drug delivery, Mater. Sci. Eng. C Mater. Biol. Appl. 81 (2017) 478–484.

[103] H.-N. Liu, N.-N. Guo, T.-T. Wang, W.-W. Guo, M.-T. Lin, M.-Y. Huang-Fu, et al., Mitochondrial targeted doxorubicin-triphenylphosphonium delivered by hyaluronic acid modified and pH responsive nanocarriers to breast tumor: in vitro and in vivo studies, Mol. Pharm. 15 (3) (2018) 882–891.

[104] G. Zhang, X. Jiang, Temperature responsive nanoparticles based on PEGylated polyaspartamide derivatives for drug delivery, Polymers 11 (2) (2019) 316.

[105] T. Hashimoto, Y. Hirai, E. Yuba, A. Harada, K. Kono, Temperature-responsive molecular assemblies using oligo(ethylene glycol)-attached polyamidoamine dendron lipids and their functions as drug carriers, J. Funct. Biomater. 11 (1) (2020) 16.

[106] J. Zong, S.L. Cobb, N.R. Cameron, Short elastin-like peptide-functionalized gold nanoparticles that are temperature responsive under near-physiological conditions, J. Mater. Chem. B Mater. Biol. Med. 6 (41) (2018) 6667–6674.

[107] M. Amoli-Diva, R. Sadighi-Bonabi, K. Pourghazi, Switchable on/off drug release from gold nanoparticles-grafted dual light- and temperature-responsive hydrogel for controlled drug delivery, Mater. Sci. Eng. C Mater. Biol. Appl. 76 (2017) 242–248.

[108] F. Farjadian, S. Rezaeifard, M. Naeimi, S. Ghasemi, S. Mohammadi-Samani, M.E. Welland, et al., Temperature and pH-responsive nano-hydrogel drug delivery system based on lysine-modified poly(vinylcaprolactam), Int. J. Nanomedicine 14 (2019) 6901–6915.

[109] J. Ma, J. Sun, L. Fan, S. Bai, H. Panezai, Y. Jiao, Fractal evolution of dual pH- and temperature-responsive P(NIPAM-co-AA)@BMMs with bimodal mesoporous silica core and coated-copolymer shell during drug delivery procedure via SAXS characterization, Arab. J. Chem. 13 (2020) 4147–4161.

[110] A. Suksuwan, L. Lomlim, T. Rungrotmongkol, T. Nakpheng, F.L. Dickert, R. Suedee, The composite nanomaterials containing (R)-thalidomide-molecularly imprinted polymers as a recognition system for enantioselective-controlled release and targeted drug delivery, J. Appl. Polym. Sci. 132 (18) (2015) 41930.

[111] K. Hemmati, R. Sahraei, M. Ghaemy, Synthesis and characterization of a novel magnetic molecularly imprinted polymer with incorporated graphene oxide for drug delivery, Polymer 101 (2016) 257–268.

[112] K. Zhang, X. Guan, Y. Qiu, D. Wang, X. Zhang, H. Zhang, A pH/glutathione double responsive drug delivery system using molecular imprint technique for drug loading, Appl. Surf. Sci. 389 (2016) 1208–1213.

[113] M. Esfandyari-Manesh, B. Darvishi, F.A. Ishkuh, E. Shahmoradi, A. Mohammadi, M. Javanbakht, et al., Paclitaxel molecularly imprinted polymer-PEG-folate nanoparticles for targeting anticancer delivery: characterization and cellular cytotoxicity, Mater. Sci. Eng. C Mater. Biol. Appl. 62 (2016) 626–633.

[114] B.T. Luk, L. Zhang, Cell membrane-camouflaged nanoparticles for drug delivery, J. Control. Release 220 (Pt B) (2015) 600–607.

[115] F.M. Veronese, G. Pasut, PEGylation, successful approach to drug delivery, Drug Discov. Today 10 (21) (2005) 1451–1458.

[116] Z. Hussain, S. Khan, M. Imran, M. Sohail, S.W.A. Shah, M. de Matas, PEGylation: a promising strategy to overcome challenges to cancer-targeted nanomedicines: a review of challenges to clinical transition and promising resolution, Drug Deliv. Transl. Res. 9 (3) (2019) 721–734.

[117] Y. Qi, A. Simakova, N.J. Ganson, X. Li, K.M. Luginbuhl, I. Ozer, et al., A brush-polymer/exendin-4 conjugate reduces blood glucose levels for up to five days and eliminates poly(ethylene glycol) antigenicity, Nat. Biomed. Eng. 1 (2016) 0002.

[118] C.-M.J. Hu, R.H. Fang, B.T. Luk, L. Zhang, Polymeric nanotherapeutics: clinical development and advances in stealth functionalization strategies, Nanoscale 6 (1) (2014) 65–75.

[119] N. Hadjesfandiari, A. Parambath, Stealth coatings for nanoparticles, in: Engineering of Biomaterials for Drug Delivery Systems, 2018, pp. 345–361.

[120] J.B. Schlenoff, Zwitteration: coating surfaces with zwitterionic functionality to reduce nonspecific adsorption, Langmuir 30 (32) (2014) 9625–9636.

[121] M. Debayle, E. Balloul, F. Dembele, X. Xu, M. Hanafi, F. Ribot, et al., Zwitterionic polymer ligands: an ideal surface coating to totally suppress protein-nanoparticle corona formation? Biomaterials 219 (2019) 119357.

[122] F. Ding, S. Yang, Z. Gao, J. Guo, P. Zhang, X. Qiu, et al., Antifouling and pH-responsive poly(Carboxybetaine)-based nanoparticles for tumor cell targeting, Front Chem. 7 (2019) 770.

[123] S. Peng, B. Ouyang, Y. Men, Y. Du, Y. Cao, R. Xie, et al., Biodegradable zwitterionic polymer membrane coating endowing nanoparticles with ultra-long circulation and enhanced tumor photothermal therapy, Biomaterials 231 (2020) 119680.

[124] A. Runser, D. Dujardin, P. Ernst, A.S. Klymchenko, A. Reisch, Zwitterionic stealth dye-loaded polymer nanoparticles for intracellular imaging, ACS Appl. Mater. Interfaces 12 (1) (2020) 117–125.

[125] D. Gromadzki, V. Tzankova, M. Kondeva, C. Gorinova, P. Rychter, M. Libera, et al., Amphiphilic core-shell nanoparticles with dimer fatty acid-based aliphatic polyester core and zwitterionic poly(sulfobetaine) shell for controlled delivery of curcumin, Int. J. Polym. Mater. Polym. Biomater. 66 (18) (2017) 915–925.

[126] S. Chen, Q. Li, H. Li, L. Yang, J.-Z. Yi, M. Xie, et al., Long-circulating zein-polysulfobetaine conjugate-based nanocarriers for enhancing the stability and pharmacokinetics of curcumin, Mater. Sci. Eng. C 109 (2020) 110636.

[127] A.C.G. Weiss, H.G. Kelly, M. Faria, Q.A. Besford, A.K. Wheatley, C.-S. Ang, et al., Link between low-fouling and stealth: a whole blood biomolecular corona and cellular association analysis on nanoengineered particles, ACS Nano 13 (5) (2019) 4980–4991.

[128] J.-M. Noy, C. Cao, M. Stenzel, Length of the stabilizing zwitterionic poly(2-methacryloyloxyethyl phosphorycholine) block influences the activity of the conjugated arsenic drug in drug-directed polymerization–induced self-assembly particles, ACS Macro Lett. 8 (2019) 57–63.

[129] R. Xie, Y. Tian, S. Peng, L. Zhang, Y. Men, W. Yang, Poly(2-methacryloyloxyethyl phosphorylcholine)-based biodegradable nanogels for controlled drug release, Polym. Chem. 9 (2018) 4556–4565.

[130] G. Liu, H.-I. Tsai, X. Zeng, W. Cheng, L. Jiang, H. Chen, et al., Phosphorylcholine-based stealthy nanocapsules decorating TPGS for combatting multi-drug-resistant cancer, ACS Biomater. Sci. Eng. 4 (5) (2018) 1679–1686.

[131] G. Liu, H.-I. Tsai, X. Zeng, Y. Zuo, W. Tao, J. Han, et al., Phosphorylcholine-based stealthy nanocapsules enabling tumor microenvironment-responsive doxorubicin release for tumor suppression, Theranostics 7 (2017) 1192–1203.

[132] S. Banskota, S. Saha, J. Bhattacharya, N. Kirmani, P. Yousefpour, M. Dzuricky, et al., Genetically encoded stealth nanoparticles of a zwitterionic polypeptide-paclitaxel conjugate have a wider therapeutic window than Abraxane in multiple tumor models, Nano Lett. 20 (4) (2020) 2396–2409.

[133] A.C. Engler, X. Ke, S. Gao, J.M.W. Chan, D.J. Coady, R.J. Ono, et al., Hydrophilic polycarbonates: promising degradable alternatives to poly(ethylene glycol)-based stealth materials, Macromolecules 48 (6) (2015) 1673–1678.

[134] H. Bludau, A.E. Czapar, A.S. Pitek, S. Shukla, R. Jordan, N.F. Steinmetz, POxylation as an alternative stealth coating for biomedical applications, Eur. Polym. J. 88 (2016) 679–688.

[135] Y. Qi, A. Chilkoti, Protein-polymer conjugation-moving beyond PEGylation, Curr. Opin. Chem. Biol. 28 (2015) 181–193.

[136] S. Lowe, N.M. O'Brien-Simpson, L.A. Connal, Antibiofouling polymer interfaces: poly(ethylene glycol) and other promising candidates, Polym. Chem. 6 (2) (2015) 198–212.

[137] G. Aldrian, A. Vaissière, K. Konate, Q. Seisel, E. Vivès, F. Fernandez, et al., PEGylation rate influences peptide-based nanoparticles mediated siRNA delivery in vitro and in vivo, J. Control. Release 256 (2017) 79–91.

[138] Y. Hu, Z. Zhao, T. Harmon, P.R. Pentel, M. Ehrich, C. Zhang, Paradox of PEGylation in fabricating hybrid nanoparticle-based nicotine vaccines, Biomaterials 182 (2018) 72–81.

[139] F.S. Mozar, E.H. Chowdhury, PEGylation of carbonate apatite nanoparticles prevents opsonin binding and enhances tumor accumulation of gemcitabine, J. Pharm. Sci. 107 (9) (2018) 2497–2508.

[140] F. Ding, X. Gao, X. Huang, H. Ge, M. Xie, J. Qian, et al., Polydopamine-coated nucleic acid nanogel for siRNA-mediated low-temperature photothermal therapy, Biomaterials 245 (2020) 119976.

[141] A. Almeida, M. Araújo, R. Novoa-Carballal, F. Andrade, H. Gonçalves, R.L. Reis, et al., Novel amphiphilic chitosan micelles as carriers for hydrophobic anticancer drugs, Mater. Sci. Eng. C 112 (2020) 110920.

[142] M. Wang, G. Siddiqui, O.J.R. Gustafsson, A. Käkinen, I. Javed, N.H. Voelcker, et al., Plasma proteome association and catalytic activity of stealth polymer-grafted iron oxide nanoparticles, Small 13 (2017) 1701528.

[143] Y. Du, Z. Wang, T. Wang, W. He, W. Zhou, M. Li, et al., Improved antitumor activity of novel redox-responsive paclitaxel-encapsulated liposomes based on disulfide phosphatidylcholine, Mol. Pharm. 17 (1) (2020) 262–273.

[144] Y. Du, W. He, Q. Xia, W. Zhou, C. Yao, X. Li, Thioether phosphatidylcholine liposomes: a novel ROS-responsive platform for drug delivery, ACS Appl. Mater. Interfaces 11 (41) (2019) 37411–37420.

[145] R.A.H. Ishak, N.M. Mostafa, A.O. Kamel, Stealth lipid polymer hybrid nanoparticles loaded with rutin for effective brain delivery - comparative study with the gold standard (Tween 80): optimization, characterization and biodistribution, Drug Deliv. 24 (1) (2017) 1874–1890.

[146] A. Ivask, E.H. Pilkington, T. Blin, A. Käkinen, H. Vija, M. Visnapuu, et al., Uptake and transcytosis of functionalized superparamagnetic iron oxide nanoparticles in an in vitro blood brain barrier model, Biomater. Sci. 6 (2018) 314–323.

[147] R.A. Meyer, R.S. Meyer, J.J. Green, An automated multidimensional thin film stretching device for the generation of anisotropic polymeric micro- and nanoparticles, J. Biomed. Mater. Res. A 103 (8) (2015) 2747–2757.

[148] X. Ju, X. Wang, Z. Liu, R. Xie, W. Wang, L. Chu, Red-blood-cell-shaped chitosan microparticles prepared by electrospraying, Particuology 30 (2017) 151–157.

[149] T.J. Merkel, S.W. Jones, K.P. Herlihy, F.R. Kersey, A.R. Shields, M. Napier, et al., Using mechanobiological mimicry of red blood cells to extend circulation times of hydrogel microparticles, Proc. Natl. Acad. Sci. U. S. A. 108 (2) (2011) 586–591.

[150] K. Chen, T.J. Merkel, A. Pandya, M.E. Napier, J.C. Luft, W. Daniel, et al., Low modulus biomimetic microgel particles with high loading of hemoglobin, Biomacromolecules 13 (9) (2012) 2748–2759.

[151] R.J.C. Bose, S.-H. Lee, H. Park, Biofunctionalized nanoparticles: an emerging drug delivery platform for various disease treatments, Drug Discov. Today 21 (8) (2016) 1303–1312.

[152] J.A. Copp, R.H. Fang, B.T. Luk, C.-M.J. Hu, W. Gao, K. Zhang, et al., Clearance of pathological antibodies using biomimetic nanoparticles, Proc. Natl. Acad. Sci. U. S. A. 111 (37) (2014) 13481–13486.

[153] C.-M.J. Hu, L. Zhang, S. Aryal, C. Cheung, R.H. Fang, L. Zhang, Erythrocyte membrane-camouflaged polymeric nanoparticles as a biomimetic delivery platform, Proc. Natl. Acad. Sci. U. S. A. 108 (27) (2011) 10980–10985.

[154] Y. Guo, D. Wang, Q. Song, T. Wu, X. Zhuang, Y. Bao, et al., Erythrocyte membrane-enveloped polymeric nanoparticles as nanovaccine for induction of antitumor immunity against melanoma, ACS Nano 9 (7) (2015) 6918–6933.

[155] L. Hui, S. Qin, L. Yang, Upper critical solution temperature polymer, photothermal agent, and erythrocyte membrane coating: an unexplored recipe for making drug carriers with spatiotemporally controlled cargo release, ACS Biomater. Sci. Eng. 2 (12) (2016) 2127–2132.

[156] B.T. Luk, R.H. Fang, C.-M.J. Hu, J.A. Copp, S. Thamphiwatana, D. Dehaini, et al., Safe and immunocompatible nanocarriers cloaked in RBC membranes for drug delivery to treat solid tumors, Theranostics 6 (7) (2016) 1004–1011.

[157] H. Zhou, Z. Fan, P.K. Lemons, H. Cheng, A facile approach to functionalize cell membrane-coated nanoparticles, Theranostics 6 (7) (2016) 1012–1022.

[158] Q. Wang, H. Cheng, H. Peng, H. Zhou, P.Y. Li, R. Langer, Non-genetic engineering of cells for drug delivery and cell-based therapy, Adv. Drug Deliv. Rev. 91 (2015) 125–140.

[159] M. Magnani, L. Rossi, Approaches to erythrocyte-mediated drug delivery, Expert Opin. Drug Deliv. 11 (5) (2014) 677–687.

[160] J.S. Brenner, D.C. Pan, J.W. Myerson, O.A. Marcos-Contreras, C.H. Villa, P. Patel, et al., Red blood cell-hitchhiking boosts delivery of nanocarriers to chosen organs by orders of magnitude, Nat. Commun. 9 (1) (2018) 2684.

[161] I.V. Zelepukin, A.V. Yaremenko, V.O. Shipunova, A.V. Babenyshev, I.V. Balalaeva, P.I. Nikitin, et al., Nanoparticle-based drug delivery via RBC-hitchhiking for the inhibition of lung metastases growth, Nanoscale 11 (4) (2019) 1636–1646.

[162] C.-M.J. Hu, R.H. Fang, B.T. Luk, K.N.H. Chen, C. Carpenter, W. Gao, et al., "Marker-of-self" functionalization of nanoscale particles through a top-down cellular membrane coating approach, Nanoscale 5 (7) (2013) 2664.

[163] L. Rao, J.-H. Xu, B. Cai, H. Liu, M. Li, Y. Jia, et al., Synthetic nanoparticles camouflaged with biomimetic erythrocyte membranes for reduced reticuloendothelial system uptake, Nanotechnology 27 (8) (2016), 085106.

[164] X. Ren, R. Zheng, X. Fang, X. Wang, X. Zhang, W. Yang, et al., Red blood cell membrane camouflaged magnetic nanoclusters for imaging-guided photothermal therapy, Biomaterials 92 (2016) 13–24.

[165] A.N. Barclay, T.K. Van den Berg, The interaction between signal regulatory protein alpha (SIRPα) and CD47: structure, function, and therapeutic target, Annu. Rev. Immunol. 32 (2014) 25–50.

[166] D. Rosenblum, N. Joshi, W. Tao, J.M. Karp, D. Peer, Progress and challenges towards targeted delivery of cancer therapeutics, Nat. Commun. 9 (1) (2018) 1410.

[167] E. Blanco, H. Shen, M. Ferrari, Principles of nanoparticle design for overcoming biological barriers to drug delivery, Nat. Biotechnol. 33 (9) (2015) 941–951.

[168] J.-W. Yoo, D.J. Irvine, D.E. Discher, S. Mitragotri, Bio-inspired, bioengineered and biomimetic drug delivery carriers, Nat. Rev. Drug Discov. 10 (7) (2011) 521–535.

[169] M.-R. Choi, K.J. Stanton-Maxey, J.K. Stanley, C.S. Levin, R. Bardhan, D. Akin, et al., A cellular Trojan Horse for delivery of therapeutic nanoparticles into tumors, Nano Lett. 7 (12) (2007) 3759–3765.

[170] Z. Li, H. Huang, S. Tang, Y. Li, X.-F. Yu, H. Wang, et al., Small gold nanorods laden macrophages for enhanced tumor coverage in photothermal therapy, Biomaterials 74 (2016) 144–154.

[171] G. Schackert, R.D. Simmons, T.M. Buzbee, D.A. Hume, I.J. Fidler, Macrophage infiltration into experimental brain metastases: occurrence through an intact blood-brain barrier, J. Natl. Cancer Inst. 80 (13) (1988) 1027–1034.

[172] M.-R. Choi, R. Bardhan, K.J. Stanton-Maxey, S. Badve, H. Nakshatri, K.M. Stantz, et al., Delivery of nanoparticles to brain metastases of breast cancer using a cellular Trojan horse, Cancer Nanotechnol. 3 (1-6) (2012) 47–54.

[173] J. Qin, X. Yang, R.-X. Zhang, Y.-X. Luo, J.-L. Li, J. Hou, et al., Monocyte mediated brain targeting delivery of macromolecular drug for the therapy of depression, Nanomedicine 11 (2) (2015) 391–400.

[174] L. Wayteck, H. Dewitte, L. De Backer, K. Breckpot, J. Demeester, S.C. De Smedt, et al., Hitchhiking nanoparticles: reversible coupling of lipid-based nanoparticles to cytotoxic T lymphocytes, Biomaterials 77 (2016) 243–254.

[175] R. Palomba, A. Parodi, M. Evangelopoulos, S. Acciardo, C. Corbo, E. de Rosa, et al., Biomimetic carriers mimicking leukocyte plasma membrane to increase tumor vasculature permeability, Sci. Rep. 6 (2016) 34422.

[176] A. Parodi, N. Quattrocchi, A.L. van de Ven, C. Chiappini, M. Evangelopoulos, J.O. Martinez, et al., Synthetic nanoparticles functionalized with biomimetic leukocyte membranes possess cell-like functions, Nat. Nanotechnol. 8 (1) (2013) 61–68.

[177] W. He, J. Frueh, Z. Wu, Q. He, How leucocyte cell membrane modified Janus microcapsules are phagocytosed by cancer cells, ACS Appl. Mater. Interfaces 8 (7) (2016) 4407–4415.

[178] W. He, J. Frueh, Z. Wu, Q. He, Leucocyte membrane-coated Janus microcapsules for enhanced photothermal cancer treatment, Langmuir 32 (15) (2016) 3637–3644.

[179] M. Xuan, J. Shao, L. Dai, J. Li, Q. He, Macrophage cell membrane camouflaged Au nanoshells for in vivo prolonged circulation life and enhanced cancer photothermal therapy, ACS Appl. Mater. Interfaces 8 (15) (2016) 9610–9618.

[180] M. Xuan, J. Shao, L. Dai, Q. He, J. Li, Macrophage cell membrane camouflaged mesoporous silica nanocapsules for in vivo cancer therapy, Adv. Healthc. Mater. 4 (11) (2015) 1645–1652.

[181] R.H. Fang, C.-M.J. Hu, B.T. Luk, W. Gao, J.A. Copp, Y. Tai, et al., Cancer cell membrane-coated nanoparticles for anticancer vaccination and drug delivery, Nano Lett. 14 (4) (2014) 2181–2188.

[182] J. Li, Y. Ai, L. Wang, P. Bu, C.C. Sharkey, Q. Wu, et al., Targeted drug delivery to circulating tumor cells via platelet membrane-functionalized particles, Biomaterials 76 (2016) 52–65.

[183] J. Li, C.C. Sharkey, B. Wun, J.L. Liesveld, M.R. King, Genetic engineering of platelets to neutralize circulating tumor cells, J. Control. Release 228 (2016) 38–47.

[184] Q. Hu, C. Qian, W. Sun, J. Wang, Z. Chen, H.N. Bomba, et al., Engineered nanoplatelets for enhanced treatment of multiple myeloma and thrombus, Adv. Mater. 28 (43) (2016) 9573–9580.

[185] A. Sahari, M.A. Traore, B.E. Scharf, B. Behkam, Directed transport of bacteria-based drug delivery vehicles: bacterial chemotaxis dominates particle shape, Biomed. Microdevices 16 (5) (2014) 717–725.

[186] A.C. Anselmo, S. Mitragotri, Nanoparticles in the clinic: an update, Bioeng. Transl. Med. 4 (3) (2019), e10143.

[187] G.M. Chertow, M. Dillon, S.K. Burke, M. Steg, A.J. Bleyer, B.N. Garrett, et al., A randomized trial of sevelamer hydrochloride (RenaGel) with and without supplemental calcium. Strategies for the control of hyperphosphatemia and hyperparathyroidism in hemodialysis patients, Clin. Nephrol. 51 (1) (1999) 18–26.

[188] G. Reid, S.C. Kao, N. Pavlakis, H. Brahmbhatt, J. MacDiarmid, S. Clarke, et al., Clinical development of TargomiRs, a miRNA mimic-based treatment for patients with recurrent thoracic cancer, Epigenomics 8 (8) (2016) 1079–1085.

[189] J.A. MacDiarmid, N.B. Mugridge, J.C. Weiss, L. Phillips, A.L. Burn, R.P. Paulin, et al., Bacterially derived 400 nm particles for encapsulation and cancer cell targeting of chemotherapeutics, Cancer Cell 11 (5) (2007) 431–445.

[190] J.A. MacDiarmid, N.B. Amaro-Mugridge, J. Madrid-Weiss, I. Sedliarou, S. Wetzel, K. Kochar, et al., Sequential treatment of drug-resistant tumors with targeted minicells containing siRNA or a cytotoxic drug, Nat. Biotechnol. 27 (7) (2009) 643–651.

[191] G. Reid, M.E. Pel, M.B. Kirschner, Y.Y. Cheng, N. Mugridge, J. Weiss, et al., Restoring expression of miR-16: a novel approach to therapy for malignant pleural mesothelioma, Ann. Oncol. 24 (12) (2013) 3128–3135.

CHAPTER 15

Radioactive nanoparticles and their biomedical application in nanobrachytherapy

Carla Daruich de Souza, Beatriz Ribeiro Nogueira, Carlos Alberto Zeituni and Maria Elisa Chuery Martins Rostelato
Laboratory for Radiation Therapy Sources Production, Nuclear and Energy Research Institute, National Nuclear Energy Commission, São Paulo, Brazil

1. Introduction

Nanotechnology is a new science. If the term nanoparticle was searched in ScienceDirect, in1996, only 623 papers were published. In 2006, the number increased to 7937. In September 2019, the number of papers published was almost 50,000 [1]. This rapid increase is certainly because of the huge potential of nanotechnology in different application fields. From biosensors to drug delivery, the nanoparticles can be applied to directly modify or receive information from systems and organs. This "new science" concentrates on the study of particles, structures, and materials that are in the nanoscale (nm) and have unique properties [2–4]. Their configuration in the nanometric size promotes alteration in their physical, chemical, and biological behaviors [3–5]. The quantum confinement effects state that the properties of materials are size-dependent in this scale range. In nanomaterials, the energy levels of the electrons are not continuous when compared to the bulk form. They are discrete due to the confinement of the electronic wave function in up to three physical dimensions, which leads to a change in surface area and electron confinement, making the changes in materials properties [3–5]. For example, properties such as melting point, fluorescence, electrical conductivity, magnetic permeability, and chemical reactivity change as a function of the size of the particle [3, 5]. Maybe the most important result of the nanoscale quantum effects is the ability to tune the properties. It means that a researcher can change the proprieties behavior to serve a determined purpose. For example, changing the fluorescence color of a particle can be used to identify diseases [4, 6, 7].

The surface chemistry of nanoparticles plays a direct role in the interactions between nanoparticles and biological systems can be altered by attaching molecules to the surface. As an example, charge, hydrophobicity, or even reactivity can be changed. Also, molecules of interest can be attached to their surface. For example, a chemotherapeutic drug can be encapsulated in a radioactive shell [8–10].

The union of brachytherapy with nanotechnology is inevitable. Brachytherapy is when a radiation source is placed in close contact or inside a patient's body. It can be left in the target area (TA) as a permanent seed implant, or it can treat the TA for a period of time. The primary use of brachytherapy is in the treatment of various types of cancers. It has the potential to be portable, due to the small size of the radiation sources. For example, iodine-125 seeds have 4.5 mm in length and 0.8 mm in diameter. Radioactive nanoparticles can be delivered by a syringe [11–13]. This is a major advantage of the method because it can be used in areas that people do not have access to large linear particle accelerators (linacs), which are expensive robust machines used in radiation therapy [14, 15].

The radiation dose received by the TA should be large enough to cause a lot of cellular damage, to the point of destruction. It needs to be focused on the TA as much as possible, saving the surrounding healthy tissues. Dose constraints are applied in all treatments and protocols are in constant review. Of course, since brachytherapy is placed inside the TA, it can focus even more on the target than EBRT (external beam radiation therapy), the other modality used for radiation therapy [13].

A radioactive nanoparticle would easily penetrate tumor vascularity and deliver radioactivity to the entire TA for enhanced effect. In the following part of the chapter, basics of radiation physics, radiological effects, factors to consider when creating a new brachytherapy modality, which nanosystems are promising, and the challenges that will be faced are explained.

2. Radiation concepts

When an atom has excess energy, that energy can be emitted in the form of radiation in several different forms. Basically, the atom nucleus will emit either particles or electromagnetic radiation. In radiation therapy, alpha, beta, gamma, and/or X-rays are used for treatment. Protons and neutrons are less used. Fig. 1 shows the alpha, beta, and X-ray/gamma radiation emission and the electromagnetic spectra [12, 16, 17].

Alpha decay typically occurs in very heavy nuclei, where electrostatic repulsion between protons in the nucleus is very large. The original radioactive isotope, also called father isotope (with an atomic number Z), emits its excess energy in the form of a helium nucleus (2 protons and 2 neutrons). The atom that remains, called daughter nuclei, has a Z-2 atomic number. The sum of the masses of the daughter nucleus and that of the alpha particle is slightly less than the mass of the father nucleus. Considering Einstein's $E = m^* c^2$, that mass difference is equivalent to the amount of energy released. They are usually very energetic and very heavy, thus have large ionization power and low penetration in matter [12, 16–19].

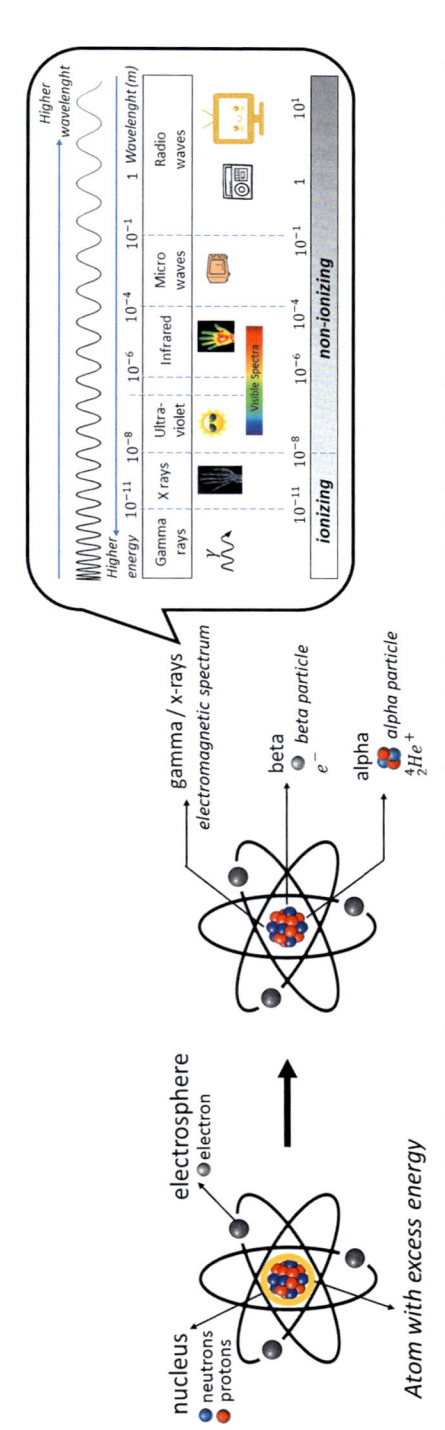

Fig. 1 Radioactive decay. When the atom has excess energy, the nucleus emits that energy in the form of radiation. Alpha emissions are actually a high-energy heavy helium atom nucleus. Beta emissions are a less heavy, less energetic electron. Gamma and X-rays are part of the electromagnetic spectra. They are from the same family as visible light, microwaves, and radio waves. The difference is in wavelength and ionization power.

In beta decay, a nucleus with too many protons or too many neutrons need to become stable. One of the protons or neutrons is transformed into the other. In beta minus decay (β^-), a neutron decays into a proton plus an electron and an antineutrino. In beta plus decay (β^+), a proton decays into a neutron plus a positron and a neutrino. To respect conservation laws, when an electrically neutral neutron becomes a positively charged proton, an electrically negative particle (electron in β^-) must be produced also. Because the electron mass is a tiny fraction of an atomic mass unit, the beta decaying nucleus mass is altered only by a very small amount. The core mass number does not change. Beta decay is, as a rule of thumb, less energetic than alpha emission and less heavy, thus have less ionization power but higher penetration in matter [12, 16, 17, 20, 21].

A gamma ray is when the nucleus changes from a higher energy state to a lower energy state through the emission of electromagnetic radiation (photons). The only thing that distinguishes a gamma ray from visible light photons emitted by a lamp is its wavelength. The gamma ray wavelength is hundreds of thousands of times smaller than that of visible light. Therefore, the frequency is hundreds of thousands of times greater. For complex nuclei of heavy elements, there are numerous different possibilities in which protons can rearrange themselves within the nucleus. Neither the mass number nor the atomic number of a nucleus changes when a gamma ray is emitted. However, the nucleus mass also decreases slightly and is converted into photon energy [12, 16, 17, 22, 23].

X-rays are produced by radiation interactions outside of the nucleus. Characteristic X-ray emission is when a moving electron hits an orbital electron with energy enough to remove it out of the inner electron shell of the target atom. After that, electrons from higher energy levels fill the vacancies, and X-ray photons are emitted. These X-rays have a discrete spectrum. Bremsstrahlung X-ray emission occurs when electrons are scattered due to the strong electric field near the nuclei. These X-rays have a continuous spectrum. The frequency of bremsstrahlung x-rays is limited by the energy of incident electrons [12, 16, 17].

Gamma decay and X-rays are, as a rule of thumb, less energetic than beta emission and they have no mass, thus have less ionization power but even higher penetration in matter [12, 16, 17].

Viewing all of this from the treatment standpoint, a few considerations need to be made. The first energy is deposited from the inside-out in brachytherapy. To concentrate the dose in the target and minimize the dose in the healthy surrounding tissues, a careful dosimetry study needs to be made. Energy should be high enough to reach the target but not as high to the point of causing necrosis to the healthy organs at risk (OAR). Fig. 2 presents radiation from both views [11, 13].

Radiation activity, half-life, and dose directly impact the treatment results. Activity is defined as the number of atoms that decay or emit their radiation, per second. This is measured in decay per second, or Becquerel (Bq). Activity after a period of time (A) depends on the initial activity (A_0) on the half-life ($T_{1/2}$) of the isotope used. Half-life

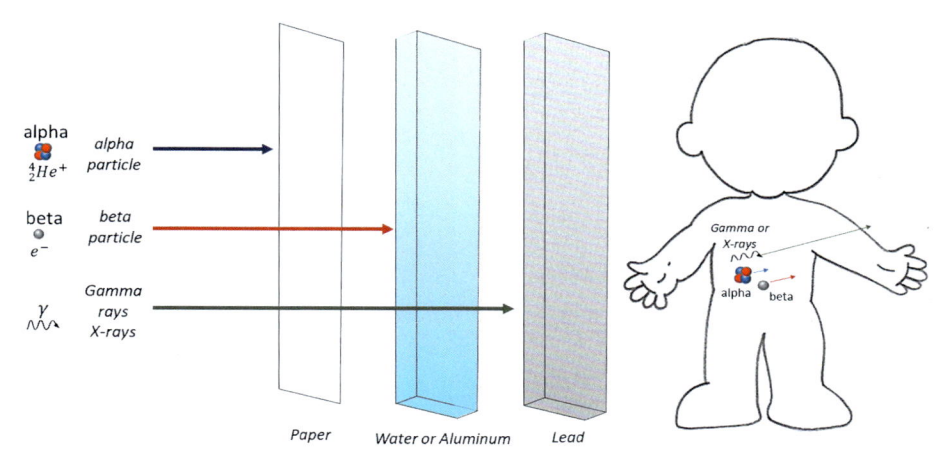

Fig. 2 Radiation penetration in matter from two standpoints: radiation source located outside the body, with examples of shielding materials; from inside the body.

is the time a radioactive source takes to emit half of the radiation. So, activity has an exponential behavior. Eqs. (1) and (2) shows the calculation [12, 16, 17].

$$A = A_0 . e^{-\lambda . t} \tag{1}$$

or

$$N = N_0 . e^{-\lambda . t} \tag{2}$$

where

A = final activity
A_0 = initial activity
N = final atoms number
N_o = initial atoms number
Decay coefficient: $\lambda = \frac{\ln 2}{T_{1/2.}}$
t = time

To know the necessary isotope mass to reach a certain activity, Eqs. (3) and (4) are used:

$$A = \lambda . N \tag{3}$$

$$m = \frac{M_a \, x \, N}{N_a} \tag{4}$$

where

A = activity needed
N = correspondent atoms number
M_a = atomic mass
N_a = Avogadro number
m = equivalent mass.

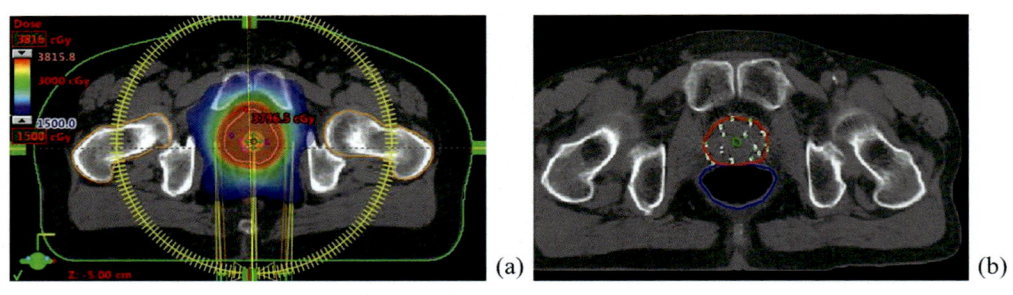

Fig. 3 Hiperfractionated EBRT (A) versus Brachytherapy planning (B) showing isodose curves. Notice that radiation deviates very little from the TA in brachytherapy.

The final dose would be calculated depending on treatment time. It is desirable that a high dose rate, higher activity, and energy in less treatment time are used. If a low dose rate is required, lower activity and energy in less treatment time is performed. The dose is the amount of energy released by the unit of mass. Usually, the unit Gy is used (equals to J/Kg). The higher the dose, the higher the effect. In a radiation therapy treatment, a dose chart is used to show where the cancer is, how much radiation is hitting it, and to quantify the dose in the OAR [11, 13]. Depending on the placement of the radiation source, the different organs with different intensities will be hit by radiation. Fig. 3 shows isodose curves for an Intensity Modulated Radiation therapy linac treatment versus a brachytherapy implant.

Radiation interacts with matter releasing the energy in the tissues. Basically, the energy is lost by two major routes: excitation or ionization. In excitation, which occurs in nonionizing radiation, the photon interacts with the atoms' electronsphere exciting electrons, but not strong enough to remove them. In ionization, energy is high enough to remove electrons. That is what occurs in ionizing radiation [24].

Since the human body is mostly water, radiation has a higher chance to hit a water molecule, creating free radicals. Those can cause cellular and DNA damage, killing the cell. There is also a chance that radiation directly hits the DNA strand causing mutations that can lead to cellular death. The amount of water directly impacts radiosensitivity, which will reflect on treatment radiation dose [24].

Another important aspect is angiogenesis. Cancer needs more blood supply to grow and, eventually, spread the damaged cells. To achieve this, it stimulates the growth of the vascular network [25]. It is proven that without good blood irrigation, tumors may become necrotic or even apoptotic [26, 27]. Radioactive nanoparticles could unite both forms of treatment: use radiation effects to kill cancer cells, use radiation effects to destroy tumor vascularity.

3. Necessary paths to develop nanobrachytherapy

To develop a new source, several factors must be considered. They will limit the radio-isotopes that can be used.

The cancer site is very important. If the target area has more water and oxygen, more free radicals and oxygen reactive species will be formed, and the radiation effects will be greater. If the cancer is large, is likely the internal cells will be in anoxia, so treatment needs to be fractioned (killing the outside cells first, allowing the internal cells to restore, and then killing them with another radiation treatment fraction). If the cancer is aggressive, a higher dose rate may be necessary, all of which is limited by cancer location. In the case of the surrounding areas being sensitive, the dose must be restricted. For example, it is "easier" to treat prostate cancer than spinal cord cancer [12, 13, 28].

The way that the radioisotope is produced impacts final price and availability. If the isotope is produced naturally as part of the decay chain of uranium-238, uranium-235, or thorium-232, it is easily found (but high values can be spent with purification). If the isotope is a by-product of fission, it can be recovered from the nuclear fuel that has already been used. Only five countries in the world have facilities to reprocess spent fuel (England, India, France, Japan, and Russia). If the isotope needs to be produced in nuclear reactor or accelerator, costs can be high (due to the high cost of enriched targets) [28, 29].

Chemical stability, form, and waste management are also key. The radioactive material must be stable inside the body or the capsule. The decay product must do so also. For example, if a solid welded source decay product is gaseous, it might damage the structure. If a radiation source has a high half-life, it might be considered as a high-level waste after being used, needing management for a long period of time [12, 13, 28].

Energy, decay mode, radioactivity, and dose rate must be adequate to deliver the maximum amount of energy to the TA and minimum to the OAR. They must also not pose a threat to medical staff and the production operators. If manipulations require heavy shielding, the cost of treatment and manufacture is higher [12, 13]. Table 1 presents a few of the radioisotopes considered for use [28].

Dosimetry is important for all radiation therapy treatments. In brachytherapy, it depends on how the sources are positioned relative to the dose calculation points and on the source strength. The use of a well-established dosimetric system for the brachytherapy treatment of cancer needs to be established for each new source format and/or radioisotope used. However, the use of a model alone is not sufficient to validate results; reliability in source strength determination is necessary in order for the dose calculation to be accurate. Thus, brachytherapy sources need to be calibrated with traceability to a national or international standards laboratory. In situations in which the system to be used is not obvious, such as in nanobrachytherapy, all dosimetric modes must be investigated, and new models proposed [28].

Table 1 Radioisotopes used in brachytherapy and characteristics.

Radioisotope	Main emission type	Energy (MeV)	Half-life	Source format
Pd–103	X-rays	0.021	17 days	Seeds and nanoparticles
I–125	γ, X-rays	0.035, 27.4	59.6 days	Seeds
Ir–192	β^-, γ	0.29 and 0.16, 0.32	74 days	Wire, pellets, and seeds
P–32	β^-	0.695	14.29 days	Plaques
Cs–131	X-rays	0.030	9.7 days	Seeds
Au–198	β^-, γ	0.31, 0.41	2.7 days	Nanoparticles
Ru–106 (decays to Rh–106)	β^- β^-	0.04 1.5	1.02 years 30 s	Plaques
Lu–177	β^-	0.149	6.64 days	Nanoparticles
Y–90	β^-	2.28	64 h	Plaques
In–111	β^- X-rays	0.245 and 0.171 0.023	2.8 days	Nanoparticle coatings

Next, the patient and source production journey will be summarized (basic steps). When a patient with cancer enters a radiation therapy center, the medical staff will evaluate a 3D computed tomography (CT). If brachytherapy is recommended, treatment planning starts. Let us suppose that iodine–125 sources will be implanted inside the prostate. The staff will perform a mock-up treatment in the planning system (TPS), a software containing all information from the patient and radiation source. This will consider the patient's prostate anatomy and will inform the staff approximatively how many sources needed to be used for treatment success (80–120). Accessories that maybe be used (such as immobilizers, extra shielding, and others) are also determined. The purchase order is relayed to the manufacturer. Inside the industrial radioactive plant, a hot cell (completely shielded and airtight led structure) is used in production. The iodine–125 seed can have many configurations that impact production. Let us consider the most used: silver core that contains the radioactive material surrounded by a titanium capsule. The production for the patient seeds starts by performing the chemical iodine–125 binding reaction to the core. Then, the radioactive core is placed inside a titanium capsule that is then welded in both ends. Leakage tests are performed to ensure weld success. Radioactivity is measured to confirm the correct value. The order is filled. The TPS already has all the information about the source dosimetry. That information was obtained in a study made before the seed even reach the medical market. Dosimetry methods, such as calculations according to protocols (AAPM TG-43 for iodine-125 [30]), thermoluminescent dosimeters (such as TLD-100), film, Monte Carlo simulations (AAPM TG-186 [30]), in vitro and in vivo dosimetry, among others, should be already performed. The seeds reach the hospital after authorized transport to the hospital. Source activity is calculated to reach the desired value at the time of the implant. Depending on distance, they might leave the manufacturing site with more radioactive activity to compensate for decay. After the implant is performed, a dose curve is calculated by the TPS, like the example shown in Fig. 3B.

This example is here to show the difficulties of developing and implementing a new form of treatment in the radiation therapy field. A lot of investment in manufacturing, calibration, quality insurance, transport logistics, software development, training, hospital operation room needs to be made.

For nanobrachytherapy, we are still in the early developing stages. Efforts are still in synthesis methods, validation in cell and animal models, and dosimetry models propositions. Mostly, gold-198 and palladium-103 nanoparticles are being explored.

The pathway to developed radioactive nanoparticles also depends on if the technique can be reproduced nonradioactively. Most of the characterization cannot be performed with the radioactive product due to the risk of contamination. Also, the equipment needs to be in a controlled area with all radiation protection measurements in place. Usually, DLS (Dynamic light sizer), Zeta Potential, FTIR (Fourier Transform Infrared Spectroscopy), and UV-vis (UV–Visible Spectroscopy) are possible to access the radioactive product (that, of course, depends on the equipment characteristics). However, TEM (Transmission electron microscopy), AFM (Atomic Force Microscope), CP-MS (Inductively coupled plasma mass spectrometry), and NMR (Nuclear Magnetic Resonance) cannot be used. The synthesis technique must also consider this contamination issue. If a unique glassware, rotary evaporator, shakers, centrifuge, HPLC (High performance liquid chromatography) is being used to produce and refine the nonradioactive product, the radioactive route might not be possible.

3.1 Nanoparticle synthesis

Nanoparticles can be synthesized by several routes (Fig. 4). Usually, nanoparticle type, final use, and equipment availability are determinant factors when choosing a method. For radioactive nanoparticles, simplicity must be pursued when performing the synthesis. Choosing a method that needs an expensive machine, high heating, or has volatilization is unfeasible.

Radioactive nanoparticles for cancer treatment are mainly being synthesized by chemical reduction (CR). It is a simple route that only needs basic lab glassware and a chemical reducing agent to form nanoparticles. In Fig. 5, the most used CR pathways are presented.

Co precipitation

- Metalic nanoparticles cores formed in aqueous solutions by: chemical reduction, electrochemical reduction, and decomposition of metallorganic precursors

- Formed by nucleation, growth, coarsening, and/or agglomeration

Microemulsion

- The mixing of microemulsion materials causes reactants exchange during the collision of water droplets in the microemulsion

- Precipitation occurs in the form of nanodroplets, followed by nucleation growth and coagulation

Hydrothermal Synthesis

- The crystal growth is performed insinde an autoclave
- the precursor is supplied in solution
- A temperature gradient is maintained at the opposite ends, so that the hotter end dissolves the precursor and the cooler end causes nanoseeds to take additional growth.

Inert gas condensation

- An ultrahigh vacuum chamber occupied with helium or argon gas at very high pressure is used to evaporate metals

- The evaporated metal atoms lose their kinetic energy by collisions with the gas, condensing into small particles.

Microwave

- Chemical reactions are often faster than traditional convection heating methods

- Microwave assisted synthesis achieve higher yields and creates less side products

Laser Ablasion

- Small parts of a solid/liquid surface of a given material can be removed by a laser beam

- At low laser flux, the material is heated by the absorbed laser energy and evaporates or sublimates in the nano range

Sol-Gel

- A solution can gradually evolves toward the formation of a gel-like diphasic system containing both a liquid phase and a solid phase

- Ultrafine and uniform ceramic nanopowder can be formed by precipitation

Ultrasound

- Ultrasonic irradiation causes ultrasonic cavitation in liquids and, maybe, forms nanomaterials

- Results in physical and chemical effects, such as high temperature, pressure, and cooling rate

Spark Discharge

- A spark is an abrupt electrical discharge

- The high electric field creates an ionized, electrically conductive channel through a normally insulating medium were nanoproducts can be formed

Biological

- Nanoparticles can be sintetize by microbes, viruses, or bacterias

- Mechanisms should be studied in detail to increase the rate of synthesis and improve the properties of the nanoparticles

- Biological nanoparticles are not monodispersed and the rate of synthesis is slow

Radiation

- Radiation acts a reducing agent, supplying electrons for the synthesis

- Different forms of radiations, dose and activity are being studied

- Usually, dose is high, demanding large irradiators or nuclear reactors

Sputtering

- Is the ejection of atoms from a target surface by bombardment

- Atoms from a cathode/target are driven off by ion collision

- Sputtered atoms travel until they strike a substrate, where they deposit to form the desired nanolayer

Template

- Uniform void spaces of porous materials are used as hosts to confine the synthesized nanoparticles as guests

- Have the role of a skeleton in order to organize the different functions of a device, the active components, and the different interfaces

Fig. 4 Summary of the main methods used in the synthesis of nanoparticles. Information from [31].

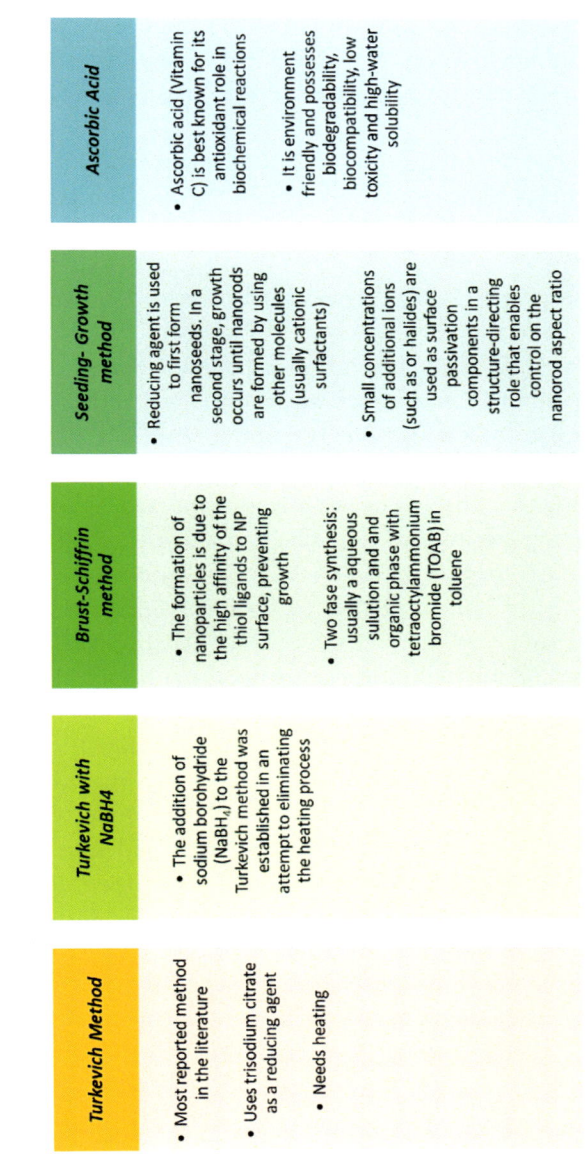

Fig. 5 Summary of the main methods used in the synthesis of gold nanoparticles. Information from [3].

Turkevich Method

- Most reported method in the literature
- Uses trisodium citrate as a reducing agent
- Needs heating

Turkevich with NaBH4

- The addition of sodium borohydride (NaBH₄) to the Turkevich method was established in an attempt to eliminating the heating process

Brust-Schiffrin method

- The formation of nanoparticles is due to the high affinity of the thiol ligands to NP surface, preventing growth
- Two fase synthesis: usually a aqueous sulution and and organic phase with tetraoctylammonium bromide (TOAB) in toluene

Seeding- Growth method

- Reducing agent is used to first form nanoseeds. In a second stage, growth occurs until nanorods are formed by using other molecules (usually cationic surfactants)
- Small concentrations of additional ions (such as or halides) are used as surface passivation components in a structure-directing role that enables control on the nanorod aspect ratio

Ascorbic Acid

- Ascorbic acid (Vitamin C) is best known for its antioxidant role in biochemical reactions
- It is environment friendly and possesses biodegradability, biocompatibility, low toxicity and high-water solubility

Green synthesis methods

- Synthesis of gold nanoparticles using plant extracts is well
- Various geometrical shapes and different sizes may be obtained impacting function and thus use
- Different plants deviced amino acids, enzymes, flavonoids, aldehydes, ketones, amines, carboxylic acids, phenols, proteins and alkaloids can provide electrons to form nanoparticles

For successfully synthesized nanoparticles, the aggregation barrier (activation energy that two particles have to overcome to merge into one) increases to an optimum particle size allowing colloidal stability to be reached. The steps are presented below [32]:

Step 1: High reduction rate increases the number of particles. Clusters with 1–2 nm are formed.

Step 2: Reduction continues at a much lower rate. Since particles have weak stabilization at this stage, coalescence processes take place (two or more droplets, bubbles or particles merge during contact to form a single daughter droplet, bubble, or particle), resulting in a decrease in the number of particles. When the radius is around 2.5 nm, the number of particles is now constant, but they keep growing in size.

Step 3: NPs grow due to the diffusion of the metal atoms reduced in the solution.

Step 4: When the particles reach a radius of around 4–5 nm, the growth rate increases drastically and the remaining metal salt is consumed. Particle size increases to the final radius.

Synthesis routes for the preparation of gold nanoparticles are presented in Fig. 5 and elaborated below:

(a) *Turkevich method:* 95 mL of $HAuCl_4$ solution (containing 5 mg Au) was heated to the boiling point and 5 mL of 1% sodium citrate solution was added to the boiling solution with good mechanical stirring. The reaction mixture was colorless for 12 s after the addition of the citrate, and then it turned purplish-blue within a fraction of a second. After 5 min, the final color was deep wine red. The best results were obtained with 5–50 mg citrate addition. The particle size achieved 20 nm. [33]

(b) *Turkevich method with $NaBH_4$:* 1 mL of 1% $HAuCl_4$ was added to 90 mL of H_2O at room temperature. After 1 min of stirring, 2.00 mL of 38.8 mM sodium citrate was added. One minute later, 1.00 mL of fresh 0.075% $NaBH_4$ in 38.8 mM sodium citrate was added. The colloidal solution was stirred for an additional 5 min and stored in a dark bottle at 4°C. Particles had 13 nm [34, 35].

(c) *Brust-Schiffrin method:* $HAuCl_4$ (30 mL, 30 mM) was mixed with a solution of tetraoctylammonium bromide (TOAB) in toluene (80 mL, 50 mM). The two–phase mixture was vigorously stirred until all the tetrachloroaurate was transferred into the organic layer and dodecanethiol (170 mg) was then added to the organic phase. A freshly prepared aqueous solution of sodium borohydride (25 mL, 0.4 M) was slowly added with vigorous stirring. After further stirring for 3 h, the organic phase was separated, evaporated to 10 mL in a rotary evaporator, and mixed with 400 mL ethanol to remove excess thiol. The mixture was kept for 4 h at −18°C, and the dark brown precipitate was filtered off and washed with ethanol. The crude product was dissolved in 10 mL toluene and again precipitated with 400 mL ethanol. Particles produced by this method reached 2.5 nm [36, 37].

(d) Seeding-growth method:

(i) Seed solution—20 mL aqueous solution containing 2.5×10^{-4} M HAuCl$_4$ and 2.5×10^{-4} M trisodium citrate was prepared in a conical flask. Next, 0.6 mL of ice-cold, freshly prepared 0.1 M NaBH$_4$ solution was added to the solution while stirring. The solution turned pink immediately after adding NaBH$_4$, indicating particle formation. The particles in this solution were used as seeds within 2–5 h after preparation. Citrate serves only as a capping agent since it cannot reduce the gold salt at room temperature (25°C). [38]

(ii) For the growth solution, a 200 mL aqueous solution of 2.5×10^{-4} M of HAuCl$_4$ was prepared in a conical flask. Next, 6 g of solid cetyltrimethylammonium bromide (CTAB- 0.08 M) was added to the solution, and the mixture was heated until the solution turned a clear orange color. The solution was cooled to room temperature and used as a stock growth solution [38].

(iii) Several routes were successful to produce various-sized gold nanoparticles by mixing different quantities of the seed and growth solutions. For example, 7.5 mL of growth solution was mixed with 0.05 mL of freshly prepared 0.1 M ascorbic acid solution. Next, 2.5 mL of seed solution was added while stirring. Stirring continued for 10 min after the solution turned wine red. Particles prepared this way were spherical with a diameter of 5.5 nm [38].

(e) *Ascorbic Acid method:* 10^{-2} M gold salt (200 mL) was mixed with 20 mM ice-cold ascorbic acid (70 mL) under continuous stirring. The temperature of the reaction synthesis was maintained to 4°C throughout. With the addition of gold salt, the colorless solution of ascorbic acid is changed to purple and finally faint blue. The average size of nanoflowers was measured to be of the order of 45 nm, the core size is around 35 nm, and the length of the petals is around 5 nm [39].

(f) *Green synthesis:* Different compositions and quantities of reducing agents are found in organic extracts can be used to produce NPs. CH-AuNPs were synthesized by using biopolymer chitosan for the reduction of gold salt. For preparation, 500 μL of a freshly prepared solution of chitosan (8 mg/mL) dissolved in 1% acetic acid solution was added to 10 mL of 1 mM tetrachloroauric acid (HAuCl$_4$) solution and stirred at 70°C until the color changed from pale yellow to red, which indicates the nanoparticle formation. Gold NPs had a narrow size distribution with an average size of 10–15 nm [40].

During the NPs fabrication process, the coordination sites of the surface atoms are not complete, allowing the binding of donor-acceptor species, or ligands. These molecules are easily bound to the surface establishing a double layer of charge that prevents nanoparticle aggregation. This coating agent can also be functionalized to provide reactive groups, such as amines, for subsequent modification [10].

Surfactants can be used to control the growth of nanomaterials to achieve desired morphologies [41]. At low concentration, those amphiphilic substances can be adsorbed on surfaces or interfaces resulting in an alteration of free energy available resulting in well-dispersed NP solutions [42].

3.2 Gold-198 nanoparticles, Au^{198}NPs

In the beginning, brachytherapy was limited to only Radium-226 and Radon-222. However, due to complications associated with high energy emissions from radium [43], gold-198 was introduced as an option by Flocks et al. from Iowa State University in 1951. They treated 400 men with inoperable prostate cancer and shown significant palliative results [44]. Today, it is still being used to treat an array of cancers, such as prostate, in different formats, such as seeds, foils, and pellets [45].

Gold nanoparticles were the first to be described in the literature. They were used as a method for staining glass to a wine-red color. The Lycurgus Cup, manufactured in 4th-century Rome, changes color depending on the location of the light source [46]. When the light source is transmitted, glass turned red by the addition of a precious metal–bearing material when the glass was molten. Now it is known that the color change is because of gold and silver nanoparticles [46].

AuNPs have small gold particles with a diameter of 1–100 nm, which, once dispersed in water, are also known as colloidal gold. Functionalized gold nanoparticles are the subject of intensive studies and biomedical applications [47], including genomics [48], biosensors [49], immunoassays [50], clinical biochemistry [51], laser phototherapy of cancer cells and tumors [52], the targeted delivery of drugs [53], optical bioimaging [39], and monitoring of cells and tissues [54].

Faraday [55] presented in 1857 the first synthesis methodology with a scientific eye. He used chloroauric acid (HAuCl$_4$) as a precursor and a phosphorus–ether solution as a reducing agent. The solution turned first brown, then gray, purple, and red, finally giving a deep-red product. He had produced extremely fine particles with a mean diameter of 50 Å (5 nm).

Oxidation states of gold are Au^{+1} (aurous), Au^{+3} (auric), and its nonoxidized state Au^0. Au^0 is the final electric state found in nanoparticles. Therefore, the synthesis of AuNPs is actually a reduction from Au^{+1} or Au^{+3} states to Au^0 using an electron donor (reduction agent). The precursor of choice for the majority of researchers is chloroauric acid, HAuCl$_4$ with gold in its Au^{+3} oxidation states. [3, 5, 56]

Gold in colloidal form has been used since the 1950s. The research group led by P. Hahn (Meharry medical college, Nashville-TN, USA) had used gold-198 in several types of cancer, such as leukemia and lung cancers [57–59]. In 1949, Miller's group located in Switzerland proposed an injection of colloidal radioactive gold directly into the body cavity as an effective way to treat cancer [60]. The research in the field became scarce with the development of EBRT cobalt units and, later, linacs.

Nanoparticles are generally produced by reducing tetrachloroauric acid with various agents such as borohydride, amines, alcohols, carboxylic acids, sodium citrate, sodium borohydride, ascorbic acid, and others (examples in [61, 62]). At the same time, small organic molecules or polymers should/can be added to the system to prevent the aggregation of the formed nanoparticles. The final shape achieved is generally spherical [3, 63].

The major question is: Can tetrachloroauric acid be placed inside a nuclear reactor and only gold be activated? If no, can it be fabricated from radioactive pure gold? The answers are: no and yes. Chlorine forms a high energy high half-life isotope making it impossible to use inside a human body. Gold foils or wires can be turned into $HAu^{198}Cl_4$ by being dissolved with aqua regia.

Aqua Regia dissolves gold, though neither HCl nor HNO_3 has the power to dissolve it alone, because, in combination, each acid performs a different task. Nitric acid is a powerful oxidizer, which dissolves a virtually undetectable amount of gold, forming ions of gold (Au^{3+}). The hydrochloric acid provides chloride ions (Cl^-), which react with the ions gold to produce tetrachlorourate (III) anions, also in solution. The reaction with hydrochloric acid is an equilibrium reaction that favors the formation of chlorourate anions ($AuCl_4^-$). This results in the removal of gold ions from the solution allowing further oxidation of the remaining gold. The gold foil then becomes chlorouronic acid [3, 64].

To produce gold–198 a nuclear reactor is necessary. Highly pure gold (99.99+%) is placed in the reactor neutron flux. The core becomes unstable, radioactive. Stable gold (197 in mass) turns into radioactive gold-198. After emitting beta radiation with 314.55 keV and gamma with 411.8 keV, the remaining atom is now stable mercury with 198 mass [65]. Eq. (5) summarizes gold–198 formation and decay.

$$^{197}_{79}Au + {}^{1}_{0}n \xrightarrow{\text{98.65 barns}} {}^{198}_{79}Au \xrightarrow[\gamma, \beta^-]{} {}^{198}_{80}Hg \tag{5}$$

In order to estimate the final activity, the following calculation is used (Eq. 6) [17]. Explanation of Eq. (6) indexes can be found in Table 2.

$$A = \frac{M.N.\theta.\sigma.\phi}{P} \left(1 - e^{-\lambda t}\right) \tag{6}$$

Next, we will present the advance in $^{198}AuNPs$ reported in the literature.

Table 2 Initials meaning for Eq. (1).

Initial	Meaning	Unity
A	Activity	Bequerel
P	Atomic weight	g/mol
M	Mass	g
N	Avogadro	6.02×10^{23}
θ	Isotopic abundance	%
λ	Decay	$time^{-1}$
ϕ	Reactor flow	$n\,cm^2/s^{-1}$
σ	Cross section	$cm^{2\ a}$
t	Irradiation time	time

$^{a}1\,barn = 10^{-24}\,cm^2$.

The group headed by K. Katti (University of Missouri, EUA) developed a gold-198 nanoparticle coated with gum arabica [62]. The authors state that their new method is better than the classic ones that consisted in the reduction of Au^{3+} in $AuCl_4^-$ with $NaBH_4$ or sodium citrate, compounds that the author claims are highly toxic. Instead of the traditional route, the researchers synthesized a novel reducing agent that was named THPAL-alanine trimeric with phosphines. The amount 0.500 g, 4.033 mmol of Tris(hydroxymethyl)phosphine (THPAL) in 5 mL of distilled water was added dropwise to (L)-Alanine/(D)-Alanine (1.077 g, 12.00 mmol) in 10 mL of distilled water at 25°C. The reaction mixture was stirred under dry nitrogen for 1 h. The solvent was removed in vacuo to obtain a white solid. The white solid was washed with methanol and dried in vacuo to give the pure product (L)- or (D)- in 90% yield. For $H^{198}AuCl_4$ fabrication, the research group conveys two routes:

(a) $^{198}AuCl_4^-$ was activated in a nuclear reactor with a flow of 8×10^{13} n/cm^2/s in an acid solution of HCl [62].

(b) Au-198 was produced by direct irradiation of natural gold foil or metal Au-197(n,γ) Au-198. Gold foil was irradiated at a neutron flux of 8×10^{13} n/cm^2/s. After irradiation, the radioactive foil was dissolved with aqua regia, dried down, and reconstituted in 0.05–1 mL of 0.05 N HCL to form $H^{198}AuCl_4$ [66].

Gum arabica was used as a coating stabilizing agent. It consists of a mixture of lower molecular weight polysaccharides and higher molecular weight hydroxyproline-rich glycoprotein. Stability studies carried out with AuNPs have demonstrated that the saccharide and protein structure provides exceptional stability for AuNPs for periods of over six months. Therefore, we have selected gum arabica stabilized ^{198}AuNPs for further in vivo pharmacokinetic studies [66].

After activation, 50–100 μL of $H^{198}AuCl_4$ is added with 6 mL of gum arabica along with 20 μL/0.0337 THPAL for each mL of solution. This reaction occurs results in radioactive gold nanoparticles in less than 5 min.

The analysis of the material, without the radioactive nucleus, was performed obtaining the following results: DLS mean diameter: 7 ± 3 nm, TEM mean diameter: 17.5 ± 2.5 nm, DLS average diameter (with gum arabica coating): 85 nm, zeta potential: -24.5 ± 1.5 mV, and UV-Vis plasmon absorption band: 540 nm. In vitro tests obtained the following results. Biological stability test performed with 10% NaCl, 0.5% cysteine, 0.2 M histidine, 0.5% HSA (human serum albumin), and various pH values were found to be stable without aggregation or decomposition. The Hemocompatibility test did not result in hemolysis and the platelet aggregation test showed that the function was inhibited or cause platelet aggregation.

The biodistribution studies were performed in normal, SCID [62] and CF-1 [66] mice. They were then euthanized, and the tissues and organs were excised from the animals following at 1 h, 4 h, and 24 h postinjection. Subsequently, the tissues and organs were weighed and counted in a NaI well counter. Two varieties were reported, yielding

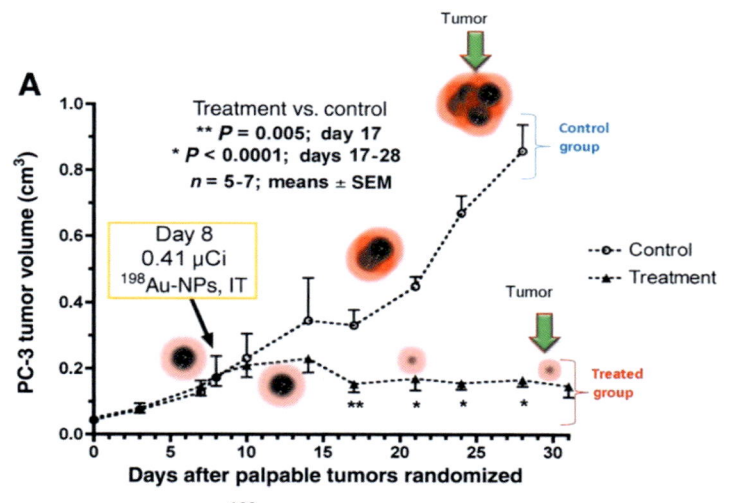

Fig. 6 Therapeutic efficacy of GA-^{198}AuNPs in prostate tumor-bearing mice. *(Reproduced from N. Chanda, et al., Radioactive gold nanoparticles in cancer therapy: therapeutic efficacy studies of GA-198AuNP nanoconstruct in prostate tumor-bearing mice. Nanomedicine 6 (2) (2010) 201–209 with permission. Elsevier RightsLink: 4,680,770,734,640.)*

the following results: 408 µCi nanoparticles were injected directly into cancer (PC3 prostate cancer cells) and remained for three weeks. After two weeks, 19.9% of the injected dose was found in the tumor, 0.91% in the liver, 0.13% in the kidney, 0.09% in the intestine, and 18.5% in the carcass [62]. On euthanasia of the animals, a reduction of up to 82% of the size of the cancer was observed (Fig. 6).

10–20 µCi nanoparticles were injected directly tail vein (20–40 µL). The pharmacokinetics and biodistribution studies of gum arabica (glycol protein) coated 198-AuNPs in mice showed >80% uptake in the liver with minimal accumulation in blood and other non-target organs. Authors suggest the use in liver cancer [66]. Red blood cell, platelet, lymphocyte, and antibody count remained the same as in healthy animals for both experiments.

The same research group developed gold-198 nanoparticles coated with epigallocatechin-gallate (EGC) as an alternative treatment for prostate cancer treatment [67]. An advantage of EGC is its ability to target the Laminin receptor (Lam 67R) that is over-expressed in human prostate cancer cells [68]. The authors initially synthesized the non-radioactive surrogate EGC–AuNP by mixing sodium tetrachloroaurate ($NaAuCl_4$) with EGC in deionized water [67]. For EGC-^{198}AuNPs, a 0.76 mg gold leaf was irradiated for 3.5 h (no extra information on the nuclear reactor characteristics was provided). The calculated activity achieved was 20.5 mCi. The now radioactive foil was dissolved in 800 µL of aqua regia. The solution was heated until the volume reaches approximately 200 µL. Then 600 µL of 0.05 M HCl was added and heating continued until most of the acid had evaporated. The solution was then allowed to cool followed by the addition of 200 µL of 0.05 M HCl. A 0.17 mg/mL solution of EGC was prepared under constant stirring at 25°C for 3 min. To this solution, 2 µL of $H^{198}AuCl_4$ (0.41–1.3 mCi) and 98 µL of 0.1 M $NaAuCl_4$ carrier solution were added. The wine-red expected color change occurred after 5 min. The reaction mixture was stirred for an additional 15 min at 25°C [67]. Probably, the carrier solution formed EGC-Au:^{198}AuNPs.

Characterization, in vitro, and in vivo studies were performed with the non-radioactive version. The nanoparticles showed particle size of 125 ± 5 nm with DLS while TEM of the core nanoparticles without the EGC layer demonstrated a mean diameter of 47.5 ± 7.5 nm and zeta potential of -37.7 mV with UV-Vis plasmon absorption band: 535 nm. The hemocompatibility assay involved direct exposure of EGC-AuNP to freshly drawn whole human blood for 2 h at 25°C. The results obtained confirmed that the EGC-AuNP remained intact. A size variation of 10 nm (monitored through changes in surface plasmon resonance—UV-Vis) when tested in presence of with 10% NaCl, 0.2 M, histidine, 0.5% human serum albumin (HSA), 0.5% Bovine Serum Albumin (BSA), and at various pH values. No detectable aggregation/decomposition was noted in EGC-AuNPs when assessed by incubating the nanoparticles in cell culture media.

SCID mice received intratumoral injections of EGC-^{198}AuNP (3.5 µCi) in DPBS (20 µL) while performed under brief inhalational anesthesia. Analysis of ^{198}Au radioactivity revealed that $72.4 \pm 5.9\%$ was retained in prostate tumors at 24 h. Slow clearance (leakage) into the blood with only 0.06% at 24 h was observed. Lungs and pancreas exhibited low uptake at 24 h, with only 0.33% and 0.22%, respectively.

Cancer size and treatment progression are shown in Fig. 7. The end-of-study biodistribution on Day 42 showed that 37.4 ± 8.1 of EGC-^{198}AuNP remained in the residual tumor, while 17.8 ± 6.1 was noted for carcass and $2.5 \pm 1.7\%$ was observed in the liver. Retention in other tissues was negligible, with radioactivity near background levels for blood, heart, lung, spleen, intestines, stomach, bone, brain, and skeletal muscle.

In comparison, EGC-^{198}AuNP was retained in tumors for more than twice time period when compared to GA-^{193}AuNPs.

Efforts continue with new research continuing being done in synthesis, validation, and dosimetry approach.

3.3 Palladium-103 nanoparticles (Pd^{103}NP)

Palladium-103 brachytherapy was introduced in 1987 as an alternative to iodine-125, also suitable for interstitial implantation. The characteristics of palladium-103 are similar to iodine-125 in that it emits a low-energy photon with an average energy of 21 KeV. The shorter half-life of palladium-103 (17 days), relative to that of iodine-125 (60 days), results in a higher initial dose rate [69]. They are being used mostly for prostate cancer in the seed form.

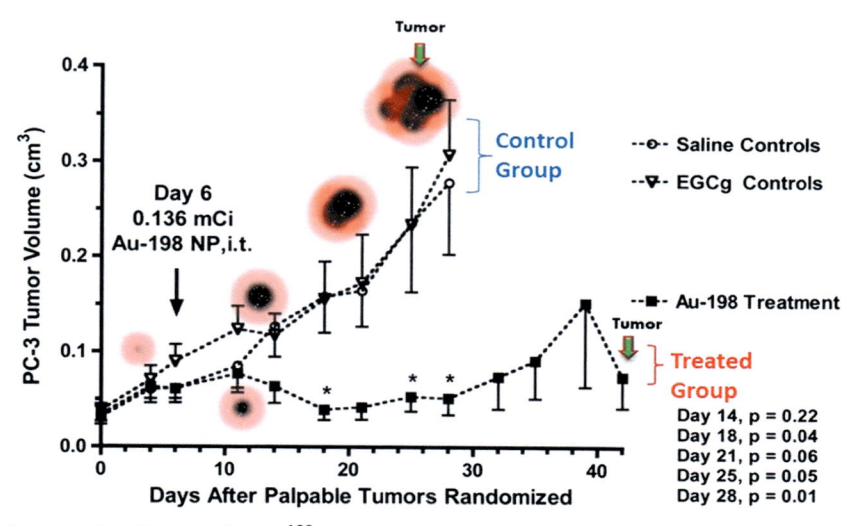

Fig. 7 Therapeutic efficacy of EGC-^{198}AuNPs in prostate tumor-bearing mice. *(Reproduced from R. Shukla, et al., Laminin receptor specific therapeutic gold nanoparticles (^{198}AuNP-EGCg) show efficacy in treating prostate cancer, Proc. Natl. Acad. Sci. U. S. A. 109 (31) (2012) 12426–31 with permission.)*

Palladium nanoparticles are used as catalytic materials [70] (due to high surface-area-to-volume ratio), hydrogen storage [71], sensing [72], and more recently as radioactive nanoparticles for cancer treatment. Their synthesis can follow chemical or electrochemical methods using a variety of stabilizers including organic ligands, salts/surfactants, polymers, and dendrimers [73]. The palladium precursor used can have an important impact on the nanoparticles formed. The Pd(acetylacetonate)$_2$ salt gave more monodispersed distributions than the corresponding nitrate or chloride salts [74]. This is maybe due to carbon monoxide molecules being formed by the decomposition of the acetylacetonate [75].

By far the most common synthesis direction is to use ligands. They are also successful in avoiding agglomeration. Sulfur-based ligands, such as thiols, acting as electron donors, are highly efficient stabilizers due to strong interactions with Palladium (and other metals). The method developed in the 1990s by Brust-Schiffrin [37] is based that metal salt is dissolved in water and first transferred to the organic phase using a suitable phase-transfer agent, such as tetraoctylammonium bromide (TOAB). Then, an aqueous sodium borohydride solution is added to the stirred biphasic system, leading to the formation of nanoparticles. The ratio of metal to organo-thiol and the reaction temperature control particle size. The nanoparticles, protected by a compact shell of organo-thiols, are stable for long periods of time, either in solution or as solids that can be readily redispersed in organic solvents [3, 37, 74]. To synthesize small nanoparticles, the use of long thiols in excess with an additional quantity of reducing agent will increase reaction yield [74].

The Brust-Schiffrin pathway has successfully been used to prepare palladium nano-particles stabilized by simple lipophilic *n*-alkanethiols. The two-phase methodology allows easy nanoparticle separation from the aqueous by-products. However, when more polar ligands are used, the purification can get more complicated [74].

An alternative synthesis route that allows a wider variety of ligands to be used was presented by Yee et al. [76]. Superhydride (lithium triethylborohydride, $LiEt_3BH$) was used as a reductant agent. The combination of palladium (II) acetate with octadeca-nethiol resulted in a soluble metal–thiolate complex in tetrahydrofuran (THF). The addi-tion of superhydride solution to this resulted in stable nanoparticles being formed with an average diameter of 2.3 nm [74, 76].

Another strategy for preparing palladium nanoparticles soluble in aqueous systems is to use ω-substituted thiol ligands (contains a charged functional group, such as ammo-nium or carboxylate salts). The sodium borohydride reduction of potassium tetrachloropalladate(II) in the presence of the chloride salt of the *N,N*-trimethyl(undecylmercapto)ammonium ligand in a one-phase system results in water-soluble PdNPs with an average diameter of 2.7 nm [74, 77].

The use of phosphine ligands has also been used for palladium nanoparticle preparation. Kim et al. [75] used thermolysis of palladium-trioctlyphosphine (TOP) complex to synthe-sized nanoparticles. Nanoparticles with 3.5 nm were formed when Palladium (acetylaceto-nate)$_2$, and TOP were heated to 300°C under an argon atmosphere. Oleylamine was used as a stabilizer solvent. Monodispersed nanoparticles up to 7.5 nm were formed [74, 75].

Adaptations from an existing route for gold nanoparticles by phase transferring were made by Tamura et al. [78]. Potassium tetrachloropalladate(II) was phase transferred into dichloromethane using tetraoctylammonium bromide before sodium borohydride reduction in the presence of optically active bidentate BINAP (2,2′-bis(-diphenylphosphino)-1,1′-binaphthyl) ligands [74, 78].

Nitrogen-based ligands, such as long-chain primary amines, are used to strongly chemically adsorb onto NPs surface. The alkyl group prevents agglomeration via steric stabilization [74]. Boron tributylamine (BTB) used in conjunction with oleylamines was presented by Sun et al. [79]. BTB was used as a coreductant and oleylamine served as a solvent, stabilizing ligand, and reductant. Several other nitrogen-based ligands have been reported, such as pyridines [80], amines [81], and imidazole derivatives [82, 83].

Surfactants have also been extensively used as stabilizers [84]. Tetra-*N*-alkylammonium halide salts are the most chosen for this purpose (examples in [85, 86]). Imidazolium-based ionic liquids are also being used, especially in biodiesel improvement. [87, 88]. The stabilization can also be achieved by incorporating PdNPs within an organic matrix, such as polymers [89], dendritic structure [90], or biomolecules (such as proteins [91], polypeptides [92], and DNA/RNA [93]).

Production of ^{103}Pd is mostly performed via ^{103}Rh(p,n)^{103}Pd nuclear reaction in a cyclotron. The highly enriched target was bombarded with protons with an electric current for a determined amount of time. A classical route is 18 MeV protons at 200 μA beam current for 15 h [94]. After irradiation, radiochemical separation of the target is necessary. The remaining rhodium is separated from palladium-103 by an exchange resin (such as Amberlite®IR-93) in acidic conditions. After emitting mostly x-rays with an average of 21 keV, the remaining atom is now stable rhodium with 103 mass [65]. Palladium-103 is mainly produced by the proton bombardment of rhodium (^{103}Rh (p, n)^{103}Pd). The electroplating of the target can be carried out by applying either a DC constant voltage (or current) or an AC constant voltage (or current) at an elevated temperature (typically 40–60°C). Eq. (7) summarizes palladium formation and decay.

$$^{103}_{45}\text{Rh} + ^{0}_{1}\text{p} \rightarrow ^{103}_{46}\text{Pd} \xrightarrow[\gamma, X^-]{} ^{103}_{45}\text{Rh} \tag{7}$$

To estimate final number of atoms formed, the following calculation is used (Eqs. 8 and 9) [17, 95] (Table 3):

$$R = n_T I \int_{E_S}^{E_0} \frac{\sigma(E)}{dE/dx} dE \tag{8}$$

$$n_T = \frac{\rho\, x_m}{A_T} N \tag{9}$$

Next, we will present the advances in ^{103}PdNPs reported in the literature.

Table 3 Initials meaning for Eqs. (8) and (9).a

Initial	Meaning	Unity
R	N	Nuclei/second Converted to activity by using Eq. (3)
n_T	The target thickness	nuclei/cm^2
I	Incident particle flux per second and is related to the beam current;	particle flux/second
N	Avogadro	6.02×10^{23}
σ	Reaction cross-section, or probability of interaction, and is a function of energy	cm^2
E	The energy of the incident particles	eV
x	The distance traveled by the particle	cm
E_0	Initial energy of the incident particle along its path	eV
E_s^a	Final energy of the incident particle along its path	eV
A_T	Atomic weight of the target material	Grams
ρ	Density	g/cm^3
x_m	The distance the particle travels through the material	cm

aAs the particle passes through the target material, it loses energy due to the interactions of the particle with the electrons of the target. This is represented in the above equation by the term dE/dx (also called the stopping power).

The group led by M.A. Fortin presented nanoparticles species fabrication by following the core-shell (denominated by @) pathway: PEG-^{103}Pd:Pd@AuNPs [96] and PEG-^{103}Pd:Pd@^{198}AuNPs [97].

Basic synthesis involved the following steps:
- 0.075 mM of H$_2$PdCl$_4$ precursor solution was mixed with ^{103}PdCl$_2$. Centrifugation (3000 × g, 2 min) to remove any aggregate was performed.
- The ^{103}Pd:PdCl$_2$ mixture was rapidly transferred into a 0.3 mM DMSA (2,3-meso–dimercaptosuccinic acid, acting a capping agent) solution and 30 mM ascorbic acid solution, under vigorous stirring (final NP concentration of 0.06 mM). A change from light yellow to dark brown indicated the formation of ultrasmall Pd cores.
- Then ^{103}Pd:PdNPs were used as seeds for the growth of a gold shell (Pd@Au).
- HAuCl$_4$ solution was added to the Pd nanocores suspension under stirring.
- NH2-PEG-SH was added to the solution to ensure colloidal stability. The reaction proceeded for 2 more hours, followed by centrifugation (16,000 × g, 15 min). The supernatant was discarded, and the nanoparticles were washed twice with nanopure water, followed by ultracentrifugation.
- NPs were concentrated by centrifugation into small volumes (typically: 30–60 μL).

Route for alginate-PEG-^{103}Pd:Pd@AuNPs synthesis is as follows:
- Same route, but with H$_2$Pd^{103}Cl$_4$ precursor in solution with 54 mL and 39.1 mCi
- The particles were centrifuge-filtered, followed by rinsing in nanopure water and two centrifugation-filtration cycles
- HAuCl$_4$ (0.5 mM) was added to Pd nanocore solution with an ascorbic acid solution (0.9 mM); Diameter was 49.5 nm
- NH2-PEG-SH was added—same procedure, $d = 84.5$ nm
- The suspension was supplemented with 20 μL of an aqueous solution of sodium alginate (2% m/v) up to a final volume of ≈55 μL (1.7 mCi/4 μL injection)

Route for alginate-PEG-^{103}Pd:Pd@^{198}AuNPs synthesis is as follows:
- The same as the first three items for alginate-PEG-^{103}Pd:Pd@AuNPs synthesis
- HAuCl$_4$ solution with H^{198}AuCl$_4$ (9.4 mCi) was mixed with 0.9 mM of ascorbic acid ($d = 58.1$ nm)
- NH2-PEG-SH was added, same procedure ($d = 86.2$ nm)
- The suspension was supplemented with 25 μL of an aqueous solution of sodium alginate (2% m/v) up to a final volume of ≈55 μL (1.6 mCi/4 μL injection).

To follow the characteristics of regular brachytherapy seeds (volume ranging from 2 to 4 mm^3 and radioactivity between 0.5 and 2 mCi), the NP solutions were concentrated by centrifugation until similar levels: 4 μL volumes, 1.7 mCi for PEG-^{103}Pd:Pd@AuNPs, and 1.6 mCi for PEG-^{103}Pd:Pd@^{198}AuNPs. The balance between both isotopes in the double radioactive species was ^{103}Pd ≈ 1.2 mCi to ^{198}Au ≈ 0.4 mCi. The injected concentration of Au was kept between 2 and 6 mg Au per kg of body weight of BALB/c mice. According to the authors, a small fraction only of the injected dose is expected to transit to vasculature over time, but this quantity is not expected to lead to acute toxicity issues in vivo. To slow this process, biocompatible alginate was used. They gradually degrade over several weeks in physiological conditions due to calcium exchange. Overall, both types of nanoparticles remained massively in the tumors (>75%). Significant fractions of nanoparticles were also found in the liver ($\approx 16\%$).

Since this was an initial study, tumor cells continue to multiply in the periphery, but tumor size diminished considerably (56% smaller for stable gold and 75% for radioactive gold). In the future, multiple injections of radioactive NPs will be tested [97].

Moeendarbari et al. [98] also used ^{103}Pd@Au nanoseeds as a brachytherapy agent. They report a Cu-mediated surface modification process to efficiently incorporate ^{103}Pd onto hollow gold nanoparticles (~120 nm in diameter). The authors used the electrochemistry route by using hydrogen nanobubbles that served as templates and reduce Au ions into metal Au^0. The high concentration of hydrogen molecules in the bubble boundary reduces the Au^+ ion to form Au^0 clusters. Subsequently, the metal clusters act as catalysts to trigger the autocatalytic disproportionation reaction of $Na_3Au(SO_3)_2$, which leads to the formation of a gold shell around the hydrogen bubble. The metallic gold covers the bubbles to form hollow Au nanoparticles. Anodic aluminum oxide (AAO) membrane with 300 nm diameter through channels was used inside the solution to collect nanoparticles (diameter larger than 100 nm). The AuNPs feature a 25 nm polycrystalline shell with a 50 nm hollow core.

The hollow Au nanoparticles were first coated with Cu by an electroless deposition process. Then, the Cu layer was partially (about 70%) replaced by Pd through a galvanic reaction. To prepare ^{103}Pd@Au nanoseeds, ^{103}Pd solution was added after the electroless plating of Cu. Copper can be readily replaced by less reactive metals, such as ^{103}Pd:PdCl$_2$, through a single displacement reaction in an aqueous solution containing metal ions without any other additives. The plating of ^{103}Pd was allowed to continue for 24 h upon the addition of the Pd plating solution. Exactly 3 mL of 0.1 M citric acid solution containing 4.37 mCi ^{103}Pd was added. Plating of ^{103}Pd on HAuNPs was then continued for 24 h, followed by the addition of cold Pd plating solution (containing 0.0025 M PdCl$_2$ in 0.4 M citric acid solution) to replace all Cu. After 1 h, 2 M NaOH was added to dissolve the membrane and the resultant ^{103}Pd@Au nanoseed suspension was washed thrice with water. The incorporation yield was >80%. The total synthesis time of ^{103}Pd@Au nanoseeds was approximately 26 h. They were found to be extremely stable and retain their original size even after being shelved for 2 months at $8 \pm 2°C$.

The resulting ^{103}Pd@Au nanoseeds as a colloidal suspension were administered by direct injection into a prostate cancer xenograft model using Severe combined immunodeficiency (SCID) mice bearing human prostate cancer tumors. PC3 cells were implanted subcutaneously; tumors were allowed to grow for 4 weeks to reach a palpable size. The injection was performed intratumorally at 6–9 locations for each tumor. The injected radioactivity was 1.5 mCi and volume under 40 μL. The gold nanoparticle concentration in each injection was 2.03×10^{10} nanoparticles/mL. The size of the nano-seeds was large enough to prevent diffusion, resulting in >95% nanoseeds being retained inside the tumor over the entire course of the 5-week treatment. A high therapeutic efficacy was observed without noticeable side effects on the liver, spleen, and other organs. Over the 5-week treatment period, the group treated with ^{103}Pd@Au nanoseeds showed a significantly retarded tumor growth or tumor size shrinkage (82.75 ± 46.25 mm^3 to 19.83 ± 20.12 mm^3).

3.4 Other nuclei

3.4.1 Indium-111

Indium-111 has a half-life of 2.8 days and emissions: γ with 245.35 KeV (94%) and 171.3 KeV (90%); and X with main emission 23 KeV [65]. It is produced by proton irradiation in a cyclotron following: (p,2n) for ^{112}Cd target or (p,n) for ^{111}Cd target [94]. Indium-111 nanoparticles have been considered as an alternative to treat HER2 breast cancer by the group led by R. M. Reilly from the University of Toronto, Canada. They have produced and tested Tras-^{111}In@AuNPs [99]. Trastuzumab (tras) is a monoclonal antibody used to treat breast cancer, specifically for HER2 receptor-positive cancer.

Briefly, 500 µg of trastuzumab in 50 µL of 100 mM NaHCO$_3$, pH 9.3 was reacted with a 25-fold molar excess of orthopyridyldisulfide-polyethyleneglycol-*N*-hydroxysuccinimide (OPSS-PEG5k-SVA) overnight at 4°C. PEGylated trastuzumab (trastuzumab-PEG-OPSS) was purified and buffer exchanged into phosphate-buffered saline (PBS), pH 7.5, by ultrafiltration. Thiolated PEG derivatized with benzylisothiocyanate diethylenetriaminepentaacetic acid (SH-PEG2k-Bn-DTPA; 10 µg) was then complexed with ^{111}InCl$_3$ in 1 M sodium acetate buffer, pH 6.0 at room temperature for 30 min. The final radiochemical purity of SH-PEG2k-Bn-DTPA-^{111}In was >95. Finally, Tras-^{111}In@AuNPs were constructed by reacting 1 mL of 30 nm AuNP (2×10^{11} particles per mL) with 0.4 µg of SHPEG2k-Bn-DTPA-^{111}In (with 0.7–0.8 MBq) at 4°C for 30 min, then with 10 µg trastuzumab-PEG5k-OPSS for 5 min. Tras-^{111}In@AuNPs were finally surface-coated with an excess (7.6 µg) of PEG2k-SH for 30 min to prevent aggregation in vitro and minimize liver and spleen uptake in vivo. The final product was then purified from excess reagents by centrifugation at $2500 \times g$ for 30 min at 4°C followed by $15,000 \times g$ for 30 min at 4°C. The supernatant was carefully removed, and trastuzumab-AuNP-^{111}In was re-suspended in 1 mL of PBS, pH 7.5 [99].

MDA-MB-361 human breast cancer (BC) cells overexpressing HER2 were injected subcutaneously into female CD1-athymic mice. 10 MBq in 100 µL of normal saline of Tras-^{111}In@AuNPs was injected directly into cancer arresting tumor growth over a 70-day observation period. In contrast, untreated mice receiving only normal saline injections exhibited rapidly increasing tumor growth reaching 8-times their initial size over 70 days [99].

3.4.2 Lutetium-177

Lutetium-177 has also being studied as a viable isotope for use in nanobrachytherapy. It has a half-life of 6.6 days and mainly β^- emissions of 149 keV [65]. It is produced by neutron irradiation in a nuclear reactor [94]. The main methods for producing ^{177}Lu of high specific activity are based on irradiation of either ^{176}Lu or ^{176}Yb with reactor neutrons following Eqs. (10) and (11) [100]:

$$^{176}_{71}Lu + {}^{1}_{0}n \rightarrow {}^{177}_{71}Lu \xrightarrow[\gamma, X, \beta^-]{} {}^{177}_{72}Hf \tag{10}$$

$$^{176}_{70}Yb + {}^{1}_{0}n \rightarrow {}^{177}_{71}Lu \xrightarrow[\gamma, X, \beta^-]{} {}^{177}_{72}Hf \tag{11}$$

Lutetium-177 gold nanoparticles were also considered by R.M. Reilly research group as an alternative to treat HER2 breast cancer. They have produced and tested Tras-^{177}Lu@AuNPs [101]. The methodology is as follows: Tras-^{177}Lu@AuNPs were constructed by stepwise incubation of 2×10^{11} AuNPs (30 nm) in 1 mL of 20 mM NaHCO$_3$, pH 9.0 at 4°C with:

- OPSS-PEG-DOTA consisted of polyethylene glycol (4 kDa) functionalized with ortho–pyridyl disulfide (OPSS) at one end to create a gold-sulfur bond and at the opposite end with a 1,4,7,10- tetraazacyclododecane-1,4,7,10-tetraacetic acid (DOTA) that acts as a chelator to complex ^{177}Lu (full description in [102]—Supporting information).
- Labeling with ^{177}Lu was carried out by incubation with ^{177}LuCl$_3$ in 1 M sodium acetate buffer, pH 4.5 at 80°C for 30 min. The final radiochemical purity of OPSS-PEG3k-DOTA-^{177}Lu was >95%.
- OPSS-PEG5k-trastuzumab was prepared by reacting trastuzumab (200 μg) with a 5-fold molar excess of OPSS-PEG5k-succinimidyl valerate (SVA) in 100 mM NaHCO$_3$ buffer, pH 9.3 overnight at 4°C. Trastuzumab-PEG5k-OPSS was purified and buffer-exchanged into Milli-Q® water by ultrafiltration.
- No details on how OPSS-PEG3k-DOTA-^{177}Lu reaction with OPSS-PEG5k-trastuzumab to result in OPSS-PEG5k-trastuzumab-DOTA-^{177}Lu were reported.
- Following each step, AuNP were centrifuged at 15,000 × g for 15 min at 4°C to remove unreacted ligands [101].

MDA-MB-361 human breast cancer (BC) cells overexpressing HER2 were injected subcutaneously into female NOD/SCID mice. 20 MBq to 1 mg AuNP in normal saline of Tras-^{111}In@AuNPs was injected directly into cancer significantly inhibiting tumor growth during the 16 days observation period [101]. However, tumor growth arrest was not achieved with Trastuzumab.

A similar study was performed by the same group but with panitumumab [102]. The mass number of NPs injected in these two studies was the same, but the radioactivity injected in the trastuzumab [101] study was one-third lower (3.0 vs. 4.5 MBq). Tumor arrest was achieved in MDA-MB-468 cells with panitumumab. The group is expanding both studies.

Vilchis-Juárez et al. demonstrated that 20 nm diameter AuNP labeled with ^{177}Lu targeted to $\alpha(v)\beta(3)$-integrin-positive C6 glioma was able to deposit high radiation doses (>60 Gy) in xenografts in mice preventing tumor progression [103]. The cysteine in DOTA-GGC (1,4,7,10-tetraazacyclododecane-N',N',N''-tetraacetic-Gly-Gly-Cys) molecule is used to interact with the gold nanoparticle surface and DOTA is used as the lutetium-177 chelator. A 5 μL aliquot of DOTA-GGC (1 mg/mL) was diluted with 40 μ of 1 M acetate buffer at pH 5, followed by the addition of 10 μL of a ^{177}LuCl$_3$ solution. The mixture was incubated at 90°C in a block heater for 30 min. Separately, to 1 mL of AuNP (20 nm), 0.025 mL of c[RGDfK(C)] (Arg-Gly-Asp cyclo peptide) was added, followed by 3 μL (40 MBq) of 177Lu-DOTA-GGC, and the mixture was stirred for 5 min to form the ^{177}Lu-DOTA-GGC-AuNP-c[RGDfK(C)]. The complex has 2 MBq/0.05 mL was injected intratumorally. High tumor retention was observed (68%). The rest of the activity occurred mainly in the kidneys and liver, as well as in the spleen, with negligible uptake in other organs. In the end, significantly decreased glioma tumor progression in mice through the effect of a combined molecular targeting therapy/radiotherapy was observed.

4. Conclusion

With the rapid increase in cancer occurrence, new forms of treatment are and must be investigated. Also, easy and portable forms are of the utmost importance. Nanoseeds-based brachytherapy fits perfectly with these conditions, with the potential of:

- bringing effective and easy treatment to remote areas
- enabling effective treatment of smaller tumors: procedure can be performed intraoperatively when optimal surgical resection is not possible
- completely halt cancer growth or significantly diminish its size
- controlling nanoseeds size to prevent diffusion, focusing radiation effects on the target

This chapter presented the basics of radiation physics and the routes being investigated to produce radioactive nanoparticles. Efforts have been made mostly with gold-198, palladium-106, indium-111, and lutetium-177. Practically, all results reported were successful in cancer progression halt or even completely cancer cure. The potential is huge since the treatment is injectable and radiation activity is considerably low.

Necrosis was found in the skin of the animals injected with alginate-PEG-^{103}Pd: Pd@^{198}AuNPs [97]. This is due to gold-198 high gamma emission energy. Necrosis confirms the impact of the treatment on tumor volume control but elucidates the difficulty of circumspect the radiotherapeutic treatment to the exact site of the tumor tissue.

There are still several steps that need to be developed so that nanobrachytherapy can be implemented on large scale. They are:

(a) Stability tests: radioactive nanoparticles need to be proven stable from the production line to the treatment place.

(b) Mercury contamination assay: gold-198 decays to mercury, which is highly toxic. Effects of this on the cells are yet to be evaluated.

(c) Nanotoxicity: evaluation of the toxicity level of nonradioactive nanoparticle counterparts needs detailed investigation.

(d) More in vitro testing: a complete assessment of inflammatory and immunity response needs further evaluation.

(e) More in vivo testing: by evaluating possible routes and understanding the nanomaterials transport it will be possible to evaluate the distribution of radioactive nanoparticles in the human body after systemic administration and reproducible demonstration of the therapeutic efficiency.

(f) Dosimetry: it is imperative that models are created that can accurately measure radiation dose delivered to the target and healthy surrounding tissues. Protocols comprising of different dosimetry studies, simulation-based dosimetry systems, and computer treatment planning systems must be investigated/created.

(g) Industrial or semi-industrial laboratory construction: tabletop research is very important because it validates the technique. However, fabrication on a large-scale, to provide to hospitals and clinics, is a different issue on its own. Methods that are too complicated may not be translated to an industrial scale. Also, these new products need to be produced with the highest level of precision under good manufacturing practices and by respecting radiation protection demands.

Even with all these challenges, some non-radioactive NPs are being investigated in clinical studies for cancer treatment. On the U.S. National Library of Medicine website (www.clinicaltrials.gov) a list of compounds in different stages (Phase I, II, and III oncology studies) can be found. As an example, AuroLase (NCT02680535) is a therapy project by Nanospectra Biosciences based on PEG-Si@AuNPs, that uses focal ablation of neoplastic prostate tissue via nanoparticle directed irradiation. Although not radioactive, these clinical studies open the door to more clinical investigations in the field, including with radioactive NPs. For nanobrachytherapy to become a viable technique, besides the higher monetary investment in this area of research, there is no doubt that interdisciplinary teams are of the utmost necessity.

References

[1] ScienceDirect, Nanoparticle search engine results, 2019 (cited 2019 Sept. 24).

[2] Alves, O.L., Cartilha sobre nanotecnologia, Agencia Brasileira de Desenvolvimento Industrial, Editor. 2011: Brasília.

[3] C. Daruich De Souza, B. Ribeiro Nogueira, M.E.C.M. Rostelato, Review of the methodologies used in the synthesis gold nanoparticles by chemical reduction, J. Alloys Compd. 798 (2019) 714–740.

[4] M. Adams, Nanoparticles Technology Handbook, first ed., NY Research Press, New York, 2015, p. 346.

[5] M.-C. Daniel, D. Astruc, Gold Nanoparticles: assembly, supramolecular chemistry, quantum-size-related properties, and applications toward biology, catalysis, and nanotechnology, Chem. Rev. 104 (1) (2004) 293–346.

[6] D.R. Baer, et al., Surface characterization of nanomaterials and nanoparticles: Important needs and challenging opportunities, J. Vacuum Sci. Technol. A, Vacuum, Surf. Films: Off. J. Am. Vacuum Soc. 31 (5) (2013) 50820.

[7] J. Zhao, M.H. Stenzel, Entry of nanoparticles into cells: the importance of nanoparticle properties, Polym. Chem. 9 (3) (2018) 259–272.

[8] M.E.C.M. Rostelato, et al., Surface coating and study of metallic cores for radioactive sources production used in cancer treatment, in: International Conference on Advanced Material Science and mechanic engineering –AMSME, Bangkok, 2016.

[9] A.S.M. Thorn, The Impact of Nanoparticle Surface Chemistry on Biological Systems, University of Iowa, 2017. p. 136.

[10] H.E. Toma, et al., The coordination chemistry at gold nanoparticles, J. Braz. Chem. Soc. 21 (2010) 1158–1176.

[11] E.B. Podgorsak, Radiation Oncology Physics: A Handbook for Teachers and Students, first ed., International Atomic Energy Agency, Viena, 2005.

[12] C.D. de Souza, Materials for the course: TNA 5744 dicipline—Applications of Intense Radiation Sources, 2019.

[13] C.D. de Souza, Materials for the course: TNA5805 Brachytherapy: Fundamentals, Production, Application, Dosimetry and Quality, 2019.

[14] C.D. de Souza, et al., New gold-198 nanoparticle synthesis to be used in cancer treatment, Brazilian J. Radiat. Sci. 9 (01-A) (2021) 18.

[15] C.D. de Souza, et al., New core configuration for the fabrication of 125I radioactive sources for cancer treatment, Appl. Radiat. Isot. 165 (2020) 109307.

[16] W.E. Meyerhof, Elements of Nuclear Physics, McGraw-Hill Book Company, New York, 1989.

[17] E.F. Pessoa, F.A.B. Coutinho, O. Sala, Introdução à Física Nuclear, Editora da Universidade de São Paulo/McGraw-Hill-Brasil, São Paulo, 1978.

[18] Departamento de Física Nuclear. Instituto de Física da Universidade de São Paulo, Decaimento Alfa, 2019 (cited 2019 Sept, 25).

[19] Berkeley Lab, Alpha Decay. Guide to Nuclear Wallchart 2000, 2000 (cited 2019 25 sept).

[20] Departamento de Física Nuclear. Instituto de Física da Universidade de São Paulo, Decaimento Beta, 2019 (cited 2019 Sept, 25).

[21] Berkeley Lab, Beta Decay. Guide to Nuclear Wallchart 2000, 2000 (cited 2019 25 sept).

[22] Departamento de Física Nuclear. Instituto de Física da Universidade de São Paulo, Decaimento Gama, 2019 (cited 2019 Sept, 25).

[23] Berkeley Lab, Gamma Decay. Guide to Nuclear Wallchart 2000, 2000 (cited 2019 Sept 25).

[24] F.A. Mettler Jr., A.C. Upton, Medical Effects of Ionizing Radiation, 2nd, Saunders, Philadelphia, PA, 1995.

[25] N. Nishida, et al., Angiogenesis in cancer, Vasc. Health Risk Manag. 2 (3) (2006) 213–219.

[26] L. Holmgren, M.S. O'Reilly, J. Folkman, Dormancy of micrometastases: balanced proliferation and apoptosis in the presence of angiogenesis suppression, Nat. Med. 1 (2) (1995) 149–153.

[27] S. Parangi, et al., Antiangiogenic therapy of transgenic mice impairs de novo tumor growth, Proc. Natl. Acad. Sci. U. S. A. 93 (5) (1996) 2002–2007.

[28] E.B. Podgorsak, Radiation Oncology Physics: A Handbook For Teachers And Students, International Atomic Energy Agency, Viena, 2005.

[29] Centro de Tecnologia das Radiações—IPEN, et al., Relatório Técnico 01—Acordo Baterias Nucleares, Instituto de Pesquisas Energéticas e Nucleares, 2019.

[30] L. Beaulieu, et al., Report of the Task Group 186 on model-based dose calculation methods in brachytherapy beyond the TG-43 formalism: Current status and recommendations for clinical implementation (Task Group No. 186), Med. Phys. 39 (10) (2012) 28.

[31] A.V. Rane, et al., Chapter 5—Methods for synthesis of nanoparticles and fabrication of nanocomposites, in: S.M. Bhagyaraj, et al. (Eds.), Synthesis of Inorganic Nanomaterials, Woodhead Publishing, 2018, pp. 121–139.

[32] J. Polte, Fundamental growth principles of colloidal metal nanoparticles—a new perspective, CrstEngComm 17 (36) (2015) 6809–6830.

[33] Turkevich, J., P.C. Stevenson, and J. Hillier, A study of the nucleation and growth processes in the synthesis of colloidal gold. Discuss. Faraday Soc., 1951. 11(0): p. 55–75.

[34] P. Kalimuthu, S.A. John, Studies on ligand exchange reaction of functionalized mercaptothiadiazole compounds onto citrate capped gold nanoparticles, Mater. Chem. Phys. 122 (2–3) (2010) 380–385.

[35] C.R. Raj, T. Okajima, T. Ohsaka, Gold nanoparticle arrays for the voltammetric sensing of dopamine, J. Electroanal. Chem. 543 (2) (2003) 127–133.

[36] M. Brust, G.J. Gordillo, Electrocatalytic hydrogen redox chemistry on gold Nanoparticles, J. Am. Chem. Soc. 134 (7) (2012) 3318–3321.

[37] M. Brust, et al., Synthesis of thiol-derivatised gold nanoparticles in a two-phase Liquid-Liquid system, J. Chem. Soc. Chem. Commun. 7 (1994) 801–802.

[38] N.R. Jana, L. Gearheart, C.J. Murphy, Seeding growth for size control of 5 − 40 nm diameter gold Nanoparticles, Langmuir 17 (22) (2001) 6782–6786.

[39] A.S. Patel, et al., Gold nanoflowers as efficient hosts for SERS based sensing and bio-imaging, Nano-Struct. Nano-Obj. 16 (2018) 329–336.

[40] Sonia, et al., Exploring the DNA damaging potential of chitosan and citrate-reduced gold nanoparticles: Physicochemical approach, Int. J. Biol. Macromol. 115 (2018) 801–810.

[41] M.S. Bakshi, How surfactants control crystal growth of nanomaterials, Cryst. Growth Des. 16 (2) (2016) 1104–1133.

[42] S. Kumar, Role of Surfactants in Synthesis and Stabilization of Nanoparticles Spectroscopic and Physicochemical Aspects, Panjab University, 2011. p. 218.

[43] S.B. Awan, et al., Historical review of interstitial prostate brachytherapy, Int. J. Radiat. Res. 5 (4) (2008) 153–168.

[44] Science and Technology Branch, in: H.G. Donald (Ed.), Bibliography from Nuclear Science Abstracts, Vol. 21, United States Atomic Energy Commission, 1968. https://books.googleusercontent.com/ books/content?req=AKW5QafMlZ0ZqnEcNom8f20jjXeJwvLqUOCfxkXl4IP8IMtStBNSsZPlZ 5Srry_zoXqQ3RjJ2pwlSXbFJxp6RuuP84wbPjjzDbhilJ6zUPAGZdd27rumDeonrsaJZaW4v sh0GNJ7Ah–C–plbQIVNJAibCe2Rf86SEfgjFDv4KuXPLV08gxKAvtr2fuJNSrdPupZRimYBhV Kbn83BLOvkwKl3gKvwcpc_TOJ5vHpzFeZAW1aOjB5QV1ulX48Dm81LZFHmBVQ0xerv AoeUhRurszB3gTte4Qdirg.

[45] C.D. Souza, et al., New Gold-198 nanoparticle synthesis to be used in cancer treatment, in: 60th AAPM Annual Meeting and Exhibition, Nashville, USA, 2018.

[46] I. Freestone, et al., The Lycurgus cup—a Roman nanotechnology, Gold Bull. 40 (4) (2007) 270–277.

[47] L.A. Dykman, N.G. Khlebtsov, Gold nanoparticles in biology and medicine: recent advances and prospects, Acta Nat. 3 (2) (2011) 34–55.

[48] J.J. Storhoff, et al., Gold nanoparticle-based detection of genomic DNA targets on microarrays using a novel optical detection system, Biosens. Bioelectron. 19 (8) (2004) 875–883.

[49] Y. Li, H.J. Schluesener, S. Xu, Gold nanoparticle-based biosensors, Gold Bull. 43 (1) (2010) 29–41.

[50] X. Liu, et al., A one-step homogeneous immunoassay for cancer biomarker detection using gold nanoparticle probes coupled with dynamic light scattering, J. Am. Chem. Soc. 130 (9) (2008) 2780–2782.

[51] P.V. Baptista, et al., Gold-nanoparticle-probe-based assay for rapid and direct detection of *Mycobacterium tuberculosis* DNA in clinical samples, Clin. Chem. 52 (7) (2006) 1433–1434.

[52] Y. Cheng, et al., Highly efficient drug delivery with gold nanoparticle vectors for in vivo photodynamic therapy of cancer, J. Am. Chem. Soc. 130 (32) (2008) 10643–10647.

[53] S. Thambiraj, S. Hema, D. Ravi Shankaran, Functionalized gold nanoparticles for drug delivery applications, Mater. Today: Proc. 5 (8, Part 3) (2018) 16763–16773.

[54] W. Lu, et al., Gold nano-popcorn-based targeted diagnosis, nanotherapy treatment, and in situ monitoring of photothermal therapy response of prostate cancer cells using surface-enhanced Raman spectroscopy, J. Am. Chem. Soc. 132 (51) (2010) 18103–18114.

[55] M. Faraday, The Bakerian lecture: experimental relations of gold (and other metals) to light, Philos. Trans. R. Soc. Lond. A 147 (1857) 145–181.

[56] S. Jain, D.G. Hirst, J.M. O'Sullivan, Gold nanoparticles as novel agents for cancer therapy, Br. J. Radiol. 85 (1010) (2012) 101–113.

[57] P.F. Hahn, et al., Intravenous radioactive gold in the treatment of chronic leukemia, Acta Radiol. 50 (6) (1958) 565–572.

[58] Hahn, P.F. and E.L. Carothers, Use of radioactive colloidal metallic gold in the treatment of malignancies. Nucleonics, 1950. 6(1): p. 54–62, illust.

[59] P.F. Hahn, J.P. Goodell, et al., Direct infiltration of radioactive isotopes as a means of delivering ionizing radiation to discrete tissues, J. Lab. Clin. Med. 32 (12) (1947) 1442–1453.

[60] J.H. Muller, Medical therapeutic use of artificial radioactivity, Bull. Schweiz. Akad. Med. Wiss. 5 (5–6) (1949) 484–510.

[61] P.C. Lee, D. Meisel, Adsorption and surface-enhanced Raman of dyes on silver and gold sols, J. Phys. Chem. 86 (17) (1982) 3391–3395.

[62] N. Chanda, et al., Radioactive gold nanoparticles in cancer therapy: therapeutic efficacy studies of GA-198AuNP nanoconstruct in prostate tumor-bearing mice, Nanomedicine 6 (2) (2010) 201–209.

[63] X. Chen, et al., Formation and catalytic activity of spherical composites with surfaces coated with gold nanoparticles, J. Colloid Interface Sci. 322 (2) (2008) 414–420.

[64] C.D. de Souza, Relatório de Pós-Doutorado, Instituto de Pesquisas Energéticas e Nucleares, 2019.

[65] International Atomic Nuclear Agency, Live Chart of Nuclides: nuclear structure and decay data, 2019 (cited 2020, 26 Feb).

[66] K.V. Katti, et al., Hybrid gold nanoparticles in molecular imaging and radiotherapy, Czechoslov. J. Phys. 56 (4) (2006) D23–D34.

[67] R. Shukla, et al., Laminin receptor specific therapeutic gold nanoparticles (198AuNP-EGCg) show efficacy in treating prostate cancer, Proc. Natl. Acad. Sci. U. S. A. 109 (31) (2012) 12426–12431.

[68] N. Rao, et al., Isolation of a tumor cell laminin receptor, Biochem. Biophys. Res. Commun. 111 (3) (1983) 804–808.

[69] J.C. Blasko, et al., Palladium-103 brachytherapy for prostate carcinoma, Int. J. Radiat. Oncol. Biol. Phys. 46 (4) (2000) 839–850.

[70] K.R. Gopidas, J.K. Whitesell, M.A. Fox, Synthesis, characterization, and catalytic applications of a palladium-nanoparticle-cored dendrimer, Nano Lett. 3 (12) (2003) 1757–1760.

[71] Y.E. Cheon, M.P. Suh, Enhanced hydrogen storage by palladium nanoparticles fabricated in a redox-active metal-organic framework, Angew. Chem. Int. Ed. 48 (16) (2009) 2899–2903.

[72] X.-M. Chen, et al., Nonenzymatic amperometric sensing of glucose by using palladium nanoparticles supported on functional carbon nanotubes, Biosens. Bioelectron. 25 (7) (2010) 1803–1808.

[73] J. Cookson, The preparation of palladium nanoparticles, Platin. Met. Rev. 56 (2) (2012) 83–98.

[74] J. Cookson, The preparation of palladium nanoparticles, Platinum Met. Rev. 56 (2) (2012) 15.

[75] S.-W. Kim, et al., Synthesis of monodisperse palladium nanoparticles, Nano Lett. 3 (9) (2003) 1289–1291.

[76] C.K. Yee, et al., Novel one-phase synthesis of thiol-functionalized gold, palladium, and iridium Nanoparticles using Superhydride, Langmuir 15 (10) (1999) 3486–3491.

[77] D.E. Cliffel, et al., Mercaptoammonium-monolayer-protected, water-soluble gold, silver, and palladium clusters, Langmuir 16 (25) (2000) 9699–9702.

[78] M. Tamura, H. Fujihara, Chiral Bisphosphine BINAP-stabilized gold and palladium nanoparticles with small size and their palladium nanoparticle-catalyzed asymmetric reaction, J. Am. Chem. Soc. 125 (51) (2003) 15742–15743.

[79] V. Mazumder, S. Sun, Oleylamine-mediated synthesis of Pd nanoparticles for catalytic formic acid oxidation, J. Am. Chem. Soc. 131 (13) (2009) 4588–4589.

[80] H. Shen, et al., Novel glycosyl pyridyl-triazole@ palladium nanoparticles: efficient and recoverable catalysts for C–C cross-coupling reactions, Cat. Sci. Technol. 5 (4) (2015) 2065–2071.

[81] R.A. Molla, et al., Mesoporous poly-melamine-formaldehyde stabilized palladium nanoparticle (Pd@mPMF) catalyzed mono and double carbonylation of aryl halides with amines, RSC Adv. 4 (89) (2014) 48177–48190.

[82] Y.M. Yamada, S.M. Sarkar, Y. Uozumi, Self-assembled poly (imidazole-palladium): highly active, reusable catalyst at parts per million to parts per billion levels, J. Am. Chem. Soc. 134 (6) (2012) 3190–3198.

[83] S.M. Sarkar, Y. Uozumi, Y.M. Yamada, A highly active and reusable self-assembled poly (imidazole/palladium) catalyst: allylic Arylation/Alkenylation, Angew. Chem. Int. Ed. 50 (40) (2011) 9437–9441.

[84] D. Astruc, Palladium nanoparticles as efficient green homogeneous and heterogeneous carbon−−carbon coupling precatalysts: a unifying view, Inorg. Chem. 46 (6) (2007) 1884–1894.

[85] W. Kleist, J.K. Lee, K. Köhler, Pd/MOx materials synthesized by sol–gel Coprecipitation as catalysts for carbon–carbon coupling reactions of aryl bromides and chlorides, Eur. J. Inorg. Chem. 2009 (2) (2009) 261–266.

[86] J. Bennett, et al., Nanoparticles of palladium supported on bacterial biomass: new re-usable heterogeneous catalyst with comparable activity to homogeneous colloidal Pd in the heck reaction, Appl. Catal. Environ. 140 (2013) 700–707.

[87] M.S. Carvalho, et al., In situ generated palladium nanoparticles in imidazolium-based ionic liquids: a versatile medium for an efficient and selective partial biodiesel hydrogenation, Cat. Sci. Technol. 1 (3) (2011) 480–488.

[88] B.S. Souza, et al., Selective partial biodiesel hydrogenation using highly active supported palladium nanoparticles in imidazolium-based ionic liquid, Appl. Catal. A. Gen. 433 (2012) 109–114.

[89] S. Jones, et al., Prominent electronic and geometric modifications of palladium nanoparticles by polymer stabilizers for hydrogen production under ambient conditions, Angew. Chem. Int. Ed. 51 (45) (2012) 11275–11278.

[90] L. Wu, et al., Phosphine dendrimer-stabilized palladium nanoparticles, a highly active and recyclable catalyst for the Suzuki-Miyaura reaction and hydrogenation, Org. Lett. 8 (16) (2006) 3605–3608.

[91] A. Prastaro, et al., Suzuki-Miyaura cross-coupling catalyzed by protein-stabilized palladium nanoparticles under aerobic conditions in water: application to a one-pot chemoenzymatic enantioselective synthesis of chiral biaryl alcohols, Green Chem. 11 (12) (2009) 1929–1932.

[92] G. Marcelo, A. Muñoz-Bonilla, M. Fernández-García, Magnetite-polypeptide hybrid materials decorated with gold nanoparticles: study of their catalytic activity in 4-nitrophenol reduction, J. Phys. Chem. C 116 (46) (2012) 24717–24725.

[93] L.A. Gugliotti, D.L. Feldheim, B.E. Eaton, RNA-mediated metal-metal bond formation in the synthesis of hexagonal palladium nanoparticles, Science 304 (5672) (2004) 850–852.

[94] International Atomic Energy Agency, Technical Reports Series No. 468: Cyclotron Produced Radionuclides: Physical Characteristics and Production Methods, 2009. https://www-pub.iaea.org/MTCD/publications/PDF/trs468_web.pdf.

[95] International Atomic Energy Agency, Cyclotron Produced Radionuclides: Physical Characteristics And Production Methods, 2009. p. 2009.

[96] D. Djoumessi, et al., Rapid, one-pot procedure to synthesise 103Pd:Pd@au nanoparticles en route for radiosensitisation and radiotherapeutic applications, J. Mater. Chem. B 3 (10) (2015) 2192–2205.

[97] M. Laprise-Pelletier, et al., Low-dose prostate cancer brachytherapy with radioactive palladium-gold nanoparticles, Adv. Healthc. Mater. 6 (4) (2017) 24.

[98] S. Moeendarbari, et al., Theranostic Nanoseeds for efficacious internal radiation therapy of Unresectable solid tumors, Sci. Rep. 6 (2016) 20614.

[99] Z. Cai, et al., 111In-labeled trastuzumab-modified gold nanoparticles are cytotoxic in vitro to HER2-positive breast cancer cells and arrest tumor growth in vivo in athymic mice after intratumoral injection, Nucl. Med. Biol. 43 (12) (2016) 818–826.

[100] R.A. Kuznetsov, et al., Production of Lutetium-177: process aspects, Radiochemistry 61 (4) (2019) 381–395.

[101] Z. Cai, et al., Local radiation treatment of HER2-positive breast cancer using Trastuzumab-modified gold Nanoparticles labeled with 177Lu, Pharm. Res. 34 (3) (2017) 579–590.

[102] S. Yook, et al., Radiation nanomedicine for EGFR-positive breast Cancer: Panitumumab-modified gold Nanoparticles complexed to the β-particle-emitter, 177Lu, Mol. Pharm. 12 (11) (2015) 3963–3972.

[103] A. Vilchis-Juarez, et al., Molecular targeting radiotherapy with cyclo-RGDFK(C) peptides conjugated to 177Lu-labeled gold nanoparticles in tumor-bearing mice, J. Biomed. Nanotechnol. 10 (3) (2014) 393–404.

PART 3

Environmental and regulatory aspects

CHAPTER 16

Regulatory pathways and federal perspectives on nanoparticles

Ajaz Hussain[a,b,c] and Sarwar Beg[d,e]
[a]Private Practice Insight, Advice, and Solutions, Frederick, MD, United States
[b]Former President of the National Institute for Pharmaceutical Technology and Education, Minneapolis, MN, United States
[c]Deputy Director, Office of Pharmaceutical Science, US FDA, Silver Spring, MD, United States
[d]Department of Pharmaceutics, School of Pharmaceutical Education and Research, Jamia Hamdard (Hamdard University), New Delhi, India
[e]School of Pharmacy and Biomedical Sciences, Faculty of Clinical and Biomedical Sciences, University of Central Lancashire, Preston, United Kingdom

1. Introduction

Why, scientists learning to reduce technical uncertainty about nanoparticles—concerning their measurements, *in vitro* and in vivo behaviors, toxicity, pharmacology, and production on an industrial scale hold valuable lessons useful beyond their "reductionist (work) compartment" in a corporation or an institution? How can they, in taking a design and systems thinking stance, better serve the world—a sociotechnical system beyond their corporate management system? This narrative spirals about these two questions to help readers recognize that emergence is a characteristic feature of a complex system, and that chaos is a system that exhibits extreme sensitivity to starting conditions and is unpredictable beyond the mean of normal and abnormal distributions. A good practice is any valid procedure consisting of an appropriately calibrated measurement system and workflow instructions to reproduce and repeatedly yield predicted or expected outcomes. In contrast to complicated and simple systems where good and best practices apply, it is not feasible to assure good practices in a complex or chaotic system. Why so and why the journey ahead is best titled "Chaos to Continual Improvement," as was the sense in the pharma sector before the COVID-19 pandemic are interesting questions in the context of nanoparticles.

The role of nanoparticles is at the center of the COVID-19 pandemic; most commonly, the world recognizes the role of lipid-nanoparticles in delivering mRNA-based vaccines, but other numerous applications are waiting to be realized. Amidst this public health emergency, nanotechnology-based vaccines and therapeutics are at the cutting-edge of solutions we seek to "get back to normal." What is "normal"—a (statistical) distribution or simply complying or conforming to a usual or a typical standard? In the more common meaning of chaos, as in making sense—economics, politics, regulations, and science, in sociotechnical systems we live and work how to anticipate (and therefore, plan to be prepared for) the emerging new "normal"? [1]. To begin our exploration to observe from a high level the federal and regulatory systems to appreciate a national perspective.

The innovations in the domain of nanotechnology are keeping high pace in the last few decades. This has led to the development of several nanotechnology-based products into existence. These primarily include nanoparticles, nanomaterials, and nanocomposites made from polymers, lipids, biomaterials, metals, and other related substances with applications in diverse fields including energy, water purification, waste management, automobiles, agriculture, medical diagnostics, and therapeutic applications, especially in drug delivery and disease treatment. Despite the benefits of nanotechnology have been accelerating at an exponential pace for benefits of mankind, yet the safety of nanopharmaceuticals is a topic of debate. As far as the medical applications of nanotechnology are concerned, the safety requirements have much higher significance. Moreover, the impact of nanopharmaceutical's use on environmental set-up also possesses equal importance, as alteration in environmental harmony also causes disturbances in the human health. In this regard, strict regulations have been adopted by the regulatory agencies in the last few years for controlled production and their usage with utmost care for avoiding human health and environmental risk. The present chapter particularly deals with the guidelines on safe and effective use of nanopharmaceutical products to make the readers attuned with the current regulations.

This chapter aims to offer an opportunity to science interns and scientists to expand their perspective and appreciate in the regulatory context the journey to be "good" practitioners. To do so, it provides an opportunity to observe efforts of a leading federal agency, the US FDA, to answer—what are traditional regulatory channels to bring nanotechnology-enabled healthcare products to market, and how might these pathways evolve, in and beyond the COVID-19 pandemic.

The US FDA is a federal agency, and federal relates to a governing system. Perspective is a viewpoint, as it is a particular attitude toward or way of regarding something. Fundamental knowledge entails how to think intrinsically about risk and uncertainty. In a dynamic and uncertain environment, as in the case of the chaos we are experiencing in the COVID-19 pandemic, there are unchanging principles and fundamental knowledge. We can and must hold on to the *first principle—first, do no harm* to find Order in chaos. This chapter hones in on the fundamentals while offering a 360-degree view of ongoing regulatory efforts with a specific focus on professional development via learning from experience. Note that it usually takes time and effort to appreciate what are "good practices" and understand why these are essential to science, broadly and specifically for therapeutic products. A "federal perspective" and how it, within the authority of existing laws, influences regulations provides a framework to understand and leverage regulatory pathways. Furthermore, understanding why it often takes a crisis to innovate and to change or establish new laws is also useful. The public health response to the COVID-19 pandemic, regardless of how history will judge it, is an experiential learning opportunity like no other. Let us leverage it to know deeply why a profession is not an occupation and why we, a community of knowledge, call our field of inquiry a discipline.

As per the US National Science Foundation (NSF) survey, it has been estimated that the market size of nanotechnology-enabled products in US market itself accounted around $92 billion in the current year and is exponentially increasing every year [2]. With the growing demand and expansion in the market size, the two major health and safety agencies of US including Environmental Protection Agency (EPA) and Food & Drug Administration (FDA) are continuously working in this area for standardizing the production and usage of nanomaterials for human use.

The major concern arises on the regulation of nanotechnology-based products that are available in market for direct human use, be it the cosmetic products for external applications or the pharmaceutical products for therapeutic benefits. In this context, agencies have already framed regulatory policies for safe use of nanoparticles, along with regulation regarding production, handling, or labeling.

2. Management of human health and environmental risks

The management of human health is considered as the foremost important part of fixing health care policies for human safety and benefits. Several studies have been published on the management of the health impact of airborne particles including smaller particles of nondimensional size, and results have shown that these are more toxic than the toxic materials [1]. This is due to the same mass per volume, the dose in terms of particle numbers increases as particle size decreases. Based upon available data, it has been argued that current risk assessment methodologies are not suited to the hazards associated with nanoparticles; in particular, existing toxicological and eco-toxicological methods are not up to the task; exposure evaluation (dose) needs to be expressed as quantity of nanoparticles and/or surface area rather than simply mass; equipment for routine detecting and measuring nanoparticles in air, water, or soil is inadequate; and very little is known about the physiological responses to nanoparticles.

Regulatory bodies in the US as well as in the EU have concluded that nanoparticles form the potential for an entirely new risk and that it is necessary to carry out an extensive analysis of the risk. The challenge for regulators is whether a matrix can be developed, which would identify nanoparticles, and more complex nanoformulations, which are likely to have special toxicological properties or whether it is more reasonable for each particle or formulation to be tested separately.

3. Regulatory considerations on nanotechnology

FDA defines nanotechnology products as the engineered material or end product has at least one dimension in the nanoscale range (approximately 1–100 nm) or exhibits properties or phenomena, including physical or chemical properties or biological effects, which are attributable to its dimension(s), even if these dimensions fall outside the nanoscale range, up to 1 µm [3]. However, European commission recommendation classify nanomaterials as means a natural, incidental or manufactured material containing particles, in an unbound state or as an aggregate or as an agglomerate and where, for 50% or more of the particles in the number size distribution, one or more external dimensions is in the size range 1–100 nm [4].

Moreover, the first question arises in this regard is how best nanotechnology should be regulated depends on whether nanotechnology represents something "new" must be answered to decide. The international groups including the Royal Society recommend that UK government assess chemicals in the form of nanoparticles or nanotubes as new substances. In 2007, the society has registered nearly 40 groups called for nanomaterials and classified them under new substances for regulating them [5].

Over the years, there is significant debate on the issue of who is responsible and accountable for the regulation of nanotechnology. Some non-nanotechnology specific regulatory agencies have currently covered some products and processes up to varying degrees—by "bolting on" nanotechnology to existing regulations. Moreover, USFDA has also reviewed immediate health effects of exposure to nanoparticles of TiO_2 nanoparticles in consumer products [6]. However, the review suggested that the impacts of such nanoparticles have been there on the aquatic ecosystems when the sunscreen rubs off and with water it goes to ecosystem. Similar to the Australian equivalent of the FDA, the Therapeutic Goods Administration (TGA) has also approved the use of nanoparticles in sunscreens after a thorough review of literature on the basis that although nanoparticles of TiO_2 and ZnO in sunscreens do produce free radicals and oxidative DNA damage in vitro, such particles were unlikely to pass the dead outer cells of the stratum corneum of human skin [7].

Moreover, above all regulatory agencies, US FDA has revealed that the agency only regulates on the basis of voluntary claims made by the product manufacturer regarding confidentiality disclosure of nature of the product containing nanotechnology elements. If no claims are made by a manufacturer, the FDA may be unaware of nanotechnology being employed. Debate is going on this to have standard guidance for making uniform regulations for controlling the nanomaterials for the purpose.

4. International laws on regulation of nanopharmaceutical products

The International Council on Nanotechnology maintains a database and Virtual Journal of scientific papers on environmental, health, and safety research on nanoparticles. The database currently has over 2000 entries indexed by particle type, exposure pathway, and other criteria. The Project on Emerging Nanotechnologies (PEN) currently lists 807 products in which the manufacturers have voluntarily identified the use of nanotechnology principles [8].

The supplier needs to provide the mandatory Material Safety Data Sheet for the materials to differentiate their nanoscale size and properties.

Despite these recommendations, chemicals comprising nanoparticles have previously been subjected to assessment and regulation may be exempt from regulation, regardless of the potential for different risks and impacts. In contrast, nanomaterials are often recognized as "new" from the perspective of intellectual property rights (IPRs) and as such are commercially protected via patenting laws.

Since products that are produced using nanotechnologies will likely enter international trade, it is argued that it will be necessary to harmonize nanotechnology standards across national borders. There is concern that some countries, most notably developing countries, will be excluded from international standards negotiations. The Institute for Food and Agricultural Standards notes that "developing countries should have a say in international nanotechnology standards development, even if they lack capacity to enforce the standards."

5. Regulation on technical aspects of nanotechnology

These existing approval frameworks almost universally use the best available science to assess safety and do not approve substances or products with an unacceptable risk benefit profile. One proposal is to simply treat particle size as one of the several parameters defining a substance to be approved, rather than creating special rules for all particles of a given size regardless of its type. A major argument against special regulation of nanotechnology is that the projected applications with the greatest impact are far in the future, and it is unclear how to regulate technologies whose feasibility is speculative at this point. In the meantime, it has been argued that the immediate applications of nanomaterials raise challenges not much different from those of introducing any other new material and can be dealt with by minor tweaks to existing regulatory schemes rather than sweeping regulation of entire scientific fields.

5.1 Size related aspects

Regulation of nanotechnology will require a definition of the size, in which particles and processes are recognized as operating at the nanoscale. The size-defining characteristic of nanotechnology is the subject of significant debate and varies to include particles and materials in the scale of at least 100–300 nm. Friends of the Earth Australia recommend defining nanoparticles up to 300 nm in size. They argue that "particles up to a few hundred nanometers in size share many of the novel biological behaviors of nanoparticles, including novel toxicity risks," and that "nanomaterials up to approximately 300 nm in size can be taken up by individual cells." The UK Soil Association defines nanotechnology to include manufactured nanoparticles where the mean particle size is 200 nm or smaller. The US National Nanotechnology Initiative defines nanotechnology as "the understanding and control of matter at dimensions of roughly 1–100 nm."

5.2 Mass volume thresholds

Regulatory frameworks for chemicals tend to be triggered by mass thresholds. This is certainly the case for the management of toxic chemicals in Australia through the National pollutant inventory. However, in the case of nanotechnology, nanoparticle applications are unlikely to exceed these thresholds (tons/kilograms) due to the size and weight of nanoparticles. As such, the Woodrow Wilson International Center for Scholars questions the usefulness of regulating nanotechnologies on the basis of their size/weight alone. They argue, for example, that the toxicity of nanoparticles is more related to surface area than weight, and that emerging regulations should also take account of such factors.

6. The US FDA's policies and practices: Product-focused and science-based

The entire gamut of the products regulated by the US FDA—foods, cosmetics, drugs, devices, veterinary products, and tobacco products can contain nanomaterials. These sectors can utilize nanotechnology in their development and manufacture. The US FDA shares with the public an overview of current FDA activities in regulating nanomaterials and nanotechnology-enabled products and guidance documents on its internet website [9].

For a nonregulatory to appreciate regulatory science-based risk assessment, let us begin with an overview of the federal perspective and its influence on regulations. The US FDA is part of the Executive branch of the US federal governance system, and it adopts and enforces regulations that impose legally binding requirements on the public. Only Congress (the Legislative branch), under the US Constitution, is vested with the legislative power. The President's authority to execute is vested in the Constitution and the prevailing laws. Regulatory agencies develop regulations based on the statutes under which they operate and also issue guidance documents to share their current thinking on specific contemporary topics. For example, the US FDA develops regulations based on the laws outlined in the Food, Drug, and Cosmetic Act (FD&C Act) or other laws—including the Public Health Service Act (PHS Act) —under which it operates.

6.1 Regulations, guidance, and executive orders

Regulations have the full force of law. Guidance is current thinking, is not binding, and one can justify alternative approaches. However, most do not or are unable to pursue alternate, more efficient, and innovative methods. Why? Because they perceive regulatory uncertainty will increase the risk of delayed time-to-market.

Presidential Executive Order 13563, dated 18 January 2011, requires all US regulatory agencies to improve regulations and regulatory review. This Order outlines the general principles of regulations. It directs the attention of regulatory agencies to the specific intent of the federal government, such as to involve public participation, integration and innovation, flexible approaches, scientific integrity, and retrospective analysis of existing rules [10].

As Presidents change, the interpretation of the spirit of laws and emphasis changes, yet it must remain within the Constitution and the letter of established acts. On 9 October 2019, an Executive Order (EO 13891) Promoting the Rule of Law Through Improved Agency Guidance Documents aimed to improve transparency in the development and utility of Guidance documents [11]. This Order came into effect a few months before the emergence of a novel coronavirus, the SARS-COV-2, and the COVID-19 pandemic.

6.2 Regulatory guidances on nanotechnology

More than a decade back, USFDA instituted the Nanotechnology Task Force to develop FDA center-specific guidance documents to help support the development of safe and effective nanomaterial-containing products [1]. To implement the recommendation, FDA need to organize data submitted on nanotechnology-based applications [1, 12]. The agency has published a manual of policies and procedures (MAPP) for submitting marketing authorization to nanotechnology-related information in CMC review. This MAPP provides a framework to CMC reviewers for reporting information about nanomaterial-containing drugs [13].

Of late, publication of several guidance documents occurred over the years on topics relating to the application of nanotechnology in FDA-regulated products [4]. One of the guidance listed on the FDA internet website—*Drug Products, Including Biological Products, that Contain Nanomaterials,* is still a "draft" guidance. It was issued in December 2017 [14]. It focuses on the deliberate and purposeful manipulation and control (i.e., intentional or by design) of dimensions to produce specific physicochemical properties, which may warrant further evaluation with regards to safety, effectiveness, performance, and quality.

Its premise is that:

(a) adequate characterization of the nanomaterials, and

(b) understanding how the attributes of nanomaterial relate to product quality, safety, and efficacy, in the context of the intended use and application, offers a sound scientific basis for evaluating its potential risk(s).

(c) biological materials that naturally occur at particle sizes ranging up to $1\,\mu m$ (1000 nm), for example, proteins, cells, viruses, nucleic acids, or other such as gene therapy or vaccine products. Admittedly, the exception being when such materials are manipulated or deliberated to have dimensions between 1 and 100 nm or to exhibit dimension dependent properties or phenomena up to 1 µm, and

(d) drug products that incidentally contain or may contain particles in the nanoscale range due to conventional manufacture or storage.

7. US FDA's risk evaluation and risk mitigation strategy in nanotechnology

It indicates that certain degree of risk evaluation and mitigation strategy is utmost required for controlling the production and evaluation of nanotechnology containing formulations. It indicates properties of and therefore the risks related to nanomaterials are to be considered unique (therefore requiring new regulations), then avenues for exposure, toxicity, or susceptibility must also be unique.

7.1 Product safety evaluation

In US, the FDA regulates cosmetic products under the Federal Food, Drug, and Cosmetic Act (FFDCA). In August 2006, the FDA Nanotechnology task force was formed to help in determining regulatory approaches for nanomaterials. This task force made various suggestions, including collecting data in early stages of the development process for products whenever nanoscale materials are used. In June 2011, the FDA has released a guidance document on its website. The FDA's risk assessment strategy includes considerations on the framed checklist, but not limited to the below mentioned points:

- whether an engineered material or end product has at least one dimension in the nanoscale range (approximately 1–100 nm); or
- whether an engineered material or end product exhibits properties or phenomena,
- including physical or chemical properties or biological effects that are attributable to its dimension(s), even if these dimensions fall outside the nanoscale range, up to 1 µm.

Thus, the regulations may hinge on demonstration that a nanoscale material imposes new properties on the drug that the bulk additive would not exhibit. This case history demonstrates that the definition of a nanomaterial may be different with different agencies. It also demonstrates that various agencies are at different stages in the road map, evolving from soft law to hard law.

8. US FDA 21st century initiatives and current nanotechnology regulations

8.1 National nanotechnology initiative

The goal of the National Nanotechnology Initiative (NNI) is "to move nanotechnology discoveries from the laboratory into new products for commercial and public benefit, encourage more students and teachers to become involved in nanotechnology education, create a skilled workforce and the supporting infrastructure and tools to advance nanotechnology and to support the responsible development of nanotechnology" [15]. The NNI consists of 25 federal agencies and has published various objectives to achieve the goals mentioned previously, which include:

- Establishing guidance standards or other methods to formulate nanotechnology-related regulatory approaches for domestic and global researchers, manufacturers, distributors, and users of nanotechnology-enabled products to ensure the protection of public health and environment.
- Obtaining stake holders perspectives by developing and using a variety of methods such as surveys, workshop, public meetings, and advisory panels; disseminating information through publicly accessible summaries of findings and developing mechanisms for integration EHS priorities and assessment methods into national and international regulatory policies.
- Building collaborations among the relevant expert communities such as regulators, ethicists, engineers, scientists including social and behavioral scientists, nongovernmental organizations, industry and consumers in order to support a rapid mobilization of stakeholders to consider the potential risk and benefits of research breakthroughs and provide perspectives on new research directions.

This body is responsible for a significant amount of the research funding pushing nano-development in the United States, and oversight is provided by a subcommittee under the Executive Office of the President of the United States. They recently issued a guidance document listing research needs in terms of risk assessment and governance. Although it would probably be easier as of this writing to list who is not influencing or lobbying the development of nanotechnology regulation in the United States, accessing the NNI Website may provide the interested reader with an idea of the breadth of the stakeholders and their agendas as regulation evolves.

9. ISO standards on nanotechnologies

As we saw earlier, regulation risks a maze of confusion if the simple definition of nano cannot be agreed upon. To ameliorate this issue, the ISO, coordinator of a systemized network of participants including more than 162 countries, issues standards guiding regulation and public policy. One new standard, ISO/TR 13121:2011, establishes and details appropriate processes for detection of nanoparticles and evaluation of nanomaterial risks [16]. The ISO/TC 229 has divided the standardization activities among four workgroups or subcommittees based on the following areas:

- Terminology and Nomenclature
- Measurement and Instrumentation
- Health Safety and Environment
- Material Specifications

10. FDA's critical path initiative for nanotechnology product regulations

Nanotechnology is an element under evaluation in FDA's initiative, and it has approved many products with particulate materials in the nanosized range. Most drugs are expected to go through a nanosized phase during the process of absorption in the body. There has been no safety concerns reported in the past because of particle size. The agency has highlighted some of the examples of nanomaterials on its current initiative that include the common carriers for "nanoscale" multifunctional therapeutics, including dendrimers, fullerenes, quantum dots, nanoshells, and liposomes [17, 18]. Key FDA-regulated products, in this regard, are expected to be impacted by nanotechnology include drugs (novel NMEs or delivery systems), medical devices, biotechnology products, tissue engineering products, vaccines, cosmetics, and combination products.

11. Other global regulatory agency initiatives

The EU has a number of key influencers, and as with the United States, the research funding body will exercise some regulatory guidance. Examples of these initiatives include:

- Norway: NANOMAT
- Finland: FinNano
- Germany: German Nanotechnology Program
- France: R3N, National Nanosciences Nanotechnology Network
- Switzerland: TopNano21
- Netherlands: NanoNed
- In Asia, these influential bodies include:
- China: Institute for NanoMaterials and Nanotechnology (INMT)
- Taiwan: National Science and Technology Program for Nanoscience and Nanotechnology
- South Korea: Nanotechnology Development Program (NDP);
- Japan: Nanotechnology Research Initiative (NRI)
- Thailand: The National Nanotechnology Center (NANOTEC)
- India: Nanomaterials Science and Technology Initiative (NSTI)

12. COVID-19-related guidance documents

The COVID-19 pandemic is an unprecedented crisis. The recurring lockdowns have dire consequences beyond the lives already lost. It is threatening the very livelihood, and it is contributing to geopolitical chaos. The SARS-CoV-2 is a novel coronavirus, and since its emergence, it is reported to have infected over 20 on million. It has been implicated in the death of over 750,000 individuals [19].

As the world learns about this novel virus, a frantic pace to understand mechanisms and disease pathology is ongoing to identify targets and develop strategies for COVID-19 diagnostics, therapeutics, and vaccine development. Potential target, for example, can be SARS-CoV-2S glycoprotein, which harbors a furin cleavage site at the boundary between the S1/S2 subunits and sets this virus apart from SARS-CoV and SARS-related corona viruses [20].

It is beyond the scope of this chapter to review or summarize ongoing efforts. To place our discussion on the regulation of nanotechnology in the context, it will be useful to glimpse the interrelationship, implicit and explicit, between regulatory guidance documents on nanotechnology and the many, about 59, guidance FDA has issued on topics related to COVID-19 [21]. For our discussion, the notion of design or intentional, deliberate, and purposeful manipulation and control (as emphasized above) can be the main conduit. Nanostrategies can overcome certain limitations of drug carriers via nanocarriers and, of course, include nanoparticle-based vaccines (e.g., a lipid nanoparticle formulation of mRNA, liposome-encapsulated mRNA, virus-like particles), vaccine adjuvant nanoparticles [22].

13. Nanoparticle therapeutics—Complex by FDA decree

The FDA considers nanoparticle therapeutics as complex dosage forms because assessments of therapeutic equivalence are still evolving. Delays in approval of complex generic drugs, drug shortages due to quality and manufacturing issues, and suboptimal competition resulting in increasing prices of generic drug products have been a dominant concern in the USA over the past two decades.

Before the COVID-19 pandemic, the affordability of healthcare and prescription medicines was reaching an intolerable level. Furthermore, evidence of an erosion of (epistemic) trust in FDA in approvals was emerging in reports of falling adherence rate upon automatic substitution when the color and shape of tablets change. We had suggested this as a "canary in the coal mine"—an early warning to the system [23]. Before the pandemic, we were struggling to provide the level of assurance in therapeutic equivalence for patients [23]. The challenge we now face, in and beyond the pandemic, is unprecedented [24].

14. Pandemic, perspectives, professionals, and pathways

The consequences of this pandemic—in lost lives, resulting in injuries and disabilities, and destruction of livelihoods are dire, and most undoubtedly unprecedented to date. As of this writing, systemic uncertainty is still increasing. If we look around without bias, we can observe simultaneously the best and worst of human nature. We see the dedication and commitment beyond the call of duty, and we see profiteering at its worst.

As science interns and scientists, we intend to be dedicated and engaged in finding solutions. However, we also fall victim to the virus real and virtual as in posts going viral on social media. We also have limited tolerance for uncertainty and ambiguity, i.e., unpredictability. Yet, around the world, amid the chaos precipitated by the COVID-19 pandemic, we must develop therapeutics and vaccines and their manufacturing processes at "warp speed." Not just to save lives of those infected, all other patients, and to end recurring lockdowns that are eroding the stability and security of economic, financial, and geopolitical systems and creating chaos.

The Emergency Use Authorization pathway became a prominent topic of intense debate and controversy in the COVID-19 pandemic. Multiple perspectives, political, public health, healthcare, scientific, and public discourse turned nasty and accusatory in the context of evidentiary standards, clinical practice, reliability of diagnostic tests, measurement uncertainty and causality of COVID-19 cases and mortality, and standards of safety and efficacy of repurposed therapeutics ad for vaccine development at "warp speed."

14.1 An unprecedented opportunity to contribute and learn experientially

Our education, training, and continuing education inform us about the facts we need to get the job done right. Information is not knowledge.

> There can be no knowledge without emotion... until we have felt the force of the knowledge, it is not ours.

"Arnold Bennett" Intelligence relates to our development, and lack thereof is evident in the gap between what we know and what we can implement. When we fail to apply what we know, it points to a gap in our ability to solve a problem while being aware of the necessary facts. We must take responsibility to fill these gaps. At the same time, we work to deliver safe, effective, and quality therapeutics and vaccines at an unprecedented scale and speed. We must recognize that we do so as chaos spread, and the nations respond to emergencies.

How we experience chaos and crises in personal and professional environments is emerging as a collective experience we share with a vast majority of the human population—this is unprecedented to learn experientially. A path forward is to determine the characteristics of chaotic and complex systems and practice to match our intelligence with the complexity of the task and commit to improving continually [24].

15. Chaos to continual improvement—A path to harmonization

Chaos and complexity theories are concerned with the behavior of dynamic systems, i.e., the systems that change in time. Chaos is a system whose outputs are unpredictable due to extreme sensitivity to starting conditions, often referred to as the "Butterfly Effect," and patterns emerge around "strange attractors." Chaos is a science of process rather than a state, i.e., of becoming rather than being [25]. In the book entitled, "Order Out of Chaos: Man's New Dialogue with Nature," Ilya Prigogine, and Isabelle Stengers trace the gradual emergence of the conception of Order and chaos and layout their argument why entropy is the price of structure, and that we grow and develop "in direct proportion to the amount of chaos we can sustain and dissipate" [26].

In 1977, Ilya Prigogine was awarded the Nobel Prize for Chemistry. It recognized his contributions to nonequilibrium thermodynamics, particularly the theory of dissipative structures [27]. Starting with a sound scientific basis to think about chaotic and complex systems, we need a practical way forward. In day-to-day practice—the process by which we can choose to understand chaos can be intricately related to what is familiar to professionals in the pharmaceutical sector—cause and effect relationships, current good (manufacturing) practices, and quality management system [28]. A high-level overview of the characteristics of systems is in Table 1. We move in and out as we work in and beyond the COVID-19 pandemic, and we should tailor our approach accordingly [29].

Prediction and empirical verification are foundational in science and engineering, and it is not possible in chaos. Chaos is unpredictability due to extreme sensitivity to initial or starting conditions. Then the solution to reduce epistemic uncertainty (i.e., lack of knowledge) to act—as in federally funding massively parallel experimentation to cover as many initial conditions as feasible.

As of this writing, it is estimated that there are 231 vaccine candidates in development The notion of "sense," in Table 1, refers to observation and assessment, and the response would be to make safe and effective vaccines available. Similar considerations are useful to consider in our day-to-day interactions.

Table 1 Systems in the context of known and unknown cause and effect relationships and stance we should take.

Type of system	Causes and their effect	Stance
Chaotic	Unknowable, not predictable due to extreme sensitivity to starting conditions (Butterfly effect)	Act, sense, respond
Complex	Unknown without research and experimentation can be predictable after knowledge acquired via experiments	Probe, sense, respond
Complicated	Known unknowns, i.e., expertise needed to understand cause-and-effect relationships. Good practices	Sense, analyze, respond
Simple	Self-evident, without specialized expertise. Known knowns. Best practices	Sense, categorize, respond

16. Awareness and self-authorship

In uncertainty are risk and opportunity. To leverage this opportunity, we must be self-authored in what we do and how we know what to do. Distinguish between informed and knowledge-based decisions is an essential consideration of the perpetual "information asymmetry" we confront in the healthcare sector. Information is not knowledge; professionals make knowledge-based decisions whereas the public has no choices to make an informed decision—based on information we provide in product package inserts.

They do so in the assurance our corporate and regulatory systems provide, i.e., they do so on trust. Making false claims, errors, and recalling products erode trust.

Experience is feeling emotions, and experiential learning is a practice of integrating our intention, emotion, observation, and cognition seamlessly. To learn experientially is not limited to "hands-on" experience; we can learn from our imagination and learn for mistakes others have made. To do so—empathizing or imagining walking in other's shoes can help to experience the consequence of errors others have made and to consider patients. The result of releasing poor quality for distributions and product recalls.

When interfacing, interrelating, and interacting with other systems such as the social system, we must recognize the asymmetries of information, knowledge, and understanding. We must expand our awareness of potential impacts rapid changes in the social, economic, political, and technological domains can have on us in the context of our professional and corporate suitability capability. It is essential to pay attention to changing patterns (e.g., expressed apprehensions of receiving a vaccine and protests against perceptions of mandated vaccines) that relate to the evolving expectations and need for assurance of quality population and healthcare providers. In chaos and complexity, cause and effects are known after experimentation, and we cannot estimate the likelihood of occurrence. Therefore, risk needs to be considered differently—as the product of Impact x (Threat x Vulnerability).

To be "good" in practice, we need to utilize fundamental knowledge that remains the same in chaos, emergency or routine operations in which "good laboratory, clinical and manufacturing practices" or GXP we commit to follow. That being careful in calibrating our measurement systems—particularly those measurements that relate to shape, size, and surface characteristics of particulate systems that relate to their stability, disposition, and pharmacodynamics. Our measurement systems must not just be reproducible in one system (e.g., R&D); they must be repeatable by others in other systems such as in commercial operations and other laboratories such as that of the US FDA. Inadequate attention to measurement system uncertainty, particularly for physical particulate systems, has been a chronic challenge in the sector for decades [30]. It should not and cannot be perpetuated in and beyond the COVID-19 pandemic. Unfortunately, we are still struggling in this context. Historically many pharma companies have been in a reactionary mode, i.e., correcting after regulatory observations and warning letters. In the COVID-19 pandemic, routine regulatory inspections have been either suspended or made virtual [30–32]. Therefore, the assurance patients derived from regulatory oversight is compromised, and the consequence of incidents of external quality failures can be expected to be more severe, if not catastrophic.

Let us remember that by decree FDA classifies nanoparticulate therapeutics as complex in the context of assessing their therapeutic equivalence—between a follow-on produce to brand reference and between clinical trial product and commercial products after postapproval changes such as scale-up. In a beyond the COVID-19 pandemic, the consequence of therapeutic inequivalence is more likely to be a national security issue.

16. Closing thoughts

The COVID-19 pandemic, the first world lockdown, and recurring lockdowns have fundamentally changed the economy, politics, and how we organize our activities and societies. As the notion of globalization had begun yielding to localization, a novel coronavirus took the crown to remind us of our obligations as we transition to the Anthropocene epoch. Now it is no longer "alarmist" to ask—will our fledgling democracies survive? How to rebuild the economy and reorganize governance of a nation, a corporation, and the public are burning issues.

We are professionals, not occupants of a job or position. To give assurance to others, we must first be self-assured. To do so, learn experientially to match our intelligence to the complexity of nanoparticle therapeutics and strive for therapeutic equivalence on the traditional and emergency regulatory paths while accounting for the evolving individual, local, and federal perspectives.

References

[1] Nanotechnology Task Force Report, 2007. https://www.fda.gov/science-research/nanotechnology-programs-fda/nanotechnology-task-force-report-2007. (Accessed 7 May 2021).

[2] Opportunities and risks of Nanotechnologies, Report in Co-Operation with the OECD International Futures Programme, 2021. https://www.oecd.org/science/nanosafety/44108334.pdf. (Accessed 7 May 2021).

[3] Guidance for Industry—Considering Whether an FDA-Regulated Product Involves the Application of Nanotechnology, 2014. June https://www.fda.gov/regulatory-information/search-fda-guidance-documents/considering-whether-fda-regulated-product-involves-application-nanotechnology. (Accessed 7 May 2021).

[4] European commission, Definition of a Nanomaterial. https://ec.europa.eu/environment/chemicals/nanotech/faq/definition_en.htm. (Accessed 7 May 2021).

[5] UK Government Response to The Royal Commission on Environmental Pollution (RCEP) Report "Novel Materials in the Environment: The Case Of Nanotechnology", 2009. https://assets.publishing.service.gov.uk/government/uploads/system/uploads/attachment_data/file/228785/7620.pdf. (Accessed 7 May 2021).

[6] Guidance for Industry: Safety of Nanomaterials in Cosmetic Products, 2014. June https://www.fda.gov/regulatory-information/search-fda-guidance-documents/guidance-industry-safety-nanomaterials-cosmetic-products. (Accessed 7 May 2021).

[7] Literature Review on the Safety of Titanium Dioxide and Zinc Oxide Nanoparticles in Sunscreens: Scientific Review Report, 2016. https://www.tga.gov.au/literature-review-safety-titanium-dioxide-and-zinc-oxide-nanoparticles-sunscreens.

[8] Analysis of Consumer Products Inventory [website] Washington, DC: The Project on Emerging Nanotechnologies, Woodrow Wilson International Center for Scholars; http://tinyurl.com/yk46ye2. (Accessed 05.07.2021).

[9] The US FDA. Nanotechnology Programs at FDA. Content Current as of 07/30/2020 2020 https://www.fda.gov/science-research/science-and-research-special-topics/nanotechnology-programs-fda (Accessed 08.15.2020).

[10] The White House, Office of the Press Secretary. Executive Order 13563—Improving Regulation and Regulatory Review, 2011. https://obamawhitehouse.archives.gov/the-press-office/2011/01/18/executive-order-13563-improving-regulation-and-regulatory-review. (Accessed 08.15.2020).

[11] Federal Register Notice, Promoting the Rule of Law Through Improved Agency Guidance Documents. FR Doc. 2020–14433 Filed 7-1-20, 9 October, 2019. https://www.federalregister.gov/documents/2020/07/02/2020-14433/promoting-the-rule-of-law-through-improved-agency-guidance-documents. (Accessed 08.15.2020).

[12] L. Fatehi, S.M. Wolf, J. McCullough, et al., Recommendations for nanomedicine human subjects research oversight: An evolutionary approach for an emerging field, J. Law Med. Ethics 40 (4) (2012) 716–750.

[13] CDER Manual of Policies & Procedures | MAPP. https://www.fda.gov/about-fda/center-drug-evaluation-and-research-cder/cder-manual-policies-procedures-mapp (Accessed 05.07.2021).

[14] The US FDA, Nanotechnology Guidance Documents, 2020. https://www.fda.gov/science-research/nanotechnology-programs-fda/nanotechnology-guidance-documents. (Accessed 08.15.2020).

[15] Nanotechnology Programs at FDA, 2021. https://www.fda.gov/science-research/science-and-research-special-topics/nanotechnology-programs-fda. (Accessed 7 May 2021).

[16] Safe Nano, International Organisation for Standardization, 2021. https://www.safenano.org/knowledgebase/standards/international-organisation-for-standardization/. (Accessed 7 May 2021).

[17] The critical path initiative of Department of Health and Human Services US Food and Drug Administration, 2009. https://www.fda.gov/media/77780/download. (Accessed 7 May 2021).

[18] The US FDA, Draft Guidance for Industry. Drug Products, Including Biological Products that Contain Nanomaterials, 2017, December https://www.fda.gov/media/109910/download. (Accessed 08.15.2020).

[19] Johns Hopkins University. Coronavirus Resouce Center. https://coronavirus.jhu.edu/map.html (Accessed 08.15.2020).

[20] A.C. Walls, Y.J. Park, M.A. Tortorici, A. Wall, A.T. McGuire, D. Veesler, Structure, function, and antigenicity of the SARS-CoV-2 spike glycoprotein, Cell 181 (2020) 281–292.

[21] The US FDA.n.d. COVID-19-Related Guidance Documents for Industry, FDA Staff, and Other Stakeholders . Content Current as of 08/10/2020. https://www.fda.gov/emergency-preparedness-and-response/coronavirus-disease-2019-covid-19/covid-19-related-guidance-documents-industry-fda-staff-and-other-stakeholders(Accessed 08.15.2020).

[22] G. Chauhan, M.J. Madou, S. Kalra, V. Chopra, D. Ghosh, S.O. Martinez-Chapa, Nanotechnology for COVID-19: therapeutics and vaccine research, ACS Nano 14 (7) (2020) 7760–7782.

[23] A.S. Hussain, V.J. Gurvich, K. Morris, Pharmaceutical "new prior knowledge": Twenty-first-century assurance of therapeutic equivalence, AAPS PharmSciTech 20 (3) (2019) 140, https://doi.org/10.1208/s12249-019-1347-6 (Accessed 08.15.2020).

[24] V.J. Gurvich, A.S. Hussain, In, and beyond COVID-19: US academic pharmaceutical science and engineering community must engage to meet critical national needs, AAPS PharmSciTech 21 (2020) 153, https://doi.org/10.1208/s12249-020-01718-9 (Accessed 08.15.2020).

[25] A.S. Hussain, Chaos to Continual Improvement: Path to Harmonization. CPhI Annual Industry Report 2019: Expert Contribution, 2019, CPhI Worldwide, November 2019, Frankfurt | Produced

by Defacto https://www.cphi.com/content/dam/Informa/cphi/en/cphi-insights/HLN19-CPhI% 20Insights-2019-Industry-Report.pdf. (Accessed 08.15.2020).

[26] D. Larsen-Freeman, Chaos/complexity science, and second language acquisition, Appl. Linguis. 18 (2) (1997) 141–165.

[27] I. Prigogine, I. Stengers, Order out of chaos: Man's New Dialogue with Nature, Verso Books, 2018.

[28] I. Prigogine, Nobel Lecture: Time, Structure, and Fluctuations, 1977. https://www.nobelprize.org/ prizes/chemistry/1977/prigogine/lecture/. (Accessed 08.15.2020).

[29] A.S. Hussain, Pharmaceutical beyond 2020: Professionals and artificial intelligence, Pharma Times 52 (06) (2020) 15–19.

[30] D.J. Snowden, M.E. Boone, A leader's framework for decision making, Harv. Bus. Rev. 85 (11) (2007) 68.

[31] COVID-19 vaccine development pipeline. Last updated on 10 August 2020. https://vac-lshtm. shinyapps.io/ncov_vaccine_landscape/ (Accessed 08.15.2020).

[32] A.S. Hussain, Past, Present, and Future of Pharmaceutical Dissolution Testing: A Disciplined Reflection and Synthesis, Society for Dissolution Science, Mumbai, India, e-Disso Newsletter, ed. Prof. Mala Menon, 2019. September.

CHAPTER 17

Fate and potential hazards of nanoparticles in the environment

Govind Sharan Gupta[a] and Alok Dhawan[b,c]
[a]Unit of Molecular Toxicology, Institute of Environmental Medicine, Karolinska Institutet, Stockholm, Sweden
[b]Nanomaterials Toxicology Group, CSIR-Indian Institute of Toxicology Research, Lucknow, Uttar Pradesh, India
[c]Centre of Biomedical Research, Lucknow, Uttar Pradesh, India

1. Background

The rapid development of nanotechnology has caused nanoparticle (NP) exposure to organisms at the size range of 1–100 nm [1]. Nanotechnology has made significant progress in the last decade and has been successfully incorporated in several industrial products. According to the inventory of consumer nanoproducts, there are 1800–3100 registered nanoproducts in the global and European markets (https://www.nanotechproject.org/cpi/products/; http://www.nanodb.dk). Most of the existing nanoproducts fall into the personal care and clothing categories ($\geq$300), followed by sporting goods and cleaning ($>$200) [2]. Such a rapid development of this new technology poses hazardous effects in addition to benefits to society. Avoiding the unintended hazards of a material can further potentiate into a long-term risk, as seen before for other technologies such as agrochemicals, petroleum, and plastic. NPs can enter environmental matrices at different stages of their life cycle from production to recycling [3–8]. The hazard of nanomaterials in the environment depends on their fate after entering the environment, as it is most likely that they behave differently after interactions with abundant biotic and abiotic components. It is well established that environmental components such as natural organic matters, clay minerals, and ions interact with NPs and alter their aggregation or agglomeration potential in the environmental matrices. Studies show a potential increase or decrease in the bioavailability and toxicity of nanoparticles depending on the types of nanoparticles and their interactions with environmental components. For example, the interaction of nanoparticles with particulate humic matter mitigates their toxicity; however, clay colloids increase the toxicity of nanoparticles in the environment [9, 10]. In this chapter, we summarize the current development so far on the fate and impact of NPs in the environment, with special consideration at the level of the individual organism to the food chain.

2. Exposure of nanoparticles in the environment

Exposure is a primary criterion of the hazard assessment of nanoparticles to humans and the environment. Environmental exposure can be described as available concentrations/doses to which biotic organisms come across at different matrices. At this stage, there is sufficient literature available on the modeling-based predictive concentration of metal NPs in the environment; however, no studies have shown the actual available concentration of NPs [11–13]. The studies predicted very low environmental concentrations (ng/L) of the three most-used NPs (TiO_2 NPs, ZnO NPs, and AgNPs) in different environmental matrices [14]. The dose response studies showed that NPs elicit toxic effects in the model organisms at doses that are multiple times higher than the available concentrations in the environment [8]. However, the NP concentration is expected to rise in the environment because of increased demands in the consumer market and improper disposal. Hence, the hazards and the risk assessment of NPs to the environment and humans are important in order to prove the safety and maintain the sustainable growth of nanotechnology in the future. The appropriate test systems to measure NP hazards in the global environment are still under discussion. However, it has been suggested that certain microcosm- or mesocosm-based studies at environmentally relevant NP dosages could help to understand their hazards in the environment.

3. Fate and behavior of nanoparticles

Significant developments in understanding the fate of nanoparticles in the environment has been recorded in the last decade, with an average of >100 peer-reviewed research articles appearing per year in PubMed over the last 5 years (Fig. 1). The fate and stability of NPs in the environmental matrices majorly depend on their interactions with

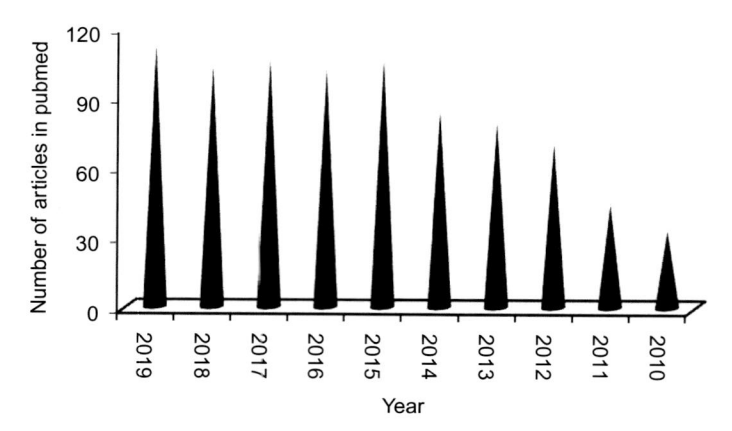

Fig. 1 Yearly distribution of research articles that appeared in PubMed in the last decade with the specific keywords "fate of nanoparticles in environment."

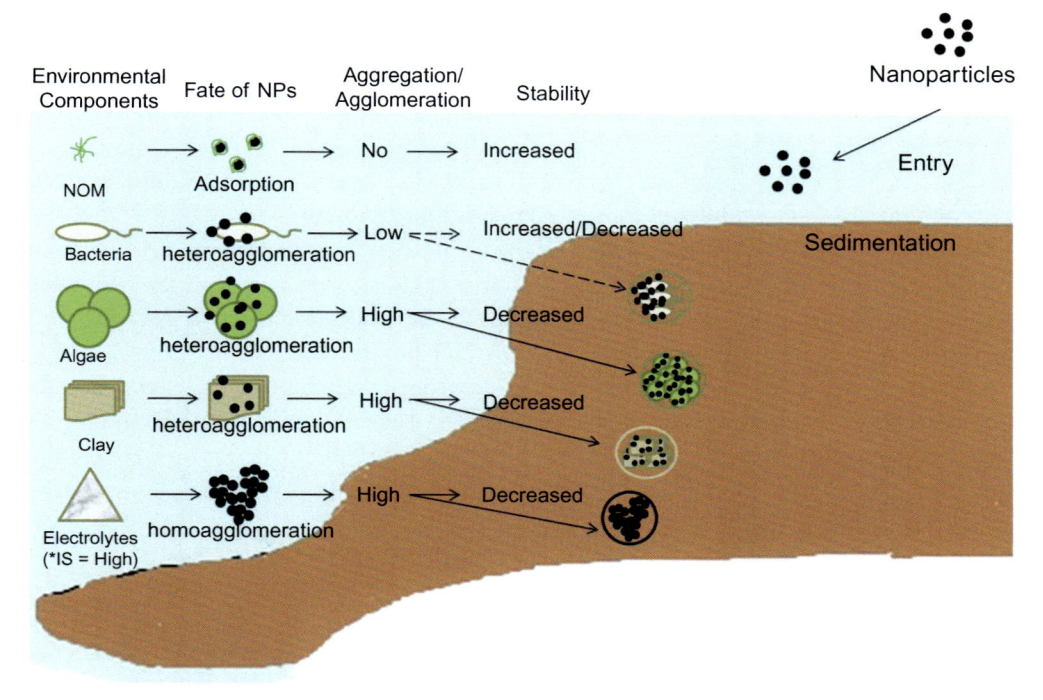

Fig. 2 The fate and stability of nanoparticles in the aquatic environment. The interaction of NPs with biotic and abiotic components leads to the surface transformation of nanoparticles, which consequently affects NP stability in environmental matrices.

environmental factors such as particulate organic and inorganic colloids. The surface charge and functional groups on the NPs determine their stability and the mechanism of interactions with the environmental components. The NP surface coating with uniform charge moieties either with positive or negative ions leads to increased dispersion in the media while opposite charges on the surface lead to NP destabilization in the media. Considering the complex natural conditions, it is difficult to achieve uniform NP dispersion due to their encounters with biotic and abiotic components upon entry into the aquatic environment (Fig. 2). The biotic components include the living items and their components such as producers, consumers, decomposers, exopolymeric substances (EPS), natural organic matter (NOM), and biomolecules. The abiotic components are nonliving entities such as inorganic colloids (e.g., clay particles) and ultraviolet (UV) radiation. Recent developments in studies on eco–nano interactions have demonstrated the interaction of NPs with these environmental factors. NPs behave differently after interactions with these factors due to changes in their physiochemical as well as surface properties.

NPs have undergone numerous modifications. The most common are homo- or heteroaggregation, sedimentation, dissolution, and speciation due to the interactions of chemical and biological moieties.

The homoagglomeration process occurs among the same types of NPs while heteroagglomeration occurs between dissimilar particles such as NPs and biotic or abiotic factors (Fig. 3). The aquatic chemistry such as pH and ionic strength (IS) can influence NP agglomeration. Earlier studies demonstrated that homoagglomeration usually occurs due to the increase in NP-NP interactions at higher concentrations of NPs or the high IS of the environmental media. The presence of electrolytes (Ca^{2+} and Na^+) can also influence the homoaggregation of NPs by compressing the electrodouble layer of NPs [15,16]. However, an increase in homoagglomeration can further decrease NP heteroagglomeration due to reduced interactions of NPs with environmental factors. The heteroagglomeration occurs due to NP binding on the surface of either biotic (bacteria and algae) or abiotic components (e.g., clay particles). These bindings usually occur by electrostatic

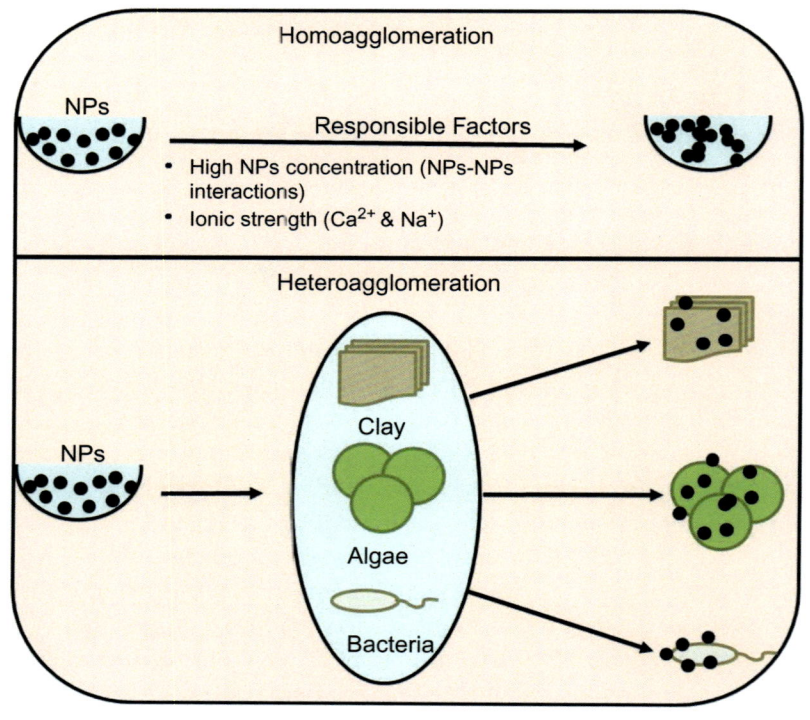

Fig. 3 Agglomeration of nanoparticles in the environment. A high concentration of NPs and the increased ionic strength of the medium lead to NP homoagglomeration. An abundance of clay minerals and/or bacterial and algal cells in the environmental medium induces NP heteroagglomeration.

as well as van der Waals interactions. Also, the interactions of NPs with the surface biomolecules of organisms have been demonstrated. It has been shown that environmentally abundant bacterial cells absorb a higher extent of released NPs because of their higher surface-area-to-volume ratio than other organisms [17]. Other studies have shown that the presence of exopolymeric substances (EPS) on the bacterial and algal cells potentiates the binding of NPs on their surface [18–20]. Furthermore, the studies by Jucker et al. [21] and Host et al. [22] show that the presence of certain exopolymers of bacteria such as lipopolysaccharides (LPS) and siderophores also supports the adsorption of metal NPs, especially TiO_2, to the bacterial cells. Interestingly, Li and Logan [23] observed that a long-chain carbon LPS in *E. coli* cells adheres more strongly to TiO_2 NPs than short chains. Ma et al. [24] showed the heteroagglomeration of metal oxide NPs with algal cells. The heteroagglomeration of NPs with clay depends on the pH and ionic strength of the media [25]. At low pH conditions, positively charged NPs interact with the face of NPs while negatively charged NPs are on the edge of the clay. At neutral pH conditions, there is a poor interaction of negatively charged NPs on the edges due to the presence of an aluminol group while positively charged NPs interact strongly on both the edges and faces of NPs. Labile et al. [26] demonstrated that the heteroagglomeration of TiO_2 NPs with clay occurs due to the electrostatic interactions of oppositely charged particles.

The sedimentation of NPs can effectively reduce their mobility in environmental matrices, which could further decrease NP bioavailability to organisms living on the surface of water systems. The generation of large clusters of NP agglomerates leads to their sedimentation in the environmental matrices. Several factors have been recognized as responsible for NP sedimentation in environmental media such as water chemistry (high ionic strength), biotic factors (bacteria, algae), and abiotic factors (clay and UV light). The sedimentation occurs for both homo- as well as heteroagglomerated nanoparticles. The sedimentation of heteroagglomerates is also defined as cosedimentation [24,27]. Ma et al. [24] and Gupta et al. [27] have shown the cosedimentation of metal oxide nanoparticles with algal and bacterial cells. In both these studies, it was demonstrated that the rate of sedimentation mainly depends on the intensity of NP homo- or heteroagglomerations. Heteroagglomerates of NPs with clay minerals also lead to the coagulation and sedimentation of the NPs [25,28].

The dissolution of metal NPs, especially Ag, CuO, and most recently Au, is the major factor that needs to be considered while assessing their fate, speciation, and toxicity in the environment. The particulate organic (humic or fulvic acid) and inorganic substances (e.g., sulfides, cyanides, phosphates, and nitrates) can influence the dissolution and speciation of NPs in the environment [29–36]. Most of the earlier studies on AgNPs demonstrated their toxicity due to the release of Ag ions. The studies demonstrated that the environmental factors and water chemistry, e.g., pH, oxygen content, complexing ions (Cl^- or SO_4^{2-}) and particulate or dissolved organic matter–can affect release of silver from AgNPs (Fig. 4). Silver nanoparticles release silver ions (Ag+) after oxidation in the

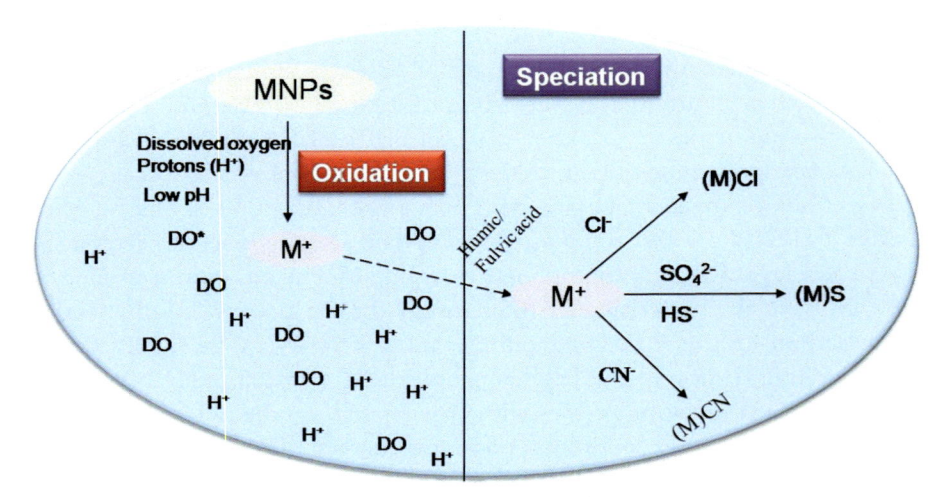

Fig. 4 Dissolution and speciation of metal nanoparticles in the environment. A high level of dissolved oxygen in water and the presence of dissolved organic matter or low pH conditions induce the dissolution of certain NPs such as Ag, ZnO, CuO, and Au. The released metal ions (M^+) from NPs undergo speciation after reaction with sulfides, cyanide, and chloride present in the environment.

presence of dissolved oxygen. The oxidation process of AgNPs in the environment is also facilitated by high proton content at low pH [37,38]. Furthermore, these Ag + can undergo a speciation process after interactions with abundant inorganic species (e.g., Cl- or HS-, SO_4^{2-}) and form secondary precipitates such as silver chloride and/or silver sulfide (Ag_2S) [39–41]. The released Ag^+ can also be readsorbed onto the surface of NPs or other reactive surfaces [37,42]. The presence of chemical or biological substances such as citric acid or natural organic matter (fulvic acid) can reduce the Ag^+ into Ag(0) [43,44]. There are contradictory observations on the dissolution of AgNPs in the presence of NOM. Some studies have shown an increase in the dissolution of AgNPs while others are in contrast. Collin et al. [45] have shown that sulfur-rich NOM (Pony Lake fulvic acid) influences the dissolution of sulfadized AgNPs while Suwanee River and Pahokee peat fulvic acid suppress the release of Ag ions from AgNPs. This was due to the sulfur ligand exchange between sulfadized AgNPs and sulfur-rich fulvic acid.

The modification in the surface properties of NPs further impacts their potential toxicity in environmental organisms and food chains. In addition, an administered versus a realistic dose of NPs is equally important to understand the toxicity of NPs from unicellular to multicellular organisms. It is essential to note that under realistic conditions, there may be some "hot spots" of NP reservoirs due to their tendency to sediment after aggregation/agglomeration; aquatic sediments and soil could be two such NP reservoirs in the environment [46,47]. In addition, during standard waste treatment, it is likely that a major extent of NPs going through the stream will end up associated with the solid phase and then potentially be deposited in certain areas of the environment. It is now acceptable that to understand the possible presence of NPs in free form, a knowledge on release situations is very important. Unfortunately, as of now, there are no suitable sources of information or studies available on this subject. Additionally, there is still a lack of understanding with respect to the environmental fate of NPs in the air [48]. NPs in air come through road traffic exhaust, combustion, explosion, and the oxidation of atmospheric gases [49].

4. Nanoparticles hazard at individual organism level

Ecotoxicity studies of NPs have been performed mainly at alternate model organisms at the unicellular and multicellular levels such as bacteria, protozoa (*Paramecium* and *Tetrahymena*), algae, crustaceans (*Daphnia* spp.), nematodes (*Caenorahbditis elegans*), and zebrafish (*Danio rerio*). There were more than 67 research articles per year published in last one decade specifically under the theme of "nanoparticle hazard in the environment" (Fig. 5). One such article by Bondarenko et al. [50] grouped the seven physicochemically well-characterized NPs based on their hazard ranking determined in model organisms. The hazard ranking of the NPs was in the following order: $Ag > ZnO > CuO > TiO_2 > MWCNTs > SiO_2 > Au$. Bondarenko et al. [51] showed that crustaceans, algae, and fish are the most sensitive organisms for Ag, CuO, and ZnO NPs among the tested organisms of algae, crustaceans, fish, bacteria, yeast, nematodes, and protozoa. Hansen et al. [2] categorized the high-risk nanoproducts using predictive modeling and indicated that NPs used in health and fitness pose the maximum risk to humans and the environment.

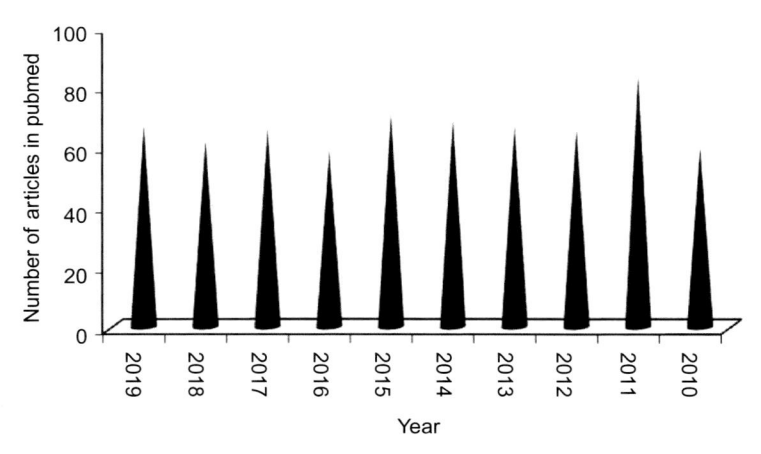

Fig. 5 Yearly distribution of research articles that appeared in PubMed in the last decade with the specific keywords "nanoparticles hazard in environment."

In aquatic life, NP toxicity could be low to high depending on the complexity of the organism. A higher exposure of metal NPs than realistic environmental doses poses toxicity to aquatic microorganisms, (e.g., *Escherichia coli* and *Aeromonas hydrophila*) [52,53], algae [54], and other model organisms. A mechanistic investigation on the toxicity of metal NPs showed that Ag, TiO_2, and ZnO NPs exert toxic effects on aquatic bacteria by damaging their cell membrane via the generation of reactive oxygen species (ROS) near the bacterial surface [53,55,56]. In the terrestrial environment, AgNPs have shown potential bactericidal activity in the rhizospheric microbial community at far below the concentration levels of any other heavy metal [57]. The toxic effects of NPs on beneficial microbes involved in nutrient recycling and plant growth–promoting activities such as nitrogen fixation and siderophore production can pose potential long-term consequences in the ecosystem, denitrification processes, and ecosystem services [58].

Tetrahymena is a model ciliated protozoan widely used in ecotoxicity assessment [59]. Mortimer et al. [60] showed a higher toxicity of ZnO NPs (EC50 value 5 mg metal/L) than CuO NPs (EC50 value 128 mg-metal/L). Furthermore, in another study, Mortimer et al. [61] showed that the toxicity of CuO NPs in *Tetrahymena thermophila* is due to ROS generation, changes in fatty acid profiles, and lipid peroxidation. Blinova et al. [62] showed that CuO NPs were toxic to *Tetrahymena thermophila* under normal exposure conditions in the experimental medium, but that toxicity can be mitigated in natural water due to the presence of dissolved organic matter. Zou et al. [63] showed that the combined exposure of TiO_2 NPs with AgNPs enhances the toxicity potential of AgNPs in *Tetrahymena* under natural light conditions; however, the toxicity was reduced under dark conditions. Gupta et al. [28] showed that ZnO NPs are toxic to *Tetrahymena* cells and the presence of clay minerals potentiated their toxicity.

Daphnia magna and *Ceriodaphnia dubia* are the majorly used crustaceous species for ecotoxicological studies on NPs in the aquatic environment [64]. The higher sensitivity of daphnids shows them to be bioindicators of NP pollution in the environment [65]. Muna et al. [66] showed that the exposure medium of daphnia plays an important role in the toxicity of NPs. In a minimal medium with algae, the 48 h exposure to CuO and ZnO NPs showed similar EC50 values to ~2 mg metal/L. However, the exposure of the same NPs in natural water with algae mitigated the toxicity by 4- to 11-fold. Das et al. [67] showed both the acute and chronic toxicity of AgNPs and TiO_2 NPs in *Daphnia magna*. The 48 h median lethal concentrations (LC_{50}) for AgNPs and bare TiO_2 NPs were 2.75 μg/L and 7.75 mg/L, respectively. In contrast, carboxy-functionalized TiO_2 NPs showed no toxicity in *Daphnia* exposed at maximum tested concentrations (30 mg/L). The chronic exposure of bare TiO_2 NPs in *Daphnia* showed negative effects on growth, reproduction, and survival at the concentration range of 4.5–7.5 mg/L. The exposure of TiO_2 NPs is also known to increase sensitivity in the next-generation offspring of *Daphnia magna*. Bundschuh et al. [68] found that the juvenile offspring of adult daphnids pre-exposed to TiO_2 NPs showed higher toxicity than the offspring of normal adults. The aging of NPs has been recognized to increase the toxicity of TiO_2 NPs in *Daphnia magna*. Seitz et al. [69] have shown that 1–3-day-old TiO_2 NPs in an exposure medium supplemented with natural organic matter showed a significantly increased acute toxicity (by approximately 30%) over 0-day-old particles while 6-day-old particles showed a reduction (60%) in the toxicity of TiO_2 NPs.

The zebrafish is an important aquatic model because of its numerous properties such as easy maintenance, high fecundity, rapid development, transparent embryos, significant association in the food web, accessibility of the complete genome, and its significant overlap (~70%) with the human genome [70–72]. The early stage embryos and larvae of zebrafish are broadly in use for the high-throughput screening of NP safety for the environment and humans, as presented in Fig. 6. The exposure of AgNPs has been shown to decrease the survival of zebrafish embryos [73]. Gupta et al. [9] also showed the toxicity of AgNPs in zebrafish eletheroembryos, which was potentiated in the presence of montmorillonite clay minerals in the exposure medium. The complexation of TiO_2 NPs with lead (Pb) significantly increased the uptake, bioavailability, and toxicity in zebrafish embryos [74]. The acute exposure of CuO NPs affects embryo growth by suppressing viability, delaying hatching, and reducing heart rate, even without entering the zebrafish embryos or larvae [75]. Additionally, CuO NPs downregulated the expressions of the genes related to the neurodevelopment (gfap, syn2α, and elavl3) of zebrafish embryos. The presence of NOMs such as humic substances has been shown to mitigate NP toxicity in zebrafish [10,71,76,77].

The hazards of NPs in the terrestrial environment have also been studied. *Caenorhabditis elegans* has been used as a model organism for the toxicity assessment of NPs in a soil environment [78,79]. Li et al. [80] showed a decrease in the survival of *C. elegans* after early stage chronic exposure of ZnO NPs at realistic NP doses (50 and 500 µg/L). Zhang et al. [78] showed that CeO_2 NPs increased ROS accumulation, which resulted in oxidative

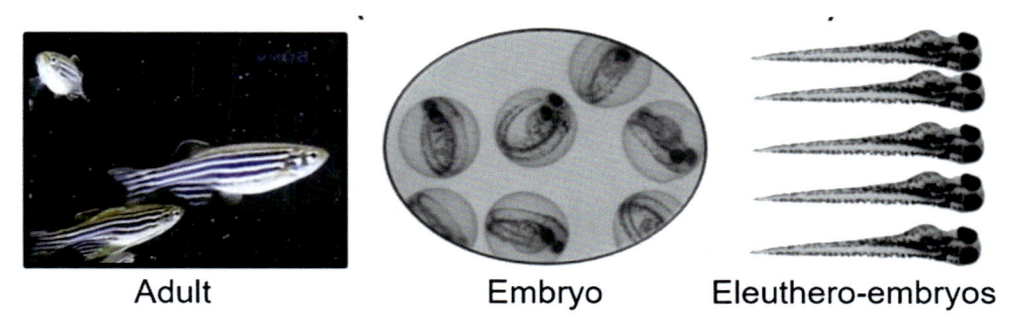

Adult   Embryo   Eleuthero-embryos

Fig. 6 Zebrafish and their different developmental stages. The images were self-captured in a zebrafish risk assessment facility at Ahmedabad University. The embryo and eleutheroembryos images were captured under a stereo-zoom microscope (Model CETi; Medline Scientific, Bangalore, India). The adult zebrafish image was captured by an SLR digital camera.

damage and the consequent death of *C. elegans*. Moon et al. [81] showed that the NP shape plays an important function in NP toxicity in soil organisms. Among different size- and shape-based AgNMs (AgNPs; 10 μm silver nanowires, 10-AgNWs; 20 μm 20-AgNWs and silver nanoplates, AgPLs), only AgNPs and AgPLs showed negative effects on the growth and reproduction of *C. elegans*, whereas AgNWs did not show any toxic effects on *C. elegans*. TiO$_2$ NP exposure showed phototoxic effects in *C elegans* and such effects were regulated through the activation of the JAK/STAT and TGF-β pathways [82].

5. Nanoparticle hazards at food chain level

So far, most studies on NP risk assessment have determined the direct exposure of the model organism to NPs for a short duration and the measurement of certain toxicity parameters representing stress or death. Only a few studies are available on the accumulation and impact of NPs on the food chain after long-term exposure.

NPs with partial water solubility and less degradability in biophysical systems can influence their biopersistence, bioaccumulation, and trophic transfer in food chains [83,84]. The potential environmental pollutants such as DDT, lead, asbestos, and mercury have been recognized as highly dangerous for humans and the environment because of their high bioaccumulative properties in cells, tissues, and entire organisms [84,85]. Most of the available studies show the trophic transfer of NPs up to 2–3 trophic levels in food chains constructed under laboratory conditions. The studies also show that NPs have the potential to biomagnify up to two trophic levels with a biomagnification factor (BMF) value >1 [85]. At the third trophic level, the NP concentration decreased with a BMF value <1, demonstrating the trophic transfer of NPs without biomagnification [84,86]. Considering the very low realistic environmental concentration of NPs, they may not be toxic to organisms at the moment, but their continuous exposure and accumulation in cells and tissues may ultimately reach a lethal concentration. Hence, it is important to study the potential accumulation and food chain impacts of NPs.

The interaction of NPs with the food chain is particularly dependent on the mode of exposure in the organisms such as through water (bioconcentration), water and diet (bioaccumulation), and diet only (trophic transfer and/or biomagnification).

Bioconcentration: When the organism's NP intake is limited through water in aquatic organisms and via air or water in terrestrial organisms and plants, it is considered bioconcentration [87]. The NP uptake mostly occur through passive diffusion mechanism during bioconcentration [88,89].

Bioaccumulation: When the NP intake is not limited to a single route and can occur through all the possible ways such as water, air, and diet, this is termed bioaccumulation [87].

Trophic Transfer and Biomagnification: This occurs when NPs move up the food chain to higher trophic levels with higher concentrations than expected. For example, >1-fold higher concentration of NPs in predators from prey is known as biomagnification, whereas if the concentration is <1-fold, then it is termed trophic transfer without bio-magnification [87,90].

In terms of quantitation, biomagnification is expressed as biomagnification or trophic magnification factor (BMF or TMF). BMF can be shown as the ratio of the pollutant between the predator and the prey [91]. It can be calculated by the following equation:

$$\text{BMF or TMF} = \text{Concentration of pollutant in predator organism} \left(C_B\right)/$$

$$\text{Concentration of pollutant in prey or diet} \left(C_D\right)$$

The impact of NPs in the food chain at different trophic levels from producers to consumers is summarized in Table 1. The studies are limited to certain organisms such

Table 1 Trophic transfer and bioaccumulation of NPs in environmental food chains.

Nanoparticles	Test species	Bioaccumulation	Food chain type	References
Ag$_2$S–NPs	Soybean *Glycine* max L. to Snail	Approx. 78% of the Ag assimilated by snails	Terrestrial	[92]
Tannate–coated Au NPs	Tobacco hornworm (*Manducasexta*) caterpillars	BAF = 0.16	Terrestrial	[93]
CeO$_2$ and ZnO NPs	Cucumber plants	1.27 mg of Ce and 110 mg Zn per kg dry weight	Terrestrial	[94]
Tyrosine (T-AgNP), epigallocatechin gallate (*E*-AgNP), and curcumin (C-AgNP)	Raphidocelis subcapitata to *Daphnia carinata*	Accumulation of T-AgNPs in daphnids via trophic transfer was 2.6 times higher than T-AgNPs	Aquatic	[95]
Al$_2$O$_3$ NPs	*Ceriodaphniadubia*	BMF = 0.19	Aquatic	[96]
AuNPs (5, 10, and 15 nm)	*Manducasexta* (tobacco hornworm)	BMF = 6.2, 11.6, and 9.6 for 5, 10, and 15 nm	Terrestrial	[97]
Au Nanorods (CTAB-coated)	Estuarine mesocosm (sea water, sediment, sea grass, microbes, bioflms, snails, clams, shrimp, and fish)	BCF (L kg^{-1} DW) Biofilm = 15,300, Sea grass = 8.21, Shrimp = 115, Fish = 474, Snail = 167, Clam = 22,800	Esturine	[98]

Continued

Table 1 Trophic transfer and bioaccumulation of NPs in environmental food chains—cont'd

Nanoparticles	Test species	Bioaccumulation	Food chain type	References
ZnO NPs	Soybean (Glycine max)	BCF Leaf = 0.69, Stem = 0.25, Root = 0.24, Nodule = 0.07	Terrestrial	[99]
ZnO NPs	Peanut (Arachishypogaea L.)	Zn bioaccumulation in leaf = 42%	Terrestrial	[100]
TiO_2 NPs	Tomato	% Accumulation Stem = 11.0%–11.4%, Leaves = 6.1%–8.4%	Terrestrial	[101]
CeO_2 NPs	Soybean	BCF Root = 0.21, Nodule = 0.011, Stem = 0.0001, Leaf = 3×10^{-7}	Terrestrial	[101]
ZnO NPs	Isopods (Porcellio scaber)	BAF (kg dry leaf per kg dry biomass) = 0.1 and 0.2	Terrestrial	[102]
CuO NPs	*Artemia salina* to Amatitlania nigrofasciata	Accumulation of Cu after the depuration phase in cichlid larvae was 25.4 ± 0.5, 29 ± 8.0, 33.9 ± 9.7, and $42.3 \pm 4.0 \mu g/g$ dry weight at 0, 1, 10, and 100 mg/L of CuO-NP-treated Artemia.	Aquatic	[103]
TiO_2 NPs	*Ceriodaphnia dubia*	BAF = 214.38 (L/Kg) BMF = 0.218	Aquatic	[104]
Anatase-TiO2 NPs, rutile + anatase-TiO2 NPs, and rutile-TiO2 NPs	Algae to Daphnia	BMF = 5.7–122	Aquatic	[105]
ZnO NPs	Algae to Daphnia	BMF = 1	Aquatic	[106]
TiO_2 nanoparticle aggregates	*Daphnia magna*	BCF = 118,062.84 L/Kg BAF = 1232.28 L/Kg	Aquatic	[107]

Table 1 Trophic transfer and bioaccumulation of NPs in environmental food chains—cont'd

Nanoparticles	Test species	Bioaccumulation	Food chain type	References
AgNP-citrate, AgNP-PVP	Polychaete (*Nereisvirens*)	Ag accumulation = 32–44%, relative to the sediments	Estuarine Sediment	[108]
CdSe core–ZnS shell carboxylated and biotinylated QDs	Rotifer (Brachionus calciflorus)	BMF (dry weight) = Carboxylated QDs–0.62 Biotinylated QDs–0.29	Aquatic	[84]
TiO$_2$ nanoparticle	*Tetrahymena thermophile*	BMF = 0.16	Aquatic	[109]
TiO$_2$ nanoparticle	Zebrafish (Deniorerio)	BMF = 0.009	Aquatic	[86]
AgNP (Coated with carbonate)	Daphnia magna	BCF = 46,000 104 (L kg^{-1} dry Weight)	Aquatic	[110]
CdSe QDs (Citrate-coated)	Ciliated protozoa (*Tetrahymena thermophila*)	BMF (DW) ¼ 5.4	Aquatic	[85]
CdTe QDs	Ciliated protozoan (*Paramecium caudatum*)	BMF– >1.0	Aquatic	[111]
TiO$_2$ NPs	Polychaets (*Arenicola marina*)	BSAF (kg OC per kg Lipid) = 0.156–0.196	Sediment	[112]
TiO$_2$ NPs	Carp (Cyprinuscarpio)	BCF (L kg^{-1} whole DW) = 495	Aquatic vertebrate	[113]
TiO$_2$ NPs	Zebrafish (*Danio rerio*)	BCF (L kg^{-1} DW) = 25 and 181	Aquatic vertebrate	[86]
CdSe-ZnS QD (coated with Poly(acrylic acid)-octylamine)	Zebrafish (*Danio rerio*)	BMF = 0.04 and 0.004 for adult and juvenile zebrafsh	Aquatic	[114]

BAF, Bioaccumulation factor; *BMF*, Biomagnification factor; *BCF*, Bioconcentration factor; *BSAF*, Biota-sediment accumulation factor.

as plants, algae, bacteria, crustaceans, ciliated protozoans, snails, fish, and earthworms. Very few studies have discussed the consequent ecological impact of bioaccumulated NPs in the food chain.

The biomagnification of CdSe QDs from bacteria to protozoa with a BMF >5 in the aqueous environment was observed in a laboratory-based simple microbial food chain established using *Pseudomonas* (bacteria) and *Tetrahymena* (protozoa) as the prey and predator species [85]. A loss in bacterial digestion and a delay in the reproduction of the *Tetrahymena* cells due to QD biomagnification was also observed. Chen et al. [115] demonstrated the biomagnification of TiO_2 NPs in *Daphnia magna* from *Scenedesmus obliquus* with a high BMF value of 7.8–2.2. Additionally, an increased accumulation of TiO_2 NPs in *Daphnia magna* was observed after surface modification with sodium dodecyl benzene sulfonate.

In the terrestrial environment, there are fewer studies on the biomagnification of metal NPs with a BMF value of more than 1. Majumdar et al. [116] indicated the biomagnification (BMF = 5.0) of CeO_2 NPs up to three trophic levels in a food chain that consisted of a kidney bean plant (*Phaseolus vulgaris* var. red hawk) to adult beetles (*Epilachnavarivestisand*) at lower trophic levels and then in bugs (*Podisusmaculiventris*), a consumer at the top of the assembled food chain. Judy et al. [93] and Unrine et al. [117] also showed the biomagnification (BMF 6.2–11.6) of Au NPs from the tomato plant (*Lycopersiconesculentum*) to the tobacco hornworm (*Manducasexta*).

There are fewer studies available on the fate, ecotoxicity, and food chain impact of carbon-based nanomaterials (e.g., fullerenes, carbon nanotubes, and graphene) due to a limitation in their quantification, as isotope labeling is the only way to quantify carbon-based nanomaterials [118]. One such study with ^{14}C-labeled multiwalled carbon nanotubes (^{14}C-MWCNT) showed bioaccumulation in the protozoan *Tetrahymena thermophila* at low to ultralow dose (0.0–1 mg/L) exposure via bacterial prey *Pseudomonas aeruginosa* and direct uptake from growth media [119].

6. International guidelines on limiting environmental exposure of nanoparticles

The newly developed technologies and products should receive a thorough investigation for their possible risks to the environmental and human health before implementation in the consumer market [48]. According to the precautionary principles introduced by Kriebel et al. [120], the following suggestions should be taken into account: (1) take preventive action in the face of uncertainty; (2) shift the burden of proof to the proponents of an activity; (3) explore a wide range of alternatives to possibly harmful actions; and (4) increase public participation in decision-making. However, such suggestions were not followed during the implementation of NP-based products in the market, leaving insecurities about NP safety in consumers. Taken together, the dispute and agreement with scientists working to assure the safety of NP exposure for a sustainable future, the government systems and the regulatory agencies are functioning to implement substantial regulation on NP use. In line with the discussion on NP safety, a European commission has acknowledged that NPs are "difficult to regulate" due to their complexity and the limited availability of knowledge in the field [121]. Organizations working on international regulatory standards have recommended careful occupational exposure limits (OELs) for NPs. For instance, the recommended exposure limits (RELs) according to the US National Institute for Occupational Safety and Health (NIOSH) for nanorange particles and pigmentary (>100 nm) TiO_2 are 0.3 and 2.4 mg-m^{-3} for 10 h TWA per day, respectively [122]. Besides NIOSH, the US Environmental Protection Agency guidelines suggest that manufacturers of nanoproducts provide detailed information (e.g., chemical nature, manufacturing process, produced volume, exposure and potential release, use, and health and safety information) on NPs to the agency for evaluation in order to ensure the safety of the products for humans and/or the environment [123]. The British Standard Institute proposed a benchmark exposure level for fibrous NPs of 0.01 fibers-mL^{-1} [124]. The German Institute for Occupational Safety and Health also recommended a benchmark exposure limit for metals, metal oxides, and bio-persistent granular NPs (density > 6000 kg-m^{-3}) of 20,000 particles-cm^{-3} and for bio-persistent granular NPs (density < 6000 kg-m^{-3}) of 40,000 partcles-cm^{-3} [125]. However, such limits on NP exposure levels may not be sufficient for safety because of the highly reactive properties of NPs. Hence, it recommends bringing the exposure limit to a maximum lower level by working on proper control measures such as safety data sheets and proper labeling.

7. Conclusions and future aspects

This chapter showed the potential hazards of metal NPs in different environmental settings such as terrestrial and aquatic systems. There are plenty of studies showing the negative effects of NPs in model organisms such as bacteria, algae, ciliated protozoa, nematodes, crustaceans, and fish. It was also observed that NPs have the potential to move up the food chains to the next trophic levels through prey and predators. Furthermore, the fate of NPs has been reviewed and it was demonstrated that NPs potentially interact with the biotic and abiotic components of the environment that affect the original properties of NPs by causing agglomeration and speciation. There are still limited or no studies available on the fate and hazards of NPs in an air atmosphere. There is also a lack of research on the bioaccumulative property of NPs under realistic environmental situations and the ultimate effects of NPs in the performance of ecosystem services such as biogeochemical cycling. Other important aspects that need to be investigated in the near future are the effect of NPs on predation and the interaction of predator–prey species and vice-versa in natural communities. At the end, omics-based NP investigations of the effects on trophic-level organisms can further strengthen understanding on the safe use and disposal of nanoproducts.

References

[1] S.H. Lee, B.-H. Jun, Silver nanoparticles: synthesis and application for nanomedicine, Int. J. Mol. Sci. 20 (2019) 865.

[2] S.F. Hansen, L.R. Heggelund, P.R. Besora, A. Mackevica, A. Boldrin, A. Baun, Nanoproducts—what is actually available to European consumers? Environ. Sci. Nano 3 (1) (2016) 169–180.

[3] T. Walser, L.K. Limbach, R. Brogioli, E. Erismann, L. Flamigni, B. Hattendorf, M. Juchli, F. Krumeich, C. Ludwig, K. Prikopsky, M. Rossier, D. Saner, A. Sigg, S. Hellweg, D. Gunther, W.J. Stark, Persistence of engineered nanoparticles in a municipal solid-waste incineration plant, Nat. Nanotechnol. 7 (2012) 520–524.

[4] A.A. Keller, A. Lazareva, Predicted releases of engineered nanomaterials: from global to regional to local, Environ. Sci. Technol. Lett. 1 (2014) 65–70.

[5] F. Gottschalk, T. Sonderer, R.W. Scholz, B. Nowack, Modeled environmental concentrations of engineered nanomaterials (TiO_2, ZnO, Ag, Cnt, fullerenes) for different regions, Environ. Sci. Technol. 43 (2009) 9216–9222.

[6] J. Lee, S. Mahendra, P.J. Alvarez, Nanomaterials in the construction industry: a review of their applications and environmental health and safety considerations, ACS Nano 4 (2010) 3580–3590.

[7] P.A. Holden, R.M. Nisbet, H.S. Lenihan, R.J. Miller, G.N. Cherr, J.P. Schimel, J.L. Gardea-Torresdey, Ecological nanotoxicology: integrating nanomaterial hazard considerations across the subcellular, population, community, and ecosystems levels, Acc. Chem. Res. 46 (2013) 813–822.

[8] E.S. Bernhardt, B.P. Colman, M.F. Hochella Jr., B.J. Cardinale, R.M. Nisbet, C.J. Richardson, L. Yin, An ecological perspective on nanomaterial impacts in the environment, J. Environ. Qual. 39 (2010) 1954–1965.

[9] G.S. Gupta, A. Dhawan, R. Shanker, Montmorillonite clay alters toxicity of silver nanoparticles in zebrafish (Danio rerio) eleutheroembryo, Chemosphere 163 (2016) 242–251.

[10] P.R. Cáceres-Vélez, M.L. Fascineli, M.H. Sousa, C.K. Grisolia, L. Yate, P.E.N. de Souza, I. Estrela-Lopis, S. Moya, R.B. Azevedo, Humic acid attenuation of silver nanoparticle toxicity by ion complexation and the formation of a Ag3+ coating, J. Hazard. Mater. 353 (2018) 173–181.

[11] F. Gottschalk, C. Lassen, J. Kjoelholt, F. Christensen, B. Nowack, Modeling flows and concentrations of nine engineered nanomaterials in the Danish environment, Int. J. Environ. Res. Public Health 12 (5) (2015) 5581–5602.

[12] A. Boxall, K. Tiede, Q. Chaudhry, R. Aitken, A.D. Jones, B. Jefferson, J. Lewis, E. Team, Current and future predicted exposure to engineered nanoparticles, Sci. Total Environ. 390 (2007) 396–409.

[13] N.C. Mueller, B. Nowack, Exposure modeling of engineered nanoparticles in the environment, Environ. Sci. Technol. 42 (12) (2008) 4447–4453.

[14] T.Y. Sun, F. Gottschalk, K. Hungerbühler, B. Nowack, Comprehensive probabilistic modelling of environmental emissions of engineered nanomaterials, Environ. Pollut. 185 (2014) 69–76.

[15] B. Mukherjee, J.W. Weaver, Aggregation and charge behaviour of metallic and nonmetallic nanoparticles in the presence of competing similarly-charged inorganic ions, Environ. Sci. Technol. 44 (2010) 3332–3338.

[16] Y.H. Shih, C.M. Zhuang, Y.H. Peng, C.H. Lin, Y.M. Tseng, The effect of inorganic ions on the aggregation kinetics of lab-made TiO2 nanoparticles in water, Sci. Total Environ. 435–436 (2012) 446–452.

[17] P.A. Holden, J.P. Schimel, H.A. Godwin, Five reasons to use bacteria when assesing manufactured nanomaterial environmental hazards and fates, Curr. Opin. Biotechnol. 27 (2014) 73–78.

[18] W. Zhang, B. Rittmann, Y. Chen, Size effects on adsorption of hematite nanoparticles on E. coli cells, Environ. Sci. Technol. 45 (6) (2011) 2172–2178.

[19] K. Chojnacka, Biosorption and bioaccumulation—the prospects for practical applications, Environ. Int. 36 (3) (2010) 299–307.

[20] S.S. Khan, P. Srivatsan, N. Vaishnavi, A. Mukherjee, N. Chandrasekaran, Interaction of silver nanoparticles (SNPs) with bacterial extracellular proteins (ECPs) and its adsorption isotherms and kinetics, J. Hazard. Mater. 192 (2011) 299–306.

[21] B.A. Jucker, A.J. Zehnder, H. Harms, Quantification of polymer interactions in bacterial adhesion, Environ. Sci. Technol. 32 (19) (1998) 2909–2915.

[22] A.M. Horst, A.C. Neal, R.E. Mielke, P.R. Sislian, W.H. Suh, L. Mädler, G.D. Stucky, P.A. Holden, Dispersion of TiO2 nanoparticle agglomerates by *Pseudomonas aeruginosa*, Appl. Environ. Microbiol. 76 (21) (2010) 7292–7298.

[23] B. Li, B.E. Logan, Bacterial adhesion to glass and metal-oxide surfaces, Colloids Surf. B: Biointerfaces 36 (2) (2004) 81–90.

[24] S. Ma, K. Zhou, K. Yang, D. Lin, Heteroagglomeration of oxide nanoparticles with algal cells: effects of particle type, ionic strength and pH, Environ. Sci. Technol. 49 (2) (2015) 932–939.

[25] D. Zhou, A.I. Abdel-Fattah, A.A. Keller, Clay particles destabilize engineered nanoparticles in aqueous environments, Environ. Sci. Technol. 46 (14) (2012) 7520–7526.

[26] J. Labille, C. Harns, J.Y. Bottero, J. Brant, Heteroaggregation of titanium dioxide nanoparticles with natural clay colloids, Environ. Sci. Technol. 49 (11) (2015) 6608–6616.

[27] G.S. Gupta, A. Kumar, R. Shanker, A. Dhawan, Assessment of agglomeration, co-sedimentation and trophic transfer of titanium dioxide nanoparticles in a laboratory-scale predator-prey model system, Sci. Rep. 6 (2016) 31422.

[28] G.S. Gupta, V.A. Senapati, A. Dhawan, R. Shanker, Heteroagglomeration of zinc oxide nanoparticles with clay mineral modulates the bioavailability and toxicity of nanoparticle in Tetrahymena pyriformis, J. Colloid Interface Sci. 495 (2017) 9–18.

[29] J.B. Glenn, S.J. Klaine, Abiotic and biotic factors that influence the bioavailability of gold nanoparticles to aquatic macrophytes, Environ. Sci. Technol. 47 (18) (2013) 10223–10230.

[30] J.A. Hernandez-Viezcas, H. Castillo-Michel, J.C. Andrews, M. Cotte, C. Rico, J.R. Peralta-Videa, Y. Ge, J.H. Priester, P.A. Holden, J.L. Gardea-Torresdey, In situ synchrotron X-ray fluorescence mapping and speciation of CeO2 and ZnO nanoparticles in soil cultivated soybean (Glycine max), ACS Nano 7 (2) (2013) 1415–1423.

[31] H. Sun, X. Zhang, Z. Zhang, Y. Chen, J.C. Crittenden, Influence of titanium dioxide nanoparticles on speciation and bioavailability of arsenite, Environ. Pollut. 157 (4) (2009) 1165–1170.

[32] G.V. Lowry, E.M. Hotze, E.S. Bernhardt, D.D. Dionysiou, J.A. Pedersen, M.R. Wiesner, B. Xing, Environmental occurrences, behavior, fate, and ecological effects of nanomaterials: an introduction to the special series, J. Environ. Qual. 39 (6) (2010) 1867–1874.

[33] S. Liu, Y. Liu, B. Pan, Y. He, B. Li, D. Zhou, Y. Xiao, H. Qiu, M.G. Vijver, W.J. Peijnenburg, The promoted dissolution of copper oxide nanoparticles by dissolved humic acid: copper complexation over particle dispersion, Chemosphere 245 (2020) 125612.

[34] E. Yin, Z. Zhao, Z. Chi, Z. Zhang, R. Jiang, L. Gao, J. Cao, X. Li, Effect of Chlamydomonas reinhardtii on the fate of CuO nanoparticles in aquatic environment, Chemosphere (2020) 125935.

[35] E. McGivney, X. Gao, Y. Liu, G.V. Lowry, E. Casman, K.B. Gregory, J.M. Van Briesen, A. Avellan, Biogenic cyanide production promotes dissolution of gold nanoparticles in soil, Environ. Sci. Technol. 53 (3) (2018) 1287–1295.

[36] A. Avellan, M. Simonin, E. McGivney, N. Bossa, E. Spielman-Sun, J.D. Rocca, E.S. Bernhardt, N.K. Geitner, J.M. Unrine, M.R. Wiesner, G.V. Lowry, Gold nanoparticle biodissolution by a freshwater macrophyte and its associated microbiome, Nat. Nanotechnol. 13 (11) (2018) 1072–1077.

[37] J.Y. Liu, R.H. Hurt, Ion release kinetics and particle persistence in aqueous nano-silver colloids, Environ. Sci. Technol. 44 (6) (2010) 2169–2175.

[38] C.N. Lok, C.M. Ho, R. Chen, Q.Y. He, W.Y. Yu, H. Sun, P.K.H. Tam, J.F. Chiu, C.M. Che, Silver nanoparticles: partial oxidation and antibacterial activities, J. Biol. Inorg. Chem. 12 (4) (2007) 527–534.

[39] O. Choi, K.K. Deng, N.J. Kim, L. Ross, R.Y. Surampalli, Z.Q. Hu, The inhibitory effects of silver nanoparticles, silver ions, and silver chloride colloids on microbial growth, Water Res. 42 (12) (2008) 3066–3074.

[40] O. Choi, T.E. Cleuenger, B.L. Deng, R.Y. Surampalli, L. Ross, Z.Q. Hu, Role of sulfide and ligand strength in controlling nanosilver toxicity, Water Res. 43 (7) (2009) 1879–1886.

[41] J. Gao, S. Youn, A. Hovsepyan, V.L. Llaneza, Y. Wang, G. Bitton, J.C.J. Bonzongo, Dispersion and toxicity of selected manufactured nanomaterials in natural river water samples: effects of water chemical composition, Environ. Sci. Technol. 43 (9) (2009) 3322–3328.

[42] J.R. Morones, J.L. Elechiguerra, A. Camacho, K. Holt, J.B. Kouri, J.T. Ramirez, M.J. Yacaman, The bactericidal effect of silver nanoparticles, Nanotechnology 16 (10) (2005) 2346–2353.

[43] J. Turkevich, P.C. Stevenson, J. Hillier, A study of the nucleation and growth processes in the synthesis of colloidal gold, Discuss. Faraday Soc. 11 (1951) 55–—75.

[44] P. Sarkar, D.K. Bhui, H. Bar, G.P. Sahoo, S. Samanta, S. Pyne, A. Misra, Aqueous-phase synthesis of silver nanodiscs and nanorods in methyl cellulose matrix: Photophysical study and simulation of UV-vis extinction spectra using DDA method, Nanoscale Res. Lett. 5 (10) (2010) 1611–1618.

[45] B. Collin, O.V. Tsyusko, D.L. Starnes, J.M. Unrine, Effect of natural organic matter on dissolution and toxicity of sulfidized silver nanoparticles to *Caenorhabditis elegans*, Environ. Sci. Nano 3 (4) (2016) 728–736.

[46] M. Baalousha, A. Manciulea, S. Cumberland, K. Kendall, J.R. Lead, Aggregation and surface properties of iron oxide nanoparticles: influence of pH and natural organic matter, Environ. Toxicol. Chem. 27 (9) (2008) 1875–1882.

[47] S.J. Klaine, P.J. Alvarez, G.E. Batley, T.F. Fernandes, R.D. Handy, D.Y. Lyon, S. Mahendra, M. J. McLaughlin, J.R. Lead, Nanomaterials in the environment: behavior, fate, bioavailability, and effects, Environ. Toxicol. Chem. 27 (9) (2008) 1825–1851.

[48] E. Kabir, V. Kumar, K.H. Kim, A.C. Yip, J.R. Sohn, Environmental impacts of nanomaterials, J. Environ. Manag. 225 (2018) 261–271.

[49] A.C. John, M. Küpper, A.M. Manders-Groot, B. Debray, J.M. Lacome, T.A. Kuhlbusch, Emissions and possible environmental implication of engineered nanomaterials (ENMs) in the atmosphere, Atmosphere 8 (5) (2017) 84.

[50] O.M. Bondarenko, M. Heinlaan, M. Sihtmäe, A. Ivask, I. Kurvet, E. Joonas, A. Jemec, M. Mannerström, T. Heinonen, R. Rekulapelly, S. Singh, Multilaboratory evaluation of 15 bioassays for (eco) toxicity screening and hazard ranking of engineered nanomaterials: FP7 project NANOVALID, Nanotoxicology 10 (9) (2016) 1229–1242.

[51] O. Bondarenko, K. Juganson, A. Ivask, K. Kasemets, M. Mortimer, A. Kahru, Toxicity of ag, CuO and ZnO nanoparticles to selected environmentally relevant test organisms and mammalian cells in vitro: a critical review, Arch. Toxicol. 87 (7) (2013) 1181–1200.

[52] T. Tong, C.M. Wilke, J. Wu, C.T.T. Binh, J.J. Kelly, J.F. Gaillard, K.A. Gray, Combined toxicity of nano-ZnO and nano-TiO2: from single-to multinanomaterial systems, Environ. Sci. Technol. 49 (13) (2015) 8113–8123.

[53] A. Kumar, A.K. Pandey, S.S. Singh, R. Shanker, A. Dhawan, Engineered ZnO and Tio(2) nanoparticles induce oxidative stress and DNA damage leading to reduced viability of Escherichia coli, Free Radic. Biol. Med. 51 (2011) 1872–1881.

[54] J. Li, S. Schiavo, G. Rametta, M.L. Miglietta, V. La Ferrara, C. Wu, S. Manzo, Comparative toxicity of nano ZnO and bulk ZnO towards marine algae Tetraselmis suecica and Phaeodactylum tricornutum, Environ. Sci. Pollut. Res. 24 (7) (2017) 6543–6553.

[55] W.R. Li, X.B. Xie, Q.S. Shi, H.Y. Zeng, O.Y. You-Sheng, Y.B. Chen, Antibacterial activity and mechanism of silver nanoparticles on Escherichia coli, Appl. Microbiol. Biotechnol. 85 (4) (2010) 1115–1122.

[56] B. Das, S.K. Dash, D. Mandal, T. Ghosh, S. Chattopadhyay, S. Tripathy, S. Das, S.K. Dey, D. Das, S. Roy, Green synthesized silver nanoparticles destroy multidrug resistant bacteria via reactive oxygen species mediated membrane damage, Arab. J. Chem. 10 (6) (2017) 862–876.

[57] I.N. Throbäck, M. Johansson, M. Rosenquist, M. Pell, M. Hansson, S. Hallin, Silver (Ag+) reduces denitrification and induces enrichment of novel nirK genotypes in soil, FEMS Microbiol. Lett. 270 (2) (2007) 189–194.

[58] N.R. Panyala, E.M. Peña-Méndez, J. Havel, Silver or silver nanoparticles: a hazardous threat to the environment and human health? J. Appl. Biomed. 6 (3) (2008) 117–129.

[59] P. Ghafari, C.H. St-Denis, M.E. Power, X. Jin, V. Tsou, H.S. Mandal, N.C. Bols, X.S. Tang, Impact of carbon nanotubes on the ingestion and digestion of bacteria by ciliated protozoa, Nat. Nanotechnol. 3 (6) (2008) 347–351.

[60] M. Mortimer, K. Kasemets, A. Kahru, Toxicity of ZnO and CuO nanoparticles to ciliated protozoa Tetrahymena thermophila, Toxicology 269 (2–3) (2010) 182–189.

[61] M. Mortimer, K. Kasemets, M. Vodovnik, R. Marinšek-Logar, A. Kahru, Exposure to CuO nanoparticles changes the fatty acid composition of protozoa Tetrahymena thermophila, Environ. Sci. Technol. 45 (15) (2011) 6617–6624.

[62] I. Blinova, A. Ivask, M. Heinlaan, M. Mortimer, A. Kahru, Ecotoxicity of nanoparticles of CuO and ZnO in natural water, Environ. Pollut. 158 (1) (2010) 41–47.

[63] X. Zou, J. Shi, H. Zhang, Coexistence of silver and titanium dioxide nanoparticles: enhancing or reducing environmental risks? Aquat. Toxicol. 154 (2014) 168–175.

[64] C. Blaise, J.F. Férard (Eds.), Small-scale Freshwater Toxicity Investigations: Volume 2—Hazard Assessment Schemes, In: vol. 2, Springer Science & Business Media, 2005.

[65] S. Pakrashi, S. Dalai, A. Humayun, S. Chakravarty, N. Chandrasekaran, A. Mukherjee, Ceriodaphnia dubia as a potential bio-indicator for assessing acute aluminum oxide nanoparticle toxicity in fresh water environment, PLoS One 8 (9) (2013) e74003.

[66] M. Muna, I. Blinova, A. Kahru, I. Vinković Vrček, B. Pem, K. Orupõld, M. Heinlaan, Combined effects of test media and dietary algae on the toxicity of CuO and ZnO nanoparticles to freshwater microcrustaceans Daphnia magna and Heterocypris incongruens: food for thought, Nano 9 (1) (2019) 23.

[67] P. Das, M.A. Xenopoulos, C.D. Metcalfe, Toxicity of silver and titanium dioxide nanoparticle suspensions to the aquatic invertebrate, Daphnia magna, Bull. Environ. Contam. Toxicol. 91 (1) (2013) 76–82.

[68] M. Bundschuh, F. Seitz, R.R. Rosenfeldt, R. Schulz, Titanium dioxide nanoparticles increase sensitivity in the next generation of the water flea daphnia magna, PLoS One 7 (11) (2012).

[69] F. Seitz, S. Lüderwald, R.R. Rosenfeldt, R. Schulz, M. Bundschuh, Aging of TiO2 nanoparticles transiently increases their toxicity to the pelagic microcrustacean Daphnia magna, PLoS One 10 (5) (2015).

[70] K. Howe, M.D. Clark, C.F. Torroja, J. Torrance, C. Berthelot, M. Muffato, J.E. Collins, S. Humphray, K. McLaren, L. Matthews, S. McLaren, The zebrafish reference genome sequence and its relationship to the human genome, Nature 496 (7446) (2013) 498–503.

[71] G.S. Gupta, K. Kansara, H. Shah, R. Rathod, D. Valecha, S. Gogisetty, P. Joshi, A. Kumar, Impact of humic acid on the fate and toxicity of titanium dioxide nanoparticles in *Tetrahymena pyriformis* and zebrafish embryos, Nanoscale Adv. 1 (1) (2019) 219–227.

[72] S. George, T. Xia, R. Rallo, Y. Zhao, Z. Ji, S. Lin, X. Wang, H. Zhang, B. France, D. Schoenfeld, R. Damoiseaux, Use of a high-throughput screening approach coupled with in vivo zebrafish embryo screening to develop hazard ranking for engineered nanomaterials, ACS Nano 5 (3) (2011) 1805–1817.

[73] C.Y. Lee, J.L. Horng, P.Y. Chen, L.Y. Lin, Silver nanoparticle exposure impairs ion regulation in zebrafish embryos, Aquat. Toxicol. 214 (2019) 105263.

[74] S. Hu, J. Han, L. Yang, S. Li, Y. Guo, B. Zhou, H. Wu, Impact of co-exposure to titanium dioxide nanoparticles and Pb on zebrafish embryos, Chemosphere 233 (2019) 579–589.

[75] F.I. Aksakal, A. Ciltas, Impact of copper oxide nanoparticles (CuO NPs) exposure on embryo development and expression of genes related to the innate immune system of zebrafish (Danio rerio), Comp. Biochem. Physiol. Part C: Toxicol. Pharmacol. 223 (2019) 78–87.

[76] T. Fang, L.P. Yu, W.C. Zhang, S.P. Bao, Effects of humic acid and ionic strength on TiO$_2$ nanoparticles sublethal toxicity to zebrafish, Ecotoxicology 24 (10) (2015) 2054–2066.

[77] S. Lin, A.A. Taylor, Z. Ji, C.H. Chang, N.M. Kinsinger, W. Ueng, S.L. Walker, A.E. Nel, Understanding the transformation, speciation, and hazard potential of copper particles in a model septic tank system using zebrafish to monitor the effluent, ACS Nano 9 (2) (2015) 2038–2048.

[78] H. Zhang, X. He, Z. Zhang, P. Zhang, Y. Li, Y. Ma, Y. Kuang, Y. Zhao, Z. Chai, Nano-CeO2 exhibits adverse effects at environmental relevant concentrations, Environ. Sci. Technol. 45 (8) (2011) 3725–3730.

[79] T. Wu, H. Xu, X. Liang, M. Tang, Caenorhabditis elegans as a complete model organism for biosafety assessments of nanoparticles, Chemosphere 221 (2019) 708–726.

[80] S.W. Li, C.W. Huang, V.H.C. Liao, Early-life long-term exposure to ZnO nanoparticles suppresses innate immunity regulated by SKN-1/Nrf and the p38 MAPK signaling pathway in *Caenorhabditis elegans*, Environ. Pollut. 256 (2020) 113382.

[81] J. Moon, J.I. Kwak, Y.J. An, The effects of silver nanomaterial shape and size on toxicity to *Caenorhabditis elegans* in soil media, Chemosphere 215 (2019) 50–56.

[82] H. Kim, J. Jeong, N. Chatterjee, C.P. Roca, D. Yoon, S. Kim, Y. Kim, J. Choi, JAK/STAT and TGF-ß activation as potential adverse outcome pathway of TiO 2 NPs phototoxicity in Caenorhabditis elegans, Sci. Rep. 7 (1) (2017) 1–12.

[83] P.H.M. Hoet, A. Nemmar, B. Nemery, Health impact of nanomaterials? Nat. Biotechnol. 22 (2004) 19.

[84] R.D. Holbrook, K.E. Murphy, J.B. Morrow, K.D. Cole, Trophic transfer of nanoparticles in a simplified invertebrate food web, Nat. Nanotechnol. 3 (2008) 352–355.

[85] R. Werlin, J.H. Priester, R.E. Mielke, S. Kramer, S. Jackson, P.K. Stoimenov, G.D. Stucky, G. N. Cherr, E. Orias, P.A. Holden, Biomagnification of cadmium selenide quantum dots in a simple experimental microbial food chain, Nat. Nanotechnol. 6 (2011) 65–71.

[86] X. Zhu, J. Wang, X. Zhang, Y. Chang, Y. Chen, Trophic transfer of Tio(2) nanoparticles from daphnia to zebrafish in a simplified freshwater food chain, Chemosphere 79 (2010) 928–933.

[87] D.E. Alexander, Bioaccumulation, bioconcentration, biomagnification, in: Environmental Geology, Springer, Netherlands, 1999, , pp. 43–44.

[88] R. van der Oost, J. Beyer, N.P. Vermeulen, Fish bioaccumulation and biomarkers in environmental risk assessment: a review, Environ. Toxicol. Pharmacol. 13 (2003) 57–149.

[89] R.V. Thomann, J.P. Conolly, T.F. Parkerton, An equilibrium model of organic chemical accumulation in aquatic food webs with sediment interaction, Environ. Toxicol. Chem. 11 (1992) 615–629.

[90] W.B. Neely, Chemicals in the Environment: Distribution, Transport, Fate, Analysis, M. Dekker, 1980.

[91] J.A. Arnot, F.A. Gobas, A review of bioconcentration factor (BCF) and bioaccumulation factor (BAF) assessments for organic chemicals in aquatic organisms. Environ. Rev. 14 (4) (2006) 257–297, https://doi.org/10.1139/a06-005.

[92] F. Dang, Y.Z. Chen, Y.N. Huang, H. Hintelmann, Y.B. Si, D.M. Zhou, Discerning the sources of silver nanoparticle in a terrestrial food chain by stable isotope tracer technique, Environ. Sci. Technol. 53 (7) (2019) 3802–3810.

[93] J.D. Judy, J.M. Unrine, W. Rao, P.M. Bertsch, Bioaccumulation of gold nanomaterials by ManducaSexta through dietary uptake of surface contaminated plant tissue, Environ. Sci. Technol. 46 (2012) 12672–12678.

[94] L. Zhao, Y. Sun, J.A. Hernandez-Viezcas, A.D. Servin, J. Hong, G. Niu, J.R. Peralta-Videa, M. Duarte-Gardea, J.L. Gardea-Torresdey, Influence of Ceo2 and Zno nanoparticles on cucumber physiological markers and bioaccumulation of Ce and Zn: a life cycle study, J. Agric. Food Chem. 61 (2013) 11945–11951.

[95] S. Lekamge, A.F. Miranda, A.S. Ball, R. Shukla, D. Nugegoda, The toxicity of coated silver nanoparticles to Daphnia carinata and trophic transfer from alga Raphidocelis subcapitata, PLoS One 14 (4) (2019).

[96] S. Pakrashi, S. Dalai, N. Chandrasekaran, A. Mukherjee, Trophic transfer potential of aluminium oxide nanoparticles using representative primary producer (chlorella Ellipsoides) and a primary consumer (*Ceriodaphnia dubia*), Aquat. Toxicol. 152 (2014) 74–81.

[97] J.D. Judy, J.M. Unrine, P.M. Bertsch, Evidence for biomagnification of gold nanoparticles within a terrestrial food chain, Environ. Sci. Technol. 45 (2011) 776–781.

[98] J.L. Ferry, P. Craig, C. Hexel, P. Sisco, R. Frey, P.L. Pennington, M.H. Fulton, I.G. Scott, A. W. Decho, S. Kashiwada, C.J. Murphy, T.J. Shaw, Transfer of gold nanoparticles from the water column to the estuarine food web, Nat. Nanotechnol. 4 (2009) 441–444.

[99] J.A. Hernandez-Viezcas, H. Castillo-Michel, A.D. Servin, J.R. Peralta-Videa, J. L. Gardea-Torresdey, Spectroscopic verification of zinc absorption and distribution in the desert plant Prosopis Juliflora-Velutina (velvet Mesquite) treated with Zno nanoparticles, Chem. Eng. J. 170 (2011) 346–352.

[100] T.N.V.K. Prasad, P. Sudhakar, Y. Sreenivasulu, P. Latha, V. Munaswamy, K.R. Reddy, T. S. Sreeprasad, P.R. Sajanlal, T. Pradeep, Effect of nanoscale zinc oxide particles on the germination, growth and yield of peanut, J. Plant Nutr. 35 (2012) 905–927.

[101] U. Song, H. Jun, B. Waldman, J. Roh, Y. Kim, J. Yi, E.J. Lee, Functional analyses of nanoparticle toxicity: a comparative study of the effects of Tio2 and ag on tomatoes (Lycopersicon Esculentum), Ecotoxicol. Environ. Saf. 93 (2013) 60–67.

[102] Z. Pipan-Tkalec, D. Drobne, A. Jemec, T. Romih, P. Zidar, M. Bele, Zinc bioaccumulation in a terrestrial invertebrate fed a diet treated with particulate ZnO or ZnCl$_2$ solution, Toxicology 269 (2010) 198–203.

[103] T. Nemati, M. Sarkheil, S.A. Johari, Trophic transfer of CuO nanoparticles from brine shrimp (Artemia salina) nauplii to convict cichlid (Amatitlania nigrofasciata) larvae: uptake, accumulation and elimination, Environ. Sci. Pollut. Res. 26 (10) (2019) 9610–9618.

[104] S. Dalai, V. Iswarya, M. Bhuvaneshwari, S. Pakrashi, N. Chandrasekaran, A. Mukherjee, Different modes of Tio2 uptake by Ceriodaphnia Dubia: relevance to toxicity and bioaccumulation, Aquat. Toxicol. 152 (2014) 139–146.

[105] X. Chen, Y. Zhu, K. Yang, L. Zhu, D. Lin, Nanoparticle TiO2 size and rutile content impact bioconcentration and biomagnification from algae to daphnia, Environ. Pollut. 247 (2019) 421–430.

[106] M. Bhuvaneshwari, V. Iswarya, S. Vishnu, N. Chandrasekaran, A. Mukherjee, Dietary transfer of zinc oxide particles from algae (Scenedesmus obliquus) to daphnia (Ceriodaphnia dubia), Environ. Res. 164 (2018) 395–404.

[107] X. Zhu, Y. Chang, Y. Chen, Toxicity and bioaccumulation of TiO2 nanoparticle aggregates in daphnia magna, Chemosphere 78 (2010) 209–215.

[108] Y. Wang, A.J. Miao, J. Luo, Z.B. Wei, J.J. Zhu, L.Y. Yang, Bioaccumulation of Cdte quantum dots in a freshwater alga Ochromonas Danica: a kinetics study, Environ. Sci. Technol. 47 (2013) 10601–10610.

[109] R.E. Mielke, J.H. Priester, R.A. Werlin, J. Gelb, A.M. Horst, E. Orias, P.A. Holden, Differential growth of and nanoscale TiO(2) accumulation in Tetrahymena Thermophila by direct feeding versus trophic transfer from pseudomonas aeruginosa, Appl. Environ. Microbiol. 79 (2013) 5616–5624.

[110] C.M. Zhao, W.X. Wang, Comparison of acute and chronic toxicity of silver nanoparticles and silver nitrate to daphnia magna, Environ. Toxicol. Chem. 30 (2011) 885–892.

[111] G.S. Gupta, A. Kumar, V.A. Senapati, A.K. Pandey, R. Shanker, A. Dhawan, Laboratory scale microbial food chain to study bioaccumulation, biomagnification, and ecotoxicity of cadmium telluride quantum dots, Environ. Sci. Technol. 51 (3) (2017) 1695–1706.

[112] T. Galloway, C. Lewis, I. Dolciotti, B.D. Johnston, J. Moger, F. Regoli, Sublethal toxicity of Nanotitanium dioxide and carbon nanotubes in a sediment dwelling marine polychaete, Environ. Pollut. 158 (2010) 1748–1755.

[113] H. Sun, X. Zhang, Q. Niu, Y. Chen, J.C. Crittenden, Enhanced accumulation of arsenate in carp in the presence of titanium dioxide nanoparticles, Water Air Soil Pollut. 178 (2007) 245–254.

[114] N.A. Lewinski, H. Zhu, C.R. Ouyang, G.P. Conner, D.S. Wagner, V.L. Colvin, R.A. Drezek, Trophic transfer of amphiphilic polymer coated Cdse/Zns quantum dots to danio Rerio, Nanoscale 3 (2011) 3080–3083.

[115] J. Chen, H. Li, X. Han, X. Wei, Transmission and accumulation of Nano-TiO2 in a 2-step food chain (Scenedesmus obliquus to Daphnia magna), Bull. Environ. Contam. Toxicol. 95 (2) (2015) 145–149.

[116] S. Majumdar, J. Trujillo-Reyes, J.A. Hernandez-Viezcas, J.C. White, J.R. Peralta-Videa, J. L. Gardea-Torresdey, Cerium biomagnification in a terrestrial food chain: influence of particle size and growth stage, Environ. Sci. Technol. (2015).

[117] J.M. Unrine, W.A. Shoults-Wilson, O. Zhurbich, P.M. Bertsch, O.V. Tsyusko, Trophic transfer of au nanoparticles from soil along a simulated terrestrial food chain, Environ. Sci. Technol. 46 (17) (2012) 9753–9760.

[118] E.J. Petersen, D.X. Flores-Cervantes, T.D. Bucheli, L.C. Elliott, J.A. Fagan, A. Gogos, S. Hanna, R. Kägi, E. Mansfield, A.R.M. Bustos, D.L. Plata, Quantification of carbon nanotubes in environmental matrices: current capabilities, case studies, and future prospects, Environ. Sci. Technol. 50 (9) (2016) 4587–4605.

[119] M. Mortimer, E.J. Petersen, B.A. Buchholz, E. Orias, P.A. Holden, Bioaccumulation of multiwall carbon nanotubes in Tetrahymena thermophila by direct feeding or trophic transfer, Environ. Sci. Technol. 50 (16) (2016) 8876–8885.

[120] D. Kriebel, J. Tickner, P. Epstein, J. Lemons, R. Levins, E.L. Loechler, The precautionary principle in environmental science, Environ. Health Perspect. 109 (2001) 871–876. Available from: http://ehp.niehs.gov/members/2001/109p871-87kriebel/kriebel-full.html.

[121] Organisation for Economic Co-cperation and Development (OECD), Categorisation of manufactured nanomaterials. Workshop Report ENV/JM/MONO (2016)9, http://go.nature.com/2uQY5lS, 2016.

[122] NIOSH, Occupational Exposure to Carbon Nanotubes and Nanofibres, National Institute for Occupational Safety and Health, 2013.

[123] USEPA (United States Environmental Protection Agency), Reviewing New Chemicals under the Toxic Substances Control Act (TSCA). Control of Nanoscale Materials under the Toxic Substances Control Act, Available from: https://www.epa.gov/reviewing-new-chemicals-under-toxic-substances-control-act-tsca/controlnanoscale-materials-under, 2017. Accessed 19 January 2017.

[124] BSI-British Standards Institute, Nanotechnologies – Part 2: Guide to Safe Handling and Disposal of Manufactured Nanomaterials. PD 6699-2, BSI, London, 2007. Available from: https://shop.bsigroup.com/Browse-By Subject/Nanotechnology/Guidance-fornanotechnology/.

[125] IFA – Institut für Arbeitsschutz der Deutschen Gesetzlichen Unfallversicherungen, Criteria for Assessment of the Effectiveness of Protective Measures, Available from: http://www.dguv.de/ifa/fachinfos/nanopartikel-am-arbeitsplatz/beurteilung-vonschutzmassnahmen/index-2.jsp, 2011.

CHAPTER 18

Biological toxicity of nanoparticles

Violina Kakoty[a], Sarathlal K.C.[a], Meghna Pandey[a], Sunil Kumar Dubey[b], Prashant Kesharwani[c], and Rajeev Taliyan[a]

[a]Department of Pharmacy, Birla Institute of Technology and Science, Pilani, Rajasthan, India
[b]R&D Healthcare Division, Emami Ltd, Belgharia, Kolkata, India
[c]Department of Pharmaceutics, School of Pharmaceutical Education and Research, Jamia Hamdard, New Delhi, India

1. Introduction

Nanotechnology is an emerging field, and it has diverse applications in the field of science. The concept of nanoscience and nanotechnology was given by the physicist Richard Feynman in the topic "There's Plenty of Room at the Bottom" at an American Physical Society meeting that was held on December 29, 1959, at the California Institute of Technology [1]. During his talk, Feynman described a process to control and manipulate the atoms and molecules, whereby the term nanotechnology was coined by an engineer Prof. Norio Taniguchi in 1974 from the University of Tokyo, in his article about semiconductors process and material handling. The word *nano* has come from a Latin word, meaning dwarf [2]. Since then, nanotechnology has proved its worth in numerous clinical uses possessing wide applications in the field of science like physics, engineering, electronics, with recent advancement in pharmaceutical and biomedical sciences. Nowadays, the term nanomedicine is used that involves the application of principles of nanotechnology for the benefits of human healthcare. Nanomedicine encompasses nanoparticles in such a way that it can be used for diagnostic, therapeutic, and research purposes [3].

Nanoparticles are generally regarded as the particulate system with a size ranging from 10 to 1000 nm, and for medical applications, the size should be preferentially less than 200 nm [4]. The drug is dissolved, dispersed, entrapped, or encapsulated to the matrix system. Depending on the type of method of preparation used, the nanosphere, nanocapsules, or nanoparticles can be obtained [5]. The nanoparticles are designed with unique physicochemical properties when compared with other materials. As the nanoparticle directly targets the tissue after the administration, therefore, it becomes necessary to evaluate its toxicological profile. During the evaluation, the toxicity of the whole formulation is determined, which involves both the nanoparticles and the drug, and discrimination between these two is not possible. So, the toxicology of the blank nanoparticles without the drug should be determined, and this becomes more critical if the polymer/material used is nonbiodegradable [6, 7].

Nanoparticle Therapeutics
https://doi.org/10.1016/B978-0-12-820757-4.00016-8

The nanoparticles owing to its size can penetrate the organs and the deepest layer of the skin easily. Therefore, toxicity testing has to be done in both the external and internal environment. The toxicity of the nanoparticles can be broadly classified into two types: environmental toxicity and biological toxicity. The biological toxicity of nanoparticles can be responsible for the oxidative stress and inflammatory responses. The toxicity in the circulatory system includes the paradox effect on heart function and the prethrombotic effects. The nanoparticles can also cause carcinogenicity, genotoxicity, and teratogenicity. Some nanoparticles have the ability to cross the blood-brain barrier (BBB) readily, so it can also slow the chances of toxicity to the brain [8, 9].

2. Routes of nanoparticle exposure and its relation to toxic effects

The nanoparticles can be ingested in the body via inhalation, oral, intravenous, and by the dermal route. The oral route is the noninvasive route for the administration of the nanoparticles, and it has some of the associated disadvantages that possess high chances to translocate into the systemic circulation, high risk of the liver toxicity due to the first-pass metabolism, and for the uptake of the nanoparticles, the intact intestinal mucosa is required. The intravenous route has the problem of systemic toxicity as the nanoparticles are directly injected into the bloodstream. Despite the BBB present in the brain, the nanoparticles have the ability to penetrate the tight junctions owing to its size, which makes it prone to cause the particle-mediated toxicity. The pulmonary route shows great potential for the local and systemic toxicity. The aggregation of the nanoparticles can lead to the inflammation in the mucosal linings. Transdermal drug delivery has the chance to cause the local irritation and has the ability to translocate to the systemic circulation. Therefore the data for nanoparticle toxicity is necessary to avoid the detrimental side effects [10].

3. Influence of the physicochemical properties on the toxicity

Though the nanoparticles are designed to reduce the toxic effects of the drugs and enhance biocompatibility, it has certain risks associated with it owing to its unique properties. Owing to this challenge, the term "nanotoxicology" has been used that is the branch of biomedicine which is specifically designed for the study and application of the toxicity of nanomaterials. These studies determine up to what extent the nanomaterials pose a risk to human health and the environment. The nanoparticles have unique properties that make it essential to study its toxicity [11].

3.1 Size and surface area

The interaction of the nanoparticles with the biological system is largely influenced by the size and surface characteristics. The nanoparticles have a large specific surface area that influences the high catalytic activity and reaction capability. Huo et al. have proved in the study that the gold nanoparticles of the size less than 6 nm can easily penetrate the nucleus, whereas the nanoparticles of the size 10 or 16 nm have the ability to penetrate only to the cytoplasm. This implies that the smaller size nanoparticles are more toxic as compared with the larger size nanoparticles [12]. Pan et al., in their study, proved that the nanoparticles of the size of 15 nm are 60 times less toxic to the macrophages, fibroblasts, melanoma cells, and epithelial cells when compared with the 1.4 nm nanoparticles. These data also reveal that the nanoparticles not only penetrate to the nucleus but also intermingle with the DNA backbone [13].

The nanoparticles of the size less than 50 nm and administered by the intravenous injection has the ability to show more toxic effects on the tissue while the size greater than the 50 nm and with the positive charge are taken up by the reticuloendothelial system (RES) and hinders their path to the tissues. However, it protects the other tissues from the toxicity but increases the chances of the oxidative stress on the liver and spleen [14].

The particles with the dimension of the <100 nm have the ability to cause severe respiratory problems in comparison with the larger nanoparticles. The particles with a size less than 100 nm have deposition in all the parts of the lungs, and the particles with less than 10 nm deposit mainly in the tracheobronchial region and the particles of the size ranging from 10 to 20 nm have the deposition in the alveolar region. The size influences the distribution of the nanoparticles that, in turn, decides the toxicity of the nanoparticles [15].

The nanoparticles have effective adsorption on the cell surface due to the large specific surface area and their smaller size. The study was conducted to understand the hemolytic activity of the mesoporous silicon particles with a size ranging from 100 to 600 nm toward the human erythrocytes. The particles of the size of 100 nm are effectively adsorbed on the erythrocytes without disturbing the membrane, whereas adsorption of the particles of 600 nm caused the deformation of the cell membrane and resulted in the erythrocyte destruction and hemolysis [16].

3.2 Shape

The nanoparticles have the characteristic shapes that are spheres, cylindrical, ellipsoids, cubes, rod, and sheets. The toxicity of the nanoparticles largely depends on its shape [17]. The nanoparticles of the spherical shape are more liable to the endocytosis as compared with the nanofibers and nanotubes [18]. This indicates that the shape impacts the membrane wrapping process in vivo during phagocytosis and endocytosis [19]. The hydroxyapatite nanoparticles were compared on the basis of the shape in the cultured BEAS-2B cells, and the result indicates that the needle shape and plate shape nanoparticles are responsible for more cell death as compared with the rod shape and spherical nanoparticles [20].

3.3 Surface charge

The interaction of the nanoparticles to the biological system largely depends on the surface charge. The surface charge also regulates colloidal behavior, selective absorption, plasma protein binding, transmembrane permeability, and BBB integrity [21, 22]. The neutral and negative charge nanoparticles show less cellular uptake as compared with the positively charged nanoparticles [23]. The positive charge nanoparticle has the ability to penetrate effectively through the membrane and bind effectively to the DNA that is a negative charge and cause damage by prolonging the G0/G1 phase of the cell cycle [24]. The positively charged particles have higher chances of the opsonization and complement with the blood and biological fluids [25].

The proteins that are absorbed on the surface of the nanoparticles have the ability to alter the surface properties of the nanoparticles. The proteins that are absorbed are known as the protein crown which is composed of the immunoglobulin proteins like the albumin, immunoglobulin G, and fibrinogen along with other functional molecules [26]. This binding alters the structure of the protein that leads to enzymatic activity loss and ultimately disturbs the biological process and can cause diseases like amyloidosis [27].

3.4 Chemical composition

Even though the toxicity of the nanoparticle largely depends on the shape and size, but the other factors like the crystal structure and chemical composition plays a role in the toxicity. The study was conducted on the mouse fibroblasts with 20 nm zinc oxide and silicon dioxide nanoparticles. The results show that the silicon dioxide nanoparticles cause the DNA structure alternation, and zinc oxide nanoparticles are responsible for the oxidative stress [28].

The crystal structure also influences the toxicity of nanoparticles. A study was conducted on the human bronchial cell line using a different type of crystal lattice of the titanium oxide nanoparticles. The octahedral (anatase)-shaped crystals of the titanium oxide is nontoxic, whereas the prism (rutile) shape crystals of the titanium oxide lead to the oxidative stress to DNA and abnormal chromosomal segregation during mitosis [29].

3.5 Shell and surface coating

The nanoparticles biocompatibility, stability, solubility in water and biological fluid, and the electrical, optic, and magnetic properties depend on the shell. The shell also impacts the nanoparticles pharmacokinetics and accumulation in the body. Therefore the shell mainly decreases the toxicity by influencing the specific interaction to the cells and biological molecules [30].

The coating of the nanoparticle surface is done with the help of inorganic or organic compounds such as lipids, poly(lactic acid), proteins, and silicon. These have the ability to change the nanoparticle properties and their transport and accumulation in the body [31]. The quantum dots toxicity is decreased due to the shell that improves the solubilization, increases the stability and prevents the oxidative and desalination. This ultimately reduces the leakage of the metal ions from the quantum dots core [32].

3.6 Aggregation and concentration

The aggregation is dependent on the shape and size of the nanoparticles. It has been reported that carbon nanotubes are generally accumulated in the lungs, spleen, and liver that induce the cytotoxic effects by accumulating at the site for a long time. The toxicity of the nanoparticles generally decreases with the increase in concentration [32].

4. Mechanism of biological toxicity

The nanoparticle toxicity is mainly influenced by its physical and chemical properties like the shape, size, surface properties, and presence or absence of the active moieties on the surface. Smaller-size nanoparticles permit entry into the epithelial and endothelial barrier into the blood and lymph and through the blood and lymph to the different tissues and organs, including the heart, brain, liver, kidneys, bone marrow, spleen, and nervous system. The nanoparticles can also be transported to the cells by simple diffusion through the cell membrane or the transcytosis mechanism [33, 34].

The nanoparticles due to its nanosize can ooze out through the endothelium in inflammatory sites, tumors and even penetrate the microcapillaries [35]. The nanoparticles can increase the plate aggregation and cause thrombosis, myocardial infarction, and inflammation at the upper and lower respiratory tracts [36, 37]. The nanoparticle can reach up to the cell organelles level like nuclei and mitochondria altering the metabolism of cells that leads to DNA mutations, lesions, and even cell death [33].

The nanoparticles with a size of 50 nm after coming in contact with the lung tissue result in the perforation of the alveolar cells and cause the entry of it into the cells. This results in cell necrosis due to the release of the lactate dehydrogenase [38]. On the other hand, the nanoparticles induce the formation of the reactive oxygen species (ROS) by peroxidation of the membrane lipids that results in the loss of the membrane flexibility and can ultimately lead to cell necrosis [39].

Cell differentiation and protein synthesis can be affected by the nanoparticle, and it can also activate the pro-inflammatory genes and synthesis of the inflammatory mediators.[40] The nanoparticle interaction with the cell increases the gene expression for the formation of lysosomes and hinders the protein synthesis [41]. Fig. 1 describes the possible mechanism(s) by which nanoparticles exert tissue toxicity after being ingested.

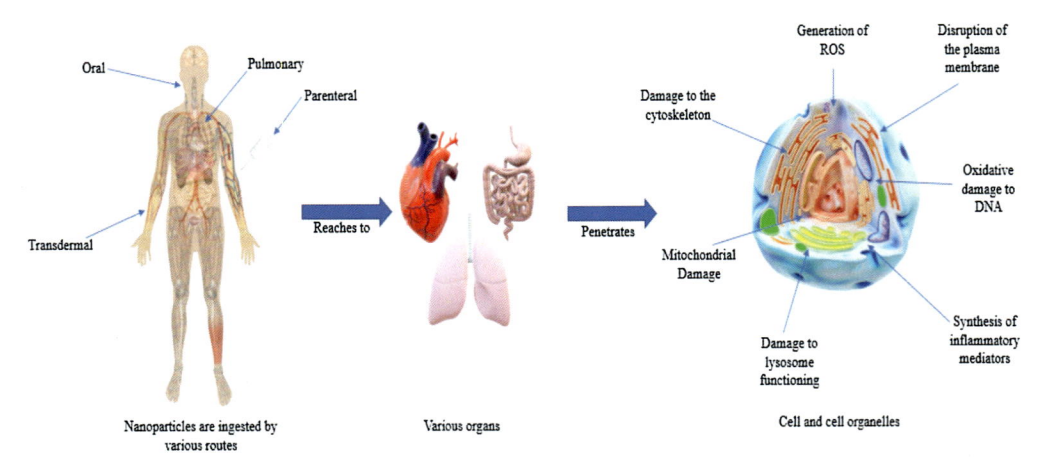

Fig. 1 Mechanism of toxicity of nanoparticles.

5. Molecular toxicity of nanomedicines

Nanoparticles interact with biological environment after entering into the circulation and the biomolecules forms a coat over the nanoparticles [42]. The coat is known as protein corona consisting of soft and hard layers. The hard layer will be tightly bound with nanoparticle whereas the soft layer is dynamic. The feature of hard corona imparts to the characteristic feature of the nanoparticle [43, 44].

The hard corona consists more than 100 types of proteins [45, 46] which influence the property of nanoparticle, including size, shape, and charge[42]. The interaction with protein can either increase or decrease the size of nanoparticle causing the zeta potential to reach to more anionic phase [46–49]. The characteristic change of nanoparticles on interaction with proteins influences the functionalities which is implicated with the safety of the nanoparticles [42, 50, 51]. Studies reported that nanoparticles could alter the structural changes in albumin [52] cytochrome [53] and ribonuclease A [54].

The larger nanoparticles with lesser surface curvature interact in ease with proteins which ultimately cause protein structure alteration. Further the irreversible changes occur due to nanoparticle and protein interaction. The transferrin when interacts with iron oxide nanoparticles, it loses its structure and function irreversibly, that causes premature release of iron [55]. The nanoparticle on interaction with fibrinogen cause its unfolding which activates inflammatory pathways [56]. Another serious issue of nanoparticle is protein aggregation on its surface. The β2-microglobulin fibrillation was increased in acidic environment in presence of nanoparticles [57].

The protein corona of nanoparticle which leads to altered protein structure and cause decreased toxicity [58, 59] may be due to the poor interaction with membrane protein and nanoparticle coated with proteins as it preserve membrane integrity [58, 60, 61].

Studies also reported that protein corona prevent the interaction of nanoparticle with platelet activation and hemolysis [46]. Hence, the formation of circulatory proteins layer over nanoparticle prevent the interaction of membrane proteins with nanoparticles. Hence nanoparticle-mediated protein corona induced damage to circulatory proteins considered to be insignificant as it reduces overall toxicity of nanomedicine.

6. Cellular toxicity of nanomedicines

The cellular toxicity of nanoparticles is due to disruption of proteins within the cellular membrane. The nanoparticle can alter the cellular integrity as it can bind and damage the cell membrane proteins which cause the leakage of cellular components. The several nanoparticle made of zinc oxide [62], titanium dioxide [63], polycation particles [64], and polystyrene [65] disrupt lysosomal membrane. The damage of lysosomal membrane cause to release iron, protons, and hydrolytic enzymes, that will induce endoplasmic reticulum stress, oxidative stress, mitochondrial dysfunction, and protein aggregation [66].

Moreover, the nanoparticles can alter the environment of lysosome without damaging its membrane. For instance, gold nanoparticles alkalize the lysosome compartment which leads to accumulation of autophagosomes [67]. The polyethylenimine (PEI) which is found in polymeric nanoparticle can damage the mitochondrial membrane, causing to ooze out the protons and inhibition of cytochrome c oxidase enzyme action [68]. Further, it impairs the electron transport system and generation of adenosine triphosphate [68]. Hence, the cytotoxicity of nanoparticle is caused by the damage to cellular membrane or disrupting endomembrane integrity.

Another mechanism of nanoparticle toxicity is nanoparticle-mediated generation of ROS which can damage DNA, proteins, and lipids. The ROS is generated either by nanoparticle catalyzed free radical reactions or by disrupting cellular homeostasis [69]. ROS further damages other cellular organelles and sensitize cells to oxidative or nitrosative stress [70]. Moreover, the nanoparticle induce apoptosis regulated necrosis and autophagic cell death [71]. The nanoparticle size, shape [13, 72], dose [73], incubation duration [73], and the associated charge [21] can influence on the cell death pathways.

7. Tissue toxicity of nanomedicines

The cellular level toxicities of nanoparticles causing alteration in endogenous biomolecules can affect the functionality of tissue system/organ. Generally, the organs which have more accumulation of nanoparticle is more prone to get affected which can be identified with the cellular or molecular level alterations prompted by nanoparticles. The accumulation of nanoparticle in organ depends on its route of administration as well. For instance, accumulation will be higher in lungs if administered by intratracheal and accumulation will be more in liver when administered intravenously. Reports reveal that the accumulation of carbon nanotubes in lungs causes lung inflammation [74, 75].

Further, on intravenous injection of positively charged nanoparticles caused hepatotoxicity as evidenced from the elevated level of hepatic enzymes [76].

8. Nanoparticle toxicity at different organs

The nanoparticle after ingestion reaches to various organs of the body, so the chance of the toxicity increases in the organ.

8.1 Liver toxicity

The liver toxicity is the first organ to be considered because the clearance of the nanoparticles depends on the liver. The BALB/c mice were injected intraperitoneally with the naked gold nanoparticles of the size ranging from 3 to 100 nm with a dose of 8 mg/kg/week. Gold nanoparticles with size ranging from 8 to 37 nm cause severe sickness in mice. The Kupffer cells in the liver increased, and the structural integrity of the lungs are lost that are toxicity signs [77]. The liver toxicity's primary mechanism arises from the acute inflammatory changes and apoptosis [78].

The cadmium selenide (CdSe) quantum dots were found to be cytotoxic because of the liberation of Cd^{+2} ions. The cytotoxic activity was measured using the macrophage cell viability test for a period of 48 h. The breakdown of the nanoparticle surface has caused the liberation of the ion and toxicity [79, 80].

8.2 Heart toxicity

The heart is one of the critical organs and the nanoparticle has shown to affect cardiac electrophysiology. The study was conducted on the mice in which the pegylated gold nanoparticles of size of 10, 30, or 50 nm have been injected. After administration of the first injection, the mice are sacrificed in 2, 4, or 12 weeks. The signs of the cardiac problem have been shown after the 2 weeks. The symptoms that are shown in the mice are an increase in the left ventricular mass and enhanced left ventricular end-diastolic inner dimension. These symptoms are absent in the 4 and 12 week mice that indicate that these symptoms are reversible [81].

8.3 Kidney toxicity

The kidney is the target of the nanoparticles with a size of less than 6 nm and is widely eliminated by this route. The quantum dots coated with the zwitterion or neutral layer of the size <6 nm are widely cleared by the kidney crossing the glomeruli. The kidney toxicity was low as there was very little interaction with the nanoparticles and the renal tissues [82].

8.4 Lung toxicity

Lungs are a critical organ that is highly prone to nanoparticle toxicity. The alveolar macrophages are exposed to the silver nanoparticles of the size 15, 30, and 55 nm in the concentration range of 10–75 µg/mL. The nanoparticles are internalized after 24 h and show the toxicity signs like an increase in the level of ROS and enhanced cytokines levels [83].

8.5 Brain toxicity

The human neural cells, like the hippocampal cells, are the most delicate and sensitive cells in the body, which plays a critical role in the brain functions, and they are highly prone to ischemia and other factors [84]. The nanoparticles have the ability to cross the BBB and produce toxic effects in the brain. The small-sized particles, because of their higher mobility, can easily cross the BBB by either carrier-mediated or passive diffusion [85]. The silver nanoparticles are exposed to the neuroendocrine (PG-12) cell lines, and the results indicate the reduction in the level of the dopamine. It has been found out that the silver nanoparticles are more toxic to the human brain as compared with the manganese nanoparticles [86].

The silica–coated cobalt ferrite nanoparticles after the intravenous injection have been found in the mice brain [87]. In another study conducted on the maternal mice that have been exposed to the titanium dioxide nanoparticles, and the result shows that the gene expression in the brain of the new-born pups has been altered. The genes that are linked to the apoptosis and brain development are altered, and also the oxidative stress has been enhanced in the mice. This indicates that the nanoparticles have potential toxicity on the developmental process of new-borns [88].

9. Immunological responses to nanomedicines

Unwanted immunological recognition is a major challenge in nanomedicine. Accumulation of nanoparticles is primarily seen in liver and spleen following intravenous injection [89, 90]. On passing through the circulation, organ macrophages recognize and engulf these nanoparticles. The reticuloendothelial system composed of phagocytic cells is responsible for eliminating foreign bodies. Opsonins and macrophages play a collaborative dual role in recognizing and clearance of nanoparticle. Opsonins are plasma proteins that bind to the surface of foreign particles through its immunostimulatory action, hence, marking them for phagocytosis. Immunoglobulins and complement proteins are the most common opsonins. Nanoparticle engulfment is activated by immunoglobulins either by direct binding to macrophages or by activating the complement system [91]. Recognition of foreign microparticles and nanoparticles is served by the complement system. Since bacteria and viruses usually express repetitive surface units the recognition by complement system majorly depends on identifying surface patterns [92, 93].

Therefore the similar principle applies to the nanoparticle-induced complement activation, suggesting an unbeneficial basis for uniform nanoparticle surface [92]. Nanoparticle-induced complement activation is also dependent on size and surface charge properties, for instance, larger size, enhanced cholesterol content, and cationic or anionic charge have been correlated with liposome-induced complement activation [94], nanoparticle hydrophobicity and immunological activation also share such type of positive correlation [95], originating from the fact that recognition of hydrophobic molecules possess damage-associated patterns [96]. In fact, release of hydrophobic cellular components due to disruption of cells warns the immune system of the damage caused. Additionally, endogenous plasma proteins also activate the complement system specifically those undergoing nanoparticle-induced conformational changes [42]. Proteins binding to the nanoparticle surface may acquire a misfolded form, rendering the immune system to misunderstand them for foreign bodies. For instance, one study speculated that due to conformational changes in albumin, polymeric nanoparticles lead to induction of the complement system [97]. Covering of all the nanoparticles by opsonins, begins the internalization by immune cells mediated via nonspecific interactions or through binding to Fc receptors [91, 98]. Moreover, interaction of nanoparticles with toll-like receptors (TLRs) can directly stimulate the immune system [99, 100]. Although immune reactions are generated independently by these receptors and the complement system; however, they might act synergistically to enhance inflammatory responses [101]. Clearance by macrophages leads to a lower level of nanoparticle accumulation in the target tissue, thereby lowering its therapeutic efficacy. However, other immune cells and vascular endothelial cells are activated by complement fragments resulting in the production of proinflammatory and vasoactive agents [102]. Therefore, serious complications like adverse inflammatory reactions might be possibly triggered by nanoparticle-induced complement activation [103]. For instance, hypersensitivity reactions like skin reactions, hypotension, hypertension, respiratory issues or pain was reported in humans [104] and pigs [105] triggered by the nanotherapeutic Doxil. All these reactions were believed to occur due to activation of the complement system. Taxol is another clinically approved nanotherapeutic, administered as a micellar formulation, is found to manifest complement-induced adverse reactions [94]. Various small molecule drugs through non-complement mediated pathways also cause immunological hypersensitivity reactions [106]. Hence, based on these observations it cannot be concluded that nanotherapeutics are more immunogenic than conventional drugs, rather a fair possibility could be that the main mechanisms for immune activation could be different. Release of proinflammatory cytokines were done by cationic liposomes in a TLR4-dependent manner [76]. Understanding the nanoparticles interaction with an in vivo environment have identified some of the major proteins that on exposure to serum or plasma bind to the surface of nanoparticles [26, 45, 46, 107]. In majority of these studies, complement proteins are identified as one of the main compositional elements of the protein corona. Nevertheless, information about the immunological activity cannot be predicted by the mere presence of complement proteins on the nanoparticle surface. Accordingly, initially binding to the nanoparticle surface activated complement proteins may be released in the fluid phase [92]. Hence, examination of the nanoparticle surface and the fluid phase should be done by a complete analysis of complement activation. Furthermore, ethylenediaminetetraacetic acid (EDTA), an inhibitor of complement activation has also been used to prevent blood coagulation in several protein corona studies [26, 45, 108, 109]. Consequently, disturbance between the experimental and in vivo conditions may arise due to the unintentional suppression of the complement system. Enzyme-linked immunosorbent assay (ELISA) can be efficiently used to measure the amount of produced complement activation products in the fluid phase followed by incubation of nanoparticles with serum or heparin-treated plasma samples. For instance, complement activation in in-vitro studies have been elicited by pegylated liposomes [105], carbon nanotubes [110], and polystyrene particles [111] when measured by ELISA. Measurement of cytokine production in human peripheral blood mononucleated cells (PBMCs) can be used to predict nanoparticle-induced immune reactions [112]. Inflammatory responses can be indicated by cytokines such as interferon gamma (IFN-γ), tumor necrosis factor (TNF-α), and interleukin-12 (IL-12). Moreover, on the genome level, immunological adverse drug reactions can be linked to distinct individual differences [113] suggesting that individual identification, for avoiding adverse reactions in the clinic sensitive would be helpful after infusion of nanoparticles [114]. Immunological activation can be partially prevented by various other ways in which the most common way is the use of antifouling agents, which reduce protein binding. For example, various nanodelivery systems uses a stealth polymer known as polyethylene glycol (PEG) [64, 115–117]. Nevertheless, PEG coating of nanoparticles does not provide complete protection from protein binding [111, 118]. More or less, regulating the density of PEG chains can optimize the stealth effect. For instance, conformation change from a mushroom to a brush structure exhibits greater protection from protein which could be achieved by increasing the density of PEG on a liposome surface [119]. PEG may also cause immunological reactions even with its benefits of pegylation. In certain cases, incubation of PEG with human serum has paradoxically been found to trigger the complement system [120]. Moreover, accelerated blood clearance (ABC) phenomenon can take place invoking a rapid clearance of particles from the circulation through the formation of anti-PEG antibodies via repeated injections of pegylated nanoparticle [121]. There are other strategies that have been developed so far to prevent immunological activation besides the use of antifouling agents. For example, lipoproteins, such as high-density lipoprotein and low-density lipoprotein through their interaction with these nanoparticles are able to prevent complement activation caused by polymeric nanostructures as evidenced by in vitro studies [122]. Likewise, cholesterol-rich liposomes induced complement activation can be suppressed by these liposomes in vitro, and also intravenous injection of these nanoparticles in vivo to reduce the adverse effects can be brought about by these liposomes [123]. Immunological recognition of nanoparticles can also be reduced by certain inhibitors of the complement system [103] administered in conjugation to the nanoparticle surface or delivered before nanoparticle injection. One study demonstrated that pretreatment with complement inhibitors could downregulate doxil-induced hypersensitivity reactions in pigs [105]. Although these strategies are in the early stages of development, they provide promising opportunities for improving the safety of nanotherapeutics. Additionally, nonimmunogenic nanoparticles can be designed by certain biomimetic approaches for immunological disguise. Complement activation could be evaded by pathogenic organisms through multiple mechanisms, such as inhibiting or enzymatically degrading complement proteins and interfering with complement regulation [124]. Hence, it is worth emphasizing that the immunological recognition of nanoparticles can also be utilized for therapeutic purposes.

10. Toxicity of different nanoparticles

10.1 Quantum dots toxicity

The quantum dots are composed of the semiconducting material with a core of cadmium selenide (inorganic) with a shell of zinc sulfide (organic) and a size of 2–10 nm, and it has improved optical properties. Quantum dots can be utilized as a diagnostic tool for the in vitro bioimaging and monitoring [125]. However, the utilization of the quantum dots has been limited due to the severe elimination problems attributing to its toxicity. There is a property of the quantum dots that, after the internalization, its size increases, and the size increases up to that extent that it is not filtered/eliminated by the renal capillaries that lead to its accumulation and results in the toxicity. The studies also show that quantum dots leach the Cd^{2+} ions and induce the formation of reactive oxygen that damages the plasma membrane, nucleus, and mitochondria [125].

10.2 Dendrimers toxicity

Dendrimers are the hyperbranched, tree-like structures polymer whose shape and size can be controlled. These are attractive drug carriers because of their capability of surface functionalization, stability, and monodispersity of size properties. Either encapsulation or complexation can accomplish the drug incorporation [126, 127]. The dendritic nature and the branching limit the extensive drug loading. The anionic dendrimer has less cytotoxicity profile as compared with cationic dendrimers. The cationic dendrimers have the ability to cause membrane instability that leads to cell lysis. The coating of the dendrimer surface by the PEG can decrease the property of the cytotoxicity [128]. However, there are very little data on the toxicity, so it has been concluded that the dendrimers are safe.

10.3 Fullerenes toxicity

The fullerenes are the carbon allotrope that is composed of the 60 carbon atoms, which is arranged in shape known as truncated icosahedrons with a diameter of 7 Å [129]. They are widely utilized for imaging and diagnostic processes [130]. Fullerenes are recently used as antimicrobial agents because of their ability to induce the ROS after photoexcitation [83]. It has also been reported that the cases of brain toxicity by dissolving lipid-rich brain tissue and even the production of oxygen radicals by microglia. The fullerenes also show them examples of environmental toxicity by causing the ecotoxicity [131].

10.4 Gold nanoparticles toxicity

The flexibility of metallic nanoparticles is more, owing to its properties of controllable size, structure, shape, composition, tunable optical properties, and encapsulation. The gold nanoparticles have shown their efficacy in cancer treatment. It can be utilized for the diagnosis of diseases due to tunable optical property. The nanoparticles formed are highly rigid three-dimensional, and ordered structures [132, 133]. The gold itself does not show any toxicity, but the use of the cetrimonium bromide (CTAB) as a stabilizer during the preparation of the nanoparticle causes the cytotoxicity. Also, the acute oral toxicity reports showed no signs of the adverse effects when the nanorod suspension is administered at the single dose for acute oral LD_{50} being greater than 5000 mg/kg body weight [134].

10.5 Silica nanoparticles

Silica nanoparticles have unique properties like high chemical stability, biocompatibility, and targeted and controlled release. The Si—O bond of the silica is responsible for the higher stability. They have shown its application in cell tracing and biosensing, diagnostic tool, and in targeting tumor-specific drugs [135]. The silica nanoparticles have caused the increased level of the ROS levels and reduction in the level of glutathione, which ultimately leads to enhancing the oxidative stress. The silica nanoparticle has also shown a decrease in cell proliferation/viability by increasing the level of lactate dehydrogenase [136]. Later we have depicted an image portraying the mechanism of tissue/cell toxicity by some of the nanoparticles (Fig. 2).

10.6 Toxicity assessment of nanoparticles

The analytical method validation that will be used to determine the toxicity of nanoparticles is of foremost importance. The evaluation of the nanoparticles comprises the characterization of the physicochemical property, biodistribution, and both in vitro and in vivo toxicity evaluation. The size and shape of nanoparticles can be measured with the help of dynamic light scattering (DLS) and transmission electron microscopy

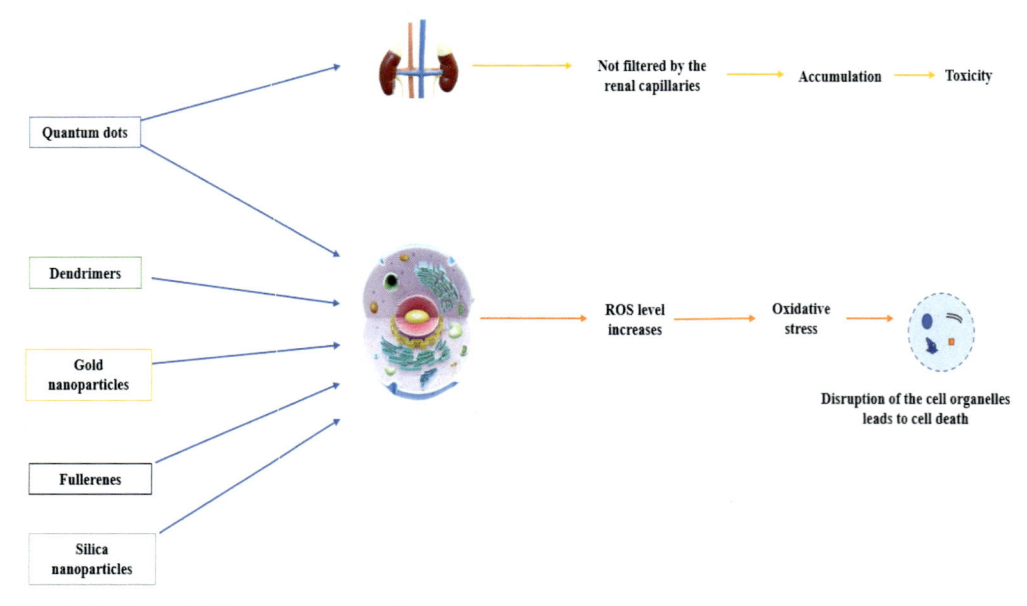

Fig. 2 Toxicity of different nanoparticles.

(TEM). The cellular viability, apoptosis, necrosis, and oxidative stress markers are measured by in vitro toxicity tests. The in vivo toxicity tests that are suggested by the Organization for Economic and Development (OECD) guidelines include oral toxicity, corrosion, dermal toxicity, eye irritation, and lethal dose (LD_{50}) [137–139].

10.7 In vitro assessment methods

In vitro assessment methods are the most crucial methods because it is faster, less expensive, and comes with minimum ethical concern.

10.7.1 Proliferation assay

Cellular metabolism is measured by evaluating the metabolically active cells. The most commonly used tetrazolium salt for the in vitro toxicity assessment is 3-(4,5-dimethylthiazol-2-yl)-2,5-diphenyltetrazolium bromide (MTT). The technique gives quicker and reproducible results, and less manipulation of the model cells.[140] The thymidine (^{3}H) incorporation method is the alternative method for evaluating cellular proliferation, but it has the limitation of the toxicity and expensive method [141]. The Alamar Blue and the Cologenic assay are the other two assays that can be used [142, 143].

10.7.2 Apoptosis assay

The major marker that has been seen in the assessment of the nanoparticle toxicity is the apoptosis. The Annexin-V assay, TdT-mediated dUTP-biotin nick-end labeling (TUNEL) assay, Comet assay, and morphological changes inspection are the different assay available [144]. The agarose gel electrophoresis can be used to differentiate the necrotic model and the apoptotic model of cell death [145]. The propidium iodide (PI) and Annexin-V are cell death markers [146]. The comet assay or single cell gel electrophoresis assay (SCGE) is used for the identification of the mutagenic potential by determining the single and double-stranded DNA break in the separate cells [147]. TUNEL assay is commonly used for the identification of the DNA damage by the use of in situ TUNEL staining. It can be also be used for the detection of DNA damage related to nonapoptotic events [148, 149].

10.7.3 Necrosis assay

The integrity of the membrane is used to determine cell viability. The Trypan Blue and Neutral Red are the most commonly used dyes for this test. The method is reliable, reproducible, quick, and inexpensive. Neutral red diffuses through the membrane and accumulate in the lysosome and alter the surface of the cell leading to lysosomal fragility [150]. It can easily distinguish between the dead and viable cells. The trypan blue only enters the dead cells leaving the viable cells [151].

10.7.4 Oxidative stress assay

The method for determination of the ROS and reactive nitrogen species (RNS) includes the reaction of the 2,2,6,6-tetramethylpiperidine (TEMP) with the oxygen radical that is detected by the use of the X-band electron paramagnetic resonance (EPR) [152]. This method has a limitation of the higher cost. The fluorescent probe molecule can be an alternative to the earlier mentioned method, and it is also cost-effective [153]. The limitation of this technique is its inability to react with the various reactive species. The different other assay method includes the 2,7-dichlorofluorescein diacetate (DCFH-DA) assay, Amplex Red assay, nitro blue tetrazolium assay, and many other assays [154].

10.8 In vivo assessment methods

10.8.1 Oral toxicity test

In the acute oral toxicity test, mice receive colloidal nanoparticle orally at the dose of 5000 mg/kg body weight (LD_{50}). For the first 3 h, the animals are observed for the toxic symptoms, and after 24 h, the numbers of mice survived are noted. The animals are observed daily for a period of 14 days for skin symptoms (erythema, edema, bloody scabs, ulcers, scars, and discoloration) and toxic symptoms (water and food consumption, behavior, weight loss). The skin biopsies and blood are performed for the histopathological and biochemical evaluation, respectively, and after 14 days, the animals are sacrificed, and the liver and skin are collected for the histopathological assessment [155].

10.8.2 Dermal toxicity test

The animals are first subdivided into the three groups that are group 1 receiving the vehicle, group 2 receives the 50-ppm colloidal suspension, and group 3 receives the 100-ppm colloidal suspension. These experiments are performed in accordance with the OECD 434 guidelines for acute dermal toxicity- fixed-dose procedure, and the colloidal suspension is applied on the shaved area of skin that has been shaved in the recommended dimension according to the proposed guidelines. The area is then covered with the non-irritating tape for 24 h. The number of survivors is noted at the end of 24 h. The animals are observed for a period of 14 days for skin symptoms and toxic symptoms. The animals are sacrificed after the 14 days period, and skin is collected for the histopathological evaluation [156].

10.8.3 Eye irritation test

The colloidal suspension was placed in the conjunctival sac in one eye, and the other eye is injected with the same volume of distilled water. The toxic symptoms in the animals are observed for a period of 1, 12, 24, 48, and 72 h after administration. The eye reactions are graded as recommended by the OECD 405 guidelines. The animals are observed and maintained daily for a period of 14 days [157].

In the following table, we have enlisted various results upon toxicity assessment of different nanoparticles (Table 1).

11. Future aspects

Presently, there are no guidelines for nanoparticle toxicity. The EPA has although listed some of the restriction and environmental hazards of the nanoparticles. However, there is still a need for the development and stringency of the guidelines.

Also, in the near future, the utilization of the properties of nanoparticles for the testing and optimization in the early developmental phase will eliminate the identified toxic characteristics. The need for the development of the more specific assays that can easily predict the toxicity in early stage and help to remove it. Although coating with hydrophilic polymers showed a positive result in reducing the toxicity, there is still a need to develop other methods to reduce the toxicity of nanoparticle and increase its efficacy.

Table 1 Result of estimation of toxicity after administering nanoparticle by different routes.

Sr. no.	Nanoparticle	Concentration	Size	Cell line	Detection method	Conclusion	References
1	Cadmium telluride quantum dots	1, 0.1 and 0.01 mg/L; 24 h	3.5–4.5 nm	Caco-2	Fluorescent microscopy; transepithelial electrical resistance measurement	Death of the cell	[158]
2	Carbon nanotubes	5–50 µg/mL; 24 h	14–170 nm	TH-1 Met5a	ELISA, trypan blue, ROS assay; flow cytometry	The cell viability has been decreased and increase in the production of the ROS and cytokines	[159]
3	Titanium nanoparticles	0.008–80 µg/ mL; 6, 24 and 48 h	124.9 nm	A431	MTT assay; comet assay; ROS assay; flow cytometry; glutathione level measurement; lipid peroxidase assay	After 48 h treatment, there is a decrease in the cell viability	[160]
4	Polyamidoamine dendrimers	0.01–21 µg/ mL; 24 h; 8 days	4.5–6.7 nm	HaCaT SW480	MTT, Alamar Blue, clonogenic, and neutral red assay	The toxicity increased linearly with the increase in the size	[161]
5	Fullerenes	1 ng/mL; 80 days	178 nm	CHO HELA HEK293	Micronucleus assay	Chromosomal damage with the DNA strand breakage	[162]
6	Silica nanoparticles	10–100 µg/ mL; 48 h	15–46 nm	Human bronchoalveolar carcinoma cells	DCFH-DA assay	The level of ROS increased	[163]

12. Conclusion

Nanotechnology is an emerging field in the field of medicine, and it has an impact on both aspects of toxicological exerting a great impact on the health and the environment. Generally the nanoparticles are used to reduce the potential side effects of the drugs by targeting it to specific sites. Until recently, the studies are showing that this carrier itself possess the threat of showing toxic effects due to its unique physicochemical properties. The toxicity that is demonstrated by the nanoparticles is far more different as compared with the conventional delivery systems. Nevertheless, still, the exact mechanism by which the nanoparticles show the toxicity is still lacking that needs to be extensively studied in the near future. Also, the collaboration of working on toxicology and drug delivery is essential for the exchange of concepts and to develop methods that will pave way for nanoparticle-mediated targeted drug delivery with minimal level of toxicity.

References

[1] R.P. Feynman, Plenty of room at the bottom, in: Feynman and Computation: Exploring the Limits of Computers, 1959.

[2] N. Taniguchi, Proc. Intl. Conf. Prod. Eng. Tokyo, Part II, J. Jpn. Soc. Precis. Eng. (1974).

[3] S. Bhatia, S. Bhatia, Nanoparticles types, classification, characterization, fabrication methods and drug delivery applications, Springer International Publishing, Natural Polymer Drug Delivery Systems, 2016, pp. 33–93, https://doi.org/10.1007/978-3-319-41129-3_2.

[4] M. Hamidi, A. Azadi, P. Rafiei, Hydrogel nanoparticles in drug delivery, Adv. Drug Deliv. Rev. 60 (2008) 1638–1649, https://doi.org/10.1016/j.addr.2008.08.002.

[5] R. Langer, Biomaterials in drug delivery and tissue engineering: one laboratory's experience, Acc. Chem. Res. 33 (2000) 94–101, https://doi.org/10.1021/ar9800993.

[6] G. Oberdörster, E. Oberdörster, J. Oberdörster, Nanotoxicology: an emerging discipline evolving from studies of ultrafine particles, Environ. Health Perspect. 113 (2005) 823–839, https://doi.org/10.1289/ehp.7339.

[7] C.S. Yah, G. Simate, S.E. Iyuke, Nanoparticles toxicity and their routes of exposures, Pak. J. Pharm. Sci. 25 (2012) 477–491, https://doi.org/10.5772/51230.

[8] B. Rothen-Rutishauser, C. Mühlfeld, F. Blank, C. Musso, P. Gehr, Translocation of particles and inflammatory responses after exposure to fine particles and nanoparticles in an epithelial airway model, Part. Fibre Toxicol. 4 (2007) 1–9, https://doi.org/10.1186/1743-8977-4-9.

[9] C. Mühlfeld, P. Gehr, B. Rothen-Rutishauser, Translocation and cellular entering mechanisms of nanoparticles in the respiratory tract, Swiss Med. Wkly. 138 (2008) 387–391, https://doi.org/10.4414/smw.2008.12153.

[10] L. Yildirimer, N.T.K. Thanh, M. Loizidou, A.M. Seifalian, Toxicological considerations of clinically applicable nanoparticles, Nano Today 6 (2011) 585–607, https://doi.org/10.1016/j.nantod.2011.10.001.

[11] W.H. De Jong, P.J.A. Borm, Drug delivery and nanoparticles:applications and hazards, Int. J. Nanomed. 3 (2008) 133–149, https://doi.org/10.2147/IJN.S596.

[12] S. Huo, S. Jin, X. Ma, X. Xue, K. Yang, A. Kumar, P.C. Wang, J. Zhang, Z. Hu, X.J. Liang, Ultrasmall gold nanoparticles as carriers for nucleus-based gene therapy due to size-dependent nuclear entry, ACS Nano 8 (2014) 5852–5862, https://doi.org/10.1021/nn5008572.

[13] Y. Pan, S. Neuss, A. Leifert, M. Fischler, F. Wen, U. Simon, G. Schmid, W. Brandau, W. Jahnen-Dechent, Size-dependent cytotoxicity of gold nanoparticles, Small 3 (2007) 1941–1949, https://doi.org/10.1002/smll.200700378.

[14] W.H. De Jong, W.I. Hagens, P. Krystek, M.C. Burger, A.J.A.M. Sips, R.E. Geertsma, Particle size-dependent organ distribution of gold nanoparticles after intravenous administration, Biomaterials 29 (2008) 1912–1919, https://doi.org/10.1016/j.biomaterials.2007.12.037.

[15] B. Asgharian, O.T. Price, Deposition of ultrafine (NANO) particles in the human lung, Inhal. Toxicol. 19 (2007) 1045–1054, https://doi.org/10.1080/08958370701626501.

[16] Y. Zhao, X. Sun, G. Zhang, B.G. Trewyn, I.I. Slowing, V.S.Y. Lin, Interaction of mesoporous silica nanoparticles with human red blood cell membranes: size and surface effects, ACS Nano 5 (2011) 1366–1375, https://doi.org/10.1021/nn103077k.

[17] B. Kong, J.H. Seog, L.M. Graham, S.B. Lee, Experimental considerations on the cytotoxicity of nanoparticles, Nanomedicine 6 (2011) 929–941, https://doi.org/10.2217/nnm.11.77.

[18] J.A. Champion, S. Mitragotri, Role of target geometry in phagocytosis, Proc. Natl. Acad. Sci. U. S. A. 103 (2006) 4930–4934, https://doi.org/10.1073/pnas.0600997103.

[19] A. Verma, F. Stellacci, Effect of surface properties on nanoparticle-cell interactions, Small 6 (2010) 12–21, https://doi.org/10.1002/smll.200901158.

[20] X. Zhao, S. Ng, B.C. Heng, J. Guo, L. Ma, T.T.Y. Tan, K.W. Ng, S.C.J. Loo, Cytotoxicity of hydroxyapatite nanoparticles is shape and cell dependent, Arch. Toxicol. 87 (2013) 1037–1052, https://doi.org/10.1007/s00204-012-0827-1.

[21] N.M. Schaeublin, L.K. Braydich-Stolle, A.M. Schrand, J.M. Miller, J. Hutchison, J.J. Schlager, S.M. Hussain, Surface charge of gold nanoparticles mediates mechanism of toxicity, Nanoscale 3 (2011) 410–420.

[22] A.M. El Badawy, R.G. Silva, B. Morris, K.G. Scheckel, M.T. Suidan, T.M. Tolaymat, Surface charge-dependent toxicity of silver nanoparticles, Environ. Sci. Technol. 45 (2011) 283–287, https://doi.org/10.1021/es1034188.

[23] C.M. Goodman, C.D. McCusker, T. Yilmaz, V.M. Rotello, Toxicity of gold nanoparticles functionalized with cationic and anionic side chains, Bioconjug. Chem. 15 (2004) 897–900, https://doi.org/10.1021/bc049951i.

[24] Y. Liu, W. Li, F. Lao, Y. Liu, L. Wang, R. Bai, Y. Zhao, C. Chen, Intracellular dynamics of cationic and anionic polystyrene nanoparticles without direct interaction with mitotic spindle and chromosomes, Biomaterials 32 (2011) 8291–8303, https://doi.org/10.1016/j.biomaterials.2011.07.037.

[25] F. Alexis, E. Pridgen, L.K. Molnar, O.C. Farokhzad, Factors affecting the clearance and biodistribution of polymeric nanoparticles, Mol. Pharm. (2008) 505–515, https://doi.org/10.1021/mp800051m.

[26] M. Lundqvist, J. Stigler, G. Elia, I. Lynch, T. Cedervall, K.A. Dawson, Nanoparticle size and surface properties determine the protein corona with possible implications for biological impacts, Proc. Natl. Acad. Sci. 105 (2008) 14265–14270, https://doi.org/10.1073/pnas.0805135105.

[27] A. Sukhanova, S. Poly, A. Shemetov, I.R. Nabiev, Quantum dots induce charge-specific amyloid-like fibrillation of insulin at physiological conditions, in: Nanosystems in Engineering and Medicine, SPIE, 2012, p. 85485F, https://doi.org/10.1117/12.946606.

[28] H. Yang, C. Liu, D. Yang, H. Zhang, Z. Xi, Comparative study of cytotoxicity, oxidative stress and genotoxicity induced by four typical nanomaterials: the role of particle size, shape and composition, J. Appl. Toxicol. 29 (2009) 69–78, https://doi.org/10.1002/jat.1385.

[29] J.-R. Gurr, A.S.S. Wang, C.-H. Chen, K.-Y. Jan, Ultrafine titanium dioxide particles in the absence of photoactivation can induce oxidative damage to human bronchial epithelial cells, Toxicology 213 (2005) 66–73, https://doi.org/10.1016/j.tox.2005.05.007.

[30] H. Arami, A. Khandhar, D. Liggitt, K.M. Krishnan, In vivo delivery, pharmacokinetics, biodistribution and toxicity of iron oxide nanoparticles, Chem. Soc. Rev. 44 (2015) 8576–8607, https://doi.org/10.1039/c5cs00541h.

[31] A.E. Gregory, R. Titball, D. Williamson, Vaccine delivery using nanoparticles, Front. Cell. Infect. Microbiol. 4 (2013) 13, https://doi.org/10.3389/fcimb.2013.00013.

[32] G. Guo, W. Liu, J. Liang, Z. He, H. Xu, X. Yang, Probing the cytotoxicity of CdSe quantum dots with surface modification, Mater. Lett. 61 (2007) 1641–1644, https://doi.org/10.1016/j.matlet.2006.07.105.

[33] G.D. Ad, 基因的改变, NIH Public Access, 2008, https://doi.org/10.1038/jid.2014.371.

[34] G. Oberdörster, A. Maynard, K. Donaldson, V. Castranova, J. Fitzpatrick, K. Ausman, J. Carter, B. Karn, W. Kreyling, D. Lai, S. Olin, N. Monteiro-Riviere, D. Warheit, H. Yang, Principles for characterizing the potential human health effects from exposure to nanomaterials: elements of a screening strategy, Part. Fibre Toxicol. 2 (2005) 1–35, https://doi.org/10.1186/1743-8977-2-8.

[35] S. Rajesh, J.W. Lillard Jr., Nanoparticle-based targeted drug delivery, Exp. Mol. Pathol. 86 (2000) 215–223, https://doi.org/10.1016/j.yexmp.2008.12.004.Nanoparticle-based.

[36] M.S. Lucia, C. Das, K.C. Hansen, J.K. Tyler, 乳鼠心肌提取 HHS Public Access, Physiol. Behav. 176 (2017) 139–148, https://doi.org/10.1016/j.physbeh.2017.03.040.

[37] A. Radomski, P. Jurasz, D. Alonso-Escolano, M. Drews, M. Morandi, T. Malinski, M.W. Radomski, Nanoparticle-induced platelet aggregation and vascular thrombosis, Br. J. Pharmacol. 146 (2005) 882–893, https://doi.org/10.1038/sj.bjp.0706386.

[38] P. Ruenraroengsak, P. Novak, D. Berhanu, A.J. Thorley, E. Valsami-Jones, J. Gorelik, Y.E. Korchev, T.D. Tetley, Respiratory epithelial cytotoxicity and membrane damage (holes) caused by amine-modified nanoparticles, Nanotoxicology 6 (2012) 94–108, https://doi.org/10.3109/17435390.2011.558643.

[39] A. Sukhanova, S. Bozrova, P. Sokolov, M. Berestovoy, A. Karaulov, I. Nabiev, Dependence of nanoparticle toxicity on their physical and chemical properties, Nanoscale Res. Lett. 13 (2018), https://doi.org/10.1186/s11671-018-2457-x.

[40] C.D. Walkey, J.B. Olsen, H. Guo, A. Emili, W.C.W. Chan, Nanoparticle size and surface chemistry determine serum protein adsorption and macrophage uptake, J. Am. Chem. Soc. 134 (2012) 2139–2147, https://doi.org/10.1021/ja2084338.

[41] J. Puppi, R.R. Mitry, M. Modo, A. Dhawan, K. Raja, R.D. Hughes, Use of a clinically approved iron oxide MRI contrast agent to label human hepatocytes, Cell Transplant. 20 (2011) 963–975, https://doi.org/10.3727/096368910X543367.

[42] J. Wolfram, Y. Yang, J. Shen, A. Moten, C. Chen, H. Shen, M. Ferrari, Y. Zhao, The nano–plasma interface: implications of the protein corona, Colloids Surf. B Biointerfaces 124 (2014) 17–24.

[43] M. Zhu, S. Perrett, G. Nie, Understanding the particokinetics of engineered nanomaterials for safe and effective therapeutic applications, Small 9 (2013) 1619–1634.

[44] M. Zhu, G. Nie, H. Meng, T. Xia, A. Nel, Y. Zhao, Physicochemical properties determine nanomaterial cellular uptake, transport, and fate, Acc. Chem. Res. 46 (2013) 622–631.

[45] S. Tenzer, D. Docter, S. Rosfa, A. Wlodarski, J. Kuharev, A. Rekik, S.K. Knauer, C. Bantz, T. Nawroth, C. Bier, Nanoparticle size is a critical physicochemical determinant of the human blood plasma corona: a comprehensive quantitative proteomic analysis, ACS Nano 5 (2011) 7155–7167.

[46] S. Tenzer, D. Docter, J. Kuharev, A. Musyanovych, V. Fetz, R. Hecht, F. Schlenk, D. Fischer, K. Kiouptsi, C. Reinhardt, et al., Rapid formation of plasma protein corona critically affects nanoparticle pathophysiology, Nat. Nanotechnol. 8 (2013) 772–781.

[47] N.P. Mortensen, G.B. Hurst, W. Wang, C.M. Foster, P.D. Nallathamby, S.T. Retterer, Dynamic development of the protein corona on silica nanoparticles: composition and role in toxicity, Nanoscale 5 (2013) 6372–6380.

[48] S. Goy-López, J. Juárez, M. Alatorre-Meda, E. Casals, V.F. Puntes, P. Taboada, V. Mosquera, Physicochemical characteristics of protein—NP bioconjugates: the role of particle curvature and solution conditions on human serum albumin conformation and fibrillogenesis inhibition, Langmuir 28 (2012) 9113–9126.

[49] J. Wolfram, K. Suri, Y. Yang, J. Shen, C. Celia, M. Fresta, Y. Zhao, H. Shen, M. Ferrari, Shrinkage of pegylated and non-pegylated liposomes in serum, Colloids Surf. B Biointerfaces 114 (2014) 294–300.

[50] B. Pelaz, G. Charron, C. Pfeiffer, Y. Zhao, J.M. De La Fuente, X.-J. Liang, W.J. Parak, P. Del Pino, Interfacing engineered nanoparticles with biological systems: anticipating adverse nano-bio interactions, Small 9 (2013) 1573–1584.

[51] G. Zuo, S. Kang, P. Xiu, Y. Zhao, R. Zhou, Interactions between proteins and carbon-based nanoparticles: exploring the origin of nanotoxicity at the molecular level, Small 9 (2013) 1546–1556.

[52] L. Shang, Y. Wang, J. Jiang, S. Dong, pH-dependent protein conformational changes in albumin: gold nanoparticle bioconjugates: a spectroscopic study, Langmuir 23 (2007) 2714–2721.

[53] W. Shang, J.H. Nuffer, V.A. Muñiz-Papandrea, W. Colón, R.W. Siegel, J.S. Dordick, Cytochrome c on silica nanoparticles: influence of nanoparticle size on protein structure, stability, and activity, Small 5 (2009) 470–476.

[54] W. Shang, J.H. Nuffer, J.S. Dordick, R.W. Siegel, Unfolding of ribonuclease a on silica nanoparticle surfaces, Nano Lett. 7 (2007) 1991–1995.

[55] M. Mahmoudi, M.A. Shokrgozar, S. Sardari, M.K. Moghadam, H. Vali, S. Laurent, P. Stroeve, Irreversible changes in protein conformation due to interaction with superparamagnetic iron oxide nanoparticles, Nanoscale 3 (3) (2011) 1127–1138, https://doi.org/10.1039/c0nr00733a.

[56] Z.J. Deng, M. Liang, M. Monteiro, I. Toth, R.F. Minchin, Nanoparticle-induced unfolding of fibrinogen promotes mac-1 receptor activation and inflammation, Nat. Nanotechnol. 6 (2011) 39–44.

[57] S. Linse, C. Cabaleiro-Lago, W.-F. Xue, I. Lynch, S. Lindman, E. Thulin, S.E. Radford, K.A. Dawson, Nucleation of protein fibrillation by nanoparticles, Proc. Natl. Acad. Sci. 104 (2007) 8691–8696.

[58] C. Ge, J. Du, L. Zhao, L. Wang, Y. Liu, D. Li, Y. Yang, R. Zhou, Y. Zhao, Z. Chai, Binding of blood proteins to carbon nanotubes reduces cytotoxicity, Proc. Natl. Acad. Sci. 108 (2011) 16968–16973.

[59] L. Wang, J. Li, J. Pan, X. Jiang, Y. Ji, Y. Li, Y. Qu, Y. Zhao, X. Wu, C. Chen, Revealing the binding structure of the protein corona on gold nanorods using synchrotron radiation-based techniques: understanding the reduced damage in cell membranes, J. Am. Chem. Soc. 135 (2013) 17359–17368.

[60] W. Hu, C. Peng, M. Lv, X. Li, Y. Zhang, N. Chen, C. Fan, Q. Huang, Protein corona-mediated mitigation of cytotoxicity of graphene oxide, ACS Nano 5 (2011) 3693–3700.

[61] A. Lesniak, F. Fenaroli, M.P. Monopoli, C. Åberg, K.A. Dawson, A. Salvati, Effects of the presence or absence of a protein corona on silica nanoparticle uptake and impact on cells, ACS Nano 6 (2012) 5845–5857.

[62] W.-S. Cho, R. Duffin, S.E.M. Howie, C.J. Scotton, W.A.H. Wallace, W. MacNee, M. Bradley, I.L. Megson, K. Donaldson, Progressive severe lung injury by zinc oxide nanoparticles; the role of Zn^{2+} dissolution inside lysosomes, Part. Fibre Toxicol. 8 (2011) 1–16.

[63] R.F. Hamilton, N. Wu, D. Porter, M. Buford, M. Wolfarth, A. Holian, Particle length-dependent titanium dioxide nanomaterials toxicity and bioactivity, Part. Fibre Toxicol. 6 (2009) 35.

[64] R. Molinaro, J. Wolfram, C. Federico, F. Cilurzo, L. Di Marzio, C.A. Ventura, M. Carafa, C. Celia, M. Fresta, Polyethylenimine and chitosan carriers for the delivery of RNA interference effectors, Expert Opin. Drug Deliv. 10 (2013) 1653–1668.

[65] O. Lunov, T. Syrovets, C. Loos, G.U. Nienhaus, V. Mailänder, K. Landfester, T. Rouis, T. Simmet, Amino-functionalized polystyrene nanoparticles activate the NLRP3 inflammasome in human macrophages, ACS Nano 5 (2011) 9648–9657.

[66] S.T. Stern, P.P. Adiseshaiah, R.M. Crist, Autophagy and lysosomal dysfunction as emerging mechanisms of nanomaterial toxicity, Part. Fibre Toxicol. 9 (2012) 20.

[67] X. Ma, Y. Wu, S. Jin, Y. Tian, X. Zhang, Y. Zhao, L. Yu, X.-J. Liang, Gold nanoparticles induce autophagosome accumulation through size-dependent nanoparticle uptake and lysosome impairment, ACS Nano 5 (2011) 8629–8639.

[68] A. Hall, A.K. Larsen, L. Parhamifar, K.D. Meyle, L.-P. Wu, S.M. Moghimi, High resolution respirometry analysis of polyethylenimine-mediated mitochondrial energy crisis and cellular stress: Mitochondrial proton leak and inhibition of the electron transport system, Biochim. Biophys. Acta Bioenergetics 1827 (2013) 1213–1225.

[69] L. Yan, Z. Gu, Y. Zhao, Chemical mechanisms of the toxicological properties of nanomaterials: generation of intracellular reactive oxygen species, Chem. Asian J. 8 (2013) 2342–2353.

[70] W. Zhang, C. Wang, Z. Li, Z. Lu, Y. Li, J.-J. Yin, Y.-T. Zhou, X. Gao, Y. Fang, G. Nie, et al., Unraveling stress-induced toxicity properties of graphene oxide and the underlying mechanism, Adv. Mater. 24 (2012) 5391–5397.

[71] F.T. Andón, B. Fadeel, Programmed cell death: molecular mechanisms and implications for safety assessment of nanomaterials, Acc. Chem. Res. 46 (2013) 733–742.

[72] T.-H. Kim, M. Kim, H.-S. Park, U.S. Shin, M.-S. Gong, H.-W. Kim, Size-dependent cellular toxicity of silver nanoparticles, J. Biomed. Mater. Res. A 100 (2012) 1033–1043.

[73] R. Foldbjerg, P. Olesen, M. Hougaard, D.A. Dang, H.J. Hoffmann, H. Autrup, PVP-coated silver nanoparticles and silver ions induce reactive oxygen species, apoptosis and necrosis in THP-1 monocytes, Toxicol. Lett. 190 (2009) 156–162, https://doi.org/10.1016/j.toxlet.2009.07.009.

[74] D.B. Warheit, B.R. Laurence, K.L. Reed, D.H. Roach, G.A.M. Reynolds, T.R. Webb, Comparative pulmonary toxicity assessment of single-wall carbon nanotubes in rats, Toxicol. Sci. 77 (2004) 117–125.

[75] J. Muller, F. Huaux, N. Moreau, P. Misson, J.-F. Heilier, M. Delos, M. Arras, A. Fonseca, J.B. Nagy, D. Lison, Respiratory toxicity of multi-wall carbon nanotubes, Toxicol. Appl. Pharmacol. 207 (2005) 221–231.

[76] R. Kedmi, N. Ben-Arie, D. Peer, The systemic toxicity of positively charged lipid nanoparticles and the role of toll-like receptor 4 in immune activation, Biomaterials 31 (2010) 6867–6875.

[77] Y.S. Chen, Y.C. Hung, I. Liau, G.S. Huang, Assessment of the in vivo toxicity of gold nanoparticles, Nanoscale Res. Lett. 4 (2009) 858–864, https://doi.org/10.1007/s11671-009-9334-6.

[78] W.S. Cho, M. Cho, J. Jeong, M. Choi, H.Y. Cho, B.S. Han, S.H. Kim, H.O. Kim, Y.T. Lim, B.H. Chung, J. Jeong, Acute toxicity and pharmacokinetics of 13 nm-sized PEG-coated gold nanoparticles, Toxicol. Appl. Pharmacol. 236 (2009) 16–24, https://doi.org/10.1016/j.taap.2008.12.023.

[79] M.J.D. Clift, J. Varet, S.M. Hankin, B. Brownlee, A.M. Davidson, C. Brandenberger, B. Rothen-Rutishauser, D.M. Brown, V. Stone, Quantum dot cytotoxicity in vitro: an investigation into the cytotoxic effects of a series of different surface chemistries and their core/shell materials, Nanotoxicology 5 (2011) 664–674, https://doi.org/10.3109/17435390.2010.534196.

[80] Y. Zhao, D. Sultan, Y. Liu, Biodistribution, excretion, and toxicity of nanoparticles, in: Theranostic Bionanomaterials, Elsevier, 2019, pp. 27–53, https://doi.org/10.1016/B978-0-12-815341-3.00002-X.

[81] X.-D. Zhang, D. Wu, X. Shen, P.-X. Liu, N. Yang, B. Zhao, H. Zhang, Y.-M. Sun, L.-A. Zhang, F.-Y. Fan, Size-dependent in vivo toxicity of PEG-coated gold nanoparticles, Int. J. Nanomed. 6 (2011) 2071–2081, https://doi.org/10.2147/IJN.S21657.

[82] H. Soo Choi, W. Liu, P. Misra, E. Tanaka, J.P. Zimmer, B. Itty Ipe, M.G. Bawendi, J.V. Frangioni, Renal clearance of quantum dots, Nat. Biotechnol. 25 (2007) 1165–1170, https://doi.org/10.1038/nbt1340.

[83] C. Carlson, S.M. Hussein, A.M. Schrand, L.K. Braydich-Stolle, K.L. Hess, R.L. Jones, J.J. Schlager, Unique cellular interaction of silver nanoparticles: size-dependent generation of reactive oxygen species, J. Phys. Chem. B 112 (2008) 13608–13619, https://doi.org/10.1021/jp712087m.

[84] J. Brooking, S.S. Davis, L. Illum, Transport of nanoparticles across the rat nasal mucosa, J. Drug Target. 9 (2001) 267–279, https://doi.org/10.3109/10611860108997935.

[85] P.H.M. Hoet, I. Brüske-Hohlfeld, O.V. Salata, Nanoparticles—known and unknown health risks, J. Nanobiotechnol. 2 (2004), https://doi.org/10.1186/1477-3155-2-12.

[86] S.M. Hussain, A.K. Javorina, A.M. Schrand, H.M.H.M. Duhart, S.F. Ali, J.J. Schlager, The interaction of manganese nanoparticles with PC-12 cells induces dopamine depletion, Toxicol. Sci. 92 (2006) 456–463, https://doi.org/10.1093/toxsci/kfl020.

[87] J.S. Kim, T.J. Yoon, K.N. Yu, B.G. Kim, S.J. Park, H.W. Kim, K.H. Lee, S.B. Park, J.K. Lee, M.H. Cho, Toxicity and tissue distribution of magnetic nanoparticles in mice, Toxicol. Sci. 89 (2006) 338–347, https://doi.org/10.1093/toxsci/kfj027.

[88] J. Wang, C. Chen, Y. Liu, F. Jiao, W. Li, F. Lao, Y. Li, B. Li, C. Ge, G. Zhou, Y. Gao, Y. Zhao, Z. Chai, Potential neurological lesion after nasal instillation of TiO_2 nanoparticles in the anatase and rutile crystal phases, Toxicol. Lett. 183 (2008) 72–80, https://doi.org/10.1016/j.toxlet.2008.10.001.

[89] R. Kumar, I. Roy, T.Y. Ohulchanskky, L.A. Vathy, E.J. Bergey, M. Sajjad, P.N. Prasad, In vivo biodistribution and clearance studies using multimodal organically modified silica nanoparticles, ACS Nano 4 (2010) 699–708.

[90] M.J.-E. Lee, O. Veiseh, N. Bhattarai, C. Sun, S.J. Hansen, S. Ditzler, S. Knoblaugh, D. Lee, R. Ellenbogen, M. Zhang, Rapid pharmacokinetic and biodistribution studies using cholorotoxin-conjugated iron oxide nanoparticles: a novel non-radioactive method, PLoS ONE 5 (2010), e9536.

[91] H.S. Goodridge, D.M. Underhill, N. Touret, Mechanisms of Fc receptor and dectin-1 activation for phagocytosis, Traffic 13 (2012) 1062–1071.

[92] S.M. Moghimi, A.J. Andersen, D. Ahmadvand, P.P. Wibroe, T.L. Andresen, A.C. Hunter, Material properties in complement activation, Adv. Drug Deliv. Rev. 63 (2011) 1000–1007.

[93] S.E. Degn, S. Thiel, Humoral pattern recognition and the complement system, Scand. J. Immunol. 78 (2013) 181–193.

[94] J. Szebeni, Complement activation-related pseudoallergy: a new class of drug-induced acute immune toxicity, Toxicology 216 (2005) 106–121.

[95] D.F. Moyano, M. Goldsmith, D.J. Solfiell, D. Landesman-Milo, O.R. Miranda, D. Peer, V.M. Rotello, Nanoparticle hydrophobicity dictates immune response, J. Am. Chem. Soc. 134 (2012) 3965–3967.

[96] S.-Y. Seong, P. Matzinger, Hydrophobicity: an ancient damage-associated molecular pattern that initiates innate immune responses, Nat. Rev. Immunol. 4 (2004) 469–478.

[97] C. Vauthier, B. Persson, P. Lindner, B. Cabane, Protein adsorption and complement activation for di-block copolymer nanoparticles, Biomaterials 32 (2011) 1646–1656.

[98] D.E. Owens, N.A. Peppas, Opsonization, biodistribution, and pharmacokinetics of polymeric nanoparticles, Int. J. Pharm. 307 (2006) 93–102, https://doi.org/10.1016/j.ijpharm.2005.10.010.

[99] Y. Yang, J. Wolfram, X. Fang, H. Shen, M. Ferrari, Polyarginine induces an antitumor immune response through binding to toll-like receptor 4, Small 10 (2014) 1250–1254.

[100] I. Tamayo, J.M. Irache, C. Mansilla, J. Ochoa-Repáraz, J.J. Lasarte, C. Gamazo, Poly (anhydride) nanoparticles act as active Th1 adjuvants through toll-like receptor exploitation, Clin. Vaccine Immunol. 17 (2010) 1356–1362.

[101] B. Holst, A.-C. Raby, J.E. Hall, M.O. Labéta, Complement takes its toll: an inflammatory crosstalk between toll-like receptors and the receptors for the complement anaphylatoxin C5a, Anaesthesia 67 (2012) 60–64.

[102] D. Ricklin, G. Hajishengallis, K. Yang, J.D. Lambris, Complement: a key system for immune surveillance and homeostasis, Nat. Immunol. 11 (2010) 785–797.

[103] S.M. Moghimi, Z.S. Farhangrazi, Nanomedicine and the complement paradigm, Nanomed.: Nanotechnol., Biol. Med. 9 (2013) 458–460.

[104] A. Chanan-Khan, J. Szebeni, S. Savay, L. Liebes, N.M. Rafique, C.R. Alving, F.M. Muggia, Complement activation following first exposure to pegylated liposomal doxorubicin (Doxil®): possible role in hypersensitivity reactions, Ann. Oncol. 14 (2003) 1430–1437.

[105] J. Szebeni, L. Baranyi, S. Sávay, M. Bodó, J. Milosevits, C.R. Alving, R. Bünger, Complement activation-related cardiac anaphylaxis in pigs: role of C5a anaphylatoxin and adenosine in liposome-induced abnormalities in ECG and heart function, Am. J. Physiol. Circ. Physiol 290 (2006) H1050–H1058.

[106] W.J. Pichler, J. Adam, B. Daubner, T. Gentinetta, M. Keller, D. Yerly, Drug hypersensitivity reactions: pathomechanism and clinical symptoms, Med. Clin. 94 (2010) 645–664.

[107] M.A. Dobrovolskaia, A.K. Patri, J. Zheng, J.D. Clogston, N. Ayub, P. Aggarwal, B.W. Neun, J.B. Hall, S.E. McNeil, Interaction of colloidal gold nanoparticles with human blood: effects on particle size and analysis of plasma protein binding profiles, Nanomed.: Nanotechnol., Biol. Med. 5 (2009) 106–117.

[108] M.P. Monopoli, D. Walczyk, A. Campbell, G. Elia, I. Lynch, F. Baldelli Bombelli, K.A. Dawson, Physical- chemical aspects of protein corona: relevance to in vitro and in vivo biological impacts of nanoparticles, J. Am. Chem. Soc. 133 (2011) 2525–2534.

[109] S. Zapf, M. Loos, Effect of EDTA and citrate on the functional activity of the first component of complement, C1, and the C1q subcomponent, Immunobiology 170 (1985) 123–132.

[110] A.J. Andersen, J.T. Robinson, H. Dai, A.C. Hunter, T.L. Andresen, S.M. Moghimi, Single-walled carbon nanotube surface control of complement recognition and activation, ACS Nano 7 (2013) 1108–1119.

[111] I. Hamad, O. Al-Hanbali, A.C. Hunter, K.J. Rutt, T.L. Andresen, S.M. Moghimi, Distinct polymer architecture mediates switching of complement activation pathways at the nanosphere-serum interface: implications for stealth nanoparticle engineering, ACS Nano 4 (2010) 6629–6638.

[112] C. Hanley, A. Thurber, C. Hanna, A. Punnoose, J. Zhang, D.G. Wingett, The influences of cell type and ZnO nanoparticle size on immune cell cytotoxicity and cytokine induction, Nanoscale Res. Lett. 4 (2009) 1409.

[113] A.K. Daly, Using genome-wide association studies to identify genes important in serious adverse drug reactions, Annu. Rev. Pharmacol. Toxicol. 52 (2012) 21–35.

[114] S.M. Moghimi, P.P. Wibroe, S.Y. Helvig, Z.S. Farhangrazi, A.C. Hunter, Genomic perspectives in inter-individual adverse responses following nanomedicine administration: the way forward, Adv. Drug Deliv. Rev. 64 (2012) 1385–1393.

[115] M.L. Schipper, G. Iyer, A.L. Koh, Z. Cheng, Y. Ebenstein, A. Aharoni, S. Keren, L.A. Bentolila, J. Li, J. Rao, et al., Particle size, surface coating, and PEGylation influence the biodistribution of quantum dots in living mice, Small 5 (2009) 126–134.

[116] J. Wolfram, K. Suri, Y. Huang, R. Molinaro, C. Borsoi, B. Scott, K. Boom, D. Paolino, M. Fresta, J. Wang, et al., Evaluation of anticancer activity of celastrol liposomes in prostate cancer cells, J. Microencapsul. 31 (2014) 501–507.

[117] C. Celia, E. Trapasso, M. Locatelli, M. Navarra, C.A. Ventura, J. Wolfram, M. Carafa, V.M. Morittu, D. Britti, L. Di Marzio, et al., Anticancer activity of liposomal bergamot essential oil (BEO) on human neuroblastoma cells, Colloids Surf. B Biointerfaces 112 (2013) 548–553.

[118] H.R. Kim, K. Andrieux, C. Delomenie, H. Chacun, M. Appel, D. Desmaële, F. Taran, D. Georgin, P. Couvreur, M. Taverna, Analysis of plasma protein adsorption onto PEGylated nanoparticles by complementary methods: 2-DE, CE and protein lab-on-chip system, Electrophoresis 28 (2007) 2252–2261.

[119] R. Bartucci, M. Pantusa, D. Marsh, L. Sportelli, Interaction of human serum albumin with membranes containing polymer-grafted lipids: spin-label ESR studies in the mushroom and brush regimes, Biochim. Biophys. Acta Biomembr. 1564 (2002) 237–242.

[120] I. Hamad, A.C. Hunter, J. Szeberi, S.M. Moghimi, Poly (ethylene glycol) s generate complement activation products in human serum through increased alternative pathway turnover and a MASP-2-dependent process, Mol. Immunol. 46 (2008) 225–232.

[121] K. Knop, R. Hoogenboom, D. Fischer, U.S. Schubert, Poly(ethylene glycol) in drug delivery: pros and cons as well as potential alternatives, Angew. Chem. Int. Ed. 49 (2010) 6288–6308, https://doi.org/10.1002/anie.200902672.

[122] I. Hamad, A.C. Hunter, S.M. Moghimi, Complement monitoring of Pluronic 127 gel and micelles: suppression of copolymer-mediated complement activation by elevated serum levels of HDL, LDL, and apolipoproteins AI and B-100, J. Control. Release 170 (2013) 167–174.

[123] S. Moein Moghimi, I. Hamad, R. Bünger, T.L. Andresen, K. Jørgensen, A. Christy Hunter, L. Baranji, L. Rosivall, J. Szebeni, Activation of the human complement system by cholesterol-rich and pegylated liposomes—modulation of cholesterol-rich liposome-mediated complement activation by elevated serum LDL and HDL levels, J. Liposome Res. 16 (2006) 167–174.

[124] J.D. Lambris, D. Ricklin, B.V. Geisbrecht, Complement evasion by human pathogens, Nat. Rev. Microbiol. 6 (2008) 132–142.

[125] R.E. Bailey, A.M. Smith, S. Nie, Quantum dots in biology and medicine, Physica E 25 (2004) 1–12, https://doi.org/10.1016/j.physe.2004.07.013.

[126] S.M. Moghimi, A.C. Hunter, J.C. Murray, Nanomedicine: current status and future prospects, FASEB J. 19 (2005) 311–330, https://doi.org/10.1096/fj.04-2747rev.

[127] J.H. Adair, M.P. Parette, E.I. Altinoğlu, M. Kester, Nanoparticulate alternatives for drug delivery, ACS Nano 4 (2010) 4967–4970, https://doi.org/10.1021/nn102324e.

[128] S. Svenson, D.A. Tomalia, Dendrimers in biomedical applications—reflections on the field, Adv. Drug Deliv. Rev. 57 (2005) 2106–2129, https://doi.org/10.1016/j.addr.2005.09.018.

[129] W. Krätschmer, L.D. Lamb, K. Fostiropoulos, D.R. Huffman, Solid C60: a new form of carbon, Nature 347 (1990) 354–358, https://doi.org/10.1038/347354a0.

[130] K.R.S. Chandrakumar, S.K. Ghosh, Alkali-metal-induced enhancement of hydrogen adsorption in C60 fullerene: an ab initio study, Nano Lett. 8 (2008) 13–19, https://doi.org/10.1021/nl071456i.

[131] B.X. Chen, S.R. Wilson, M. Das, D.J. Coughlin, B.F. Erlanger, Antigenicity of fullerenes: antibodies specific for fullerenes and their characteristics, Proc. Natl. Acad. Sci. U. S. A. 95 (1998) 10809–10813, https://doi.org/10.1073/pnas.95.18.10809.

[132] X. Huang, P.K. Jain, I.H. El-Sayed, M.A. El-Sayed, Gold nanoparticles: interesting optical properties and recent applications in cancer diagnostics and therapy, Nanomedicine 2 (2007) 681–693, https://doi.org/10.2217/17435889.2.5.681.

[133] E. Boisselier, D. Astruc, Gold nanoparticles in nanomedicine: preparations, imaging, diagnostics, therapies and toxicity, Chem. Soc. Rev. 38 (2009) 1759, https://doi.org/10.1039/b806051g.

[134] J. Conde, G. Doria, P. Baptista, Noble metal nanoparticles applications in cancer, J. Drug Deliv. 2012 (2012), https://doi.org/10.1155/2012/751075.

[135] B. Charu, U. Nagaich, A.K. Pal, N. Gulati, Mesoporous silica nanoparticles in target drug delivery system: a review, Int. J. Pharm. Investig. 5124 (2015), https://doi.org/10.4103/2230-973X.160844.

[136] Y. Jin, S. Kannan, M. Wu, J.X. Zhao, Toxicity of luminescent silica nanoparticles to living cells, Chem. Res. Toxicol. 20 (2007) 1126–1133, https://doi.org/10.1021/tx7001959.

[137] S.I. Adiloğlu, C. Yu, R. Chen, J.J. Li, J.J. Li, M. Drahansky, M. Paridah, A. Moradbak, A. Mohamed, H.A.T. Owolabi, FolaLi, M. Asniza, S.H.A. Khalid, T. Sharma, N. Dohare, M. Kumari, U.K. Singh, A.B. Khan, M.S. Borse, R. Patel, A. Paez, A. Howe, D. Goldschmidt, C. Corporation, J. Coates, F. Reading, We are IntechOpen, the world's leading publisher of Open Access books Built by scientists, for scientists TOP 1%, Intech i (2012) 13, https://doi.org/10.1016/j.colsurfa.2011.12.014.

[138] J.B. Hall, M.A. Dobrovolskaia, A.K. Patri, S.E. Mcneil, Characterization of nanoparticles for therapeutics: physicochemical characterization, Futur. Med. 2 (2007) 789–803, https://doi.org/10.1007/s00406-008-2003-4.

[139] K.W. Powers, M. Palazuelos, B.M. Moudgil, S.M. Roberts, Characterization of the size, shape, and state of dispersion of nanoparticles for toxicological studies, Nanotoxicology 1 (2007) 42–51, https://doi.org/10.1080/17435390701314902.

[140] N.J. Marshall, C.J. Goodwin, S.J. Holt, A critical assessment of the use of microculture tetrazolium assays to measure cell growth and function, Growth Regul. 5 (1995) 69–84.

[141] S.M. Hussain, J.M. Frazier, Cellular toxicity of hydrazine in primary rat hepatocytes, Toxicol. Sci. 69 (2002) 424–432, https://doi.org/10.1093/toxsci/69.2.424.

[142] A. Casey, E. Herzog, M. Davoren, F.M. Lyng, H.J. Byrne, G. Chambers, Spectroscopic analysis confirms the interactions between single walled carbon nanotubes and various dyes commonly used to assess cytotoxicity, Carbon N.Y. 45 (2007) 1425–1432, https://doi.org/10.1016/j.carbon.2007.03.033.

[143] S.P. Low, K.A. Williams, L.T. Canham, N.H. Voelcker, Evaluation of mammalian cell adhesion on surface-modified porous silicon, Biomaterials 27 (2006) 4538–4546, https://doi.org/10.1016/j.biomaterials.2006.04.015.

[144] V. Kumar, N. Sharma, S.S. Maitra, In vitro and in vivo toxicity assessment of nanoparticles, Int. Nano Lett. 7 (2017) 243–256, https://doi.org/10.1007/s40089-017-0221-3.

[145] "Apoptosis: a basic biological phenomenon with wide-ranging implications in tissue kinetics" (1972), by John F. R. Kerr, Andrew H. Wyllie and Alastair R. Currie | The Embryo Project Encyclopedia.

[146] M.T. Silva, Secondary necrosis: the natural outcome of the complete apoptotic program, FEBS Lett. 584 (2010) 4491–4499, https://doi.org/10.1016/j.febslet.2010.10.046.

[147] E.R. Kisin, A.R. Murray, M.J. Keane, X.C. Shi, D. Schwegler-Berry, O. Gorelik, S. Arepalli, V. Castranova, W.E. Wallace, V.E. Kagan, A.A. Shvedova, Single-walled carbon nanotubes: Geno- and cytotoxic effects in lung fibroblast V79 cells, J. Toxicol. Environ. Health, Part A 70 (2007) 2071–2079, https://doi.org/10.1080/15287390701601251.

[148] Y. Gavrieli, Y. Sherman, S.A. Ben-Sasson, Identification of programmed cell death in situ via specific labeling of nuclear DNA fragmentation, J. Cell Biol. 119 (1992) 493–501, https://doi.org/10.1083/jcb.119.3.493.

[149] D.T. Loo, TUNEL assay, An overview of techniques, Methods Mol. Biol. 203 (2002) 21–30, https://doi.org/10.1385/1-59259-179-5:21.

[150] E. Borenfreund, J.A. Puerner, Toxicity determined in vitro by morphological alterations and neutral red absorption, Toxicol. Lett. 24 (1985) 119–124, https://doi.org/10.1016/0378-4274(85)90046-3.

[151] W. Strober, Trypan blue exclusion test of cell viability, Curr. Protoc. Immunol. 111 (2015) A3.B.1–A3.B.3, https://doi.org/10.1002/0471142735.ima03bs111.

[152] T. Takajo, Y. Kurihara, K. Iwase, D. Miyake, K. Tsuchida, K. Anzai, Basic investigations of singlet oxygen detection systems with ESR for the measurement of singlet oxygen quenching activities, Chem. Pharm. Bull. 68 (2020) 150–154, https://doi.org/10.1248/cpb.c19-00770.

[153] A. Gomes, E. Fernandes, J.L.F.C. Lima, Fluorescence probes used for detection of reactive oxygen species, J. Biochem. Biophys. Methods 65 (2005) 45–80, https://doi.org/10.1016/j.jbbm.2005.10.003.

[154] S.M. Hussain, A.K. Javorina, A.M. Schrand, H.M. Duhart, S.F. Ali, J.J. Schlager, The interaction of manganese nanoparticles with PC-12 cells induces dopamine depletion, Toxicol. Sci. 92 (2006) 456–463, https://doi.org/10.1093/toxsci/kfl020.

[155] S. Clichici, A. Filip, In vivo assessment of nanomaterials toxicity, in: Nanomaterials—Toxicity and Risk Assessment, InTech, 2015, https://doi.org/10.5772/60707.

[156] OECD, Guideline for Testing of Chemicals Proposal for a New Draft Guideline 434: Acute Dermal Toxicity-Fixed Dose Procedure, 2004.

[157] OECD, OECD/OCDE 405 OECD Guideline for the Testing of Chemicals Acute Eye Irritation/Corrosion Introduction, 2012.

[158] B.A. Koeneman, Y. Zhang, K. Hristovski, P. Westerhoff, Y. Chen, J.C. Crittenden, D.G. Capco, Experimental approach for an in vitro toxicity assay with non-aggregated quantum dots, Toxicol. In Vitro 23 (2009) 955–962, https://doi.org/10.1016/j.tiv.2009.05.007.

[159] F.A. Murphy, A. Schinwald, C.A. Poland, K. Donaldson, The mechanism of pleural inflammation by long carbon nanotubes: interaction of long fibres with macrophages stimulates them to amplify pro-inflammatory responses in mesothelial cells, Part. Fibre. Toxicol. (2012), https://doi.org/10.1186/1743-8977-9-8.

[160] R.K. Shukla, V. Sharma, A.K. Pandey, S. Singh, S. Sultana, A. Dhawan, ROS-mediated genotoxicity induced by titanium dioxide nanoparticles in human epidermal cells, Toxicol. In Vitro 25 (2011) 231–241, https://doi.org/10.1016/j.tiv.2010.11.008.

[161] S.P. Mukherjee, M. Davoren, H.J. Byrne, In vitro mammalian cytotoxicological study of PAMAM dendrimers—towards quantitative structure activity relationships, Toxicol. In Vitro 24 (2010) 169–177, https://doi.org/10.1016/j.tiv.2009.09.014.

[162] A. Dhawan, J.S. Taurozzi, A.K. Pandey, W. Shan, S.M. Miller, S.A. Hashsham, V.V. Tarabara, Stable colloidal dispersions of C60 fullerenes in water: evidence for genotoxicity, Environ. Sci. Technol. 40 (2006) 7394–7401, https://doi.org/10.1021/es0609708.

[163] W. Lin, Y.W. Huang, X.D. Zhou, Y. Ma, In vitro toxicity of silica nanoparticles in human lung cancer cells, Toxicol. Appl. Pharmacol. 217 (2006) 252–259, https://doi.org/10.1016/j.taap.2006.10.004.

Index

Note: Page numbers followed by *f* indicate figures and *t* indicate tables.

N

Nab-technology, 331
NanoActivator magnetic field applicator, 482–483
Nanobiotechnology, 377
Nanobrachytherapy, biomedical application in, 529–530, 535–554
    development of, 535–554
    radiation concepts, 530–534
    radiation penetration, 533f
    radioactive decay, 531f
    radioisotopes, in brachytherapy and characteristics, 536t
        gold-198 nanoparticles, $Au^{198}NPs$, 539f, 542–546
        nanoparticle synthesis, 537–541, 538f
        nuclei, 552–554
        palladium-103 nanoparticles, $Pd^{103}NP$, 546–552
Nanocapsules, 11, 136
Nanocarriers (NCs), 133–134, 294–296
    characterization of, 137–172, 138f, 139–141t
    classification of, 134–137, 135f
        inorganic nanocarriers, 137
        particulate carriers, 136–137
        vesicular nanocarriers, 134–135
    dry states, characterization in, 154–162
        crystallinity, 160–162
        differential scanning calorimetry (DSC), 160
        high-resolution transmission electron microscopy (HRTEM), 157–158
        polarized light microscopy (PLM), 159
        polymorphism, 160–162
        porosity determination, 159
        scanning electron microscopy (SEM), 155–156
        scanning tunneling microscopy (STM), 158
        surface morphology, 155–159
        transmission electron microscopy (TEM), 156–157
        ultrastructure electron microscopy, 155–159
        X-ray diffraction, 161–162
    internal structure and interaction mechanism, 167–171
    lamellarity, 164–166, 165f
    liquid states, characterization in, 137–154
        analytical ultracentrifugation (AUC), 145
        asymmetrical flow field fractionation (AF4), 146, 147f
        differential centrifugal sedimentation (DCS), 144–145
        drug release, 154

dynamic light scattering (DLS), 142–143, 142f
        electroacoustic approach, 152
        electrophoresis approach, 151–152
        entrapment efficiency (EE), 152–154
        field flow fractionation (FFF), 145–146
        fluorescence correlation spectroscopy (FCS), 149–150
        inductively coupled plasma mass spectrometry (ICP-MS), 153–154
        laser diffraction, 143–144
        loading capacity (LC), 152–154
        nanoparticle tracking analysis, 146–149
        particle size determination, 141–150
        polydispersity index (PDI) and surface charge, 150–152, 151t
        size exclusion chromatography, 149
        spectroscopic analysis (UV-visible spectroscopy), 153
        static laser light scattering (SLS), 143–144
        tunable resistive pulse sensing (TRPS) methodology, 150
    molecular weight analysis, 163–164
    pore size distribution analysis, 171–172
    porosity determination, 171–172
    regulatory and safety aspects of, 172–173
    size analysis, 162–163
    surface topography and size analysis, 166–167
Nanocrystals, 85–97, 86f
    advantages of, 86
    architecture of, 88, 91t
    characteristics, 85–86
    formulations, 87–88, 89–90t
    gel (nanogel) of rutin, 440
    hydrogel, 440–441
    limitations, 86
    methods of preparation, 88–97, 91f
    oral solid dosage forms (OSDs), 433f
    products, 448t
    properties of, 86–87
    technique, 94
    topical formulations, 441
Nanocrystals, as drug delivery systems, 413–415
    formulation space, 416
    functionalized nanocrystals, 443–447
        targeting using, 444–446
    nanosuspension to oral solid dosage forms (OSDs)
        conversion of, 421–422
        development of, 421
        solidified nanosuspension, 432–434
        techniques of, 422–431, 423t

Printed in the United States
by Baker & Taylor Publisher Services